Handbook of Circuit Analysis Languages and Techniques

HANDBOOK OF

EDITORS

Randall W. Jensen
Senior Systems Engineer
Hughes Aircraft Company

Lawrence P. McNamee
Computer Science Department
University of California, Los Angeles

PRENTICE-HALL, INC.
Englewood Cliffs, New Jersey

CIRCUIT ANALYSIS LANGUAGES AND TECHNIQUES

Library of Congress Cataloging in Publication Data

JENSEN, RANDALL W.
  Handbook of circuit analysis languages and techniques.

  Includes bibliographical references.
  1. Electronic circuit design—Data processing.
  2. Programming languages (Electronic computers)
  I. McNamee, Lawrence P.   joint author
  II. Title
  TK7867.J39   621.3815'3   75-23235
  ISBN 0-13-372649-5

Handbook of Circuit Analysis Languages and Techniques
Jensen / McNamee

© 1976 by Prentice-Hall, Inc.
Englewood Cliffs, N. J.

All rights reserved. No part of this book may be reproduced in
in any form or by any means without permission in writing
from the publisher.

10 9 8 7 6 5 4 3 2 1

Printed in the United States of America

PRENTICE-HALL INTERNATIONAL, INC., *London*
PRENTICE-HALL OF AUSTRALIA, PTY, LTD., *Sydney*
PRENTICE-HALL OF CANADA, LTD., *Toronto*
PRENTICE-HALL OF INDIA PRIVATE LIMITED, *New Delhi*
PRENTICE-HALL OF JAPAN, INC., *Tokyo*
PRENTICE-HALL OF SOUTHEAST ASIA (PTE.) LTD., *Singapore*

To

MARGE	**KATE**
Kathy	*Michele*
Lori	*Lauren*
Jenny	*Kevin*
	Colleen
	Lawrence
	Cathy
	Brian

CONTENTS

Preface xxi

1
Introduction 2

    1.1 Computer-Aided Design/Analysis (CAD)   3
    1.2 Characteristics of Analog Circuit Analysis Programs   6
        1. Analysis Capabilities   6
            dc analysis   6
            ac analysis   6
            Transient Analysis   6
            Sensitivity Analysis   6
            Worst Case   7
            Monte Carlo Analysis   7
        2. Models   7
        3. Circuit Formulation Techniques   7
            Tableau   8
            State Variable Method   8
            Hybrid Formulation   9
            Nodal Method   9
            Topological Method   9
            Ports Method   10
    1.3 Analog Circuit Analysis Design Procedure   10
    1.4 Historical Development of Computer-Aided Circuit Analysis   12
        Nodal Analysis   14
        State Variables   17
        Topological (Signal Flow)   18
        Hybrid Formulation   19

		Ports	20
		Tableau Method	20
	1.5	Organization of the Handbook	21
	1.6	References	26

2
ASTAP: Advanced Statistical Analysis Program 34

	2.1	Introduction	35
		General Description	35
		Capabilities	36
		dc Analysis	36
		Transient Analysis	38
		ac Analysis	38
		Statistical Analysis	39
		Circuit Failure Analysis	41
		Circuit Modification and Rerun	41
		Run Controls	42
	2.2	Program Structure	42
	2.3	Input Language	44
		Sample Problem: Digital Logic Inverter	45
		Describing Element and Parameter Dependencies	51
		Equations	52
		Summary of System-Supplied Functions	52
		Fortran-Supplied Functions	53
		User-Written Functions	54
		Nested Models	54
		Model References and Changes	56
		Global Variables	56
		Model Library Facilities	57
		Diagnostic Messages	58
	2.4	Language Statement Descriptions	59
		Heading Statements	59
		MODEL DESCRIPTION Heading	59
		MODEL Heading	60
		ELEMENTS Heading	60
		Element Statement	60
		Parameter Statement	67
		Model Reference Statement	68
		Mutual Inductance Statement	69
		FUNCTIONS Heading	69
		TABLE Statement	69
		EQUATION Statement	70
		DISTRIBUTION Statement	70
		STATISTICAL TABLE Statement	72

	FEATURES Heading	73
	GROUND Node Statement	73
	GLOBAL Statement	73
	PORTS Statement	73
	NORMALIZE Statement	75
	EXECUTION CONTROLS Heading	76
	ANALYZE Heading	76
	RUN CONTROLS Heading	76
	Run Control Statement	77
	INITIAL CONDITIONS Heading	78
	OPERATING POINT Heading	79
	OPERATING POINT Statement	79
	OUTPUTS Heading	80
	PRINT Statement	81
	PLOT Statement	81
	HISTOGRAM Statement	84
	SCATTERGRAM Statement	84
	TEST Statement	85
	RERUNS Heading	86
	RERUN Statement	86
2.5	**Examples**	86
	Current Switch Emitter Follower	86
	Statistical Description of CSEF Circuit	89
	Transistor Model	90
	Statistical Output	91
	Crystal Oscillator	92
	Acknowledgments	95
2.6	**References**	97

3 BELAC 98

3.1	**Introduction**		99
	3.1.1	Routines	104
	3.1.2	Examples	105
	3.1.3	Example of BELAC'S Editing Features	115
	3.1.4	Transformer-Coupled Circuit	117
	3.1.5	Filter Design	120
	3.1.6	Amplifier Example	121
	3.1.7	Nonlinear Optimization Example	126
3.2	**Input Data Format and Miscellaneous Input Statements**		131
	3.2.1	Deck Setup	132
	3.2.2	Simplified BELAC Job Flow Diagram	134
	3.2.3	Input Statement Description	135

	3.2.4 Input Statement Format Conventions	136
	3.2.5 General Input Statement Format	137
	3.2.6 Arithmetic Expressions	138
	3.2.7 Modification of Stored Parameters	140
	3.2.8 Miscellaneous Input Statements	141
3.3	**Model Library**	142
	3.3.1 Library Input Statements	144
	3.3.2 Example of Model Library Use	145
3.4	**Internal Generated Entries**	149
	3.4.1 Butterworth and Chebyshev Filter Design	149
3.5	**Standard Network Analysis**	150
	3.5.1 Index Minimization	151
	3.5.2 Sweep Response	154
	3.5.3 Sensitivities	157
	3.5.4 Monte Carlo	158
	3.5.5 Worst Case	160
	3.5.6 Time Response	160
	3.5.7 Poles/Zeros	161
3.6	**Standard Network Elements**	162
	3.6.1 Short	163
	3.6.2 Linear Resistor	163
	3.6.3 Linear Capacitor	163
	3.6.4 Linear Inductor	164
	3.6.5 Mutual Inductance	164
	3.6.6 Two-Winding Transformer	164
	3.6.7 Independent Voltage Generator	165
	3.6.8 Independent Current Generator	165
	3.6.9 Voltage-Dependent Current Generator	166
	3.6.10 Voltage-Dependent Voltage Generator	167
	3.6.11 Current-Dependent Current Generator	167
	3.6.12 Current-Dependent Voltage Generator	168
	3.6.13 Voltage Battery	168
	3.6.14 Current Battery	169
	3.6.15 Transfer Function	170
	3.6.16 Time Delay	170
	3.6.17 Transmission Line	171
	3.6.18 Switch Current Generator (Nonlinear)	172
	3.6.19 Saturation Current Generator (Nonlinear)	173
	3.6.20 Logarithmic Current Generator (Nonlinear)	173
	3.6.21 Arctangent Current Generator (Nonlinear)	174
	3.6.22 Cosine Current Generator (Nonlinear)	175
	3.6.23 Rational Function Current Generator (Nonlinear)	175
	3.6.24 Product Current Generator (Nonlinear)	176

Contents

3.6.25	Ratio Current Generator (Nonlinear)	177
3.6.26	Generalized Diode Current Generator (Nonlinear)	177
3.6.27	Voltage Generator (Nonlinear)	178
3.6.28	Diode (Nonlinear)	179
3.6.29	Transistion Capacitance (Nonlinear)	180
3.6.30	Diffusion Capacitance (Nonlinear)	180

4 CIRC 182

4.1 Introduction	183
General Description	183
Summary of Salient Features	185
General	185
CIRC-DC	186
CIRC-AC	186
CIRC-TR	187
Circuit Size Capability	187
Major Feature Capability	188
Worst-Case Analysis	189
Open-Loop Analysis Technique	205
4.2 Circuit Description Elements	210
Introduction	210
Junction Temperature	210
Voltage Supplies	211
ac Voltage Supplies—Complex Number Usage	211
TR Voltage Supplies	213
Parameter Control Tables	216
CIRC-TR (CIRC-DC) Network Elements	217
Resistor	217
Voltage-Resistance Source	218
Current Source	219
Diode	220
Transistor	224
Inductor	230
Capacitor	231
CIRC-AC Network Elements	231
Inductor	232
Capacitor	234
Transistor	234
Current-Admittance Source	238
Voltage-Impedance Source	240
4.3 Input to CIRC	243
Control Messages	243

		Data Input Conventions	246
		Data Format Conventions	246
		Parameter Data Entry	246
		Special Data for CIRC-TR	248
		Connection and Control Data	250
		Special Circuit Description Termination Data	255
		Frequency Data Concludes ac Circuit Description	255
		Plot Variables for CIRC-AC	256
		Analysis Control for CIRC-TR	256
		Output Interval Controls for CIRC-TR	258
		Initial Conditions Option for CIRC-TR	258
		Plotting Selection and Control	259
	4.4	**CIRC Models**	260
		CIRC-TR Diode and Transistor Models	260
		CIRC-TR Diode Model	261
		CIRC-TR Transistor Model	265
		CIRC-TR Diode and Transistor Time Domain Methods	273
		Special Comments About XDS Models	275
		Warning and Information Messages for CIRC-TR Diodes and Transistors	277
		CIRC-TR L and C Models	280
		CIRC-AC Models	282
		CIRC-AC Inductor	282
		CIRC-AC Transistor Model	283
		Emitter Resistance	288
		Mutual Inductance Model	288
		Information and Error Messages for CIRC-AC Inductors and Mutuals	289
	4.5	**Examples**	290
		Worst-Case Analysis	291
		AC Nominal and Open-Loop Analysis	298
	4.6	**References**	305

5
CIRCUS 2   307

	5.1	Introduction	307
	5.2	**Program Structure**	308
		5.2.1 Topological Matrix Generator	309
		5.2.2 Sparse Matrix Processor	310
		5.2.3 Steady-State Analysis	310
		5.2.4 Transient Analysis	311
		5.2.5 Output Processor	311
		5.2.6 Model Processor	312
		5.2.7 Device Processor	313

5.3	Network Elements	313
5.4	**Input Language**	314
	5.4.1 Key Words	314
	5.4.2 Names	315
	5.4.3 Numbers	316
	5.4.4 Scaling Factors	316
	5.4.5 Data Card Format	317
	5.4.6 Input Preparation	317
	5.4.7 Circuit Description Rules	319
	5.4.8 Circuit Description Data Statements	320
	5.4.8.1 Fixed-Value Element Input Statements	320
	5.4.8.2 Mutual Coupling	320
	5.4.8.3 Pulsed Sources	321
	5.4.8.4 Sine-Wave Sources	322
	5.4.8.5 Tabular Sources	323
	5.4.8.6 Rational Impedances and Admittances	324
	5.4.8.7 Devices	325
	5.4.9 Basic Control Statements for Transient Analysis	326
	5.4.10 Naming Conventions	327
	5.4.11 Functions	328
	5.4.12 Parameter Variation	329
	5.4.12.1 Circuit Element Modification	329
	5.4.12.2 Device Replacement	330
	5.4.12.3 Device Parameter Modification	331
	5.4.12.4 Global Change	332
	5.4.13 Additional Steady-State Control Statements	332
	5.4.13.1 Specifying Initial Conditions	332
	5.4.13.2 Steady-State Guess	333
	5.4.13.3 Multiple Steady-State Analysis	333
	5.4.13.4 Selection of Optimum Parameter Values	334
	5.4.14 Multiple Transient Analysis	335
	5.4.15 Utility Statements	335
	5.4.16 Input Order	336
5.5	**Output Specifications**	337
5.6	**Models**	339
	5.6.1 Models and Devices in CIRCUS	339
	5.6.2 Basic Modeling Rules	341
	5.6.3 Model Description Data Statement Formats	342
	5.6.3.1 Model Name Statement	343
	5.6.3.2 External Nodes Statement	343
	5.6.3.3 Topology	344
	5.6.3.4 Equations	344
	5.6.3.5 Default Device Parameter Statement	352
	5.6.3.6 Fixed Parameters	352

xiv Contents

		5.6.3.7 Arrays	353
		5.6.3.8 Convolution Integrals in CIRCUS Models	353
		5.6.3.9 Predefined Variables	355
	5.6.4	Model Input	355
	5.6.5	Device Definition	356
	5.6.6	Device Description Data Statements	356
		5.6.6.1 Device Name/Model Name Statement	356
		5.6.6.2 Single-Valued Device Parameters	357
		5.6.6.3 One-Dimensional Tabular Device Parameters	357
		5.6.6.4 Two-Dimensional Tabular Device Parameters	358
		5.6.6.5 Arrays	358
		5.6.6.6 Device Parameter Input	359
	5.6.7	Global Device Parameters	359
5.7	Sample Circuits		360
	5.7.1	Digital Logic Inverter	360
	5.7.2	18-Volt Regulator	364
5.8	Program Limitations		371
5.9	Error Diagnostics		372
5.10	References		380

6 ECAP II 382

6.1	Introduction	383
	Major Features	385
6.2	Analysis Facilities	387
	dc Analysis	387
	dc Parameter Studies	387
	Initial Response Estimates	388
	dc Analysis Run Controls	388
	Transient Analysis	388
	Transient Analysis Initial Conditions	389
	Run Controls for Transient Analysis	389
6.3	Processing Description	390
	Program Structure and Processing Phases	391
	Interaction with the Circuit Description	392
	Automatic Replace Feature	394
	Interaction with the Model Library	394
	Program Control Commands	394
	Examples of Typical ECAP II Runs	395
6.4	Network Elements	397
	Basic Elements	397
	Elements	398
	Element Couplings	402
	Model References	402

Contents   xv

	Auxiliary Components	406
6.5	**The ECAP II Language**	406
	Statement Conventions	406
	Format	406
	Literals	408
	Basic Statement Format	408
	Statement Name	409
	Data Section	410
	Comment Section	411
	Network Nomenclature	411
	Element, Element Coupling, and Model Reference Names	411
	Element Voltage and Current Names	412
	Node Names	412
	Node Voltage Names	412
	Parameter Names	413
	Qualified Names	413
	External References	415
	Internal References	416
	Value Specification	416
	Numeric Constants	417
	Scale Factors	417
	Units	418
	Values Expressed as Function References	419
	Primary and Secondary Response Variables	421
6.6	**Statement Descriptions**	423
	Circuit Description Statements	424
	Run Specification Statements	429
	Output Description Statements	431
	Commands	434
	Miscellaneous Control Statements	435
6.7	**Sample Problems**	436
	Digital Logic Inverter	436
	Running the Analysis	440
	Application to Nonelectrical System Analysis	445
6.8	**Error Messages**	449
	Messages Formats	449
	PLAN Messages	450
6.9	**References**	450

7
LISA: Linear Systems Analysis 452

7.1	**Introduction**	453
7.2	**Program Structure**	455
7.3	**Network Elements**	455

	Components	455
	Component Names	455
	Component Values	457
	Node Labeling	457
	Component Card Format	457
7.4	**Input Language and Output Specifications**	460
7.5	**Models**	471
7.6	**Examples**	471
7.7	**Limitations**	489
	Rules	489
	Restrictions	489
7.8	**Error Diagnostics**	491
	Input Phase	491
	Compute Phase	493
7.9	**References**	494

8
MARTHA 496

8.1	**Introduction**	497
	General Description	497
	Capabilities	498
	Documentation	499
	Availability	500
8.2	**Program Structure**	500
8.3	**Network Elements**	501
8.4	**Input Language**	504
8.5	**Output Specifications**	510
	Response Functions	510
	Modifiers	512
	Formats	513
8.6	**Models**	514
	Types of Models Allowed	514
	Built-in Models	515
	Input Techniques and Format	515
8.7	**Examples**	519
	Transistor Amplifier	519
	Crystal Filter	523
	Coaxial Low-Pass Filter	529
8.8	**Limitations**	533
	Resource Limitations	533
	Ill-Conditioned Networks	533
8.9	**Error Diagnostics**	534
8.10	**References**	535

9 SCEPTRE 536

9.1 Introduction 537
Brief History 537
Basic Features of SCEPTRE 538
Program Availability 539

9.2 Program Operation 539
Transient Formulation 540
dc, Monte Carlo, Optimization, Sensitivity, Worst-Case, and ac Solution 543
   dc Solution 543
   Monte Carlo 543
   Sensitivity, Worst Case, and Optimization 543
   ac Calculation 544
SCEPTRE Execution 544

9.3 Network Elements and Units 545

9.4 Input Language 546
Modes or Major Headings 547
Groups or Subheadings 548
Element Definition 552
   Defined Parameter Definition 555
   Optimization, Worst Case, Sensitivity, and Monte Carlo 557
Initial Conditions Specifications 558
Functions Specification 559
Run Control Specification 561
Run Controls to Specify Mode of Analysis 563
Special Run Controls for dc Options 567
Special Run Controls for ac Analysis 570
Special Run Controls for Diagnostics 570
Program Data Limits 571
Vectorized Notation 572
Model Description and Storage of Models 572
Reruns 576
User-Supplied FORTRAN Subprograms 578
Use of CONTINUE Heading 578
RE-OUTPUT Heading 579

9.5 Output Definition 579
General Output Specifications 580
ac Outputs 581

9.6 Models for Use in SCEPTRE 582
Diode Models 583
Transistor Models 584

9.7	**Examples**	589
	Example 9.1. Transient Analysis of a Control System	590
	Transistor Models	594
	Actuator Transfer Function	595
	Transfer Function Implementation Using Defined Parameters	595
	Transfer Function Implementation Using Network Synthesis	595
	Control System SCEPTRE Circuit Description	601
	Example 9.2. Inverter Circuit Loades with *RC* Network	610
	Example 9.3. Transformer-Coupled Amplifier	612
	Example 9.4. Darlington Pair	621
	Example 9.5. Use of Small-Signal Equivalent Circuit	629
	Example 9.6. Solution of Simultaneous Differential Equations	631
	Example 9.7. Use of Monte Carlo	634
	Example 9.8. Use of Sensitivity	638
	Example 9.9. Use of Worst Case	640
	Example 9.10. Use of Optimization	643
	Example 9.11. Use of ac Analysis	645
	Example 9.12. Transfer Function Simulation	648
9.8	**Pathological Problems**	655
	Topological Restrictions	655
	Computational Delays	657
	Program Limits	658
	Convergence Problems on Initial Conditions	658
9.9	**Error Diagnostics**	659
	9.10 **References**	666

10
SYSCAP 670

10.1	**Introduction**	671
	Historical Background	672
	Mathematical Approach	673
	General Capability	674
	DICAP	674
	ALCAP	679
	TRACAP	679
	Engineering-Oriented Features	680
	SYSCAP Application Areas	682
	SYSCAP Availability	683
10.2	**Program Structure**	685
	DICAP	685
	ALCAP	687
	TRACAP	689

Contents   xix

10.3	SYSCAP Models	692
	DIODE-Zener	692
	Transistor	694
	Transformer	697
	Field Effect Transistor	697
	General-Purpose Amplifier (AMP/AMP WC Model)	705
10.4	Input Language	706
	General Ground Rules for Data Entries	710
	Input Coding	710
	Control Card	711
	Title Card	712
	Topology Cards	713
	Solution Control Cards	725
	Special Control Cards	725
10.5	Output Options	728
10.6	SYSCAP Examples	728
	Transient Simulation	728
	ac Simulation	731
	General Simulation	732
10.7	Special User Modeling	769
	FUNC Option	769
	MACRO Models	771
10.8	Limitations	773
10.9	References	773
	Appendix 10.1 Major SYSCAP Developers/Contributors	775
	Appendix 10.2 Radiation Simulation	775

Appendix A
Diode and Transistor Equivalent Circuits   778

A.1	Diode Equivalent Circuit	779
A.2	Transistor Equivalent Circuit	781
A.3	References	783

Appendix B
**Derivation of the Ebers-Moll
and Charge-Control Transistor Model
from the Two-lump Intrinsic Drift Model   784**

B.1	Ebers-Moll Transistor Model	785
B.2	Charge-Control Transistor Model	789

Appendix C
Diode Model Parameter Tables   794

Appendix D
Transistor Model Parameter Tables   798

PREFACE

The application of the digital computer to electrical engineering has had a profound effect on the methods of electronic circuit design and analysis. A new discipline, referred to as computer-aided design (CAD), began to emerge and gain widespread acceptance in about 1965. Since that time the power and flexibilities of the CAD programs, driven by the increased complexity and development costs of electronic circuits, have constantly improved. New circuit analysis and design programs and extensions of existing programs are continually being introduced at a rate which makes it almost impossible to remain aware of the new advances.

We have certainly not included in this handbook all the popular CAD programs available. Some obvious omissions from the text include NET-2 and the old standard ECAP, just to mention a couple. The list of omissions would probably comprise a chapter by itself. However, after weighing this volume it is obvious that a line had to be drawn somewhere. A great deal of time was spent determining which programs should be included in this work based on capability, structure, analysis approach and options, modeling techniques, and availability. Besides these criteria, we also discovered that legal problems encountered involved in publishing company-proprietary programs served to further shorten the list of available programs.

We did attempt to include a subset of the CAD programs which would provide a broad perspective of the capabilities and analysis techniques available today. In each case we included the most recent version of the program. In a few cases the version was not released at the time of delivery to Prentice-Hall.

The material contained on each program in this handbook is intended to be used as a user's reference guide. The information is concise by necessity, and the user is referred to program documentation where completeness in the handbook is impossible. In some programs the list of error messages alone would fill many pages.

We express our gratitude to Jo Anne Fox, Joanne Hawk, Irene Wilson, and Kate McNamee for their assistance in preparing the manuscript and to our wives and families for their support throughout the ordeal. We also wish to thank the contributing authors for their diligent effort, without which there would be no *Handbook of Circuit Analysis Languages and Techniques*.

RANDALL W. JENSEN
LAWRENCE P. MCNAMEE

Handbook of Circuit Analysis Languages and Techniques

1
INTRODUCTION

RANDALL W. JENSEN
LAWRENCE P. McNAMEE

1.1 COMPUTER-AIDED DESIGN/ANALYSIS (CAD)

During the past decade the role of the digital computer in the electronics industry had expanded from a relatively limited member of simulation applications programs to elaborate design automation systems, reaching into all phases of the operation—from the conceptual stage, through fabrication and testing, to manufacturing and packaging. For most companies the competitive edge one enjoys over another can be traced directly to the sophistication and cost effectiveness of their computer-aided design facility and software.

In logic applications, the success and acceptance of CAD problems by designers have been virtually universal. Software has been developed to simulate and synthesize logic networks, route and place components, and generate fault test procedures. In most companies CAD for logic design is considered the standard mode of operation; manual methods are almost nonexistent.

In analog circuit design, the utilization of software by engineers has not been a widely accepted practice, even though large sums of money have been spent in the development of a variety of circuit analysis programs. There are a couple reasons for this. First, and foremost, the original CAD programs were oversold; because of their limitations, many of the problems confronting the circuit designer at that time were either unsolved or solved with great difficulty. A considerable amount of user sophistication was required to make effective use of the programs.

The initial programs were very limited. The number of circuit elements which could be handled was small (50–100 elements). The integration routines, which were based on the classical Runge-Kutta, predictor-corrector, or implicit schemes, produced results that were suspect, or, even worse, would not converge. Modeling capability and device libraries were not generally available [27].

Moreover, virtually all the programs were designed to operate in batch mode (because time sharing was still in a primitive state), typically requiring a day or more for turnaround. This usually frustrated the circuit analyst, particularly when he discovered that a small coding or key-punching error aborted the run. On subsequent runs additional errors emerged, caused either by the user or quite often by the program.

To most circuit designers this was unacceptable, particularly when a manager could not understand why a "straightforward" circuit analysis problem required so much time and expense. As a result circuit analysts found that they could solve their problems quicker using a variety of approximations learned from textbooks and from their own experience. In fact, many circuit analysts competed with the program to prove that a "human" was superior to the machine. This attitude still prevails to a large extent throughout the industry.

A second reason that many circuit designers did not readily accept analog analysis programs was that the training in school was much more intense in circuit theory and electronics than in logic design and the associated routing and test methods. Thus, when a new employee was hired by a company and asked to design logic circuits, he automatically grasped for the in-house software packages even though they may have suffered from the same turnaround problems as the analog software. If he had to design analog circuits, he used the techniques taught in school.

The evolution of CAD over the years has been both blessed and frustrated by technological developments. When the circuit analysis programs were first being written, circuit designers were dealing with discrete elements and relatively small circuits. It was difficult to analyze those discrete circuits by first-generation programs since they could accommodate only a relatively few numbers and types of circuit elements and employed inadequate numerical and programming techniques. Two other serious deficiencies hampered this capability and acceptance of the first-generation programs. First, the level of sophistication and/or experience of the user was minimal. Comparing first- and second-generation circuit analysts is akin to comparing the flying skills of the Wright brothers to those of Manfred Von Richthofen. Second, the applicability of automated circuit analysis could progress no faster than the advances in device and circuit modeling. There had been no developments in transistor models from the mid-1950s until the pressures placed on the modeling fraternity by the circuit analysts in the mid-1960s produced many rapid advances in the field. In turn the model advances placed additional requirements on the type and complexity of circuit elements available in the analysis programs.

As the device and circuit technology improved to produce faster computers with a larger storage capacity, the same technology increased the complexity of the typical circuits to be analyzed from discrete components through integrated circuits to large-scale integration (LSI), imposing an order-of-magnitude increase on the size of the circuit.

The first-generation programs proved to be quite successful in many respects. Many different network formulation and numerical techniques were attempted and evaluated, thereby providing great insight into what the characteristics of the most efficient and accurate methods should be. They also allowed the user to try them out and let his opinions be heard. More important, the experience gained set the stage for the development of the new techniques for analyzing the large networks encountered because of the LSI technology.

All the first-generation programs were limited to some degree whether they lacked an analysis capability or modeling facility. It very quickly became apparent that the circuit analyst wanted as many of the following features as possible: a user-oriented input language; nonlinear dc, nonlinear transient, and linear (small-signal) ac analysis capabilities with the ac and transient solutions taken about the dc (or steady-state) operating conditions; built-in models of diodes, zeners, and bipolar transistors for both small-signal and large-signal analyses, FETs, unijunction, SCRs, and system-supplied functions such as the black-box models; voltage and current sources of sine, exponential, step, and ramp functions and user-defined tables; current/voltage-dependent current/voltage sources; a system-supplied model library of available solid-state devices; a user-derived library capability; improved accuracy and stability of solutions; the ability to analyze large circuits; fast turnaround; and reduced costs.

Most all the second-generation programs have provisions for the user to virtually define almost any functional dependencies imaginable and to define models and store them in a library for subsequent recall. Accuracy and computational stability are no longer serious problems what with implicit integration and the introduction of Gear's method [47], the adjoint method [12], and sparse matrices [16, 17, 33]. The circuit size that the program can accommodate is no longer specified to an element or node limit but to the size of the computer.

The second-generation programs have made it possible to analyze networks of size and complexity that were considered impossible to analyze only a few years ago. The programs, the device model and modeling techniques, and the analyst's sophistication have each developed at the precise time where all three capabilities are necessary to design and analyze the second-generation class of circuits (MSI and LSI). This handbook contains a powerful group of programs which can, in most cases, analyze current contemporary circuit analysis problems.

1.2 CHARACTERISTICS OF ANALOG CIRCUIT ANALYSIS PROGRAMS

Circuit analysis programs are usually classified according to the features incorporated in them. Since some of these characteristics have been used in the literature with conflicting meanings, the context with which they are employed throughout the handbook are defined in order to clarify their usage.

1. Analysis Capabilities

dc Analysis

The dc analysis capability in a CAD program refers to the dc steady-state solution of the circuit. The dc solution is usually transferred to other analysis routines, such as ac and transient, to compute the responses taken about the dc operating point.

ac Analysis

An ac analysis means a small-signal steady-state solution of a network subjected to a sinusoidal excitation at a fixed frequency. All nonlinearities are linearized about its operating point. A frequency sweep can also be employed to obtain a frequency spectrum of the circuit performance.

Transient Analysis

A transient analysis capability of a program computes the time response solution of a network subjected to user-specified driving functions. The solution is obtained starting at a set of initial conditions input by a user or obtained from the dc analysis. Nonlinear elements in some programs can be incorporated in the network by functional relationships or by built-in models.

Sensitivity Analysis

A sensitivity analysis determines the change obtained in an output variable (such as a node voltage, e_n) by a corresponding change of a network parameters (such as P); i.e.,

$$S = \frac{\partial e_n}{\partial P}$$

Sensitivities are based on small variations computed in the neighborhood of the given parameter values. Sensitivities can be computed in dc, ac, and transient modes of operation.

Worst Case

A worst-case analysis is obtained for a circuit by computing the extremes of the circuit response of a network composed of elements which are set to their extreme tolerance limits. A band of the upper and lower response limits is given as output.

Monte Carlo Analysis

A Monte Carlo analysis is a series of analyses of a circuit whose parameter values are randomly chosen within their tolerance limits. The output is commonly expressed by histograms.

2. Models

Models used in CAD programs are classified as (1) built-in, (2) user-defined, and (3) library-stored. A built-in model is a model that can be called by a user by supplying parameters as arguments to a reserved input language format. Usually some standard form of a diode and transistor such as that of Ebers and Moll [59] is made available to the user as built-in models.

Models which can be defined by a user for immediate application in a program are classified as a user-defined model. On time-shared systems and in most batch-oriented programs user-defined models can be stored and recalled for subsequent use in other analyses. In this manner a designer can create, modify, and delete models as desired.

Library-stored models are similar to user-stored models except that library models are not controlled by users but by a group of specialists who have exclusive access to the library. A model library usually allows the designer to insert models of devices by defining them by their manufacturer numbers or some other designators without needing to input a large number of parameters.

3. Circuit Formulation Techniques

The network equations that are ultimately solved by an analysis program are usually formulated from a primitive set of component-nodal descriptions of a network. The manner in which the network parameters are stored and operated on are dictated by the numerical methods employed. Of all the formulations that have been developed over the years, the following six fundamental approaches can be identified in modern CAD programs.

Tableau [16, 17]

The tableau method of formulation is based on the representation of Kirchhoff's voltage law [KVL], Kirchhoff's current law [KCL], and the functional relationships [FR] of components; i.e.,

$$\text{KVL:} \quad A^T v_n = v_b$$
$$\text{KCL:} \quad A i_b = 0$$
$$\text{FR:} \quad Y v_b + Z i_b = S$$

where

i_b = branch current vector
v_n = node voltage vector
v_b = branch voltage vector
A = incidence matrix
Z = impedance matrix
Y = admittance matrix
S = matrix of sources and nonlinear relationships

Operation on the tableau is carried out in its primitive (sparse matrix) form. Reordering of terms is done prior to calculation to retain as much of the sparsity of the original matrices as possible [4, 28] This has the effect of reducing the number of calculations required to solve the network equations as well as reducing the computational errors normally encountered in the traditional full matrix approach.

State Variable Method [3, 19, 23]

The state variable method is based on the formulation of a set of first-order differential equations of capacitor voltages and inductor currents taken as dependent variables. Adjunct algebraic equations, generated during the formulation process, are eliminated by substitution. The state equations take the form

$$\dot{x} = Ax + Bu$$

where

x = state vector of inductor currents and capacitor voltages
u = vector of independent current and voltage sources

The output variables are given by

$$Y = Cx + Du$$

since the state variables may not necessarily be the output variables desired by a user. The A, B, C, and D matrices incorporate the topological and relational properties of the circuit components.

Hybrid Formulation [7]

The hybrid formulation is a modification of the state space method except that the algebraic relationships are not eliminated. That is, the state vector, x, contains variables of the algebraic equations, and no attempt is made to eliminate them in subsequent calculations. The hybrid formulation, using Branin's notation [7], is given by

$$\begin{bmatrix} V_{TK} \\ V_{LK} \\ I_{TL} \\ I_{LL} \\ V_{TG} \\ V_{LG} \\ I_{TR} \\ I_{LR} \end{bmatrix} = \begin{bmatrix} U_{TK} & 0 & 0 & 0 \\ -C^t_{TKK} & 0 & 0 & 0 \\ 0 & C_{TLL} & 0 & 0 \\ 0 & U_{LL} & 0 & 0 \\ 0 & 0 & U_{TG} & 0 \\ -C^t_{TKG} & 0 & -C^t_{TGG} & 0 \\ 0 & C_{TRL} & 0 & C_{TRR} \\ 0 & 0 & 0 & U_{LR} \end{bmatrix} \begin{bmatrix} V_{TK} \\ I_{LL} \\ V_{TG} \\ I_{LR} \end{bmatrix} + \begin{bmatrix} 0 & 0 \\ -C^t_{TEK} & 0 \\ 0 & C_{TLJ} \\ 0 & 0 \\ 0 & 0 \\ -C^t_{TEG} & 0 \\ 0 & C_{TRJ} \\ 0 & 0 \end{bmatrix} \begin{bmatrix} E \\ J \end{bmatrix}$$

where the first subscript T or L refers to a tree or link element, respectively, and the second subscript E, K, G, R, L, or J denotes a voltage source, capacitor (since C is used for a cutset matrix), conductance, resistance, inductance, or current source, respectively.

Nodal Method [6, 19]

In the nodal analysis method Kirchhoff's current law is applied at each node to form a set of n independent equations which can be written as

$$Yv_n = I$$

where Y is the nodal admittance matrix, v_n is the node voltage vector, and I is the current source vector. Currents through inductors and capacitors are approximated by difference equations. The unknown voltages are solved by means of some form of the Gaussian elimination method, possibly incorporating some of the sparse matrices techniques.

Topological Method [18, 25, 34, 35]

The topological methods are based on Mason's formula [25], which constructs the transfer function of a signal flow graph representation of a source node x_0 to any other nonsource node x_v. The transfer function can be determined by

$$T = \frac{X_v}{X_0} = \frac{\sum_i P_i \Delta_i}{\Delta}$$

where

$$\Delta = 1 + \sum (-1)^j \sum_k L_{k,j} = \text{determinant of the graph}$$

$L_{k,j}$ = product of the transmittance of the kth set of j nonintersecting loops (product)

P_i = transmittance product of the ith path between nodes

Δ_i = partial determinant obtained from after the removal of all loops intersecting the ith path between x_0 and x_v

The tree-enumeration methods [34], which are similar to the signal flow method, are also classified as a topological method. Both classes of topological methods suffer from a numerical accuracy problem and excess storage requirements which force the number of networks that can be handled to be very small (50 elements or less).

Ports Method [29]

The ports method is based on the interconnection and reduction of one- and two-port networks. The networks are characterized by their impedance, admittance, hybrid, *ABCD*, etc., matrices commonly described in textbooks.

1.3 ANALOG CIRCUIT ANALYSIS DESIGN PROCEDURE

The process of designing an electronic circuit by means of circuit analysis programs follows a relatively straightforward procedure (Fig.1.1). First, one must start with a complete description of the circuit specifications, such as input and output waveforms, frequency of operation, power dissipation, tolerances, and environmental effects (block 1). Once the specifications have been set, the initial design can be formulated (block 2). This is usually done by the designer drawing upon his own experience or from previous designs by others.

Components, nominal values, and tolerances are then selected. Next a circuit analysis program is chosen which can best analyze the circuit to verify the accuracy of the design (block 3). Sometimes a group of programs is required since a single program may not have all the computational capabilities to verify the circuit design.

The circuit is then prepared for computer analysis. All components and devices are modeled using elements acceptable to the analysis program (block 4). This may require some judicious substitution of components to

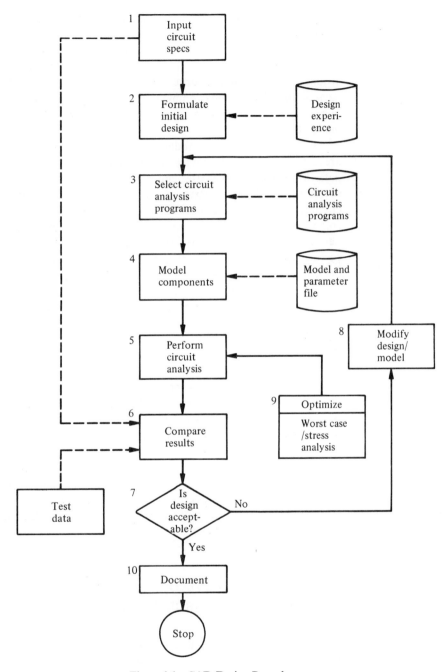

Figure 1.1 CAD Design Procedure.

perform the analysis without substantially changing the performance of the original circuit.

With the circuit modeled, computer runs are made (block 5) either by batch or on-line operation, the difference between the two being the turnaround time. The batch turnaround time is typically a matter of hours or days in an industrial environment, depending on the work load at the computation center. For on-line operation the analysis is usually completed in a matter of minutes.

Once all coding and modeling errors have been eliminated, the results are analyzed by comparing them with the design specifications and/or test data (block 6). In some cases engineering insight must override any discrepancies found in the comparison. If the design is acceptable (block 7), the circuit performance characteristic is documented for the fabrication and test engineers (block 10). If the performance is unacceptable, the circuit and the device models must be modified (block 8). The modifications are usually performed by the circuit designer, who knows which changes should be considered. The analysis cycle is then started again. It is here the circuit design/analysis programs are useful.

The circuit analysis programs permit the designer to use the capabilities of the digital computer in several areas. First, the effects of a parameter variation can be much more easily observed using a digital computer program than using breadboard evaluation. For example, the effects of changing only the collector capacitance or the alpha cutoff frequency of a transistor are very difficult to determine experimentally, but they are easy to ascertain when using the digital computer. Second, the circuit designer can analyze the effects of a component failure on circuit performance without destroying the component involved. Third, the computer allows the designer to study combinations of component parameters, such as worst case, which are difficult if not almost impossible to achieve in the laboratory. Fourth, the program makes it possible to simulate the effects of expensive or hard-to-obtain components. Finally, the program makes it possible to perform measurements on a circuit model that would be difficult to make or time-consuming and expensive to instrument in the laboratory.

1.4 HISTORICAL DEVELOPMENT OF COMPUTER-AIDED CIRCUIT ANALYSIS

General-purpose circuit analysis programs were not made available to the engineering community until the early 1960s, even though some companies developed programs for internal use. Prior to 1960 virtually all the circuit analysis programs were written ad hoc to solve specific problems at hand or restricted classes of problems such as filters [14]. When the first general-purpose programs emerged, it was apparent that no common approach was employed, even though the basic underlying theory was known for quite some time.

The basic concepts behind computer-aided design and analysis of analog circuits, as currently employed in virtually all general-porpose programs, were first presented by Gabriel Kron in his book *Tensor Analysis of Networks* [21]. Kron recognized that tensors could be employed to do all the bookkeeping chores required by Kirchhoff's laws to formulate network equations from the primitive impedance/admittance matrices and their topological matrices used as operators. His procedure was ideal for computer implementation, even though his book was published before the first digital computers were invented.

In 1953 Kron extended his basic approach to large-scale network problems by "tearing" a network into smaller manageable parts, solving them separately, and then operating on each subnetwork solution matrix with the appropriate topological matrices to obtain the overall solution [22]. The time of solution and the accuracy of the results improved by an order of magnitude over the standard Gaussian elimination methods.

Kron's methods did not achieve wide acceptance by practicing engineers because they found the tensor concept too complicated to understand and apply. Kron actually needed the tensor formalism for his book in order to describe the operation of motors and generators in electrical networks. It was shown later that the simpler matrix theory was sufficient to analyze stationary (no moving parts) networks—the area of interest to electronic engineers.

It was not until the early 1960s that the matrix algebraic topological techniques expounded by Kron caught on. This was due to a large measure to Branin [5, 6], who advocated and extended Kron's work. At just about the same time Seshu and Reed [34] published a book presenting a matrix approach to network analysis that was palatable to engineers.

In 1962 Branin [5] developed TAP (Transistor Analysis Program), employing Kron's methods to formulate the mesh and nodal equations and to obtain the inverse of these matrices during the solution process. The transient portion of the program incorporated for the first time a form of Bashkow's A matrix [3] to derive the differential equations—but without topological matrices.

In 1961 Ashcraft and Hochwald specified the requirements for the design of circuits by computer taking into account reliability (worst-case) considerations [2]. These specifications provided the basis for the SPARC program written at Autonetics (now Rockwell International). SPARC was actually an equation solver which required that the circuit designer write the equations by hand and then input them into the program for solution. In 1962 a pre-processor, called SCAN, performed the equation-forming task automatically from an element-nodal description of the circuit. The combination of SPARC and SCAN resulted in the first version of TRAC [88], one of the first general-purpose circuit analysis programs written.

With the limited success of TAP (it had an inflexible transistor model embedded in it, and the dc portion could not be run in conjunction with the transient portion) and SPARC, the computer-aided circuit analysis field took a dynamic turn—to what appears as a proliferation of programs. Figure 1.2 depicts the evolution of the major programs from the time of TAP and SPARC to the present. The nodes shown list the release date for each program, and the branches indicate the influence each program had on the development of subsequent programs. It should be noted that the influence that a program had on another varies considerably, for a free exchange of ideas was practiced and incorporated, either consciously or unconsciously, into subsequent programs.

The rectangular boxes in Fig. 1.2 indicate the major work or research project on which the formulation of the program is based. Six network formulation schemes were employed: nodal, state variable, hybrid, tableau, topological (signal flow), and ports. The evolution of each of the six program groups is described in the following paragraphs. We have not attempted to include all the hundreds of programs available in the discussion but have limited our discussion to those which have had widespread acceptance or have furnished a substantial framework from which major programs have evolved.

Nodal Analysis

The TRAC [88], ECAP [82, 87], LISA [78], and ASAP* programs constitute the first group of major programs to employ the nodal method of analysis. Each program assumed its own special flavor in the nature of the output a circuit analyst could request. LISA provided a transfer function, pole-zero type of analysis; ECAP was the first major program to provide dc, ac, and transient analysis capabilities with a common language; TRAC was written to analyze circuits for transient radiation effects, and ASAP was the first program to provide a statistical (Monte Carlo) analysis capability.

TRAC, perhaps more than any other program at that time, influenced much of the subsequent computer-aided circuit analysis development. It was the first program to employ implicit integration, include built-in models, and use sparse matrices for compact storage. TRAC also was one of the first programs to be placed at the disposal of a large group of circuit analysts operating at remote terminals.

*A. G. Kennard, "Automatic Statistical Analysis Program," *IBM Technical Report TR*00.896, Aug. 3, 1962 (not generally available).

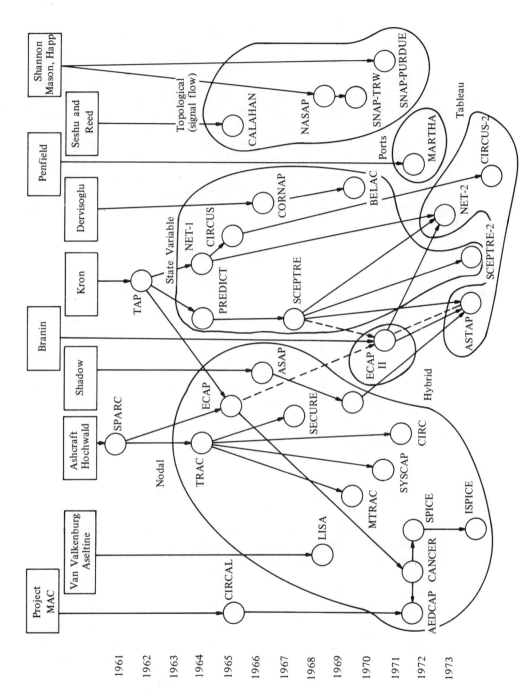

Figure 1.2 Genealogy of Circuit Analysis Programs.

The programs MTRAC [97], SECURE, SYSCAP [102, 103], and CIRC [77] were spawned from TRAC. MTRAC, developed by Nitzan at Stanford, is an extended version of TRAC which includes magnetic devices. SECURE is an improved version of TRAC which could handle much larger circuits, allowed "black-box"-type modeling, and processed internally stored functional blocks or subsystems. SECURE is available only on a need-to-know basis. SYSCAP, a streamlined version of TRAC which was released in 1970, was made available on the time-share Cybernet system by CDC through a licensing arrangement with Rockwell. SYSCAP now includes nonlinear dc, small-signal ac, and nonlinear transient modes of analyses. CIRC was developed at Xerox by Richard McNair, who was a member of the team that developed the original version of TRAC. CIRC runs on the SIGMA series machines of Xerox and can perform nonlinear dc, small-signal ac, and nonlinear transient analysis.

The ECAP program was developed by IBM in Los Angeles under the direction of Howell Tyson, who incorporated the dc portion of TAP, an ac analysis program written by Herb Wall in Washington, D.C., and his own transient analysis program to form an integrated circuit analysis program [36]. The transient portion of ECAP included the implicit integration technique first introduced in SPARC.

ECAP became the most widely used circuit analysis program; it is estimated that the number of copies of the program is in the thousands, being implemented on virtually every make of digital computer around. It is still being used in a number of academic institutions and industrial companies.

ECAP became the starting point for the development of the CANCER program at Berkeley under the direction of Rohrer in the late 1960s [96]. Later CANCER was combined with CIRCAL [20, 32, 80] to form AEDCAP [70], a popular interactive circuit analysis program. CIRCAL, developed as part of project MAC [15] at M.I.T., established the basic data structure principles underlying the interactive circuit analysis programs.

After the development of CANCER at Berkeley, another group of CAD programs were developed for instructional purposes [84, 92, 95]. One in particular, SPICE [95], proved to be very popular and has been widely distributed throughout the country. It has also recently been converted into an interactive program, called ISPICE [85], which has been made available through National CSS.

The LISA program was developed at the IBM San Jose facility by Decker and Johnson [78] to satisfy the needs of circuit analysts who were trained in the traditional Laplace transform theory and application approach taught in the electrical engineering departments at that time. LISA drew upon the fundamental work of Van Valkenburg [37] and Aseltine [1] and the numerical techniques of Crout [42] and Muller [50]. LISA represents an alternative approach to the computation of transfer functions and poles and zeros expounded by the topological signal flow methods.

ASAP was the first program written to solve circuit equations symbolically and to compute tolerance bands by Monte Carlo "simulation." The development of ASAP was originally started by Art Kennard in 1962 to perform dc Monte Carlo analysis on an IBM 7094.

ASAP was based on the work done at M.I.T. in the Shadow Project [9], which was devoted to the recognition of symbol patterns within the FORTRAN language. ASAP was probably the first program to allow a free-format input language. Later, under the direction of Hooks, ASAP was converted to the IBM 360 and rewritten to include ac analysis [99]. This expanded program, called PANE [99], was released in 1969. PANE is still being maintained and improved by IBM at their Boulder, Colorado facility. ASAP and PANE paved the way for adoption of statistical techniques in other circuit analysis programs.

State Variables

Three circuit analysis programs based on the state variable formulation first appeared in the mid-1960s. Two of them, PREDICT [71] and NET-1 [91], which were influenced by the TAP program, were funded by the government to model transient radiation effects in electronic circuits. Another program, CORNAP, was developed at Cornell University by Pottle [101] to provide both the frequency domain characteristics of a network and the transient response.

PREDICT and NET-1 did not receive wide acceptance by the engineering community because both programs had a large portion of their code written in machine assembly language, thereby restricting their operation to IBM 7090/94 computers. NET-1 moreover required that the user control the program at the operator's console, a practice contrary to the closed-shop philosophy of most computing centers. PREDICT, on the other hand, suffered from being too large to fit into the core, which was typically 32,000 words at that time. Both programs started to tackle the large circuit analysis problem with the associated large computation time, providing great insight as to what modifications and improvements had to be made.

CORNAP, which was developed from the work by Dervisoglu [11] at the Coordinated Science Laboratory at the University of Illinois, was initially extremely fast computationally but was fraught with roundoff error problems. This was later corrected by the addition of the LU decomposition method, a scheme that proved to be so successful that it was adopted by virtually all subsequent CAD programs [104].

CORNAP was initially taken in its entirety and embedded into a BELAC program [83], developed at General Electric, to perform part of their analysis requirements. In recent years BELAC has been modified considerably to include many high-frequency features, such as four-terminal characteristics and transmission line modeling. BELAC, like many other circuit analysis programs, is constantly being changed and improved.

When the difficulties of PREDICT were realized, it was decided to completely revamp it to make it transportable to other machines. The SCEPTRE program was the result [72, 73, 74].

SCEPTRE, like PREDICT, was designed to model transient radiation effects on circuits. It was written entirely in FORTRAN IV and had an elaborate overlay structure to assure that it could fit into the available computer. It also provided the user with a choice of three integration routines, an implicit integration scheme, an exponential integration method to handle circuits with widely varying time constants, and the traditional Runge-Kutta constant step integration method [44, 54]. SCEPTRE allowed FORTRAN statements to be embedded into the input language, which allowed, for the first time, great flexibility in nonlinear modeling.

Other features were added to SCEPTRE over the years to improve it. The additional features have resulted in a much more powerful program, called SCEPTRE-2, which was released in 1973 as a new version of SCEPTRE.

SCEPTRE-2 includes a model library capability (although the models that cannot be called directly by the program, model parameters are available for a number of diode and transistors and can be input to the program), Gear's method of integration [47], and a small-signal ac analysis capability. Block-box modeling and transfer function relationships can also be accommodated by the program.

Another popular state variable program that was developed by the Boeing Company during this period was the CIRCUS program [94]. CIRCUS was also funded to model radiation effects in networks. It was written entirely in FORTRAN, which made it very easy to adapt to almost any computer system. CIRCUS provided a dc initial condition solution and transient analysis of a discrete component network. The program could also handle four-region devices and accommodate a device library. It used the exponential integration method to compute transient analysis [51].

Topological (Signal Flow)

One of the first circuit analysis programs written based on a topological formulation was called CALAHAN [76], named after its developer Donald Calahan of the University of Michigan. Calahan used the tree-enumeration approach, described by Seshu and Reed [34], as the theoretical basis of his program.

Although Calahan's program did receive wide distribution, it did not receive wide acceptance except as a pedagogical program for teaching circuit theory. It could accommodate no more than 20 linear circuit elements because the cost of analyzing larger circuits was prohibitive.

A short time after Calahan's program appeared, the NASAP program was developed as a cooperative effort of universities and institutions across the country [93, 98]. NASAP, which was funded by NASA Electronics Research Center in Cambridge, Massachusetts, was based on a signal (flowgraph) formulation first proposed by Shannon [35] and Mason [25]—and later Happ [18], who developed the algorithm for its implementation from a node-element description. The major difficulty with NASAP was that, like CALAHAN, it could solve only small circuits—consisting of 30 linear elements or less.

In 1969 NASAP was converted to provide an interactive mode of operation at TRW to allow a user to reduce a circuit flowgraph to a simpler form before the application of the transfer function formula is made [10]. This program was renamed SNAP (Symbolic Network Analysis Program). Professor P. M. Lin, at Purdue University, also developed a program, called SNAP, which utilized the signal flowgraph method to formulate a transfer function [89].

The topological signal flowgraph approaches have not gained wide acceptance because of their inability to handle large networks and nonlinear elements [24].

Hybrid Formulation

ECAP-II was developed from the hybrid formulation presented by Branin et al. [7, 75, 81]. It has no relationship to ECAP other than its name and parent company. ECAP-II drew upon some of the techniques which were being developed at IBM and which later appeared in the ASTAP program.

ECAP-II can perform both a nonlinear dc and nonlinear transient analysis of a circuit of very large size. Two thousand element problems have been solved by ECAP-II with respectable execution times and accuracy. Branin incorporated Gear's method for the integration routines to improve the accuracy of solution and to overcome computational instability difficulties in conventional integration methods. Sparse matrices were also incorporated to reduce run time (and thus accuracy) and storage. Branin also developed a highly efficient and accurate method for solving nonlinear equations [40].

A number of convenient data management techniques were incorporated into ECAP-II to provide for a great deal of its flexibility. Dynamic storage allocation capabilities were employed to eliminate the restrictions placed on the number of circuit elements allowed. The limits were determined by the available storage of the computing system, not by the number of elements or nodes. Models could also be defined by a user and stored in a library. They could also be modified or deleted from the library. All these operations could be requested by a user with the input language of ECAP-II without knowing data management techniques.

Ports

MARTHA is written in APL [100]. It has been designed for high-frequency circuits, which for the most part are ignored by the other CAD programs with the possible exception of BELAC. MARTHA does, however, offer a new interactive approach to computing. As with any APL-based program, the solution of a large problem can be expensive.

An APL version of ECAP, called ECAPL [86], is the only other general-purpose APL-based program reported in the literature.

Tableau Method

In 1969 the paper presented by Hachtel et al. introduced the tableau sparse matrix approach [16, 17] to solving large network problems. This paper has had a major inpact on the development of the second-generation programs. Sparse matrix techniques have been employed in the past in a number of circuit analysis programs [33, 53, 77, 88, 102], principally to reduce the core requirements by storing only the nonzero terms. What made the tableau approach significantly different from the previous hybrid or state variable approach is that the matrices are not reduced in size by combining the algebraic equations with the differential equations but are kept in their original sparse form. It was found that the number of subsequent operations on the orginal matrices was less than those resulting from first reducing the matrices to a lower order and then performing the calculations required. It is the tableau/sparse matrix approach which has made it possible to solve the large LSI network problems. The tableau method is employed in ASTAP [69]. NET-2 [90] and CIRCUS-2 [79] have also incorporated some form of the tableau or sparse matrix methods in their formulations.

The ASTAP is probably the most powerful program in use today. It can provide nonlinear dc, small-signal ac, and nonlinear transient analysis of very large circuits whose elements can be a function of any other voltage, current or user-defined parameter. A 3500-element problem has been solved with ASTAP.

NET-2 was written by Malmberg, who developed the original version of NET-1. NET-2 provides nonlinear dc, small-signal ac, and nonlinear transient analyses of circuits consisting of no more than 500 elements or 500 nodes. It also allows a user to conduct optimization analysis in all three modes of analysis. NET-2, like NET-1, models radiation effects in solid-state devices.

CIRCUS-2 is another radiation effects program which can perform nonlinear dc and transient analysis of large networks. It can also provide a convolution of rational functions in the transient mode of operation. Its circuit size is limited by the computer system. CIRCUS-2, like NET-2 and ASTAP, has provisions for the establishment of a model library by a user.

1.5 ORGANIZATION OF THE HANDBOOK

Nine circuit analysis programs are described in this handbook. The programs presented are ASTAP, BELAC, CIRC, CIRCUS-2, ECAP-II, LISA, MARTHA, SCEPTRE-2, and SYSCAP. Each program description is self-contained, in that most of the circuit analysis problems that can be handled by a given program can be coded from the descriptions contained herein, run, and debugged and the results interpreted without any reference to outside sources. Each chapter contains a complete program description, which, in most cases, includes the program structure, a comprehensive description of the program's allowable network elements, the input language, the output specifications, modeling techniques, program application examples, limitations, error diagnostics, and the theoretical references upon which the program was developed. Because of space limitations, the error diagnostics sections for some programs have been eliminated from the handbook. In those cases, reference to the diagnostic sections of the program documentation has been included.

In using the analysis programs, a single program is usually adequate to obtain the desired results. If this proves not to be the case, a judicious combination of programs presented within the handbook should be sufficient. A summary of the characteristics of the nine programs is given in Table 1.1.

The table lists the analysis capabilities of each program and states whether they can perform linear or nonlinear (NL) analyses. The basic network formulations and computational techniques employed in the analysis phases are also included in the table. The program restrictions presented have been extracted from user manuals or from original program documentation. The modeling capabilities are grouped into two categories: models already built into the program, and those which allow a user to define models and, in most cases, store them in a library for recall.

Table 1.1 *Summary of Program Characteristics*

Program (Issue Date)	Analysis Capability	Network Formulation	Computational Techniques	Models BUILT-IN	Models USER-DERIVED

ASTAP (1973)	dc (NL)
	  Monte Carlo
	ac
	  Monte Carlo
	  Transfer functions
	    ABCD
	    Scattering
	    Hybrid (h)
	    Hybrid (g)
	    Admittance (y)
	    Impedance (z)
	  Bode response
	tr (NL)
	  Monte Carlo
		Hybrid tableau	Sparse matrices
			Gear's method
				Diode function
				Equation can be defined
					Library can be formed
					Nesting is allowed

BELAC	dc (NL)
	  Transfer characteristics
	  Driving point characteristics
	  Monte Carlo
	  Sensitivities
	  Worst case
	ac
	  Sensitivities
	  Worst case
	  Monte Carlo
	  Poles/zeros
	  Root locus
	  Bode response
	  Optimization
	tr (NL)
		State variable	Embedded CORNAP
			Gear's method
				CIRCUS diode and transistor models	Yes

CIRC (1970–1971)	dc (NL)
	  Worst case
	  Sensitivity
	ac
	  Real & imaginary
	  Bode response
	  Open loop
	tr
	  Auto initial conditions
		Nodal	Newton-Raphson
			Sparse matrices
				Ebers-Moll diode and transistor	Yes—not stored in library

CIRCUS–2	dc (NL)
	  Parameter variation
	tr (NL)
	  Convolution of rational functions
		Tableau	Sparse matrices
			LU decomposition
			Newton's method
			Rosenbroch's method
				None	Model library can be built

Program	Program Restrictions	Comments
ASTAP	System restricted	Turn-on transient analysis can be obtained Failure modes can be determined dc operating point can be passed to ac and transient analysis Over 3500-element problem has been solved
BELAC	64 elements 51 nodes 31 reactive elements	Time delays can be modeled Has been converted to conversational mode
CIRC	500 nodes and 100 transistors 630 nodes in ac 400 nodes and 80 transistors in dc worst case	dc, ac, and tr calculations are separate Initial conditions cannot be input into tr analysis
CIRCUS–2	System limited	Subroutines can be called Functions of network behavior can be defined Symbolic variables can be defined in a model Initial conditions can be user-supplied or computed

Table 1.1—*Cont.*

Program (Issue Date)	Analysis Capability	Network Formulation	Computational Techniques	Models Built-in	Models User-derived
ECAP–II (1970)	dc (NL) Parameter variation tr (NL)	Hybrid	Gear's method Sparse matrices Newton's method	Diode	Model library can be built Models can be nested
LISA (1968)	Poles/zeros Variation of poles and zeros to component variation Root locus Time response Bode response Function sensitivities	Nodal	Cramer's rule Muller's method Gaussian elimination	None	None

MARTHA (1971)	ac
	  Real & imaginary
	  Nichols and Smith chart
	  Bode response
	  Impedance functions
	  Admittance functions
		Ports	Adjoint method
			Tellegen's theorem
				Bipolar transistor
				FET
				Operational amplifier
					Yes
					A library can be constructed

SCEPTRE–2 (1973)	dc (NL)
	  Optimization
	  Worst case
	  Monte Carlo
	  Sensitivity
	ac
	  Optimization
	  Worst case
	  Monte Carlo
	  Sensitivity
	  Nyquist
	  Bode response
	tr (NL)
	  Convolution
		State space	Sparse matrices
			Newton-Raphson
			Gear's method
			Exponential, Range-Kutta, and trapezoidal integration
			Adjoint method
				Diode equation	Device library can be user-derived

SYSCAP (1970)	dc (NL)
	  Parameter variation
	  Circuit failure
	  Worst case
	  Sensitivity
	  Monte Carlo
	  Power supply turn-on
	ac
	  Bode response
	  Nichols plots
	tr (NL)
	  Fourier analysis
		Nodal	Sparse matrices
			Implicit integration
			Newton-Raphson
				Ebers-Moll diode and transistor
				Zener diode
				FET
				Transformer
					Data bank is available

Program	Program Restrictions	Comments
ECAP–II	Over 2000 element problems have been solved System limited	No ac analysis Dynamic storage allocation is used Initial conditions can be user-supplied or computed
LISA	50 nodes 50 equations	Input may include transfer functions, matrix equations, and block diagrams
MARTHA	System limited	An APL-based program—especially adept at handling high-frequency and microwave problems
SCEPTRE–2	300 elements 301 nodes 50 mutual inductances 50 convolution kernels	Elements as functions of frequency are not permitted Initial conditions can be input or computed by either Newton-Raphson or Gear's method
SYSCAP	255 nodes 500 elements	Initial conditions can be inserted by user or computed

1.6 REFERENCES

Network Theory and Formulation Techniques

1. Aseltine, John, *Transform Method in Linear System Analysis*, McGraw-Hill, New York, 1958.
2. Ashcraft, W. D., and W. Hochwald, "Design by Worst-Case Analysis: A Systematic Method to Approach Specified Reliability Requirements," *IRE Trans. Reliability and Quality Control*, RQC-10, Nov. 1961, pp. 15–21.
3. Bashkow, T. R., "The A Matrix, New Network Concept," *IRE Trans. Circuit Theory*, CT-4, Sept. 1957, pp. 117–119.
4. Berry, R. D., "An Optimal Ordering of Electronic Circuit Equations for a Sparse Matrix Solution," *IEEE Trans. Circuit Theory*, Col. CT-18, Jan. 1971, pp. 40–50.
5. Branin, Jr., F. H., "dc and Transient Analysis of Networks Using a Digital Computer," *IRE International Convention Record*, 10, Part 2, 1962, pp. 236–256.
6. Branin, Jr., F. H., "Computer Methods of Network Analysis," *Proc. IEEE*, 55, Nov. 1967, pp. 1787–1801.
7. Branin, Jr., F. H., and L. E. Kugel, "The Hybrid Method of Network Analysis," *IEEE Symp. Circuit Theory*, San Francisco, Calif., Dec. 1969.
8. Bryant, P. R., "The Explicit Form of Bashkow's A Matrix," *IRE Trans. Circuit Theory*, CT-9, Sept. 1962, pp. 303–306.
9. Carter, E. J., and R. P. Futrelle, *Symbol Pattern Recognition with FORTRAN—The Shadow III System*, M.I.T. SSMTG Program Note No. 39, 1962.
10. Demari, A., "On-Line Computer Active Network Analysis and Design in Symbolic Form," *Proc. 2nd Cornell Electrical Engineering Conference*, Aug. 1969, pp. 94–106.
11. Dervisoglu, A. "State Models of Active RLC Networks," *Coordinated Science Laboratory Report R-237*, Univ. of Illinois, Urbana, Ill., 1964.
12. Director, S. W., and R. A. Rohrer, "The Generalized Adjoint Network and Network Sensitivities," *IEEE Trans. Circuit Theory*, CT-16, Aug. 1969, pp. 318–323.
13. Director, S. W., and R. A. Rohrer, "Automated Network Design: The Frequency-Domain Case," *IEEE Trans. Circuit Theory*, CT-16, Aug. 1969, pp. 330–336.
14. Falk, H., "Computer Programs for Circuit Design," *Electro-Technology*, March 1964, pp. 101–104, 162–164.
15. Fano, R. M., "The MAC System: The Computer Utility Approach," *IEEE Spectrum*, 2, Jan. 1965, pp. 55–64.

16. HACHTEL, G. D., R. BRAYTON, and F. GUSTAVSON, "A Sparse Matrix Approach to Network Analysis," *Proc. 2nd Biennial Cornell Electrical Engineering Conference,* Cornell Univ., Ithaca, N.Y., Aug. 26–28, 1969, pp. 68–82.

17. HACHTEL, G. D., R. K. BRAYTON, and F. G. GUSTAVSON, "The Sparse Tableau Approach to Network Analysis and Design," *IEEE Trans. Circuit Theory,* CT-18, No. 1, Jan. 1971, pp. 101–113.

18. HAPP, W. W., "Flowgraph Techniques for Closer Systems," *IEEE Trans. Aerospace and Electronics Systems,* AES-2, May 1966, pp. 252–264.

19. JENSEN, R. W., and B. O. WATKINS, *Network Analysis: Theory and Computer Methods,* Prentice-Hall, Englewood Cliffs, N.J., 1974.

20. JESSEL, G. P., and J. R. STINGER, "CIRCAL-2: A General Approach to On-line Circuit Analysis," *1969 NEREM Record,* Vol. 11, Nov. 1969, pp. 30–31.

21. KRON, G., *Tensor Analysis of Networks,* Wiley, New York, 1939.

22. KRON, G., "A Set of Principles To Interconnect the Solutions of Physical Systems," *J. Applied Physics,* 24, 1953, pp. 965–980.

23. KUH, E. S., and R. A. ROHRER, "The State-Variable Approach to Network Analysis," *Proc. IEEE,* 53, July 1965, pp. 672–686.

24. LIN, P. M., "A Survey of Applications of Symbolic Network Functions," *IEEE Trans. Circuit Theory,* CT-20, No. 6., Nov. 1973, pp. 732–737.

25. MASON, S. J., "Feedback Theory: Further Properties of Signal Flow Graphs," *Proc. IRE,* 44, No. 7, July 1956, pp. 920–926.

26. MASON, S. J., "Topological Analysis of Linear Non-Reciprocal Methods," *Proc. IRE,* 45, June 1957, pp. 829–838.

27. MEYERHOFF, D. E., "Investigation and Evaluation of General Purpose Digital Computer Circuit/Analysis Programs," Master's Thesis, Engineering Department, Univ. of California, Los Angeles, Calif., 1969.

28. NORIN, R. S., and C. POTTLE, "Effective Ordering Of Sparse Matrices Arising from Nonlinear Electrical Networks," *IEEE Trans. Circuit Theory,* CT-18, No. 1, Jan. 1971, 139–145.

29. PENFIELD, JR., P., "Description of Electrical Networks Using Wiring Operators," *Proc. IEEE,* 60, No. 1, Jan. 1972, pp. 49–53.

30. PENFIELD, JR., P., R. SPENCE, and S. DUINKER, *Tellegen's Theorem and Electrical Networks,* M.I.T. Press, Cambridge, Mass., 1970, Appendix D.

31. POTTLE, C., "State-Space Techniques for General Active Network Analysis," in *System Analysis by Digital Computer,* F. F. Kuo and J. F. Kaiser, eds., Wiley, New York, 1966, Chap. 3.

32. RODRIGUEZ, J. E., and R. A. ROHRER, "CIRCAL-2: A Computer-Aided Environment for the Design of Electrical Networks," *1970 NEREM Record,* 12, Nov. 1970, pp. 185–189.

33. SATO, N., and W. F. TINNEY, "Techniques for Exploiting the Sparsity of the Network Admittance Matrix," *IEEE Trans. Power and Apparatus,* 82, Dec. 1963, pp. 944–950.

34. SESHU, S., and M. B. REED, *Linear Graphs and Electrical Networks*, Addison-Wesley, Reading, Mass., 1961.

35. SHANNON, C. E., "The Theory and Design of Linear Differential Equation Machines," National Defense Research Committee, *OSRD Report 411*, Jan. 1942.

36. TYSON, H. N., G. R. HOGSETT, and D. A. NISEWANGER, "The IBM Electronic Circuit Analysis Program (ECAP)," *Proc. 1966 Annual Symposium on Reliability*, Jan. 1966, pp. 45–65.

37. VAN VALKENBURG, M. E., *Introduction to Modern Network Synthesis*, Wiley, New York, 1960.

38. WEEKS, W. T., et al., "Algorithms for ASTAP: A Network Analysis Program," *IEEE Trans. Circuit Theory*, CT-20, No. 6, Nov. 1973, pp. 628–634.

39. WILSON, R. L., and W. A. MASSENA, "An Extension of Bryant-Bashkow A Matrix," *IEEE Trans. Circuit Theory*, CT-12, March 1965, pp. 120–122.

Numerical Techniques

40. BRANIN, JR., F. H., "A Controlled-Step Newton Method for Solving Nonlinear Equations," *Digest Record, ACM-SIAM-IEEE Conference on Mathematical and Computer Aids to Design*, Anaheim, Calif., Oct. 1969, p. 378.

41. CERTAINE, J., "The Solution of Ordinary Differential Equations with Large Time Constants," in *Mathematical Methods for Digital Computers*, A. Ralston and H. S. Wolf, eds., Wiley, New York, 1960, Chap. 11.

42. CROUT, P. D., "A Short Method for Evaluation, Determinants and Solving Systems of Linear Equations with Real or Complex Coefficients," *AIEE Trans.*, 60, 1941, pp. 1235–1241.

43. FORSYTHE, G., and C. B. MOLER, *Computer Solutions of Linear Algebraic Systems*, Prentice-Hall, Englewood Cliffs, N.J., pp. 27–33.

44. FOWLER, M. E., and R. M. WARTEN, "A Numerical Technique for Ordinary Differential Equations with Widely Separated Eigen Values," *IBM J. Research and Development*, 11, No. 5, Sept. 1967, pp. 537–543.

45. FRAME, J. S., "Matrix Functions and Applications, Part IV: Matrix Functions and Constituent Matrices," *IEEE Spectrum*, June 1964, pp. 123–131.

46. FRANCIS, J. G. F., "The Q-R Transformation—I," *Computer J.*, 4, Oct. 1961, pp. 265–271; "The Q-R Transformation—II" *Computer J.*, 4, Jan. 1962, pp. 332–345.

47. GEAR, C. W., "The Automatic Integration of Stiff Ordinary Differential Equations," in *Information Processing 68*, A. H. H. Morell, ed., North-Holland, Amsterdam, 1969, pp. 186–193.

48. HOUSEHOLDER, A. S., *Principles of Numerical Analysis*, McGraw-Hill, New York, 1963.

49. KAPLAN, W., *Ordinary Differential Equations*, Addison-Wesley, Reading, Mass., 1958, pp. 402–403 (Heun's method).

50. MULLER, D. E., "A Method for Solving Algebraic Equations Using An Automated Computer," *MTAC*, 1956, pp. 208–215.

51. POPE, D. A., "An Exponential Method of Numerical Integration of Ordinary Differential Equations," *Commun. ACM*, 6, No. 8, Aug. 1963, pp. 491–493.

52. ROSENBROCK, H. H., "Some General Processes for the Numerical Solution of Differential Equations," *Computer J.*, 5, No. 4, Jan. 1963, pp. 329–330.

53. TINNEY, W. F., and J. W. WALKER, "Direct Solutions of Sparse Network Equations by Optimally Ordered Triangular Factorization," *Proc. IEEE*, 55, Nov. 1967, pp. 1801–1809.

54. WARTEN, R. M., "Automatic Step-Size Control for Runge-Kutta Integrater," *IBM J.*, 7, Oct. 1963, pp. 340–341.

Modeling

55. BEAUFOY, R., and J. J. SPARKES, "The Junction Transistor as a Charge-Controlled Device," *Automatic Telephone and Electric Co. J.*, 13, Oct. 1957, pp. 310–327.

56. CORDWELL, W. A., "Transistor and Diode Model Handbook," *USAF Technical Report AFWL-TR-69-44*, Air Force Weapons Laboratory, Kirkland AFB, N.M., Oct. 1969.

57. DANIEL, M. E., "Development of Mathematical Models of Semiconductor Devices for Computer-Aided Circuit Analysis," *Proc. IEEE*, 55, No. 11, Nov. 1967, pp. 1913–1920.

58. DIERKING, W. H., and C. T. KLEINER, "Phenomenological Magnetic Core Model for Circuit Analysis Programs," *IEEE Trans. Magnetics*, Sept. 1972, pp. 594–596.

59. EBERS, J. J., and J. L. MOLL, "Large Signal Behavior of Junction Transistors," *Proc. IRE*, 42, Dec. 1954, pp. 1761–1772.

60. FROHMAN-BENTCHKOWSKY, D., and L. VADASZ, "Computer-Aided Design and Characterization of Digital MosIntegrated Circuits," *IEEE J. Solid State Circuits*, SC-9, April 1969, pp. 57–64.

61. GUMMEL, H. K., and H. C. POON, "An Integral Charge Control Model of Bipolar Transistors," *Bell System Technical J.*, 49, No. 5, May-June 1970, pp. 827–852.

62. GUMMEL, H. K., and H. C. POON, "Modeling of Emitter Capacitance," *Proc. IEEE*, 57, No. 12, Dec. 1969, pp. 2181–2182.

63. LINVILL, J. G., "Lumped Models of Transistors and Diodes," *Proc. IRE*, 46, No. 6, June 1958, pp. 1141–1152.

64. LYNN, D. K., C. S. MEYER, and D. J. HAMILTON, *Analysis and Design of Integrated Circuits*, McGraw-Hill, New York, 1967.

65. PHILLIPS, A. B., *Transistor Engineering*, McGraw-Hill, New York, 1962, p. 298.

66. POCOCK, D. N., M. G. KREBS, and C. W. PERKINS, "Simplified Microcircuit Modeling," *Air Force Weapons Laboratory Technical Report AFWL-TR-73-272*, Air Force Weapons Laboratory, Kirkland AFB, N.M., 1973.

67. RAYMOND, J. P., and R. E. JOHNSON, "Study of Generalized, Lumped Transistor Models for Use with SCEPTRE," *USAF Technical Report AFWL-TR-68-86*, Air Force Weapons Laboratory, Kirkland AFB, N.M., March 1969.

68. SCHICHMAN, H., and D. A. HODGES, "Modeling and Simulation of Insulated-Gate-Field-Effect Transistor Switching Circuits," *IEEE J. Solid State Circuits*, SC-3, Sept. 1968, pp. 285–289.

Program Documentation

69. "Advanced Statistical Analysis Program (ASTAP)," IBM Corp-Data Processing Division, White Plains, N.Y., Installed User Program, SH 20-1118-X.

70. "AEDCAP User's Guide," *Report A-CIR-000-4*, SofTech, Inc., Waltham, Mass., 1973.

71. "Automated Digital Computer Program for Determining Responses of Electronic Systems to Transient Nuclear Radiation (PREDICT)," IBM Space Guidance Center, Oswego, N.Y., IBM File 64-521-5, July 1964.

72. "Automated Digital Computer Program for Determining the Response of Electronic Systems for Transient Nuclear Radiation (SCEPTRE)," *Air Force Weapons Laboratory Technical Report AFWL-TR-69-77*, Kirkland AFB, N.M., 1969.

73. BECKER, David, et al., "SCEPTRE (Improved)," *USAF Technical Report AFWL-TR-73-75*, Air Force Weapons Laboratory, Kirkland AFB, N.M. 1973.

74. BOWERS, J. C., and S. R. SEDORE, *SCEPTRE: A Computer Program for Circuit and Systems Analysis*, Prentice-Hall, Englewood Cliffs, N.J., 1971.

75. BRANIN, Jr., F. H., "ECAP-II—A New Electronic Circuit Analysis Program," *IEEE J. Solid State Circuits*, SC-6, Aug. 1971, pp. 146–166.

76. CALAHAN, D. A., "Linear Network Analysis and Realization Digital Computer Programs and Instruction Manual," *Univ. of Illinois Bulletin*, 62, No. 58, Feb. 1965.

77. "CIRC-DC Reference Manual and User's Guide," *No. 901697*; CIRC-AC Reference Manual and User's Guide, *No. 901698*; CIRC-TR Reference Manual and User's Guide, *No. 901786*, Xerox Corporation, El Segundo, Calif.

78. DECKER, K. L., and E. T. JOHNSON, "User's Guide for LISA/360, A Program for Linear System Analysis," *IBM Technical Report TR 02.432*, San Jose, Calif., July 31, 1968.

79. DEMBART, B., and L. MILLIMAN, "CIRCUS-2—A Digital Computer Program for Transient Analysis of Electronics Circuits—User's Guide," Boeing Co., Harry Diamond Laboratory, July 1971.

80. DERTOUZOS, M. L., "CIRCAL: On-Line Circuit Design," *Proc. IEEE*, 55, May 1967, pp. 637–654.

81. "ECAP-II Application Description Manual" (GH20-0983), IBM Corp., Mechanicsburg, Pa, 1971.

82. "1620 Electronic Circuit Analysis Program (ECAP)," IBM Corporation Data Processing Division, White Plains, N.Y., Application Program 1620-EE-02X, 1965.

83. GEORGE, C. A., "BELAC User's Manual," *Technical Information Service R69 EMLII*, General Electric Co., Utica, N.Y., 1969; also updated R72ELI, 1972.

84. IDLEMAN, T. E., F. S. JENKINS, W. J. MCCALLA, and D. O. PEDERSON, "SLIC—A Simulator for Linear Integrated Circuits," *IEEE J. Solid State Circuits*, SC-6, Aug. 1971, pp. 188–203.

85. *ISPICE User's Manual*, National CSS, Stamford, Conn.

86. JENSEN, R. W., T. D. HIGBEE, and P. M. HANSEN, "ECAPL: An APL Electronic Circuit Analysis Program," *Proc. 4th International APL User's Conference*, June 15–16, 1972, Atlanta, Ga., pp. 161–190.

87. JENSEN, R. W., and M. D. LIEBERMAN, *IBM Electronic Circuit Analysis Program: Techniques and Applications*, Prentice-Hall, Englewood Cliffs, N.J., 1968.

88. JOHNSON, E. D., C. T. KLEINER, L. R. MCMURRAY, E. L. STEELE, and F. A. VASSALLO, *Transient Radiation Analysis by Computer Program* (*TRAC*), Autonetics Div., Rockwell International, Harry Diamond Laboratory Technical Report, June 1968.

89. LIN, P. M., and G. E. ALDERSON, "SNAP—A Computer Program for Generating Symbolic Network Functions," *School of Electrical Engineering Report TR-EE70-16*, Purdue Univ., Aug. 1970.

90. MALMBERG, A. F., "User's Manual—NET-2 Network Analysis Program," *Harry Diamond Laboratory Report AD-752-600*, Braddock, Dunn, and McDonald, Sept. 1972.

91. MALMBERG, A. F., F. L. CORNWALL, and F. N. HOFNER, "NET-1 Network Analysis Program," *Rept. LA-3119*, Los Alamos, N.M., 1964.

92. MCCALLA, W. J., and W. G. HOWARD, JR., "BIAS-3—A Program for the Nonlinear dc Analysis of Bipolar Transistor Circuits," *IEEE J. Solid State Circuits*, SC-6, Feb. 1971, pp. 14–19.

93. MCNAMEE, L. P., and H. POTASH, "A User's and Programmer's Manual for NASAP," *Report 68–38*, Univ. of California, Los Angeles, Aug. 1968.

94. MILLIMAN, L. D., W. A. MASSENA, and R. H. Dickhaut, "CIRCUS—A Digital Computer Program for Transient Analysis of Electronic Circuits-User's Guide," Boeing Co., Harry Diamond Laboratory, July 1971.

95. NAGEL, L. W., and D. O. PEDERSON, "SPICE Electronics Research Laboratory Report," College of Engineering, Univ. of California, Berkeley, Ca., April 1973.

96. NAGEL, L., and R. ROHRER, "CANCER: Computer Analysis of Nonlinear Circuits Excluding Radiation," *IEEE J. Solid State Circuits*, SC-6, Aug. 1971, pp. 166–182.

97. NITZAN, D., and J. R. HERNDON, "MTRAC—A Computer Program for Analysis of Circuits Including Magnetic Cones," *SRI Project 6408, Report 6*, Stanford Research Institute, Menlo Park, Calif., June 1969.

98. OKRENT, H., and L. P. MCNAMEE, "NASAP-70 User's and Programmer's Manual," *Technical Report No. 7044*, Univ. of California, Los Angeles, School of Engineering and Applied Science, Contract NAS 12-2138, Sept. 1970.

99. "PANE-Performance Analysis of Electrical Networks," *No. 360–16.4.003*, IBM Corp., 1969, Hawthorne, New York.

100. PENFIELD, JR., P. "MARTHA User's Manual," M.I.T. Press, Cambridge, Mass., 1971; addendum, 1973.

101. POTTLE, C., "CORNAP User Manual," School of Electrical Engineering, Cornell Univ., Ithaca, N.Y., 1968.

102. "SYSCAP—Teletype User Guide Preliminary," Control Data Corp., Feb. 1973.

103. "SYSCAP: User Information Manual," *Nos. D000135/9012/22/32*, 3 Volts Publications, Control Data Corp., Sept. 1970.

Textbooks

104. CALAHAN, D. A., *Computer-Aided Network Design*, rev. ed., McGraw-Hill, New York, 1972.

105. HERSKOWITZ, G. J., ed., *Computer-Aided Integrated Circuit Design*, McGraw-Hill, New York, 1968.

106. HERSKOWITZ, G. J., and R. B. SCHILLING, eds., *Semiconductor Device Modeling for Computer-Aided Design*, McGraw-Hill, New York, 1972.

107. KUO, F. F., and J. F. KAISER, eds., *System Analysis by Digital Computer*, Wiley, New York, 1966.

108. KUO, F. F., and W. G. MAGNUSON, JR., eds., *Computer Oriented Circuit Design*, Prentice-Hall, Englewood Cliffs, N.J., 1969.

109. WOLFENDALE, E., ed., *Computer-Aided Design Techniques*, Butterworth's, London, 1970.

110. ZOBRIST, G. W., ed., *Network Computer Analysis*, Boston Technical Publishers, Cambridge, Mass., 1969.

Special Proceedings

111. *Biennial Conference Proceedings on Computerized Electronics*, Cornell Electrical Engineering Conference, Aug. 1969.

112. Special Issue on Computer-Aided Circuit Analysis and Device Modeling, *IEEE J. Solid State Circuits*, SC-6, Aug. 1971.

113. Special Issue on Computer-Aided Design, *Proc. IEEE*, 55, Nov. 1967.
114. Special Issue on Computer-Aided Design, *IEEE Trans. Circuit Theory*, CT-20, Dec. 1973.
115. Special Issue on Computer-Aided Network Design, *IEEE Trans. Circuit Theory*, CT-18, Jan. 1971.
116. Special Issue on Computer-Oriented Microwave Practices, *IEEE Trans. Microwave Theory*, MTT-17, Aug. 1969.
117. Special Issue on Computers in Design, *Proc. IEEE*, 60, Jan. 1972.
118. Special Issues on Educational Aspects of Circuit Design, *Computer IEEE Trans. Education*, E-12, Sept. and Dec. 1969.

2

ASTAP
Advanced Statistical Analysis Program

GERALD W. MAHONEY
MARTIN J. GOLDBERG

IBM Corporation
Data Processing Division
White Plains, New York

2.1 INTRODUCTION*

During the past decade IBM has developed numerous programs for computer-aided circuit design (CACD) to satisfy technology requirements. The programs developed were aimed at specific needs or were limited to a particular mode of analysis. The program described in this chapter is the culmination of a development effort within IBM to provide a general CACD program which meets the needs of CACD within a single system.

General Description

The Advanced Statistical Analysis Program (ASTAP) was developed by IBM to satisfy the circuit analysis requirements of large-scale integrated circuit technology [1, 2]. ASTAP is able to handle networks of up to thousands of elements while efficiently using machine resources.

ASTAP offers the capability of performing dc, ac, and transient analyses using a single program with a common input language. Furthermore, it has the facility for automatically passing the results of a prior dc or transient analysis to the small-signal ac analysis to establish the operating point of the nonlinear devices.

In addition, ASTAP provides a Monte Carlo statistical analysis facility which can be used in all three basic analysis modes. Distributions of circuit outputs due to input parameter variations such as component tolerances, aging, or temperature effects can be determined. With this capability, ASTAP more fully serves the needs of reliability- and quality-assurance engineers and

---

*Portions hereof reprinted by permission from *Advanced Statistical Analysis Program* (ASTAP), Program Reference Manual (SH20–1118–0), 1973, International Business Machines Corporation.

others concerned with production and field maintenance costs. Since production yields for a given set of component tolerances can be accurately predicted, it is not necessary to overdesign circuits, so that reliable circuits can be produced at lower cost.

The performance advantages and large problem size capability of the program stem in large part from the use of advanced mathematical and programming techniques. These include the use of implicit integration, sparse matrix methods, a modified tableau formulation of the network equations, and dynamic storage allocation. Some of these techniques are advances developed at the IBM Thomas J. Watson Research Center [3, 4].

Besides the major analysis facilities mentioned above, ASTAP contains many other facilities designed to provide ease and convenience to the user. These include a user-oriented input language, model library facilities, an in-line FORTRAN-like expression capability, and several system-supplied standard functions. With these facilities, the user does not need extensive knowledge of programming or the mathematics of circuit theory.

Because of its advanced functional and operational capabilities, the program can often be used to advantage in simulating network problems in other disciplines. This is facilitated through the ASTAP input language and model library, which permit electrical analogs for nonelectrical components to be stored and used in problem descriptions. The program can be of particular advantage in simulating mixed systems described by both differential equations (Laplace transfer functions) and electrical components, such as found in electromechanical systems.

Capabilities

ASTAP provides three major analysis modes: dc steady-state, ac steady-state, and transient analysis. A Monte Carlo simulation may be performed in any of the analysis modes. The input language integrates these facilities so that any type of analysis may be performed with a single circuit description. In addition, certain analysis modes may be combined so that required data from one analysis are automatically passed to another analysis. Figure 2.1 depicts the analysis combinations supported, with the arrows indicating the control paths and direction of data transfer.

dc Analysis

A dc analysis on a model named AMPLIFIER is requested using the following statement:

ANALYZE AMPLIFIER (DC)

The dc analysis finds the equilibrium or steady-state operating point of a linear or nonlinear circuit. For a circuit which contains storage elements (Cs and Ls) and so would exhibit a transient response to time-varying driving functions, the dc solution corresponds to the steady-state condition which

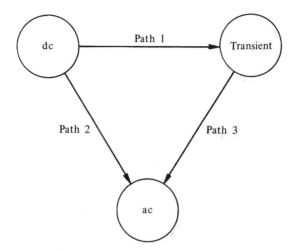

Figure 2.1 Analysis Combinations.

exists after all transients have decayed—that is, when the driving functions are removed or become constant and the response voltages and currents have settled to constant values with respect to time.

Typical outputs of a dc analysis are the steady-state levels of element voltages and currents associated with nonlinear devices (which define their operating points), node voltages, and user-defined quantities such as power dissipations.

In the input for the digital logic inverter of Fig. 2.4, which is discussed in subsequent sections, the TRANSIENT option in statement 29 requests a combined dc and transient analysis (corresponding to path 1 of Fig. 2.1). Here the initial conditions obtained by the dc analysis is automatically passed to the succeeding transient analysis. The dc analysis computes the steady-state response prior to the applications of the driving function EIN. The dc solution corresponds to the circuit state in which the power supplies, ES1 and ES2, are turned on and the transistor is at the operating point. If a dc analysis had not been performed prior to the transient analysis, the transient response would have included the turn-on transients of the circuit as well as the response to EIN. The TRANSIENT ONLY option may be used to obtain the circuit response to turn-on of the power supplies.

As indicated above, a dc solution can be obtained by performing a transient analysis and letting the solution continue until steady-state conditions are reached. In fact, the program obtains a dc solution in just this way by performing a pseudotransient analysis which is essentially transparent to the user. This approach has exhibited better reliability than the direct technique of iteratively solving only the algebraic network equations. Such purely iterative techniques may not converge to a solution on certain types of circuits, whereas in such cases the pseudotransient approach rarely fails to produce a solution.

Transient Analysis

A transient analysis produces a time domain solution, providing the transient response of a linear or nonlinear network to driving functions specified by the user. The driving functions may be time-varying or constant.

Typical outputs of a transient analysis are plots of element voltages or currents and node voltages as a function of time.

A transient analysis on a model named AMPLIFIER may be requested in one of two ways:

1. ANALYZE AMPLIFIER (TRANSIENT).
2. ANALYZE AMPLIFIER (TRANSIENT ONLY).

As discussed in the previous section, the TRANSIENT option in the first example requests a dc analysis followed by a transient analysis, where the dc analysis provides the initial conditions from the beginning (time = 0) of the transient solution. The TRANSIENT ONLY option requests only a transient analysis to be performed, in which case a particular set of initial conditions desired by the user may be specified.

A variable-step, variable-order, implicit integration algorithm (see References 4 and 5) is used to obtain a transient solution. The prime advantage of the use of implicit integration is the elimination of the so-called time constant barrier, since the size of the time step is not limited by the minimum time constant of a circuit. With explicit integration schemes, which have been used in the past, the integration step size must generally be less than the smallest circuit time constant in order to maintain numerical stability. This often gives rise to exceedingly long running times in circuits with a wide variation between the largest and smallest time constants.

The ASTAP implicit integration algorithm is much less subject to such stability problems. The program automatically adjusts the time step and the integration order so that the step size used in advancing the solution from one point in time to the next is as large as possible, consistent with the desired solution accuracy.

The implicit integration technique is combined with the use of a modified tableau formulation (see References 2 and 3) of the network equations, which fully exploits the use of sparse matrix techniques for the solution of the network equations at each time step. This combination accounts in large part for the faster transient execution speeds of ASTAP, making statistical transient analysis economically feasible for many circuits.

ac Analysis

An ac analysis produces a frequency domain solution, providing the small-signal frequency response of a linear network. Nonlinear networks are linearized at a given operating point, and the frequency response of the linearized network is computed.

The operating point for a nonlinear network may be supplied by the user, or it may be obtained automatically from a dc or transient analysis, as illustrated by paths 2 and 3 of Fig. 2.1. When a transient analysis is used to obtain the operating point, the user may specify a particular time at which the operating point is to be computed.

When the operating point is obtained automatically, the topological structure of the network for the ac analysis must be the same as that used for the dc or transient analysis. This restriction is commonly found in programs which automatically compute an operating point for ac analysis. However, with ASTAP the user may also manually input an operating point for the ac analysis through use of special language statements. The operating point data might be obtained from a prior dc analysis. With this approach, there is no restriction on the topological structure of the ac problem, permitting elements or entire sections of circuitry to be added or deleted at will.

Output which may be requested for a complex parameter or response variable includes its real component, imaginary component, magnitude, phase in degrees, and magnitude in decibels.

In addition, a user may request output of any of the transfer functions associated with seven types of N-port network parameters:

A: $ABCD$ chain parameters

F: transfer scattering parameters

H: hybrid h parameters

K: hybrid g parameters

S: scattering parameters

Y: admittance parameters

Z: impedance parameters

The user may specify an arbitrary number of ports (points of access to the circuit), except for the A, F, H, and K parameters, which are inherently two-port parameters, and each port may be associated with any pair of nodes. For example, if port 1 were specified between node IN and ground and port 2 between node OUT and ground, an output request for K21 would give the output/input voltage transfer ratio, NOUT/NIN. (The prefix N signifies a node voltage.)

Statistical Analysis

Statistical analysis simulates circuit performance variations caused by tolerance variations and other actual circuit production factors. The procedure is to simulate the selection of many sample circuits and tabulate the results, which are expected to correspond to the actual results which would be obtained from testing many physical samples. ASTAP accepts probability distributions on element values as input and yields statistical outputs such as histogram and scattergram plots of circuit performance parameters. The method used for the statistical analysis is the Monte Carlo method.

Monte Carlo Method

The basic Monte Carlo method produces the probability distributions of the performance parameters of a system from known probability distributions on the input parameters. It involves two ingredients: random element selection and performance simulation relating the inputs to the outputs. In ASTAP, circuit analysis provides the functional relationship between the input parameters and the output parameters. All the modes of analysis are available: dc, transient, and ac analysis. The procedure is simply to reanalyze the circuit many times, each time letting the varying elements (input parameters) assume random values according to their prescribed distributions.

The Monte Carlo method makes no assumptions concerning the distribution shapes themselves or the linearity of the input/output functional relations. It does, however, assume that all input parameters are statistically independent. Correlations which exist between circuit elements, resulting from the manufacturing process, for example, can be accounted for explicitly by entering these correlations, or tracking relationships, into the random selection process. The ASTAP language accepts quite general expressions for simulating tracking between elements.

The distributions themselves are input to the program as tables of points and can therefore be any arbitrary shape. The program adjusts the end points of any general distribution to the range limits for each element. The range limits are specified with the distribution associated with the element. The program performs any normalization which may be required.

Describing a Circuit for Statistical Analysis

Describing a circuit for statistical analysis is similar to describing a circuit for a nominal analysis, that is, an analysis using the specified nominal value for each input element. For a statistical analysis, distributions for the statistically varying input quantities must be supplied.

For example, under the ELEMENTS heading, circuit element values may be expressed in terms of a range and statistical distribution in addition to the other options for element values. For an example of this, see the section entitled "Global Variables." Statistical tables giving a spread to the curves with some distribution can be used to define nonlinearities. The statistical table reference is similar to a normal TABLE reference with the addition of the word STATISTICAL. The table name and arguments follow TABLE. The distribution shapes themselves, together with the statistical table definitions, are included under the FUNCTIONS heading in each model. Some additions are made also to the RUN CONTROLS and OUTPUTS to control the analysis and display the statistical results.

Further information in regard to statistical analysis input language details can be found in the sections entitled "Distribution Reference," "Statistical Table Reference," and "Statistical Run Controls."

Statistical Outputs

The Monte Carlo method can be used with any of the modes of analysis available in the program. Essentially all variables which are available as outputs from these analyses can be displayed as histograms in a statistical run. Output in the form of waveforms is displayed as envelope plots. These plots are the upper and lower envelopes of all the waveforms produced during the statistical run, together with the mean value curve plotted between them. Histograms can be obtained at sampled points of the waveform. A histogram tabulates the frequency of occurrence of the values obtained for a response variable over all cases.

Another statistical output which may be requested is called a scattergram plot. A scattergram can be requested at a given sample point on a frequency or transient response curve, equivalent to a "snapshot" at a given frequency or point in time. It produces a plot (scatter diagram) of one or more variables versus another variable over all statistical cases and can be used to visually determine correlation between response variables.

For each mode of analysis, a complete circuit solution is performed for each randomly selected case. For statistical transient runs, therefore, a complete run is made each time. A nominal run is performed prior to each statistical run.

Circuit Failure Analysis

ASTAP permits the user to specify a failure criterion for one or more output quantities, such as voltages, currents, element values, and parameter values. If any of the output quantities exceeds user-specified bounds during an analysis (a failure), the values of the output quantities under test are printed. In addition, during a statistical analysis, the values of all statistical elements and parameters are also printed when a failure occurs.

Failure criteria are specified using the TEST statement as illustrated below:

TEST IRLOAD (4, 9.9), POWER (8.6, 9.4)

For the above example, the program prints a failure report whenever IRLOAD, the current through RLOAD, exceeds the range (4, 9.9) or POWER exceeds the range (8.6, 9.4).

Circuit Modification and Rerun

Using the rerun facility, the input to a circuit may be modified and reanalyzed. In this way, parameter studies may be carried out under any of the analysis modes. The RERUNS heading is used to indicate the desired input modifications, as illustrated in the example below:

$$\text{RERUNS}$$
$$P1 = (10, 20, 30)$$
$$P2 = (30, 25)$$

Three reruns are requested above, where the first rerun changes parameter P1 to 10 and P2 to 30 (from their original values as specified under ELEMENTS). The second rerun changes P1 to 20 and P2 to 25. The third rerun changes P1 to 30 and leaves P2 set at 25. Any elements dependent on these parameters are modified accordingly for the reruns.

Run Controls

ASTAP contains over 40 run controls which can be used to control such items as the duration of an analysis (for example, STOP TIME and STOP FREQ), the resolution of the output (PLOT INTERVAL), and the accuracy of the results.

All the controls affecting the accuracy of the results have default values which for most problems provide a reasonable trade-off between solution accuracy and the computation time required. Therefore, the user normally sets only a few controls to specify the desired analysis duration and output resolution and relies on the default settings for those controls affecting the accuracy of the solution.

2.2 PROGRAM STRUCTURE

ASTAP consists of two phases: a setup phase and an execution phase. The setup phase may be thought of as a compiler which transforms the high-level ASTAP language statements into executable machine instructions. This "compilation" is done in two stages. First the ASTAP statements are translated into FORTRAN statements, which in turn are compiled into executable code. The execution phase then executes the code produced by the setup phase to perform an analysis.

The setup phase consists of the following steps:

1. Input scan. The circuit input data are read and scanned for errors. Models are retrieved or retained in the model libraries.

2. Dissolving of models. From the description of the network and desired outputs, lists containing interconnections of network elements, the type and value of each element, and the voltages and currents which are to be observed are generated.

3. Forming of matrices. A tree of the network is selected, the incidence matrix is generated in sparse form from the interconnection list, and the cutset matrix is generated from the incidence matrix corresponding to the selected tree. Kirchhoff's voltage and current law equations are formed and fill the top half of the tableau matrix.

4. Generation of subroutines. The rest of the tableau is filled in. This bottom half contains the current/voltage relationships of each element, including the differential and algebraic equations. Derivatives are replaced by terms corresponding to an implicit integration scheme, and the nonlinear equations are linearized in a Taylor series expansion. The elements in the tableau may be variable, and FORTRAN subroutines which evaluate these elements are written and saved. These subroutines will be called repetitively during the execution phase.

5. Formation of solution matrix. Gaussian elimination is performed to solve the entire system of equations of the tableau. Symbolic pivoting, a prescan of the equations to select pivots, is performed under control of algorithms to minimize growth of nonzero elements.

6. Generation of solution code. Three solution methods are available in ASTAP. Each takes proportionately more core storage but produces faster execution times. The user may trade speed for storage by choosing from the different solution methods. Method 1, which takes the least storage, consists of the execution of a preassembled subroutine which uses data written in this step in the setup phase. Method 2 consists of an interpretive execution of pseudocode produced in this step. Method 3, the fastest, consists of executing a stream of floating-point operations produced in this step which solve the system of equations. Method 3 is the standard default method.

7. Compilation of the subroutines. The subroutines produced in step 3 together with user-supplied subroutines are compiled and combined with the analysis and output subroutines of the execution phase.

The execution phase consists of the following steps:

1. Reading of data. The data produced by the setup phase are read in.

2. Analysis. The solution of the circuit is produced, and the solution points are stored on disk. This consists of iterations through the solution code calling the subroutines each time to evaluate the variable elements.

3. Output. The solution points are printed and/or plotted according to the user's output requests.

Since the code generated by the setup phase is executed repetitively, typically several hundred times, to obtain the solution of a single problem, ASTAP contains a relatively elaborate setup phase to produce efficient executable code. Machine time required to obtain a solution during the execution phase is thereby reduced. This is most apparent in the transient analysis of large problems and problems in which the user requests a long simulation time (STOP TIME).

Another advantage of this approach is that user-written FORTRAN subprograms may be submitted with the normal ASTAP language statements describing a circuit. The user subprograms are automatically passed to the FORTRAN compiler, together with the circuit description statements, so that it is not necessary to precompile user subprograms in a separate run.

By the use of dynamic storage allocation in the management of the program data, there are no internal restrictions on the various dimensions of the problem such as the number of elements or the number of nodes, etc. The network size that the program can analyze is limited only by the amount of main storage available under OS/MFT or OS/MVT. Under VS (virtual storage) on IBM S/370 even this restriction is removed, and large problems which could not otherwise be run in smaller machines can be accommodated. Furthermore, the program is automatically extendable to any amount of storage specified for each job. By structure, the program is thus unlimited in capacity. Networks of up to about 200 elements can be analyzed using 220K bytes of storage on an IBM S/360. Networks of up to 3000 elements have been analyzed in less than 1.5 megabytes. Problems can be run in as little as 100K bytes of real storage on a virtual storage system.

2.3 INPUT LANGUAGE

The ASTAP input language consists of a set of statements for describing circuit elements (resistors, capacitors, etc.), for storing and recalling models, for specifying special element dependencies (tables or equations), for selecting output variables to be printed or plotted, for specifying run controls, and for selecting analysis options.

The language has been specifically designed for use by circuit designers. It permits completely general nonlinear models of circuit devices to be described with ease. This is due to the flexibility permitted in describing nonlinear elements and other dependencies, such as statistical tracking between elements. For example, element values may be specified as dependent on any of the following program variables:

- Any element voltage
- Any element current
- Any element value
- User-defined parameters
- Time or frequency

In addition to the variables used in describing various dependencies, any of the above variables as well as node voltages may be requested as outputs. The dependencies themselves may be expressed in terms of the above variables using the following forms:

- FORTRAN-like expressions defining a single element value directly
- General equations, which can be defined once using a FORTRAN-like expression and then referenced as many times as necessary to define individual elements
- Tables containing pairs of values of the dependent and independent variables
- References to system-supplied functions
- References to user-written functions

Sample Problem: Digital Logic Inverter

A sample problem* is presented at this point to illustrate the basic structure and overall function of the input language and to set the stage for later discussions of the program's facilities. Figure 2.2(a) is the circuit schematic of the digital logic inverter to be coded. The transient response of the output for the input signal of Fig. 2.2(b) is to be obtained.

The first step in preparing a circuit for analysis is to develop an equivalent circuit diagram. Models for the nonlinear devices must be selected from the model library or developed. In addition, models for any parasitic effects are included. An equivalent circuit diagram suitable for coding the digital logic inverter circuit in the ASTAP language is shown in Fig. 2.3(a).

The parasitics which most affect the output response are the stray capacitances from the transistor base and collector to ground. These are modeled by C1 and C2 in Fig. 2.3a. The equivalent circuit diagram is annotated with unique names for all components and nodes. These names (labels) may contain any of the alphameric characters A–Z and 0–9, and they may be of any length. The first letter of a component name indicates the element type. Arrows are inserted to show the assumed direction for positive current flow through elements. This labeling procedure facilitates preparation of the input statements and later interpretation of the output generated by the program.

Figure 2.3(b) shows an equivalent circuit diagram for a modified Ebers-Moll model of the 2N2369A transistor, together with the equations defining the dependent elements of the model.

---

*Adapted with permission from *IBM Electronic Circuit Analysis Program Techniques and Applications* by R. W. Jensen and M. D. Lieberman, Prentice-Hall, Englewood Cliffs, N.J., 1968.

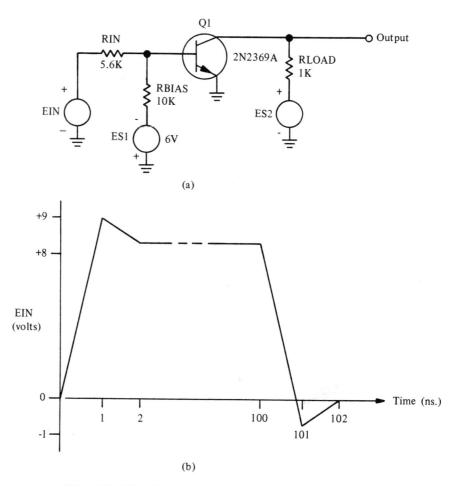

Figure 2.2 Digital Logic Inverter: (a) Circuit Schematic; (b) Input Signal.

The input language statements for the inverter circuit and the transistor model might be coded as shown in Fig. 2.4. Statement numbers have been added for reference purposes.

The statements in Fig. 2.4 have been purposely indented to show the various levels of heading statements contained in the language. Such indentation is not necessary in actual practice. Statements 1, 3, 5, 15, 18, 20, 28, 29, 31, and 33 are heading statements which serve to identify the type of data following the heading. The two major or highest-level headings are MODEL DESCRIPTION and EXECUTION CONTROLS. The MODEL DESCRIPTION heading precedes all statements describing models (or circuits), and the EXECUTION CONTROLS heading precedes all statements controlling the execution of the analysis.

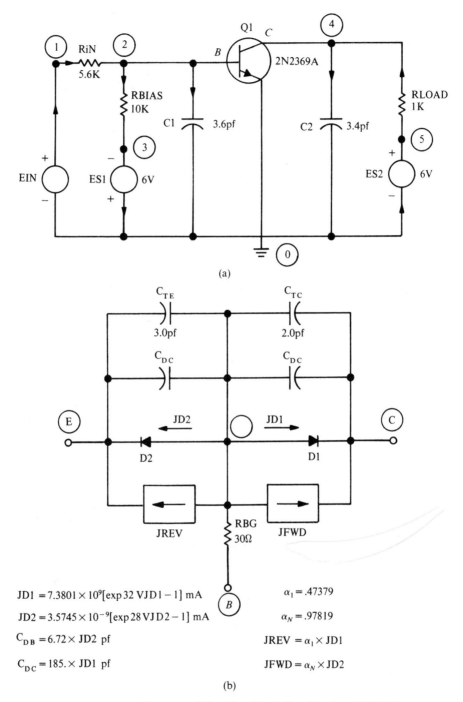

Figure 2.3 Inverter and Transistor Equivalent Circuits: (a) Digital Logic Inverter Equivalent Circuit Diagram; (b) Transistor Equivalent Circuit Diagram.

```
 1          MODEL DESCRIPTION
 2                    CODING EXAMPLE - ASTAP LANGUAGE
 3              MODEL DIGITAL LOGIC INVERTER ( )
 4                    THE FOLLOWING DESCRIBES THE INVERTER CIRCUIT
 5                 ELEMENTS
 6                       ES1,3-0 = 6
 7                       ES2,0-5 = 6
 8                       EIN,0-1 = TABLE1(TIME)
 9                       C1,2-0 = 3.6
10                       C2,4-0 = 3.4
11                       RIN,1-2 = 5.6
12                       RBIAS,2-3 = 10
13                       RLOAD,5-4 = 1
14                       Q1=MODEL 2N2369A(0-2-4)(RBB=.02)
15                 FUNCTIONS
16                       TABLE1,0,0,1.9,2,8,100,8,101,-1,
17                              102,0,110,0
18              MODEL 2N2369A (RETAIN) (E-B-C)
19                    THE FOLLOWING DESCRIBES THE TRANSISTOR MODEL
20                 ELEMENTS
21                       RBB,B-BP = .03
22                       CEMIT,BP-E = (3 + 6.72 * JD2)
23                       CCOL,BP-C = (2 + 185 * JD1)
24                       JD1,BP-C = (DIODEQ(7.3801E-9,32,VJD1))
25                       JD2,BP-E = (DIODEQ(3.5745E-9,28,VJD2))
26                       JFWD,C-BP = (.97819*JD2)
27                       JREV,E-BP = (.47379*JD1)
28          EXECUTION CONTROLS
29              ANALYZE DIGITAL LOGIC INVERTER (TRANSIENT)
30                    RUN NUMBER 1, 1/6/73
31                 RUN CONTROLS
32                       STOP TIME = 200
33                 OUTPUTS
34                       PRINT,VC1,VC2,CEMIT.Q1,CCOL.Q1
35                       PLOT,VC1(BASE VOLTS), VC2(OUTPUT VOLTS)
```

Figure 2.4 Digital Logic Inverter Circuit Input Data.

Statements 2, 4, 19, and 30 are examples of comment statements which can be inserted after certain heading statements to clarify the input. Comments inserted after the ANALYZE heading also appear in the output and can be used to insert a run title (see statement 30).

The MODEL heading, statement 3, signals the beginning of the inverter circuit description and supplies the name of the model (all networks are designated as models, whether they are complete circuits, subcircuits, or device models). The parentheses following the model name are required to delimit the name. This heading is followed by statements 6–14 describing the elements of the inverter circuit. The general form for describing an element is

element name, from node-to node = element value

The element name is arbitrary except that it must begin with one of the following key letters to identify the element type:

Key Letter	Type
R	Resistance
C	Capacitance
L	Inductance
M	Mutual inductance
G	Conductance
J	Current source
E	Voltage source

The FROM node and TO node are the names of the nodes to which the element is connected. Positive current is assumed to flow from the FROM node to the TO node. This convention applies consistently to all elements. Voltage is assumed positive at the FROM node with respect to the TO node in all elements except the voltage source. The above element conventions, as well as the nomenclature for referring to an element voltage and current, are summarized in Fig. 2.5.

Node voltages are referred to by prefixing the letter N to the node name. For example, NFROM and NTO refer to the node voltages associated with the FROM and TO nodes of Fig. 2.5.

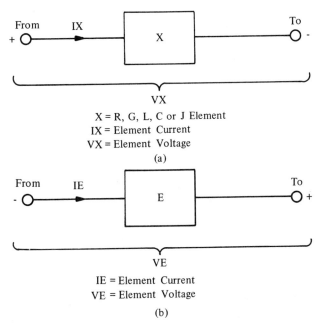

Figure 2.5 ASTAP Element Conventions: (a) R, G, L, C, J Element Conventions; (b) Element Convention.

In Fig. 2.4 all the elements of the inverter except for the independent voltage source EIN are specified with values equal to numeric constants. Although any consistent set of units may be used, those used in this example are generally the most convenient: volts, milliamps, kilohms, microhenries, picofarads, nanoseconds, and gigahertz. The value of EIN is specified as a tabular function of time given by the TABLE 1 statement. The table itself is described under the FUNCTIONS heading and supplies the data of Fig. 2.2(b) in the form of time/voltage pairs of values. As illustrated by the TABLE 1 statement, any statement may be continued onto the next card as many times as necessary. No special continuation symbol is required.

Statement 14 is called a model reference statement. It references (or calls) the 2N2369A transistor model, which may be retrieved from the model library; or, as in this example, the model description may be supplied directly in the input deck with the main circuit description. This illustrates the model-nesting feature of the program, which permits one model to call another.

The model reference name Q1 is arbitrary, being used to label the model for this given instance of its use in the inverter circuit. The Q1 statement causes all the elements of the model to be inserted in the circuit between nodes 0, 2, and 4 [see Fig. 2.3(a)]. The composite circuit consisting only of basic elements is actually analyzed. In the Q1 statement, element RBB in the model is changed to 0.02 kilohms for this use of the model only. This illustrates the model change feature of the program, which permits the same standard model to be used for more than one device by suitable changes in the model parameters.

The MODEL heading, statement 18, signals the beginning of the description of the transistor model and supplies the model name 2N2369A. The keyword RETAIN is used to cause the model to be stored in the model library with the model name 2N2369A. Once it has been stored, the model can be referenced in later runs of the inverter circuit without supplying the statements describing the model. The external nodes of the model are labeled E—B—C. These nodes are replaced by circuit nodes 0—2—4, supplied on the Q1 statement, when the model is inserted in the circuit.

Following the MODEL heading are the statements describing the elements constituting the transistor model of Fig. 2.3(b). The RBB element with a nominal value of 0.03 kilohms is changed to 0.02 kilohms by the model change on the Q1 statement. Element CEMIT models the charge storage effects represented in Fig. 2.3(b) by the constant transition capacitance CTE and the diffusion capacitance CDE, which is a function of the junction diode current JD2. While these are shown as separate capacitors in Fig. 2.3(b), they have been combined into a single element CEMIT whose value is defined by a FORTRAN-like expression. As illustrated, expressions are enclosed in parentheses. In a similar manner, CCOL models the storage effects represented by CTC and CDC in Fig. 2.3(b). While only simple arithmetic expressions have been used in these examples, the full set of FORTRAN arithmetic operators is available. These are +, −, *, /, and ** for addition, subtraction, multiplication, division, and exponentiation, respectively. In addition, all the standard subroutines in the FORTRAN IV library, such as DSIN, DCOS, and DSQRT, may be used in writing expressions. Expressions may also contain references to user-written and system-supplied functions, as discussed below.

Element JD1 is a nonlinear current source which is a function of its own voltage VJD1. It models the transistor collector-base junction diode shown as D1 in Fig. 2.3(b). Its value is defined by reference to the DIODEQ function, which is one of the ASTAP system-supplied functions. The DIODEQ function solves the diode characteristic equation shown for JD1 in Fig. 2.3(b). Since a function reference must appear within an expression, the outermost parentheses are required. Current source JD2 similarly models the emitter-base junction. The forward and reverse gains of the transistor are modeled by elements JFWD and JREV, respectively. These are linear functions of the diode current JD2 and JD1, respectively.

The ANALYZE statement requests a transient analysis of the inverter. The keyword TRANSIENT implies that a dc analysis is to be performed first, followed by a transient analysis. The dc analysis supplies the steady-state conditions for the start of the transient analysis. The ASTAP run controls contain default values which are adequate for most problems. The only control which must be set by the user is the STOP TIME, which is set at 200 nanoseconds of simulation time in this example.

The PRINT statement requests a tabular printout of the voltages across capacitors C1 and C2 and the capacitances CEMIT and CCOL within transistor Q1, all as a function of time. CEMIT.Q1 and CCOL.Q1 are examples of qualified names used to refer to quantities within a lower-level model. Such names are discussed later, in the section entitled "Qualified Names." The PLOT statement requests a printer plot of VC1 and VC2, which are renamed BASE VOLTS and OUTPUT VOLTS, respectively. These new names appear as labels for the variables on the printer plot.

The coding example in Fig. 2.4 illustrates many of the facilities of the input language. Some additional language facilities are discussed below.

Describing Element and Parameter Dependencies

As mentioned previously, the values of dependent elements and parameters may be described using expressions, equations, tables, and references to system-supplied or user-written functions. Discussion of the use of some of these facilities is provided in the following sections.

Equations

The coding example illustrated the use of expressions in defining element values. However, when the same mathematical relationship is used in defining several elements or parameters, it is more convenient to define the mathematical relationship only once, using an equation.

For example, statements 22 and 23 of Fig. 2.4 might be defined using an equation reference such as:

CEMIT, BP−E = EQUATION 1 (3,6.72,JD2)
CCOL, BP−C = EQUATION 1 (2,185,JD1)

For this example, the following two statements would be added to the 2N2369A model description under the FUNCTIONS heading:

FUNCTIONS
EQUATION 1 (A,B,X) = (A+B*X)

A, B, and X are dummy arguments used to define the variables in the equation. They are replaced by the corresponding constants or variables enclosed in parentheses in the equation reference.

Generally, an equation is used when many elements can be specified by the same equation and the expression defining the equation is complicated, as in the following example:

EQUATION2 (A,B,C,Y) = (A/B**.8+DLOG(C/(Y+.1))/AVEVAL(Y))

In the above expression, DLOG is a FORTRAN library function and AVEVAL is an ASTAP system-supplied function.

Summary of System-Supplied Functions

The ASTAP input language contains several built-in functions to facilitate input preparation. They are generally used for describing nonlinear elements, as illustrated by the use of the DIODEQ function to define the values of elements JD1 and JD2 in Fig. 2.4. They may also be used to obtain special outputs. Brief descriptions of some of the system-supplied functions are given below:

TRISE. Provides the time at which a specified variable first rises through a specified level. For example, TRISE might be used to compute the delay between an input and output as follows:

PIN = (TRISE(VIN,0))
POUT = (TRISE(VOUT,0))
DELAY = (PIN-POUT)

Parameters PIN and POUT provide the times at which the input VIN and the output VOUT first cross (rise) through the zero level.

TFALL. Provides the time at which a specified variable first falls through a specified level.

DIODEQ. Used to model a diode or transistor junction characteristic equation. For an example, see statement 24 of Fig. 2.4.

FETN. Used to model the drain-to-source static characteristic equation of an N-type field-effect transistor (FET).

FETNT. Used in conjunction with the FETN function to obtain the threshold voltage of an N-type FET device during the course of an analysis.

FETP. Similar to FETN, but for P-type FET devices.

FETPT. Similar to FETNT, but for P-type FET devices.

TRAP. Used to specify a time-dependent voltage or current source as a train of pulses with a trapezoidal wave shape. By adjustment of the arguments of the function, pulse trains with other wave shapes can be defined, such as square waves and sawtooths.

SINSQ. Used similarly to TRAP, but specifies a pulse train with a sine-squared shape for the rise and fall portions of the pulse.

AVEVAL. Used in a transient analysis to specify an output parameter which is the average value of a specified network variable. For example, it might be used to obtain the average power dissipated in a resistor RLOAD:

POWER = (AVEVAL(VRLOAD∗IRLOAD))

In a tabular printout of POWER versus TIME, the value of POWER at, say, 100 nanoseconds is the average power over the range TIME = 0 to TIME = 100.

AREA. Used similarly to AVEVAL, but obtains the area under a plot of a variable as a function of time. For example, it might be used to obtain the energy dissipated in a resistor.

PTIME. Used in conjunction with a time-dependent element described by a table to indicate that the table is periodic. For example, element EIN in Fig. 2.4 might be specified as a periodic function of time, starting at TIME = 0 and with a period of 110 nanoseconds, using the following statement:

EIN,0−1 = TABLE1 (PTIME(0,110))

Fortran-Supplied Functions

All the FORTRAN-supplied functions may be used in ASTAP expressions. The double-precision form must be used. A description of all the FORTRAN functions available can be found in the following publications:

IBM System/360 and System/370 FORTRAN IV Language (GC28-6515) and IBM System/360 Operating System FORTRAN IV Library—Mathematical and Service Subprograms (GC28-6816).

A listing of some of the functions which are commonly used is given below:

Function	Definition
DEXP (x)	Exponential function
DLOG (x)	Natural logarithm
DLOG10 (x)	Logarithm to the base 10
DSQRT (x)	Square root
DSIN (x)	Sine function
DCOS (x)	Cosine function
DTAN (x)	Tangent function
DABS (x)	Absolute value
DMAX1 (x1, ..., xn)	Algebraic maximum of arguments
DMIN1 (x1, ..., xn)	Algebraic minimum of arguments

User-Written Functions

Most circuits can be described without the need for special user-written functions due to the flexibility offered by the use of FORTRAN-like expressions, which may include arithmetic operations involving circuit variables, FORTRAN library functions, and system-supplied functions.

The user may write his own functions, which may then be used in the same way as the system-supplied functions. The user-written functions are prepared as FORTRAN function subprograms which are inserted in the input deck following the normal ASTAP language statements for the circuit. The ASTAP program automatically invokes the FORTRAN G or H compiler during the circuit input phase to compile the ASTAP language statements together with any user-written subprograms. With this approach, it is not necessary for the user to precompile his subprograms in a separate job before executing an ASTAP job. Also, the user is not required to insert special calling statements into the ASTAP program itself in order to invoke his subprograms.

Alternatively, the user's subprograms may be added permanently to the system so that they do not have to be submitted with the input deck each time they are used. With either approach, the user is in effect extending the set of system-supplied functions, so that the total set of functions is open-ended.

Nested Models

A nested model is a model which references (or calls) one or more lower-level models. In the coding example of Fig. 2.4, only one level of nesting was illustrated: The Q1 model reference statement in the inverter circuit (main model) references the 2N2369A transistor model (first level).

In addition to the main model referencing other models, any of those models may reference another model, and so on to any level of nesting. This model-nesting feature facilitates the description of larger networks, especially those with a repetitive structure.

Figure 2.6 further illustrates the concept of nested models. Each box represents a model. The corresponding model names are given inside each box. In ASTAP, even the main model is given a model name. There are two levels of model nesting, since main model CHIP contains model CELL (first level), which contains model TRA (second level).

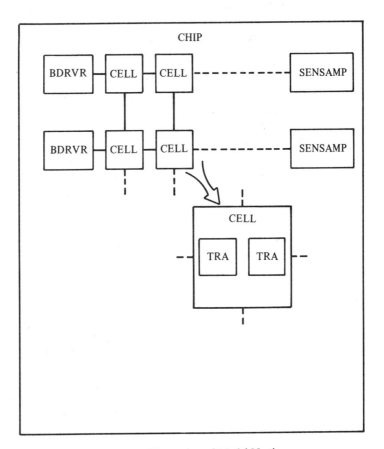

Figure 2.6 Illustration of Model Nesting.

The main model is the one which is not referenced by any higher-level models; however, it may be made part of a larger network in a subsequent analysis. Usually the main model is the one which is analyzed, although any model in the nested structure may be selected for analysis.

Model References and Changes

The mechanism for referencing a lower-level model and connecting it to another model or circuit is the model reference statement. Statement 14 of Fig. 2.4 illustrates such a model reference, where Q1 is the model reference name assigned by the user. A different model reference name is assigned for each use of the same model in a given circuit.

The model change facility permits changes to be applied to a given model reference or use of a model. The model as it exists in the model library is not affected. The value of any element or parameter within the model may be changed to a different constant value. A change to an element is illustrated in the Q1 statement of Fig. 2.4 (statement 14) by the change RBB = .02. In addition, using the model change feature, a new table may be substituted for a table defined within a model.

Global Variables

This language facility permits parameters, tables, and other quantities which are defined in a main model to be used in defining quantities in other models referenced by the main model. This is a convenient feature for the user in preparing input data, and it economizes storage in the program as well.

Without the global facility, it would be very difficult to simulate such effects as the tracking of several devices with respect to a common circuit parameter, such as temperature or sheet resistivity. This facility is illustrated below:

```
            MODEL MAIN
                .
                .
                .
            PSHEET = 20
                .
                .
                .
            Q1 = MODEL XTR (9−2−1)
            Q2 = MODEL XTR (4−3−7)
            Q3 = MODEL XTR (5−8−6)
                .
                .
                .
            MODEL XTR
                .
                .
                .
            R1,B−2 = (5*PSHEET)
            R2,E−3 = (8*PSHEET)
                .
                .
                .
            FEATURES
            GLOBAL = (PSHEET)
```

The MAIN model contains three devices (Q1, Q2, and Q3) which track each other with respect to parameter PSHEET. Although the parameter itself is defined in the MAIN model, it may be used in defining elements within the XTR model because it is specified as a global parameter under the FEATURES heading.

This facility assumes added importance in a statistical analysis, since it may be used to define statistical tracking between devices. For example, parameter PSHEET might be specified with a 5% tolerance and with a normal (Gaussian) distribution about a nominal value of 20:

PSHEET = DISTRIBUTION NORMAL(20,5%)

Correspondingly, R1 and R2 within model XTR might be defined as

R1,B−2 = DISTRIBUTION NORMAL(5∗PSHEET,2%)
R2,E−3 = DISTRIBUTION NORMAL(8∗PSHEET,2%)

In the above, R1 and R2 are assigned independent normal distributions of 2%, while their nominal values are correlated with the statistical variation of the PSHEET parameter. (The letter D is permitted as an abbreviation for DISTRIBUTION.)

During a statistical analysis of the above example, devices Q1, Q2, and Q3 would all track each other with respect to the PSHEET parameter. A statistical transient analysis on the circuit above is requested with the following statements:

EXECUTION CONTROLS
ANALYZE MODEL MAIN (TRANSIENT)
RUN CONTROLS
CASES=100

where CASES indicates the number of separate transient runs to be made, with the values for each run selected randomly according to their associated distributions.

Model Library Facilities

ASTAP provides two libraries for storage of models. The first is called the user model library; This is the library normally used by circuit designers for storage and retrieval of models. A user model library may be utilized in common by all users, or private user model libraries may be assigned to individual users.

The second library is called the program model library, and it can be accessed only through use of a special key. It is intended for use by modeling specialists or quality assurance personnel at a given installation to store standard models for use by circuit designers. Any model in the program library can be referenced (retrieved) for use in defining other models or circuits; however, a model cannot be stored, replaced, or deleted without use of the special key. This key is described in the *ASTAP Operations Guide*.

The input language provides the same facilities for accessing either model library. To avoid confusion, the general term *model library* is used to refer to the user model library.

Normal operations on the model library are carried out through use of the MODEL heading statement. Statements 3 and 18 of Fig. 2.4 are specific examples of its use. The operations which can be performed on the model library with this statement can best be described by referring to its general form:

MODEL model name (disposition)(node list)

where

model name = the name of the model

node list = a list of the external nodes of the model, separated by hyphens

disposition = the desired operation on the library

The disposition options are

RETAIN. Causes the model to be retained in the model library, provided a model with the same name is not already in the library.

REPLACE. Causes the model to replace an existing model with the same name.

DELETE. Causes the model to be deleted from the library.

If no disposition is explicitly specified, the model is temporary and exists only for the current execution of the program. This default can be used to avoid cluttering of the model library with untested models.

The language statements describing any model in the library may be printed using the MODEL heading statement, as illustrated below for a model named SCR:

MODEL SCR (PRINT)

Additional operations on the model library can be carried out through the use of utility control statements. These provide for printing of the names of all models in the library, printing of the statements defining all models, and compression of the model library to eliminate holes left by deleted models.

Diagnostic Messages

ASTAP contains over 300 diagnostic messages designed to assist the user in correcting input errors. Special care has been taken to make the messages meaningful to the user. The message text is designed to be complete and self-contained, so that it is not necessary for the user to refer to a manual for a more detailed explanation of the message. For example, when it is not self-evident, the message states the action taken by the program as well as the cause of the error.

2.4 LANGUAGE STATEMENT DESCRIPTIONS

This section gives a description of all the statements which comprise the ASTAP language. Statements are entered in free format, with blanks ignored, in columns 1–72 of a card. Statements may be continued on any number of cards. Multiple statements per card may be entered separated by a semicolon, excepting headings.

Heading Statements

The complete list of the ASTAP language heading statements, organized in a typical input order, is shown below:

```
            MODEL DESCRIPTION
                     (comments)
            MODEL model name (disposition) (nodes)
                     (comments)
                ELEMENTS
                FUNCTIONS
                FEATURES
            EXECUTION CONTROLS
                     (comments)
            ANALYZE model name (mode of analysis)
                     (comments)
                RUN CONTROLS
                INITIAL CONDITIONS
                OPERATING POINT
                OUTPUTS
                RERUNS
            UTILITY CONTROLS
```

MODEL DESCRIPTION Heading

The MODEL DESCRIPTION heading introduces one of the three major sections of data. The descriptions of each of the models of the circuit including the main model—that model designated for analysis—are placed in this section.

MODEL Heading

> MODEL model name (disposition) (node-node- . . . -node)
> or
> MODEL model name (node-node . . . -node)
> or
> MODEL model name ( )

where

MODEL is a key word.

model name is an arbitrary name uniquely identifying the model.

disposition is one of the three key words RETAIN, REPLACE, or DELETE which indicates the status of the model with respect to the model libraries.

(*node-node-* . . . *-node*) are the external nodes of the model.

Example.

MODEL TRANS (B—C—E)

Nodes B, C, and E are the nodes in the model, TRANS, which are designated as the external nodes. Any subset of the nodes in a model may be declared external. These are the nodes to which the model which references the model TRANS connect. It is not necessary to have any nodes designated as external, but the open and close parentheses must always be included both in the model heading statement and in the referencing statement.

ELEMENTS Heading

ELEMENTS is one of the three headings which may appear under the MODEL heading. Under this heading appear the statements which describe the elements of a model.

Element Statement

> element name, from node-to node = element value

where

element name is the unique name assigned to the element.

The first letter of the name must correspond to the type of the element as shown in the table:

Key Letter	Type
R	Resistance
C	Capacitance
L	Inductance
G	Conductance
J	Current source
E	Voltage source
M	Mutual inductance

from node-to node are the names of the nodes to which the element connects. The order in which the node names are placed indicate the assumed polarity of the element. Refer to Fig. 2.5 for the definition of the ASTAP element conventions.

element value is the field in which the constant or dependent value of the element is given. The element types, R, C, L, and G must *never* have a zero value during any part of the analysis. Care should be exercised when defining dependent element values (e.g., tables and expressions) to ensure that this does not happen. An element value may be specified using one of the general forms listed below:

1. Numeric constant
2. TABLE table name (argument)
3. EQUATION equation name (argument list)
4. (FORTRAN-like expression)
5. DISTRIBUTION distribution name (nominal value, tolerance %)
6. STATISTICAL TABLE table name (arguments)
7. COMPLEX (real argument, imaginary argument)
8. COMPLEX MAGPH (magnitude, phase in degrees)
9. FOURIER (argument list)

ac analysis sources are described using the COMPLEX, COMPLEX MAGPH, and FOURIER forms. These sources are set equal to zero during a dc or transient analysis. Passive elements, defined by one of the three above forms, are evaluated using the real part with FREQ equal to zero during a dc or transient analysis.

Detailed Description of Element Value Forms

Constant Element. A constant element may be specified using a single numeric constant. Numeric constants may be entered with optional signs, decimal points, and exponents.

Examples.

1. RLOAD, C–GND = 5
2. CE, B–E = .3
3. L6, 5–7 = 1.56E–2 (means 1.56×10^{-2})

Table Reference.

> TABLE table name (argument)

where

TABLE is a key word. It may be abbreviated with the letter T.

table name refers to a table specified under the FUNCTIONS heading.

argument may be any of the following:
- Numeric constant
- Element voltage or current
- Element or parameter name
- TIME or FREQ
- Function reference
- Expression

Examples.

1. EIN, 0–1 = TABLE 1 (TIME)
2. JE, B1–E = TABLE JE (VJE)

Equation Reference.

> EQUATION equation name (argument list)

where

EQUATION is a key word. It may be abbreviated with the letter Q.

equation name is a unique name among equation names in the model. It refers to an equation defined under the FUNCTIONS heading.

argument list is a list consisting of any of the following arguments of the type listed under the TABLE form.

Examples.

1. CE, P–N = EQUATION 1 (PCE, 12,1, JE)

   The arguments that appear in the argument list are used to evaluate the equation which is defined using dummy arguments under FUNCTIONS.

2. CC, 1−2 = Q2(1.275, .9, VCC, .2556, 20.559, PTBL4, JC, 2.8367E−6)

FORTRAN-Like Expression. A full mathematical expression may be used to define an element (or parameter) value. It may be continued on as many lines as necessary, like any statement. The expression may be formed from any of the following:

- Numeric constant
- Element voltage or current
- Element or parameter name
- TIME or FREQ
- Function reference

The expressions are formed using the FORTRAN rules for arithmetic expressions. The expressions in ASTAP may contain variables, as listed above, which consist of more than six characters. Also, the numeric constants may be integer, single- or double-precision real constants, because ASTAP prescans all expressions and converts external variables to internal variable names and all numeric constants to double-precision real constants. FORTRAN-supplied, ASTAP system-supplied, and user-supplied functions may be used in the expressions. These functions must always be double-precision functions. The expression must always be enclosed in parentheses.

Examples.
  1. JC, BB−C = (.98*JE)
  2. CE, 4−6 = (.4 + (2.6*PIDB**0.464*JE) + DMAX1 (0.01, JE/5)**4.4)
     The FORTRAN-supplied function DMAX1 is used to ensure that the exponential always operates on a positive number no matter what the value of JE is.
  3. JE, BB−C = (DIODEQ (2.16E−12, 38.37, VCE))

DISTRIBUTION Reference. Such a reference is used to specify the value of an element (or parameter) which varies according to a statistical distribution.

| DISTRIBUTION distribution name (nominal value, tolerance %) |

where

  DISTRIBUTION is a key word which may be abbreviated by the letter D.

  distribution name may be a name identifying a user-defined distribution, or the key word GAUSSIAN (or NORMAL); or the distribution name may be omitted, indicating a uniform distribution. The user-defined distribution is described under the FUNCTIONS heading. The end points of the distribution are associated with the element value range limits and normalization is automatic during the Monte Carlo selection. The key word GAUSSIAN (or NORMAL) indicates that a symmetric Gaussian distribution truncated at the plus and minus three sigma points should be used. Omitting the distribution name indicates that a uniform distribution be used.

nominal value is the base value to which the tolerance is applied to determine the range of the element value. It is also the value which is used during the nominal run which precedes the statistical cases.

tolerance % is the value of the symmetric tolerance, which must be expressed as a percentage. The element value range limits are computed using the nominal value and plus and minus this tolerance.

Range Field Options.

a. DISTRIBUTION distribution name (nominal value, tolerance %)

b. DISTRIBUTION distribution name (low tol %, nominal value, high tol %)

c. DISTRIBUTION distribution name (min value, nominal value, max value)

d. DISTRIBUTION distribution name (min value, max value)

These are the optional forms for the range field. In option b, nonsymmetric tolerances are assigned. In options c and d, the actual range limits are entered. In option d, since a nominal value is not indicated, the value corresponding to a random number of .5 is used for the nominal value. The values which appear in the range field may be numeric constants, parameters, or expressions in terms of constants or constant parameters.

Examples.

1. CI, V10−V7 = D (.01, 10%)
2. R6, GND−V6 = D NORMAL (.34, 20%)
3. RE, EP−E = D (PRE−.0001, PRE + .0001)
4. E15, GND−V15 = D3 (−4.25, 4%)
5. RBIAS, 4−5 = D RBIAS (−2%, .56, 10%)
6. R5, V5−V15 = D (1.4, 1.82, 2.18)
7. RA, V1−VN = D GAUSSIAN (2.98∗PX + 1.04∗PX∗PRP, 1%)

STATISTICAL TABLE Reference. A tabular dependent element (or parameter) can be made to vary statistically using a statistical table reference.

> STATISTICAL TABLE table name (argument, DISTRIBUTION distribution name)
>
> or
>
> STATISTICAL TABLE table name (argument, parameter)
>
> or
>
> STATISTICAL TABLE table name (argument)

where

STATISTICAL TABLE is a key word. It may be abbreviated by STABLE, ST, or S.

table name is an arbitrary name of the statistical table.

argument may be any of the types listed under table arguments.

DISTRIBUTION *distribution name* (optional) is a field where DISTRIBUTION is a key word and *distribution name* may take the same forms as described in the section entitled "DISTRIBUTION Statement." When this field appears, it overrides the corresponding field in the statistical table statement.

parameter (optional) is the name of a parameter which must have a value between 0 and 1. When it is used, its value is used to pick the statistical curve from within the statistical table instead of the random number for the statistical case. The curve selection process is discussed in the section entitled "STATISTICAL TABLE Statement."

Examples.

1. JET1, V1−EP = STATISTICAL TABLE 1 (VC1)
2. J1, V1−1 = STABLE SJ1 (VJ1, PT)
   J2, V1−2 = STABLE SJ1 (VJ2, PT)

COMPLEX Values—Rectangular Form. A complex element (or parameter) for ac analysis may be described using the COMPLEX form.

> COMPLEX (real argument, imaginary argument)

where

COMPLEX is a key word.

real argument and *imaginary argument* may be

- Numeric constant
- Parameter (real-valued)
- FREQ or TIME
- Function reference
- Expression (using only the above types of quantities)

Examples.

1. EIN, 0−1 = COMPLEX (1, 0)
    This source has unit value at all frequencies of evaluation. Both the real and imaginary arguments must be defined in terms of real quantities.

2. RS, V7−GND = COMPLEX (PTABR, PTABI)
    PTABR = TABLE REAL (FREQ)
    PTABI = TABLE IMAG (FREQ)
    Here both the real and imaginary parts are given by tables which are functions of frequency and defined indirectly using parameters. This represents the skin effect dependence of RS on frequency.

E and J sources defined with the COMPLEX form are set to zero value during the dc and transient modes of analysis. Passive elements (R, G, L, and C) are set equal to the real part of the COMPLEX value evaluated at FREQ equal to zero. The real part of the passive element value must be nonzero during a dc or transient analysis.

COMPLEX Sources—Polar Form.

> COMPLEX MAGPH (magnitude, phase)

where

COMPLEX MAGPH is a key word.

magnitude and *phase* are arguments which may be any of the types listed for the COMPLEX value form.

Example.

> EIN, 0−1 = COMPLEX MAGPH (1, 0)

Both the magnitude and phase arguments must be real quantities. In this example, the source EIN has unit magnitude at zero phase for all frequencies. This form can be used only for sources.

FOURIER Source Element Value. This form of element value may be used for E and J sources in the ac (frequency domain) mode of analysis.

> FOURIER (argument 1, argument 2)

where

FOURIER is a key word.

argument 1 is a parameter which is defined by a table of points describing a time-dependent waveform.

argument 2 is the number of harmonics which will be computed for the waveform and must be a numeric constant or a parameter defined as a numeric constant.

Example.

JSOURCE, 0−1 = FOURIER (PTABFR, 10)
PTABFR = TABLE FOUR (TIME)

The waveform described by a table must consist of at least $2N + 1$ points where N is the number of harmonics. In this example the table must consist of at least 21 points. The analysis is carried out for each of the harmonic frequencies and the standard ac outputs are provided.

Parameter Statement

$$\boxed{\text{parameter name} = \text{parameter value}}$$

where

parameter name is the unique name of the parameter, which must begin with one of the following letters: A, B, F, H, K, O, P, U, W, X, Y, Z. The letter P is most often used.

parameter value may be expressed using one of the following forms:

1. Numeric constant
2. EQUATION equation name (argument list)
3. TABLE table name (argument)
4. (FORTRAN-like expression)
5. DISTRIBUTION distribution name (nominal value, tolerance %)
6. STATISTICAL TABLE-table name (arguments)
7. COMPLEX (real argument, imaginary argument)
8. COMPLEX EXP (FORTRAN-like complex expression)

Examples.

1. P6 = 7.5
2. PB = TABLE BETA (VJE)
   PA = (PB/(1 + PB))
   Parameters may be included anywhere under ELEMENTS and may be used to define auxiliary quantities for use in expressions or as outputs. In this example a calculation with a table value is carried out using the parameter PB.
3. PT = D GAUSSIAN (PSBD−.1, PSBD + .1)
   Parameters are useful for evaluating statistical variables and in defining tracking relationships.

4. PK21 = COMPLEX EXP (VJOUT/VJIN)
COMPLEX EXP is a special value form to be used only with parameters which are used as outputs. It allows expressions involving the complex-valued variables of the problem to be calculated. In this example the voltage transfer ratio of VJOUT to VJIN is computed for outputting.

Model Reference Statement

> reference name = MODEL model name (node-node-... -node)
> or
> reference name = MODEL model name (node-node-... -node)(changes)

where

reference name is an arbitrary name which uniquely identifies the particular model reference. It corresponds to the name assigned to a particular device in the circuit.

MODEL is a key word.

model name is the name of the referenced model to be inserted in the calling circuit.

(*node-node-...-node*) are the nodes of the calling circuit to which the referenced model is connected. They must correspond in order and number with the external node list of the referenced model. No common ground node between models is assumed by the program. All nodes which are common between models must be passed using the node lists. The parentheses must appear even if there are no nodes in the node list.

(*changes*) is an optional field in which changes to the referenced model are specified for this particular model reference (or use of the reference model). Any number of changes may be indicated, separated by commas. Element and parameter values in the model may be changed from whatever value form they have to new numeric constants. A new table can also be substituted.

Examples.

1. Q2 = MODEL 2N2369 (GND−2−4)
The node designated as GND is intended as the ground node but must be passed through the node list.

2. TOC = MODEL TX (VB−VRI−VOC−VN) (PWE = .2, PLE = .5, B1 = 1.5, TABLE 1 = T2)
Several changes are indicated in this model reference including replacing TABLE 1 of model TX by TABLE 2, which is defined in the model in which the reference appears.

Mutual Inductance Statement

> mutual name, element 1 — element 2 = mutual inductance value

where

mutual name is an arbitrary unique name which must begin with the letter M.

element 1 — element 2 are the names of the two inductance elements (self-inductances) between which the mutual inductance exists.

mutual inductance value may be specified using any of the following forms:

- Numeric constant
- TABLE table name (argument)
- EQUATION equation name (argument list)
- DISTRIBUTION distribution name (nominal value, tolerance %)
- STATISTICAL TABLE table name (arguments)

The mutual inductance value is constrained such that the coefficient of coupling must always be strictly less than unity. For ideal transformers controlled sources rather than mutual inductance should be used. For three or more mutually coupled inductances, the constraint is that the inductance matrix of these inductances must be positive-definite.

The dot convention is followed in assigning polarity to the mutual inductance. It states that if both the assumed directions of current, as fixed by the self-inductances, enter or leave the dotted ends, the mutual inductance value is positive; otherwise it is negative.

Example.

A two winding transformer with the dots placed at nodes A and C:

$$L1, A-B = 1$$
$$L2, C-D = 1$$
$$M12, L1-L2 = .998$$

FUNCTIONS Heading

The FUNCTIONS heading is used when it is necessary to supply definitions for certain items used in specifying elements and parameters. The following are placed under the FUNCTIONS heading.

TABLE Statement

> TABLE table name, x,y,x,y, ... x,y

where

> TABLE is a key word. It may be abbreviated with the letter T.
>
> *table name* is the unique name assigned to the table. This name also appears in a table reference used to specify an element or parameter value.
>
> $x,y,x,y,\ldots x,y$ form the table description and are the coordinate points of the tabular function being described. The x's are the values of the independent variable and the y's are the corresponding function values. The x's must be in algebraically increasing order. Step functions may be described using two points having the same x value.

Example.

> TABLE 1, 0, 1.023, 5, 1.023, 10, 1.545, 60, 1.545, 65, 1.023, 100, 1.023

EQUATION Statement

> EQUATION equation name (dummy argument list)
>   = (FORTRAN-like expression)

where

> EQUATION is a key word. It may be abbreviated with the letter Q.
>
> *equation name* is a unique name assigned to the equation. This name also appears in the corresponding equation reference used to specify an element or parameter value.
>
> (*dummy argument list*) is a list of dummy variables separated by commas. These variables must be no longer than six characters and must not begin with the letters I, J, K, L, M, or N.
>
> (*FORTRAN-like expression*) is an expression written with FORTRAN arithmetic operators using the variables in the dummy argument list.

Examples.

1. EQUATION 2(A, B, C, D, E) = (A*B/((C + D)**E))
2. Q CAP (A, B, C) = (A + DMAX1 (.01, B*C))
   The expressions defining the equations may include reference to FORTRAN, system-supplied, or user-supplied functions.

DISTRIBUTION Statement

This statement is used to define shapes of the statistical distributions used in specifying element or parameter values.

General Distributions

> DISTRIBUTION distribution name(type)x,y,x,y, ... x,y
> or
> DISTRIBUTION distribution name, x,y,x,y, ... x,y

where

DISTRIBUTION is a key word which may be abbreviated with the letter D.

distribution name is the unique name assigned to the distribution. This name also appears on the distribution reference used in specifying an element or parameter value.

type is either DENSITY or CUMULATIVE, indicating the distribution type. If no type is entered (using the second form above), density type is assumed.

$x,y,x,y, \ldots x,y$ are the coordinate points of the distribution. The x values are the independent variable values to any scale. The y values are the density distribution or CDF values. The x values must be in algebraically increasing order. The y values for the CDF must be in algebraically increasing order.

Example.

```
         ELEMENTS
            .
            .
            .
         PB = DISTRIBUTION BETA (10, 25, 40)
            .
         FUNCTIONS
            .
            .
            .
         D BETA (DENSITY) 10, 0,  20, 1,  40, 0
```

Special Gaussian Definition

> DISTRIBUTION distribution name (GAUSSIAN) MIN=w, MAX=x, MEAN=y, SIGMA=z

where

MIN, MAX, MEAN, and SIGMA are key words and may appear in any order.

w, x, y, and z are the values on any scale of values, and $|w-y| \leq 4z$ and $|x-y| \leq 4z$ must hold.

Example.

D 6 (GAUSSIAN) MIN = −1, MAX = 2.5, MEAN = 0, SIGMA = 1

As in this example, this form is used to describe a nonsymmetrically truncated Gaussian distribution.

STATISTICAL TABLE Statement

A curve which has statistical variation can be described by a statistical table which gives inner and outer bounding curves of the variation. On each statistical case, an intermediate curve is chosen.

> STATISTICAL TABLE table name (DEP=DISTRIBUTION name)
> $x,y1,y2,x,y1,y2$ ...
>
>   or
>
> STATISTICAL TABLE table name (IND=DISTRIBUTION name)
> $x1,x2,y,x1,x2,y$ ...

where

STATISTICAL TABLE is a key word. It may be abbreviated by the words STABLE, ST, or S.

table name is the unique name assigned to the statistical table. This name is used in a statistical table reference to specify an element or parameter value.

DEP is a key word indicating that the spread is on the dependent variable.

IND is a key word indicating that the spread is on the independent variable.

=*DISTRIBUTION name* is an optional field which includes the equals sign (=). It specifies a distribution to be applied to the spread of either the dependent or independent variable. The default is the uniform distribution.

$x,y1,y2,x,y1,y2$ ... is a list of coordinates with two dependent variable values corresponding to each independent variable value.

$x1,x2,y,x1,x2,y$ ... is a list of coordinates with two independent variable values corresponding to each dependent variable value.

Example.

STABLE SBD (IND = D GAUSSIAN) −5., −5., −.001, 0, 0, 0,
.207, .257, .001, .277, .327, .01, .350, .400, .1,
.370, .420, .2, .412, .462, .5, .464, .514, 1, .535, .585, 2

On each statistical case, a single random number is used to select the intermediate curve. At each breakpoint, the proportionate distance produced by the random number and the specified distribution determines the coordinate of the curve being selected.

FEATURES Heading

Under the FEATURES heading, the following statements which provide additional specifications for the model description may be included.

GROUND Node Statement

Node voltages may not be requested as outputs unless a ground node is defined.

$$\boxed{\text{GROUND} = (\text{node name})}$$

where

GROUND is a key word.

node name is the name of any node in the main model (the circuit to be analyzed). Lower-level models may not specify a ground node.

GLOBAL Statement

The quantities listed below may be used as global quantities, that is, may be referenced in models other than the model in which they were defined by making them entries in the GLOBAL statement in the model where they are referenced. To be used in a GLOBAL statement, the quantities must be defined in the main model.

$$\boxed{\text{GLOBAL} = (\text{entry,entry}, \ldots, \text{entry})}$$

Each entry in the GLOBAL statement can be one of the following:

- Parameter name
- TABLE table name
- DISTRIBUTION distribution name
- STATISTICAL TABLE table name

Example.

   GLOBAL = (PRE, PCI, S1, PB)

PORTS Statement

In ac analysis, the PORTS statement is used to designate node pairs as ports of a network for transfer function parameter calculations. The PORTS statement must appear in the main model.

$$\boxed{\text{PORTS}=(\text{node-node},\text{node-node},\ldots,\text{node-node})}$$

The nodes in the PORTS statement are any node names in the model. Each port is specified by a pair of nodes. The first pair is identified as port 1, the second as port 2, etc.

Transfer Functions

In the ac or frequency mode of analysis, seven sets of transfer function parameters are available for calculation. The transfer functions available and the associated key letters are

Key Letter	Transfer Function	Port Limit
Z	Impedance parameters	N ports
Y	Admittance parameters	N ports
H	Hybrid h parameters	2 ports
K	Hybrid g parameters	2 ports
S	Scattering parameters	N ports
F	Transfer-scattering parameters	2 ports
A	ABCD or chain parameters	2 ports

The transfer functions are defined as follows using ASTAP nomenclature:

1. Impedance parameters (Z):

$$\begin{bmatrix} v_1 \\ v_2 \end{bmatrix} = \begin{bmatrix} z_{11} & z_{12} \\ z_{21} & z_{22} \end{bmatrix} \begin{bmatrix} i_1 \\ i_2 \end{bmatrix}$$

2. Admittance parameters (Y):

$$\begin{bmatrix} i_1 \\ i_2 \end{bmatrix} = \begin{bmatrix} y_{11} & y_{12} \\ y_{21} & y_{22} \end{bmatrix} \begin{bmatrix} v_1 \\ v_2 \end{bmatrix}$$

3. Hybrid h parameters (H):

$$\begin{bmatrix} v_1 \\ i_2 \end{bmatrix} = \begin{bmatrix} h_{11} & h_{12} \\ h_{21} & h_{22} \end{bmatrix} \begin{bmatrix} i_1 \\ v_2 \end{bmatrix}$$

4. Hybrid g parameters (K):

$$\begin{bmatrix} i_1 \\ v_2 \end{bmatrix} = \begin{bmatrix} k_{11} & k_{12} \\ k_{21} & k_{22} \end{bmatrix} \begin{bmatrix} v_1 \\ i_2 \end{bmatrix}$$

5. ABCD or chain parameters (A):

$$\begin{bmatrix} v_1 \\ i_1 \end{bmatrix} = \begin{bmatrix} a_{11} & a_{12} \\ a_{21} & a_{22} \end{bmatrix} \begin{bmatrix} v_1 \\ i_2 \end{bmatrix}$$

6. Scattering parameters (S):

$$\begin{bmatrix} b_1 \\ b_2 \end{bmatrix} = \begin{bmatrix} s_{11} & s_{12} \\ s_{21} & s_{22} \end{bmatrix} \begin{bmatrix} a_1 \\ a_2 \end{bmatrix}$$

where the as are the incident wave amplitudes and the bs are the reflected wave amplitudes. In terms of voltages and currents and the normalizing numbers,

$$a_1 = \frac{E_1 + J_1}{2} \qquad b_1 = \frac{E_1 - J_1}{2}$$

and

$$E_1 = \frac{v_1}{\sqrt{SNUM1}} \qquad J_1 = \frac{i_1}{\sqrt{SNUM1}}$$

where SNUM1 is the normalizing number for port 1 and v_1 and i_1 are the port 1 voltage and current. Corresponding definitions apply to port 2.

7. Transfer-scattering parameters (F):

$$\begin{bmatrix} a_1 \\ b_1 \end{bmatrix} = \begin{bmatrix} f_{11} & f_{12} \\ f_{21} & f_{22} \end{bmatrix} \begin{bmatrix} a_2 \\ b_2 \end{bmatrix}$$

where the definitions for a_1, b_1, a_2, and b_2 are the same as for the S parameters.

Parameter Names. The parameter names are formed by using the key letter and adding the appropriate numbers of ports as determined from the PORTS statement. In a two-port example, Z11 is the impedance at port 1 with port 2 open-circuited.

Example.

$$\text{PORTS} = (0-1, 0-3)$$

Nodes 0 and 1 define port 1 and nodes 0 and 3 define port 2.

NORMALIZE Statement

Scattering parameters (S) and transfer-scattering parameters (F) can be normalized with respect to user-specified real or complex normalizing numbers. These are specified using the NORMALIZE statement. The normalizing numbers must be numeric constants.

```
NORMALIZE = S(SNUM1,SNUM2,...)
NORMALIZE = S(SNUM1R,SNUM1I,SNUM2R,SNUM2I,...)
NORMALIZE = F(FNUM1,FNUM2,...)
NORMALIZE = F(FNUM1R,FNUM1I,FNUM2R,FNUM2I,...)
```

EXECUTION CONTROLS Heading

The headings and statement under the EXECUTION CONTROLS heading describe the manner in which the analysis is to be carried out.

A list of the heading statements follow:

> EXECUTION CONTROLS
> (comments)
> ANALYZE model name (mode of analysis)
> (comments)
> RUN CONTROLS
> INITIAL CONDITIONS
> OPERATING POINT
> OUTPUTS
> RERUNS

ANALYZE Heading

This heading must be the first heading under EXECUTION CONTROLS.

> ANALYZE model name (mode of analysis)

where

ANALYZE is a key word.

model name is the model name of the main model to be analyzed.

mode of analysis can be one of the following key words:

DC: Determines the dc levels.

TRANSIENT: Determines the dc levels, or initial conditions, followed by the transient response.

TRANSIENT ONLY: Determines the transient response starting from zero or specified initial conditions.

FREQUENCY or AC: Determines the frequency response over a range of frequencies of a linear circuit or a nonlinear circuit which is automatically linearized about a specified operating point (see the section entitled "OPERATING POINT Heading").

RUN CONTROLS Heading

The RUN CONTROLS heading signals the beginning of the run control statements, which provide facilities for controlling an analysis run with respect to solution accuracy, duration, increments, choice of methods, etc.

Run Control Statement

The run control statement can assume one of the two following forms:

> run control name = numeric constant
> or
> run control name

Transient and dc Analysis Run Controls

Run Control Name	Default Value
STOP TIME =	100
START TIME =	0
PRINT INTERVAL =	(STOP TIME − START TIME)/100
PLOT INTERVAL =	(STOP TIME − START TIME)/100
MAXIMUM STEP SIZE =	PRINT INTERVAL or PLOT INTERVAL, whichever is smaller.
MINIMUM STEP SIZE =	(STOP TIME − START TIME)/10000
CONTINUE AT MINIMUM STEP SIZE	Do not continue
STARTING STEP SIZE =	$(MAX \times MIN)^{1/2}$, where MAX and MIN are the maximum and minimum step sizes
SOLUTION METHOD =	3, 2, or 1 depending on available real storage
INTEGRATION ORDER =	2 (maximum order)
SETTLING TIME =	0
CHARGE UP REACTANCE =	1
NO REACTANCE SUBSTITUTION	Substitute unit value reactances
MAXIMUM PASSES =	10000
MAXIMUM INITIAL CONDITION PASSES =	1000

Transient and dc Analysis Error Controls

Error Control	Default
ITERATION RELATIVE ERROR =	See note below
ITERATION ABSOLUTE ERROR =	See note below
TRUNCATION RELATIVE ERROR =	See note below
TRUNCATION ABSOLUTE ERROR =	See note below
DC LEVEL RELATIVE ERROR =	See note below
DC LEVEL ABSOLUTE ERROR =	See note below
DC ITERATION RELATIVE ERROR =	See note below
DC ITERATION ABSOLUTE ERROR =	See note below

Note: The above defaults are determined by the values of the following two controls: RELATIVE ERROR determines the defaults for the above four relative controls, and ABSOLUTE ERROR acts correspondingly on the four absolute controls.

RELATIVE ERROR =	.005
ABSOLUTE ERROR =	.0005

Frequency (ac) Run Controls

Run Control	Default
START FREQ =	None
STOP FREQ =	None
POINTS PER DECADE =	10

The START FREQ must not equal the STOP FREQ.

Statistical Run Controls

Run Controls	Default
CASES =	0
STARTING RANDOM NUMBER SEED =	Random number
NOMINAL RANDOM NUMBER =	.5
PRINT ALL FAILURES	Print first failure in case

The use of the CASES control indicates a statistical run.

General Run Controls

Run Control	Default
COMPUTER TIME LIMIT =	No limit (CPU minutes)
OUTPUT TIME =	Not in effect (CPU minutes)
COMMON SCALES	Not in effect (applies to all plots)
TOPOLOGY	No topology output
TOPOLOGY ONLY	No topology output (used for debugging)
INPUT ONLY	Complete execution (used for debugging)
FULL PIVOT SELECTION	Diagonal pivot selection
DISK ACTIVITY =	200

Example.

```
        RUN CONTROLS
          TOPOLOGY
          OUTPUT TIME  =    .8
          STOP TIME    =    20
          CASES        =    1000
          PRINT INTERVAL =  .1
```

INITIAL CONDITIONS Heading

Under the INITIAL CONDITIONS heading, the user may specify the initial state of a network at the start of an analysis performed in the transient-only, transient, or dc mode. Capacitor voltages and inductor currents are used to specify initial conditions. Parameters and element values can also be initialized under this heading. Problem variables default to zero if not specified. The initial condition statements assume the following forms:

> V capacitance name = numeric constant
> or
> I inductance name = numeric constant

where

V is a key letter which indicates the voltage across an element.

I is a key letter which indicates the current through an element.

Example.
```
           INITIAL CONDITIONS
                VCE = .2
                IL4 = 7.5
                PB = 1
                VCC.T4 = −.8
```

OPERATING POINT Heading

The operating point of a nonlinear circuit must be specified prior to an ac (or frequency) analysis of the circuit. To specify the operating point it is necessary to specify the voltage or current variable on which the nonlinear elements fundamentally depend. Variables which are not specified are assumed to be zero. Differentiation is performed on the J and E source elements to determine the small-signal values at the operating point. The other elements are simply evaluated at the operating point. For example, if it is desired to have a small-signal resistance used during ac analysis, the nonlinear resistance should be entered as a dependent voltage or current source. Nonlinear capacitances and inductances which are entered as differential values are handled correctly in this way.

OPERATING POINT Statement

In specifying the operating point, it is only necessary to specify the variables on which the nonlinear elements depend. For elements and parameters defined indirectly in terms of other elements and parameters, only the lowest level of dependency in terms of the program's variables must be specified. Variables used in nonlinear dependencies and not specified under the OPERATING POINT heading are assumed to have zero value at the operating point.

General Form.

> V element name = numeric constant
> I element name = numeric constant

where

V prefacing an element name denotes the voltage across the element.

I prefacing an element name denotes the current through the element.

Example.

```
ELEMENTS
    CE, 1-E =   Q1 (4.289, .9, VCE, .3708, 38.3, PTBL3, JE, 2.16E-12)
    PTBL3 =     T3 (JE)
    CC, 1-2 =   Q1 (1.28, .9, VCC, .256, 20.6, PTBL4, JC, 2.84E-6)
    PTBL4 =     T4 (JC)
    JE, 1-E =   (DIODEQ (2.16E-12, 38.4, VCE))
    JC, 1-2 =   (DIODEQ (2.84E-6, 20.6, VCC))
        .
        .
        .
OPERATING POINT
    VCE =       .742
    VCC =       -11.
```

OUTPUTS Heading

The request for the display of the results of the analysis is made under the OUTPUTS heading.

Output variables can be any of the following:

1. *Voltage across an element.* The voltage name is formed by prefixing the letter V to the element name. All element voltages are available as outputs. Voltages in lower-level models are referred to by their qualified names.

2. *Current through an element.* The current name is formed by prefixing the letter I to the element name. All element currents are available as outputs.

3. *Node voltage.* A node voltage name is formed by prefixing the letter N to any node name. Node voltages may be used only in output statements. The node voltage is given with respect to the specified ground node. See "GROUND Node Statement" in the section entitled "FEATURES Heading."

4. *Element value.* Asking for element values as outputs is a useful aid in problem verifications.

5. *Parameter value.* Parameters can be used for outputting quantities defined by expressions more general than the basic voltage and current outputs.

6. *Transfer function parameter* (X, Y, H, K, S, F, A). These are output variables available as a result of an ac analysis. They are requested in conjuction with the use of a PORTS statement. See "PORTS Statement" in the section entitled "FEATURES Heading."

7. *TIME or FREQ.* In the respective mode of analysis the running variable time or frequency may be used in output statements. They are referred to as TIME and FREQ, respectively.

8. *Internal control variable.* FORTRAN variables internal to ASTAP can be requested as outputs. In some cases they can be useful in monitoring the operation of the program. A partial list of them is

   NPASS: Count of iteration passes

   NPASSI: Count of initialization passes

   NSTEP: Running count of time steps

   DELT: Time step of integration

9. *Partial derivatives of dependent sources.* In ac analysis these partial derivatives, which are automatically computed as part of the linearization procedure, can be requested as outputs. Each partial derivative name is formed by the dependent variable prefixed by the letter D, followed by a /, followed by the independent variable prefixed by D.

PRINT Statement

Output variables are added to the print list, printed at every time step, by listing them on a PRINT statement. The general form of this statement is

> PRINT variable list

where

variable list is a list of the individual output variables of the general form

> variable(rename), variable (rename), . . .

The renaming of each variable is optional. If used, the rename replaces the variable name in the tabular listing. The rename must be unique among all other renames and must not contain any commas or parentheses. (Other special and alphabetic characters may be used.)

Example.

PRINT VRC, VRIN2, JDIODE.T1, JETRAN.T2 (IFWRD)

PLOT Statement

Each PLOT statement generates a separate plot. Each variable on the PLOT statement is plotted with a separate alphabetic character; thus, there is a maximum of 26 curves per plot.

> PLOT (plot options) variable list, VS ind. variable

Plot Options

1. LABEL = (label text)
   The label text is used to title the plot.
2. COMMON SCALES or COMMON SCALES (min,max)
   This option causes all curves to be plotted to the same scale. Normally all curves are scaled independently.
3. INTERVAL = x
   With this option the plot interval on the independent variable is specified, overriding the default.
4. DIVISIONS = n
   The number of plot divisions can be specified with this option. This option and the INTERVAL option cannot be used simultaneously.
5. BODE
   In ac analysis this option designates a log magnitude and phase plot for each variable on a log scale.
6. REAL: Real component.
7. IMAG: Imaginary component.
8. MAG: Magnitude.
9. PHASE: Phase in degrees.
10. LOGM: logarithm of the magnitude (20 log(MAG)).

One of the above five options (6–10) causes the respective component of all the variables in the output variable list to be used on the plot in ac analysis. Only one option can be used in any one PLOT statement.

11. ENVELOPE
    This option causes an envelope plot to be plotted for each variable in the variable list in the transient or ac nodes under statistical analysis. It is a plot of the upper and lower envelope of all the curves of response for all cases of the statistical run.

Variable List

The general form for variable list is

> variable(rename)(min,max)(variable option), . . .

where

Each of the quantities in parentheses following the variable is optional for each variable.

(*rename*) allows a name other than the output variable name to be used on the plot for the variable. See the description of renaming for the PRINT statement above.

(*min,max*) specifies the coordinate scale on the dependent axis for the particular variable it follows, overriding the automatic scaling.

(*variable option*) specifies the single type of component to be plotted for the complex-valued ac output variable it follows. This option applies only in an ac analysis. The variable option may be one of the following:

REAL: Real component

IMAG: Imaginary component

MAG: Magnitude

PHASE: Phase in degrees

LOGM: Logarithm of the magnitude (20 log(MAG))

Optional Independent Variable

Any output variable may be specified as the independent variable on the plot. If none is specified, the independent variable is assumed to be TIME for a transient analysis and FREQ (on a logarithmic scale) for an ac analysis.

```
VS independent variable(rename)(min,max)(variable option)
```

where

VS is a key word which must be preceded by a comma, and the independent variable may have the same modifying options following it as the dependent variables.

Examples.

1. PLOT N1, N2, VCIN, IRIN

2. PRINT, PLOT (COMMON SCALES,
   LABEL = (PROPAGATION DELAY)) EIN,
   VRL2 (IN-PHASE), VRL1 (OUT OF-PHASE)
   PRINT and PLOT statements may be combined. Both labeling of the plot and renaming of the variables are illustrated in this example.

3. PLOT   JE, VS   VJE
   This calls for a parametric plot of current versus voltage.

4. PLOT (INTERVAL = 20.0, ENVELOPE) IJC.TE.PHSA, IJC.TE.PHSG, IJC.TC5

5. PLOT   Z21 (IMAG), VS   Z21 (REAL)
   This statement results in a Nyquist plot.

HISTOGRAM Statement

For each variable in a HISTOGRAM statement a separate histogram is plotted. A histogram is a plot of the number of cases which fall into each of the class intervals over the range of variable values. The value of the variable used in the histogram is the value it has at the STOP TIME or STOP FREQ of each case, whether dc, transient, or ac analysis is performed. The SAMPLE option can be used to obtain variable values during the course of an analysis. The general form is

> HISTOGRAM (histogram options) variable list

where

> HISTOGRAM is a key word which may be abbreviated by HIST. In the variable list, the *rename* and *variable option* options can be used. These are explained in the section entitled "PLOT Statement."

Histogram Options

1. SAMPLE = (T1,T2,...)
   T1, T2,... are the sample times or frequencies at which a histogram is produced.

2. INTERVAL = class interval value
   This is the actual class interval to be used when overriding the automatic default.

3. DIVISIONS = number of class intervals
   This specifies the number of class intervals in the histogram. The default is 50. Do not use this option if INTERVAL is used.

Examples.

1. HISTOGRAM   N1, VR6, PTOFF
2. PRINT, HIST (SAMPLE = (2, 4, 10)) VRLOAD
   The PRINT and HIST statements may be combined. VLOAD is sampled at three times (or frequencies) and three histograms result.

SCATTERGRAM Statement

A scattergram can be requested as a result of a statistical analysis. It is a plot of one output variable versus another over all the statistical cases. The general form is

> SCATTERGRAM(sample option)variable list, VS ind. variable

where

> SCATTERGRAM is a key word which may be abbreviated with SCAT.
>
> *sample option* has the form SAMPLE = (T1,T2, . . .), where T1,T2, . . . are the sample times or frequencies at which a scattergram is generated for each of the variables.
>
> *variable list* is a list of output variables, which cannot be TIME or FREQ.
>
> *ind. variable* is any output variable except TIME or FREQ. It must be specified.

The SCATTERGRAM request cannot be lumped with other output requests. Any number of scattergram statements can be included under OUTPUTS. If the run control, CASES, is not specified, a scattergram is not produced.

Example.

> SCATTERGRAM   VOUT, VS PDOPING

The scatter diagram request allows a visual picture of the correlation of VOUT with the input process parameter PDOPING.

TEST Statement

During a statistical run, any variable can be tested for a failure condition. A failure is defined as the failure of the variable under test to stay within specified upper and lower bounds. When a failure occurs on any case, a printout occurs which contains the values of the statistical variables for that case plus the outputs under test. This output can be used to help diagnose the failure condition. The general form of the TEST statement is

> TEST variable (min,max), variable(min,max), . . .

where

> *variable* is the name of any output variable which is designated to be tested.
>
> (*min,max*) are the specified bounds for the variable they follow expressed as numeric constants. If the variable is less than min or greater than max, the case is designated a failure.

Examples.

1. TEST   VR7 (0.4, 0.6)
2. TEST   VC1 (.8, 1.2), IR2 (1.9, 2.1), VR8 ($-1E9$, .6)
   A failure will register if any one or more of the variables listed fail. The test on VR8 illustrates a one-sided test.

RERUNS Heading

A RERUNS heading is placed under the ANALYZE heading when a reanalysis, or a rerun, of the specified circuit with parameter modification is desired. Modification between reruns is limited to parameters which are defined as constants. They may be modified to other constants. Elements may be modified by defining them in terms of parameters which are modifiable. Any number of reruns are allowed, and any number of RERUN statements are possible. The output for each rerun is the same as for the basic run. The general RERUN statement is defined as follows:

RERUN Statement

$$\boxed{\text{parameter name} = (\text{constant1}, \text{constant2}, \ldots)}$$

where

parameter name designates the parameter to be modified. It may be a qualified name.

constant1, constant2, ... are numeric constants.

Example.

```
RERUNS
   PSET1 = (4, 5, 6)
   PSET2 = (10)
```

In this example, PSET1 takes the value 4 on the first rerun, 5 on the second, and 6 on the third. There will be as many reruns as parameter changes indicated on all RERUN statements. PSET2 takes the value 10 on the first rerun and retains that value for all subsequent reruns.

2.5 EXAMPLES

Current Switch Emitter Follower

This sample problem illustrates a statistical transient analysis used to investigate the propagation delay in a current switch emitter follower (CSEF) logic circuit [also known as emitter-coupled logic (ECL)]. Figure 2.7 shows a CSEF circuit which is used as a basic building block for a family of logic circuits employing the monolithic system technology used in certain IBM 370 systems.

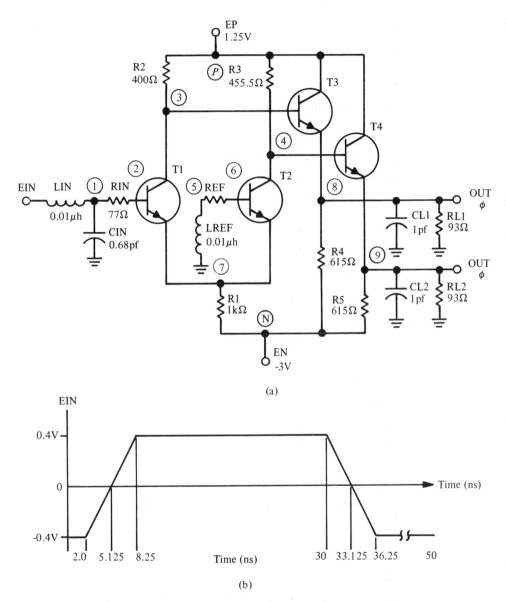

Figure 2.7 Current Switch Emitter Follower: (a) Equivalent Circuit Diagram; (b) Input Signal.

A key measurement of the performance of the circuit is the propagation delay between the input and output. The propagation delays are defined as the time differences between zero crossings of the input signal and the corresponding zero crossings of the output signals.

```
 1          MODEL DESCRIPTION
 2          MODEL CURRENT SWITCH ( )
 3
 4              CURRENT SWITCH EMITTER FOLLOWER CIRCUIT.  STATISTICAL DESCRIPTION
 5
 6          ELEMENTS
 7              PTRACK = DISTRIBUTION NORMAL (.077, 15%)
 8              RIN, 1-2 = D NORMAL (PTRACK, 5%)
 9              REF, 5-6 = D NORMAL (PTRACK, 5%)
10
11              PS = D NORMAL (.15, 8%)
12              R1, 7-N = D NORMAL (6.67*PS, 3.6%)
13              R2, P-3 = D NORMAL (2.66*PS, 3.6%)
14              R3, P-4 = D NORMAL (2.97*PS, 3.6%)
15              R4, 8-N = D NORMAL (4.1*PS, 3.6%)
16              R5, 9-N = D NORMAL (4.1*PS, 3.6%)
17              RL1, 8-GND = .093
18              RL2, 9-GND = .093
19
20              LIN, EIN-1 = .01
21              LREF, 5-GND = .01
22
23              CIN, 1-GND = .68
24              CL1, 8-GND = 1
25              CL2, 9-GND = 1
26
27              T1 = MODEL JCNTRAN (2-3-7)
28              T2 = MODEL JCNTRAN (6-4-7)
29              T3 = MODEL JCNTRAN (3-P-8)
30              T4 = MODEL JCNTRAN (4-P-9)
31
32              EP, GND-P = 1.25
33              EN, GND-N = -3
34              EIN, GND-EIN = TABLE EIN(TIME)
35
36              PTONDIN  =   (TRISE(VRL2,0) - 5.125)
37              PTONDOUT =   (TFALL(VRL1,0) - 5.125)
38              PTOFFDIN =   (TFALL(VRL2,0) - 33.125)
39              PTOFFDOUT =  (TRISE(VRL1,0) - 33.125)
40
41              PIDBG = DISTRIBUTION 3 (4, 20, 47.3)
42
43          FUNCTIONS
44              TABLE EIN,  0,-.4, 2,-.4, 8.25,.4, 30,.4, 36.25,-.4, 50,-.4
45              DISTRIBUTION 3 (NORMAL) MIN=4, MEAN=20, MAX=47.3, SIGMA=9.1
46
47          FEATURES
48              GROUND = (GND)
49
50          MODEL JCNTRAN (B-C-E)
51
52              DIFFUSED NPN JUNCTION TRANSISTOR MODEL.  STATISTICAL DESCRIPTION.
53
54          ELEMENTS
55              RE, 6-E = .0005
56              RC, C-5 = .008
57              PIDB = D NORMAL (PIDBG, 20%)
58              RB,B-4=(2/PIDB**.8+DLOG(3/(JE+.1))/PIDB)
59              CRB, B-4 = TABLE 1 (PIDB)
60              PY = D NORMAL (0.27, 20%)
61              CC1,B-5=(2.32*PY/(.8-DMIN1(0,VCC1))**.33333)
```

Figure 2.8 CSEF Circuit Input for Statistical Transient Analysis.

```
62        CC2,4-5=(0.4*PY/(.8-DMIN1(0,VCC1))**.33333)
63        CE, 4-6 = (0.4 + (2.6*PIDB**0.464*JE) + DMAX1(0.01, JE/5)**4.4)
64        PISAT = (2.5E-13*(20/PIDB)**.84)
65        JE, 4-6 = (DIODEQ(PISAT, 38.1, VJE))
66        PBETA = (1200*(1 - 1/PIDB)/PIDB)
67        PALPHA = (PBETA/(1 + PBETA))
68        JC, 5-4 = (PALPHA*JE)
69
70    FUNCTIONS
71        TABLE 1, 5,3.5,10,3.5,20,2.5,40,.1,50,.1
72    FEATURES
73        GLOBAL = (PIDBG)
74    EXECUTION CONTROLS
75    ANALYZE CURRENT SWITCH (TRANSIENT)
76        CURRENT SWITCH EMITTER FOLLOWER USING BIPLOAR JUNCTION TRANSISTORS.
77    RUN CONTROLS
78        STOP TIME = 50
79        MAXIMUM STEP SIZE = .1
80        CASES = 100
81    OUTPUTS
82        PRINT    EIN,VRL1,VRL2
83        PRINT    IRE.T1,IRE.T2,IRE.T3,IRE.T4
84        PLOT    N1,N2,N3,N4
85        HISTOGRAM    PTONDIN,PTONDCUT,PTOFFDIN,PTOFFDCUT
86        HIST (SAMPLE=(0,30,50))    VRL2
87        PLOT(ENVELOPE,COMMON SCALES,LABEL=(IN PHASE DELAY))    EIN,VRL2
88        PLOT(ENVELOPE,COMMON SCALES,LABEL=(OUT OF PHASE DELAY))    EIN,VRL1
89        SCATTERGRAM    PTONDIN, VS PIDBG
90        SCATTERGRAM    PTOFFDIN, VS PIDBG
91    END
```

Figure 2.8—*Cont.*

Statistical Description of CSEF Circuit

In the CSEF circuit, resistors R1 through R5 are dependent on the sheet resistivity of the material used for the resistors. In addition, variations within the tolerances of their dimensions give an actual value within the statistical spreads about their respective nominal values. Such a situation is modeled as shown in the input data for the CSEF circuit in Fig. 2.8. The parameter PS represents the sheet resistivity. It is normally distributed about a nominal value of 0.15 and varies $\pm 8\%$ from the nominal at the ± 3 sigma points of the distribution. The nominal values of R1-R5 are each defined as some factor of PS to model the tracking between these elements as a function of sheet resistivity. In addition, each resistor has an independent normal distribution of $\pm 3.6\%$ about its nominal value. Data to determine the distributions are gathered from measurements on actual samples of circuits manufactured using the monolithic system technology. The units used for this problem are volts, milliamps, kilohms, microhenries, picofarads, nanoseconds, and gigahertz.

The lead resistors, RIN and REF, form a separate, independent tracking group with respect to parameter PTRACK.

Transistor Model

The transistor model and the equations defining the values for its elements are shown in Fig. 2.9. The equations for the element values in Fig. 2.9 may be translated directly into mathematical expressions (FORTRAN-

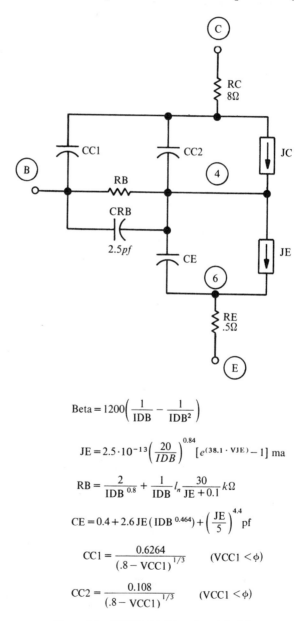

$$\text{Beta} = 1200\left(\frac{1}{\text{IDB}} - \frac{1}{\text{IDB}^2}\right)$$

$$\text{JE} = 2.5 \cdot 10^{-13}\left(\frac{20}{\text{IDB}}\right)^{0.84} [e^{(38.1 \cdot \text{VJE})} - 1] \text{ ma}$$

$$\text{RB} = \frac{2}{\text{IDB}^{0.8}} + \frac{1}{\text{IDB}} l_n \frac{30}{\text{JE}+0.1} k\Omega$$

$$\text{CE} = 0.4 + 2.6\,\text{JE}\,(\text{IDB}^{0.464}) + \left(\frac{\text{JE}}{5}\right)^{4.4} \text{pf}$$

$$\text{CC1} = \frac{0.6264}{(.8 - \text{VCC1})^{1/3}} \quad (\text{VCC1} < \phi)$$

$$\text{CC2} = \frac{0.108}{(.8 - \text{VCC1})^{1/3}} \quad (\text{VCC1} < \phi)$$

Figure 2.9 JCNTRAN Transistor Model.

like) in the ASTAP language. The expressions may be as general as one wishes as long as they represent physically meaningful relationships. Standard FORTRAN library functions, ASTAP system-supplied functions, or user-supplied functions may be used in these expressions.

The statistical description of the transistor model illustrates additional capabilities of the language. Many of the elements of the transistor model are nonlinear. Furthermore, they are correlated with each other. To describe this correlation to the program, independent parameters are found on which the parameters in questions are dependent. Beta, the JE versus VJE curve, RB, CRB, and CE are considered to be dependent on total base doping. An additional measurement is made for each device which measured IDB, a current proportional to the sheet conductivity of the active base. IDB is then closely correlated with total base doping. The dependent parameters are plotted against IDB for a large number of samples, and curve-fitting techniques are used to determine their relationships with IDB. The CRB dependency is described by a table, as shown in Fig. 2.8. The mathematical relationships describing the other parameters dependent on IDB are shown in Fig. 2.9.

These relationships are used in the element value definitions of the transistor model in Fig. 2.8. The parameter PIDB represents the IDB and is defined also in the model. This value varies from device to device and thus has a $\pm 20\%$ spread around a nominal value chosen for the whole circuit (or chip), PIDBG. PIDBG must have the same value for all devices in the circuit. Since any parameter is normally defined only for the model in which it appears, PIDBG is declared a global parameter under the FEATURES heading of each model in which it is used. The definition of this global parameter appears in the model designated for analysis, the highest-level model of the circuit description. In this case the global parameter PIDBG is defined by a distribution which represents the circuit-to-circuit variation of IDB. Tables and statistical distributions are other quantities which may be declared global.

The collector capacitors, CC1 and CC2, depend on another independent parameter, the P to N capacitance, which is represented by PY in the model. These form another tracking group.

Further details on the statistical model of the transistor may be found in Reference 6.

Statistical Output

In general, the HISTOGRAM statement requests the program to plot histograms of the single-valued variables listed after the key word HISTOGRAM or HIST. The first HISTOGRAM statement in the example requests histograms of the four delay parameters. For variables which are functions of time, the SAMPLE option in the HISTOGRAM statement requests histograms of the variables at those values of time listed following SAMPLE =. In the second HISTOGRAM statement, the variable VRL2 is sampled at 0, 30, and 50 nanoseconds.

Another statistical output for variables which are functions of time (or frequency) is requested by the ENVELOPE option in the standard plot statement. This produces an envelope of all the responses of each variable over all the cases in the statistical simulation. The mean value at each time point is also plotted with the upper and lower envelope values. Envelope plots for the in-phase and out-of-phase outputs, VRL2 and VRL1, are requested in the example.

The SCATTERGRAM statement requests another form of statistical output. It produces a scatter diagram of one or more variables versus another over all the statistical cases and can be used to visually determine correlation of the variables. In the example, the turn-on and turn-off in-phase delays are plotted versus IDB (PIDBG) to assess the correlation of the delay with the IDB measurement.

Crystal Oscillator

This example* demonstrates some of the ac analysis features of ASTAP. The analysis is performed on a two-transistor 10-megahertz crystal oscillator circuit. The frequency response of the complete oscillator is calculated to determine the oscillator frequency. A detailed discussion of a similar analysis of this circuit using the ECAP program is described on page 200 of Reference 7.

The circuit diagram for the 10-megahertz crystal oscillator is shown in Fig. 2.10. This is the main circuit or model. The data for this circuit are listed as the first model in the MODEL DESCRIPTION section of the input data listing (see Fig. 2.11, statements 2–15). The units for this circuit are volts, milliamps, kilohms, picofarads, microhenries, nanoseconds, and gigahertz.

The model for the crystal is shown in Fig. 2.12. The derivation of the model can be found in Reference 7. The model consists of four linear elements and could have been included as part of the main model, but by defining it as a seperate entity it can easily be added to or removed from the circuit.

The model used here for the 2N2369 transistor is a full nonlinear Ebers-Moll model.† Since ac analysis requires a linear circuit, the nonlinear model

---

*Adapted with permission from *IBM Electronic Circuit Analysis Program Techniques and Applications* by R. W. Jensen and M. D. Lieberman, Prentice-Hall, Englewood Cliffs, N.J., 1968.

†Adapted with permission from *SCEPTRE: A Computer Program for Circuit and System Analysis* by J. C. Bowers and S. R. Sedore, Prentice-Hall, Englewood Cliffs, N.J., 1971.

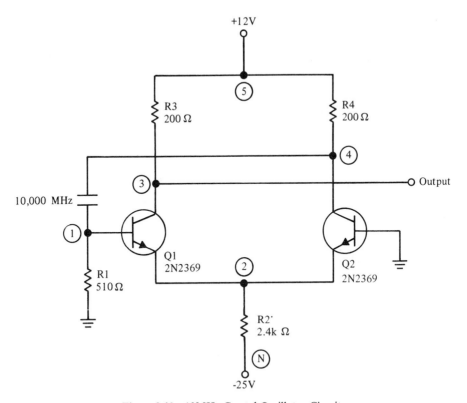

Figure 2.10 10MHz Crystal Oscillator Circuit.

is linearized automatically by the program about a specified operating point, illustrating an important ac analysis feature. The equivalent circuit diagram for the model is shown in Fig. 2.13.

In this problem, the operating point was found by a prior dc analysis of the circuit in a balanced condition with the crystal removed and the base of Q1 grounded. The operating point must be specified by the user when the topology for the dc analysis is different from that used for the ac analysis. This is done under the OPERATING POINT heading in the EXECUTION CONTROLS section. It is not necessary to define the complete state of the network. Only the primary voltages or currents on which the nonlinearities depend have to be specified, so that in Fig. 2.11 only the junction voltage of each of the transistors is specified (see statements 66–70).

The program then automatically linearizes the circuit by evaluating partial derivatives of the nonlinear quantities at the operating point. Although not illustrated in the sample problem, the program has an option for printing the linearized element and parameter values.

Note that no complex ac driving function is specified in the input data.

```
 1          MODEL DESCRIPTION
 2          MODEL OSCILLATOR ( )
 3          FEATURES
 4          PORTS = (GND-1, GND-4)
 5          ELEMENTS
 6          R1, 1-GND = .510
 7          R2, 2-N = 2.4
 8          R3, 5-3 = .2
 9          R4, 5-4 = .2
10          E25, GND-N = -25
11          E12, GND-5 = 12
12          Q1 = MODEL 2N2369 (1-2-3)
13          Q2 = MODEL 2N2369 (GND-2-4)
14          CRYSTAL = MODEL CRYSTAL (1-4)
15          PFREQ = (FREQ)
16          MODEL 2N2369 (B-E-C)
17          ELEMENTS
18          CE, 1-E = Q1(4.289,.9,VCE,.3708,38.376,PTBL3,JE,2.1596E-12)
19          PTBL3 = T3(JE)
20          CC, 1-2 = Q1(1.275,.9,VCC,.2556,20.559,PTBL4,JC,2.8367E-6)
21          PTBL4 = T4(JC)
22          RB, B-1 = .0498
23          RC, C-2 = .00274
24          R1, 1-E = 8.33E6
25          R2, 1-2 = 4.16E6
26          JE, 1-E = (DIODEQ (2.1596E-12,38.376,VCE))
27          JC, 1-2 = (DIODEQ (2.8367E-6,20.559,VCC))
28          JN, 2-1 = (PTBL1*JE)
29          PTBL1 = TABLE 1 (JE)
30          JI, E-1 = (PTBL2*JC)
31          PTBL2 = TABLE 2 (JC)
32          FUNCTIONS
33          Q1(A,B,C,D,E,F,G,H) = (A/(B-C)**D+E*F*(G+H))
34          TABLE 1,
35          .001,.97276,.0514,.97276,.1026,.97466,.2047,.97704,.5103,
36          .97982,1.0195,.98087,2.0353,.98265,5.0832,.98363,10.165,
37          .98376,20.325,.98400,30.510,.98328,40.697,.98287,50.92,
38          .98193,61.111,.98182,71.396,.98058,81.65,.97979,91.963,
39          .97865,102.28,.97771,500,.97771
40          TABLE 2,
41          .1,.02045,.489,.02045,.727,.02751,1.25,.04,1.92,.05208,2.95,
42          .06779,3.88,.07731,5.38,.09293,9.03,.11074,15.0,.1333,25,.1333
43          TABLE 3,
44          .1,.4597,1.02,.4597,2.54,.3521,5.108,.2823,10.16,.2177,
45          25.41,.1803,50.92,.1441,100,.1441
46          TABLE 4,
47          .1,8.556,.7935,8.556,1.845,6.559,2.932,6.228,4.031,6.277,
48          5.155,6.036,10,6.036
49          MODEL CRYSTAL (1-4)
50          ELEMENTS
51          L, 1-2 = 1E5
52          C2, 2-3 = 2.53299E-3
53          R, 3-4 = .025
54          C1, 1-4 = 5
55          EXECUTION CONTROLS
56          ANALYZE OSCILLATOR (AC)
57          RUN CONTROLS
58          START FREQ = .009998
59          STOP FREQ = .010002
60          POINTS PER DECADE = 1E5
```

Figure 2.11 Input for 10 MHz Oscillator.

```
61      OUTPUTS
62      PRINT    K21(N4/N1)
63      PLOT     (LABEL=(FEEDBACK-NETWORK-TERMINALS)) K21(N4/N1)
64      PLOT     (LABEL=(FEEDBACK-NETWORK-TERMINALS)) K21(N4/N1)(LOGM),
65               K21(N4/N1)(PHASE),  VS PFREQ(.C09995,.C10005)(MAG)
66      OPERATING POINT
67      VCE.Q1 = .74216
68      VCE.Q2 = .74216
69      VCC.Q1 = -10.996
70      VCC.Q2 = -10.996
71      END
```

Figure 2.11—*Cont.*

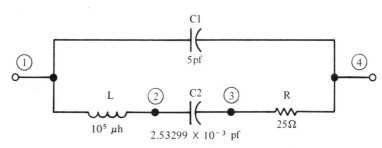

Figure 2.12 Equivalent Circuit for Crystal.

In this problem, the only output of interest is a voltage transfer ratio used to determine the frequency of oscillation. (Output requests are found in statements 61–65 in Fig. 2.11.) This voltage transfer ratio is obtained by requesting the appropriate two-port matrix parameter as an output, in this case K21, one of the G parameters (the letter K is used for the G parameters since the letter G is used for conductance). In this case no external source need be specified, since the program automatically accounts for it during the internal calculation of the requested matrix parameter.

When calling for any of the matrix parameters as outputs, PORTS must be specified under the FEATURES heading in the main model (see statements 3 and 4). In the sample problem nodes GND to 1 form the first port and GND to 4 the second. K21 (G21) is defined as the ratio of the voltage at port 2 to the voltage at port 1, or, in this case, the voltage at node 4 to the voltage at node 1.

Acknowledgments

ASTAP is the result of the efforts of many people. The authors of this chapter which to recognize and express their appreciation to the principal developers of ASTAP: A. J. Jimenez, G. W. Mahoney, D. A. Mehta, H.

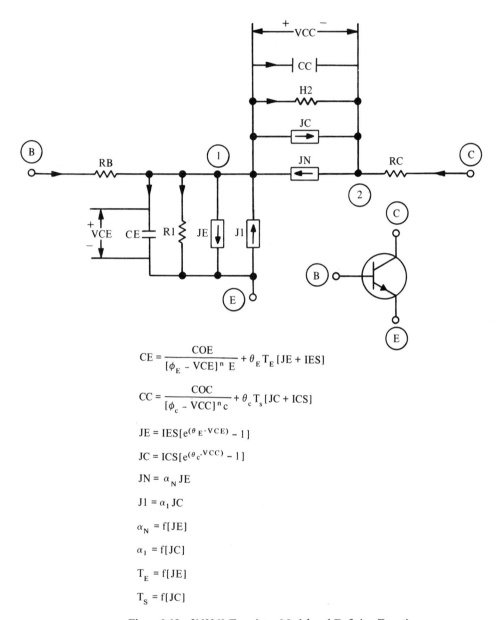

Figure 2.13 2N2369 Transistor Model and Defining Equations.

Qassemzadeh, T. R. Scott, and W. T. Weeks. They also wish to acknowledge the contributions of H. Spiro to the initial definition and implementation of the input language and the efforts of M. J. Goldberg and W. R. Reini for the preparation of ASTAP as a released product.

2.6 REFERENCES

1. MAHONEY, G. W., A. J. JIMENEZ, H. QASSEMZADEH, T. R. SCOTT, W. T. WEEKS, and D. A. MEHTA, "A Statistical Analysis Program for Computer-Aided Circuit Design," *Digest of Technical Papers*, 1973 IEEE Southeast Conference, April 30–May 2, 1973, pp. c-1-1, c-1-6.

2. WEEKS, W. T., A. J. JIMENEZ, G. W. MAHONEY, H. QASSEMZADEH, and T. R. SCOTT, "Algorithms for ASTAP—A Network-Analysis Program"; *IEEE Trans. Circuit Theory*, CT20, No. 6, Nov. 1973, pp. 628–634.

3. HACHTEL, G. D., R. D. BRAYTON, and F. G. GUSTAVSON, "The Sparse Tableau Approach to Network Analysis and Design", *IEEE Trans. Circuit Theory*, CT-18, No. 1, Jan. 1971, p. 101.

4. BRAYTON, R. D., F. G. GUSTAVSON, and G. D. HACHTEL, "A New Efficient Algorithm for Solving Differential-Algebraic Systems Using Implicit Backward Differentiation Formulae," *IBM Research Report RC 3381*, IBM Thomas J. Watson Research Center, Yorktown Heights, N.Y., June 1, 1971.

5. GEAR, C. W., "The Automatic Integration of Stiff Ordinary Differential Equations," *Proc. 1968 IFIPS Congress*, Edinburgh, Aug. 1968, pp. A81–A85.

6. FOX, P. E., and F. C. WERNICKE, "Statistical Analysis of Propagation Delay in Digital Integrated Circuits," *Digest of Technical Papers*, 1972 IEEE International Solid-State Circuits Conference, p. 66, Philadelphia, Penn., Feb. 1972.

7. JENSEN, R. W., and M. D. LIEBERMAN, *IBM Electronic Circuit Analysis Program Techniques and Applications*, Prentice-Hall, Englewood Cliffs, N.J., 1968.

8. BOWERS, J. C., and S. R. SEDORE, *SCEPTRE: A Computer Program for Circuit and System Analysis*, Prentice-Hall, Englewood Cliffs, N.J., 1971.

3

BELAC

C. A. GEORGE

General Electric Company
Aerospace Electronic Systems
Utica, New York

3.1 INTRODUCTION

BELAC is a digital computer program which was developed as a tool for computer-aided network design. Since many physical systems admit to a network model representation, this program has application in many seemingly unrelated physical disciplines. BELAC has seen its greatest applications in the analysis of electrical circuits but has also been applied to mechanical, thermal,* and servomechanism systems. It was written for the user who is qualified in his area but unfamiliar with numerical analysis and digital computer programming. However, provisions have been made so that those familiar with FORTRAN programming can add routines that model special components and/or add special analysis capabilities. The standard routines in BELAC are capable of analyzing both linear and nonlinear networks, performing ac, dc, and transient analyses, and have a multitude of significant features which makes BELAC exceptionally versatile, powerful, and convenient to use. Using the easy-to-learn input format, it is only necessary to describe the network to be analyzed, indicate the analyses desired, and state the output that is to be generated. The results are printed neatly in tabular form, are plotted if desired, and are quite suitable for inclusion into records and reports. If the model is nonlinear and a small-signal analysis is desired, the nonlinear dc quiescent solution will be calculated and the nonlinearities automatically linearized about that solution so that the small-signal analysis can then be performed. The small-signal analysis could be, for example, the calculation of the poles and zeros, or the determination of the input impedance. The input consists of a description of the components and

---

*T. J. Rossiter, "THEROS, A Computer Program For Heat Flow Analysis", RADC-TR-74-113 Rome Air Development Center, Griffiss Air Force Base, N.Y., April 1974.

their interconnections. Some of the elements recognized by BELAC are

- Resistors, capacitors, inductors
- Ideal independent voltage and current generators
- Ideal current- or voltage-controlled dependent current generators
- Integrators and differentiators
- Transformers
- Time delay (presently in the frequency domain only)
- Two-port parameters (h, y, z, a, g, s)
- Transfer functions
- Transmission lines (presently in the frequency domain only)
- Nonlinear voltage-controlled voltage and current generators
- Nonlinear capacitors
- Elements defined by user-supplied FORTRAN routines in the input data (may be nonlinear, time-dependent, or frequency-dependent)

A FORTRAN built-in compiler allows the user to add code to the program at execution time. This code is interspersed in the input data and is identified by the fact that "line numbers" appear at the extreme left of the input statement. BELAC will collect all input statements with line numbers and put them in sequence. It will then convert the FORTRAN to machine code and place the code in the dynamic memory area. Such coding can be used to define new elements. For example, the complete Ebers-Moll model has been defined as an element so that the model can be invoked by one input statement that contains the values for a particular device. This is not to be confused with putting a model which contains many elements on the model library (which is also allowed in BELAC). Rather, the use of FORTRAN coding allows the defining of a new element (e.g., a resistor with a resistance that is a function of time or frequency). Thus, the user is not confined by the program's repertoire of elements when describing his circuit or when building models. If he needs the program to recognize a transformer with hysteresis, he can add such an element so that he can then reference it in the same manner as he references elements intrinsic to BELAC.

FORTRAN code has been used to add an element which models a transmission line specified by a real characteristic impedance, a length, an effective dielectric constant, a dielectric loss in dB/inch, and a conductor loss in dB/inch. The effective dielectric constant may be a tabular function of frequency. An element which models a pair of couple lines specified in a similar manner, except that there are even and odd mode impedances, effective dielectric constants, and losses, has also been coded.

Element parameter values may be correlated (e.g., three transmission line lengths may be declared to be equal during an optimization which will vary them).

The model library may be used to store BELAC input data that is expected to be used repeatedly. Most often the library is used to store device models each composed of many interconnected elements. This allows one BELAC input statement to cause the entire set of interconnected elements that models some device to be read from the library and placed in the BELAC input data stream. However, the library may be used to store other types of data. For example, FORTRAN coding defining an element may be stored and then retrieved with one input statement.

One of BELAC's most important assets is its simple and flexible input format.

The input language spares the user most of the preliminary calculations and modeling that the computer can do. For example, a resistance of 1000 ohms may be entered as 1K, as engineers are prone to do. A conductance of $1/\sqrt{.7}$ mhos can be entered by the expression 1/SQRT(0.7). The engineering-oriented input language can be learned in functional groups, so that the user need not be familiar with all the features of the program to use it. It is possible to learn to solve some types of problems in a few minutes.

The possible outputs include the value of node and branch voltages, small-signal ac gains, dc transfer characteristics, charge, flux, impedances, and currents—plotted and tabulated. BELAC also can calculate the sensitivities of these outputs to parameter value variations, optimize these outputs relative to a desired response function by varying the parameter values, perform Monte Carlo simulations, and do worst-case analyses. These capabilities exist for both small-signal ac and nonlinear dc networks. In addition, BELAC has transient, and pole-zero capabilities. This is outlined in Fig. 3.1.

The addition of new routines (including complete programs) to BELAC can be easily accomplished due to BELAC's structure, which allows such

Large signal dc	*Small signal ac**
Optimum parameter values	Optimum parameter values
Transfer characteristic plot	Transfer function plot
Driving point characteristic	Driving point function plot
Sensitivity to parameter values	Sensitivity to parameter values
Monte Carlo results	Monte Carlo results
Worst-case values	Worst-case values

Other analyses	
Time domain	*If the network is nonlinear
Pole/zero*	BELAC will linearize it before doing these calculations

Figure 3.1 BELAC Results.

routines to be added as new links. The present input link would be common to any new links; therefore, the features of the new links would become available in the present BELAC input language. These new links could also use the equation writing and solving routines already in BELAC, as well as its plotting routine. Moreover, they can operate on data produced by some of the present routines to produce additional output. An FFT routine, for example, has been added so that it receives as input the frequency response generated by BELAC and produces a tabulation of the time response, it can also transform a BELAC-produced time response into the frequency domain. The entire CORNAP* program has been incorporated into BELAC in this manner and is presently operable. NASAP† had been incorporated into an earlier version of BELAC but the interface would require some updating to make it operable with the present version.

Every digital computer program contains the results of trade-offs made by the author. We invariably tried to make BELAC convenient to use by making it flexible. For example, it is our desire to be able to answer "yes" to questions such as

- I want to remove a capacitor from my input data and make another computer run. Can I do it without relabeling nodes?
- Can the computer tune my IF strip before doing the rest of the analysis?
- I know nothing about numerical integration. Can the computer pick an integration step size?
- I know the transfer function of a black box in my circuit. Can I enter it directly?
- I work in radians per second. Can BELAC?
- I have common base y-parameters. Can I use them?
- I have common emitter h-parameters. Can I use them?
- Can the computer linearize the network before doing an ac optimization?
- Can BELAC optimize for small-signal ac and nonlinear dc simultaneously?
- Can I use y-parameters that are a function of frequency?
- Can I use frequency-dependent scattering parameters?
- Can coupled microstrip lines have different frequency-dependent effective constants for the even and odd modes.

The program is simple to use. Its use may appear complex during the following attempt to display so many details; however, *very little of the*

---

*C. Pottle, "State-Space Techniques for General Active Network Analysis," *System Analysis by Digital Computer*, F. F. Kuo and J. F. Kaiser, eds., Wiley, New York, 1966.

†L. P. McNamee and H. Potash, "A User's and Programmer's Manual for NASAP," *Report 68-38*, UCLA, Los Angeles, Aug. 1968.

information in this chapter will pertain to any one computer run, since only a few features will be used in any given run. It is easy to learn the features needed for one type of analysis. Only when almost the entire input language is presented, as in this chapter, might the use of the program appear less than simple. The reader should not assume BELACs flexibility preclude ease of use.

All the flexibility of BELAC is not described here. Many types of input statements are not mentioned, and many possible analyses are left unillustrated: calculating sensitivities using the adjoint network technique, for example, or producing plots of the *v-i* characteristics of a transistor model (just as pretty as from transistor specification sheets), a Monte Carlo of a distributed network, the sensitivity of dc bias conditions to element parameter changes, optimization of microwave networks, Fast Fourier Transforms, FORTRAN code in the input data, ad infinitum.

Sections 3.2 through 3.6 define how to prepare input data for BELAC. They are written so that they form a user's reference guide. *This introductory section is to acquaint readers with the general features of BELAC, details being deferred to the later sections.* During the first reading, the reader should read this section rather quickly and should not be concerned if many questions occur to him for which answers do not immediately appear in the text.

It is important to note that very good correlation between calculated values and measured values have been obtained, even at times when the transistor model was only approximately known. This may be because an engineer attempts to design for small sensitivities to parameters that are not accurately known or tightly controlled.

BELAC can be run as a nonconversational program in which case all the input data are submitted to the computer and all the results are calculated and printed and then presented to the user simultaneously; or BELAC can be run as a conversational program in which case it will obtain the input data from a remote terminal as they are needed and print the results on the terminal as they are calculated. This allows the user to interact with BELAC so as to be able to accomplish tasks that would require much more elapsed time in the non-conversational mode. For example, if the user wanted to select the resistance value that produces an optimum time response, he could submit a nonconversational computer run which will produce the time responses for a few different values of resistance. After he receives the results (say a few hours later) he could pick the response closest to his needs and submit a new run with additional resistance values to determine a still-improved response. In this manner he could, after a few runs, "home-in" on the desired resistance value and the corresponding time response. Alternatively, he could run BELAC in a conversational mode. In this case he would submit a resistance value through a remote terminal and receive back, almost at once, the time response. He could then submit a new value and again receive the corresponding response. In this fashion he could iterate to the desired time response with much less lapse time (say an hour instead of a couple of days).

3.1.1 Routines

The basic routines of BELAC include a routine that solves a set of linear algebraic equations with complex coefficients, an eigenvalue determination routine, and a differential-equation-solving routine. These are standard routines and are available at most computer installations. However, as is well known, the effort required to produce the network's equation often precludes the use of these routines to simulate anything but very small networks. This effort is minimized by BELAC routines which generate the equations from a description of the network. These equation-generating routines and the previously mentioned equation-solving routines are used by other routines that produce a variety of information about the network operation. The functions of some of these BELAC routines are described next.

The *input routine* determines if a run is conversational, reads input data and stores it in appropriate arrays, performs diagnostic tests on the input data, senses input instructions to make changes or deletions in data previously read and operated upon, makes note of the types of analyses that are to be performed, allows jobs to be stacked, performs network designs if desired, terminates the run after the last job, generates and uses the model library, and performs other similar organizational functions.

The *index minimization* routine varies the network parameters to minimize a performance index and can therefore be used to optimize the network parameters, to find the worst-case network parameter values, or to perform operations within the computer that are analogous to tuning components in the physical system. The user specifies the nominal, the maximum, and the minimum network parameter values of the variable elements.

The *sweep routine* calculates values of the output variables over a range of complex frequencies, dc voltages, dc currents, or parameter values. This routine allows the generation of Bode, Nyquist, and root locus plots and dc characteristic curves. For example, it can produce a plot of some dc voltage versus some resistance value, or the dc characteristic curve relating collector current to collector voltage for various values of base current, or the small-signal ac gain of a network.

The *sensitivities routine* calculates the sensitivities of the output variables to changes in the network parameter values, i.e., the amount an output variable will change if a network parameter value changes 1%. Calculations are performed for selected network parameters.

The *Monte Carlo routine* is used to determine the statistics of the output variables from the statistics of the network parameters. This routine simulates the network many times. Each time the network is simulated, network parameter values are randomly chosen (as when components are picked from bins). Each simulated network is analyzed, and the results of all the analyses are used to calculate the statistics of the output variables.

The *worst-case routine* determines the maximum and/or the minimum values the output variables can take on when the networks parameter values are constrained to remain between given minimum and maximum values. This routine assumes that the minimum and/or the maximum values of the output variables occur when the network parameter values are at their extreme allowable values. (The routine checks this and prints a warning if it is not true.) When it is not true the index minimization routine should be used to find the worst-case values. Although the index minimization routine is more general than this routine, it is usually more expensive to use when there is more than one output variable and/or more than one frequency of interest.

The *poles and zeros routine* calculates the poles and zeros of the output variables' frequency functions. Therefore, this routine can be used to determine expressions for driving point and transfer functions in terms of the complex frequency variable.

The *time response routine* calculates values of the output variables as functions of time. The method used to integrate the differential equation is given in C. W. Gear, "Simultaneous Numerical Solution of Differential-Algebraic Equations" *IEEE Trans. Circuit Theory*, Jan. 1971.

3.1.2 Examples

Examples of networks and their analysis are presented next. These examples are expected to raise many questions in the reader's mind since he has not yet read the sections which define BELAC's input data format. However, reading these examples will make those sections more meaningful. It will undoubtedly be desirable to refer to these examples when reading later sections. We shall start with the analysis of the low-pass filter shown in Fig. 3.2.

Two equivalent sets of input data, describing the circuit and requesting a frequency response are given in Fig. 3.3. Either of the sets may be used. *Note that every input statement (line of data) is a special case of a general form in which data are separated by a comma, a left parenthesis*, a slash, a right parenthesis, a plus sign, an asterisk, or an equal sign. *To the right of an equal sign, only the comma can be used to separate data. To the left of an equal sign, the above-mentioned symbols (except the equal sign) can be used interchangeably.*

The leftmost characters in an input statement are called the identifier, since they identify the data in the input statement. For example, the **VG** identifies the second input statement in Fig. 3.3 as describing an independent

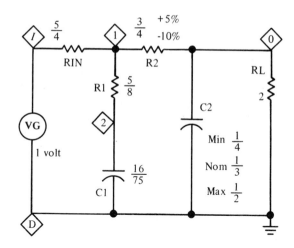

Figure 3.2 Low-pass Filter.

START	START
VG(I−D)=1	**VG*I,D**=1
RIN(I,1)=5/4	**RIN,I,1**=5/4
R1(1,2)=15/8	**R1,1,2**=15/8
C1(2,D)=16/75	**C1,2,D**=16/75
R2(1,0)=3/4	**R2−1,0**=3/4
C2(0,D)=1/3	**C2,0−D**=1/3
RL(0,D)=2	**RL*0*D**=2
SENSITIVITY=0,7,9	**SENSITIVITY**=0,7,9
RUN	**RUN**

Figure 3.3 BELAC Input Data.

voltage generator. This is a sinusoidal generator during a small-signal ac analysis, a wave-shape generator during a time domain analysis, and a short circuit during a nonlinear dc analysis. In addition to describing the circuit to BELAC, the user must command it to do an analysis. The **SENSITIVITY** input statement is a command to calculate the small-signal frequency response at dc, 7 hertz, and 9 hertz. The results are shown in Fig. 3.4. (The reason the word sensitivity is used will become clearer in the sequel, as will the boldface.)

Note that the voltages are identified by node labels. When the letter D appears in a numeral it indicates that the numeral to the left of the D is to be multiplied by 10 raised to the numeral to the right of the D. As shown in the previous example, the input data can be written in more than one format. Although BELAC is not sensitive to the symbol used to separate the two node labels in the **VG** entry, the minus sign between them reminds the user that BELAC was programmed so that **VG(I−D)**=1 states that $V_I - V_D = 1$. What do you think **VV(1−2/3−4)**=7 models? It models a voltage-controlled

HERTZ	OUTPUT	MAGNITUDE	DB	ANGLE
0.0	I	1.0000000D 00	0.	0.
	1	6.8750000D−01	−3.2545	0.
	2	6.8749999D−01	−3.2545	0.
	0	5.0000000D−01	−6.0206	0.
7.0	I	1.0000000D 00	0.	0.
	1	3.0140469D−01	−10.4170	−3.2313
	2	1.7104550D−02	−35.3378	−89.9781
	0	2.7199612D−02	−31.3087	−86.1034
9.0	I	1.0000000D 00	0.	0.
	1	3.0085223D−01	−10.4329	−2.5210
	2	1.3287610D−02	−37.5311	−89.9896
	0	2.1180994D−02	−33.4811	−86.9658

Figure 3.4 Computer Output.

voltage generator connected between the nodes labeled 1 and 2 such that

$$\frac{V_1-V_2}{V_3-V_4}=7$$

Likewise, **VS**(A−B/C−D)=1,2,3,4,5,6 models an element with a voltage transfer function

$$\frac{V_A-V_B}{V_C-V_D}=\frac{1+3s+5s^2}{2+4s+6s^2}$$

However, **VS**,A,B,C,D=1,2,3,4,5,6 would be interpreted by BELAC in exactly the same way (so would **VS**,A∗B−C∗D=1,2,3,4,5,6).

The identifier of an input statement that represents a network element usually indicates the type element by its first one or two letters. The identifier of an input statement that represents a command usually does so by its first eight characters. Other sections in this chapter contain descriptions of some of the identifiers recognized by BELAC. In those sections and elsewhere, the symbol # is used to indicate that the character replacing it will not be used by BELAC to identify the type of element or the command represented by the input statement. The reader should briefly look at Sec. 3.6 before continuing to read this section.

Recall that the output produced by the input data in Fig. 3.3 included the voltage at every node. For a large circuit, this could be an overwhelming amount of data. Assume that in our example we were only interested in the voltage at node O and the current in RIN. We obtained the voltages at all the nodes because we did not specify the network variables of interest; if we had, only those variables would have been printed. This is done by including an **OA** input statement in the data, as shown in Fig. 3.5.

START, LOW PASS FILTER
RIN(I,1)=5/4
R1(1,2)=15/8
C1(2,D)=15/75
R2(1,0)=3/4
RL(0,D)=2
C2(0,D). 1/3
TITLE,FREQUENCY RESPONSE
VG(I−D)=1
SWEEP RANGE=0,.02,.6
OA, VOLTAGE,O
OA(*CURRENT*,RIN)*PRINTER*,PLOT
RUN

Figure 3.5 Input Data.

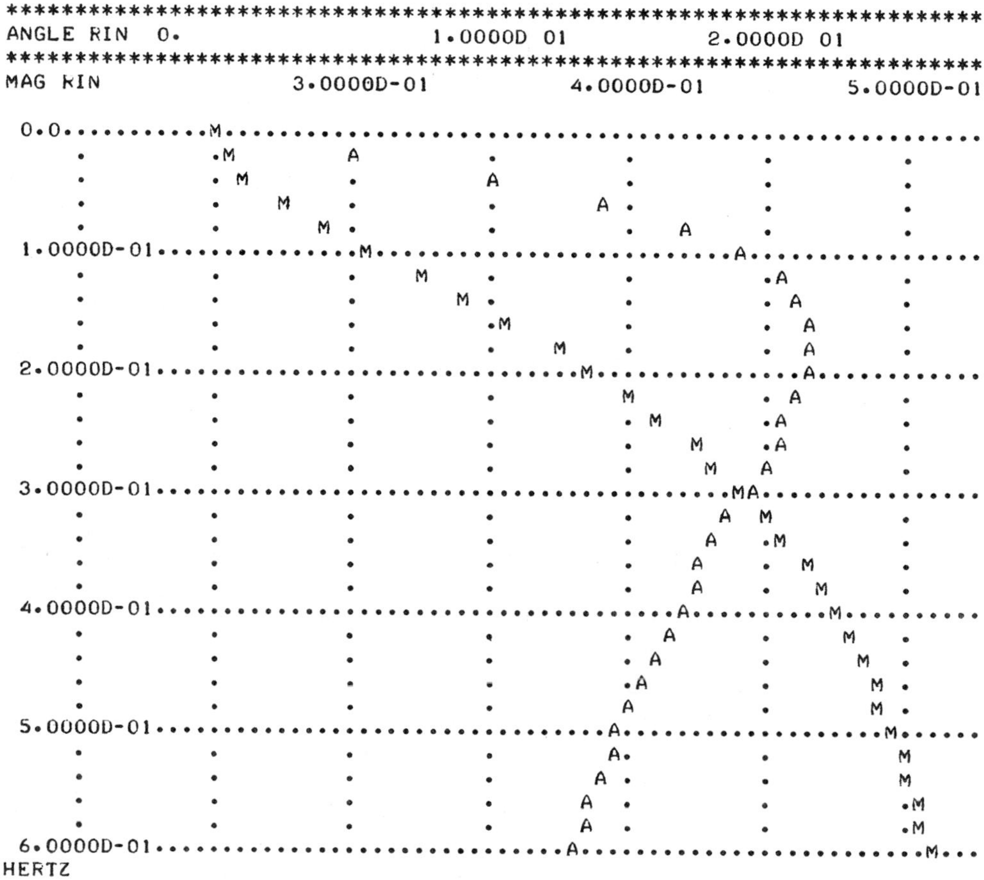

```
****************************************************************
ANGLE RIN  0.               1.0000D 01           2.0000D 01
****************************************************************
MAG RIN         3.0000D-01          4.0000D-01          5.0000D-01

 0.0..........M.................................................
    .         .M        A       .         .         .         .
    .         . M       .       A         .         .         .
    .         .   M     .       .         A .       .         .
    .         .     M   .       .         .   A     .         .
 1.0000D-01..............M..........................A..........
    .         .         . M     .         .         .A        .
    .         .         .    M  .         .         . A       .
    .         .         .       .M        .         .  A      .
    .         .         .       .    M    .         .   A     .
 2.0000D-01.....................................M........A.....
    .         .         .       .         .   M     . A       .
    .         .         .       .         .     M   .A        .
    .         .         .       .         .       M A         .
 3.0000D-01.................................................MA.
    .         .         .       .         .         A   M     .
    .         .         .       .         .         A    .M   .
    .         .         .       .         .         A    .  M .
    .         .         .       .         .         A    .   M.
 4.0000D-01..........................................A........M.
    .         .         .       .         .       A       . M .
    .         .         .       .         .      . A      .  M.
    .         .         .       .         .     .A        .   M.
    .         .         .       .         .     A         .    M.
 5.0000D-01................................................A.........M....
    .         .         .       .         .       A .     .      M
    .         .         .       .         .      A .      .      M
    .         .         .       .         .     A .       .      .M
    .         .         .       .         .    A .        .      .M
 6.0000D-01....................................A.................M...
HERTZ
```

Figure 3.6 Frequency Response.

The **SWEEP RANGE** input statement causes a frequency sweep from dc to 0.6 hertz in steps of 0.02 hertz. The current in RIN and the node voltage at node O are requested as output variables by the **OA** input statements. The pages are to be titled FREQUENCY RESPONSE, and a printer plot of the current in RIN is to be generated. It is shown in Fig. 3.6.

Recall that in the output tabulations previously shown the outputs were identified by node labels. At times, it may be preferable to label the outputs with words, such as GAIN, VIN, VOUT, ZOUT, etc. This is done by including these characters (no more than six) in the **O** input statement as shown in Fig. 3.7.

```
START,LOW PASS FILTER DESIGN
RIN(I,1)=5/4,PERCENT=1
R1(1,2)=15/8
TITLE2,DESIGN NUMBER ONE
TITLE1,LOW PASS FILTER
C1(2,D)=16/75
SENSITIVITY,RPS=2,5
R2,(1,0)=3/4
C2(O,D)=1/3
RL(O,D)=2,PERCENT=1
VG(I−D)=1
SWEEP RANGE(RPS,LOG)=.1,20,30
OAVOUT,VOLTAGE,O,PLOTTER PLOT
OEV1,VOLTAGE,1
OSIRIN(CURRENT,RIN)
OTIRIN(CURRENT,RIN)PLOTTER PLOT
TIME RESPONSE=0,.025,2
RUN
```

Figure 3.7 Input Data.

The output, that is, the result of the sensitivity input statement and the percentages that appear in the input statements describing RIN and RL, is shown in Fig. 3.8. Three sets of data appear in the output. The first is the value of VOUT and IRIN at 2 and 5 rps. These data were generated because of the presence of the **SENSITIVITY** input statement. The next set of data is the sensitivity of VOUT and IRIN to changes in the value of RIN. The last set of data gives the sensitivities to changes in RL. The sensitivities were generated because the input data contained both the **SENSITIVITY** input statement and the percentages on the **RIN** and **RL** input statement. The **TIME RESPONSE** input statement causes the step responses to be calculated. Similarly, the **SWEEP RANGE** input statement causes a frequency response to be generated (in terms of rps).

DESIGN NUMBER ONE

FOR NOMINAL VALUES OF THE PARAMETERS

RPS	OUTPUT	MAGNITUDE	DB	ANGLE
2.0000D 00	VOUT	3.4362771D–01	−9.2782	−41.0353
	IRIN	4.4854610D–01	−6.9639	18.7904
5.0000D 00	VOUT	2.0069247D–01	−13.9494	−61.2060
	IRIN	5.2675986D–01	−5.5677	10.5892

SENSITIVITIES WITH RESPECT TO RIN PARAMETER VALUE = 1.250000D 00

OUTPUT CHANGE/PERCENT PARAMETER CHANGE

RPS	OUTPUT	MAGNITUDE	DB	ANGLE
2.0000D 00	VOUT	−1.8148969D–03	−4.5875D–02	−1.0293D–01
	IRIN	−2.3690316D–03	−4.5875D–02	−1.0293D–01
5.0000D 00	VOUT	−1.2907458D–03	−5.5863D–02	−6.8882D–02
	IRIN	−3.3878355D–03	−5.5863D–02	−6.8882D–02

SENSITIVITIES WITH RESPECT TO RL PARAMETER = 2.000000D 00

OUTPUT CHANGE/PERCENT PARAMETER CHANGE

RPS	OUTPUT	MAGNITUDE	DB	ANGLE
2.0000D 00	VOUT	1.0639995D–03	2.6895D–02	−1.2291D–01
	IRIN	1.1209234D–04	2.1706D–03	7.3510D–02
5.0000D 00	VOUT	2.6539548D–04	1.1486D–02	−1.1440D–01
	IRIN	1.3648016D–04	2.2505D–03	1.5849D–02

Figure 3.8 Sensitivities of VOUT and IRIN to RIN and RL.

The user may sometimes want a circuit variable to be an output during one of the analyses but not during another. This is accomplished by using a letter other than **A** after the **O** in the output statement requesting the output. In the present example, IRIN will be an output in the sensitivity and time domain analysis only; V1 will be printed during the frequency sweep only; VOUT will be an output in all the analyses performed. The time response plot appears in Fig. 3.9.

The philosophy used is to build into the program parameters which the user can override if he wishes. For example, the datum node is assumed to be labeled D, but any one of the nodes can be made the datum node by use of a **DATUM** input statement. As another example of overriding an assumed option, consider the independent generator which was assumed to be a unit

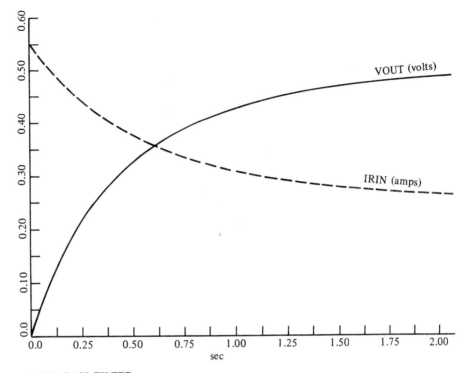

LOW - PASS FILTER
DESIGN NUMBER ONE

Figure 3.9 Plotter Plot of Time Response.

step generator in the time domain analysis. By use of a **DRIVE** input statement, functions other than a step can be specified. For example,

$$\textbf{DRIVE } (\textbf{\textit{TRAP}}) = 0,1,2,2.5$$

causes the voltage across the generator during a time domain analysis to be as shown below.

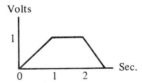

The options are too numerous to be listed here, but it is important to state that they are not frills. Many were added to the program in response to a need discovered during a program application. Their use can often reduce the time required to prepare the input data and/or the computer time used. Thus, the time spent learning to use them can be considered to be an investment.

Previous analysis produced the gain from node I to node O. Suppose that we also need the gain from 1 to O, in which case we could ask for the voltages at O and 1 and manually calculate the gain. More simply, **OAGAIN**, *RATIO* (O/1) will cause the node voltage at O to be divided by the node voltage at 1 and the results printed. Likewise, **OAPOW**, *PRODUCTS* (O∗IRL) will cause the power in RL to be calculated and printed. Consider now the data in Fig. 3.10.

```
START,LOW PASS FILTER DESIGN
RIN(1,1)=5/4
R1,1,2=15/8
C1,2,D=16/75
R2,1,0=3/4
C2(O,D)=1/3
RL(O,D)=2
VG,I,D=1
SWEEP RANGE(RPS)=0,4,18
OAVOUT,VOLTAGE,O
OAV1,VOLTAGE,1
OAIRL,CURRENT,RL
OFGAIN,RATIO(O/1)
OFPOW,PRODUCTS(O)/IRL)
NATURAL,POLES,ZEROS
ALL
RUN
```

Figure 3.10 Input Data.

In Fig. 3.10, the **ALL** input statement is a command to print all node voltages in addition to any other outputs requested by **O** entries. The **NATURAL** input statement will cause the poles and zeros of the frequency functions to be calculated and printed. For the input data shown in Fig. 3.10,

$$\text{VOUT} = \frac{6(s+2.5)}{5(s+4)(s+1.5)}$$

The constant $\frac{6}{5}$ has to be manually calculated (unfortunately) as follows: Because of the **SWEEP RANGE** input statement we learned that $\text{VOUT} = \frac{1}{2}$ when $s=0$. Therefore,

$$\text{VOUT} = \frac{K(s+2.5)}{(s+4)(s+1.5)}\bigg|_{s=0} = \frac{1}{2}$$

from which $K = \frac{6}{5}$.

The values of VOUT, V1, and IRL; the gain from node 1 to O; and the power in RL were calculated by BELAC at 0, 4, 8, 12, 16, and 20 rps. For example, at 8 rps

$$\text{VOUT} = .138\underline{/-70}$$
$$\text{V1} = .335\underline{/-15}$$
$$\text{IRL} = .069\underline{/-70}$$
$$\text{GAIN} = .412\underline{/-55}$$
$$\text{POW} = .009\underline{/\ 0}$$

For illustrative purposes, this section describes many sets of input data relating to one circuit. However, *all the computer output described could be obtained with one set of input data and one computer run*. For example, a worst-case analysis could have been obtained during one of the previous analyses by slightly modifying the input data. For clarity, a worst-case analysis is described in Fig. 3.11 as a separate example.

```
START, WORST CASE EXAMPLE
RIN(I,1)=5/4
R1(1,2)=15/8
C1(2,D)=16/75
R2(1,O)=3/4,TOLERANCE=10
C2(O,D)=1/3,MIN=.25,MAX=.5
RL(O,D)=2
OAIRL,CURRENT,RL
OAVOUT,VOLTAGE,O
OA,VOLTAGE,1
WORST CASE=1
RUN
```

Figure 3.11 Worst-case Analysis.

The worst-case analysis, produced as a result of the **WORST CASE** input statement, shows that the current in RL is nominally 0.0843 ampere (i.e., when $R2 = \frac{3}{4}$ ohm and $C2 = \frac{1}{3}$ farad). The current is a minimum of 0.0574 ampere when R2=0.8250 and C2=0.5 and is a maximum of 0.1073 ampere when R2=0.675 and C2=0.25. Similar information pertaining to V1 and VOUT was also calculated and is summarized in the table in Fig. 3.12.

As a summary, consider the following problem. The circuit is as shown in Fig. 3.2. The desired outputs are the current through, the power dissipated in, and the voltage across RL; the voltage gain from node 1 to node O; and the voltage at node 1—for frequencies ranging from dc to 5 rps in steps of 1 rps. In addition, poles, zeros, and worst-case values are desired. The input data are shown in Fig. 3.13. The use of two slashes, as shown in Fig. 3.13, generally allows the inclusion of remarks.

Magnitude of	Value at 1 Hz	R2	C2
Current in RL IRL	Nominal 0.0843 Minimum 0.0579 Maximum 0.1073	0.7500 0.8250 0.6750	0.333 0.500 0.250
Voltage across RL VOUT	Nominal 0.1687 Minimum 0.1158 Maximum 0.2146	0.7600 0.8250 0.6750	0.333 0.500 0.250
Voltage between 1 and D V1	Nominal 0.3521 Minimum 0.3173 Maximum 0.3814	0.7500 0.6750 0.8250	0.333 0.500 0.250

Figure 3.12 Worst-case Results.

```
START, LOW PASS FILTER ANALYSIS EXAMPLE
RIN(I,1)=5/4                              //     THESE ENTRIES
R1(1,2)=15/8                              //     DESCRIBE
C1(2,D)=16/75                             //     THE
R2(1,O)=3/4,MIN=.9*3/4MAX=1.05*3/4        //     COMPONENTS
C2(O,D)=1/3,MIN=.25,MAX=.5                //     IN THE
RL(O,D)=2                                 //     CIRCUIT
VG(I-D)=1       // VOLTAGE AT I MINUS THAT AT D IS 1 VOLT
OFPOW,PRODUCTS(OIRL)                      //     THESE
OAIRL,CURRENT,RL                          //     ENTRIES
OAVOUT,VOLTAGE(O)                         //     DESCRIBE
OFGAIN,RATIO(O/1)                         //     THE DESIRED
OAV1,VOLTAGE,1                            //     OUTPUTS
NATURAL,POLES,ZEROS               //     NATURAL FREQUENCIES
INDEX MIN=1                       //     WORST CASE FREQUENCY
SWEEP RANGE,RPS=0,1,5             //     FREQUENCY SWEEP RANGE
RUN
```

Figure 3.13 Input Data.

The computer results include a tabulation of POW, IRL, VOUT, GAIN, and V1 versus frequency; poles and zeros of IRL, VOUT, and V1; and the worst-case values shown in Fig. 3.12. The significance of the input statements of Fig. 3.13 are described below:

Identifier	Significance of the Input Statement
START	Defines the start of an analysis
RIN	A $\frac{5}{4}$-ohm resistor is between nodes I and 1.
R1	A $\frac{15}{8}$-ohm resistor is between nodes 1 and 2.
C1	A $\frac{16}{75}$-farad capacitor is between nodes 2 and D.

Identifier	Significance of the Input Statement
R2	A nominally $\frac{3}{4}$-ohm resistor is between nodes 1 and 0; it has a resistance range of 0.675 to 0.825 ohms.
C2	A $\frac{1}{3}$-farad capacitor is connected between nodes O and D; it has a capacitance range of $\frac{1}{4}$ to $\frac{1}{2}$ farad.
RL	A 2-ohm resistor is connected between nodes O and D.
VG	A 1-volt signal generator is connected between nodes I and D with the plus terminal at node I.
Blank	A blank line for readability
OFPOW	The power dissipated in RL is requested as the voltage at node 0, times the conjugate of current in RL.
OAIRL	The current in RL is requested and is to be printed under the title "IRL."
OAVOUT	The voltage between nodes and 0 and D is requested and is to be printed under the title "VOUT."
OFGAIN	The voltage gain (the voltage at node O divided by the voltage at node 1) of the circuit is requested and will be printed under the title "GAIN."
OAV1	The voltage between nodes 1 and D is requested and will be printed under the title "V1."
NATURAL	The poles and zeros of the output variables IRL, VOUT, and V1 are requested.
INDEX MIN	A worst-case analysis is requested at a frequency of 1 hertz; worst case can be obtained using either the **WORST CASE** input statement or the **INDEX MIN** input statement as explained in Sec. 3.1.1.
SWEEP RANGE	A frequency sweep is desired from 0 to 5 rps in steps of 1 rps.
RUN	Indicates the end of the data for this activity

3.1.3 Example of BELAC'S Editing Features

Assume for a moment that BELAC had only simple editing capabilities; that is, a user could instruct BELAC to do an analysis, make some network changes, and repeat the analysis. Further, assume that each change in an element parameter value requires an input statement to describe the change. Then BELAC could conveniently be used to analyze a circuit, change the load resistor, and reanalyze the circuit. But how would the user study the effect of temperature changes on the circuit? A change in temperature could very likely affect the values of most of the parameters. In fact, the parameter values are typically not all similarly dependent on temperature. The input statements defining the changes could be almost as numerous as the original deck. Therefore, an input statement is needed that can change an entire set of parameter values by a given factor. (It is possible to group element parameter values so that they are correlated; however, due to space considerations the method is not described in this chapter.) For example, if all resistors with a given temperature coefficient are labeled such that HP is their third and fourth characters (R1HP, R2HP, R3HP, etc.), a method of changing all resistors so labeled by a given percentage could be used to reflect a temperature change. Figure 3.15 contains a set of input data that illustrates some of BELAC's editing features. The circuit is described in Fig. 3.14.

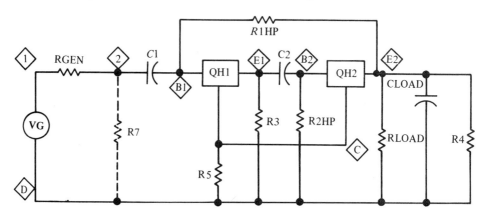

Figure 3.14 Network Schematic Diagram.

START, EXAMPLE FOR SECTION 3.1.3
VG(1,D) 1
R1HP(B1,E2)=22K
C1(2,B1)=.01U
RGEN(1,2)=100
R5(C,D)=200
R3(E1,D)=24K
R4(E2,D)=24K
R2HP(B2,D)=22K
C2(E1,B2)=.01U
RLOAD(E2,D)=50K
CLOAD(E2,D)=.01U
QH1(E1,C,B1)=10K,1,−360,45U   ///USING LOW FREQ H-PARM
QH2(E2,C,B2)=10K,1,−360−45U   //FOR TRANSISTORS
      //A FREQUENCY SWEEP IS CALLED FOR
      //FROM DC TO 700 HZ IN 20 HZ STEPS
SWEEP RANGE=0,20,700
OF,*VOLTAGE*,E2,*PRINTER* PLOT // NODE E2 VOLTAGE REQUEST

Figure 3.15 Example of BELAC's Editing Capability.

```
RUN,LIST /////////////////////////////////////////////////
        ////BELAC CALCULATES THE FREQUENCY SWEEP /////////
        ///// TABULATIONS AND PLOTS ARE PRODUCED ///////////
        ///////////////////////////////////////////////////
RLOAD(E2,D)-100K      // CHANGES RLOAD FROM 50K TO 100K
RUN,LIST/////////////////////////////////////////////////
        /////TABULAR AND PLOTTED INFORMATION AGAIN //////
CLOAD,DELETE       //CLOAD IS DELETED FROM THE CIRCUIT
R7(2,D)=5.1K       // R7 IS ADDED TO THE CIRCUIT
    ///OTHER EDITING CAPABILITIES EXIST LIKE
    ///THE MODIFY INSTRUCTION USED NEXT
MODIFY(..HP)=200
    ///ALL COMPONENTS HAVING HP IN THEIR 3TH AND 4TH
    ///IDENTIFIER CHARACTER POSITIONS ARE INCREASED
    ///BY 200 PERCENT
    ///THE PREVIOUS ANALYSIS WILL BE REPEATED
    ///WITH THE TWO 22K RESISTORS CHANGED TO 66K
    ///A (200 PERCENT INCREASE)
RUN,LIST/////////////////////////////////////////////////
        ////TABULAR AND PLOTTED INFORMATION AGAIN /////////
        ///////////////////////////////////////////////
MODIFY,DELETE
    ///THE 2 RESISTORS THAT WERE CHANGED IN THE
    ///LAST ACTIVITY ARE NOW 22K AGAIN
MODIFY (R. . . .)=20
    ///ALL RESISTANCES ARE INCREASED BY 20 PERCENT
RUN,LIST/////////////////////////////////////////////////
        /////TABULAR AND PLOTTED INFORMATION AGAIN/////////
        ///////////////////////////////////////////////
```

Figure 3.15—*Cont.*

3.1.4 Transformer-Coupled Circuit

Consider as another example the network and its model shown in Fig. 3.16. The input data are given in Fig. 3.17, and a plot of the output is shown in Fig. 3.18. The output shows the manner in which the gain of the circuit varies with the coupling coefficient of the transformer. The frequency response plot shown in Fig. 3.18 was produced with somewhat different input data than are shown in Fig. 3.17. The data described in Fig. 3.17 produce three separate plots. The plot in Fig. 3.18 was generated with input data structured like those used in the worst-case analysis of Sec. 3.3.

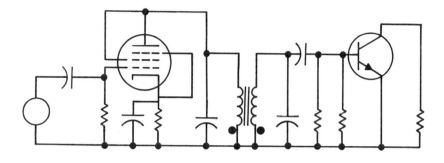

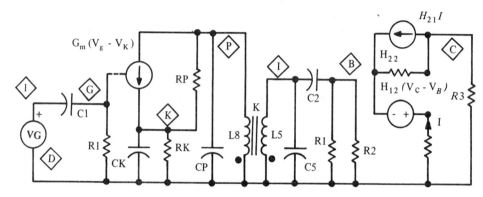

Figure 3.16 Network Schematic and Model.

```
START, TRANSFORMER COUPLED AMPLIFIER
VG(1−D)=1
CI(1,G)=9.E−8
RI(D,G)=.1M
IVTUBE(P,K/G−K)=1.65E−3        //GM OF TUBE
CK(D,K)=2U
RK(D,K)=500
RP(K,P)=4500                   //PLATE RESISTANCE
CP(D,P)=.1U
CS(D,I)=2U
C2(I,B)=50U
R1(B,D)=10K
R2(B,D)=.1M
QHCB(C,B,D)=3.3,3.E−5,−.98,6.E−7
R3(C,D)=5K
XF(D,P,D,I)=.001,50U,,.1
```

Figure 3.17 Input Data for Transformer Coupled Circuit.

```
FPLOT=1
OFGAIN,VOLTAGE(C)PLOTTER PLOT
SWEEP RANGE(RPS)=60K,500,150K
RUN
XF(D,P,D,I)=.001,5.E−5,,.3
RUN
XF(D,P,D,I)=.001,5.E−5,,.5
RUN
XF(D,P,D,I)=.001,5.E−5,,1.
RUN
```

Figure 3.17—*Cont*.

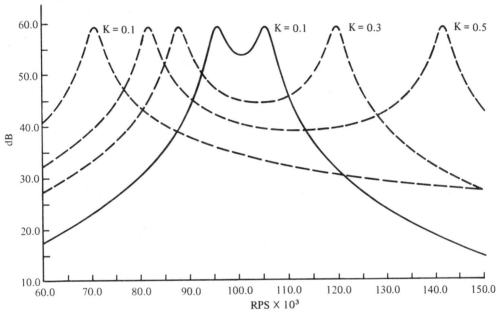

Figure 3.18 Plot of the Transformer-Coupled Amplifier Frequency Response.

The following description of the input data of Fig. 3.17 explains the meaning of each statement:

Identifier	Significance of the Input Statement
START	Defines the start of an analysis
VG	A 1-volt voltage generator is connected between nodes D and 1
CI	An 0.09-microfarad capacitor is connected between nodes 1 and G.

Identifier	Significance of the Input Statement
RI	An 0.1-megohm resistor is connected between nodes D and G.
IVTUBE	A current generator is connected between nodes and K and G; the current through it equals 0.00165 times the grid-to-cathode voltage.
CK	A 500-ohm resistor is connected between nodes D and K.
RK	A 2-microfarad capacitor is connected between nodes D and K.
RP	A 4500-ohm resistor is connected to nodes K and P.
CP through R2	Defines values and connections of CP, CS, C2, R1, and R2
QHCB	A transistor is connected to nodes C, B, and D; common base base h-parameters are specified.
R3	R3 is 5000 ohms and connects to nodes C and D.
XF	A transformer (XF) is connected to nodes D, P, and I; primary inductance is 1 millihenry, secondary inductance is 50 microhenries, and the coupling coefficient is 0.1.
FPLOT	Plot dB.
OFGAIN	The voltage at node C is desired; on the output sheet, this voltage will be identified by the title "GAIN."
SWEEP RANGE	The frequency sweep is to be produced at frequencies from 60 to 150 krps in steps of 500 rps.
RUN	The run input statement signals the program to perform whatever instructions it had received; in this case, only a frequency response has been requested.

After the frequency response is produced, the next input statement is read. Because this input statement has the same identifier (**XF**) as a previous input statement, the program will replace the previous **XF** parameters with the new ones. Then the next input statement is read. Since it is a **RUN** input statement, the program performs the calculations requested. That is, it produces a frequency response with a coupling coefficient of 0.3. The remaining input statements produce frequency responses for coupling coefficients of 0.5 and 1.0.

3.1.5 Filter Design

This example introduces BELAC's synthesis capability. The input data request BELAC to design a filter and then analyze it in a circuit described by the input data. The test circuit specified is very simple in this instance, but the filter could be analyzed while connected to a more complex circuit (e.g., a complete transistor amplifier). The test circuit and the designed filter are shown in Fig. 3.19. The frequency and time responses are shown in Fig. 3.20.
BELAC designed the Butterworth filter by use of closed-form expressions for the component values. Chebyshev filters can be similarly designed

Ch. 3 / George

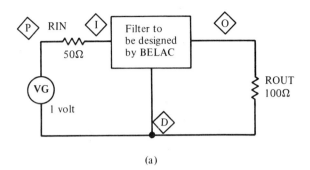

(a)

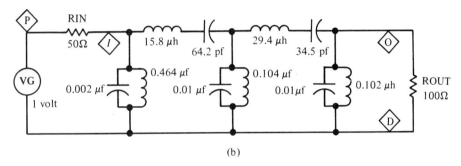

(b)

Figure 3.19 BELAC Designed Butterworth Filter: (a) Network Described by Input data; (b) Filter.

by BELAC. Design routines for other types of circuits can be added to BELAC by the user.

3.1.6 Amplifier Example

Consider the example network shown in Fig. 3.21.

It is desired to use BELAC to calculate the amplitude and phase of the node voltage at the nodes labeled V0 and V2 for frequencies in the 30- to 50-Hz range. The encircled **VG** in the diagram models a zero output impedance sinusoidal signal generator. The encircled **VV**, along with **RP**, models the electronic tube used in the physical network. The symbol **VG** indicates an ideal independent voltage generator, while the symbol **VV** indicates a generator that has a potential across it which is proportional to a potential difference elsewhere in the network, e.g., between the nodes labeled V2 and VK.

Since each element has been assigned a unique parameter value, it is possible to relate the elements to the input statements in the input data in Fig. 3.22. The first input statement is a **START** input statement, which indicates the start of a new job (i.e., the data that follow it are not related to any

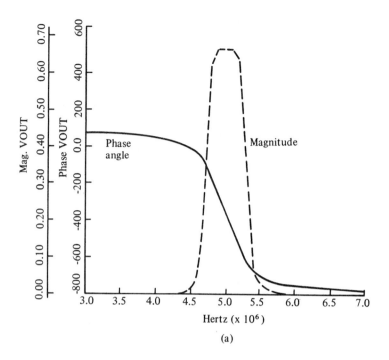

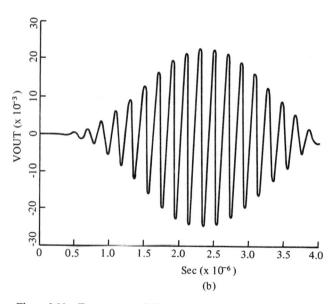

Figure 3.20 Frequency and Time Responses of BELAC Designed Butterworth Filter: (a) Plotter Plot of the Frequency Response; (b) Plotter Plot of the Step Response.

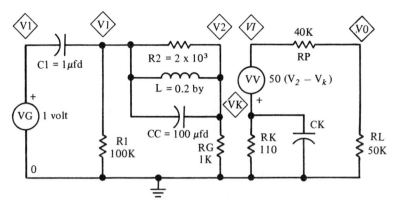

Figure 3.21 Network Schematic Diagram.

```
START, AMPLIFIER ANALYSIS
DATUM(0)
VG(VI−0)=1
CI(VI,V1)=1U
R1(0,V1)=100K
RC(V1,V2)=2K
LC(V1,V2)=.2
QC(V1,V2)=100U
RG(0,V2)=1000
RK(0,VK)=1100
CK(0,VK)=10U
VV(VK−VT/V2−VK)=50
RP(VT,V0)=40K
RL(0,V0)=50000
      //////////////////////////////////////////////////
      //THE CIRCUIT IS DESCRIBED ABOVE
      //THE DESIRED OUTPUT IS DESCRIBED BELOW
      //////////////////////////////////////////////////
OA,VOLTAGE(V0)
OA,VOLTAGE(V2)
SWEEP RANGE=30,2,50
RUN
FPRINT=1,1,1  //  PRINT DB,PHASE, AND MAGNITUDE
RL(0,V0)=5000
RUN
R1,DELETE
CF(V0,V2)=10U,PERCENT=3
SENSITIVITY(RPS)=5,6,7
RUN
```

Figure 3.22 BELAC Input Data for Amplifier Example.

previous input data that might have been read by BELAC). The **DATUM** input statement identifies the node that is to be used as the datum node (when this input statement is not present, D is assumed to be the datum node label). The next 12 entries describe the network. The **OA** input statements are instructions to tabulate the voltages at nodes V0 and V2. The **SWEEP RANGE** input statement defines the frequencies of interest to be 30, 32, 34, ..., 50 hertz. The **RUN** input statement defines the end of a set of data and signals BELAC to calculate the requested frequency sweep and print the desired data. The table shown in Fig. 3.23 contains a portion of the frequency

FREQUENCY RESPONSE FOR NOMINAL PARAMETER VALUES

HERTZ	OUTPUT	ANGLE	MAGNITUDE
3.000000D 01	V0	−88.678	4.559D 00
	V2	79.011	1.877D−01
3.200000D 01	V0	−90.147	4.989D−00
	V2	77.905	2.030D−01
3.400000D 01	V0	−93.293	5.620D 00
	V2	75.122	2.263D−01
3.600000D 01	V0	−98.058	4.297D 00
	V2	70.716	1.714D−01

Figure 3.23 Partial Tabulation of Node Voltages Versus Frequency.

response data. After BELAC completes its assigned task, the next input statement in the input data will be considered. If this were a **START** input statement, all previous input statements would be forgotten. In the present situation, however, the next input statement is a new instruction and is stored by BELAC. The **RL** input statement has the same identifier (**RL**) as a previous input statement which signals BELAC to change the stored value of **RL** (from 50,000 to 5000). Actually, the entire stored **RL** input statement is replaced with the new input statement. The **RUN** input statement will cause BELAC to again calculate and print the desired node voltages (nodes V0 and V2) for the desired frequency range (30–50 hertz). A portion of the output is shown in Fig. 3.24. The **R1** input statement then removes R1 from the stored network, and the **CF** input statement adds a new capacitor in a manner such that the sensitivity of the outputs to it can be calculated. The **SENSITIVITY** input statement is a command to calculate these sensitivities at the discrete frequencies listed. The **RUN** input statement again commands BELAC to generate the outputs requested, which now include the calculation of sensitivities as well as a frequency sweep. The sensitivity outputs are shown in Fig. 3.25 with RL equal to 5000, RL removed, and CF added. These sensitivities represent the amount the item would change if CF changed by 1 %.

FREQUENCY RESPONSE FOR NOMINAL PARAMETER VALUES

HERTZ	OUTPUT	DB	ANGLE	MAGNITUDE
3.000000D 01	V0	−2.099	−79.440	7.853D−01
	V2	−14.529	79.011	1.877D−01
3.200000D 01	V0	−1.227	−80.975	8.683D−01
	V2	−13.850	77.905	2.030D−01
3.400000D 01	V0	−0.109	−84.217	9.875D−01
	V2	−12.907	75.122	2.263D−01
3.600000D 01	V0	−2.366	−89.099	7.616D−01
	V2	−15.321	70.716	1.714D−01

Figure 3.24 Partial Tabulation of Node Voltages Versus Frequency.

FOR NOMINAL VALUES OF THE PARAMETERS

RPS	OUTPUT	MAGNITUDE	DB	ANGLE
5.0000E 00	V0	1.1490369D−02	−38.7933	−116.4492
	V2	4.7647353D−03	−46.4392	80.6638
6.0000E 00	V0	1.3395283D−02	−37.4610	−121.1995
	V2	5.6136505D−03	−45.0151	79.2049
7.0000E 00	V0	1.5135422D−02	−36.4001	−125.7103
	V2	6.4203818D−03	−43.8488	77.9195

SENSITIVITIES WITH RESPECT TO CF PARAMETER
VALUE=1.000000D=−05

OUTPUT CHANGE/PERCENT PARAMETER CHANGE

RPS	OUTPUT	MAGNITUDE	DB	ANGLE
5.0000E 00	V0	−1.5880536D−05	−1.2005D−02	−2.5637D−01
	V2	−4.2952784D−06	−7.8301D−03	−7.3014D−02
6.0000E 00	V0	−2.4989427D−05	−1.6198D−02	−2.9303D−01
	V2	−6.6919477D−06	−1.0354D−02	−7.6837D−02
7.0000E 00	V0	−3.5743016D−05	−2.0512D−02	−3.2415D−01
	V2	−9.4704478D−06	−1.2812D−02	−7.6950D−02

Figure 3.25 Tabulation of Sensitivities of Node Voltage to Changes in the Value of CF.

For example, a 1% increase in the value of CF will cause V0 at 5 rps to decrease from −38.7933 decibels (dB) to −38.7933−0.012005=−38.805305.

The numbers in Fig. 3.23 relate to a physical network for which Fig. 3.21 is a good model. A 1-volt sinusoidal generator is connected to terminals VI and O. The magnitude of the voltage at a node is measured, as is the amount by which the phase at that node *leads* the phase at node V1. These two values correspond to the numbers labeled MAGNITUDE and ANGLE, respectively.

Rather than the obvious interpretation that the generator is sinusoidal, another interpretation is that the generator is a white noise generator with a one-sided power density spectrum equal to unity. The numbers in the columns of Figs. 3.23 and 3.24 labeled MAGNITUDE represent the square root of the power density spectrum at two of the nodes.

The significance of the numbers in the columns labeled MAGNITUDE and ANGLE has been considered. If these numbers are considered to be polar coordinates of points in a plane, then the numbers in additional columns labeled REAL PART and IMAG PART would be Cartesian coordinates of the same points; these columns were not requested in this example and do not appear in the figures. The numbers in the column labeled DECIBEL are 20 times the logarithm of the corresponding magnitude and can be considered to be the gain referenced to the input if there is only one independent generator (input) and it has unity strength (as in this example). The columns labeled HERTZ (would be RPS, if radians were being used) contain the frequency at which the calculations were performed.

3.1.7 Nonlinear Optimization Example

This example serves to illustrate the solution of nonlinear networks. The information required about a nonlinear network might be the dc voltage at a node at various values of supply voltage, or it might be the small-signal gain of an amplifier at various values of supply voltage, or a histogram of the small-signal gain given the statistics of the parameters in a nonlinear transistor model, ad infinitum. This example illustrates the manner in which BELAC is commanded to produce such information. The network is described by Fig. 3.26.

Consider the input statements in Fig. 3.27 between the **START** and the first **RUN** input statements. The **PN** input statement models a diode, and the **VB** input statement models a battery. A method is needed to vary network

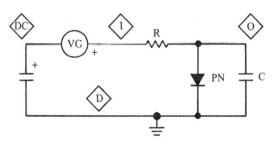

Figure 3.26 Nonlinear Example Network.

```
START
VB(DC−D)=0,1
VG(I−DC)=1
R,I,0=1/(40*U*EXP(20))
PN(O,D)=U,40
C(O,D)=2*40*U*EXP(20)
OA,VOLTAGE(O)PRINTER PLOT
INDEPENDENT VARIABLE=DC
INDEXMIN=.5,.52,.54,.56,.58,.6
RUN
INDEXMIN,DELETE
SWEEP RANGE=0,05,1
RUN
INDEPENDENT,DELETE
SWEEP RANGE, RPS=0,1.,2
INDEXMIN,RPS=0,.5,1,1.5,2
VB(DC−D)=0.5+(1−EXP(−20))/40
RUN
VB(DC−D)=.2,MIN=0,MAX=2
C(O,D)=E+3,MIN=E+2,MAX=E+5
SWEEPRANGE,DELETE
INDEXOPT=,,1,2,,3,200
INDEXMIN,RPS=DC,0,1
MAG=.5,.5,.5/SQRT(2),SIG=1,2,2
ANGLE=,,−45
RUN
```

Figure 3.27 Input Data.

parameters in a dc solution. In a time domain solution, the independent variable is time, and as time varies the value of a capacitor or the voltage across a signal generator might vary. Similarly in a dc solution we define an independent variable (δ). As δ varies, resistance values, battery voltages, or other parameters might vary. The **INDEPENDENT VARIABLE**=DC input statement identifies the values on the **INDEXMIN** statement to be values of the independent variable δ (rather than ω). The voltage across the battery modeled by the **VB** input statement equals $0+1\delta$ or simply δ (see Sec. 3.6.13). The effect is that during the index minimization effort the battery voltage will take on the values. .5, .52, .54, .56, .58, and .6 as BELAC assigns these values to δ (analogous to a 1-henry inductor taking on the impedance levels .5j, .52j, ... when ω takes on the values .5, .52, ...).

The index minimization will not actually take place since there are no parameters that are adjustable (see Sec. 3.5.1); however, the presence of the **INDEXMIN** input statement will cause the nominal response to be printed as shown in Fig. 3.28.

DELT	MAGNITUDE	DECIBEL	ANGLE
5.000D–01	4.8582D–01	−6.2705	0.
5.200D–01	4.9744D–01	−6.0652	0.
5.400D–01	5.0697D–01	−5.9004	0.
5.600D–01	5.1480D–01	−5.7672	0.
5.800D–01	5.2133D–01	−5.6578	0.
6.000D–01	5.2684D–01	−5.5664	0.

Figure 3.28 Index Minimization dc Response.

Similar information can be obtained by use of the **INDEPENDENT VARIABLE=DC** statement along with a **SWEEP RANGE** statement. To illustrate this, the input statements after the first **RUN** input statement delete the **INDEXMIN** input statement and add a **SWEEP RANGE** statement which causes δ to vary from zero to unity in steps of 0.05. The plot produced is shown in Fig. 3.29. (No plot was produced from the index minimization routine even though **PRINTER PLOT** was requested because that request applies only to the **SWEEP RANGE** and **TIME RESPONSE** input statements.) Note that the presence of the capacitor during a dc analysis causes no problems. Neither would the presence of inductors, differentiators, etc. The reader has probably noticed that the removal of the first **RUN** input statement and the **INDEXMIN,** *DELETE* input statement would affect the results only to the extent that both the index minimization and the sweep range analyses

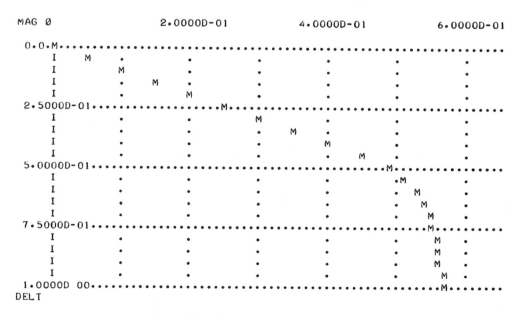

Figure 3.29 Sweep Range dc Response vs. Battery Voltage.

will take place in the first analysis set (see Sec. 3.2.1) rather than each having an analysis set devoted exclusively to a single analysis.

In the third analysis the **INDEPENDENT VARIABLE**=DC input statement is deleted, and the **SWEEP RANGE**, **INDEXMIN**, and **VB** input statements are replaced with new ones. Since there is no longer the **INDEPENDENT VARIABLE**=DC input statement, the numbers on the **SWEEP RANGE** and **INDEXMIN** input statements are values of ω (i.e., not δ). Therefore, the frequency sweep and the index minimization will be performed in the frequency domain. That is, in both analyses, the nonlinear network will be solved for the dc solution about which the network will then be linearized. Finally a frequency domain solution will be effected to produce the results shown in Figs. 3.30 and 3.31.

RPS	MAGNITUDE	DECIBEL	ANGLE
0.	5.0000D–01	−6.0206	0.
5.000D–01	4.4721D–01	−6.9897	−26.5651
1.000D 00	3.5355D–01	−9.0309	−45.0000
1.500D 00	2.7735D–01	−11.1394	−56.3099
2.000D 00	2.2361D–01	−13.0103	−63.4349

Figure 3.30   Index Minimization ac Response.

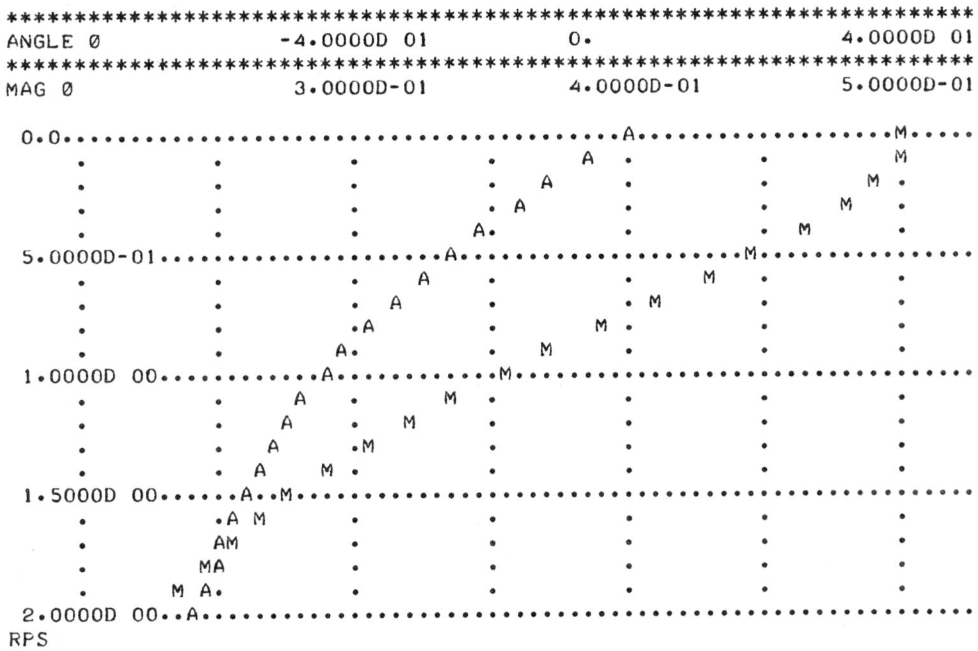

```
************************************************************
ANGLE 0         -4.0000D 01          0.              4.0000D 01
************************************************************
MAG 0            3.0000D-01       4.0000D-01         5.0000D-01

 0.0............................A...............M.....
        .          .         .     A    .       .      M
        .          .         .    . A   .       .    M .
        .          .         .    . A   .       .    M .
        .          .         .    A.    .       .  M   .
 5.0000D-01..................A............M.............
        .          .         A    .       . M   .      .
        .          .        . A   .       M .   .      .
        .          .        .A    .     M   .   .      .
        .          .       A.     . M       .   .      .
 1.0000D 00...............A........M...................
        .          . A     .  M    .        .    .     .
        .          .A     .M        .        .    .     .
        .          . A  M .         .        .    .     .
 1.5000D 00......A..M..................................
        .        .A M      .         .        .    .     .
        .         AM        .         .        .    .     .
        .         MA        .         .        .    .     .
        .        M A.       .         .        .    .     .
 2.0000D 00..A..........................................
RPS
```

Figure 3.31   Sweep Range ac Response.

BELAC recognizes that the network is nonlinear and automatically linearizes it before performing an ac analysis. However, there are times when the dc solution itself is important. For example, it is possible to optimize simultaneously to a given dc value and given ac values. If DC is present on the **INDEXMIN** input statement as in the fourth and last analysis, then the dc solution becomes part of the index minimization effort. Note that when DC was not present on the **INDEXMIN** input statement the dc values did not appear in the tabulation of output voltages in Fig. 3.30. However, when DC is present it is included (in Fig. 3.32 DC is indicated, as always, by 7.622D−06). During this last analysis BELAC is asked to find a battery voltage between 0 and 2 volts and a capacitance between 1000 and 100,000 farads (not a very useful result) which will minimize

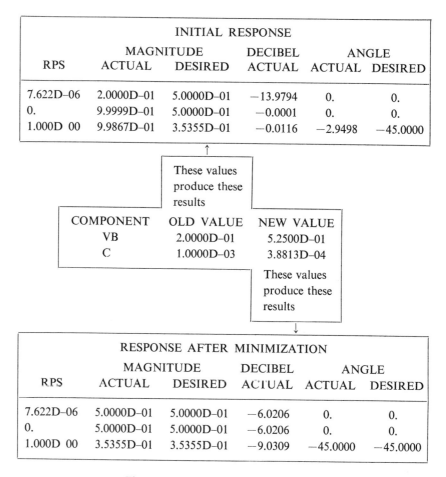

Figure 3.32 Index Minimization Results.

$$2\{1[.5-\mathrm{M(DC)}]^2+2[.5-\mathrm{M}(0)]^2+2\left[\frac{0.5}{\sqrt{2}}-\mathrm{M}(1)\right]^2\}$$
$$+1\{1[0-\mathrm{A(DC)}]^2+1[0-\mathrm{A}(0)]^2+1[-45-\mathrm{A}(1)]^2\}$$

where

The 1 and 2 in front of the braces are the weights from the **INDEXOPT** input statement.

The 1, 2, 2, in front of the first three brackets are the weights (*SIG*) on the **MAG** input statement and the three 1s in front of the last three brackets are assigned to BELAC since no weights appear on the **ANGLE** input statement.

The numbers inside the first three brackets are the desired quiescent dc voltage, and the magnitudes of the ac gain at 0 and 1 rps. The numbers inside the last three brackets are the desired angles.

M(DC) is the value of the quiescent voltage at node O. M(O) and M(1) are the magnitudes of the ac voltage at node O at zero and 1 rps., respectively. A(DC) is the angle of the dc voltage (equals zero in this case, might equal 180 in other networks). A(O) and A(1) are the angles of the ac voltage at node O. [A(O) is also zero in this case.]

M(DC), which is the *value* of the voltage at node O, is positive in this case, but in other networks it might be negative (and the angle would be 180°). However, for ac domain solutions such as M(O) and M(1), the values represent *magnitudes* and are always positive. This is a good place to point out the difference between a large-signal solution such as M(DC) and a small-signal solution such as M(0). M(DC) is the value of the quiescent voltage at node O. M(0) is the magnitude of the ratio of the change in M(DC) to a (small) change in the **VG** generator's voltage. If M(DC)=0.2, then the dc voltage at node O is 0.2. If M(0)=1, then changing the voltage generators (**VG**) voltage from zero dc to 0.001 dc will cause M(DC) to change from 0.2 to 0.201.

The results of the last analysis are given in Fig. 3.32. A different set of weights would give different (and possibly better) results.

3.2 INPUT DATA FORMAT AND MISCELLANEOUS INPUT STATEMENTS

Input data are read by the input routine and stored in memory for use by the rest of the program. These data can be considered to be composed of input statements. If computer cards are used as the computer input medium, one input statement will usually appear on each card; if a teletypewriter is used, one input statement will usually appear on each line. If a program or a graphic terminal builds a file, each record might be an input statement. However, an input statement may require more than one card, line, or record. Furthermore, a card, line, or record may contain more than one input statement. In this section the general format of the input data will be described along with certain input statements that are used by the input routine itself.

The first eight or less characters of an input statement identify the type of information contained in the input statement; therefore, they are called the input statements' identifier. The identifier (minus any characters assigned to a pound signs as described below) is used in the sequel to characterize the input statement in which it appears. For example, an input statement with the identifier **RUN** will be referred to as a **RUN** input statement.

3.2.1 Deck Setup

When computer cards are used the input deck will be as described in Fig. 3.33. If some other input medium is used, the concepts are the same and the description that follows applies if the words "card equivalent" is replaced with the appropriate word (e.g., with "line for teletypewriter input).

One computer run may be used to develop solutions to a series of *independent* problems, called jobs. A job is composed of one or more *dependent* problems, called *analysis sets*. The data pertinent to a computer run are inserted into card equivalents, which are then assembled into a deck equivalent, as shown in Fig. 3.33(a).

A job starts with a **START** input statement and end with a **RUN** input statement, followed by either a **START** input statement, a **FINISH** input

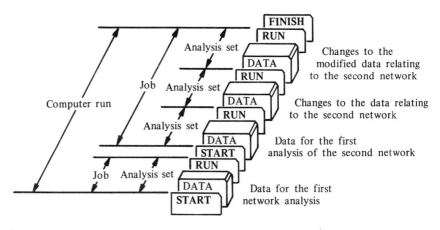

Figure 3.33 (a) Typical Input Card Deck.

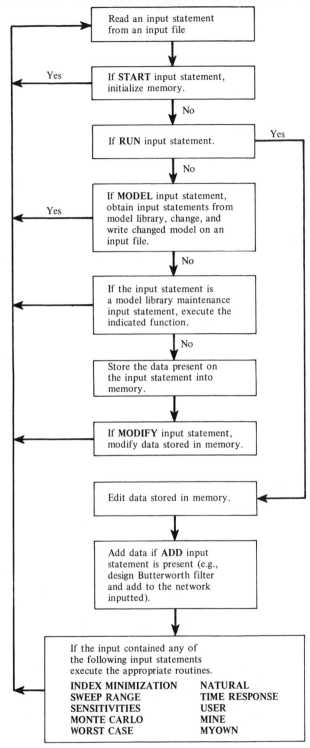

Figure 3.33—*Cont.* (b) Simplified Overall Flow Diagram.

Is the identifier stored in memory?	Does the second field contain the word *DELETE*	Action
Yes due to a previous analysis set	No	Replace old input statement with new
Yes due to this analysis set	No	Add this input statement to memory
Yes due to this or a previous analysis set	Yes	Delete this input statement from memory
No	No	Add this input statement to memory
No	Yes	Deleting a non-existing input statement

Figure 3.33—*Cont.* (c) Processing of Input Statements.

statement, or an end-of-file mark. The data for an analysis set are preceded by either a **START** or a **RUN** input statement and are followed by a **RUN** input statement. There is no limit to the number of analysis sets that can be included in a job. The last **RUN** input statement is followed by a **FINISH** input statement or an end-of-file mark. The sequence in which the input statements appear within the data is immaterial (except for very unusual usage of the **MODIFY** and library-generating input statements, to be described later in this section).

3.2.2 Simplified BELAC Job Flow Diagram

The analysis of a network starts with the computer reading input data that includes a topological description of the circuit and other pertinent data. For example, these data may include initial conditions for a time domain analysis. The data that are read by the input routine are stored in memory for use by the other routines.

Data are read and stored until a **RUN** input statement is encountered, at which time the **ADD** input statements that are stored, if any, will cause routines to be executed. These routines add data to memory in such a fashion that to the rest of the program these data appear no differently than they would appear if they had been included in the input. For example, a routine could realize a filter with a Butterworth amplitude characteristic and add the network description to memory. After the requested analyses are complete the data added to memory due to **ADD** input statements are forgotten. However, the **ADD** input statements are not forgotten. Additional data are then read. (See Figs. 3.33(b) and 3.33(c).) These additional reads give the user the option of changing any of the data stored in memory (including the data on an **ADD** input statement) and then performing additional analyses. One of the ways to change data is by the **MODIFY** input statement described later;

another way is to instruct BELAC to delete data or add new data, as described next.

When a **START** input statement encountered, BELAC discards previously stored data. The data on subsequent input statements are stored in memory until a **RUN** input statement is encountered, at which time BELAC follows the instructions contained in the stored data. When this effort is completed, additional input statements are processed as defined in Fig. 3.33 (this comment does not apply to the input statements **START**, **FINISH**, and **RUN** and all the model library-related input statements except **MODEL**).

3.2.3 Input Statement Description

An input statement can be considered to be composed of data separated by separators. To the left of the first equal sign a separator may be an asterisk, a minus sign, a plus sign, a slash, a left parenthesis, a right parenthesis, or a comma. To the right of an equal sign only commas and additional equal signs may be used as separators (not interchangeably). Although the equal sign is a separator, it has a different connotation than the other separators. The identifier is the first eight or less characters to the left of the first separator. If there are more than eight characters before the first separator the extra ones are ignored by BELAC. Only the characters in columns 1 through 72 are used by BELAC.

Comments may be inserted into an input statement by enclosing them in apostrophes or preceding them with two adjacent slashes. Such comments will be printed but otherwise ignored by BELAC. If only one apostrophe appears in the input statement, the comment consists of the characters between it and the end of the card equivalent. Likewise, when two adjacent slashes are used, the comment consists of the characters between the slashes and the end of the card equivalent. Semicolons may not be used as part of a comment.

Blanks may not separate the two adjacent slashes described in the preceding paragraph, nor may they be used arbitrarily in the **MODEL NAME IS** input statement described in the model library section. Otherwise, blanks are ignored by BELAC and can therefore be used to increase the readability of the input. An input statement containing only blanks causes a blank line to appear in the output.

The data for one input statement may appear on more than one card equivalent. This is done by placing the card equivalents in sequence and by using a comma or an equal sign as the last (noncomment) character on each card equivalent in the series except the last. Such a comma or equal sign serves its normal function, (i.e., a separator), as well as an indicator that the input statement is continued on the next card equivalent. No input statement can contain more than 2000 nonblank characters not counting those in comments.

More than one input statement may be placed in a card equivalent by separating them with semicolons. However, a semicolon may not separate a **RUN** or a **START** input statement from an input statement following it, nor can it be used to separate input statements obtained from a model library.

Some of the above statements are illustrated by the following example:

Card equivalents like the following can be used to input one input statement:

>XP2(A,9,C,D)=1,
>2, // MORE TO COME
>3,4 'ALMOST', 5,
>6, 'NOW' 7 // DONE

A card equivalent like the following is equivalent to those described above:

>XP2(A,9,C,D)=1,2,3,4,5,6,7

A card equivalent like the following can be used to input three input statements:

>**VG**(1,D)=1; **R**(2,1)=1; **C**, 'FIRST NODE='2, 'SECOND NODE=' D 'VALUE'=1

The following input statements are equivalent:

>**VG**(1,D)=1
>**R**(2,1)=1
>**C**,2,D=1

3.2.4 Input Statement Format Conventions

Every acceptable BELAC input statement has a format similar to one of those described in this chapter. To aid the reader in understanding the description of these input statements, conventions are adhered to whenever the format of an input statement is presented. In this section we shall describe these conventions and define some of the terms used.

An *alphanumeric character* is either a letter or a numeral. An *alphanumeric value* is a given combination of alphanumeric characters. An *alphanumeric constant* has a fixed alphanumeric value, while an *alphanumeric parameter* takes on various alphanumeric values. These definitions have completely analogous to numerical constants and parameters. It is often necessary when describing input statements to use a symbol to represent a numerical parameter, an alphanumeric constant, or an alphanumeric parameter. The following conventions are used in this chapter:

A lowercase italicized letter or word represents a numerical parameter to which the user must assign a numerical value.

A lowercase boldface letter or word represents an alphanumeric parameter to which the user must assign an alphanumeric value.

A pound sign (#) represents an alphanumeric parameter to which the user may, if he wishes, assign an alphanumeric character. The pound sign indicates that the character assigned to it will not affect the interpretation of the identifier of the input statement.

An uppercase boldface letter or word represents an alphanumeric constant whose value is simply the uppercase letters themselves.

Uppercase boldface italicized words are optional alphanumeric constants that may appear in an input stantement.

The first example in Sec. 3.2.3 might be an input statement corresponding to a format such as

$$\mathbf{XP}\#\#\#\#\#\#(\mathbf{i,j,k,l})=a,b,c,d,e,f,g,h$$

In the example the first pound sign was assigned a 2, and the others were ignored (BELAC will assign blanks to them). The uppercase boldface alphanumeric constant **XP** was used unchanged (and is the part of the identifier that identifies the input statement). This is an **XP** input statement. The lowercase boldface letters **i**, **j**, **k**, and **l**, represent alphanumeric parameters and were assigned alphanumeric values (the character 9 assigned to **j** does not represent the number 9 but rather the numeral). Nothing was assigned to **l**; therefore, a blank would be assigned to it by BELAC. Numerical values were assigned to the lowercase letters a through g. Since nothing was assigned to h, zero would be assigned to it by BELAC.

In general, if the input statement does not assign a value to an alphanumeric parameter, BELAC will assign a blank to it; likewise, BELAC will assign a zero to any numerical parameters that are not assigned a value by the input statement.

3.2.5 General Input Statement Format

All the input statement formats are special cases of a general format. This format is as follows:

$$\mathbf{identifier}, n1,n2,n3\ldots,=a,b,c,\ldots,$$
$$\mathit{TOL}\mathbf{ERANCE}=a_0,b_0,c_0,\ldots,$$
$$\mathit{DEV}\mathbf{IATION}=a_1,b_1,c_1,\ldots,$$
$$\mathit{PER}\mathbf{CENT}=a_2,b_2,c_2,\ldots,$$
$$\mathit{MIN}\mathbf{IMUM}=a_3,b_3,c_3,\ldots,$$
$$\mathit{MAX}\mathbf{IMUM}=a_4,b_4,c_4,\ldots,$$
$$\mathit{SIG}\mathbf{MA}=a_5,b_5,c_5,\ldots,$$
$$\mathit{TAB}\mathbf{LE}=a_6,b_6,c_6,\ldots$$

This general format was printed on eight lines for readability only. It would be just as correct had we printed the entire format on one line. This is the general format; it is very unlikely that any input statement will contain everything in this format. However, all input statements must have an identifier. Whenever values are assigned to the subscripted parameters, the equal sign shown to the left of a must appear, even if no values are assigned to $a,b,c,\ldots$. The words PERCENT, MINIMUM, etc, may be abbreviated to their first three letters; they may, but need not be, preceded by a comma. The word *GAM*MA may be used instead of *SIG*MA. *TOL*ERANCE may be used in place of *MIN*IMUM and *MAX*IMUM. When it is used, *MINI*MUM and *MAX*IMUM values are calculated by BELAC. For example, $b_3 = b(1-b_0/100)$.

3.2.6 Arithmetic Expressions

The user assigns numerical values to the parameters $a,b,c,\ldots, a_0,b_0, c_0,\ldots$. This is done by including an arithmetic expression in the input statement. Arithmetic expressions may simply be constants or may be complicated expressions.

Let n and m be any numerals, possibly preceded by a sign and possibly containing a decimal point (e.g., 3, 7.2, -43, $+.77$). Let λ be one of the following letters:

λ	Value	Name
G	10^9	Giga
M	10^6	Mega
K	10^3	Kilo
T	10^{-3}	Mili
U	10^{-6}	Micro
N	10^{-9}	Nano
P	10^{-12}	Pico
A	3.14159	π
B	2.71828	ϵ

Then, any of the following forms are acceptable:

Form	Value	Example
λ	Define in above table	K
n	n	1000
nEm	10 raised to the mth power times n	1E3
nλ	n times the value associated with λ	1K
En	10 raised to the nth power	E3
nDm	10 raised to the mth power times n	1D3
Dn	10 raised to the nth power	D3

It is also permissible to use numbers such as 5KM to represent 5×10^9. As used above, λ is part of the constant; however, it may be used to operate on an expression enclosed in parentheses; i.e., (5+1)K is allowable and has the value 6000. To remember $A=\pi$, think of "apple pie," and to remember $T=.001$, think of thousandths. Examples of constants are

Example	Value
7	7.
7E3	7000.
7.E3	7000.
7E3.	7000.
7K	7000.
7.4	7.4
2A	6.2831854
A	3.1415927
E2	100.

A special constant has the form DC. This is usually used in an input statement where a frequency would normally appear to indicate that a large signal dc solution is desired. It may be used only in the following input statements: **INDEX MINIMIZATION, SENSITIVITY, MONTE CARLO, WORST CASE, USER, MINE, MYOWN,** and **INDEPENDENT VARIABLE**. It should not appear in any other input statements, in particular, it should not appear in the following input statements: **SWEEP RANGE, NATURAL,** and **TIME RESPONSE**.

The simplest expressions consist of a single constant. Two expressions separated by one of the following binary operators is also an expression:

Operator	Operation
**	Exponentiation
*	Multiplication
/	Division
+	Addition
−	Subtraction

Expressions may be connected by the arithmetic operators given above to form more complicated expressions. Operators higher in the table have a higher priority than operators lower in the table. When expressions are evaluated, these priorities dictate the order of the calculations (i.e., exponentiation first, multiplications and divisions next, and, finally, additions and subtractions).

Expressions may be enclosed in parentheses to change the order of calculation. The expression within the innermost parenthesis will be evaluated first. If the order of calculations is not otherwise determined, they will be performed left to right. Arithmetic operators must be separated by expressions.

Function names may also be included within expressions. Let FUN be a valid function name; then a valid expression is FUN (expression).

Example.

3*SIN(7**2+1) has a value that is three times the sine of 50 radians.

Valid function names are given in the table below. The letter ϵ is used there to indicate an expression. Circular function angles are in radians.

Name	Definition
EXP(ϵ)	exp(ϵ)
SQRT(ϵ)	$\sqrt{\epsilon}$
LN(ϵ)	$Ln_e(\epsilon)$
LOG(ϵ)	$Log_{10}(\epsilon)$
SIN(ϵ)	Sine(ϵ)
COS(ϵ)	Cosine(ϵ)
ATAN(ϵ)	Arc tangent(ϵ)
TANH(ϵ)	Hyperbolic tangent(ϵ)

3.2.7 Modification of Stored Parameters

> **MODIFY##,name**=$a,b,c,\ldots$,**PER**CENT=$a_2,b_2,c_2,\ldots$,
> **SIG**MA=$a_5,b_5,c_5,\ldots$

This input statement will cause parameter values stored in memory from a previous activity to be permanently varied. (The word *permanent* implies that BELAC will not restore these values, although the user may.) Consider the special case when ....2. is assigned to **name**. This will cause input statements stored in memory to be varied if the fifth character of their identifier is a 2 and the identifier has six characters or less. Let a',b',c'... be the parameter values stored in memory; then they will be varied as follows:

a' is changed a percent

b' is changed b percent

c' is changed c percent

.
.
.

a_1' is changed a_1 percent

b_1' is changed b_1 percent

c_1' is changed c_1 percent

.
.
.

In general, then, name is composed of n characters (n not greater than 8). If there are identifiers stored in memory with n or less characters and the identifiers are identical to name, except possibly in those positions in which name has a period, the values on those stored input statements will be permanently changed in the computer's memory. Any number of **MODIFY** input statements may appear in an activity. This input statement can be deleted by use of the word ***DELETE*** (i.e., **MODIFY**##,***DELETE***).

3.2.8 Miscellaneous Input Statements

TITLE, title1
TITLE1, title1
TITLE2, title2
TITLE3, title3

If any or all of the **TITLE** input statements are present, the titles will be printed at the top of each page of the computer output. Like all input statements, they may be deleted and need not appear in any special order. Starting at the top of the computer output page, the titles will be printed in the order title1, title2, title3, regardless of the actual ordering of the input statements. Note that the identifier **TITLE** and **TITLE1** are equivalent.

DEPENDENT VARIABLE$=n$

BELAC assumes that the analysis will not require more than 20 dependent variables. If more are required, BELAC automatically compensates and obtains sufficient computer memory to perform the analysis. The process of determining that more memory is needed is somewhat time-consuming and can be avoided by including this input statement with n set equal to the estimate of the number of dependent variables.

> **TIMEMAX**=*minutes*

BELAC has a built-in limit of 10 minutes of processor time per analysis per activity. This limit can be changed with this input statement. The value assigned to minutes can be less than unity.

> **INDEPENDENT VARIABLE**=DC

As described in Sec. 3.5 many of the analyses will be performed as small-signal ac analyses unless this input statement is present. When this statement is present, they will be performed as large-signal dc analyses.

> **START**,label

This input statement defines the beginning of a job and an optional label that identifies the job.

> **RUN**,*LIST*

Each analysis set must be terminated by a **RUN** input statement. If the word *LIST* appears in the input statement, the input data stored in memory will be printed. This printout is especially useful when an **ADD** input statement is used since it shows the user the input statements that have been added to his input. This input statement causes BELAC to perform the analyses, if any, indicated by input statements stored in memory.

3.3 MODEL LIBRARY

The need for a model library stems from the fact that a device model may contain many parameters and have a topology containing a dozen or more branches. If this device is used more than once, the clerical operations of repeatedly inputting the data are tedious, error-prone, and unnecessary. After all, this is the very type of operations at which the computer excels. Therefore, provisions were provided that allow the user to describe a model to the computer and have the computer add the model to a storage file. Thereafter, the user need only reference the device by name and indicate any changes he wants made to the model as it is copied from the file and added to the input data stored in memory. The model can be a generic model, which requires the user to supply parameter values when it is used, or a device model already containing the parameter values. The model need not actually represent a device at all but may, for instance, represent an entire network. For example, to do an analysis of the effect of temperature on a network, the network may be stored as a temperature-dependent model in which temperature is represented by the letter T. All the input statements necessary for the analysis can also be stored in the model (including the **RUN** input statement). An analysis could then be requested with a **START** input statement and a **MODEL** input statement that assigns to T the value of the temperature of interest. An analysis at a second temperature would require a second pair of these input statements, and so on. In a similar fashion, models dependent on parameters other than temperature, such as process parameters, age, manufacturing tolerances, and other considerations, may be stored.

The utility routines in BELAC allow model libraries to be created, maintained, and used. These routines allow models to contain models (which may be on the same or a different library). An input statement added to the library becomes one record in the library. These records may contain any BELAC input statement.

Below it will be necessary to indicate a string of eight characters. Such a string will be denoted by the word *name*. If less than eight characters are assigned to name by the user, sufficient blanks will be assigned by BELAC and so name will represent eight characters. Any valid character except an equals sign may be used in these strings; that is, they are not restricted to letters and numbers.

To facilitate the use of the model library, one record format in the library is given special attention by the utility routines:

MODEL NAME IS name

The word **MODEL** starts in column 1 and the first character of **name** is in column 15. The blanks shown in the above format are important. This type of record precedes each model on the library, and **name** should equal the model's name.

BELAC starts reading input data from the card reader and continues to read from this reader until a **MODEL** input statement is read (this input statement is described below). The **MODEL** input statement causes a model to be placed onto an input scratch file. BELAC then reads the input scratch file rather than cards, treating the data exactly as it would have if the data came from the card reader.

Once the entire input scratch file is read, BELAC reverts to reading cards. If, while reading the input scratch file, a **MODEL** input statement is detected, that model is added to the end of the scratch file. This gives this routine the ability to handle models which contain models (which in turn, may contain models, etc.). If a **START** or a **RUN** input statement is encountered on the scratch file, any data remaining on that file are ignored.

3.3.1 Library Input Statements

The utility routines are executed when the BELAC input routine reads one of the following input statements:

> REWIND(n)

Causes file **n** to be rewound. An end-of-file record will be written on the file before it is rewound if the preceding reference to this file caused it to be left in the output mode (by a copy input statement for example).

> BACKSPACE(n)

Causes the file **n** to be backspaced one record. This leaves the file in the mode so that following a **BACKSPACE** with a **REWIND** will not cause an end-of-file to be written before the file is rewound.

> END OF FILE(n)

Causes an end-of-file record to be written on file **n**.

> COPY(n)ONTO(m)THRU(name)
> COPY(n)ONTO(m)

Causes records to be copied from file **n** onto file **m** until an end-of-file record is reached or the first eight characters of the record copied equals **name**, except possibly in those positions in which name has a period. The files are not moved before the copying starts. Similarly, after the last record is read from **n** and written on **m** the files are not moved. An end-of-file record read from **n** will not be copied onto **m**; furthermore, **n** will be positioned at the end-of-file record, not past it. An example to illustrate the use of the period: To copy through the first record that has an R as the leftmost character, name would have R. . . . . . . assigned to it; to copy through the first record that has R as the leftmost character and has blanks in columns 2 through 8, name would simply equal R.

> MOVE(n)THRU(name)

Causes file **n** to be forward until a record is reached (and passed) whose first eight characters equal **name** or until an end-of-file record is read (the comments associated with reading an end-of-file record and the COPY input statement apply here also). This command is used to position **n**. A period in name has the same significance as indicated in the description of the copy option.

> **LIST MODELS ON(n)**

Causes the model names on file **n** to be printed. File **n** is left rewound.

> **MODEL(name)FROM(n)*CHANGES*(a=b,c=d,e=f,$=g, . . .)**

Causes the model on file **n** with the appropriate **name** to be located and copied, with changes, onto the input scratch file. The letters, **a,b,c**, etc. denote strings (of any length) of characters (no commas or equal signs). If while copying a record onto the input, scratch file a string of characters delineated by the special symbols defined below is found to be equal to **a,c,e**, etc., it will be replaced with the appropriate string, i.e., **b,d,f,** etc. The delineators are the characters (),—*/=+ and a blank space. Any dollar signs ($) found in the model will be replaced with the string **g**. If there is a dollar sign in the model and the user does not assign a value to it via the ***CHANGE*** option, BELAC will assign characters to it, different, of course, for each model. This entry repositions file **n**. If no changes are to be made, **MODEL (name) FROM (n)** is sufficient.

3.3.2 Example of Model Library Use

A worst-case analysis is used to illustrate the use of the model library. In this example, the change in the frequency response of a low-pass active filter as the temperature is changed is to be determined. This can be accomplished in many ways. The one illustrated in Fig. 3.34 creates a temperature-

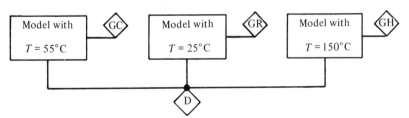

Figure 3.34 Low-pass Filter Analysis.

dependent model of the circuit including the signal generator, the request for an output, and the filter. It then analyzes one network composed of three of these models, each representing a different temperature. BELAC is commanded to calculate and plot the voltages at nodes GL, GR, and GH as well as the voltage between nodes GL and GR and between nodes GH and GR.

The input data are shown in Figs. 3.35 and 3.36. The plotter plot is shown in Fig. 3.37.

START,MAKE LIBRARY ON UNIT 3
REWIND,3
LIST WHILE COPYING (5) ON TO (3) THRU (EOF)
   MODEL NAME IS LOWPASS
     VG,I$−D=1
     R6,I$−A$=336.17K*(1.+.001*(T))
     R4,A$−B$=336.17K*(1.+.001*(T))
     R2,B$−C$=336.17K*(1.+.001*(T))
     C1,C$−D=7.94P*(1.+25.U*(T))
     C3,B$−E$=2000.P*(1.+25.U*(T))
     C5,A$−D=317.15P*(1.=25.U*(T))
     VV,E$−D/C$−D=1
     R1,E$−H$=142.89K*(1.+.001*(T))
     R5,H$−O$=142.89K*(1.+.001*(T))
     C4,H$−G$=540P*(1.25.U*(T))
     R3,G$−F$=71.45K*(1.+.001*(T))
     C6,E*−F$=270P*(1.+25.U*(T))
     C2,F$−O$=270P*(1.+25.U*(T))
     W,G$−D/O$−D=1.15
   OALOW$,*VOLTAGE*,G$,*PLOTTER* PLOT,*PRINTER* PLOT
   EOF
BACKSPACE(3)
END OF FILE(3)
RUN

           Figure 3.35 Input Statements to Create Model.

START, TEST DEPENDENT MODEL LOWPASS
MODEL(LOWPASS)FROM(3)*CHANGE*(T=−55,$=L)
   MODEL NAME IS LOWPASS
     VG,IL−D=1
     R6,IL−AL=336.17K*(1.+.001*(−55−25))
     R4,AL−BL=336.17K*(1.+.001*(−55−25))
     R2,BL−CL=336.17K*(1.+.001*(−55−25))
     C1,CL−D=7.94P*(1.+25.U*(−55−25))
     C3,BL−EL=2000.P*(1.+25.U*(−55−25))
     C5,AL−D=317.15P*(1.+25.U*(−55−25))
     VV,EL−D/CL−D=1
     R1,EL−DL=142.89K*(1.+.001*(−55−25))
     R5,HL−OL=142.89K*(1.+.001*(−55−25))
     C4,HL−GL=540P*(1.+25.U*(−55−25))
     R3,GL−FL=71.45K*(1.+.001*(−55−25))
     C6,EL−FL=270P*(1.+25.U*(−55−25))
     C2,FL−OL=270P*(1.+25.U*(−55−25))
     VV,GL−D/OL−D=1.15
   OAGAINL,*VOLTAGE*,GL,*PLOTTER* PLOT,*PRINTER* PLOT

The three indented sections are not input data. They are the model called from the library and modified.

           Figure 3.36 Data Used to Analyze Network.

MODEL(LOWPASS)FROM(3)*CHANGE*(T=25,$=$)
  MODEL NAME IS LOWPASS
    VG,IR−D=1
    R6,IR−AB=336.17K*(1.+.001*(25−25))
    R4,AR−BR=336.17K*(1.+.001*(25−25))
    R2,BR−CR=336.17K*(1.+.001*(25−25))
    C1,CR−D=7.94P*(1.+25.U*(25−25))
    C3,BR−ER=2000.P*(1.+25.U*(25−25))
    C5,AR−D=317.15P*(1.+25.U*(25−25))
    VV,ER−D/CR−D=1
    R1,ER−BR=142.89K*(1.+.001*(25−25))
    R5,HR−OR=142.89K*(1.+.001*(25−25))
    C4,HR−GR=540P*(1.+25.U*(25−25))
    R3,GR−FR=71.45K*(1.+.001*(25−25))
    C6,ER−FR=270P*(1.+25.U*(25−25))
    C2,FR−OR=270P*(1.+25.U*(25−25))
    VV,GR−D/OR−D=1.15
    OAGAINR,VOLTAGE,GR,*PLOTTER* PLOT, *PRINTER* PLOT
MODEL(LOWPASS)FROM(3)*CHANGE*(T=150,$=H)
  MODEL NAME IS LOWPASS
    VG,IH−D=1
    R6,IH−AH=336.17K*(1.+.001*(150−25))
    R4,AH−BH=336.17K*(1.+.001*(150−25))
    R2,BH−CH=336.17K*(1.+.001*(150−25))
    C1,CH−D=7.94P*(1.+25.U*(150−25))
    C3,BH−EH=2000.P*(1.+25.U*(150−25))
    C5,AH−D=317.15P*(1.+25.U*(150−25))
    VV,EH−D/CH−D=1
    R1,EH−HH=142.89K*(1.+.001*(150−25))
    R5,HH−OH=142.89K*(1.+.001*(150−25))
    C4,HH−GH=540P*(1.+25.U*(150−25))
    R3,GH−RH=71.45K*(1.+.001*(150−25))
    C6,EH−RH=270P*(1.+25.U*(150−25))
    C2,RH−OH=270P*(1.+25.U*(150−25))
    VV,GH−D/OH−D=1.15
    OAGAINH,*VOLTAGE*,GH,*PLOTTER* PLOT,*PRINTER* PLOT
DEPENDENT=40
SWEEP RANGE(*LOG*)=300,50,10K
SWEEPOPT=132,,,,,1,1
TITLE1,   LOWPASS ACTIVE FILTER DESIGN
TITLE2,   EXAMPLE OF TEMPERATURE DEPENDENT MODEL
TITLE3,   C.A.GEORGE GE−AESD UTICA N Y
OAL.R(*DIFFEREN*CE)GL,GR(*PLOTTER* PLOT,*PRINTER* PLOT)
OAH.R(*DIFFEREN*CE)GH,GR(*PLOTTER* PLOT,*PRINTER* PLOT)
RUN,*LIST*

Figure 3.36—*Cont.*

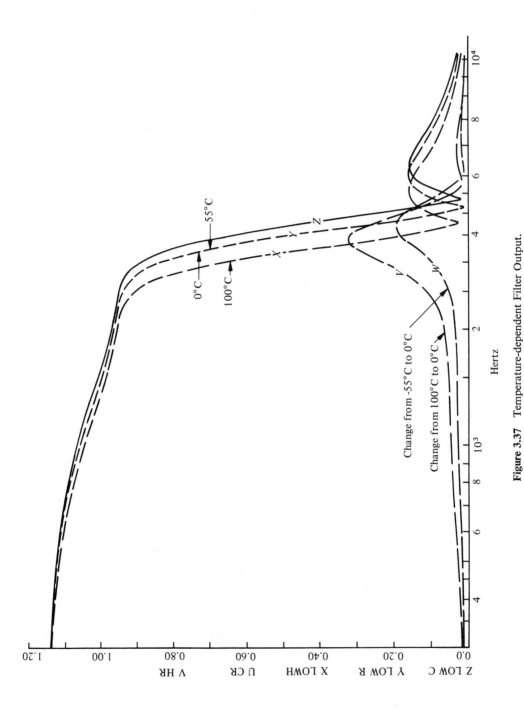

Figure 3.37 Temperature-dependent Filter Output.

3.4 INTERNALLY GENERATED ENTRIES

BELAC contains subroutines that add to the data stored in memory. This is done in such a way that the rest of the program cannot tell that these data were not part of the input deck. There is also provision for the user to add his own subroutine, called MYADD. Such a subroutine could design active filters, model some component or subsystem, read non-BELAC data and convert it to the appropriate BELAC parameters, or any one of many other possibilities. One of the subroutines that is a permanent part of BELAC is described in Sec. 3.4.1.

3.4.1 Butterworth and Chebyshev Filter Design

ADD#####,**type**(n1,n2,n3)=$n,ro,ri,fo,bw,fl,fh,f,db,rip,per$

Description

This entry causes BELAC to design a passive ladder network which, when driven with a generator with an output impedance of *ri* and loaded with an impedance of *ro*, will have a transfer characteristic given in terms of a Butterworth or Chebyshev polynomial. The synthesis is done by the use of closed-form expressions for the element values. The alphanumeric parameter **type** is composed of two characters and defines the magnitude characteristic as follows:

    First character: C for Chebyshev
                                    B for Butterworth

    Second character: H for high-pass
                                      L for low-pass
                                      P for band-pass
                                      R for band-reject

The numerical parameters which characterize the filter shown in Fig. 3.38 are described below:

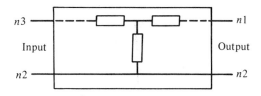

Figure 3.38 Filter Block Port Designation.

Symbol	Description	Remarks
n	Total number of poles in filter (equals the number of reactive elements)	Do not specify if f is.
ro	Output resistance of network	
ri	Input resistance of network	
fo	Filter cutoff frequency of low-pass or high-pass, or geometrical center frequency of band-pass or band reject	Do not specify if fh and fl are.
bw	Bandwidth for band-pass or band reject filter; the equivalent fh and fl are such that $bw = fh - fl$ and $fo = \sqrt{fh \cdot fl}$.	Specify if fo is specified.
fl	Lower cutoff frequency for band-pass or band reject filter	Do not specify if fo is.
fh	Upper cutoff frequency for band-pass reject filter	Do not specify if fo is.
f, db	At the frequency f, the voltage is down db decibels relative to its maximum value.	Do not specify if n is.
rip	For Chebyshev filters; the pass-band (or stop-band) peak-to-peak ripple in db	Do not specify if per is.
per	For Chebyshev filters; the pass-band (or stop-band) peak-to-peak ripple in percent of peak amplitude	Do not specify if rip is.

All frequencies are in hertz. For Chebyshev filters, f should be in the stop-band.

3.5 STANDARD NETWORK ANALYSIS

In this section we shall describe the entries that are used to instruct BELAC to perform analyses. There are five basic types of analyses that can be performed. They are identified in Fig. 3.39.

Notwithstanding the presence of other input statements related to a particular type of analysis, the analysis will be performed only if the ap-

This will be performed	If this statement is present
Index minimization	**INDEX MINIMIZATION**
Swept response	**SWEEP RANGE**
Sensitivities	**SENSITIVITY**
Monte Carlo	**MONTE CARLO**
Worst Case	**WORST CASE**
Poles/Zeros	**NATURAL**
Time Response	**TIME RESPONSE**

Figure 3.39 Five Basic Types of Analyses.

propriate input statement described in Fig. 3.39 appear in the input data. More than one of these input statements may be present within an activity; however, each such input statement must refer to a different type of analysis. For example, there can be both a **SWEEP RANGE** and a **SENSITIVITY** input statement present, but there should not be two **SENSITIVITY** input statements. The analyses are executed in the order in which the input statements appear in Fig. 3.39. For example, an optimization would be performed prior to a frequency response analysis.

3.5.1 Index Minimization

BELAC may be used to vary network parameter values to meet a small-signal ac or a large-signal dc specification on the output variables. Either worst-case analysis or network parameter optimization may be effected using this feature. Note that varying a network parameter value to minimize a voltage, for example, can be considered to be finding either the worst-case or the optimum values, depending on your point of view. The output variables are identified by **OA** or **OG** input statements.

Elements of the network are identified by input statements described in subsequent sections. In short, each element is identified by an input statement that contains element parameter values and interconnection data. These input statements have the form

$$\text{identifier}(\ ) = a, b, c, \ldots$$

The values assigned to $a, b, c, \ldots$ are the elements nominal* network parameter values. If it is desired to vary one or more of them to meet a specification, the allowable range of those parameters must be identified by specifying the lower and upper limits of the range. (See Sec. 3.2.5.) The result of the analysis will be a value of the parameter within the specified range. Those parameter values for which a range is not specified will remain at their nominal values. The limits of the range are included in the input statement as follows:

$$\text{identifier}(\ ) = a, b, c, \ldots, MIN = a_3, b_3, c_3, \ldots, MAX = a_4, b_4, c_4, \ldots$$

For example, if b is to be varied and c is not to be varied, b_2 and b_3 must be assigned but no values for c_2 and c_3 should be assigned.

Discrete values of the independent variables are specified by an input statement of the following form:

INDEX MINIMIZATION, $RPS = a, b, c, \ldots, SIGMA = a_5, b_5, c_5, \ldots$
---

*The user will usually enter typical parameter values in these input statements for worst-case analysis and "ball-park" values for an optimization. However, in this section these parameter values will be referred to simply as "nominal network parameter values."

If the input data does not contain the input statement **INDEPENDENT VARIABLE=DC**, numerical values on the **INDEX MINIMIZATION** input statement will be treated as the frequencies to be used during a small-signal ac analysis. During such an analysis, the discrete values that will be assigned to the Laplace transform variable, $s = \sigma + j\omega$, are $a_s + ja, b_s + jb, c_s + jc, \ldots$. If the optional RPS is not present in the input statement, BELAC will assume that $a,b,c,\ldots$ represent hertz and will convert them to radians per second before assigning them to s. Usually, in a small-signal ac analysis, only complex values with a zero real part are to be assigned to s, that is, only values that lie on the j axis in the complex s-plane. In this case the following simplified input statement form may be used:

$$\boxed{\textbf{INDEX MINIMIZATION,RPS}=a,b,c,\ldots}$$

If the network is nonlinear, DC has not been assigned to any of the frequencies, $a,b,c,\ldots$ and the input does not contain the statement **INDEPENDENT VARIABLE=DC**, BELAC will determine the quiescent operating point and linearize the network about that point automatically before the small-signal ac analysis is performed. However, the user can control this linearization himself by assigning DC to one of the frequencies (usually to a). Suppose that DC is assigned to c. Then the large-signal elements will act as defined in Sec. 3.6 when the small-signal response at a and b is calculated. The response at c will be the large-signal dc response. The responses at $d,e,\ldots$ will be the response of the linearized network. The large-signal dc value calculated at c will be treated in the rest of the calculations as if it were an output of the small-signal analysis. For example, by assigning DC to a and various values of frequency to $b,c,d,\ldots$ it is possible to optimize for a desired quiescent value and small-signal ac response simultaneously.

Assigning DC to one of the frequencies is quite different from assigning zero to it. DC implies a large-signal analysis of a network (possibly nonlinear), while zero implies a small-signal analysis of a linear network.

If the input data contains the input statement **INDEPENDENT VARIABLE=DC**, then no small-signal ac calculations will be performed; all calculations will be large-signal dc. The option RPS will be ignored if present. During a large-signal dc calculation some of the elements may depend on the value of the pair (γ,δ). This pair will take on the values $(a_s,a),(b_s,b),(c_s,c).\ldots$ This allows BELAC to perform large-signal dc calculations for discrete values of (γ,δ) in a manner analogous to its small-signal ac capability. Whereas in a small-signal ac calculation the value assigned to $\sigma+j\omega$ affects such items as the impedance of an inductor, in a large-signal dc analysis the value assigned to (γ,δ) affects such items as the voltage of a battery or the resistance of a resistor. The battery voltage in the example of Sec. 3.17 varied with δ, for example; see also Sec. 3.6.13.

Some of the performance indices described below contain the desired values of the outputs' magnitude, dB, phase, and weights. These are entered by input statements of the form:

$$\textbf{MAG} = m_1, m_2, m_3 \ldots, \textbf{SIGMA} = \bar{m}_1, \bar{m}_2, \bar{m}_3 \ldots,$$
$$\textbf{MIN} = \tilde{m}_1, \tilde{m}_2, \tilde{m}_3 \ldots, \textbf{MAX} = \hat{m}_1, \hat{m}_2, \hat{m}_3$$

$$\textbf{DB} = d_1, d_2, d_3 \ldots, \textbf{SIGMA} = \bar{d}_1, \bar{d}_2, \bar{d}_3 \ldots,$$
$$\textbf{MIN} = \tilde{d}_1, \tilde{d}_2, \tilde{d}_3 \ldots, \textbf{MAX} = \hat{d}_1, \hat{d}_2, \hat{d}_3$$

$$\textbf{ANGLE} = a_1, a_2, a_3 \ldots, \textbf{SIGMA} = \bar{a}_1, \bar{a}_2, \bar{a}_3 \ldots,$$
$$\textbf{MIN} = \tilde{a}_1, \tilde{a}_2, \tilde{a}_3 \ldots, \textbf{MAX} = \hat{a}_1, \hat{a}_2, \hat{a}_3$$

Values need only be entered if they are necessary for the performance index used (for example the *MAX* and *MIN* values are not always needed).

The user selectable options are defined by an input statement of the form:

$$\textbf{INDEX OPT} = strategy, d, a, m, step, pi, n, smin, change$$

If this input statement is not present in the input data, zero will be assigned to all of these parameters except as indicated below.

The value of *strategy* identifies the minimization strategy BELAC is to use.

Value	Strategy
0	Fletcher-Powell
1	Zangwill-Powell
2	Steepest Descent (old)
3	Steepest Descent (new)
4	Cyclic Coordinate Descent

The performance index, PI, which BELAC minimizes is controlled by the value of *pi* as follows:

If $pi=1$; $\text{PI} = \sum_{s_i} |m(m_i - M_i)\bar{m}_i| + |d(d_i - D_i)\bar{d}_i| + |a(a_i - A_i)\bar{a}_i|$ Minimize

If $pi=2$; $\text{PI} = \sum_{s_i} -|m(m_i - M_i)\bar{m}_i| - |d(d_i - D)\bar{d}_i| - |a(a_i - A_i)\bar{a}_i|$ Maximize

If $pi=3$; $\text{PI} = \sum_{s_i} \{m[m_i - M_i]\bar{m}_i\}^2 + \{d[d_i - D_i]\bar{d}_i\}^2 + \{a[a_i - A_i]\bar{a}_i\}^2$

If $pi=4$; $\text{PI} = \sum_{s_i} \{m[m_i - M_i]\bar{m}_i/m_i\}^2 + \{d[d_i - D_i]\bar{d}_i/d_i\}^2 + \{a[a_i - A_i]\bar{a}_i/a_i\}^2$

where

M_i is the calculated magnitude of the output at s_i, A_i is the calculated value of angle and D_i is the calculated value of dB.

The values taken on by s_i are specified on the **INDEX MINIMIZATION** input statement. The desired values of the magnitude, m_i, are identified on the **MAG** input statement as $m_1, m_2, m_3, \ldots$. Likewise the desired values of angle and dB are identified on the **ANGLE** and **DB** input statements respectively.

The weight corresponding to m_i is $\bar{m}_i$ as specified on the **MAG** input statement. Likewise, $d_i, \bar{d}_i, a_i$ and $\bar{a}_i$ are the desired dB, its weight, the desired angle in degrees, and its weight. If all the weights $\bar{m}_i$ equal zero, they will all be set equal to unity. Likewise if all $\bar{d}_i$ equal zero, or all $\bar{a}_i$ equal zero.

The weights d, a, m, and the parameter pi is specified on the **INDEXOPT** entry. If d, a, and m all equal zero, m will be set equal to unity. If pi equals zero the analysis will be first performed with pi equal to one and then with pi equal to two. That is, BELAC will find both the minimum and the maximum values the performance index can take on.

If pi equals four and any m_i equal zero for a value of s_i for which the product $m\bar{m}_i$ does not equal zero, the value of pi will be changed to three. Likewise for d_i, $d\bar{d}_i$, a_i and $a\bar{a}_i$.

The step size is the largest of the percentage changes in the parameter values due to a step in a one dimension search. The step size will be adjusted during the hill climbing process. Its initial value is *step*, or sixteen if *step* is zero.

The hill-climbing process will be terminated after n iterations, the performance index becomes zero, the processor time limit is exceeded, the step size becomes less than *smin*, or one of the other similar criterion is satisfied. If *smin* is assigned zero, 0.1 will be used and if n is assigned zero, three will be used.

If *change* equals zero, the values of the parameters will be restored to the values they had prior to the hill-climbing operation. That is, any subsequent analysis will be performed using the nominal values entered by the user. If *change* does not equal zero the parameters will not be restored; therefore, any subsequent analysis will be performed using the parameters calculated by the minimization process.

3.5.2 Sweep Response

Small-signal ac or large-signal dc solutions can be obtained by specifying ranges for the independent variables. The output variables are identified by either **OA** or **OF** input statements. The values of the independent variables are specified by an input statement of the following form:

$$\boxed{\textbf{SWEEP RANGE}, RPS, LOG = a, b, c, d, e, \ldots, \text{SIGMA} = a_5, b_5, c_5, \ldots}$$

If the input does not contain the input statement **INDEPENDENT VARIABLE=DC**, numerical values on the **SWEEP RANGE** input statement will be treated as the frequencies to be used during small-signal ac analysis. If the option **RPS** is not present in the input statement, BELAC will assume that $a,b,c,\ldots$ represent hertz and will convert them to radians per second before assigning them to s. If the option **LOG** is present, the imaginary part of s will be swept logarithmically, in which case b (and $d,f,\ldots$, if they are present) is the number of frequencies per decade. That is, the imaginary part of the $(b+1)$th frequency will be $10a$ (assuming that $10a$ is less than c). If the option **LOG** is not present, b (and $d,f,\ldots$, if they are present) is a linear increment and the imaginary part of s will take on the values $a, a+b, a+2b, \ldots a+nb, a+nb+c, a+nb+2c, \ldots$, where n is such that $a+(n+1)b<c$ and $a+nb\geq c$.

This input statement in its most general form allows the user to request calculations at the values of s that lie on the intersection of lines that form a rectangular grid in the s plane. The grid can be considered to be formed of bands of straight lines as depicted in Fig. 3.40.

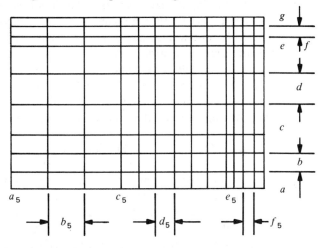

Figure 3.40 Grid in s-plane Defining Independent Variables.

From Fig. 3.40 it is evident that there are three vertical bands (there could be more). Within each band the lines are equally spaced, although the spacing may vary from band to band—similarly for the horizontal bands. If the option **LOG** were present, the spacing within the horizontal bands would be logarithmic. Examples in Fig. 3.41 serve to illustrate the above facts. Each example displays an input statement and the values of s for which calculations will be performed.

If the network is nonlinear, BELAC will determine the quiescent operating point and linearize the network about this point before the small-signal ac solutions are obtained.

SWEEP RANGE,*RPS*=0,1,2,2,5,SIGMA=0,1,2,3,5
(0,0)(0,1)(0,2)(0,4)(0,6)(1,0)(1,1) ... (1,6)(2,0) ... (2,6)(5,0) ... (5,6)

SWEEP RANGE,*RPS*=3,2,8	SWEEP RANGE=3,1,4 *SIG*MA=2,1,3
(0,3)(0,5)(0,7)(0,9)	(2,2π3)(2,2π4)(3,2π3)(3,2π4)
SWEEP RANGE,*RPS*=3	SWEEP RANGE,*LOG*=1,2,10,*SIG*MA=2
(0,3)	(2,2π)(2,2π 10)(2,2π10)
SWEEP RANGE=*SIG*MA=2	SWEEP RANGE,*LOG*,*RPS*=1,0.5,10K
(2,0)	(0,1)(0,100)(0,10000)
SWEEP RANGE=*SIG*MA=2,1,3	SWEEP RANGE,*RPS*=3,2,8,*SIG*MA=2
(2,0)(3,0)	(2,3)(2,5)(2,7)(2,9)
SWEEP RANGE=3=*SIG*MA=2	SWEEP RANGE,*RPS*=3 *SIG*MA=2,1,4
(2,3)	(2,3)(3,3)(4,3)

Figure 3.41 Values assigned to $s = (\sigma + j\omega)$ for Various SWEEP RANGE Input Statements.

If the input data contain the input statement INDEPENDENT VARIABLE=DC, then no small-signal ac analysis will be performed and all calculations will be large-signal dc. The option ***RPS*** will be ignored if present. During a large-signal dc calculation, some of the elements may depend on the value of the pair (γ,δ). This pair will take on values determined from the data on the **SWEEP RANGE** input statement in the same way (σ,ω) takes values, except that ***RPS*** will be ignored if present. Therefore, Fig. 3.41 applies if (σ,ω) is changed to (γ,δ) and the 2πs are deleted.

$$\boxed{\textbf{FPLOT}=a,b,c,d,e,f,g}$$

This input statement can be used to control the amount of information plotted. It applies to both printer plots and plotter plots. If one of the constants $a,b,c,\ldots$ is not equal to zero, the information it controls will be plotted; if it is equal to zero, the information will not be plotted. The controlled information is defined in Fig. 3.42, along with the option BELAC assumes if this input statement is not present.

Symbol	Controlled information	Small signal ac	Large signal dc
a	DB	Not desired	Not desired
b	Phase	Desired	Not desired
c	Magnitude	Desired	Desired
d	Phase delay	Not desired	Not desired
e	Group Delay	Not desired	Not desired
f	Real Part	Not desired	Not desired
	Imaginary Part	Not desired	Not desired

Figure 3.42 Controlled Information and Assumed Options.

Ch. 3 / George

> **FPRINT**=a,b,c,d,e,f,g

This input statement can be used to control the amount of information printed. If one of the constants $a,b,c\ldots$ is not equal to zero, the information it controls will be printed; if it is equal to zero, the information will not be printed. The controlled information is defined in Fig. 3.42, along with the option assumed by BELAC if this input statement is not present. Note that the information that is printed may be different from that plotted, since the plotted information is controlled by an **FPLOT** input statement.

3.5.3 Sensitivities

BELAC can compute the sensitivities of the output variables to changes in values of the network parameters. The output variables are specified by either **OA** or **OS** input statements. The network parameters of interest are identified by including percentages in their input statements in a manner similar to that used to define minimum and maximum values in Sec. 3.5.1. The statements in Sec. 3.5.1 about small-signal ac and large-signal dc analyses also apply here except for the obvious differences due to the different analysis being considered here. The discrete values of the independent variables are specified by an input statement of one of the following forms:

> **SENSITIVITY**,***RPS***=$a,b,c,\ldots,$ ***SIGMA***=$a_5,b_5,c_5,\ldots$

> **SENSITIVITY**,***RPS***=a,b,c

Sensitivities will be computed only for those network parameters for which percentages have been assigned. The sensitivities are calculated according to the following formula:

$$S_P^V = \frac{\Delta V}{\%P}$$

where

$\quad$ V is the magnitude, dB, or phase associated with some variable

$\quad$ P is the value of some network parameter

$\quad$ S_P^V is approximately equal to the change in V due to a 1% change in P

$\quad$ %P is the nonzero number inputted with the network parameter (e.g., %P=b_5 if P=b)

$\quad$ ΔV is the change observed in V due to varying P $\#$ %P.

That is, the calculated sensitivity of V with respect to P is the change in V divided by the percent change in P that caused V to vary. The more common definition of sensitivity requires that %P be infinitesimal. Since the

BELAC approximation approaches this definition as %P approaches zero, the user can calculate sensitivity by making the variation in P essentially infinitesimal. However some considerations will be pointed out next. Consider the circuit in Fig. 3.43.

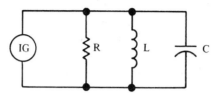

Figure 3.43 Sample Circuit for Sensitivity Calculation.

At the resonant frequency, the partial derivative of the magnitude of the voltage across the inductor with respect to the inductance equals zero. It is therefore correct to equate the sensitivity to zero. However, a finite change in the inductance *will* cause the voltage to vary. (Consider Fig. 3.44.)

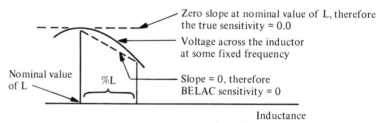

Figure 3.44 BELAC's Secant Approximation of Sensitivity.

The sensitivity calculated by BELAC is proportional to the slope of the secant shown and is probably a better measure of how the voltage will vary as the inductance varies between L and L+%L than is the slope of the tangent at the nominal inductance. The region between L and L−%L might also be considered. The hazard associated with using a very small but finite value for %P stems from the fact that %P must be large enough so that sufficient figures are retained when ΔV is calculated [$\Delta V = V(P(1+P/100)) - V(P)$], and if the true sensitivity is desired, small enough so that the secant accurately approximates the tangent. The more sensitive network parameters can use a smaller percentage than less sensitive parameters; however, the selection of %P has never appeared critical and 1% is often used.

3.5.4 Monte Carlo

Monte Carlo analysis is used to determine the statistics of the output variables from the statistics of the network parameters. The output variables are identified by **OA** or **OM** input statements. The network parameters to which statistics are assigned are identified by including the standard deviation value and/or the minimum and maximum values or tabular values of the cumulative distribution function, in a manner similar to that used to define minimum and maximum values in Sec. 3.5.1. See Fig. 3.45. The statements

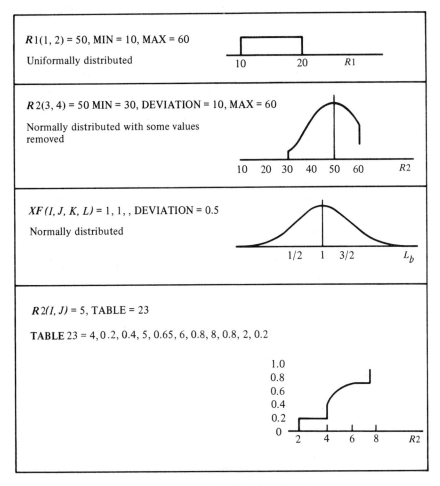

Figure 3.45 Examples of Random Parameters.

in Sec. 3.5.1 about small-signal ac and large-signal dc analyses also apply here except for the obvious differences due to the different analysis being considered here. The discrete values of the independent variables are specified by an input statement of one of the following forms:

```
MONTE CARLO,RPS=a,b,c, ... , SIGMA=a₅,b₅,c₅ ...
MONTE CARLO,RPS=a,b,c, ...
```

A network parameter can be made to have a uniform probability density function by assigning minimum and maximum values to it.

A network parameter can be made to have a normal probability density function by assigning a standard deviation to it. If a standard deviation, a minimum value, and a maximum value are all assigned to a network parameter, it will have a truncated normal probability density function.

Other probability density functions can be approximated by a tabular description of the cumulative distribution which can be assigned to a network parameter by assigning a table number to that parameter and including the tabular values on an input statement of the following form:

$$\text{TABLE}\#\#\# = f_1, p_1, f_2, p_2, f_3, \ldots$$

A numeral between 1 and 999 should be assigned to the symbol $\#\#\#$. This same numeral should appear in the TABLE part of the input statement describing the statistical network parameter as shown in Fig. 3.45. The points may appear in any order in the table. The smallest and the largest abscissa in the table are the minimum and maximum values taken on by the parameter. In those areas in which there are three or more points with distinct abscissa values and distinct ordinate values, BELAC will approximate the curve by a quadratic curve; elsewhere, BELAC will approximate the curve by straight-line segments.

3.5.5 Worst Case

The worst-case analysis is used to determine the maximum and the minimum values output variables take on as the element parameters vary over their tolerance range. The output variables are identified by **OA** or **OW** input statements. The network parameters that are variable are identified by including their minimum and maximum values in the manner described in Sec. 3.5.1. The statements in Sec. 3.5.1 about small-signal ac and large-signal dc analyses also apply here except for the obvious differences due to the different analysis being considered here. The discrete values of the independent variable are specified by an input statement of one of the following forms:

$$\text{WORST CASE}, RPS = a,b,c\ldots, SIG\text{MA} = a_5, b_5, c_5 \ldots$$
$$\text{WORST CASE}, RPS = a,b,c, \ldots$$

3.5.6 Time Response

The time response of linear and nonlinear networks can be obtained by specifying the range of time for which the response is desired. The output variables are identified by **OA** or **OT** input statements. The values of the

Ch. 3 / George

independent variable are specified by an input statement of the following form:

> **TIME RESPONSE**=*pstart,pdelta,pstop,instart*

The output variables will be printed for those values of time that are less than or equal to *pstop* and are equal to *pstart* + k · *pdelta* where k = 0, 1, 2, ... The initial value of time for which the time response is calculated (not necessarily printed) is equal to *instart*. If *pstart* were larger than *instart*, the early values of the response would be calculated but not printed. That is, the printing of the turn-on transient would be suppressed. If *instart* is greater than *pstart*, the early values would be calculated for printing by extrapolating from the values at *pstart* and could be inaccurate if *pstart* is much larger than *instart*.

3.5.7 Poles/Zeros

The poles (eigenvalues) and the eigenvectors of network equation can be calculated by BELAC. The zeros associated with an output-dependent variable can also be calculated. The output-dependent variables are identified by **OA** or **OZ** input statements.

> **NATURAL,*POLES,ZEROS,VECTORS***

If the input data contained the following input statements,

> **OA,*VOLTAGE*,X**
> **NATURAL,*POLES,ZEROS***
> **SWEEP RANGE**=*SIG*MA=1

it would be possible to express the frequency function of the node voltage at X as

$$v(s) = \frac{k(s - z_1)(s - z_2)(s - z_3)\ldots}{(s - p_1)(s - p_2)(s - p_3)\ldots}$$

where

$z_1, z_2, z_3, \ldots$ will be calculated because of the **ZEROS** option.

$p_1, p_2, p_3, \ldots$ will be calculated because of the **POLES** option.

k can be calculated by the user using the results of the **SWEEP RANGE** analysis i.e.,

$$k = \frac{v(1)(1 - p_1)(1 - p_2)(1 - p_3)\ldots}{(1 - z_1)(1 - z_2)(1 - z_3)\ldots}$$

Of course, if there is a pole or a zero at s equal to 1, a different value must be used for s. In fact, if there is neither a pole nor a zero at s equal to zero, it would be more convenient to use zero and the input statement **SWEEP RANGE=0**.

3.6 STANDARD NETWORK ELEMENTS

Physical systems that can be analyzed by BELAC are analogous to networks which have inputs that cause responses. Responses that are to be observed are called output-dependent variables. The usual method used to simulate a system is to consider the system as an interconnection of ideal elements such that the topology of the system and its network model are similar. Each device in the physical system is represented in the network model by one or more ideal elements which model the device such that the network model is expected to be adequate for the analyses contemplated. It is not possible to create one model of a device that would be adequate for all possible analyses. The mathematical relationships that define the ideal elements are well defined, and these elements are not burdened with such annoyances as parasitics, failure modes, nonlinearities, and the like, unless they are part of the mathematical relationships.

A BELAC analysis of a physical system starts with the construction of an iconic system model made up of interconnections of some of the ideal elements described in this section. This is a crucial step and represents the engineer's contribution to the analysis. It is the engineer who must decide whether a diode can be modeled by a resistor in the series with a battery, by an open circuit, by the so-called diode equation, or by a complex model containing inductance and capacitance. This must be done by the engineer (often subconsciously). The computer cannot relieve him of this resposibility. Once the iconic system model is drawn, the nodes must be labeled with names composed of one to eight characters. The general input statement format used to represent these elements has the following form:

$$\text{identifier}(n1,n2,n3,n4)=a,b,c,d, \ldots$$

The following conventions are used in this section:

The type of element is indicated by the characters assigned to the identifier, usually by the first character or the first two characters.

Characters assigned to **n1, n2, n3**, and **n4** are node labels, or identifiers of other elements. They may be eight alphanumeric characters or less.

V_i is the voltage between the node labeled **ni** and the datum node.

The letter s is used to indicate a complex frequency in small-signal ac analysis (i.e., $\sigma+j\omega$), d/dt in time domain analyses, and zero in a large-signal dc analyses. In the time domain, 1/s is used to indicate integration with respect to time.

A prime is used to indicate that the item primed relates to an element different from the one modeled by the input statement being described.

E_i is the dc voltage which has been *previously calculated* during the analysis being performed between the node labeled **ni** and the datum node.

The words *after large-signal dc analysis* is a contraction of *after a large-signal dc analysis that is performed during the analysis presently being performed*. For example, a large-signal dc analysis performed during an index minimization analysis will in no way effect a sensitivity analysis.

3.6.1 Short

$$SH\#\#\#\#\#\#(n1,n2)$$

Description

This input statement is used to short node **n1** to node **n2**.

Symbol

n1 ————▶———— n2
         I

3.6.2 Linear Resistor

$$R\#\#\#\#\#\#(n1,n2)=a$$

Description

The input statement is used to model a linear time-invariant resistor.

Symbol

n1 ———⧸⋀⋀⋀⧵——▶—— n2
         $R = a$ I

3.6.3 Linear Capacitor

$$C\#\#\#\#\#\#(n1,n2)=a$$

Description

This input statement is used to model a linear time-invariant capacitor.

Symbol

n1 ————⊣⊢————▶—— n2
         $C = a$ I

3.6.4 Linear Inductor

$$L\#\#\#\#\#\#\#(n1,n2)=a$$

Description

This input statement is used to model a linear time-invariant inductor with possible mutual coupling to other inductors. The summation is over all the other inductors that are identified as coupled to this inductor by an **MU** input statement. **MU** input statements, which are described next, contain the values of the mutual inductance.

Symbol

3.6.5 Mutual Inductance

$$MU\#\#\#\#\#\#\#(ident',ident'')=a$$

Description

This input statement is used to indicate that the two inductors with identifiers, **ident'** and **ident''**, are coupled with a mutual inductance of M. The sign of M is positive if the voltage self-induced in one of the inductors due to sinusoidal current flowing into the dotted node of that inductor is in phase with the voltage mutually induced in that inductor by the same current flowing into the dotted node of the other inductor.

Symbol

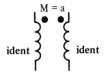

3.6.6 Two-Winding Transformer

$$XF\#\#\#\#\#\#(n1,n2,n3,n4)=a,b,c,d$$

Description

This input statement is used to model a transformer with a self-inductance equal to L_a between the nodes labeled **n1** and **n2**, a self-inductance equal to L_b between the nodes **n1** and **n2**, and a mutual inductance equal to M. The user may specify the coefficient of coupling (d) instead of M.

Symbol

$L_a = a$ $L_b = b$ ⌐2⌐ If $c \neq 0$, M = c
If $c = 0$, M = $d\sqrt{ab}$

with nodes n1, n3 (dots) and n2, n4

3.6.7 Independent Voltage Generator

$$VG\#\#\#\#\#\#(n1,n2)=a$$

Description

This input statement is used to model a generator that maintains a specified voltage between nodes **n1** and **n2**. This generator is a short circuit during a large-signal dc analysis and a signal source during a small-signal ac or a time domain analysis. The function f(t) is a unit step function unless it is defined otherwise on a **DRIVE** input statement.

Symbol

Mathematical Model

   Frequency Domain

$$V_1-V_2=a$$

   Time Domain

$$V_1-V_2=af(t)$$

   Large-Signal dc

$$V_1-V_2=0$$

3.6.8 Independent Current Generator

$$IG\#\#\#\#\#\#(n1,n2)=a$$

Description

This input statement is used to model a generator that maintains a specified current through it. This generator is an open circuit during a large-signal dc analysis and a signal source during a small-signal ac or a time domain analysis. The function f(t) is a unit step function unless it is defined otherwise on a **DRIVE** input statement.

Symbol

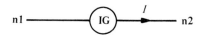

Mathematical Model

    Frequency Domain

$$I=a$$

    Time Domain

$$I=af(t)$$

    Large-Signal dc

$$I=0$$

3.6.9 Voltage-Dependent Current Generator

$$IV\#\#\#\#\#\#(n1,n2,n3,n4)=a,b,c$$

Description

    This input statement is used to model a generator that maintains a current through it to be a linear function of the difference between the voltages at nodes **n3** and **n4**. If *b* is the only nonzero parameter, this input statement models an integrator. If *c* is the only nonzero parameter, this input statement models a differentiator. Most often only *a* nonzero.

Symbol

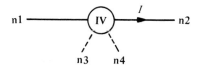

Mathematical Model

$$I=(a+b/s+cs)(V_3-V_4)$$

3.6.10 Voltage-Dependent Voltage Generator

$$VV\#\#\#\#\#\#(n1,n2,n3,n4)=a,b,c,$$

Description

This input statement is used to model a generator that maintains the voltage difference between nodes **n1** and **n2** to be a linear function of the difference between the voltages at nodes **n3** and **n4**. If b is the only nonzero parameter, this input statement models an integrator. If c is the only nonzero parameter, this input statement models a differentiator. Most often only a is nonzero.

Symbol

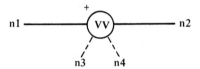

Mathematical Model

$$V_1 - V_2 = (a + b/s + cs)(V_3 - V_4)$$

3.6.11 Current-Dependent Current Generator

$$II\#\#\#\#\#\#(n1,n2,\text{ident})=a,b,c$$

Description

This input statement is used to model a generator that maintains a current through it that depends on the current in a resistor, capacitor, inductor, or short. If b is the only nonzero parameter, this input statement models an integrator. If c is the only nonzero parameter, this input statement models a differentiator. Most often only a is nonzero.

Symbol

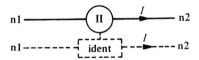

Mathematical Model

$$I=(a+b/s+cs)(V_1-V_2)/R' \quad \text{if ident}=\text{R}\#\#\#\#\#\#\#, \quad R'\neq 0$$
$$I=(a+b/s+cs)sC(V_1-V_2) \quad \text{if ident}=\text{C}\#\#\#\#\#\#\#$$
$$I=(a+b/s+cs)I' \quad \text{if ident}=\begin{cases}\text{L}\#\#\#\#\#\#\# \\ \text{R}\#\#\#\#\#\#\#, \\ \text{SH}\#\#\#\#\#\#\end{cases} \quad R'=0$$

3.6.12 Current-Dependent Voltage Generator

$$\text{VI}\#\#\#\#\#\#(\text{n1,n2,ident})=a,b,c$$

Description

This input statement is used to model a generator that defines the voltage difference between nodes **n1** and **n2** independent of the generators interconnections to other elements but dependent on the current in a resistor, capacitor, inductor, or short. If b is the only nonzero parameter, this input statement models an integrator. If c is the only nonzero parameter, this input statement models a differentiator. Most often only a is nonzero.

Symbol

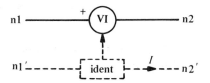

Mathematical Model

$$V_1-V_2=(a+b/s+cs)(V'_1-V'_2)/R' \quad \text{if ident}=\text{R}\#\#\#\#\#\#\#, \quad R'\neq 0$$
$$V_1-V_2=(a+b/s+cs)sC'(V'_1-V'_2) \quad \text{if ident}=\text{C}\#\#\#\#\#\#\#,$$
$$V_1-V_2=(a+b/s+cs)I' \quad \text{if ident}=\begin{cases}\text{L}\#\#\#\#\#\#\# \\ \text{R}\#\#\#\#\#\#\#, \\ \text{SH}\#\#\#\#\#\#\end{cases} \quad R'=0$$

3.6.13 Voltage Battery

$$\text{VB}\#\#\#\#\#\#(\text{n1,n2})=a,b$$

Description

This input statement is used to model a generator that maintains a specified voltage difference between nodes **n1** and **n2**. It is dc source during a large-signal dc or a time domain analysis and a short circuit during a frequency domain analysis.

Symbol

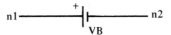

*Mathematical Model**

Large-Signal dc Analyses
$$V_1 - V_2 = a + b\delta$$

Time Domain Analyses
$$V_1 - V_2 = a$$

Frequency Domain and Pole/Zero Analyses
$$V_1 - V_2 = 0$$

3.6.14 Current Battery

$$\text{IB}\#\#\#\#\#\#(n1,n2) = a,b$$

Description

This input statement is used to model a generator that maintains a specified current through it. It is a dc source during a large-signal dc or time domain analysis and an open circuit during a frequency domain analysis.

Symbol

*Mathematical Model**

Large-Signal dc Analyses
$$I = a + b\delta$$

Time Domain Analyses
$$I = a$$

*A definition of δ appears in Sec. 3.5.2.

Frequency Domain and Pole/Zero Analyses
$$I=0$$

3.6.15 Transfer Function

$$YS\#\#\#\#\#\#(n1,n2,n3,n4)=a,b,c,\ldots$$
$$VS\#\#\#\#\#\#(n1,n2,n3,n4)=a,b,c,\ldots$$
$$IS\#\#\#\#\#\#(n1,n2,ident')=a,b,c,\ldots$$
$$ZS\#\#\#\#\#\#(n1,n2,ident')=a,b,c,\ldots$$

Description

This input statement is used to model a black box whose transfer function is known. It does not load the controlling item, nor is it affected by the load it drives.

Symbol

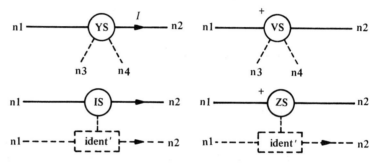

Mathematical Model

$$T=\frac{a+cs+es^2+\ldots}{b+ds+fs^2+\ldots}$$

Symbol	T Equals	Definition
YS	$I/(V_3-V_4)$	Transconductance
VS	$(V_1-V_2)/(V_3-V_4)$	Transpotential
IS	I/I'	Transfluence
ZS	$(V_1-V_2)/I'$	Transresistance

3.6.16 Time Delay

$$YD\#\#\#\#\#\#(n1,n2,n3,n4)=a,b$$
$$VD\#\#\#\#\#\#(n1,n2,n3,n4)=a,b$$
$$ID\#\#\#\#\#\#(n1,n2,ident')=a,b$$
$$ZD\#\#\#\#\#\#(n1,n2,ident')=a,b$$

Description

This input statement is used to model a black box whose output is a function of its input delayed in time. However, BELAC is presently programmed only to recognize this element as a time delay in the frequency domain.

Symbol

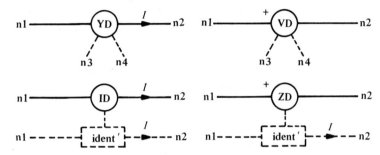

Mathematical Model

$$T = b \cdot \exp(-as)$$

Symbol	T Equals	Definition
YD	$I/(V_3 - V_4)$	Transconductance
VD	$(V_1 - V_2)/(V_3 - V_4)$	Transpotential
ID	I/I'	Transfluence
D	$(V_1 - V_2)/I'$	Transresistance

3.6.17 Transmission Line

$$TL\#\#\#\#\#\#(n1,n2,n3) = a,b,c,d,e$$

Symbol

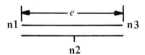

Mathematical Model

$R = a \triangleq$ series resistance per unit length

$L = b \triangleq$ series inductance per unit length

$G = c \triangleq$ leakage conductance per unit length

$C = d \triangleq$ shunt capacitance per unit length

$X = e \triangleq$ length of line, $\quad e \neq 0$

$X = 1, \quad e = 0$

$Z_0 = \sqrt{(R+sL)/G+sC)} \triangleq$ characteristic impedance

$\theta = \gamma X = X\sqrt{(R+sL)(G+sC)} \triangleq$ length times propagation constant

Frequency Domain Analyses

$$\begin{bmatrix} I_1 \\ I_3 \end{bmatrix} = \frac{1}{Z_0 \sinh \theta} \begin{bmatrix} \cosh \theta & -1 \\ -1 & \cosh \theta \end{bmatrix} \begin{bmatrix} V_1 - V_2 \\ V_3 - V_2 \end{bmatrix}$$

3.6.18 Switch Current Generator (Nonlinear)

$$I2\#\#\#\#\#\#(n1,n2,n3,n4) = a,b,c,d$$

Description

This input statement is used to model a nonlinear voltage-dependent current generator whose conductance can be switched from one value to another.

Symbol

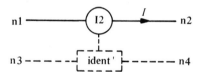

Mathematical Model

Large-Signal dc and Time Domain Analysis

$$I = a(V_3 - V_4) + d \quad V_3 - V_4 \geq 0$$
$$I = b(V_3 - V_4) + d \quad V_3 - V_4 < 0$$

Frequency Domain and Pole/Zero Analysis Before Large-Signal dc Analysis

$$I = 0$$

Frequency Domain and Pole/Zero Analysis After Large-Signal dc Analysis

$$I = aV_3 \quad E_3 - E_4 \geq 0$$
$$I = bV_3 \quad E_3 - E_4 < 0$$

3.6.19 Saturation Current Generator (Nonlinear)

$$\text{IM}\#\#\#\#\#\#(n1,n2,n3,n4)=a,b,c,d$$

Description

This input statement is used to model a nonlinear voltage dependent current generator whose current saturates.

Symbol

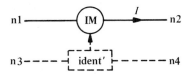

Mathematical Model

Large-Signal dc and Time Domain Analyses

$I=(b(V_3-V_4)+d-a)10^{-9}+a \qquad b(V_3-V_4)+d \leq a$
$I=b(V_3-V_4)+d \qquad a \leq b(V_3-V_4)+d \leq c$
$I=(b(V_3-V_4)+d-c)10^9+c \qquad b(V_3-V_4)+d \geq c$

Frequency Domain and Pole/Zero Analyses
Before Large-Signal dc Analysis

$$I=0$$

Frequency Domain and Pole/Zero Analyses
After Large-Signal dc Analysis

$I=b(V_3-V_4)10^{-9} \qquad b(E_3-E_4)+d<a$
$I=b(V_3-V_4) \qquad a \leq b(E_3-E_4)+b \leq c$
$I=b(V_3-V_4)10^9 \qquad b(E_3-E_4)+d>c$

3.6.20 Logarithmic Current Generator (Nonlinear)

$$\text{IL}\#\#\#\#\#\#(n1,n2,n3,n4)=a,b,c$$

Description

This input statement is used to model a nonlinear current generator whose current is the logarithm of the difference between the voltages at nodes **n3** and **n4**.

Symbol

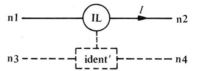

Mathematical Model

Large-Signal dc and Time Domain Analyses
$$I = a \cdot \ln(b(V_3 - V_4)) + c$$

Frequency Domain and Pole/Zero Analyses
Before Large-Signal dc Analysis
$$I = 0$$

Frequency Domain and Pole/Zero Analysis
After Large-Signal ac Analysis
$$I = a(V_3 - V_4)/(E_3 - E_4)$$

3.6.21 Arctangent Current Generator (Nonlinear)

$$\mathbf{IA\#\#\#\#\#\#(n1,n2,n3,n4) = a,b,c}$$

Description

This input statement is used to model a nonlinear current generator whose current is the arctangent of the difference between the voltages at nodes **n3** and **n4**.

Symbol

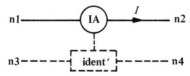

Mathematical Model

Large-Signal dc and Time Domain Analyses
$$I = a \cdot \mathrm{atan}(b(V_3 - V_4)) + c$$

Frequency Domain and Pole/Zero Analyses
Before Large-Signal dc Analysis
$$I = 0$$

*Frequency Domain and Pole/Zero Analyses
After Large-Signal dc Analysis*

$$I = ab(V_3 - V_4)/(1 + b^2(E_3 - E_4)^2)$$

3.6.22 Cosine Current Generator (Nonlinear)

$$IC\#\#\#\#\#\#(n1,n2,n3,n4) = a,b,c,d$$

Description

This input statement is used to model a nonlinear current generator whose current is the cosine of the difference between the voltages at nodes **n3** and **n4**.

Symbol

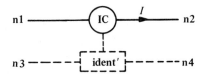

Mathematical Model

Large-Signal dc and Time Domain Analyses

$$I = a \cdot \cos(b(V_3 - V_4) + c) + d$$

*Frequency Domain and Pole/Zero Analyses
Before Large-Signal dc Analysis*

$$I = 0$$

*Frequency Domain and Pole/Zero Analyses
After Large-Signal dc Analysis*

$$I = -ab(V_3 - V_4)\sin(b(E_3 - E_4) + c)$$

3.6.23 Rational Function Current Generator (Nonlinear)

$$IF\#\#\#\#\#\#(n1,n2,n3,n4) = a,b,c,d,e,f$$

Description

This input statement is used to model a nonlinear current generator whose current is a rational function of the difference between the voltages at nodes **n3** and **n4**.

Symbol

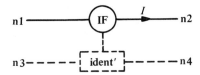

Mathematical Model

Large-Signal dc and Time-Domain Analyses

$$I = \frac{a + c(V_3 - V_4) + e(V_3 - V_4)^2}{b + d(V_3 - V_4) + f(V_3 - V_4)^2}$$

Frequency Domain and Pole/Zero Analyses Before Large-Signal dc Analysis

$$I = 0$$

Frequency Domain and Pole/Zero Analyses After Large-Signal dc Analysis

$$I = \frac{a + c(E_3 - E_4) + e(E_3 - E_4)^2}{b + d(V_3 - V_4) + f(V_3 - V_4)^2}$$

3.6.24 Product Current Generator (Nonlinear)

$$\text{IP}\#\#\#\#\#\#(\text{n1,n2,n3,n4}) = a,b$$

Description

This input statement is used to model a nonlinear current generator whose current is the product of the voltages at nodes **n3** and **n4**.

Symbol

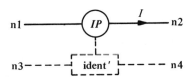

Mathematical Model

Large-Signal dc and Time Domain Analyses

$$I = aV_3V_4 + b$$

*Frequency Domain and Pole/Zero Analyses
Before Large-Signal dc Analysis*
$$I=0$$

*Frequency Domain and Pole/Zero Analyses
After Large-Signal dc Analysis*
$$I=aV_3E_4+aV_4E_3$$

3.6.25 Ratio Current Generator (Nonlinear)

$$\text{IR}\#\#\#\#\#\#(\text{n1,n2,n3,n4})=a,b$$

Description

This input statement is used to model a nonlinear current generator whose current is the ratio of the voltages at nodes **n3** and **n4**.

Symbol

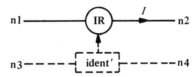

Mathematical Model

Large-Signal dc and Time Domain Analyses
$$I=aV_3/V_4+b$$

*Frequency Domain and Pole/Zero Analyses
Before Large-Signal dc Analysis*
$$I=0$$

*Frequency Domain and Pole/Zero Analyses
After Large-Signal dc Analysis*
$$I=aV_3/E_4-aV_4E_3/E_4^2$$

3.6.26 Generalized Diode Current Generator (Nonlinear)

$$\text{IE}\#\#\#\#\#\#(\text{n1,n2,n3,n4})=a,b$$

Description

This input statement is used to model a nonlinear current generator whose current is a function of the difference between the voltages at nodes

n3 and **n4**. It is sometimes used to model the alpha generator in a bipolar transistor model.

Symbol

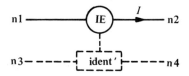

Mathematical Model

Large-Signal dc and Time Domain Analyses

$$I=(\exp(b(V_3-V_4))-1)a$$

Frequency Domain and Pole/Zero Analyses Before Large-Signal dc Analysis

$$I=0$$

Frequency Domain and Pole/Zero Analyses After Large-Signal dc Analysis

$$I=\exp(b(E_3-E_4))ab(V_3-V_4)$$

3.6.27 Voltage Generator (Nonlinear)

Voltage Generator	Current Generator	Voltage Generator	Current Generator
V2	I2	VC	IC
VM	IM	VF	IF
VL	IL	VP	IP
VA	IA	VE	IE

Description

These input statements are used to model nonlinear constraints on $V_1 - V_2$ as a function of V_3 and V_4. They are used to define a branch voltage to be a function of some other branch voltage. The input statements are identical to the input statements for the corresponding current generators except that the **I** is replaced with a **V**.

Symbol

The symbol for each of these voltage generators is the same as the symbol for a **VV** input statement except that **VV** is replaced with the appropriate letters.

Mathematical Model

The mathematical model for each of these voltage generators is the same as the mathematical model of the corresponding current generator except that I is replaced with V_1-V_2.

3.6.28 Diode (Nonlinear)

$$PN\#\#\#\#\#\#(n1,n2,OFF)=a,b$$
$$NP\#\#\#\#\#\#(n2,n1,OFF)=a,b$$

Description

This input statement is used to model a nonlinear conductor and is usually used to model semiconductor junctions. If **OFF** is included as shown above, the input statement will be ignored during large-signal dc analyses until the Newton-Raphson process converges. After convergence is accomplished the Newton-Raphson process will be repeated with this input statement being included. This option is useful in that it allows the user to cause the Newton-Raphson process to converge to a given solution when there is more than one solution.

Symbol

Mathematical Model

Large-Signal dc and Time Domain Analyses

$$I=\exp(b(V_1-V_2)-1)a$$

Frequency Domain and Pole/Zero Analyses
Before Large-Signal dc Analysis

$$I=0$$

*Frequency Domain and Pole/Zero Analyses
After Large-Signal dc Analysis*

$$I = \exp(b(E_1 - E_2))ab(V_1 - V_2)$$

3.6.29 Transition Capacitance (Nonlinear)

$$\mathbf{TC\#\#\#\#\#\#(n1,n2,n3,n4)} = a,b$$

Description

This input statement is used to model a nonlinear capacitor and is sometimes used to model semiconductor junction transition capacitors.

Symbol

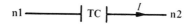

Mathematical Model

Large-Signal dc Analyses

$$I = 0$$

Time Domain Analyses

$$I = a(sV_3 - sV_4)/(b - V_3 + V_4)^2$$

*Frequency Domain and Pole/Zero Analyses
Before Large-Signal dc Analysis*

$$I = 0$$

*Frequency Domain and Pole/Zero Analyses
After Large-Signal dc Analysis*

$$I = a(sV_3 - sV_4)/(b - E_3 + E_4)^2$$

3.6.30 Diffusion Capacitance (Nonlinear)

$$\mathbf{TC\#\#\#\#\#\#(n1,n2,n3,n4)} = a,b$$

Description

This input statement is used to model a nonlinear capacitor and is sometimes used to model semiconductor junction diffusion capacitors.

Symbol

n1 ──────┤ dc ├──*I*──── n2

Mathematical Model

   Large-Signal dc Analyses
$$I=0$$

   Time Domain Analyses
$$I = a \cdot \exp(b(V_3 - V_4))(sV_3 - sV_4)$$

   Frequency Domain and Pole/Zero Analyses
   Before Large-Signal dc Analysis
$$I=0$$

   Frequency Domain and Pole/Zero Analyses
   After Large-Signal dc Analysis
$$I = a \cdot \exp(b(E_3 - E_4))(sV_3 - V_4)$$

4
CIRC

RICHARD McNAIR

Xerox Corporation
El Segundo, California

4.1 INTRODUCTION*

General Description

CIRC is a set of computer programs for the analysis of electronic circuits. CIRC is offered for use on a number of Xerox computers as three related programs: CIRC-DC, CIRC-AC, and CIRC-TR.

CIRC-DC is for dc analysis of circuits. The program contains nonlinear Ebers-Moll models of transistors and diodes that generally require only specification sheet data. CIRC-DC contains an extensive worst-case analysis capability, as well as features for parameter iteration studies with plotted results.

CIRC-AC is for ac analysis of circuits. The program handles both passive and active components. A transistor model, stored within the program, implements a two-pole modeling sophistication that includes the often-overlooked "excess phase" characteristic of transistors. The program allows tentative circuit designs to be evaluated over an automatically scanned frequency range. The program also performs open-loop analyses with proper loading handled automatically. This is a necessary feature for performing accurate stability analyses.

CIRC-TR is for transient analysis of electronic circuits. Extended Ebers-Moll transistor (and diode) models, built into the program, provide for all regions of operation and complete modeling of time dependence. CIRC-TR allows circuit designs to be evaluated over an automatically incremented interval of time. In addition, the solution increment, or time step, is selected and varied by the program in order to minimize the solution run time. The

---

*Portions of the material in this chapter are taken from Xerox's CIRC reference manuals with the permission of Xerox Corporation.

program also performs initial condition solutions to determine the state of the circuit at time zero. The initial condition solutions provide a starting point for subsequent time domain solutions.

Although electronic engineers are generally familiar with ac analysis as used by CIRC, there is a tendency to confuse ac analysis with transient (time domain) analysis programs. CIRC-AC performs small-signal, linear, frequency domain analysis. It operates on the classic complex-number-system basis that is representative of a circuit's performance when driven by *sinusoidal* (alternating current, ac) voltage or current sources. For example, complex numbers are used to represent the impedance of a resistor and inductor in series: $Z = R + j\omega L$. On the other hand, CIRC-TR provides nonlinear (and linear) circuit analysis in the time domain. It writes simultaneous, ordinary differential equations for a prescribed network of circuit elements and solves these equations as a function of time (using difference equation and sparse matrix techniques).

CIRC* is flexible with respect to circuit size and complexity in allowing a range of operational options, and it offers an efficient conversational structure that directs the user to every detail of data input. Used intensively for many years at Xerox as an in-house circuit design package, the program has proved itself as a reliable and useful design tool.

The program and its reference manuals are designed for use by a circuit engineer with no knowledge of computer programming. Running an analysis to obtain nominal values or worst-case, transient, or open-loop stability solutions is an extremely simple operation. The task becomes somewhat more complex when greater analysis flexibility or experimentation is needed, but program operation remains essentially simple throughout.

CIRC can be run under either batch-processing monitors or time-sharing monitors. Much of the power of CIRC is derived from having three modes of operation: conversational, batch, and terminal batch entry. The type of computer installation and the decisions of the computer installation management will determine which of these modes of use are available at any given installation.

The conversational mode of operation uses a remote terminal for both input and output. This mode provides true "conversational" interaction between user and computer. Following simple log-on and CIRC start-up procedures, CIRC starts the conversation by typing out the request: PROVIDE CONTROL MESSAGE. A control message is an instruction to the computer to perform a specified function, e.g., to accept circuit data, or to change a parameter, or to perform an analysis, or to iterate a cycle of calculations with changed parameters, etc. On receiving the control message, the computer—via CIRC—responds by typing out detailed requests for applicable data. For example, CIRC will request the type of data and the exact order

---

*The name CIRC, when not qualified as DC, AC, or TR, is used to refer to all three CIRC programs.

in which the data are to be input for a circuit description. The requests for data are sufficiently detailed to nearly eliminate the possibility of user error due to misunderstanding.

CIRC also automatically creates a file of every circuit description input to the computer via the conversational mode. These "filed" data can be subsequently used for input.

Another major advantage of the conversational mode is the capability of making circuit changes and getting immediate results for either the entire circuit or selected unknowns. This mode is particularly useful in a highly interactive environment producing a low volume of output and requiring limited computer central processor time.

The terminal batch entry mode preserves the convenience of the terminal as an input device while removing its constraints on output volume and CPU time. Output is on the high-speed printer, which is orders of magnitude faster than the terminal typewriter. To use terminal entry batch, two files are prepared, one containing all CIRC input including a circuit description and control messages; the other is used to direct the execution of CIRC. With these two files the job is entered from the terminal into the batch queue where it is then treated as a batch job. For jobs involving large volumes of computations and outputs, batch processing should generally be considered.

Batch processing enables the CIRC user to concentrate on data preparation for CIRC with virtually no involvement in the mechanics of computer operation. The management of the computer installation can provide a set of "start-up" cards that never change. These are placed before the deck containing the circuit information and the control messages directing the analysis and constitute the entire interface between the user and the executive software. But batch processing offers less flexibility than the conversational mode in experimenting with a circuit and requires adherence to a prescribed input sequence without CIRC providing directions. It also requires key-punch cards for input via a card reader and offers slower turnaround (time to obtain answers) than conversational use where answers start coming out immediately.

Material about CIRC presented in this chapter provides a complete overview of the features and usage of CIRC, but considerable additional detail and more examples are presented in the three CIRC manuals (listed in the References at the end of this chapter).

Summary of Salient Features

General

- Built-in models are included for transistors and diodes. For DC and TR, these models are nonlinear, Ebers-Moll type and require mainly data that are available on manufacturers' specification sheets. CIRC-TR transistor (and diode) models have complete modeling of time dependence.

- A diode or a transistor (including its nonlinear model and all necessary parameters) can often be entered into a circuit description with one or two lines of data.
- Use of equivalent circuits is generally unnecessary.
- CIRC offers true conversational structure designed for efficient man-machine interaction.
- Results of conversational circuit data entries are reusable in subsequent batch runs, offering the advantages of high-speed line printer output and economical use of larger amounts of computer time.
- CIRC automatically uses all available memory to accommodate the largest possible circuit, with complexity rather than node count being the limiting factor.
- Output of results is automatically annotated.
- Output volume is automatically adjusted to mode of usage.
- User may select output volume and unknowns to be analyzed.
- Standard analyses are extremely simple to obtain. For nonstandard analyses, an extensive set of controls is available.
- CIRC has a detailed set of reference manuals.
- Parameters and circuit configurations can be experimentally varied in both batch and conversational modes.
- Supplementary coding in FORTRAN is permitted.

CIRC-DC

- Automatic worst-case analysis can be performed with adequate output for detailed evaluation of results.
- Circuit sensitivity information is displayed, allowing effective evaluation of tolerance effects on circuit performance.
- Even if only nominal data are input, 1% parameter variations can be used for sensitivity studies.
- Parameters can be iterated with different values, and the effects on the circuit of the changes may be plotted automatically on a line printer.
- A worst-case analysis may be repeated automatically for each parameter value selected during an iteration—then nominal, minimum, and maximum curves can be plotted on a line printer.

CIRC-AC

- Automatic open-loop analysis can be performed.
- Analyses can be performed over an automatically scanned range of frequencies and plots of results obtained on the line printer.

- Parameters can be iterated with different values, and the effects on the circuit of the changes may be plotted automatically on a line printer.

CIRC-TR

- A unique, easy-to-use specification-oriented data set may be used to set up the Ebers-Moll model of transistors and diodes.
- A more conventional "classic" data set may be used to set up the Ebers-Moll model—much like the data used in SCEPTRE, NET-1, etc.
- A special Xerox data set for the extended Ebers-Moll models is provided which allows recombination diodes as well as diffusion diodes, nonlinear beta with current dependence and high current roll-off, and individual device temperatures.
- Output features are provided for economical display of selected variables versus time, allowing at the same time occasional "in-depth snapshots" of circuit performance. Since a selected variable's performance versus time is often best viewed as an effect of circuitry interaction, the snapshot allows determining the cause be displaying the interaction of all circuit elements.
- Analyses can be performed over an automatically incremented interval of time and plots of results obtained on the line printer.
- The integration techniques that are used to solve the transient performance problem are controlled by the smallest significant time constant in the circuit—not by the smallest absolute time constant. This often results in computer-time savings of 30:1 to 100:1 over techniques tied exclusively to smallest absolute time constants.
- The time step used in the integration technique can be automatically adjusted for diode and transistor circuits (1) to ensure that larger time steps are taken during periods of inactivity and (2) to reduce the time step when the circuit devices start changing rapidly.
- Automatic initial condition solutions are provided when desired.
- Analyses can be performed over an automatically incremented interval of time, with plots of results obtained on the line printer.
- Circuit parameters may be conveniently controlled and varied with tables of values. This feature permits simulation of such things as transient radiation effects, time-dependent resistors, etc.
- An analysis in progress can be dumped and the status saved so that a subsequent job can load the information and continue the analysis ("checkpointing" feature).

Circuit Size Capability

CIRC automatically uses all available memory to accommodate the largest possible circuit. The optimum use of memory is based on two special characteristics:

- Memory space allocated to data and equation storage is dynamic (i.e., variable in volume and structure).
- CIRC uses sparse matrix techniques (i.e., in its processing of node equations, it stores only nonzero numbers).

The first of these characteristics results in CIRC not having fixed limits on the size of circuits that may be analyzed. CIRC is not limited to a fixed node count, resistor count, transistor count, etc. Instead, memory space is expanded in exact proportion to the actual number of elements and nodes that exist in the circuit.

The second of these characteristics allows efficient solution of very large problems by minimizing both memory size requirements and solution time.

The combination of these two characteristics allows gradual reduction of analysis capability as circuit size becomes larger. For example, a 50 node/12 transistor circuit might be too large for a conversational worst-case analysis, but CIRC might allow a nominal analysis.

To allow memory usage to vary and expand without catastrophic effects, CIRC performs checks on the expansion. The mechanism that performs the checks informs the CIRC user of the status of the analysis capability and, at times, takes other appropriate action. Specific messages used by CIRC are presented in References 1–3. Brief discussion of the operational significance of a limiting memory environment for CIRC-DC was given earlier in the presentation of worst-case analysis of large circuits.

CIRC uses its dynamic memory feature to such an extent that memory for an analysis is sized to the specific problem. Some of Xerox's time-sharing monitors are thus allowed to make optimum use of physical memory because no significant amount of unused memory is assigned to CIRC problem solving. This optimum memory usage can be passed on as a user cost saving when the cost accounting is refined enough to have the cost reflect the amount of CPU time used *weighted* by the amount of core used.

The effectiveness of CIRCs memory handling can be seen in several examples of large analyses handled by Xerox. A nonlinear worst-case analysis has been performed of a 245-node circuit containing 54 transistors, 19 diodes, 227 resistors, and 2 power supplies. This analysis can be done on a 64K-word (32-bit) memory Sigma computer.

A transient analysis has been performed on a 204-node circuit containing 173 resistors, 245 diodes, and 36 transistors. The entire analysis can be handled in an 80K-word (32-bit) memory Sigma computer.

Major Feature Capability

The CIRC package has two analysis features which are relatively unique in their capability and which will, therefore, be discussed in some detail. The first is Worst-Case DC Analysis and the second is AC Open Loop Analysis.

Worst-Case Analysis

CIRC-DC has a worst-case analysis capability that is very complete and which makes the DC program much different from just an initial condition (dc) calculation from a transient analysis program.

The worst-case analysis discussion presented now requires several preliminary comments. First, the presentation will use output from CIRC-DC that relates to the example problem presented later in the chapter. The discussion is related to the output figures, and the annotation on the output allows understanding relative to the worst-case presentation. However, readers preferring to understand the output figures more fully may want to study the example problem before reading the worst-case presentation. It is also noted that worst-case analysis requires nominal, minimum, and maximum parameter values. The presence of all three parameter values is assumed in the worst-case discussion. Again, the example problem may be reviewed to see the method for entering the three values.

Finally, one term definition is appropriate. The term "dependent unknown" is used to describe the special information generated for each type of network element. CIRC writes a set of independent, simultaneous node equations, solves these equations, and produces its prime set of unknowns—the node voltages. CIRC then calculates the value of a new group of unknowns, such as element currents and power dissipations. The values of these new unknowns depend on the node voltages and element parameters. Hence, these unknowns are referred to as dependent unknowns.

Worst-case design and analysis may be viewed as involving two tasks:

- Determining the parameter values that create worst-case performance.

- Setting up these parameter values and solving the circuit equations to determine worst-case performance.

Sensitivities

Sensitivities indicate the influence that parameters have on circuit performance. Figure 4.1 shows the effect of a parameter on a circuit unknown. The curvature is exaggerated for illustrative purposes.

CIRC determines the sensitivity by comparing two different solutions of the circuit equations. One is the nominal solution. The other is a special solution obtained with one parameter changed from nominal—normally to its maximum. Once the solution with the incremented parameter is obtained, CIRC determines and records in memory the algebraic sign ($+$ or $-$) of the change of all circuit unknowns with respect to the one parameter that has been changed.

This process must be performed once for each parameter that is to have other than nominal value for the worst-case analysis. No sensitivity is calculated if a parameter has maximum and minimum values that are equal.

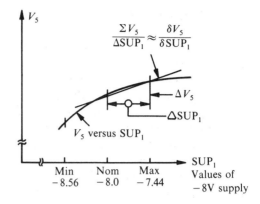

Figure 4.1 Determining Parameter Sensitivity.

Because the sensitivity information that CIRC creates for its own use is also valuable for the circuit designer, CIRC provides a number of ways for obtaining this information. Generally, the most useful output is a ranking of sensitivities. Figure 4.2 shows a typical output.

```
RANKING OF SENSITIVITIES OF NODE     5
```

RANK	PAR.		% PAR VAR.	DIFFERENCES	% OF TOTAL DIFF.
1	BN	1	100.00	-.3034E 01	63.6607
2	BN	2	100.00	.7068E 00	14.8297
3	VRV	1	5.00	.4311E 00	9.0442
4	CUR	1	5.00	-.2156E 00	4.5223
5	RES	2	3.00	-.1687E 00	3.5401
6	VBE	2	14.29	-.8622E-01	1.8089
7	SUP	1	7.00	.7720E-01	1.6197
8	RES	3	3.00	.3972E-01	.8332
9	ICBO	1	150.00	-.6596E-02	.1384
10	ICBO	2	150.00	.1291E-03	.0027
C1	C2		C3	C4	C5

Figure 4.2 Ranking of Sensitivities.

To adequately understand Fig. 4.2, the entire concept of sensitivities analysis should be understood. Information that may be obtained while the sensitivity analysis is being performed is used to explain the sensitivity analysis concept. The information is shown in Fig. 4.3. This information display has good tutorial value but little practical value and is, therefore, not normally output. However, if desired, CIRC's control message feature (discussed later) may be used to obtain it.

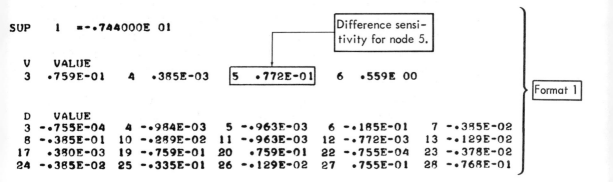

```
DIFFERENCES WITH RESPECT TO:
PAR.          VALUE


SUP    1   =-.744000E 01

  V    VALUE
  3   .759E-01      4   .385E-03      5   .772E-01      6   .559E 00

  D    VALUE
  3  -.755E-04    4  -.984E-03    5  -.963E-03    6  -.185E-01    7  -.385E-02
  8  -.385E-01   10  -.289E-02   11  -.963E-03   12  -.772E-03   13  -.129E-02
 17   .380E-03   19  -.759E-01   20   .759E-01   22  -.755E-04   23  -.378E-02
 24  -.385E-02   25  -.335E-01   26  -.129E-02   27   .755E-01   28  -.768E-01


    SPECIAL SOLUTIONS       8

** NODE VOLTAGES **

  1   .7519017E 00    2   .7419017E 00    3   .3516989E 01    4   .9992003E 01
  5   .2706031E 01    6  -.6664673E 01



** RESISTORS **

    #      IRES              PRES

    1   .1000000E-03      .1000000E-05
    2   .6475014E-02      .4192580E-01
    3   .1874141E-01      .1756202E 00
```

Figure 4.3 Sensitivities, Differences with Unknowns.

The sensitivity calculations are based on the example circuit (Fig. 4.10). The first few lines of output identify the parameter (supply 1) that has been varied from nominal and the actual value used (maximum: −7.44 volt).

The special solutions in Fig. 4.3 show that node 5 is at 2.706031 volts. The nominal solutions, presented with the example problem, show a value of 2.628831 volts. The difference is an increase of 0.0772 volt ($\Delta V_5 =$

0.772E−01 = 77.22 millivolts). Figure 4.3 displays this difference in the sensitivity table that precedes the special solutions. The sensitivity tables of Fig. 4.3 contain only four node voltages and 20 dependent unknown sensitivities (even though full output options were used). Two node voltages and eight dependent unknown sensitivities were left out because CIRC outputs only significant sensitivities. If the value of an unknown changes less than 0.001% of nominal value, then the sensitivity of that unknown is normally considered to be insignificant. This automatic editing reduces output volume and allows the circuit designer to determine quickly which unknowns are not affected by the parameter being treated.

Partial derivatives may also be obtained from CIRC-DC. They are calculated with a difference ratio as shown in Fig. 4.1; i.e.,

$$\frac{\delta U_i}{\delta P_i} \approx \frac{\Delta U_i}{\Delta P_i}$$

where

U_i = node or dependent unknown.

P_i = parameter.

Let us consider the node 5 sensitivity with respect to supply 1. Since the supply is nominally at −8 volts,

$$\Delta P = -7.44 - (-8) = 0.56 \text{ volt}$$

and previous discussion showed that

$$\Delta V_5 = 0.772 \times 10^{-1}$$

Therefore,

$$\frac{\delta V_5}{\delta S_1} \approx \frac{0.772 \times 10^{-1}}{0.56} = 0.138$$

The relative merit of differences versus partials can be appreciated best when a parameter is either very large or very small. A hypothetical case involving ICBO will illustrate the point.

Suppose that a voltage is nominally 1.0 volt and that ICBO maximum causes the voltage to become 1.005 volts. Then the difference is

$$\Delta V = 0.005 \text{ V} = 0.5 \times 10^{-2}$$

Suppose that ICBO NOM $= 0.5 \times 10^{-6}$ and that ICBO MAX $= 2.5 \times 10^{-6}$; then

$$\Delta \text{ICBO} = 2 \times 10^{-6}$$

Therefore,

$$\frac{\delta V}{\delta P} \approx \frac{0.5 \times 10^{-2}}{2.0 \times 10^{-6}} = 0.25E + 4 = 2500 \text{ volts/amp}$$

Notice that the value of the partial is quite large, even though the voltage is changed by only 5 millivolts when ICBO has its maximum value. The conclusion is that when sensitivity information is output as partial derivatives, the CIRC user must consider a parameter's value when the sensitivity to that parameter is studied. Therefore, CIRC does not normally output partial derivatives and no example printout is shown.

On the other hand, partials have the advantage that they can be used directly in estimating the magnitude of unknown change that a parameter change will cause. For example, suppose that the user wanted to know the effect of ICBO becoming 200 microamperes. The change in the voltage would be

$$\Delta V = \frac{\delta V}{\delta P} \times \Delta P = \frac{\delta V}{\delta \text{ICBO}} \Delta \text{ICBO} \approx 2500(200.0 \times 10^{-6})$$

$$\Delta V \approx 0.5 \text{ volt}$$

Sensitivity Output as Ranking

Consider again the sensitivities ranking shown in Fig. 4.2. It offers major advantages over the information in Fig. 4.3, which was obtained while the sensitivity run was in progress. First, the information is organized by unknown (node 5 in Fig. 4.10) rather than by parameter (supply 1 in Fig. 4.3). Second, the information displayed is much more refined and comprehensive, as will be seen from the discussion that follows.

The first line of Fig. 4.2 indicates that the unknown for which the information is supplied is node 5 voltage. The first column (C1) shows the ranking of the effect of each parameter. This information is obtained by CIRC's sorting of the magnitude of the differences (column C4). Column C2 shows the parameter name, and column C3 shows the percentage parameter change that was used to calculate differences. (When a nominal parameter value is zero, the percentage change would have no meaning, and CIRC, therefore, shows an S and the actual value setting used to obtain the difference calculation.) Column 3 reminds the engineer of the magnitude of parameter tolerance, while at the same time the importance of the tolerance is being displayed by columns 1, 4, and 5. Column C4 displays that amount of unknown change caused by the parameter change. Finally, column 5 shows the percentage change in the unknown contributed by each parameter that is in the table.

For example, changing SUP1 input by 7% (C3) causes a change in V5 of 0.0772V (C4), representing 1.6197% (C5) of the sum of all magnitudes of differences listed in C4 (4.76607V).

Ranking of sensitivities will be done only for unknowns that have been selected for printout. The sensitivities ranking is normally done with the parameter tolerances input for a circuit. This has the advantage of giving the circuit designer the best feel for the parameter tolerances as they are anticipated in actual production. However, there are times when a designer may prefer to know which parameters affect his circuit most—but without including predetermined tolerances. CIRC, therefore, allows the sensitivities ranking to be done with all parameters having a 1% change. The information thus obtained may be used to help the designer set tolerance limit controls over the circuit components.

CIRC Setup of a Worst-Case Analysis

A worst-case analysis uses the sentivity data to set up the appropriate parameter value conditions. The circuit equations are then solved, and the worst-case analysis is produced. The example problem, Fig. 4.10, is again used to study CIRC's worst-case analysis procedure.

Table 4.1 shows a summary of the sensitivity data as they are used in determining the worst-case maximum* for node 5. The (+) or (−) sign

Table 4.1 Sensitivity Data Interpretation for Node 5 Worst Case

Parameter		Sensitivity Results	Significant Worst-Case Parameter Values	Other Parameter Values	Nominal Values	Units
SUP	1	Positive (+)	−7.44		−8.0	Volts
TEMP	1	Equal values		25.0	25.0	Degree C
RES	1	Insignificant		103.0	100.0	Ohms
RES	2	Negative (−)	980.0		1000.0	Ohms
RES	3	Positive (+)	515.0		500.0	Ohms
VRV	1	Positive (+)	10.5		10.0	Volts
VRR	1	Equal values		0.1	0.1	Ohms
CUR	1	Negative (−)	0.095		0.1	Milliamps
CUR	2	Equal values		3.0	3.0	—
VD	1	Equal values		0.7	0.7	Volts
IREV	1	Equal values		100.0	100.0	Nanoamps
VBF	1	Insignificant		0.8	0.7	Volts
VSAT	1	Insignificant		0.2	0.3	Volts
BN	1	Negative (−)	25.0		50.0	—
BI	1	Equal values		1.0	1.0	—
ICBO	1	Negative (−)	10.0		100.0	Nanoamps
VBE	2	Negative (−)	0.6		0.7	Volts
VSAT	2	Insignificant		0.4	0.3	Volts
BN	2	Positive (+)	100.0		50.0	—
BI	2	Equal values		1.0	1.0	—
ICBO	2	Positive (+)		250.0	100.0	Nanoamps
ID	1	Condition		1.0	1.0	Milliamps
ICS	1	Condition		10.0	10.0	Milliamps
IBS	1	Condition		1.0	1.0	Milliamps
IEA	1	Condition		1.0	1.0	Milliamps
ICS	2	Condition		10.0	10.0	Milliamps
IBS	2	Condition		1.0	1.0	Milliamps
IEA	2	Condition		1.0	1.0	Milliamps

*Maximum and minimum are used in an algebraic sense; i.e., −9.0 is greater than −10.0. This applies to both input data and unknowns that are to be worst-case-analyzed.

shown in Table 4.1 is obtained from the sensitivity ranking information, Fig. 4.2. The sensitivity has the sign of the product of the percent-parameter variation (C3) times the difference (C4). This will normally turn out to be just the sign of the difference (C4) except in the rare case when a parameter's maximum and nominal values are the same.

For each parameter with a (+) or (−) showing in column 2, the corresponding parameter values (used to maximize node 5) are shown in column 3 of Table 4.1. These parameter values are at their maximum limit when the sensitivity data indicate plus effect; i.e., a parameter value increase causes an unknown value increase. When the sensitivities are minus, the minimum parameter value is selected to maximize the node voltage.

"Other Parameter Values," column 4 of Table 4.1, are the parameters whose selections were not directly based on the sensitivity data displayed by CIRC. These parameters fall into three categories: (1) parameters with "insignificant" sensitivity set to maximum or minimum values in the same way as for "significant" parameters*; (2) parameters with equal nominal, minimum, and maximum values set to their only existent value; and (3) "condition" parameters (parameters ID1 through IEA2). How the condition parameter values are selected is explained below.

Worst-Case Analysis Results

Figure 4.4 shows the actual performance of worst-case analysis for the maximum value of the node 5 voltage. The parameter values used to create this worst-case analysis are shown as an integral part of the output. The parameter values are in agreement with Table 4.1. This analysis has determined the worst-case maximum of one unknown, i.e., node 5 voltage. All the other unknown values that appear are the values that exist for the parameter set that created worst-case maximum for node 5. An entirely new analysis with the parameters at opposite limits must be used to determine the minimum worst-case value of the node 5 voltage.

Furthermore, full worst-case analysis of all the unknowns would require two solutions (one minimum and one maximum) for all nodes plus all dependent unknowns. For the example problem, there are 6 nodes and 29 dependent unknowns. Therefore, a full worst-case analysis would involve $(6 + 29) \times 2 = 70$ individual solutions. Thus, even the simple example problem involves a large volume of calculations for a full worst-case analysis.

CIRC, therefore, provides capability to select only those unknowns that are to be worst-case-analyzed and exclude other unknowns.

Figure 4.4 shows one type of output that may be obtained during worst-case analysis. This figure contains one worst-case node voltage value and a large amount of data that are not worse-case results, per se, namely, five

---

*Even sensitivities that are so small as to make printout inappropriate may be valid, and their signs are therefore stored in memory. For example, $VSAT_1$ has to have a (−) sensitivity to get the minimum value selection shown in Table 4.1.

W.C. MAX   NODE   5

** NODE VOLTAGES **

```
1  .8325546E 00   2  .8232446E 00   3  .7173819E 01   4  .1048958E 02
5  .6454986E 01   6 -.6656799E 01
```

** RESISTORS **

```
#       IRES              PRES
1   .9500000E-04      .8844500E-06
2   .3383429E-02      .1121864E-01
3   .2545978E-01      .3338231E 00
```

** VOLTAGE-RESISTANCE SOURCES **

```
#       IVRS              PVRS
1   .1042145E 00      .1093166E 01
```

** CURRENT SOURCES **

```
#     PCUR (ICUR FOR CONTROLLED SOURCES)
1   .7909269E-04
2   .7637933E-01
```

** DIODES **

```
#       IDIO              PDIO              VDIO
1   .2545978E-01      .1994013E-01      .7832014E 00
```

** TRANSISTORS **

```
#       IB                IC                IE                PQ
        VBE(VEB)          VBC(VCB)          VCE(VEC)          IC/IB

1   .9500000E-04      .2375415E-02      .2470405E-02      .1711900E-01
    .8232446E 00     -.6350574E 01      .7173819E 01      .2500436E 02

2   .1008024E-02      .1008313E 00      .1018391E 00      .4075379E 00
    .7188326E 00     -.3315760E 01      .4034593E 01      .1000287E 03
```

PARAMETERS

```
SUP   1  -.744000E 01   TEMP  1  .250000E 02   RES   1  .980000E 02
RES   2   .980000E 03   RES   3  .515000E 03   VRV   1  .105000E 02
VRR   1   .100000E 00   CUR   1  .950000E-04   CUR   2  .300000E 01
VD    1   .700000E 00   IREV  1  .100000E-06   VBE   1  .800000E 00
VSAT  1   .200000E 00   BN    1  .250000E 02   BI    1  .100000E 01
ICBO  1   .100000E-07   VBE   2  .600000E 00   VSAT  2  .400000E 00
BN    2   .100000E 03   BI    2  .100000E 01   ICBO  2  .250000E-06
ID    1   .100000E-02   ICS   1  .100000E-01   IBS   1  .100000E-02
IEA   1   .100000E-02   ICS   2  .100000E-01   IBS   2  .100000E-02
IEA   2   .100000E-02
```

Figure 4.4 Worst-Case Maximum—Node 1.

node voltages, 29 dependent unknowns, and 28 parameter values. It is, therefore, desirable to have a summary of the worst-case analysis results. CIRC provides this summary in the format shown in Fig. 4.5. For the example problem this table provides a summary of the data that could also be obtained from the nominal solution output page plus 70 individual solution outputs of the form shown in Fig. 4.4.

Both of these output formats, Figs. 4.4 and 4.5, have their advantages and disadvantages. CIRC, therefore, provides three options to select the

W.C. RESULTS

NODE			MINIMUM	NOMINAL	MAXIMUM
V	1		.63255402E 00	.75190160E 00	.86940615E 00
V	2		.62324402E 00	.74190160E 00	.85859115E 00
V	3		.39731327E 00	.34410767E 01	.71878883E 01
V	4		.94910079E 01	.99916177E 01	.10494179E 02
V	5		-.44075321E 00	.26288306E 01	.64549860E 01
V	6		-.77903976E 01	-.72233850E 01	-.66567986E 01

DEP.	ID	NO.	MINIMUM	NOMINAL	MAXIMUM
IRES	1	1	.95000000E-04	.10000000E-03	.10500000E-03
PRES	1	2	.88445000E-06	.10000000E-05	.11355750E-05
IRES	2	3	.33692185E-02	.65505410E-02	.10183070E-01
PRES	2	4	.11124601E-01	.42909587E-01	.10162101E 00
IRES	3	5	.12126488E-01	.19704431E-01	.28766937E-01
PRES	3	6	.75731631E-01	.19413230E 00	.40549298E 00
IVRS	1	7	.55402892E-01	.83822926E-01	.11744315E 00
PVRS	1	8	.52602053E 00	.83752663E 00	.12317738E 01
PCUR	1	9	.60092632E-04	.75190160E-04	.91287646E-04
ICUR	2	10	.36379464E-01	.59113294E-01	.86300812E-01
IDIO	1	11	.12126488E-01	.19704431E-01	.28766937E-01
PDIO	1	12	.92663115E-02	.15302757E-01	.22620604E-01
VDIO	1	13	.76413810E 00	.77661502E 00	.78634037E 00
IB	1	14	.95000000E-04	.10000000E-03	.10500000E-03
IC	1	15	.23753365E-02	.50053012E-02	.96593895E-02
IE	1	16	.24703265E-02	.51052012E-02	.97641395E-02
PQ	1	17	.30352875E-02	.17297816E-01	.20844842E-01
VBE	1	18	.62324402E 00	.74190160E 00	.85859115E 00
VBC	1	19	-.63646437E 01	-.26991751E 01	.35811192E 00
VCE	1	20	.39731327E 00	.34410767E 01	.71878883E 01
IC/B	1	21	.25003013E 02	.50053012E 02	.10028191E 03
IB	2	22	.48127073E-03	.15453398E-02	.36551982E-02
IC	2	23	.46640367E-01	.77272485E-01	.11392897E 00
IE	2	24	.48505952E-01	.78817725E-01	.11506775E 00
PQ	2	25	.40753793E 00	.57019605E 00	.60819680E 00
VBE	2	26	.70202014E 00	.81224612E 00	.91846687E 00
VBC	2	27	-.99794083E 01	-.65505410E 01	-.33018341E 01
VCE	2	28	.40345926E 01	.73627871E 01	.10857632E 02
IC/B	2	29	.25000111E 02	.50003555E 02	.10005650E 03

Figure 4.5 "Summary Table" Output for Worst-Case Results.

desired outputs:

1. Summary table only.
2. Individual worst-case solutions.
3. Individual worst-case solutions with parameters.

Option 1 results in no individual solutions being output during the worst-case analysis. Only the summary table is obtained (the summary table is automatically output at the end of a worst-case analysis). The disadvantage of option 1 is that no data are obtained about the overall conditions that exist when a particular worst-case result is obtained. On the other hand, with option 2 or 3, a great deal of data about overall conditions may be determined, for example, all node voltages, device currents, transistor states, etc. Or, options 2 and 3 can just show selected voltages, currents, etc.

Option 2 differs from option 3 only in that the parameters shown in Fig. 4.4 appear for option 3 but not for option 2. CIRC is initialized in option 3 for batch used and in option 2 for conversational use.

Special Parameter Considerations for Worst-Case Analysis

CIRC's acceptance of standard component models makes the program much easier to use than would be the case if the engineer had to create and input his own linear model composed of a collection of resistors, independent and controlled current, and/or voltages sources, but, there are some component subtleties that must be accommodated. Consider the diode performance data required by CIRC. A diode is described primarily by a current and voltage point from the nonlinear performance characteristic.

Three classes of diode performance—minimum, nominal, and maximum—should be described with the three pairs of data numbered 1, 2, and 3 in Fig. 4.6. CIRC is normally allowed to select each parameter value inde-

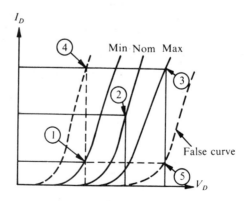

Figure 4.6 Minimum, Nominal, and Maximum Diode Performance.

pendently. If CIRC selected a minimum value of current and a maximum value of voltage, points on the false diode performance curve shown in Fig. 4.6 would be created. This curve is not acceptable.

To avoid the false curve, CIRC treats the diode voltage as an independent ("normal") parameter and the current as a dependent parameter that specifies the "condition" under which the voltage is measured. The transistor also has two cases of condition parameters.

In CIRC operation, the condition parameters are treated in the following manner. If the normal parameter of a "normal-condition" data set has the value that is less than nominal (presumably minimum) during an analysis, then the condition parameter is set up with the data directly from their minimum data storage. The same type of procedure is used for maximum parameter value conditions. This means that the present value of condition parameters is controlled by CIRC after a specific analysis is begun. This results in the following parameter change restrictions.

The present values of condition parameters (e.g., I_D, I_{CS}, I_{BS}, I_{EA}) *cannot* be changed, but the parameters associated with the condition parameters (e.g., V_D, V_{SAT}, V_{BE}) may be changed. However, there may be unexpected changes in the condition parameters themselves (if nominal, minimum, and maximum values of the condition parameters are not equal).

Consider as an example the following hypothetical diode data:

	Min.	Nom.	Max.
I_D (CONDITION)	0.1 mA	1.0 mA	10.0 mA
V_D (NORMAL)	0.6 V	0.7 V	0.8 V

If during an analysis $V_D = 0.6$ V, CIRC selects $I_D = 0.1$ mA. If $V_D = 0.7$ V, CIRC selects $I_D = 1.0$ mA. If $V_D = 0.8$ V, CIRC selects $I_D = 10.0$ mA. If the user inputs the present value of $I_D = 5$ mA and if $V_D = 0.7$ V, CIRC resets the present value of I_D to 1.0 mA. Finally, if the user sets the present value of $V_D = 0.65$ V, CIRC selects $I_D - 0.1$ mA.

Deficiencies of Worst-Case Mechanisms

Worst-case analysis mechanisms are dependent on sensitivity data. Some worst-case programs perform worst-case analysis in a manner that is dependent on the *value* of the sensitivity data (partial derivatives). These methods sacrifice accuracy to gain speed. The accuracy loss is quite significant for large problems with data variations that typically range from 10 to 100% or more (e.g., BETA = 50 nominal to 150 maximum: 300%). The loss of accuracy inherent in these methods is not compatible with the potentially high accuracy of worst-case analysis using nonlinear models.

Therefore, as previously described, CIRC uses a worst-case mechanism that depends *only* on the *sign* (+ or −) of the sensitivity data. But two sources of error remain even when this mechanism is used. They are most conveniently considered with the sketch shown in Fig. 4.7. Curve (a) shows a partial sensitivity that has an inflection between nominal and maximum parameter values. The worst-case maximum of the unknown, therefore, does *not* occur with the parameter at a maximum limit. CIRC selects a limit (maximum or minimum) value for each parameter and may, therefore, not actually find the absolute maximum for the unknown. A very good example of this type of performance is the power dissipation of a transistor (the unknown) versus the value of a resistor (the parameter) that determines the device drive. If the resistor variation changed the transistor operating point from below midrange active to above midrange active, the peak power dissipation would not be determined, since it occurs precisely at midrange active.

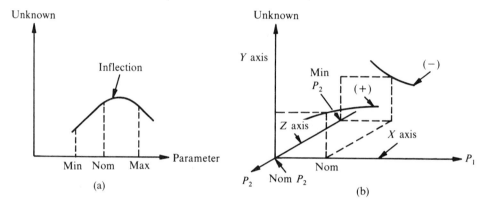

Figure 4.7 Sensitivity Difficulties: (a) Inflection of Sensitivity; (b) Partial with Secondary Parameter Effect.

Curve (b) of Fig. 4.7 displays the characteristics that may result because an unknown actually depends on a number of parameters. The sensitivity data are obtained when one parameter is varied and the other parameters are nominal (a basic characteristic of partial derivatives). The curve shows a two-parameter, three-dimensional system in which the sensitivity being studied, that of P_1, is influenced to go from (+) to (−) because of the secondary effect of parameter P_2. Actual circuits are $(n + 1)$-dimensional, where n may be a large number of parameters.

The network shown in Fig. 4.8 demonstrates a special case in the class of curve (b).

Assume that nominal conditions cause

$$V_1 - V_2 = V_D < 0.2 \text{ V}$$

The diode would be OFF and the sensitivity of node 1 with respect to R_3 and R_4 would be approximately zero. However, worst-case selections of the network parameters could turn the diode ON. Then the resistors R_3 and R_4

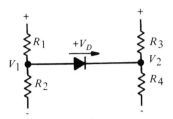

Figure 4.8 Example for Sensitivity Study.

would have an effect on node 1, but the sensitivity data would *not* be known. Therefore, R_3 and R_4 could not be reliably selected to maximize node 1.

These two sources of error are not a matter of major concern. Worse-case analysis consistently gives a valid appraisal of circuit performance under adverse parameter value conditions. The major reason that these two deficiencies do not seriously influence the validity of a worst-case analysis is this: Worst-case conditions should not normally cause devices to change bias states (e.g., diodes changing from OFF to ON, etc.). If bias states do not change, then the worst-case analysis is usually precise. If bias states do change, the parameter selection may fail slightly in creating absolute worst case, but the fact that states changed is usually sufficient to indicate a need for circuit redesign.

State Change Messages

State changes are of enough significance that CIRC-DC checks and tabulates all diode and transistor state changes. This is done by having CIRC display all nominal states just before the actual worst-case analysis is begun. CIRC then records any change of state from nominal. A typical message is STATE CHANGES: TRAN 1 SAT.

The state definitions used by CIRC are shown below.

Diode:

$$\text{FOR: } V_{AC} = +$$
$$\text{REV: } V_{AC} = -$$

where V_{AC} = anode to cathode voltage.

Transistor:

$$\text{ACT: } V_{BE} = +; \quad V_{BC} = -$$
$$\text{SAT: } V_{BE} = +; \quad V_{BC} = +$$
$$\text{INV: } V_{BE} = -; \quad V_{BC} = +$$
$$\text{OFF: } V_{BE} = -; \quad V_{BC} = -$$

where, in an *NPN* device,

$$V_{BE} = \text{base to emitter voltage}$$
$$V_{BC} = \text{base to collector voltage}$$

CIRC defines all states purely on a voltage basis. This may at times cause a device to be shown as "active" or "forward" when no significant current is flowing and the device may be correctly considered to be off. But even in such cases the state is worth pointing out.

Sorting Improves Worst-Case Analysis

Any procedure that would totally eliminate these two deficiencies (if such a procedure could be found) would require a prohibitive amount of calculation. However, CIRC has one feature that improves worst-case analysis with very little increase in analysis time (and one discussed later, PCWC, that requires considerably more analysis time). The first procedure is a simple sort. A sort is made every time a sensitivity evaluation is made or a worst case is calculated for a specific unknown.

The sort operates as follows: Each time a solution is completed, the resulting (present) unknown values (all node voltages and all dependent unknowns) are individually compared to the previously recorded minimum and maximum values of that unknown. If the present value is less than the previous minimum, the present value replaces the previous minimum. If the present value is greater than the previous maximum, the previous maximum is replaced. This procedure makes maximum use of all the calculations that are performed.

The worst-case conditions for every unknown may be unique (i.e., a different set of parameter value conditions may be required for each worst case of each unknown). However, many times there is a strong similarity (if not equality) between the worst-case conditions for two different variables. The result is that each time a specific worst-case analysis is performed, a "near" worst case may be determined for other unknowns. Because sorting is used, all of the "near" worst-case values are recorded. Therefore, performing specific worst-case analysis for a selected group of unknowns produces a near worst-case for the other unknowns. The worst-case summary table shows the worst-case and the near worst-case data.

Because of the extremely large volume of calculations required to worst-case-analyze all unknowns in large circuits, the near worst-case feature should be of considerable value.

Worst-Case Analysis of Large Circuits

Worst-case analysis of large circuits has two fairly obvious but important characteristics: Considerable time is required to perform the necessary calculation, and considerable memory is required to store the data associated with the worst-case analysis. These characteristics make special user action and special CIRC features appropriate when treating large circuits.

The time required to do the sensitivity evaluation for a given circuit need be expended only once if a permanent record of the sensitivity data is taken. The record is normally put on a standard file (DCSR) and is needed only if it is desirable to be able to restart a worst-case analysis (see Reference 1 for details). The sensitivity data should allow a valid restart of worst-case analysis provided that (1) the circuit configuration, (2) the volume of parameter data, and (3) the diode and transistor bias states all remain the same as they were when the sensitivity data were originally obtained.

Worst-case analysis has an inherent need for computer memory space to accommodate data storage associated with the worst-case analysis. However, this need is flexible, and CIRC is structured to adjust analysis capabilities to be compatible with the amount of "available" memory. CIRC automatically determines the amount of available memory.

Four levels of worst-case capability are controlled by the amount of available memory:

1. Full, unrestricted worst-case analysis may be performed.

2. Automatic worst-case analysis may be performed but only for a selected group of unknowns.

3. Automatic worst-case analysis may be performed, but CIRC cannot save information for the summary table.

4. No worst-case analysis or sensitivity data may be obtained (i.e., nominal analysis only).

CIRC itself indicates the level of capability that exists for a given circuit size (see Reference 1 for details). The present discussion about a specific level of capability simply assumes the existence of that level.

When the analysis capability is restricted to level 2, CIRC will automatically store the sensitivity information in memory and perform worst-case analysis *only* for the selected unknowns. This greatly increases the circuit size that can be handled for worst-case analysis. The only loss of flexibility is that the sensitivity data exist only for the selected unknowns. The selection tables may *not* be changed to perform a worst-case analysis of an unknown not previously selected without first redoing the sensitivity evaluation.

When analysis capability is restricted to level 3, the advantages of sorting and the worst-case summary (see Fig. 4.5) are lost. However, the essential ability for worst-case analysis is retained.

Contribution Study Improves Worst-Case Analysis (PCWC)

The CIRC PCWC feature (percent contribution to worst-case study) aids in overcoming deficiencies in worst-case analysis methods. The feature requires a relatively large amount of computer time because a new sensitivity study is performed for every worst-case analysis. This requires one complete solution for each parameter having other values than nominal only.

CIRC uses the new sensitivity information for two purposes: (1) to display the information in a special "Percent Contribution to Worst Case" format (see Reference 1 for an example printout), and (2) to perform a new worst-case analysis. This new analysis is based on sensitivity data specially tailored to the specific worst case and will generally be somewhat more accurate than the first worst-case analysis.

The use of PCWC causes no change in the way the analysis proceeds until after the first worst-case analysis has been output. At that time, CIRC starts to determine the contribution of each parameter to worst case by using the following procedure:

1. After each worst-case calculation (minimum node, maximum node, minimum dependent, or maximum dependent), the parameters are all at the appropriate values to give the worst-case condition.

2. The current worst-case value of the unknown is saved.

3. One of the parameters is set to nominal, and a new solution is calculated. By comparing the variation of this new solution from the worst case to the variation of the nominal solution from the worst case, the percentage contribution is calculated using the following formula:

$$\%C_{pi} = \text{absolute value of } \frac{U_{wc} - U_{pi}}{U_{wc} - U_{nom}} \times 100$$

where

U_{wc} = unknown value at all parameters set for the the worst case (from step 2 above).

U_{pi} = unknown value where the ith parameter is set to nominal while all other parameters are set for the worst case

U_{nom} = unknown value at nominal

4. If the percentage contribution is significant (greater than 2%), then its value is printed out. Therefore, only major contributions appear in the output. Note that on the basis of the definition of "contribution," the numbers may exceed 100%. Circuits with state changes may show very large percentage changes.

5. The new solution is then checked against the current minimum and maximum values of the unknowns in the worst-case summary table to determine if a value is worse than the value currently stored.

6. The parameter is then reset to its worst-case value.

7. Steps 3–6 are repeated for every parameter which had been set to some value other than nominal. The "contribution output" is for each parameter treated.

8. Also as steps 3–6 are performed, a new set of sensitivity data (signs) is determined and stored.

9. When all the parameters have been treated, a new second worst-case analysis is performed and output and the results are sorted into the worst-case summary table.

10. As indicated in step 1, this procedure is repeated for every unknown that is to be worst-case-analyzed.

Several comments about PCWC are appropriate. For a large circuit, it will require a large amount of computer time. When a circuit has sensitivities that are well behaved, the contribution output will be similar to sensitivity ranking output and the second worst-case analysis will be the same as the first analysis or very similar to it. In short, the feature provides little information in a well-behaved circuit. However, in circuits that are not well behaved, this feature can provide a great deal of useful information and a more accurate worst-case analysis.

Worst-Case Analysis Under Parameter Iteration Control

In this mode an entire worst-case analysis is done for each parameter value used during the iteration. Plots of the nominal, minimum, and maximum performance curves may be obtained automatically. Such an analysis will normally use a great deal of time doing sensitivity calculations. This time can be reduced if the sensitivities are reused. Therefore, CIRC provides a feature for reusing sensitivities.

Open-Loop Analysis Technique

A major capability of the CIRC-AC program is its open-loop analysis feature. The results of the application of the technique are values for open-loop gain at a number of points over a frequency range. These results can then be used to plot a stability diagram (the Bode plot can be generated directly on the line printer).

Figure 4.9 is a block diagram of a feedback amplifier in its simplest form. The overall transfer function (closed-loop gain) is

$$A_f = \frac{E_0}{E_s} = \frac{A}{1 - AB}$$

If the amplifier circuit is cut at any of the points marked X in the figure, the gain between the two ends of the cut point (open-loop gain) is given by

$$A_0 = AB$$

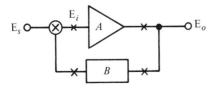

Figure 4.9 Simple Feedback Amplifier.

If a unit signal is inserted at the cut point, the return signal at the other side of the cut is itself equal to the open-loop gain.

The CIRC-AC open-loop analysis technique is based on this elementary theory, complicated somewhat by the fact that, in an actual circuit, impedance must be taken into consideration. When the loop is opened, it is essential to provide the proper load impedance. The load impedance should equal the impedance looking into the open loop from the cut. At times this impedance condition may be easy to approximate, but, in general, the impedance will be a complex, frequency-dependent quantity that is very difficult to handle properly.

Therefore, CIRC-AC provides the capability to open a loop and set up the proper load impedance automatically. The user need only indicate the point at which the loop is to be opened and the direction in which to drive it.

The designer's part in the CIRC-AC open-loop analysis technique will be easier to understand in the context of a circuit example.

Figure 4.10 shows the circuit as prepared by CIRC for open-loop analysis. The dashed line traces the path of the loop. Detail of the analysis procedures for this circuit is given in the example presentation later in this chapter. Here we are concerned only with its value in demonstrating the open-loop analysis technique.

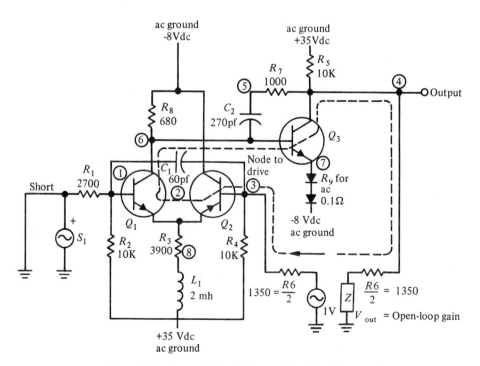

Figure 4.10 Example Circuit Prepared for Open-loop Analysis.

To make an open-loop analysis, the circuit designer has to specify two items to CIRC-AC:

1. The resistor at which the loop is to be opened.
2. The direction in which the unit drive signal is to be applied.

The open-loop analysis program has been set up to open the loop in the middle of a resistor. The resistor must meet the following criteria:

1. It must be on the loop, and all the current in the loop must flow through the selected resistor; i.e., there can be no other parallel path.
2. It must have both terminals connected to nodes; neither terminal can be connected to a supply.

It does not matter where on the loop the selected resistor is. If there is no resistor in the circuit meeting these criteria, one should be inserted. The resistor could be small in relation to other resistance values in the circuit, so as not to affect the solutions appreciably. This resistor should not be smaller than 0.001 ohm, and a larger value is preferred.

The CIRC-AC program automatically sets up a unit voltage signal to drive the loop, but the user must specify the direction of the drive (by indicating the node to drive). In an active circuit, there will normally be only one direction in which the signal can be transmitted, and this direction must be the one selected. For example, in Fig. 4.10 selection of the opposite direction (to that shown) would result in trying to drive a signal through the collector of an active transistor. This attempt would be improper and would result in calculation of a meaningless open-loop gain.

The designer should also give consideration to the question of whether the circuit he is analyzing has more than one significant loop. In one sense, because of parasitic capacitive coupling, interwiring capacitances, and internal resistive coupling in transistors, there are always feedback paths in addition to the main loop, particularly at high frequency. Also, local feedback around each transistor may be part of the design, and other local subloops may be present, too. In many cases, if the feedback is small or is itself stabilized, it is possible to ignore these loops. The CIRC-AC open-loop analysis technique provides solutions, regardless of the number of loops, but the proper meaning of the answers, if there is more than one significant loop, may be difficult to determine.*

After receiving the number of the resistor to open and the number of the node to drive, CIRC-AC takes the following steps:

1. Removes all sources—in effect, shorts all voltage supplies and opens all current supplies. (This step includes voltage supplies, fixed voltage-impedance sources, and fixed current-admittance sources.)

*Richard F. Shea, *Amplifier Handbook*, McGraw-Hill, New York, 1966, pp. 6–23 to 6–25.

2. Divides the selected resistor into two equal parts and adds the necessary circuit nodes. (One node is added to get the actual open-loop results, but two nodes are added to determine the impedance.)

3. Determines the complex impedance Z looking into the open loop in the direction of the node to be driven.

4. Sets up this value of Z as the load impedance.

5. Drives the circuit with 1 volt and calculates the voltage across the load impedance Z; with unit input voltage this output is equivalent to the open-loop gain.

These five steps are automatically iterated at each frequency specified by the user. Because Z is frequency-dependent, its value, both real and imaginary, will in general be different at each frequency.

Furthermore, since the value of Z—considered as the open-loop input impedance—is affected by the appearance of Z as the load impedance, it is necessary to determine the value of Z by a nonlinear, iterative procedure. In general, this procedure consists of applying a unit *current* drive signal to the open-loop input and calculating Z from the following relationship:

$$Z = \frac{V}{1} = V$$

where V is the voltage resulting from the current ($I = 1$ amp). This calculation is iterated until

$$|\Delta Z| \leq 0.01\% \, Z_n$$

where, for this test, the program defines Z as the sum of the real and imaginary parts:

$$Z = Z_R + Z_I$$

Z_n is the value of Z obtained on the most recent iteration, the nth iteration. And

$$\Delta Z = Z_n - Z_{n-1}$$

The normal open-loop solution output becomes self-explanatory when the titles are related to the output numbers by the arrows shown in Fig. 4.11. Note that GAIN is the open-loop gain and is referred to (for plotting) as VOL. The impedance, ZL, is the load impedance calculated by CIRC-AC. It is referred to for plotting as ZOL.

The open-loop analysis technique is of value in the design of amplifiers, nonswitching power supplies, servo systems, oscillators, etc.

With the magnitude (DB) and phase of the open-loop gain, the circuit designer can plot the Bode gain magnitude and phase diagram and assure circuit stability. The user analyzing more complex circuits will often want to do Nyquist plots that show direction of phase change and encirclement of the minus one (-1) point.

PROVIDE CONTROL MESSAGES

:OPEN

SPECIFY:
ID. OF R TO OPEN, ID. OF TERM. TO DRIVE, 1 FOR FULL OUTPUT,

:RES6,N3

FREQ(OPEN LOOP)		
GAIN(MAG)	PHASE	GAIN(D B)
ZL REAL	ZL IMAG	
.1000000E 01		
.1048327E 04	-.1799847E 03	.6040994E 02
.5204951E 04	-.3081732E-02	
.1000000E 02		
.1048323E 04	.1798467E 03	.6040991E 02
.5204951E 04	-.3081735E-01	
.1000000E 03		
.1047950E 04	.1784675E 03	.6040681E 02
.5204952E 04	-.3081995E 00	
.1000000E 04		
.1012547E 04	.1650245E 03	.6010830E 02
.5205038E 04	-.3106232E 01	
.1000000E 05		
.3662324E 03	.1108332E 03	.5127514E 02
.5205687E 04	-.3401803E 02	
.1000000E 06		
.3943576E 02	.9595559E 02	.3191780E 02
.5161060E 04	-.3386112E 03	
.1000000E 07		
.5494388E 01	.1026886E 03	.1479839E 02
.3649897E 04	-.1252126E 04	
.1000000E 08		
.5726740E 00	.7110547E 02	-.4841850E 01
.1995082E 04	-.8841680E 03	

SPECIFY:
VARIABLE TO PLOT (0 TO END ENTRIES), TYPE, MIN, MAX, TYPE, MIN, MAX

:VOL

Figure 4.11 Open-loop Analysis.

4.2 CIRCUIT DESCRIPTION ELEMENTS

Introduction

The network elements presently in the CIRC programs are

DC	AC	TR
1. Resistor (RES)	1. Resistor (RES)	1. Resistor (RES)
2. Voltage-resistor source (VRS)	2. Inductor (IND)	2. Voltage-resistor source (VRS)
3. Current source (CUR)	3. Capacitor (CAP)	3. Current source (CUR)
4. Diode (DIO)	4. Transistor (TRN)	4. Diode (DIO)
5. Transistor (TRAN)	5. Current-admittance source (CYS)	5. Transistor (TRN)
	6. Voltage-impedance source (VZS)	6. Inductor (IND)
	7. Mutual inductance (MUT)	7. Capacitor (CAP)

The resistor, inductor, capacitor, mutual inductance, diode, and transistor elements are normally used to represent actual hardware components. The voltage-impedance source (VRS or VZS) and current source elements (CUR or CYS) are used to create equivalent circuits or to represent electrical energy sources (voltage supplies, see below, may also be used for energy sources). Three aspects of each element are discussed: (1) connection information, (2) parameter information, and (3) dependent unknowns. No direct discussion of the DC network elements is given because the first three elements are the same in both the DC and TR programs and DC diode and transistor elements are a subset of the transient elements.

Before discussing the network elements in detail, let us consider the circuit description elements other than the network elements.

Junction Temperature

If diode or transistor network elements are used, CIRC will request a value for junction temperature. This parameter is needed in the classic nonlinear diode equation that CIRC uses for both diodes and transistors.

The fact that CIRC requests junction temperature does not mean that it automatically handles the predominant effects of temperature on a circuit. Actually, temperature effects on a circuit include effect on junction saturation current, leakage current (recombination), and thermal voltage, as well as effects on transisitor betas and circuit resistances.

Temperature dependence to any required degree of accuracy may be included through the use of special equations that allow CIRC to automatically handle temperature even in a worst-case analysis. A simple example is given in Chapter 10 of Reference 1.

CIRC could easily be modified to include temperature effects as a part of the standard program; however, the memory requirements would be very damaging to the size of circuits CIRC can handle in a small memory environment. It would also require a fixed set of rules for temperature dependence for which it would be difficult to get universal acceptance.

Voltage Supplies

Within CIRC there are two ways to supply voltage to a circuit—using voltage supplies or the voltage-impedance source network element. The easiest way is to use voltage supplies and simply assign a "supply" number to each point in the circuit at which a unique value of voltage is input. Then, in the connection descriptions for network elements, the supplies are referenced simply as Sn (e.g., S1). CIRC's second question during acceptance of circuit description asks how many of these voltage supplies are used; then a little later the values of the supplies are accepted. The use of voltage supplies is shown in the example analyses presented later in the chapter.

A voltage supply is inherently referenced to ground. This inherent ground reference can be avoided by the use of the voltage-impedance source (VRS or VZS) to supply voltage.

ac Voltage Supplies—Complex Number Usage

Voltage supply values are input as a magnitude and phase angle, whereas the voltage of a voltage-impedance source is input as a real and imaginary part. In both cases, complex numbers are used. Indeed, since CIRC-AC is concerned with the small-signal ac operation of the circuit, voltage and current are, in general, represented by complex numbers in polar form, that is, an amplitude and a phase angle. In all equations, V and I are understood to be complex numbers except where another interpretation is specifically indicated. Similarly, power dissipation is given in terms of complex power, that is, the product of the voltage and the current, both rms, with angle equal to the phase angle by which current lags voltage (VI/ϕ). A more convenient formula for complex power is VI∗, where I∗ is the conjugate of I.

It may be useful, also, to define the significance of current direction in ac circuits. The reference direction of current is taken to be the direction in which current flows during positive cycles. The interpretation of phase angle also needs some discussion. The phase angles as presented by CIRC can be sketched as in Fig. 4.12(a). The most significant point is that complex numbers in quadrants I and II are displayed with positive angles between 0 and

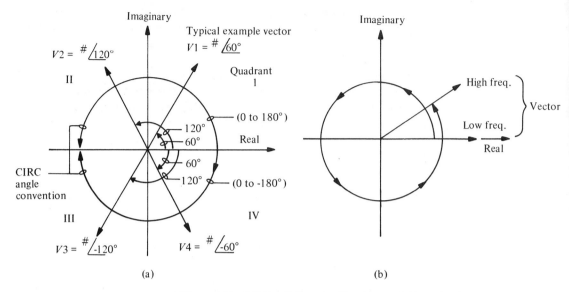

Figure 4.12 CIRC-AC Complex Numbers and Phase Angles: (a) Standard Angle Convention; (b) Direction of Change Determination.

+180 degrees. Angles in quadrants III and IV are displayed with negative angles between 0 and −180 degrees. Four example vectors (phasors) are shown that demonstrate the angle convention. Furthermore, phase angle and phase-angle change in the system using complex numbers can be definitely identified only by starting at a very low frequency where angles are known to be either 0 or ±180 degrees. Then, if frequency is increased slowly, the direction of change can be observed and the value of an angle can be identified as being more than 360 degrees because its rotation has been tracked through a smooth transition beyond 360 degrees. This idea is sketched in Fig. 4.12(b). It is also noted that for stability studies it is at times necessary to know the direction of phase-angle change. In such a case, frequency should be changed and then angle changes will be small and the direction of changes will be predictable.

Finally, it is also noted that a small-frequency shift may produce a phase-angle change that at first appears to be very large. For example, an angle may also from +178 degrees to −170 degrees. The change would probably be only 12 degrees. The appearance of a large angle change was due to the fact that the angle system must, at some point, make a transition* that appears to be an abrupt 360-degree change. In CIRC the change is

---

*This requirement for a 360-degree change is inherent in any program that operates in the complex number plane and performs a numerical analysis. There are two reasonable choices: Change from 0 to 360 degrees, or change from −180 degrees to +180 degrees.

from +180 to −180 degrees. The change is described as being "probably" 12 degrees because the most likely change was a relatively small counterclockwise angle change of 12 degrees as opposed to a very large 348-degree clockwise change. In most cases, such angles can be easily interpreted without the special attention described above. It is, however, advisable not to take large frequency steps without some care to interpret angles correctly.

TR Voltage Supplies

In CIRC-TR, voltage is supplied to a circuit by either voltage supplies or voltage-resistance source network elements. Often the easiest way to use voltage supplies is simply to assign a "supply" number to each point in the circuit at which a constant or time-dependent voltage is input. Then, in the connection input data for network elements, the supplies are referenced simply as Sn (e.g., S4). During acceptance of a circuit description, CIRC asks how many of these voltage supplies are used and what supply types are requested. Allowable supply types are dc, pulse, sine, cosine, and a table of values; the user defines the type by entering

$$\text{DC, PULSE, SIN, COS, or TABLE}$$

The first letter of each type is sufficient.

Immediately following the entry of the type information, CIRC-TR requests the parameter data for each supply. The nature of these supplies is as follows.

dc Supplies

A dc steady-state supply requires only one data entry, the dc value in volts.

Pulse Supplies

See Fig. 4.13(a) for definitions of pulse supply characteristics. Data are entered in the following order:

$$\text{V-ORG, V-PEAK}$$
$$\text{TRISE, TFALL}$$
$$\text{TWIDTH, TREPEAT}$$
$$\text{TDELAY}$$

While the time delay (TDELAY) may be set to zero, rise and fall times (TRISE and TFALL) must be nonzero. In addition, the pulse width (TWIDTH) and the repetition time (TREPEAT) must also be nonzero to be meaningful. By properly defining the parameter values, the pulse supply can be used to approximate step inputs, ramp inputs, pulses, rectangular and sawtooth periodic functions, etc.

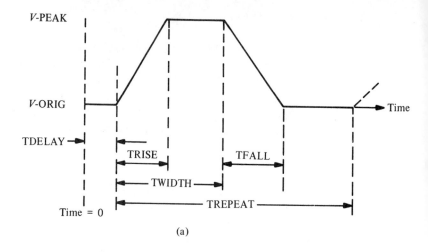

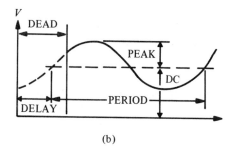

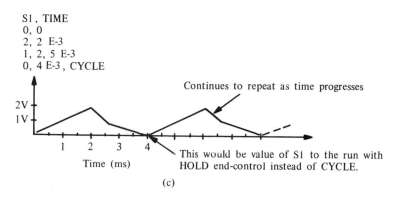

Figure 4.13 CIRC-TR Voltage Wave Shapes: (a) Pulse Supply Parameters; (b) Sine/Cosine Supply; (c) Waveform Controlled by "Table" Input Shown.

Sine and Cosine Supplies

Sine and cosine supplies are defined with the following data set:

> PEAK, DC
> PERIOD, DELAY
> DEAD

The terms are defined in Fig. 4.13(b). The supply waveform will stay at zero until t = DEAD; then it will follow the equation

$$V = DC + PEAK * SINE\left[\frac{2\pi(t-DELAY)}{PERIOD}\right]$$

COSINE is used in place of SINE when so specified.

If the values entered for PERIOD and DELAY are greater than unit, they will be assumed to be frequency and phase, respectively, rather than time intervals and will be converted as

> PERIOD = 1/VALUE ENTERED
> DELAY = PERIOD*VALUE ENTERED/360°

where the value entered for frequency is in hertz and the value entered for phase is in degrees.

If PEAK is entered as a negative value, it will be treated as the rms amplitude of the sinusoidal component. This value will be converted to peak amplitude as

> PEAK = VALUE ENTERED * $\sqrt{2}$

A typical data set for a simple sine (or cosine) is

> −1E−3
> 100E3
> 0

where the amplitude in rms is ±1 mV (1.414 millivolts peak), the dc level is zero, the frequency is 100 kHz, phase is zero, and dead time is zero.

Another typical data set is shown below:

> 1,1
> 2E−3, .2E−3
> .4E−3

where the peak amplitude of the sine or cosine wave is 1 volt, the dc level is 1 volt, the period is 2 milliseconds, the delay time is 0.2 millisecond, and the dead time is 0.4 millisecond.

Table Supplies

A "table" supply is a supply whose voltage waveform is determined by a list of voltage-time coordinates in a parameter control table. The waveform can thus be of any desired shape. CIRC-TR treats the data for these tables separately from the supply data, requesting input under the heading TABLE# after input for all other supplies have been completed.

Parameter Control Tables

Parameter control tables are used to control the behavior of a parameter with respect to time. The parameter to be controlled is typically a supply, and the controlling parameter is time. However, almost any other parameter may be controlled, and the controlling parameter can be time or a supply.

The parameters that may not be controlled are certain specification diode and transistor parameters used to calculate information for CIRC that is *not* reevaluated at each time step. These parameters not subject to proper control are specification-diode parameters 1, 2, and 4–6 (discussed later) and specification transistor parameters 1–7 and 10–15 (discussed later).

It should be noted that the original entry for a parameter that is controlled by a table is overridden by the table value. Thus, the original value is not used.

Parameter control tables are entered immediately after the supplies are entered, just as the number of tables is entered immediately after the number of supplies. CIRC-TR's data request is

PARAMETER TO CONTROL, CONTROL VARIABLE

where the control variable may be TIME or a supply (Sn). An example response would be

:S1, TIME

Next, CIRC requests

PAR. COORDINATE, CONTROL COORD.,
END CONTROL

PAR. COORDINATE refers to the value of the parameter to be controlled, and CONTROL COORD. to that of the controlling variable. For the example response given, the first value would be in volts, and the second in units of time. END CONTROL requests one of two options: HOLD or CYCLE. HOLD means that the last value input for the controlled parameter is to be held for the remainder of the run, that is, beyond the time specified by the time coordinate. CYCLE means that the range of parameter values entered in the table is to be cycled.

The supply waveform shown in Fig. 4.13(c) would be described to CIRC with the data sequence shown on the figure.

The end control (CYCLE) shown in Fig. 4.13(c) indicates that S1 is periodic. If HOLD were specified, the supply value of 0 volt at $5 = 4$ milliseconds would be used for time greater than or equal to 4 milliseconds.

The primary purpose of controlling a parameter by a supply (rather than by time, as shown) would be to expand the use of pulse, sine, and cosine supplies to include all parameters, particularly CUR and VRS sources.

This type of control is useful for a VRS to vary the resistance from open to short or to provide a time-dependent voltage not referenced to ground.

An example of a supply (S1) controlling a current source (CUR1) as a direct function of its output would result from the following data input:

$$\text{CUR1, S1}$$
$$2$$

where the current flowing in CUR1 would be calculated as twice the voltage of S1; e.g., a one-unit peak sine input from S1 to CUR1 would produce 2 amperes peak of sinusoidal current. Note that the format for this type of parameter control table is simplified in that only one data value is entered and that it is used as a multiplier of the controlling supply value.

CIRC-TR (CIRC-DC) Network Elements

Resistor

Connection Information

A schematic of the resistor element is shown in Fig. 4.14.

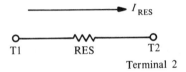

Figure 4.14 Resistor Element (RES).

The two terminals of this component are simply identified as terminal 1 and terminal 2. Since the element is bilateral, the choice of which is considered to be terminal 1 and which is terminal 2 is arbitrary, although the choice does establish a direction of current flow (from T1 to T2).

Parameter Information

This element has just one parameter: the resistance value, RES, which should never be input as exact zero. A small value (e.g., 0.001 ohm) is allowed.

Dependent Unknowns

Recall the meaning of the term *dependent unknown*. It is used to describe the special information generated for each type of network element. CIRC writes a set of independent, simultaneous node equations, solves these equations, and produces its prime set of unknowns—the node voltages. CIRC then calculates the value of a new group of unknowns, such as element currents and power dissipations. The values of these new unknowns depend on the node voltages and element parameters. Hence, these unknowns are referred to as dependent unknowns.

This element has two dependent unknowns:

1. Resistor current (from T1 to T2).
2. Resistor power dissipation.

The equations defining these unknowns are

1. $$I_{\text{RES}} = \frac{V_{T1} - V_{T2}}{R}$$

2. $$P_{\text{RES}} = I_{\text{RES}}(V_{T1} - V_{T2})$$

Voltage-Resistance Source

Connection Information

A schematic of the voltage-resistance source is shown in Fig. 4.15.

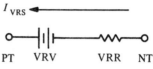

PT   VRV   VRR   NT

Positive terminal   Negative terminal

Figure 4.15 Voltage-Resistance Source Element (VRS).

The two terminals of the element are the positive and the negative. The element has an inherent polarity and the method of connecting it into a circuit (network) is not arbitrary. For details, see the discussion of dependent unknowns below.

Parameter Information

This element has two parameters:

1. Voltage value (VRV).
2. Resistance value (VRR).

VRR should never be input as exact zero. A small value (e.g., 0.001 ohm) is allowed.

Dependent Unknowns

This element has two dependent unknowns:

1. Current delivered (from NT to PT).
2. Power delivered.

The equations defining these unknowns are

1. $$I_{\text{VRS}} = \frac{V_{\text{NT}} - V_{\text{PT}} + E}{R} \qquad \text{where } E = \text{VRV and } R = \text{VRR}$$

2. $$P_{\text{VRS}} = I_{\text{VRS}}(V_{\text{PT}} - V_{\text{NT}})$$

Although this is not normally necessary, the user may control the algebraic sign of I_{VRS} by selecting the sign of the voltage, VRV. For example, a negative voltage may be coupled to a node by either

1. Connecting the "negative terminal" to the node and using a positive value of VRV, or
2. Connecting the "positive terminal" to the node that is to be negative and using a negative value of VRV.

One method calculates current flow from, and the other current flow to, the negative voltage.

Current Source

Connection and Control Information

A schematic of the current source is shown in Fig. 4.16. The two terminals are identified by the nodes to which the element connects, the receiving node (RN) and the exit node (EN). The element has an inherent polarity and must be properly connected into a network.

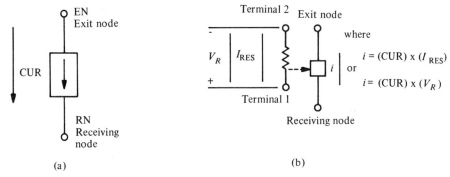

Figure 4.16 Current Source Element (CUR): (a) Fixed Current; (b) Current Source Element (CUR) as a Controlled Source.

CIRC also provides capability for controlling current sources in terms of either voltage across or current through a specified resistor. Resistor number and type of control desired (current, I, or voltage, V) are specified along with the connection data. The schematic in Fig. 4.16(b) shows polarity consideration for a controlled current source.

Examples.

G,N3	fixed current source
N2,G,R2,I	current-controlled current source
G,N5,R5,V	voltage-controlled current source

Parameter Information

The element itself has only one parameter, CUR. For a fixed source it is the current value. If a controlled current source is used, CUR represents the current gain—beta (β)—for a current-controlled source, and transconductance (GM) for a voltage-controlled source (see the equations in Fig. 4.16(b)).

Dependent Unknown

This element has one dependent unknown. For a fixed source, this is the instantaneous power delivered. The equation defining this unknown is

$$P_{CUR} = CUR(V_{RN} - V_{EN})$$

For a controlled source, the dependent unknown is the instantaneous current actually flowing. The equations are

$$I_{CUR} = (CUR)*(I_{RES}) \quad \text{for } I \text{ control}$$

or

$$I_{CUR} = (CUR)*(V_R) \quad \text{for } V \text{ control}$$

Diode

Three mutually exclusive, nonlinear Ebers-Moll-type diode models with complete modeling of time dependence are built into the CIRC-TR program. The models, referred to as specification, classic, and Xerox DIO elements, differ in the parameters used to set up the models.

The first model is oriented toward manufacturer's specification data; the second is a classic model using a more conventional set of parameters; and the third is an extended Ebers-Moll model.

Figure 4.17 shows the diode element as the user sees it and schematic drawings for each of the three models, identifying the parameters for each model.

Connection Information

The schematic for the diode element as the user sees it is shown in Fig. 4.17. Anode and cathode must be properly connected into a network since this element is unidirectional. The connection information is unaffected by whether a diode is spec, classic, or Xerox type.

Specification Model Parameter Information

The specification diode model has 12 parameters. Three are preset by the program but may be optionally entered with the input data, and two are calculated by the program. Hence, only 7 of the 12 parameters in the data set need usually be supplied.

The basic set of seven specification parameters comprises

1. ID   current level at which VD occurs ⎫
2. VD   dc voltage across diode for current ID ⎬ CIRC-DC Parameters
3. IREV   reverse (leakage) current ⎭
4. ION   forward (ON) current at which charge is stored

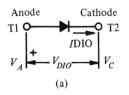

(a)

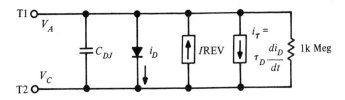

where

$C_{DJ} = f(DFJZ, VCBI, PWRD, CDJX)$ $\qquad I_S = f(VD, ID)$

$i_D = I_S (e^{qV_{DIO}/kT} - 1)$ $\qquad \tau_D = fi(ION, IOFF, TOFF)$

(b)

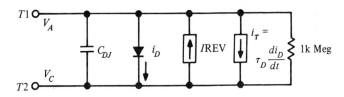

where

$C_{DJ} = f(CDFZ, VDBI, PWRD, CDJX)$

$i_D = ISD(e^{qV_{DIO}/k \cdot T \cdot MD} - 1)$

$\tau_D = TD$

(c)

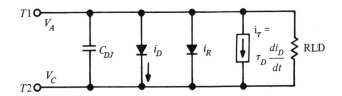

where

$C_{DJ} = f(CDJZ, VDBI, PWRD, CDJX)$ $\qquad \tau_D = TD$

$i_D = I_{DD} (e^{qV_{DIO}/kT} - 1)$ (Diffusion diode) $\qquad T = $ TMPD if TMPD $\neq 0$

$i_R = I_{DD} (e^{qV_{DIO}/k \cdot T \cdot MRD} - 1)$ $\qquad T = $ "JUNCTION TEMP."
   if TMPD = 0.
(Recombination diode) $\qquad\qquad\qquad$ as normally used.

(d)

Figure 4.17 Diode (DIO) Element: (a) Specification DIO Element as Implemented by CIRC-TR; (c) Classic DIO Element as Implemented by CIRC-TR; (d) XDS DIO Element as Implemented by CIRC-TR.

5. IOFF   reverse (off) current that is to remove stored charge
6. TOFF   turn-off time
7. CDJZ   transition capacitance at zero junction voltage

Note: If parameters 4 to 6 are entered as zero, then

$$\tau_D = 0$$

This completes the basic data set; the next three parameters are requested only when CDJZ is entered as a negative number. (The minus sign is merely a signal to CIRC that the user wants to input his own values; it does not affect the actual value of transition capacitance.)

8. VDBI   junction contact potential (default value = 0.8)
9. PWRD   junction grading constant (default value = 0.5)
10. CDJX   multiplier that limits the transition capacitance to values less than or equal to CDJX; CDJX (default value = 50)

In addition, two "hidden" parameters are calculated by the program. They are

11. ISD   dc reverse (saturation) current of the intrinsic diode
12. TD   diode time constant

The parameters ID and VD describe one point on the nonlinear dc performance characteristic of the diode. CIRC sets up its dc model characteristic to pass through this point. This is controlled by the value calculated for ISD. The parameters ION, IOFF, and TOFF describe the result of a time domain performance test and are the basis on which CIRC-TR generalizes the one test into a TD value.

Classic Model Parameter Information

The classic diode model has eight input parameters, three of which are preset by the program but may be optionally entered with the input data. Hence, only five of the required eight parameter values need usually be entered.

The basic set of five classic model parameters comprises

1. MD   junction slope factor; or the product of Kelvin temperature and the junction slope factor, if a negative value is entered
2. ISD   dc reverse (saturation) current of the intrinsic diode
3. IREV   reverse (leakage) current of the extrinsic diode
4. TD   diode time constant
5. CDJZ   transition capacitance of zero junction voltage

This completes the basic set of measurement parameters; the following three parameters are requested only if CDJZ is entered as a negative number. (The minus sign is merely a signal to CIRC that the user wants to input his own values; it does not affect the actual value of the transition capacitance.)

6. VDBI   junction contact potential (default value = 0.8)
7. PWRD   junction grading constant (default value = 0.5)
8. CDJX   multiplier that limits the transition capacitance to values less than or equal to CDJX; CDJX (default value = 50)

XDS Model Parameter Information

The XDS diode model has 10 parameters, 3 of which are preset by the program but may be optionally entered with the input data. Hence, only 7 of the 10 parameters in the data set need usually be supplied.

The basic set of specification parameters comprises

1. MRD   junction slope factor for the recombination diode; a zero entry is given the default value of 2
2. IDD   dc reverse (saturation) current of the intrinsic diffusion diode
3. IRD   dc reverse (saturation) current of the recombination diode
4. RLD   shunt (leakage) resistance
5. TMPD   diode junction temperature (allows individual device temperatures)
6. TD   diode time constant
7. CDJZ   transition capacitance at zero junction voltage

This completes the basic data set; the next three parameters are requested only when CDJZ is entered as a negative number. (The minus sign is merely a signal to CIRC that the user wants to input his own values; it does not affect the actual value of transition capacitance.

8. VDBI   junction contact potential (default value = 0.8)
9. PWRD   junction grading constant (default value = 0.5)
10. CDJX   multiplier that limits the transition capacitance to values less than or equal to CDJX; CDJZ (default value = 50)

Dependent Unknowns

The diode model has three dependent unknowns:

1. Diode current:

    I_{DIO} = total diode current   (sum of currents flowing in all elements of the appropriate equivalent circuit)

2. Instantaneous power dissipated:

$$P_{DIO} = I_{DIO} * (V_A - V_C)$$

3. Junction voltage drop:

$$V_{DIO} = V_A - V_C$$

The terms in the equations are defined in Fig. 4.17.

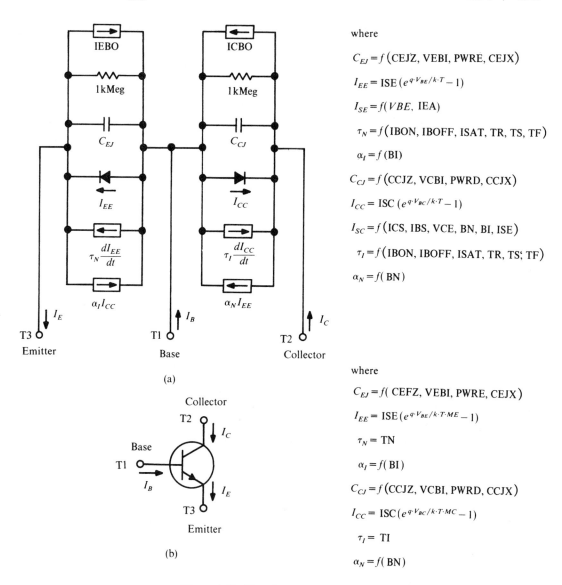

Figure 4.18 Transistor Model (TRN), SPEC and CLASSIC Types: (a) Specification Model; (b) Classic Model.

Transistor

Three mutually exclusive, nonlinear Ebers-Moll-type transistor models with complete modeling of time dependence are built into the CIRC-TR program. The models—specification, classic, and XDS—differ primarily in the parameters used to set up the models.

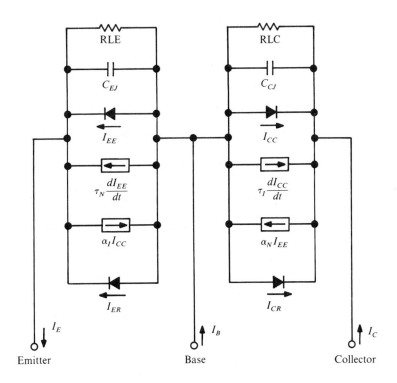

where

$C_{EJ} = f(\text{CEJZ, VEBI, PWRE, CEJX})$ $C_{CJ} = f(\text{CCJZ, VEBI, PWRC, CCJX})$

$I_{EE} = \text{IDE}(e^{q \cdot V_{BE}/k \cdot T} - 1)$ ← "Diffusion" → $I_{CC} = \text{IDC}(e^{q \cdot V_{BC}/k \cdot T} - 1)$

$\tau_N = \text{TI}$ $\tau_I = \text{TI}$

$\alpha_I = f(\text{BI, KI})$ $\alpha_N = f(\text{BN, KN})$

$I_{EE} = \text{IRE}(e^{q \cdot V_{BE}/k \cdot T \cdot 2} - 1)$ ← "Recombination" → $I_{CR} = \text{IRC}(e^{q \cdot V_{BC}/k \cdot T \cdot 2} - 1)$

$T = \text{TMPT}$ if $\text{TMPT} \neq 0$; $T = $ "JUNCTION TEMP." if $\text{TMPT} = 0$ as normally used.

Figure 4.19 Transistor Model (TRN) XDS Type.

The first model is oriented toward manufacturers' specification data; the second is a classic model using a more conventional set of parameters; and the third is an extended Ebers-Moll model.

Figure 4.18 shows the transistor element as the user views it (inset in box), and it shows schematic drawing and parameter summary for the SPEC and CLASSIC model types. Figure 4.19 shows the XDS model schematic. Additional details about the models are presented later in this chapter.

Connection Information

The three terminals CIRC requires to be specified are base, collector, and emitter, in that order. The connection information is unaffected by the type of transistor used.

Specification Model Parameter Information

The specification transistor model has 27 parameters, of which 6 are preset by the program but may be optionally entered with the input data. Four parameters are calculated by the program. (When the data input operation is finished, the calculated parameters may be accessed and modified if desired.) Hence, only 17 of the 27 parameters in the data set need usually be supplied.

The basic set of 17 specification parameters comprises

1. ICS	collector dc saturation current		⎫
2. IBS	base dc saturation current		⎪
3. IEA	active region dc emitter current		⎪
4. VBE	base-emitter voltage for current IEA		CIRC-DC
5. VCE	collector-emitter saturation voltage at currents IBS and ICS		Parameters
6. BN	normal mode common emitter current gain		⎪
7. BI	inverted mode common collector current gain		⎭
8. ICBO	collector-base reverse leakage current		
9. IEBO	emitter-base reverse leakage current		
10. IBON	base pulse turn-on current		
11. IBOFF	base pulse turn-off current		
12. ISAT	collector saturation at IBON		
13. TR	time to rise to 90% of ISAT due to IBON (a minus entry has special significance—see the discussion following parameters 26 and 27)		
14. TS	storage time that ISAT flows when the base current is switched from IBON to IBOFF		
15. TF	time to fall to 10% of ISAT, measured from the end of the storage time period (a minus entry has special significance—see the discussion following parameters 26 and 27)		
16. CCJZ	collector-base transition capacitance at $V_{BC} = 0$		
17. CEJZ	emitter-base transition capacitance at $V_{BE} = 0$		

This completes the basic data set; the following six parameters are requested only when CCJZ and/or CEJZ are entered as negative numbers. (CIRC uses absolute magnitudes as CCJZ and CEJZ parameter values.)

18. VCBI collector-base contact potential (default value = 0.8)
19. VEBI emitter-base contact potential (default value = 0.8)
20. PWRC collector-base grading constant (default value = 0.5)
21. PWRE emitter-base grading constant (default value = 0.5)
22. CCJX multiplier that limits the collector-base capacitance to values less than or equal to CCJX; CCJZ (default value = 50)
23. CEJX multiplier that limits the emitter-base capacitance to values less than or equal to CEJX; CEJZ (default value = 50)

The following four "hidden" parameters are calculated by the program:

24. ISE dc reverse current of the intrinsic base-emitter junction
25. ISC dc reverse current of the intrinsic base-collector junction

The values calculated for ISE and ISC depend on the parameter pair I_{EA}, V_{BE}, and the parameter triplet, I_{CS}, I_{BS}, and V_{SAT}. These are points on the dc performance curves of a transistor that CIRC sets up for its nonlinear Ebers-Moll-type model to pass through.

26. TN base-collector junction time constant
27. TI base-emitter junction time constant

Values for TN and TI are not normally specified for a transistor, and they are not parameters that are familiar to most engineers. Therefore, CIRC provides for specification sheet data to be used to set up the TN and TI parameters automatically. The specification data used are primarily pulse response test data: IBON, IBOFF, ISAT, TR, TS, and TF. The use of these data is explained more fully in the model discussion.

The most important aspect of this parameter calculation procedure is that it is essentially a curve (model) fit procedure. It is a "fit" procedure in that the response times measured for one condition are used to calculate TN and TI so that the model has the same performance as that actually measured. Because only two parameters are being determined, only two of the three pulse response times (TR, TF, and TS) can be made to exactly agree with measured values. The third response time will agree with the measured value to the extent that the model truly represents the device.

Usually this will be a good agreement. Because the model response has been forced to agree with the actual device response for one pulse test condition, there will be good agreement for many different variations in pulse conditions, e.g., conditions that are not ideal current drives, waveforms that are not drive steps, circuits that are not single stages, etc.

In resolving the three measurements (TR, TS, and TF) versus the derived parameters (TN and TI), CIRC-TR defines three values of TN and TI based on (1) storage and rise time, (2) storage and fall time, or (3) and average of the first two. To have CIRC-TR use TN and TI based on TR and TS, TF is input as a negative number. To have CIRC-TR use TN and TI based on TF and TS, TR is input as a negative number. If both TR and TF are positive, CIRC-TR will use an average for TN and TI. This is the normal usage.

The parameter names apply to *NPN* transistors, but all parameters are positive quantities for either type of transistor. This means that the first five specification parameters refer to normal forward bias, and the eighth and ninth parameters refer to normal reverse leakage currents. All current and voltage directions are reversed for *PNP* transistors. That is, VBE becomes VEB, VCB becomes VBC, etc.

Classic Model Parameter Information

The classic transistor model has 18 parameters, of which 6 are preset by the program but may be optionally entered with the input data. Hence, only 12 of the required 18 parameter values need usually be entered.

The basic set of 12 measurement parameters comprises

1. MC — base-collector junction slope factor, or the product of Kelvin temperature and the junction slope factor if a negative value is entered
2. ISC — dc reverse current of the intrinsic base-collector junction
3. ME — base-emitter junction slope factor with same alternative option as for MC above
4. ISE — dc reverse current of the intrinsic base-emitter junction
5. BN — normal mode common emitter current gain
6. BI — inverted mode common collector current gain
7. ICBO — collector-base reverse leakage current
8. IEBO — emitter-base reverse leakage current
9. TN — base-collector junction time constant
10. TI — base-emitter junction time constant
11. CCJZ — collector-base transition capacitance at $V_{BC} = 0$
12. CEJZ — emitter-base transition capacitance at $V_{BE} = 0$

This completes the basic set of measurement (classic) parameters. The 6 additional parameters are the capacitance parameters (13–18). Their entry is optional as described for parameters 18–23 of the SPEC model.

XDS Model Parameter Information

The XDS transistor model has 22 parameters, 6 of which are preset by the program but may be optionally entered with the input data. One is a "hidden" parameter." Only 15 of the required 22 parameter values need usually be entered.

The basic set of 15 parameters comprises

1. IDC   dc reverse saturation current of the intrinsic base-collector junction (diffusion)
2. IRC   dc reverse saturation current of the base-collector junction (recombination)
3. IDE   dc reverse saturation current of the intrinsic base-emitter junction (diffusion).
4. IRE   dc reverse saturation current of the base-emitter junction (recombination)
5. BN   normal mode common emitter current gain
6. BI   inverted mode common collector current gain
7. RLC   collector-base reverse leakage resistance (default value = 1000 megohms)
8. RLE   emitter-base reverse leakage resistance (default value = 1000 megohms)
9. KN   normal alpha high-current roll-off factor
10. KI   inverted alpha high-current roll-off factor
11. TMPT transistor junction temperature (allows individual device temperatures)
12. TN   base-collector junction time constant
13. TI   base-emitter junction time constant
14. CCJZ collector-base transition capacitance at $V_{BC} = 0$
15. CEJZ emitter-base transition capacitance at $V_{BE} = 0$

This completes the basic set of XDS model parameters. There are also 6 additional capacitance parameters (parameters 16–21) of the type previously described.

There is one "hidden" parameter. It is calculated by the program as

22. BNC   beta normal calculated—the value of the actual normal beta operative in the device, taking into account current loss in the recombination diode and high-current roll-off; this parameter may be plotted.

Dependent Unknowns

There are eight dependent unknowns for the transistor element:

1. Base current, I_B
2. Collector current, I_C See inset in Fig. 4.18 for direction of current.
3. Emitter current, I_E
4. Power dissipation (total and instantaneous): $P_Q = V_{BE}I_B + V_{CE}I_C$
5. Base-emitter drop: V_{BE}
6. Base-collector drop: V_{BC}
7. Collector-emitter drop: V_{CE}
8. Circuit gain: I_C/I_B

For a *PNP* device the voltages and currents are calculated for the opposite directions, e.g., I_B out of base, I_E into emitter, V_{EB}, V_{CB}, V_{EC}, etc.

The equations defining these unknowns are too complex to be presented here. For now, it is sufficient to know that CIRC calculates terminal currents and voltages.

Inductor

Connection Information

The inductor element is diagrammed in Fig. 4.20.

The two terminals (node connections) of this component are identified as L terminal 1 and L terminal 2. Since inductors are bilateral, the order of assignment of terminals is arbitrary. Positive current flow, however, is assumed to be from terminal 1 to terminal 2.

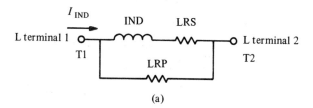

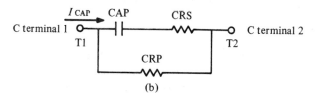

Figure 4.20 The Inductor Element (IND) and Capacitor Element (CAP) for CIRC-TR: (a) Inductor; (b) Capacitor.

Parameter Information

This element has three parameters: IND, LRS, and LRP. IND is the value of inductance (L) in henries; LRS is the value of the series resistance; and LRP is the value of the shunt (parallel) resistance. LRS and LRP are treated by CIRC-TR as 0.01 ohm and 1000 megohms, respectively, whenever zero values are entered.

Dependent Unknown

This element has one dependent unknown, IIND, the total instantaneous current from T1 to T2. This is the sum of the currents in the series and shunt resistors.

Capacitor

Connection Information

The capacitor element is diagrammed in Fig. 4.20.

The two terminals (node connections) of this component are identified as C terminal 1 and C terminal 2. Since capacitors are bilateral, assignment of terminals is arbitrary (polarized capacitors are bilateral for analysis purposes). Positive current flow, however, is taken to be from terminal 1 to terminal 2.

Parameter Information

This element has three parameters: CAP, CRS, and CRP. CAP is the value of capacitance (C) in farads; CRS is the value of the series resistance; and CRP is the value of the shunt (parallel) resistance. CRS and CRP are treated by CIRC-TR as 0.01 ohm and 1000 megohms, respectively, whenever zero values are entered.

Dependent Unknown

This element has one dependent unknown, the total instantaneous current, ICAP, from T1 to T2. This is the sum of the currents in the series and shunt resistors.

CIRC-AC Network Elements

The resistor element for CIRC-AC is effectively the same as for CIRC-DC and CIRC-TR. There is a slight difference in the nature of the dependent unknowns. The equations defining these unknowns are

$$1. \quad I_{\text{RES}} = \frac{V_{T1} - V_{T2}}{R}$$

(the values of I and V in this equation are complex) and

2. $\quad P_{\text{RES}} = (V_{T1} - V_{T2})I^*_{\text{RES}}$

Note that I^*_{RES} is the *conjugate* of I_{RES}.

Again, these terms are complex, and complex power P_{RES} is presented in units of volt-amperes at an angle. Power in a resistor is entirely real (active); however, the angle is always zero (or approximately so, within the limits of computational precision). Consequently, the unit of power for the resistor is effectively watts.

Inductor

Connection and Control Information

The inductor element is diagrammed in Fig. 4.21(a). The two terminals of this component are identified as L terminal 1 and L terminal 2. Since it is bilateral, the choice of which is to be considered L terminal 1 is completely arbitrary.

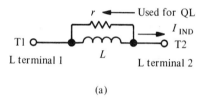

(a)

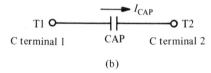

(b)

Figure 4.21 Inductor Element–IND and Capacitor Element–CAP for CIRC-AC: (a) Inductor; (b) Capacitor.

This element also allows a reference to a capacitor with which it is to be tuned. For example, in the following entry of connection and control data, N1, N2, C1, the inductor would be connected between nodes 1 and 2, and the inductance value would be calculated by CIRC as

$$L = \frac{1}{(2\pi f_r)^2 C_1}$$

where

L = inductance value used by CIRC

f_r = frequency of resonance

C_1 = value of the capacitor referenced

Values for these quantities are entered into CIRC as discussed below.

Parameter Information

This element has two parameters: IND and QL. IND is normally the value (L) of the inductance in henries. However, the value entered for IND will be treated as the resonant frequency, f_r, whenever the inductor connection and control data show a nonzero capacitor reference.

See the examples below. This IND value should never be input as exact zero. A small value is allowed, but it should keep the reactance, XL, from becoming much smaller than 0.0001. Both parameters IND and QL are input on the same data line.

QL is the value of the inductor's Q. Based on Q1, CIRC-AC determines r, shown in Fig. 4.21(a), as follows: If

$QL = 0,$ $\quad r = \infty \quad$ (i.e., r is out of the circuit completely)

$QL = -N,$ $\quad r = |N| \quad$ (i.e., the minus sign indicates that r itself is being input with a value equal to the magnitude of the number entered)

$QL = +N,$ $\quad r = N \cdot L \cdot 2\pi f_r$

Since Q is of specialized interest, the calculation of r is not discussed until the model discussion presented later.

Examples of several inductor data options follow. Note that in each row the first of the two entries indicates the connection; the second indicates the parameter.

Input Data	Comments
N1,N2 1E–3	Simple 1-millihenry inductor from node 1 to node 2.
N1,N2,C5 1E6	Example of tuned inductor: Assume that C5 is 100 pF; then $L = 1/[10^{-10}(2\pi 10^6)^2] = 0.253$ millihenry.
N1,N2,C5 1E6,100	With C and L as above and at the operating frequency of 1 MHz, but QL = 100. $r = 100(0.253 \times 10^{-3})2\pi \times 10^6 = 159 \text{ k}\Omega$
N5,N14 1E–3, –1E6	IND = 1 millihenry $r = +1 \text{ M}\Omega$

Dependent Unknowns

This element has two dependent unknowns:

1. Inductor current (from T1 to T2).

2. Inductor reactive power.

The equations defining these unknowns are

1. $\quad I_{\text{IND}} = (V_{T1} - V_{T2})\left(\dfrac{1}{r} + \dfrac{1}{j \times 2\pi fL}\right)$

2. $\quad P_{\text{IND}} = (V_{T1} - V_{T2})I^*_{\text{IND}}$

Since power to an ideal inductor is entirely *reactive*, the angle is $+90$ degrees. Consequently, the units for the ideal inductor's power (PIND) are volt-amperes reactive, or vars. CIRC-AC's inductor will be nonideal only when Q is not zero, in which case PIND is complex.

The algebraic sign of the current can be controlled by appropriate selection of the terminal to be considered terminal 1.

Capacitor

Connection Information

The capacitor element is diagrammed in Fig. 4.21(b).

This element is bilateral, and the terminals are identified as C terminal 1 and C terminal 2. (Polar capacitors must be installed in a specified orientation but, once installed, are bilateral for the purpose of ac analysis.)

Parameter Information

This element has one parameter: the capacitance value, CAP, which may be input as exact zero.

Dependent Unknowns

This element has two dependent unknowns:

1. Capacitor current (from T1 to T2)
2. Capacitor reactive power

The equations defining these unknowns are

1. $\quad I_{\text{CAP}} = (V_{T1} - V_{T2})j \times 2\pi fC$

2. $\quad P_{\text{CAP}} = (V_{T1} - V_{T2})I^*_{\text{CAP}}$

P_C is complex power. Since power to a capacitor is entirely *reactive*, the angle is always -90 degrees (or approximately so). Consequently, the unit for capacitive power is volt-amperes reactive, or vars.

The algebraic sign of the current can be controlled by appropriate selection of the terminal to be considered terminal 1.

Transistor

Connection and Control Information

The transistor model, as the user sees it, is inset in Fig. 4.22. The three terminals of this element are defined within CIRC-AC as base, collector, and emitter. CIRC-AC requires the terminals to be specified in the above

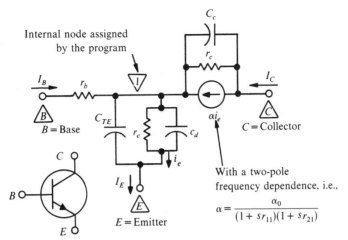

Figure 4.22 Transistor Model.

order. At least two of the three terminals must connect to nodes; that is, only one terminal may connect to a forced voltage (both supplies and ground are forced voltages).

Parameter Information

The transistor element, as the program implements it, is shown by an equivalent circuit in Fig. 4.22. This three-terminal transistor model utilizes a two-pole alpha to implement the classic one-pole current gain attenuation model, plus "excess phase." the model is discussed further later.

Shown on the model diagram are five (of the eight) parameters which must be supplied by the circuit designer:

1. I_E — dc emitter current at operating point (may be obtained from CIRC-DC analysis)
2. r_b — Bulk base resistance (refers to the ohmic drop associated with the flow of majority carriers—electrons or holes—from the base terminal to the active region of the transistor)
3. C_{TE} — Emitter transition capacitance (capacitance across the emitter-base junction due to the high electric field in the depletion region caused by the voltage across the barrier)
4. C_c — Collector capacitance (capacitance across the collector-base junction, primarily made up of transition capacitance, as diffusion capacitance is small; typically $C_c \cong C_{ob}$, as shown on specification sheets)
5. r_c — Reverse collector resistance (ac resistance of reverse-biased collector-base junction)

Three of the parameters shown on the model diagram are calculated by the

1. r_e Dynamic impedance of the emitter junction (ac resistance of forward-biased emitter-base junction); this parameter may be supplied as an alternate parameter in place of I_E

2. C_d Emitter diffusion capacitance (capacitance across the emitter-base junction due to the current flowing through the depletion region)

3. α_o Common base forward active-region current gain

For CIRC-AC to calculate r_e, C_d, and α_o, three other parameters or their alternates must be supplied:

1. βLF   Common-emitter current gain at low frequency, typically 1000 Hz

2a. EPH   Excess phase at the frequency $f_{\alpha co}$ (difference between the actual measured phase shift and the theoretical phase shift of the base-transport factor at the alpha cutoff frequency); excess phase is about 12 degrees for uniform-base transistors, more for a graded-base transistor* *or*

2b. K_θ Scale factor for excess phase† (represents an excess-phase correction factor which can vary from 0.5 to 1.0, depending on the transistor type)

3a. f_T Common emitter gain-bandwidth product (defined as the frequency at which the small-signal common emitter short-circuit forward-current transfer ration (h_{fe} or β) is unity, *or*

3b. $f_{\beta co}$ Common emitter beta cutoff frequency, *or*

3c. $f_{\alpha co}$ Common base cutoff frequency

The eight parameters or alternate parameters required by the transistor model are summarized in Table 4.2. The "notes column" of Table 4.2 discusses the method of selecting alternate parameters.

Table 4.2 Transistor Data Options

Request Number	Parameter	Symbol as Printed	Units	Notes
1	DC emitter current (Dynamic impedance of emitter junction)	IE	Amperes (Ohms)	If a minus sign is entered in front of the parameter value, the program interprets the entry as the positive resistance, r_e.
2	CE current gain (beta) at low frequency	BLF	—	

*Alvin B. Phillips, *Transistor Engineering*, McGraw-Hill, New York, 1962, p. 298.
†*Ibid.*, p. 299.

Table 4.2—*Cont.*

Request Number	Parameter	Symbol as Printed	Units	Notes
	Excess phase	EPH	Degrees	Often there are no available data on a transistor's excess-phase characteristic. In this case a value of zero should be entered. Zero is automatically entered if only one value (BLF) is provided on this second line of parameter data.
	(Scale factor for excess phase)		—	If a minus sign is entered in front of the second value on line 2, the program interprets the entry as a positive scale factor for excess phase, K_θ.
3	CE gain-bandwidth product	FTB	Hertz	If this parameter value is given as 0, the transistor is considered to have $f_{\alpha co} = 10^{30}$ Hz.
	(CE beta cutoff frequency, $f_{\beta co}$)		(Hertz)	If a minus sign is entered in front of the parameter value, the program interprets the entry as a positive beta cutoff frequency.
	(CE cutoff frequency)	FA	(Hertz)	If the user has entered in the control data that the frequency parameter is to be the alpha cutoff frequency, the symbol FA will appear in request 3 and a positive parameter value should be provided.
4	Bulk base resistance	RB	Ohms	A zero value should never be used. A small value (e.g., 0.001) may be used.
5	Emitter transition capacitance	CTE	Farads	
6	Collector capacitance	CC	Farads	0 is allowed.
	Reverse collector resistance	RC	Ohms	Often there are no data on r_c. If request 6 is not answered with two parameter values, RC is entered as zero and the model then treats RC as infinity; i.e., the conductance, $g_c = 1/\text{RC}$, is set to zero.

Dependent Unknowns

There are four dependent unknowns for the transistor element:

1. Base current, I_B.
2. Collector current, I_C.
3. Emitter current, I_E.
4. Power, P_Q.

The equations defining the three current unknowns are as follows (refer to the transistor equivalent circuit shown in Fig. 4.22):

$$I_B = \frac{V_B - V_I}{r_b}$$

$$I_E = (V_I - V_E)\left(\frac{1}{r_c} + j\omega[C_{TE} + C_d]\right)$$

$$I_C = (V_C - V_I)\left(\frac{1}{r_c} + j\omega C_c\right) + \alpha(V_I - V_E)\left(\frac{1}{r_e} + j\omega C_d\right)$$

or

$$I_C = I_E - I_B \quad \text{(used by CIRC)}$$

CIRC-AC first sets up node equations and solves for the node voltages. V_B, V_C, and V_E are the voltage phasors obtained for the three terminals (nodes) of the transistor. V_I is the voltage phasor at the internal node, within the transistor, automatically established by the program; α is a complex, frequency-dependent parameter, as defined in Fig. 4.22.

The transistor power is $P_Q = (V_B - V_E)I_B^* + (V_C - V_E)I_C^*$. Power is a complex quantity.

Current-Admittance Source

CYS Connection and Control Information

A current-admittance source is diagrammed in Fig. 4.23(a).

The two terminals of this element are identified as the exit node and the receiving node, based on the reference direction established for current flow *i*. This polarity convention is convenient for application in small-signal *y*- and *h*-equivalent circuits. At least one terminal of a CYS must be connected to a circuit node. If the current-admittance source is fixed, the data for the two terminals constitute the whole of the connection data. The element has an inherent polarity and must be properly connected into a network.

If the source is controlled, with the control being supplied by a voltage or current from some other element of the network as diagrammed in Fig. 4.23(b), then two additional facts must be provided:

1. The designator of the controlling element, which may be a resistor (for example, R2), inductor, capacitor, voltage-impedance source, or current-admittance source.
2. An indication of whether the control is by current or voltage.

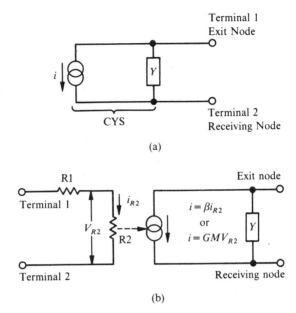

Figure 4.23 Current-Admittance Source–CYS: (a) Element; (b) Controlled Current-Admittance Source Application.

If the current source is to be voltage-controlled, control may be from any one of the five two-terminal elements—R, L, C, CYS, or VZS. If the current source is to be current-controlled, then, in general, only resistance-, inductance-, or capacitance-controlling elements are permitted. One exception has been provided: A current-admittance source may be controlled by the current in a *controlled* voltage-impedance source. This exception allows the small-signal h equivalent circuit to be directly implemented.*

Examples of CYS connection and control data are

N1,N2	fixed current source
N1,N2,R3,V	source controlled by the voltage across R3
N1,N2,L3,I	source controlled by the current through L3

Parameter Information

This element has four parameters:

1. Current transfer factor, real (CTR).

2. Current transfer factor, imaginary (CTI).

3. Admittance, real (YR).

4. Admittance, imaginary (YI).

---

*The exact nature of the allowed exception is that controlled-current source A may be controlled by the current in a controlled-voltage source B, provided that B is voltage-controlled by A (this restriction is satisfied by the h-equivalent circuit).

For fixed sources the current transfer factor is actually the value of the complex current (the value of the real part and the value of the imaginary part). For controlled sources the current transfer factor is multiplied (using complex number arithmetic) by the appropriate control variable, as shown in Fig. 4.23(b). Complex admittance is specified in mhos (real part = conductance, and imaginary part = suseptance).

Two of the most common controlled-current parameters can be related to the CIRC-AC current-admittance source model as follows:

1. When a current source is current-controlled, the current transfer factor is often referred to as beta (or, in some applications, alpha); i.e., $\beta = \text{CTR} + j\text{CTI}$.

2. When the current source is voltage-controlled, the current transfer factor is often referred to as transconductance, gm; i.e., $gm = \text{CTR} + j\text{CTI}$.

Dependent Unknowns

No dependent unknowns are evaluated within CIRC-AC for a current-admittance source.

Voltage-Impedance Source

VZS Connection and Control Information

A voltage-impedance source is diagrammed in Fig. 4.24. The two terminals of this element are identified as the positive node and the negative node, based on the reference polarity assigned to the voltage source. Although the word positive is literally correct only for dc, this is a convenient word to use in defining polarity, meaning zero degree phase shift. Negative refers to ± 180 degrees of phase shift. At least one terminal of a VZS must be connected to a circuit node. If the voltage-impedance source is fixed, the data for the two terminals constitute the whole of the connection data.

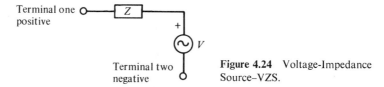

Figure 4.24 Voltage-Impedance Source–VZS.

If the source is controlled, with the control being supplied by a voltage or current from some other element of the network, then two additional facts must be provided:

1. The designator of the controlling element, which may be a resistor, inductor, capacitor, voltage-impedance source, or current-admittance source.

2. An indication of whether the control is by current or voltage.

If the voltage source is to be voltage-controlled, control may be from any one of the five two-terminal elements—R, L, C, CYS, or VZS. If the voltage source is to be current-controlled, then only resistance-, inductance-, or capacitance-controlling elements are permitted.

Parameter Information

This element has four parameters:

1. Voltage transfer factor, real (VTR).
2. Voltage transfer factor, imaginary (VTI).
3. Impedance, real (ZR).
4. Impedance, imaginary (ZI).

For fixed sources the voltage transfer factor is actually the value of the complex voltage (the value of the real part and the value of the imaginary part). For controlled sources the voltage transfer factor is multiplied by the appropriate control variable. Complex impedance is specified in ohms (real part = resistance, and imaginary part = reactance).

The inductor association data for a single MUT element would be as follows. Assume inductor X of Fig. 4.25 to be entered into CIRC as inductor element 2, and inductor Y as element 3. Then the association data will be L2, L3.

By simply adding a K to the data entry, the same line of data that describes "association" may also be used to indicate that the mutual inductance will be described using a coupling coefficient (discussed below). For example, L2,L3,K.

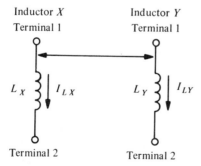

where:

L_X is the self-inductance of inductor X.

L_Y is the self-inductance of inductor Y.

M_{XY} is the mutual inductance between X and Y.

Figure 4.25 Mutual Inductance Element—MUT.

Mutual pairs must be defined when more than two inductors are involved in mutual coupling. In general, if N inductors are all mutually coupled, there will be $(N-1)N/2$ mutual inductance terms (i.e., pairs). This is illustrated in the examples below:

N (Number of Inductors)	$(N-1)N/2$ (Number of Pairs)	Appropriate Association Statements
3	3	L1,L2
		L1,L3
		L2,L3
4	6	L1,L2
		L1,L3
		L1,L4
		L2,L3
		L2,L4
		L3,L4
5	10	L1,L2; L1,L3; L1,L4; L1,L5
		L2,L3; L2,L4; L2,L5
		L3,L4; L3,L5; L4,L5

Normally each mutual inductance term will have a finite value, and the number of MUT elements input to CIRC will agree with the number in column 2 above. The association data are shown in column 2 above. The association data shown in column 3 use consecutive inductor numbers starting with 1. However, they would be equally valid to have a group with inductors 2, 6, and 7, etc.

The direction of positive current in each inductor in a group of mutually coupled inductors is defined in the same way as in an inductor that is not coupled. That is, a positive current flows from L terminal 1 to L terminal 2 as shown in Fig. 4.25.

A mutually induced voltage across the terminals of an inductor is defined as positive if it produces an increase in voltage from L terminal 2 to L terminal 1 (i.e., L terminal 1 will be more positive than 2). The sign of the mutual inductance M for a pair of mutually coupled inductors is defined as follows: If M is positive, then a positive current flowing in one of a pair of mutually coupled inductors produces a positive induced voltage across the other inductor. However, if M is negative, then a positive current flowing in one of the pair produces a negative induced voltage across the other inductor. These statements are applicable to both inductors in every coupled pair.

Parameter Data

This element has one parameter: The mutual inductance value, MUT. Optionally, a coupling coefficient may be specified as alternate data if K is indicated as described earlier. For example,

L1,L2	mutual inductance
1E−3	
L1,L2,K	coupling coefficient
.9	

Mutual inductance and the coupling coefficient (K_{XY}) are related as follows:

$$K_{XY} = \frac{M_{XY}}{\sqrt{L_X \cdot L_Y}}$$

The value of K_{XY} is restricted by CIRC to the range $-0.9999 \leq K \leq 0.9999$.

Perfect coupling occurs when $K_{XY} = \pm 1$; perfectly coupled inductors may be simulated with an ideal transformer equivalent circuit, if desired.

Dependent Unknowns

This element has no dependent unknowns; however, the effect or influence of mutual inductance is included in the inductor current and power solutions.

This effect on inductor current is explained in the modeling section of this chapter.

4.3 INPUT TO CIRC

Input to CIRC consists of (1) control messages and (2) data. Data are always associated with a preceding control message, including the important special case of data describing the circuit to be analyzed.

There is no absolute sequence in which control messages are to be used. CIRC, therefore, gives no specific directions as to what control messages are to be input at any point in an analysis—rather, CIRC simply states PROVIDE CONTROL MESSAGE.

Data, on the other hand, are always required in a specific predetermined sequence. CIRC will always direct the conversational user as to what data are required—resulting in the question and answer appearance seen in the examples presented later. The CIRC manuals, referenced at the end of this chapter, provide a sequence guide for the user who is not using CIRC conversationally and who, therefore, must prepare data in the proper sequence without the directions normally supplied by CIRC itself.

Control Messages

Control messages give the CIRC user control over the program. The first four characters alone of a control message determine the message—the remainder of the 72 characters allowed can be used as desired, either to make the message more readable or for comment. The following two messages are equivalent:

>NOMINAL SOLUTIONS
>NOMI-COMMENTS CAN BE HERE

CIRC requires only two control messages to perform a nominal analysis of a circuit: CIRCUIT DATA and NOMINAL. However, for comprehensive and economical use of the program, four additional control messages are needed. Two are used to request the special analyses:

>WORST   for dc worst-case analysis
>OPEN   for dc open-loop analysis

The other two are used to select output variables.

These six messages are the only ones discussed in this chapter. Many other control messages are allowed with CIRC. They are not essential for running an efficient analysis. All the control messages are documented in the CIRC manuals referenced at the end of this chapter.

CIRCUIT DATA is the control message required to start CIRC operations. It causes CIRC to request data describing the circuit to be analyzed. Data-input format and sequence are discussed in subsequent paragraphs. NOMINAL is the message that causes CIRC to perform the analysis on the circuit described. An example of a minimum deck structure is

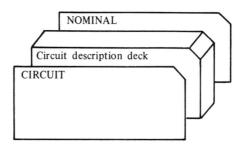

Circuit analyses in general involve a large amount of calculations and generate a large output volume. They are, therefore, uneconomical and relatively inefficient when they are applied to every variable in a circuit. It is generally more rational to seek solutions and result displays only for selected nodes and dependent unknowns. Two control messages are available for inputting such selected nodes and dependent unknowns: NODE SELECTION and DEPENDENT UNKNOWN SELECTION.

In the example sketched below, CIRC's worst-case analysis will ignore all nodes except nodes 1, 5, and 9 and all dependent unknowns except current through resistor 5, base-to-emitter voltage of transistor 3, and circuit gain for transistor 4. The symbols (e.g., IRES) to be used in selecting dependent unknowns are the same as CIRC uses in annotating its output.

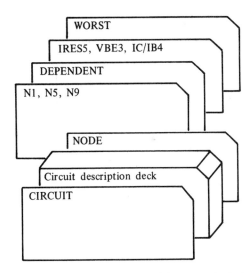

The control messages requesting OPEN-loop analysis has one associate data line:

Form:
> OPEN
> element identifier, node identifier, k

where

element identifier refers to a resistor; the symbols (RES or R) are optional (see example below).

node identifier refers to the node to be driven; again, the symbols (N or V) are optional.

k is the output control: $k = 0$—no special output; and $k = 1$—"full" output (see the comment below).

Examples.

> OPEN LOOP
> RES6,N3,1

The OPEN message causes CIRC-AC to take the necessary identifier information and then to proceed immediately with the solution. After the solution, it asks for variables to plot. The standard open-loop output previously presented is normally adequate, but the user can ask for the full solution output (nodes, currents, powers) as well ($k = 1$).

The output will be titled Open-Loop Solutions. It will be provided at each frequency just following the standard output.

If desired, the user can limit the volume of the full output by use of the control messages NODE and DEPENDENT.

Data Input Conventions

Data input conventions include rules governing entry of parameter data, connection and control data, and format conventions. The format conventions are treated first since they apply to all data.

Data Format Conventions

Data must begin in the first allowed character position. A comment may be placed in a data line following the completed data entry. It must be preceded by an asterisk (*) or a colon (:) and be contained within the allowed 72 character positions. Four special characters, colon (:), asterisk (*), slash (/), and semicolon (;), may not be used in a comment. Commas and periods are used naturally. A comma is required to terminate any number that is followed by other symbols or numbers; that is, the comma is required as a separator. Examples of these rules are

```
          N1,N2              *COMMENT
          10
          .01:ANOTHER        COMMENT
```

Exponential notion may be used but is not required. Examples of exponential notation related to common engineering units are

$$E6 = \text{mega}, \quad \text{e.g., } 1.1E6 = 1.1 \text{ M}\Omega$$
$$E3 = \text{kilo}, \quad \text{e.g., } 27E3 = 27 \text{ k}\Omega$$
$$E-3 = \text{milli}, \quad \text{e.g., } 10E-3 = 10 \text{ mA}$$
$$E-6 = \text{micro}, \quad \text{e.g., } 0.15E-6 = 0.15 \text{ }\mu\text{A}$$
$$E-9 = \text{nano}, \quad \text{e.g., } 10.E-9 = 10 \text{ nA}$$

When a single line of data (either on a card, a type line, or a line in a file) is used to input one or more numerical values, zero values may be implied by simply not completing the entries. For example, if values are to be entered for A, B, and C, then

```
   1,2,3                  yields A = 1, B = 2, C = 3
   1,2                           A = 1, B = 2, C = 0
   1,                            A = 1, B = 0, C = 0
   0                             A = 0, B = 0, C = 0
   "Blank" card or line          A = 0, B = 0, C = 0
```

If input is prepared on a typewriter terminal, pressing the RETURN key (N/L on some teletype models) is equivalent to a blank card. CIRC makes extensive use of incomplete (blank) entries for zeros.

Parameter Data Entry

CIRC-AC and CIRC-TR have a simple parameter entry system. It requests standard element parameter data entries with the message

```
          SPECIFY:
          NOM VAL (FOR PAR.S INDICATED)
```

This request is then followed by further directions to input a data line for one parameter, such as

$$\text{RES } 10 = \text{PAR. } 14$$
$$:10E3 \quad \text{(one parameter)}$$

or more than one parameter, such as

$$\text{CAP , CRS , CRP } 1 = \text{PAR.S } 58\text{–}60$$
$$:150E\text{–}12, .1, 1E6 \quad \text{(three parameters)}$$

On the other hand, CIRC-DC needs minimum and maximum parameter values for worst-case analysis. It requests parameter data entry with the message

SPECIFY:
NOM VAL, MIN INFO, MAX INFO, (FOR PARA.S INDICATED)

Minimum and maximum information is required for worst-case analysis only. A typical entry with no max/min data values is simply 10, and an entry with max/min values is 10,S5,S10. For circuit analysis purposes, minimum and maximum values are conveniently thought of in two ways: value or percent. CIRC, therefore, allows a flexible parameter data entry which has the form shown and explained in Table 4.3.

Table 4.3 Conventions for Inputing Minimum and Maximum Parameter Data

Form of parameter data entry: $N_1,\text{key}N_2,\text{key}N_3$
Example (also see and of table): 20,D5,U10
where

N_1 is the parameter's nominal value.

N_2 is the number used to determine the parameter's minimum value.

N_3 is the number used to determine the parameter's maximum value.

key indicates how N_2,N_3 are to be used, as follows:

Key	Meaning
M	The numerical value that follows is to be used directly as a minimum or maximum value.
$	The purpose is the same as for "M" but the $ symbol is desired for greater convenience in keypunching. M and $ are interchangeable and may be used on the same line (M is more convenient on a teletype).
S	(For "save"); the number that follows is a percentage change from nominal (decrease for minimum and increase for maximum) that is to be reused (R below) by subsequent parameters. Both the minimum and maximum values must be in the same line.
R	(For "reuse"); the previously saved (see S and T) percentage change from nominal is to be used for the parameter now being entered. In this case, N_2,N_3, and the second key need not be entered. R refers back to the most recently used S or T, but R need *not* immediately follow S or T. Also, the S or T used during circuit description input should *not* be relied upon during parameter changes.

Table 4.3—*Cont.*

Key	Meaning
T	(For "tolerance"); the tolerance is symmetrical, e.g., $+5\%$, -5%. In this case, the second key and N_3 need not be entered. The tolerance entered using T is also "saved" for reuse (R above).
% D U	The numerical value that follows is a percentage change from nominal. Recommended symbols for key punching are D for decrease, down, and U for increase, up. The recommended symbol for a teletype is %. Actually any symbol not in the list above (M,$,S,R,T) will be considered as a percent indication.

Examples.

		Resulting Data in Computer		
Desired Entry	DATA ENTERED	NOMINAL	MINIMUM	MAXIMUM
a. Nominal only	10	10.0	10.0	10.0
b. Direct value	10,M9,M11	10.0	9.0	11.0
c. Direct value	10,$9,$11	10.0	9.0	11.0
d. Save percentage for reuse	10,S10,S5	10.0	9.0	10.5
e. Local use of percentage	10,%3,%2	10.0	9.7	10.2
f. Reuse of saved percentage (d)	20,R	20.0	18.0	21.0
g. Local use	1E3,D20,U5	1000.0	800.0	1050.0
h. Symmetrical tolerance	10,T20	10.0	8.0	12.0
i. Reuse of symmetrical tolerance	20,R	20.0	16.0	24.0

Special Data for CIRC-TR

CIRC-TR's second request for data is

NUMBER OF VOLTAGE SUPPLIES

Later in the program CIRC requests data, first on the type and then on the characteristics of each set. A typical sequence is as follows:

VOLTAGE SUPPLIES (TYPE)
SPECIFY:
DC, PULSE, SINE, COS OR TABLE
SUP1 (USES PAR. NO.S 1, 2)
:(Response)
SUP2
 .
 .
 .
SUPn (USES PAR. NO.S N, N+1)
:(Response)

```
                VOLTAGE SUPPLIES (DATA)
                SPECIFY:
                VALUE (FOR DATA INDICATED)
                SUP1
                V-ORIG, V PEAK
                :(Response)
                TRISE, TFALL
                :(Response)
                TWIDTH, TREPEAT
                :(Response)
                TDELAY
                :(Response)
                SUP2
                    .
                    .
                    .
```

Details on voltage supply elements themselves were presented earlier. The parameter numbers associated with each supply are used by the program to identify the values it stores at each time increment for each supply. The first (lower) parameter number refers to the present supply value, and the second (higher) parameter number refers to the value of the supply at the previous time point. CIRC-TR itself moves supply values into these parameter storage positions—it is not the user's responsibility.

If the supply type is table, the program does not request table data under VOLTAGE SUPPLIES (DATA) but under the heading TABLES.

The TABLE heading requesting parameter control table data is output after all other supply data have been input. The sequence is as follows:

```
                TABLES
                SPECIFY:
                PARAMETER TO CONTROL, CONTROL VARIABLE PAR.
                    COORDINATE, CONTROL COORDINATE,
                    END CONTROL
                TABL 1
                    .
                    .
                    .
                TABL 2
                    .
                    .
                    .
                TABL n
```

The parameter to control is usually a supply and the control variable, time; but it may be almost any CIRC parameter controlled by time or a supply. The requested coordinates would be voltage and time for time control and a multiplication factor for supply control. Details on the tables themselves were presented earlier.

As previously discussed, the diode has three mutually exclusive sets of parameters for specification, classic, and Xerox-type diodes. If diodes are used, CIRC determines which model to use by the request:

SPECIFY:
DIODE MODEL TYPE(SPEC(S),CLASSIC(C), OR XDS(X))

Like the diode model, the transistor model has three mutually exclusive sets of parameters. If transistors are used, CIRC determines which model to use by the request

SPECIFY:
TRANSISTOR MODEL TYPE(SPEC(S),CLASSIC(C), OR XDS(X))

All diodes must be of one type, and all transistors must be of one type; however, diodes and transistors need not be of the same type. For example, diodes could be classic type, and the transistor, Xerox type.

Connection and Control Data

Connection data for the network elements include connections from node to node, node to ground, and node to supply. Each connection is normally specified by a letter (N for node, G for ground, S for supply) and an integer identifying the specific node or supply to which the element terminal is connected. A typical entry would be S2,N7.

Note that any two-terminal element must connect to at least one node and any three-terminal element must connect to at least two nodes (for this requirement, node 0, ground, is not considered a node).

Ground would normally be followed by zero, i.e., G0. However, because zero and alphabetic O are often interchanged and to simplify typing and improve appearance, it is recommended that G be followed immediately by a comma (G,) which has the effect of supplying a zero. At the very end of a data line, "G" may simply be omitted.

Control data are input on the same card (or line) as connection data. Their purpose varies according to the element with which they are used. The purposes are summarized below:

Elements	*Comment*
RESISTOR	None
INDUCTOR	For CIRC-AC, indicates the capacitor reference for resonant frequency option
CAPACITOR	None
TRANSISTOR	Indicates either device from which data are to be copied, or that $f_{\alpha co}$ is to be used as an input data option in CIRC-DC or CIRC-AC
CYS (CVR) & VZS sources	Indicate which element (if any) is to control a source and what type of control is to be used
MUTUAL	Indicates that a coupling coefficient is to be used to describe the mutual inductance

The copy data control saves time and trouble. If the parameter values for a diode or a transistor are the same as previously input (for another diode or transistor), the letter "C" instructs CIRC to copy such values. An integer following "C" identifies the element that is to be copied.

A summary of connection and control information is presented in Table 4.4. The table also presents examples, showing CIRC requests for the various element types and typical answers.

Table 4.4 Summary of Element Connection and Control Information

Form of Connection and Control entry: key‡,key‡,key‡,key‡,key‡
Example (also see end of table): N1,S1,N5,C0,A
where key can be any of the following:

Key	Meaning
N	Terminal is connected to the node specified by the number that follows.
S	Terminal is connected to the supply specified in the number that follows.
G	Terminal is connected to ground. G may be followed by a zero, which is an indication that ground is the datum node, node 0. Because of the tendency to mistake the alphabetic O for the desired numeric 0 (zero), it is recommended that G be followed immediately by a comma, i.e., G, rather than G0.
L	Mutual inductance exists between inductors specified by number on the same data line, e.g., L2, L3. Applies to MUT elements only and to CIRC-AC only.
INDUCTOR C	CIRC is to calculate the inductance value that gives resonance with the capacitor specified by the number that follows. Applies to CIRC-AC only.
TRANSISTOR C	CIRC is to copy (C) the parameter data of a previously treated transistor and thereby to set up the parameter data automatically for the present transistor. Most data for the present device are copied from some previously treated device specified by the integer that follows C. For CIRC-AC and CIRC-TR, all data will be copied, the IE parameter will be requested even when all other data are copied. For CIRC-DC.
P (for CIRC-DC and TR)	Indicates that a transistor is *PNP*. An *NPN* transistor is assumed if P is not used. The "copy" information must be filled in to use P. Therefore, C0 must be used if data is to be specified, since the "copy data" information is ignored if a zero or negative number is used, e.g., N5,N7,N9,C0,P.
A (for CIRC-AC)	The frequency parameter will be the alpha cutoff frequency rather than the normal frequency. Normal data are assumed if A is not used. The copy information must be filled in to use A. Use (C0) if data are to be specified; the copy data information is ignored if a zero or negative number is used, e.g., N5,N7,N9, C0,A.
CURRENT	For CIRC-DC and CIRC-TR

Table 4.4—*Cont.*

Key	Meaning
R	(For "resistor"); indicates that a current source is controlled by the resistor specified by the number that follows. (In CIRC-TR a current may be controlled only by a resistor.)
I	Indicates that control is by the current (I) in the controlling resistor.
V	Indicates that control is by the voltage (V) across the controlling resistor.
CYS and VZS sources R,L,C,Z, and Y	Indicates the type of control element as R,L,C,VZS, or CYS. Any of the five may be used for voltage control. The VZS and CYS may not be used for current control with one exception: A *controlled* voltage source may control a current source.
V or I (for CIRC-AC)	Indicates whether a controlled source is voltage (V) or current (I) controlled. If either is specified, V is assumed.
MUTUALS K	Mutual inductance will be specified using a coupling coefficient.

Examples.

RESISTORS:
SPECIFY: CIRC-DC, AC, TR
R TERM 1 INFO, TERM 2 INFO ← Typical indication of CIRC-AC's data request.
N1,G,   (Node 1, ground shows preferred G, notation.)

VOLTAGE-RESISTANCE SOURCES

SPECIFY:

POS TERM INFO, NEG TERM INFO

N1,N3   (Node 1, Node 3)

CURRENT SOURCES
SPECIFY:
EXIT NODE INFO, RECV NODE INFO, ELEMENT INFO, V OR I CONTROL

G0,N1   (Ground, Node 1. Fixed current.)

N5,G,R3,I   (Node 5, Ground. Controlled current—R3 is the controlling element, I is the type of control.)

DIODES
SPECIFY:
ANODE INFO, CATHODE INFO, COPY DATA INFO,

Table 4.4—*Cont.*

N6,S1   (Node 6, Supply 1)

TRANSISTORS

SPECIFY:
BASE INFO, COLL. INFO, EMIT, INFO, COPY DATA INFO, P IF PNP

N2,N3   (Node 2, Node 3, Ground. Shows ground being implied by not directly supplying data for the emitter terminal.)

N3,N4,N5,C1,P   (Node 3, Node 4, Node 5. C1 instructs CIRC to copy the parameter data input for transistor 1.)

INDUCTORS:
SPECIFY:
L TERM 1 INFO, 1 TERM 2 INFO,   CIRC-TR
N1,N2   (Node 1, Node 2)
CAPACITORS:
SPECIFY:
C TERM 1 INFO, C TERM 2 INFO   CIRC-AC, TR
N6,S1   (Node 6, Supply 1)
INDUCTORS:
SPECIFY: CIRC-AC
LTERM 1 INFO, L TERM 2 INFO, TUNED CAP. REF.

N1,N2 or

N1,N2,C1

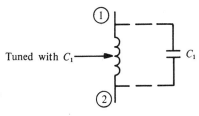

CAPACITORS:
SPECIFY:
C TERM 1 INFO, C TERM 2 INFO,
N6,S1   (Node 6, Supply 1.)
TRANSISTORS:
SPECIFY:
BASE INFO, COLL. INFO, EMIT, INFO, COPY DATA INFO, A IF FACO

N2,N3   (Node 2, node 3, ground—Shows ground being implied by not directly supplying data for the emitter terminal.)   npn vs. pnp need not be distinguished for ac analysis.

Table 4.4—*Cont.*

N3,N4,N5,C1   (Node 3, node 4, node 5—C1 instructs CIRC to copy the parameter data input for transistor 1.)

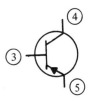

1E–3   (IE must still be input.)
N1,N2,G0,C0,A   (Transistor without data copy and with $f_{\alpha co}$ specified.)

CURRENT-ADMITTANCE SOURCES—CYS:
SPECIFY:
EXIT NODE INFO, RECV NODE INFO, ELEMENT INFO, V OR I CONTROL

G0,N1   (Ground, node 1—Fixed current.)

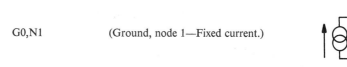

N5,G,N1,R3,I   (Node 5, ground, node 1—Controlled current—R3 is the controlling element, I the type of control.)

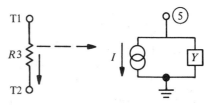

VOLTAGE-IMPEDANCE SOURCES—VZS:
SPECIFY:
POS TERM INFO, NEG TERM INFO, ELEMENT INFO, V OR I, CONTROL

N1,N3   (Node 1, node 3.)

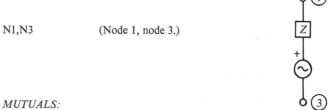

MUTUALS:
SPECIFY:
FIRST IND. REF., SECOND IND. REF., K IF COUPLING COEF.

L2,L3
L2,L3,K

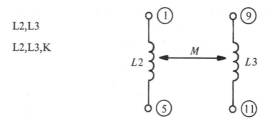

Special Circuit Description Termination Data

CIRC-AC and CIRC-TR have special analysis control data that follow the basic circuit description.

Frequency Data Concludes ac Circuit Description

In ac circuit analysis the problem is solved repetitively at various frequencies over which the circuit is expected to operate. After all the parameters have been entered, the program makes the request

FREQ DATA
START

The user enters the beginning, or lowest, frequency. CIRC then makes the request

CONTROL (A, M, L, N)

A, M, L, and N are four ways of selecting frequency analysis points between START and END (highest frequency specified):

- A   Adds the INCREMENT frequency to the START frequency and then to successive intermediate frequencies.

- M   Multiplies the START frequency by the INCREMENT; then the first intermediate frequency, etc.

- L   Logarithmically multiplies the START frequency, i.e., by 10 raised to the reciprocal of the INCREMENT:

$$F2 = FS \times 10^{1/F_I}$$

   Successive intermediate frequencies are obtained by multiplying by $10^{1/F_I}$. This method provides solutions at the rate of F_I analysis points per decade.

- N   Names the number of points to be plotted. In this case the frequency is iterated with added increments, and with the increment calculated as follows:

$$F_I = \frac{F_E - F_S}{N - 1}$$

where

F_I is originally input as N, the number of frequency values to be used and is then converted to the actual increment of frequency. After F_I is calculated, the iteration proceeds as for Add.

F_E is the END frequency (discussed below).

F_S is the START frequency (discussed above).

After the increment controls, CIRC requests

INCREMENT

Enter the appropriate INCREMENT number, as defined above. Then CIRC requests

END

Enter the final or higher frequency. A typical set of frequency data would be

      10   Start at 10 Hz
      L4   Four points per decade
      1E6   End at 1 MHz

Plot Variables for CIRC-AC

After CIRC completes a NOMINAL or OPEN-loop solution, it requests variables to plot (0 may be used to stop all plots). Allowed variables for plotting include two special open-loop variables, VOL and ZOL, all node voltages, and all dependent unknowns. VOL is the voltage gain, and ZOL, the load impedance (Z) used during the open-loop analysis.

In this chapter and the examples presented later, it is assumed that plots are restricted to these basic variables and that CIRC's standard plotting conventions are adequate. Actually CIRC provides much more plotting flexibility, as described in CIRC manuals referenced at the end of this chapter.

Analysis Control for CIRC-TR

In addition to the circuit description, CIRC-TR needs basic control information before it can do a valid analysis. This basic control information is requested and accepted as an integral part of the circuit description and consists of time controls (time step and end time), output interval controls, initial-condition control, and plot selection controls.

To obtain the necessary information for time control, CIRC requests

SPECIFY:
DELTA TIME, END TIME

Delta time serves two purposes:

1. To control the numerical accuracy of the differential equation solution procedure (by which CIRC-TR produces its time domain analysis).

2. To define the time interval that is to be used as a base point in conjunction with output interval controls (explained below).

It is necessary to pick a value for delta time that is not too large, since this gives inaccurate results, nor too small, since this requires large computation time and results in high program running costs.

One selection criterion is to pick a value of delta time that is one-tenth the smallest *significant* time constant in the circuit. CIRC-TR uses integration techniques that are not subject to control by the smallest absolute time constant but rather by significant time constants. Thus, CIRC's integration technique produces program execution efficiencies often ranging from 30 to 100 to 1 over integration techniques that are controlled by the smallest absolute time constant.

Often the switching speed of the transistors can be used to determine the delta time selection. If one is looking at circuits in which the transistors are to determine the time response of interest, it is normally a valid procedure to base delta time selection on the rule of thumb that

$$\text{DELT TIME} = \tfrac{1}{20}T$$

where T is the smallest of T_{RISE}, T_{FALL}, or T_{STORAGE}.

On the other hand, when the transistors are not controlling the time response of interest, they can normally be ignored in the delta time selection.

Actually, the value of delta time input normally is the largest value to be used by CIRC-TR in the computations. CIRC-TR automatically contracts and expands its internal Δt value as needed to achieve efficiency and accuracy in solving the nonlinear equations associated with the diodes and transistors. (The control is not based on time constants and will have no effect when neither diodes nor transistors are in the circuit.) This automatic time step control can be very effective in decreasing computer costs in circuits when there are periods of relative inactivity. This is often true of larger transistor-circuitry functional groupings.

If the user does not want CIRC-TR to use its Δt change algorithm, he may input delta time as a negative number, in which case CIRC-TR will not change Δt but will use the input value (ignoring the minus sign).

The aspect of delta time being the basic minimum time interval between outputs does not normally constrain selection of its value, since the desired output interval is usually specified to be several to many times greater than the delta time value.

The second part of the time control request is for end time. This is the tentative point at which the circuit simulation (in the time domain) is to end. CIRC-TR then requests instructions about plotting and does plots as directed (see below). Thereafter, CIRC-TR would ask the user for additional control instructions (or possibly this request will be made with the plot feature completely skipped). At this point there are several major choices. One typical choice would be to stop the analysis. Another choice would be to make some analysis control changes (using control messages) and then go on with the analysis starting at the time point now reached. Still another choice would be to make some parameter changes and redo the analysis.

Output Interval Controls for CIRC-TR

CIRC-TR requests three output interval controls:

BATCH OUTPUT INTERVAL (BOI)
CONVERSATIONAL OUTPUT INTERVAL (COI)
PLOTTING OUTPUT INTERVAL (POI)

BOI and COI are not restricted to their respective environments but are so named because of their relative importance in those environments. The following summary of rules and examples should be adequate for present purposes:

1. Numbers whose magnitudes are greater than 1 are considered as ratios (of calculations made to solutions output).

2. Numbers whose magnitudes are less than 1 are taken as the time interval between outputs.

3. Positive numbers are based on delta time, causing output to be evenly spaced in time even when CIRC-TR's Δt is changing.

4. Negative numbers (when they are ratios) are based on calculated points, not fixed time increments, resulting in more output when CIRC-TR selects a shorter solution Δt.

5. If BOI is input a 0 or blank, the value is set to 10,000.

6. If COI is input as 0 or blank, the value is set to 1000.

7. If POI is input as 0, the value is calculated as that ratio between 1 and 20 that comes as near to 150 points for plotting as an integer ratio will allow; i.e., (number of points calculated/150) = POI. Plot time is from 0 to end time.

Example A.

Symbol	Input	Results and Comments
BOI	50E–9	Typical batch run. Assuming end time = 100E–9 and NODE and DEPE selections were made, the two full outputs will be printed, but COI causes selected output to be printed every 10 nsec. Thus, if the job stopped, the user would have a good idea of the general answer quality and of how much progress was made. This run would rely heavily on the plotted output for major information display. With $\Delta t = 0.1\text{E}{-}9$ (assumed), there will then be 1000 calculations and POI = (1000/150) = 6.6, rounded to 7.
COI	10E–9	
POI	0	

Example B.

Symbol	Input	Results and Comments
BOI	0	Typical conversational run. Output is taken each nanosecond
COI	1E–9	as solution progresses. No information of practical value
POI	1E6	would be input on the plot file.

Example C.

Symbol	Input	Results and Comments
BOI	0	This is the most convenient way for the user to rely on
COI	0	CIRC for providing all output in plots and/or associated
POI	0	tabulations.

The user may also wish to see the example and the more detailed discussions in the CIRC manuals referenced at the end of this chapter.

Initial Conditions Option for CIRC-TR

When CIRC requests the initial conditions option the response must be either C for "calculate" or Z for "zero" initial conditions. (C is the default option.) C causes CIRC to find the steady-state dc voltages, currents, and powers in a circuit at time zero, whereas Z sets the initial conditions to zero. The results from calculated initial conditions may be used to automatically set up the starting point for a transient solution.

CIRC-TR does not allow the user to input initial conditions. The major reason is that in larger circuits it is better to have the program do the calculations since all inductors, capacitors, diodes, and transistors should be initialized accurately for both voltage and current. If the user wants to set some special initial conditions, this can usually be done with some small auxiliary circuit that is used as CIRC calculates initial conditions. The auxiliary circuit is then removed and the transient analysis is done.

Plotting Selection and Control

The CIRC-TR manual referenced at the end of this chapter gives a detailed description of plotting procedures and considerations. The summary given here should be sufficient for simple applications.

All plots are of variables versus time. After CIRC receives the initial condition option, it next requests plotting information. Consider this example:

```
PLOT GROUP TITLE (0 TO STOP PLOTS)
  :MY OP. AMP. RUN 3 ←──────────── A 0 entry here would have ended
                                    the circuit description and would
                                    have meant no plots were to be done.
SPECIFY:
VARIABLE TO PLOT,MIN,MAX (0 TO END)
  :V1                              ─── Not normally needed. Controls
  :V5                                  amplitude axis on plots—see CIRC-TR
                                       manual for details.
  :IRES1
  :IE1
  :0 ←─────────────────────────── This ends the list of variables
                                    to plot and completes the circuit
                                    description.
```

The symbols (e.g., IRES) to be used in selecting plot variables are the same as CIRC uses in annotating its output. Actually, the variables to plot may consist of expressions; this detail is explained in the CIRC-TR manual.

During the run, CIRC stores the plot solutions in a special file. After all solutions have been calculated (end time has been reached), the program asks for plot confirmation in the form of plot control, whereupon the user may simply input a zero and all the variables requested will be plotted.

Despite the limitations of line printer resolution, CIRC-TR's plot output is one of the most useful features of the program, and the user should study the capability with care to obtain maximum benefit of the program.

4.4 CIRC MODELS

The total network model content of the CIRC package was presented earlier. Emphasis was on the schematics of the network elements, their parameters and related input description techniques, and the definitions of related output variables. The present section is to display the necessary model equations to document the program's implementation of a model (as opposed to theoretical model development). Also, the electrical performance demonstrated by the models themselves are discussed, and model-related warning and information messages are presented.

The dc models are not directly discussed, since the dc equation and performance aspects are included in the CIRC-TR model discussions. The CIRC-TR model discussion will be followed by CIRC-AC model discussion.

CIRC-TR Diode and Transistor Models

Both the diode and transistor are nonlinear and time-dependent and can be generally classed as Ebers-Moll models, although Ebers-Moll models have various forms. The specific forms used by CIRC-TR are best described as Linvill-derived Ebers-Moll models. This model form came about as follows:

The fact that Linvill models [4] for diodes and transistors were network-oriented led to an investigation of nonlinear models by R. McNair [5] showing that literal implementation in a general-purpose circuit analysis program was practical. However, literal implementation adds excess density unknowns to node voltage unknowns, thereby increasing analysis cost. Therefore, C. Kleiner [6, 7] derived from Linvill models a form of the transistor (and diode) model that depended only on circuit voltages and currents. This model form, used in CIRC-TR, has the major unique characteristic of handling the diffusion of minority carriers with a current source that is a function of a time constant, τ, and a current derivative, di/dt. This model form, shown in Figs. 4.17, 4.18, and 4.19, is conveniently considered to be an Ebers-Moll model since it depends on voltage and current and can thus be easily related to their modeling work, as is clearly seen in the later discussion of the basis for CIRC-TR's "specification" transistor model.

The term Linvill-derived Ebers-Moll models is, however, not adequate to communicate the modeling options available in CIRC-TR. Therefore, CIRC-TR uses six model designations: specification, classic, and XDS diodes; and specification, classic, and XDS transistors. Actually, specification and classic models differ only in the parameter data used to describe them and in one true (but minor) modeling difference—inclusion or exclusion of a slope factor in the diode equation. XDS models use the same basic time domain modeling as the others, but different data. They also have appended to the basic nonlinear dc modeling some additions affecting dc junction drops, junction leakage currents, and device current gains (betas). More detail is presented later.

CIRC-TR Diode Model

The diode element consists of the five parts shown in Fig. 4.17(b): an ideal diode, a reverse (leakage) current, a high-impedance shunt, a junction transition capacitance, and a current proportional to the time rate of change of the ideal diode current. The proportionality constant, T_D, is the charge factor or diode time constant. The ideal diode is based on the diode equation:

$$I_D = I_S(e^{Vq/kTM_D} - 1)$$

where

I_S = saturation current

V = junction voltage (CIRC symbol V_{DIO})

q = electron charge

k = Boltzmann constant

T = temperature of junction (°Kelvin)

M_D = slope factor by which the junction voltage and the dynamic resistance are larger than the ideal values calculated using Vq/kT. The specification model and CIRC-DC treat M_D as equal to 1 without provision for a user input value.

When specification diodes are used, the CIRC diode element routine calculates I_S as a function of the input parameter pair (I_D, V_D), the parameter T, and the constants q and k. At times the calculated value of I_S is poor, and a warning message results. The message is presented later. The calculation is a simple curve fit procedure that causes the diode to fit the point I_D, V_D exactly. Because the diode element is based on the nonlinear diode equation, there is a close fit for all other diode operating points (see Fig. 4.26).

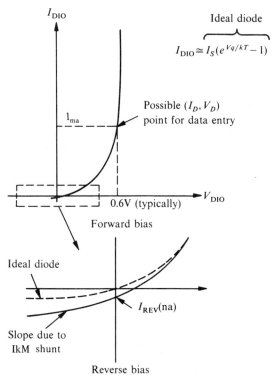

Figure 4.26 DC Electrical Characteristics of Diode Element. Region of Forward Bias is Predominated by Ideal Diode.

The temperature, T, is part of the data required by CIRC and is requested by

JUNCTION TEMPERATURE (DEG. CENT.)

as was discussed earlier.

The temperature applies to all junctions for both diodes and transistors (except for XDS diodes or transistors with specific nonzero temperature entries, or classic models, that use the special slope factor option).

The reverse current, I_{REV}, provides for simulating diode leakage, which is normally much larger than the saturation current of the ideal diode. Saturation currents are typically picoamperes (10^{-12}) and leakage currents are typically nanoamperes (10^{-9}) and larger.

The shunt impedance is present for CIRC mechanization purposes only and normally should not be important, since its value is 10^9 ohms.

The capacitance, C_{DJ}, is the diode junction transition layer (space-charge-layer) capacitance. It is normally considered to follow the equation

$$C_{DJ} \cong \text{CDJZ} \frac{1}{(1 - V_D/\text{VDBI})^{\text{PWRD}}}$$

(*note*: subscripts are reserved for symbols that are not CIRC parameters) where

CDJZ = transition capacitance at zero junction voltage

VDBI = junction contact potential (a positive quantity)

PWRD = junction grading constant

In a general-purpose transient analysis program, such as CIRC-TR, the voltage, V_D, may reach or exceed the value of VDBI specified by the user. When this happens, the above equation is invalid. CIRC-TR uses an approach based on work done by Poon and Gummel* to circumvent this problem.

In CIRC's diode implementation, the equation used for transition capacitance is

$$C_{DJ} = \text{CDJZ} \frac{1}{(X^2 + b)^{\text{PWRD}/2}} \left(1 + \frac{\text{PWRD}}{1 - \text{PWRD}} \frac{b}{X^2 + b}\right)$$

where

$$X = \frac{V_D - \text{VDBI}}{\text{VDBI}}$$

and

$$b = [(1 - \text{PWRD})\text{CDJX}]^{-2/\text{PWRD}}$$

CDJX represents the ratio of the maximum capacitance to the zero bias capacitance. The maximum capacitance occurs when $V_D = \text{VDBI}$. For voltages above VDBI the capacitance decreases. Since CDJX is a parameter that has meaning to a user, he may want to enter it with his diode input data. (The program default value is taken as 50.) This results in the capacitance following the classic equation closely (see Fig. 4.27) until V_D is very near VDBI. A peak is reached when $V_D = \text{VDBI}$ ($C_{DJ} = 50 * \text{CDJZ}$) and the capacitance starts decreasing above VDBI.

The transition capacitance model described has been validated with measurements that indicate a maximum capacitance in the vicinity of VDBI. In addition, the capacitance model is free of singularities for all voltages.

*H. C. Poon and H. K. Gummel, "Modeling of Emitter Capacitance," *Proceedings of the IEEE*, Dec. 1969, p. 2182.

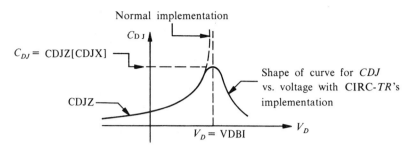

Figure 4.27 Junction Capacitance versus Voltage.

The time constant, TD, and the dependence on the current derivative, dI_D/dt, is the major charge storage mechanism of the diode model. It is CIRC's method of handling the diffusion characteristics of minority carriers. The charge stored at a forward current, I_D, is simply $I_D * TD$.

The classic and XDS diode parameter data sets include TD in the list of items to be entered by the user.

For the specification diode, the program calculates the diode time constant by using three input parameters: ION, IOFF, and TOFF. The equation used is

$$TD = TOFF/ln(1 + ION/IOFF)$$

The diode storage time, TOFF, can often be determined from a device data sheet or from a measurement where a diode is switched from a forward current, ION, to a reverse current, IOFF, as illustrated in Fig. 4.28.

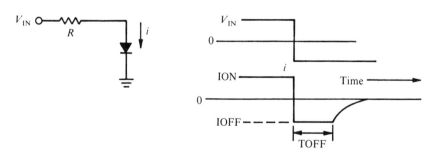

Figure 4.28 Test Circuit and Response for Measuring Diode Storage Time.

The test current is assumed to switch instantaneously from ION to IOFF. TOFF is measured from the time of the current transition to the time when the magnitude of the reverse current starts to decrease.

The performance of a diode in the test configuration can be described by the following equation when leakage current and transition capacitance are ignored:

$$i = I_D + TD\frac{dI_D}{dt}$$

The solution of this equation for the ideal diode current I_D is

$$I_D = \text{ION} + (\text{ION} + \text{IOFF})(e^{-t/\text{TD}} - 1)$$

The ideal diode current never goes negative; instead, the diode turns off. When this happens, the rate of change of I_D with respect to time also goes to zero. The storage time, TOFF, may therefore be solved for by setting $I_D = 0$ and solving for $t = \text{TOFF}$ in the above equation. The result is

$$\text{TOFF} = \text{TD} \, ln\left(1 + \frac{\text{ION}}{\text{IOFF}}\right)$$

which in turn may be solved for TD, resulting in the earlier equation showing the TD value.

It is important to emphasize that this method of determining TD is completely valid only for a perfect step current drive and a test impedance that makes the effect of the transition capacitance negligible. Therefore, the engineer may want to adjust the switching time slightly to account for not being able to achieve the ideal test conditions. This would usually be done by using a TOFF value slightly below the measured or specified value of by taking C_{DJ} to be zero.

The most important aspect of the diode element is that it possesses the nonlinear transient characteristic normally displayed by actual diodes.

Some diode device characteristics are not included in the diode element. Consider, for instance, the following three characteristics:

1. Bulk resistance (may be added by the use of a normal resistor element, if desired).
2. Breakdown voltage (reverse bias).
3. Voltage dependence for IREV.

CIRC-TR Transistor Model

The transistor element used in CIRC-TR is valid for both large-signal nonlinear and small-signal linear operation. It includes both dc and transient components. Transition capacitors are varied automatically as a function of voltage. The element covers all four modes of operation: off, active normal, active inverted, and saturated.

Both the base-emitter and base-collector junctions have a dc characteristic similar to the diode element characteristics shown in Fig. 4.26. The base-emitter junction of the specification model is described with a data set (IEA,VBE). This data set represents one point on the nonlinear base-emitter junction characteristic. The base-collector and the base-emitter leakage current are usually much larger than for an ideal junction and are, therefore, handled by separate parameters, ICBO and IEBO.

The base-collector junction of the specification model is described to CIRC by specifying the saturation performance of the transistor. There are two condition parameters, ICS and IBS, the collector current and base current conditions at which a saturation voltage, V_{SAT}, is measured (see Fig. 4.29). Part (a) highlights the saturation characteristics of the (mathematical) element and shows a possible data set for the base-collector junction. Part (b) shows all regions of the element's collector characteristic. In performing an analysis, CIRC places transistors in any appropriate region of the characteristic.

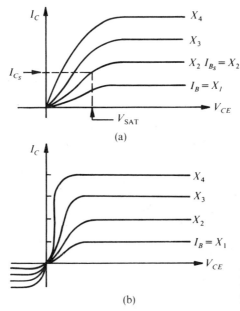

Figure 4.29 DC Collector Characteristics of Transistor Element: (a) Saturation Region; (b) All Regions.

The equivalent circuit for the CIRC-TR specification transistor model is shown in Fig. 4.18. It shows that the transient portion of the model consists of each junction being treated in a way analogous to the diode junction; i.e., a current source of the form $\tau \, dI/dt$ and a capacitance are added. The capacitance for the base-emitter junction is C_{EJ}, and for the base-collector junction it is C_{CJ}. The handling of these capacitances and the meaning and options for related parameters are exactly analogous to the information presented for the diode, and, therefore, no new discussion is provided.

The derivative current sources are associated with two time constants, τ_N and τ_I. These two current sources are CIRC-TR's method for implementing the basic Ebers-Moll time-dependent transistor model, and they alone provide the model with finite rise, fall, and storage time characteristics.

Additional explanation of τ_N and τ_I is given in the "specification" model discussion that follows.

τ_N and τ_I may be directly entered into CIRC-TR for classic and XDS models. However, values for these two constants are not normally specified for a transistor, and they are not parameters that are familiar to most engineers. Therefore, CIRC provides for "specification sheet" data to be used to set up the parameters automatically. The specification data used are primarily pulse response test data—IBON, IBOFF, ISAT, TR, TS, and TF—that are explained later.

The most important aspect of this parameter calculation procedure is that it is essentially a curve (model) fit. It is a "fit" procedure in that the response times measured for one condition are used to calculate τ_N and τ_I, so that the model has the same performance as that actually measured. Because only two parameters are being determined, only two of the three pulse response times (TR, TS, and TF) can be made to agree exactly with measured values. The third response time agrees with the measured value to the extent that the model truly represents the device. Usually this is in fairly good agreement. Because the model response has been forced to agree with the actual device response for one pulse test condition, there is good agreement for many variations in pulse conditions, e.g., conditions that are not ideal current drives, waveforms that are not steps, and circuits that are not single stages.

The τ_N and τ_I calculations are based on a transistor model that is simplified by removing leakage currents, resistances, and junction capacitances with the result shown in Fig. 4.30.

$$I_E = I_{EE} - \alpha_I I_{CC} + \tau_N \frac{dI_{EE}}{dt}$$

$$I_C = I_{CC} - \alpha_N I_{EE} = \tau_I \frac{dI_{CC}}{dt}$$

where

BN = normal β

$\alpha_N = \dfrac{BN}{BN+1}$

BI = inverted β

$\alpha_I = \dfrac{BI}{BI+1}$

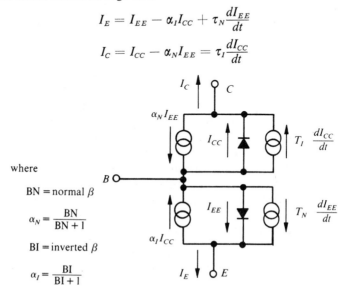

Figure 4.30 Calculation for τ_N and τ_I.

where I_{EE} and I_{CC} are currents determined by solving the appropriate diode equations. The proper solution of these equations will display finite pulse response rise time, TR; fall time, TF; and storage time, TS. These times are dependent on TN and TI.

Actually, the equations for I_E and I_C (see Fig. 4.30) represent a form of the CIRC-TR model that is exactly analogous to the Ebers-Moll model. Therefore, the solutions for τ_N and τ_I are based on the classic Ebers-Moll equations. A discussion of the theoretical basis for τ_N and τ_I calculations as done in CIRC-TR is given a little later. For now we concern ourselves with proper interpretation of CIRC's display of calculated values for (and associated with) these constants and with the proper interpretation of the parameters needed.

The data used to calculate τ_N and τ_I are the transistor current gains, normal and inverted, the pulse response time shown in Fig. 4.31, the transistor base turn-on and turn-off currents, and collector saturation current.

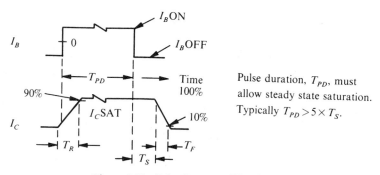

Figure 4.31 Pulse Response Waveforms.

The normal and inverse current gains are to be the dc current gains (if possible at a current level approximately equal to one-half the saturation current). As has been noted, inverse current gain is measured with the collector and emitter roles interchanged. Figure 4.32 shows a circuit that may be used for obtaining the pulse response data and the appropriate equations to get the necessary data from this circuit. The circuit is presented for tutorial purposes and is not sufficiently sophisticated to test high-speed devices. The drive currents are supposed to be ideal step currents and should, therefore, be independent of V_{BE}. Also, I_C should be an ideal current that is independent of V_{CE} SAT. Therefore, the restrictions shown in Fig. 4.32 are necessary. On the other hand, it is desired to use low resistor values to isolate basic semiconductor response characteristics from the time constant effects caused by the presence of junction and parasitic capacitance. A compromise of resistor and voltage source values that will yield good test data can usually be reached.

With this procedure there are two main quantities being determined, TN and TI. They are used by CIRC and are output for the user to see. How-

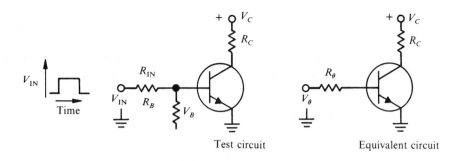

Figure 4.32 A Typical Pulse Response Test Circuit and Appropriate Data Equations.

The appropriate equations, shown below, are obtained by Thevenin's Theorem and Ohm's Law:
$$V_\theta = V_{IN} - (V_{IN} + V_B)R_B/(R_{IN} + R_B) \text{ and } R_\theta = R_{IN}R_B/(R_{IN} + R_B)$$
where it is nessessary that $|V_\theta| \gg V_{BE}\text{SAT}$ and $|V_C| \gg V_{CE}\text{SAT}$, then $I_C\text{SAT} = V_C/R_C$.

For I_BON (Npn)
——————————————
$V_0 \equiv (+)$ and $I_B\text{ON} = V_\theta/R_\theta$

For I_BOFF (Npn)
——————————————
$V_\theta \equiv (-)$ and $I_B\text{OFF} = |V_\theta|/R_\theta$

ever, as mentioned previously, there are three input quantities, one of which is redundant; i.e., rise time or fall time alone is sufficient to determine TN. Therefore, TN is calculated independently from rise time data and fall time data. Then the arithmetic average is taken, and all three values for TN are output. The value for TI is then calculated, using each of the three TN values previously calculated, and all three values for TI are output. This output follows entry of data for each transistor using the specification philosophy. The output is described later.

Because there are three values available for TN and TI, the options discussed earlier allow the user to select those he would like to use. Typical CIRC usage would result in using average values.

Figure 4.29 also shows that the transistor element is not very sensitive to the V_{CE} voltage; it also has no breakdown voltage, little reverse feedback, and little voltage effect on I_{CBO}. Slight voltage effects due to the 1.0kMΩ shunts are included for CIRC mechanization purposes, but these effects are usually negligible.

The transistor element is set up without bulk resistors in base, emitter, or collector leads. Bulk resistances, when known, may be added with normal resistor elements. Since there is no collector bulk resistor, it follows that the saturation conditions are set up with the ideal transistor saturation characteristics. This may result in a different dynamic saturation impedance in the element than is appropriate for certain high-power devices. Therefore, for high-power devices, it may be particularly desirable to use a separate resistor element in the collector path.

Basis for dc Specification Transistor Model

The basic transistor element (excluding I_{CBO}, I_{EBO}, and the 1kMΩ shunts) consists of a CIRC program implementation of the Ebers-Moll transistor model. The theory of the model is not presented here, but the equations inherent in the CIRC implementation are reported. The current directions used in the equations are as seen in the equivalent circuit, Fig. 4.30:

$$I_{EE} = I_{SE}(e^{(V_B-V_E)q/kT} - 1) \tag{4.1}$$

where CIRC calculates

$$I_{SE} = f(V_{BE}, I_{EA}) \tag{4.2}$$

$$I_{CC} = I_{SC}(e^{(V_B-V_C)q/kT} - 1) \tag{4.3}$$

where CIRC calculates

$$I_{SC} = f(I_{SE}, I_{BS}, I_{CS}, V_{SAT}) \tag{4.4}$$

$$I_E = I_{EE} - \alpha_I I_{CC} + \frac{V_B - V_E}{10^9} \tag{4.5}$$

$$I_C = -[I_{CC} - \alpha_N I_{EE}] + I_{CBO} + \frac{V_C - V_B}{10^9} \tag{4.6}$$

$$I_B = I_E - I_C \tag{4.7}$$

I_{SE} is found as

$$I_{SE} = \frac{I_{EA}}{e^{V_{BE}q/kT} - 1} \tag{4.8}$$

where I_{EA} and V_{BE} are the input data point taken from an active operating condition.

I_{SC} is found as follows. The shunt impedances and I_{CBO} are ignored and I_C of the model is considered to flow in the opposite direction from that shown in the equivalent circuit. Equations (4.5) and (4.6) then become the basic Ebers-Moll equations:

$$I_E^M = I_{EE} - \alpha_I I_{CC} \tag{4.9}$$

$$I_C^M = I_{CC} - \alpha_N I_{EE} \tag{4.10}$$

where

M indicates model terminal current.

α_N and α_I are known from input beta data.

I_{SE} has been determined by Eq. (4.8).

The input currents at saturation are used as follows:

$$I_E^M = I_{CS} + I_{BS} \tag{4.11}$$

and

$$I_C^M = -I_{CS} \tag{4.12}$$

Using Eqs. (4.11) and (4.12), Eqs. (4.9) and (4.10) may be solved as

$$I_{EE} = \frac{\alpha_I I_C^M + I_E^M}{1 - \alpha_N \alpha_I} \tag{4.13}$$

and

$$I_{CC} = \frac{\alpha_N I_E^M + I_C^M}{1 - \alpha_N \alpha_I} \tag{4.14}$$

From Eq. (4.1) we may determine

$$V_{BE} = \frac{kT}{q} \log_e \left[\frac{I_{EE}}{I_{SE}} + 1 \right] \qquad (4.15)$$

The terminal voltage relationships determine

$$V_{BC} = V_{BE} - V_{CE} \qquad (4.16)$$

where

$$V_{CE} = V_{SAT} \qquad \text{(the input parameter)}$$

Finally, then, I_{SC} is determined as

$$I_{SC} = \frac{I_{CC}}{e^{V_{BC}q/kT} - 1} \qquad (4.17)$$

Basis for TR Specification Model

The procedure for τ_N and τ_I Calculations is based on the performance equations developed by Moll [8] for relating "step" current drives to a transistor's pulse response times (i.e., rise time, storage time, and fall time). Moll's equations are in terms of circuit current levels, device gains, and alpha cutoff frequencies. However, it is desired to make measurements of device current gains, circuit current levels, and pulse response times and to use these data to determine the cutoff frequencies. Therefore, new equations are derived from those presented by Moll in order to obtain the cutoff frequencies. To convert the alpha cutoff frequencies to the necessary data for CIRC-TR, they are simply inverted (i.e., $\tau_N = 1/\omega_N$ and $\tau_I = 1/\omega_I$).

The equations for the pulse response of a common emitter amplifier operating from cutoff to saturation were developed by Moll and are presented below. The terms used in the equations that follow are defined as:

$\omega_N =$ alpha cutoff frequency in normal operation—for CIRC-TR, $\tau_N = 1/\omega_N$

$\omega_I =$ alpha cutoff frequency in inverted operation—for CIRC-TR, $\tau_I = 1/\omega_I$

$\alpha_N =$ common base current gain in normal operation

$\alpha_I =$ common base current gain in inverted operation

$I_{B1} =$ current* in base in forward direction; CIRC symbol: IBON.

$I_{B2} =$ Current* in base (reverse direction) when the transistor is first biased from the saturation condition to the cutoff condition (opposite direction from Moll's current convention); CIRC symbol: IBOFF

$I_C =$ Collector current* in saturation; CIRC symbol: JSAT.

$$\omega_B = \frac{\omega_N \omega_I (1 - \alpha_N \alpha_I)}{\omega_N + \omega_I} \qquad \text{(by definition)} \qquad (4.18)$$

*Currents I_{B1}, I_{B2}, and I_C are positive numbers for both *PNP* and *NPN* transistors.

$$\omega_A = \omega_N + \omega_I - \omega_B \quad \text{(by definition)} \tag{4.19}$$

T_R = rise time = time for collector current to rise from 0 to 90% of its saturation value (Moll's symbol is T_0)

T_S = storage time = time from point where base current is reversed (to I_{B2}) until the collector current just starts to drop from its saturation value (Moll's symbol is T_1)

T_F = fall time = time for collector current to go from just below its saturation value to 10% of its cutoff value (Moll's symbol is T_2)

The three basic equations are shown below:

$$T_R = \frac{1}{\omega_N(1 - \alpha_N)} \ln\left[\frac{I_{B1}}{I_{B1} - 0.9[(1 - \alpha_N)/\alpha_N] \cdot I_C}\right] \tag{4.20}$$

$$T_S = \frac{1}{\omega_B} \ln\left[\frac{I_{B2} + I_{B1}}{I_{B2} + I_C[(1 - \alpha_N)/\alpha_N]} \cdot \frac{1}{1 - (\omega_B/\omega_A)}\right] \tag{4.21}$$

The term

$$\frac{1}{1 - \omega_B/\omega_A} \tag{4.22}$$

does not appear in Moll's paper because it was assumed that $\omega_B \ll \omega_A$. It does appear elsewhere [9], and experimentation shows it to be useful in some applications.

$$T_F = \frac{1}{\omega_N(1 - \alpha_N)} \ln\left[\frac{I_C + [\alpha_N/(1 - \alpha_N)]I_{B2}}{0.1I_C + [\alpha_N/(1 - \alpha_N)]I_{B2}}\right] \tag{4.23}$$

Equations (4.20) or (4.23) can be easily solved for ω_N. Using Eq. (4.20), we get

$$\omega_N = \frac{1}{(1 - \alpha_N)T_R} \ln\left[\frac{I_{B1}\alpha_N}{I_{B1} \cdot \alpha_N - 0.9(1 - \alpha_N)I_C}\right] \tag{4.24}$$

Using Eq. (4.23) we get

$$\omega_N(\text{fall}) = \frac{1}{(1 - \alpha_N)T_F} \ln\left[\frac{I_C(1 - \alpha_N) + \alpha_N I_{B2}}{0.1I_C(1 - \alpha_N) + \alpha_N I_{B2}}\right] \tag{4.25}$$

It still remains to find ω_I. To do this, we start with Eq. (4.21) where we replace the ω_B term [Eq. (4.18)] that stands outside the natural log term with its full definition:

$$T_S = \frac{\omega_N + \omega_I}{\omega_N \omega_I (1 - \alpha_N \alpha_I)} \cdot \ln\left[\frac{I_{B2} + I_{B1}}{I_{B2} + I_C[(1 - \alpha_N)/\alpha_N]} \cdot \frac{1}{1 - (\omega_B/\omega_A)}\right] \tag{4.26}$$

$$A = 1 - \alpha_N \alpha_I$$

and

B = the entire natural log term in Eq. (4.26)

Then

$$T_S \omega_N \omega_I A = (\omega_N + \omega_I)B \tag{4.27}$$

and
$$\omega_I(T_s\omega_N A - B) = \omega_N B \tag{4.28}$$
and
$$\omega_I = \frac{\omega_N B}{T_s\omega_N A - B} \tag{4.29}$$

It should be noted that term B contains ω_A [Eq. (4.19)] and ω_B [Eq. (4.18)], both of which contain ω_I. Therefore, Eq. (4.29) is not an exact solution of Eq. (4.26) for ω_I. However, the effect of ω_A and ω_B is quite small, as noted before, and, therefore, Eq. (4.29) is solved ignoring these terms. Then ω_I obtained from Eq. (4.29) can be improved by substituting the first estimate of ω_I into the ω_A and ω_B equations. This process normally will quickly iterate to a sufficiently accurate result. This iteration process is carried out in CIRC-TR. Should it fail to converge, it will be reported, as discussed later.

If the model were to apply perfectly to the transistor tested and the data were taken accurately, then ω_N obtained from Eqs. (4.24) and (4.25) would be the same. In practice they will not be exactly the same. Therefore, ω_I [Eq. (4.29)] may be solved for using ω_N obtained by either Eq. (4.24) or Eq. (4.25) or from an average of the two different values. In general, the value of ω_I, obtained from Eq. (4.29); ω_N (rise), obtained from Eq. (4.24); and ω_N (fall) obtained from Eq. (4.25) can be selected depending on whether storage and rise times or storage and fall times are more important—or an average of values may be selected.

In concluding the mathematical discussion, two checks on the method are presented. The first check is to calculate the term ω_B/ω_A. For Eq. (4.21) to be exact, independent of the iteration performed, ω_B/ω_A should be zero. In general this will not be the case, and, therefore, CIRC-TR prints out this ω_B/ω_A quantity. The second check is to calculate the term

$$(\omega_I/\omega_N)\cdot(1 - \alpha_N\alpha_I) \tag{4.30}$$

It is indicated on page 1784 of Reference 8 that Eq. (4.30) is to be less than one-tenth in order to obtain good results with Moll's storage equation [Eq. (4.21)]. Because Eq. (4.21) contains the correction term $1/(1 - \omega_A/\omega_B)$, it is expected that the need for Eq. (4.30) to be less than one-tenth is not as important as indicated by Moll. Because Eq. (4.30) is important in Moll's storage equation, it may be printed out as a check when using CIRC-TR.

CIRC-TR Diode and Transistor Time Domain Methods

CIRC-TR's diode and transistor models have two fundamental time domain modeling needs:

$$C_J\frac{dv}{dt} \quad \text{and} \quad \tau\frac{di}{dt}$$

C_J is a transistion layer capacitance, and the current derivative is of a diffusion junction current.

The difference equations used to provide implicit integration for both terms is a first-order difference formula. First-order methods are used because they have proved themselves in giving efficient, accurate time domain analyses. Also, the second-order methods described for inductors and capacitors will oscillate when directly applied to diodes and transistors where the voltages and currents go through very rapid nonlinear changes.

The difference equation used for the capacitance is

$$C_j \frac{dv}{dt} = C_j \frac{v(t_n) - v(t_{n-1})}{\Delta t_n}$$

where the time-related subscript notation is as defined as

$\Delta t_n =$ time step to be taken

$t_n =$ time to be reached by the solution to be performed

$t_{n-1} =$ time back one Δt_n

The following sketch illustrates the symbols' meaning:

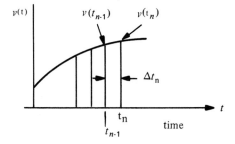

The difference equation for the current is of the same general form:

$$\tau \frac{di}{dt} = \tau \frac{i(t_n) - i(t_{n-1})}{\Delta t_n}$$

However, current is not the unknown for CIRC-TR, since node equations are used. Consequently, formulation involving voltage is needed. The formulation must also resolve the basic nonlinear aspect of the diode equation with a diode linearization based on a tangent line and an intercept voltage, as shown below:

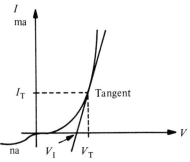

Diode characteristic curve

The current equation for such a linearized model is

$$I = \frac{v - v_I}{R_T}$$

However, the quantity v_I is not directly obtained from the node equations and is not directly useful; therefore, CIRC-TR defines

$$\left.\begin{aligned} I &= \frac{v - v_T}{R_T} = I_T \\ &= (v - v_T)G_T + I_T \end{aligned}\right\} \quad \text{where } I_T = \frac{v_T - v_I}{R_T}$$

As each solution for a time point is obtained, CIRC-TR iteratively updates v_T, G_T, and I_T. Their values at an accepted time point t_n are used to determine

$$i(t_{n-1}) = [v(t_n) - v_T(t_{n-1})]G_T(t_n) + I_T(t_n)$$

where this evaluation must be made with values at t_n but before the time step to the next point is begun.

Also,

$$i(t_n) = [v(t_n) - v_T]G_T + I_T$$

where v_T, G_T, and I_T are iteratively updated until the solution for a specific time point is obtained.

These formulations for $i(t_{n-1})$ and $i(t_n)$ are then used in

$$\tau \frac{i(t_n) - i(t_{n-1})}{\Delta t_n}$$

to obtain the specific difference formula for

$$\tau \frac{di}{dt}$$

Special Comments About the XDS Models

The XDS diode model provides a two-terminal diode model containing recombination effects in the junction so that the same model approach allowed by the XDS transistor may be used for a diode. Some details on the recombination effects are provided in the transistor discussion below.

The specification, classic, and XDS transistor models all have the same basic time-dependence modeling except the transient response may be affected by the different nonlinear dc performance of the XDS model. The specification and classic transistor models have the same basic dc modeling, but the XDS model extends the dc modeling.

The basic dc equations of CIRC's transistor models were shown in the philosophy of the dc specification model presented earlier. The same model that is set up by CIRC from specification data is directly set up from the user's data in the case of the classic model (with a minor additional data entry added—the function slope factor parameters ME and MC).

The XDS transistor model adds minority carrier recombination junctions across both the base-emitter and base-collector. The use of recombination junctions to model nonlinear beta was suggested by Autonetics in Reference 10. However, the specific form used in the XDS models, including a separate equation for high current roll-off, is different and was derived by Dr. W. E. Drobish.

The recombination junctions have four basic effects:

1. dc voltage drops for given current level are affected since the current now flows in two junctions (it will be helpful to refer to the equivalent circuit in Fig. 4.19).

2. Since the recombination junctions have a slope factor of 2 and the diffusion diodes have a slope factor of 1, the result of the parallel junctions is an intermediate slope factor.

3. The ideal saturation current of both the recombination junction and the diffusion junction combine to give larger leakage currents. Also, temperature effects can be handled differently for diffusion junctions than for recombination junctions, thereby giving superior modeling of temperature effects.

4. Since the current flow carried by the recombinations junctions is not included in the alpha-controlled current sources, the effective α current gain is changed. As α (common base) current gain is changed, so is beta (common emitter) current gain. The familiar result is that nonlinear beta current gain increases as collector current increases.

However, the beta current gain increase eventually gives way to a roll-off and decrease. The roll-off is handled with the following equations:

$$\alpha_N = \frac{\alpha_{NO}}{1 + KN \cdot I_{EE}^2}$$

$$\alpha_I = \frac{\alpha_{IO}}{1 + KI \cdot I_{CC}^2}$$

$$\alpha_{NO} = \frac{BN}{BN + 1}$$

$$\alpha_{IO} = \frac{BI}{BI + 1}$$

where KN, KI, BN, and BI are input parameters and I_{EE} and I_{CC} are currents defined in Fig. 4.19.

α_N and α_I in Fig. 4.19 follow the equations shown above.

The XDS transistor model has a "hidden" parameter, BNC, which shows that the effective normal beta (see the earlier parameter discussion) is determined by the equation

$$BNC = \frac{\alpha_N I_{EE}}{I_{ER} + I_{EE} - \alpha_N I_{EE}}$$

where the currents are as shown in Fig. 4.19. This equation serves as a good summary on the nonlinear beta. We relate it to Fig. 4.19 by assuming an active device, which means that I_{CR}, I_{CC}, and $\alpha_I I_{CC}$ are effectively zero, and by ignoring all time-dependent currents. The equation shows how the base-emitter recombination junction current I_{ER} affects the current gain. It also shows that if I_{ER} were zero, this would result in the familiar equation

$$\text{BNC} = \frac{\alpha_N}{1 - \alpha_N}$$

It should also be noted that α_N in this BNC equation is the nonlinear, roll-off form shown above.

Warning and Information Messages for CIRC-TR Diodes and Transistors

Careful consideration of the data that define the saturation performance of the specification transistor element shows two facts that require special discussion:

1. The I_{CB}/I_{BS} ratio implies a minimum device gain, normal beta.
2. The voltage difference $V_{BE} - V_{SAT}$ should be equal to V_{BC} for a forward-biased base-collector junction.

If the ratio of I_{CS}/I_{BS} is larger than the value of normal beta (β_N), improper data have been specified. To avoid problems of exact equality, the actual test is $I_{CS}/I_{BS} > 0.96\beta_N$. CIRC automatically notes this fact with the message

$$\boxed{\text{SAT ERROR } \#1 \text{ IB--S } \#2}$$

The $\#1$ identifies the transistor that has the data error, and the $\#2$ shows the value calculated for I_{BS} to allow the solution to be completed. The value of I_{BS} is determined as

$$I_{BS} = 1.05\left(\frac{I_{CS}}{\beta_N}\right)$$

This change in device performance data is minor. Usually, no attempt to correct the data and perform a new analysis is necessary.

If the voltage difference $V_{BE} - V_{SAT}$ is too small or is negative, improper data have been specified or an attempt is being made to accommodate bulk resistance voltage drop in the collector through an improper mechanism, i.e., the ideal transistor without collector bulk resistance. CIRC automatically checks for this error condition, calculates V_{BE} to agree with the actual specified saturation operating point, and uses this value when calculating $V_{BC} = V_{BE} - V_{SAT}$. CIRC requires that $V_{BC} \geq 0.1$ volt. If this requirement is not met, CIRC notes this error with the message

$$\boxed{\text{SAT ERROR } \#1 \text{ VSAT } \#2}$$

The $\#1$ identifies the transistor that has the data error, and the $\#2$ shows the value calculated for V_{SAT} to allow the solution to be completed. The value of V_{SAT} is determined as

$$V_{SAT} = V_{BE} - 0.1$$

This change in transistor data may be significant if the transistor is saturated. At times it may be desirable to correct the data and repeat a specific analysis.

The quantity V_{BE} directly controls the ideal diode saturation (leakage) current of the base-collector junction. If large collector current is required at a low V_{BE} voltage, a high saturation (leakage) current results. Even the limit value of 0.1 volt for V_{BC} may allow this condition.

A high saturation current, such as just discussed, would be reported by a separate error message. In fact, there are error messages that advise the CIRC user of both extremely large and extremely small diode and transistor saturation currents that may result from certain data inputs. Specifically, CIRC-TR now provides a warning message whenever a device saturation current exceeds the range of

$$10^{-20} \leq I_S \leq 5 \times 10^{-6} \text{ amperes}$$

The magnitude of the current and the device type and number are also printed out. The analysis is not aborted.

Before seeing examples of the WARNING messages, let us briefly consider the motivation for the messages.

When the user, either unintentionally or intentionally, specifies a very large junction voltage (e.g., greater than 1.2 volts), I_S tends to be very small, at times so small that analysis accuracy is adversely affected. Three typical reasons for using very large junction voltage input values are

1. The data include a voltage drop across bulk resistance in the device.
2. The data represent several devices in series.
3. The data represent a zener diode.

If a WARNING is related to reason 1, the user should consider putting an external resistor in series with the junction and reducing the voltage value for the junction.

If a WARNING is related to reason 2, the user is attempting a shortcut that cannot be done without slope factors. Each specification diode should be separately represented or a classic diode can be used with slope factors. For example, three diodes in series can be modeled by one classic diode with a slope factor of 3.

If a WARNING is related to reason 3, the user is attempting an unadvisable shortcut; the zener should be modeled with a VRS voltage element in series with the diode.

Likewise, when V_D tends to be small, I_S becomes large; thus, under reverse bias conditions, unrealistically large leakage currents may appear.

The most common reason for CIRC to see a very small voltage drop is that V_{SAT} has been input as a large value. This type of transistor data set was discussed earlier in this section.

Example warning messages are

> WARNING
> IS OF DIO 1 IS .16071304E−35

> WARNING
> ISE OF TRN 1 IS .16071304E−36

> WARNING
> ISC OF TRN 1 IS .42063649E−36

> WARNING
> ISC OF TRN 1 IS .19649251E−3

CIRC-TR also has two messages associated with its SPEC treatment of transient parameter determination. The first of these is an information message with the typical appearance shown:

Tran. No. = 1	Rise	Fall	Ave.
TN	.19761E−08	.24130E−08	.21728E−08
TI	.26034E−07	.25186E−07	.25651E−07
WB/WA	.34585E−01	.42400E−01	.38100E−01
(WI/WN)∗	.38696E−01	.48842E−01	.43183E−01
(1−AN∗AI)			

∗∗∗AVE. VALUES USED∗∗∗

The seond message associated with SPEC treatment of transient transistor performance is

```
****ATTEMPT TO CONVERGE TN(M) AND TI(M)
FAILED****
****SEE MANUAL DISCUSSION OF WI,WA,AND
WB****
```

where M is 1 for rise, 2 for fall, 3 for ave.

The message is immediately followed by abortion of the run; however, the message is very rare. If it should occur, input transient data for the SPEC transistor should be checked.

CIRC-TR L and C Models

The purpose of a transient circuit analysis program is to determine the time domain response of a circuit to one or more stimulus changes with respect to time. If the circuit is composed of lumped elements (as opposed to distributed elements), then the equations describing the circuit are ordinary differential (or integro-differential) equations. If all elements are linear, then a Laplace approach to the analysis is possible. However, if some of the elements are nonlinear as is the case in CIRC-TR, then numerical methods for the equation solution are normally required.

The selection of a specific numerical technique depends on many factors. There should be good engineering accuracy but with reasonable computer time requirements even for circuits with widely separated time constants (sometimes called the "stiff time constant problem"). The technique should provide for automatic control of time step size. The inductor and capacitor models' numerical techniques should work well with the diode and transistor models' numerical methods and should fit effectively into the network formulation problem. Particularly, the resultant network formulation should be capable of handling large circuits in a reasonably sized main memory (without prohibitively expensive dependence on low-speed storage in the frequently-repeated problem-solving operations). The difference-equation approach of implicit integration pioneered and refined for circuit analysis programs at Autonetics [10, 11] meets all these objectives and, therefore, serves as CIRC-TR's numerical method base for time domain modeling.

Consider the specific integro-differential equations for inductors and capacitors. The inductor model is

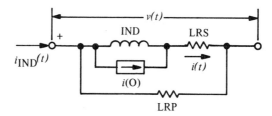

where

$$I_{\text{IND}}(t) = i(t) + \frac{v(t)}{\text{LRP}}$$

$$i(t) = \frac{1}{\text{IND}} \int_0^t [v(t) - i(t)\text{LRS}]\, dt + i(0)$$

To define the difference equation that represents this model, we introduce the following symbols:

Δt_n = time step to be taken

t_n = time to be reached by the solution to be performed

t_{n-1} = time back one Δt_n

The following sketch illustrates the symbols' meaning:

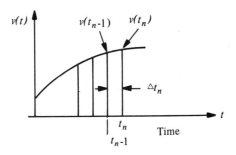

By appropriate application of the trapezoidal integration rule [10], a second-order difference equation is obtained. Its form is

$$i_{\text{IND}}(t_n) = \left\{\frac{1}{2\cdot\text{IND}/\Delta t_n + \text{LRS}} + \frac{1}{\text{LRP}}\right\} v(t_n)$$

$$\cdot \left\{+\frac{1}{2\cdot\text{IND}/\Delta t_n + \text{LRS}}\right\} v(t_{n-1})$$

$$+ \left\{\frac{2\cdot\text{IND}/\Delta t_n - \text{LRS}}{2\cdot\text{IND}/\Delta t_n + \text{LRS}}\right\} i(t_{n-1})$$

The capacitor model is

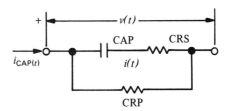

where

$$i_{\text{CAP}}(t) = i(t) + \frac{v(t)}{\text{CRP}}$$

$$i(t) = \text{CAP}\left[\frac{[dv(t) - i_n(t)\text{CRS}]}{dt}\right]$$

By appropriate trapezoidal integration of both sides of the above equation [10], a second-order difference equation is obtained. Its form is (using symbols previously defined)

$$i_{CAP}(t_n) = \left\{\frac{1}{\Delta t_n/(2 \cdot CAP) + CRS} + \frac{1}{CRP}\right\} v(t_n)$$
$$- \left\{\frac{1}{\Delta t_n/(2 \cdot CAP) + CRS}\right\} v(t_{n-1})$$
$$- \left\{\frac{\Delta t_n/(2 \cdot CAP) - CRS}{\Delta t_n/(2 \cdot CAP) + CRS}\right\} i(t_{n-1})$$

CIRC-AC Models

Special discussion of the equations implemented and electrical performance is presented for three CIRC-AC models: the inductor, transistor, and mutual inductance.

CIRC-AC Inductor

The basic inductor model implements Ohm's law for alternating current; i.e.,

$$V = IZ = Ij\omega L = Ij2\pi fL$$

In this equation, L is the inductance and f is the frequency at which the analysis is being performed.

Many applications of inductors have a practical equivalent circuit that also includes a Q—not just pure inductance. This is particularly true of inductor applications in the communications industry. Therefore, CIRC allows a Q value to be specified and then calculates a parallel resistance as

$$r = Q \times 2\pi fL$$

The significance of the equation is best seen in a sketch of resistance versus frequency:

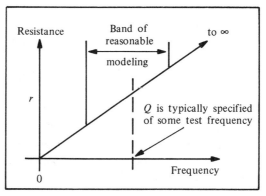

Resistance vs. frequency

Since r varies as a function of frequency, it is desirable to have the program handle r automatically. Furthermore, the CIRC Q model creates a natural band of reasonable operation for the user. Several program output messages related to the Q modeling are discussed later. The reason the range for the Q model is restricted is that r becomes unrealistically small as frequency becomes very low, and at zero frequency r becomes zero, which is not allowed in CIRC.

Similarly, at high frequency the model would show r approaching infinity (which is unrealistic but not a problem to CIRC). The actual Q of a hardware component could be modeled more perfectly with an elaborate structure having both series and parallel resistances that are mathematically related to a test frequency as well as operating frequency. However, such modeling is more complicated than required by most CIRC users and is, therefore, not implemented. Special equations can be written and coded to use such a model. Special coding can supplement any of CIRC programs and is documented in References 1–3.

CIRC-AC Transistor Model

For use in computer-aided design, a transistor model should ideally meet three objectives:

1. The parameters to be supplied should be available on published specifications or capable of being readily estimated.

2. The model should be valid to high frequencies such as may normally be encountered, even though it may be difficult to solve by manual computation standards.

3. The model should function without regard to the circuit configuration—common base, common collector, or common emitter—in which the transistor is placed.

The CIRC-AC transistor model meets these objectives. The bulk base resistance r_b is the only important parameter that is often unspecified, and it may be estimated fairly easily, for example, from the h parameters. The excess-phase parameters become significant only at high frequencies and usually can be approximated well enough, if not available in the specifications.

This model is usable to high frequencies because emitter and collector capacitances are included and because the alpha current gain employed has been modeled to be sensitive to frequency. In fact, by the use of a two-pole alpha expression the calculated value of alpha comes very close to the actual value of alpha over the entire range of frequencies at which the device would be used.

Transistor Model Performance

Before the equations for the transistor model are presented, it will be informative to observe the performance of the model itself. A one-transistor common emitter configuration is analyzed. The transistor has the collector capacitance C_c and the emitter transition capacitance C_{TE} set to 0 and the alpha cutoff frequency set at 1000 MHz. With the excess phase of the transistor assigned values of 0 and 12 degrees, a series of nominal solutions is obtained for a frequency range of 1 to 10,000 MHz. For these solutions, common base current gain ($\alpha = I_C/I_E$) is calculated, and both magnitude and phase are plotted in Fig. 4.33. Note that the phase shift with excess phase included in the model becomes asymptotic to $-180°$ beyond 10,000 MHz, indicative of a two-pole model.

The addition of excess phase to the transistor model also has an effect on the common emitter current gain, beta. This effect ($\beta = I_C/I_B$) is plotted in Fig. 4.33(b). Without excess phase the beta cutoff frequency [calculated from the alpha cutoff frequency (calculated from the alpha cutoff frequency by the program)] is 10^7 Hz. Beta gain is down to 70.8% of its maximum value and phase shift at this point is $-45°$.

With the addition of excess phase to the model, beta cutoff frequency (calculated by means of an equation given later) is shifted to a lower frequency: 8.3 MHz. At this frequency beta gain is down to 70.7% and phase shift is $-45°$. There is a phase difference of about 5.5° at this point between the two phase-shift curves, and this difference increases to 10° at the alpha cutoff frequency.

With this representation of the performance of the transistor model, the discussion of the model equations will be more meaningful.

Three parameters (α, C_d, and r_e) are calculated by CIRC-AC from the transistor element input data. The equations for the calculation of these three parameters are presented briefly here. References to supporting theoretical documentation permit the reader who is interested to study the CIRC-AC transistor model more fully.

Alpha Equation

The alpha parameter is calculated from a two-pole equation, permitting accurate representation of the semiconductor phenomena:

$$\alpha = \frac{\alpha_0}{(1 + s\tau_{11})(1 + s\tau_{21})} \quad * \tag{4.31}$$

---

*Appears as Eq. 12-13, page 369, in D. K. Lynn, C. S. Meyer, and D. J. Hamilton, *Analysis and Design of Integrated Circuits*, McGraw-Hill, New York, 1967. Further references to this work will be identified simply by Lynn and the equation number.

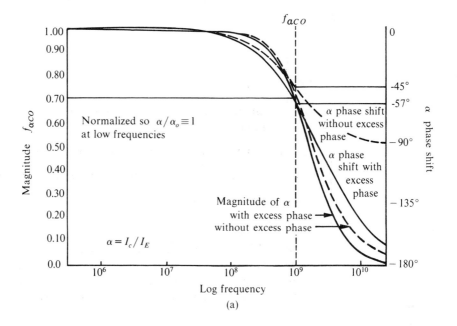

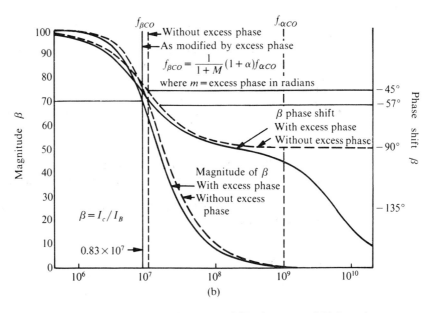

Figure 4.33 Transistor Element Model Performance of Alpha and Beta CIRC-AC: (a) Magnitude and Phase Shift of Alpha With and Without Excess Phase; (b) Magnitude and Phase Shift of Beta With and Without Excess Phase.

where

$$\alpha_0 = \frac{\beta_{LF}}{1 + \beta_{LF}} \tag{4.32}$$

β_{LF} = common-emitter current gain at low frequency—one of the input parameters

τ_{11} = predominant time constant (to be calculated by the program)

τ_{21} = secondary time constant (to be calculated by the program)

s = complex frequency variable; $s = j\omega$ in CIRC-AC

ω = radian frequency

For the CIRC-AC model, the time constants τ_{11} and τ_{21} are calculated to implement alpha with both a one-pole roll-off and excess-phase characteristics. For the one-pole roll-off (where the frequency is less than the magnitude of the first pole), alpha can be approximated by a single-pole expression:

$$\alpha = \frac{\alpha_0}{1 + j\omega/\omega_\alpha} \quad \text{(Lynn, Eq. 4-113)} \tag{4.33}$$

where

$$\ddot{\omega}_\alpha = 2\pi f_{\alpha co} \quad (f_{\alpha co} \text{ is one of the alternative input parameters}) \tag{4.34}$$

The excess-base characteristics can be modeled reasonably well up to frequencies around $f_{\alpha co}$ by adding an expression for excess phase to the one-pole equation:

$$\alpha = \frac{\alpha_0}{1 + j\omega/\omega_\alpha} \cdot e^{-jm\omega/\omega_\alpha} \quad \text{(Lynn, Eq. 4-115)} \tag{4.35}$$

where

m = the excess phase in radians measured at ω_α (excess phase EPH, measured in degrees, is one of the input parameters)

For the alternative excess-phase parameters, scale factor K_θ, m is related to K_θ as follows:

$$-m = \frac{K_\theta - 1}{K_\theta} \tag{4.36}$$

$$K_\theta = \frac{1}{1 + m} \tag{4.37}$$

A typical value of m is 0.20943951 radian, corresponding to 12° of excess phase. Corresponding to this m is $K_\theta = 0.8268294$.

The frequency parameter ω_α is defined in Eq. (4.34) in terms of $f_{\alpha co}$, an alternative input parameter. This frequency parameter is also calculated from other input parameters by the following relations [where α and β are *low-frequency* parameters related by Eq. (4.32)].*

$$f_\alpha = f_{\alpha co} \tag{4.38}$$

$$f_\tau = \beta f_{\beta co} \quad \text{(Phillips, Eq. 14–27)} \tag{4.39}$$

$$f_{\alpha co} = (\beta + 1)\frac{f_{\beta co}}{K_\theta} \tag{4.40}$$

$$= (\beta + 1)(m + 1)f_{\beta co} \tag{4.41}$$

$$= \frac{(1 + m)f_\tau}{\alpha} \tag{4.42}$$

Equations (4.40), (4.41), and (4.42) are based on Eqs. (4.32), (4.36), (4.37), and (4.39) and upon a relationship developed by Phillips:

$$f_{\beta co} = K_\theta(1 - \alpha)f_\alpha \quad \text{(Phillips, Eq. 14–23)} \tag{4.43}$$

All the parameters contained in Eq. (4.35) have now been discussed. It remains to define the time constants τ_{11} and τ_{21} of Eq. (4.31) in relation to Eq. (4.35).

Treating the two-pole model, Eq. (4.31), as a one-pole expression plus excess phase, Eq. (4.35) implies that

$$\tau_{11} = \frac{1}{\omega_\alpha} \quad \text{(Lynn, Eq. 4-116)} \tag{4.44}$$

$$\frac{1}{1 + s\tau_{21}} \longrightarrow e^{-jm\omega/\omega_\alpha} \quad \text{(Lynn, Eq. 4-117)} \tag{4.45}$$

This relationship involving τ_{21} [Eq. (4.45)] can be reduced to the expression

$$\tau_{21} = \frac{\tan(m)}{\omega_\alpha} \tag{4.46}$$

To summarize, the alpha used in the CIRC-AC transistor model is a complex, frequency-dependent parameter calculated from Eq. (4.31), which, in turn, is related to Eq. (4.35), using a dominant time constant τ_{11} based on the alpha cutoff frequency [Eq. (4.44)] and a secondary time constant τ_{21} embodying the influence of excess phase [Eq. (4.46)].

A two-pole model [Eq. (4.31)] using τ_{21} is used rather than a simple excess-phase term [as in Eq. (4.35)] so that the α will be correctly modeled at frequencies well beyond $f_{\alpha co}$.

*Alvin B. Phillips, *Transistor Engineering*, McGraw-Hill, New York, 1962, Eqs. 14–18, p. 301.

Emitter Resistance

The emitter resistance r_e is calculated for the model by the relationship

$$r_e = \frac{kT}{qI_E} \quad \text{(Phillips, Eq. 14-5)}$$

where

k = Boltzmann's constant

q = electron charge

T = temperature in degrees Kelvin (an input parameter)

I_E = dc emitter current (may be obtained from CIRC-DC) (an input parameter)

Diffusion Capacitance

The diffusion capacitance C_d is calculated for the model by the relationship

$$C_d = \frac{1}{r_e \omega_\alpha} \quad \text{(Lynn, Eq. 12-16)}$$

Mutual Inductance Model

The mutual inductance model is linear and is not temperature-dependent. The model makes use of Ohm's law for alternating current, i.e., $V_{XM} = j\omega I_Y M_{XY}$, where V_{XM} is the voltage induced in inductor X by a current in inductor Y when there is mutual coupling, M_{XY}, between the two inductors. The above equation is a reasonably complete picture of mutual inductance for a general model for loop equations.

For CIRC-AC, however, the fact that the mutual inductance is to be modeled in a general way for node equations introduces a great deal of complexity. This is so particularly with three or more inductors in a mutually coupled group. Because of this complexity and because the details of the mutual inductance model are of limited interest to the average CIRC user, they are not presented here.

Following is an overview of the MUT element model for node equations. The current contributions due to mutual inductance are generally of the form

$$I_{MY} = A_{XY} \cdot \frac{V_X}{j2\pi f}, \qquad A_{XY} \neq \frac{1}{M_{XY}}$$

where A_{XY} represents an inverse mutual inductance coefficient evaluated from an inverse matrix calculated by CIRC-AC.* The coefficient is a function of magnetic mutual coupling between inductor X and inductor Y and self-inductances of X and Y. V_X represents the voltage across inductor X. Thus, I_{MY} is a current contribution to inductor Y as a result of the voltage across inductor X. In general there will be one such contribution *from each inductor* that is magnetically coupled with another inductor. Just as inverse mutual inductances are needed for coupled current contributions, inverse self-inductances (which differ from the self-inductances themselves) are also needed and used in the actual node equations. This may be represented as

$$I_Y = A_{YY} \cdot \frac{V_Y}{j2\pi f}, \qquad A_{YY} \neq \frac{I}{L_Y}$$

where A_{YY} is inverse self-inductance. The total current in an inductor is then the result of all coupled contributions and a "self"-contribution based on inverse self-inductance.

Information and Error Messages for CIRC-AC Inductors and Mutuals

After completing acceptance of a circuit description, CIRC-AC puts out information for tuned inductors and inductors with a positive value for Q. The messages are

```
TUNED IND n = value
```

```
R(QL) FREQ = value   FREQ = value
    n        value          value
```

The first message identifies the inductor with a number, n, and shows the value obtained from the solution of the equation discussed in the ac inductor element presentation.

The second message identifies two frequencies at which r is calculated. The first frequency is the START frequency, and the second the END frequency. For both of these frequencies, it shows an inductor number, n, and the two corresponding r values.

If at any time during an analysis r, as calculated, is less than 100, CIRC-AC puts out the message

```
**WARNING – R FOR QL n = value**
```

*Details are not presented here. See M. F. Gardner and J. L. Barnes, *Transients in Linear Systems*, Wiley, New York, 1942, for the principles on which the calculations are based.

For each MUTUAL inductance, CIRC checks the coupling coefficient K_{XY} and, when appropriate, reports

> **WARNING: ABS VAL OF COUPLING COEFF. NOT L. T. I. — RESET TO .9999**

One other message may appear when mutual inductances are used:

> **ERROR — MUTUAL GROUP # PRODUCES A SINGULAR MATRIX**

In this case CIRC cannot perform the circuit analysis. The user should check the validity of his mutual inductance and self-inductance data.

CIRC-AC CYS and VZS Source Uses

Although CIRC-AC has a strong equivalent circuit modeling capability in its CYS and VZS models, its high-performance transistor model makes it unnecessary for the average CIRC-AC user to create equivalent circuits.

The detailed discussion of the CYS and VZS models is, therefore, not presented in this book. However, the CIRC-AC manual [2] has a chapter with a tutorial example on CYS and VZS source use. It also contains a discussion of

1. Transistor models.
    a. h Parameter.
    b. y Parameter.
    c. Hybrid π.

2. A junction FET.

3. A voltage-sensing circuit.

4. An ideal transformer.

5. Amplifiers.

4.5 EXAMPLES

The application of the CIRC program is demonstrated with two examples. The first example demonstrates the DC program and the worst-case analysis feature. The second example demonstrates the AC program and the open-loop analysis feature.

Worst-Case Analysis

A worst-case analysis is one that is based on the worst combination of minimum and maximum parameter values. The circuit to be analyzed is shown in Fig. 4.34. The question and answer dialog to describe the circuit

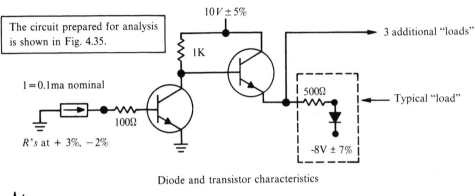

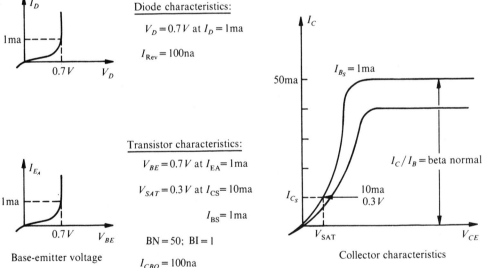

Figure 4.34 Circuit and Sketches of Nominal Diode and Transistor Data.

and request a worst-case analysis of selected unknowns is shown in Fig. 4.35.

Figure 4.35 concludes with the results of the conversational run of the worst-case analysis. The analysis itself begins with the nominal solution. Next CIRC performs a sensitivity evaluation of the circuit unknowns (node

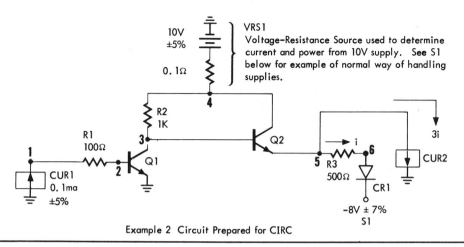

Example 2 Circuit Prepared for CIRC

CONVERSATIONAL CIRC RUN		COMMENTS
(0)	!RUN LOAD MODULE FID: DC(CIRC7) ;G EXECUTION 18:20	(0) This shows a typical procedure to access CIRC under a BTM account. Other XDS Sigma computer installation may have different procedures. Underlining shows user typing.
	PROVIDE CONTROL MESSAGES	
(1)	:CIRCUIT DATA Only lines preceded by a colon are user-input.	(1) Control message: "Accept circuit description".
	** CIRC-DC **	
(2)	NUMBER OF NODES :6	(2) Circuit has 6 nodes.
(3)	NUMBER OF VOLTAGE SUPPLIES :1	(3) Circuit has 1 voltage supply.
(4)	ARE SPECIAL EQS USED :N	(4) No special equations are used.
(5)	SPECIFY THE QUANTITY OF: RES ,VRS ,CUR ,DIO ,TRN , :3,1,2,1,2,	(5) Circuit contains 3 resistors, 1 voltage resistance source, 2 current sources, 1 diode, 2 transistors.
	VOLTAGE SUPPLIES SPECIFY: NOM VAL,MIN INFO,MAX INFO, (FOR PAR.S INDICATED)	
(6)	SUP 1 = PAR. 1 :-8,T7	(6) Supply 1 is at −8 volts, ±7 percent (Tn = ±n percent).

Figure 4.35 Conversational CIRC Run.

| CONVERSATIONAL CIRC RUN (cont.) | COMMENTS |

```
       JUNCTION TEMPERATURE   (DEG. CENT.)
       SPECIFY:
       NOM VAL,MIN INFO,MAX INFO, (FOR PAR.S INDICATED)


          TEMP   1 = PAR.    2
 (7) :25
```

(7) Junction temperature of diode and transistor models is 25 C. The single entry makes nominal, minimum, and maximum values the same, i.e., 25.

```
       RESISTORS

       SPECIFY:
       TERM 1 INFO,TERM 2 INFO,
       NOM VAL,MIN INFO,MAX INFO, (FOR PAR.S INDICATED)


          RES    1
 (8) :N1,N2
          RES    1 = PAR.    3
 (9) :100,S2,S3
```

(8) Resistor 1 is connected between nodes 1 and 2.

(9) Resistor 1 has a value of 100 ohms, with tolerances of 2% minimum and 3% maximum. (Sn = tolerance percentage to be used by CIRC and stored for reuse. Minimum and maximum are distinguished by order of input — minimum is first.)

```
          RES    2
(10) :N4,N3
          RES    2 = PAR.    4
(11) :1000,R
```

(10) Resistor 2 is connected between nodes 3 and 4.

```
          RES    3
(12) :N5,N6
          RES    3 = PAR.    5
(13) :500,R
```

(11) Resistor 2 has a nominal value of 1000 ohms and the same percentage tolerances as resistor 1. (R instructs CIRC to use the previously stored tolerances.)

(12) Resistor 3 is connected between nodes 5 and 6.

```
       VOLTAGE-RESISTANCE SOURCES
```

(13) Resistor 3 has a nominal value of 500 ohms with the same percentage tolerances as resistors 1 and 2.

```
       SPECIFY:
       POS TERM INFO,NEG TERM INFO,
       NOM VAL,MIN INFO,MAX INFO, (FOR PAR.S INDICATED)


          VRS    1
(14) :N4,G,
          VRV    1 = PAR.    6
(15) :10,T5
          VRR    1 = PAR.    7
(16) :.1
```

(14) VRS 1 is connected between node 4 and ground (indicated by G,).

(15) Voltage of VRS 1 (VRV) is 10 volts nominal, ±5%.

(16) Resistance of VRS 1 (VRR) is 0.1 ohm nominal.

```
       CURRENT SOURCES

       SPECIFY:
       EXIT NODE INFO, RECV NODE INFO, ELEMENT INFO, V OR I CONTROL
       NOM VAL,MIN INFO,MAX INFO, (FOR PAR.S INDICATED)
```

Figure 4.35—*Cont.*

CONVERSATIONAL CIRC RUN (cont.)   COMMENTS

```
         CUR    1
(17)  : G,N1
         CUR    1  =  PAR.    8
(18)  : .0001,R

         CUR    2
(19)  : N5,G,R3,I
         CUR    2  =  PAR.    9
(20)  : 3
```

(17) Current source 1 is connected between ground and node 1.

(18) Current source 1 has a nominal value of 0.0001 ampere, ±5% (R in this case refers to tolerance of VRV of VRS1, the last tolerance stored.)

(19) Current source 2 is connected between node 5 and ground, and is controlled by the current (i) in R3 (resistor 3.)

(20) The current of source 2 is three times that flowing in R3.

DIODES

SPECIFY:
ANODE INFO, CATHODE INFO, COPY DATA INFO,
NOM VAL, MIN INFO, MAX INFO, (FOR PAR.S INDICATED)

```
         DIO    1
(21)  : N6,S1
         ID     1  =  PAR.    22
(22)  : .001
         VD     1  =  PAR.    10
(23)  : .7
         IREV   1  =  PAR.    11
(24)  : 100E-9
```

(21) Diode 1 is connected between node 6 and supply 1.

(22) Diode 1 forward-bias current is 1 na.

(23) Drop across diode 1 is 0.7 volt, nominal at 1 ma.

(24) Diode 1 leakage current (IREV) is 100 nanoamperes nominal.

TRANSISTORS

SPECIFY:
BASE INFO, COLL. INFO, EMIT. INFO, COPY DATA INFO, P IF PNP,
NOM VAL, MIN INFO, MAX INFO, (FOR PAR.S INDICATED)

```
         TRN    1
(25)  : N2,N3,G
         ICS    1  =  PAR.    23
(26)  : .01
         IBS    1  =  PAR.    24
(27)  : .001
         IEA    1  =  PAR.    25
(28)  : 1E-3
         VBE    1  =  PAR.    12
(29)  : .7,M.6,M.8
         VSAT   1  =  PAR.    13
(30)  : .3,M.2,M.4
         BN     1  =  PAR.    14
(31)  : 50,M25,M100
         BI     1  =  PAR.    15
(32)  : 1
         ICBO   1  =  PAR.    16
(33)  : 100E-9.,M10E-9,M250E-9
```

(25) Transistor 1 is connected to nodes 2 and 3 and ground.

(26) ICS is 0.01 ampere nominal.

(27) IBS is 0.001 ampere nominal.

(28) IEA is 0.001 ampere nominal.

(29) VBE is at 0.7 volt nominal, 0.6 volt minimum, 0.8 volt maximum. (Mn = actual value of n.)

(30) VSAT is at 0.3 volt nominal, 0.2 volt minimum, and 0.4 volt maximum.

(31) Normal beta is 50 nominal, 25 minimum, 100 maximum.

(32) Inverted beta is 1.

(33) ICBO is 0.1 microampere nominal, 0.01 microampere minimum, and 0.25 microampere maximum.

Figure 4.35—*Cont.*

CONVERSATIONAL CIRC RUN (cont.)

  TRN 2
(34) :N3,N4,N5,C1

   SAME DATA AS TRN 1

   ICS = PAR. 26
   IBS = PAR. 27
   IEA = PAR. 28
   VBE = PAR. 17
   VSAT = PAR. 18
   BN = PAR. 19
   BI = PAR. 20
   ICBO = PAR. 21

  PROVIDE CONTROL MESSAGES

(35) :NODE SELECTION

   SPECIFY:

  NODE NUMBERS (0 TO END ENTRIES)
  (LIMIT - 12 ENTRIES PER LINE, 15 NODES
           MAXIMUM)

(36) :N1,N3,N5

  PROVIDE CONTROL MESSAGES

(37) :DEPENDENT UNKNOWN SELECTION

   SPECIFY:

  DEP. UNK. ID'S (0 TO END ENTRIES)
  (LIMIT - 12 ENTRIES PER LINE, 35 DEP. UNKNOWNS MAXIMUM)

(38) :IRES1,IRES3,ICUR2,IC/IB1,IC/IB2

  PROVIDE CONTROL MESSAGES

(39) :WORST CASE

If the "RANKING OF SENSITIVITIES" output shown on the
next page is not desired, request this analysis as follows:

 :SSOU (stop sensitivity output)
 :WORST CASE

COMMENTS

(34) Transistor 2 is connected to nodes 3, 4, and 5.
Data from transistor 1 is to be copied. (Cn means
"copy data of transistor n".)

(35) Control message: accept node selection.

When CIRC is operating conversationally,
these selections automatically limit both
output volume and the unknowns to be
worst case analyzed.

(36) Node voltages selected are V_1, V_3, and V_5.

(37) Control message: accept selection of dependent
unknowns. When CIRC is operating conversationally, these selections automatically limit
both output volume and the unknowns to be worst
case analyzed.

(38) Selected dependent unknowns are currents for
resistors 1 and 3, current source 2, IC/IB1, and
IC/IB2. IC/IB is the "circuit gain" of a transistor.

(39) Control message: perform worst case calculations.
Note that the one command WORST CASE causes
CIRC to do the entire worst case analysis with results
shown.

Figure 4.35—*Cont.*

CONVERSATIONAL CIRC RUN (cont.)   COMMENTS

NOMINAL SOLUTIONS   13

> The worst case analysis begins with a nominal solution.

** NODE VOLTAGES **

1 .7519016E 00   3 .3441077E 01   5 .2628831E 01

** DEPENDENT UNKNOWNS **

```
IRES  1  .1000000E-03    IRES  3  .1970443E-01    ICUR  2  .5911329E-01
IC/B  1  .5005301E 02    IC/B  2  .5000356E 02
```

> Compare the limited output obtained in this conversational worst case analysis to the full output of the "batch" results, Figures 2-10 to 2-13.

DIFFERENCES WITH RESPECT TO:
PAR.   VALUE

```
SUP    1  = -.744000E 01
RES    1  =  .103000E 03
RES    2  =  .103000E 04
RES    3  =  .515000E 03
VRV    1  =  .105000E 02
CUR    1  =  .105000E-03
VBE    1  =  .800000E 00
VSAT   1  =  .400000E 00
BN     1  =  .100000E 03
ICBO   1  =  .250000E-06
VBE    2  =  .800000E 00
VSAT   2  =  .400000E 00
BN     2  =  .100000E 03
ICBO   2  =  .250000E-06
```

> This output is provided primarily to show the CIRC user that calculation of sensitivity information is in progress. The parameter value (normally equal to the maximum value) helps detect incorrect input data.

RANKING OF SENSITIVITIES OF NODE   1

RANK	PAR.		% PAR VAR.	DIFFERENCES	% OF TOTAL DIFF.
1	VBE	1	14.29	.1000E 00	86.4943
2	BN	1	100.00	.1352E-01	11.6972
3	CUR	1	5.00	.1753E-02	1.5159

> Ranking of sensitivities for the other seven selected unknowns would require about 50 typed lines and is not shown.

Figure 4.35—*Cont.*

CONVERSATIONAL CIRC RUN (cont.)   COMMENTS

NOMINAL CONDITIONS OF DIODES AND TRANSISTORS | States different than these will be reported during worst case evaluations. |

DEVICE   STATE | State definitions are based on voltage alone. |

DIO   1   FOR | Forward, i.e., anode to cathode voltage is plus (+). |

TRN   1   ACT | Active, i.e., base to emitter voltage is plus (+) and base to collector voltage is negative (−). |
TRN   2   ACT

W.C. MAX   NODE   1

** NODE VOLTAGES **

  1 .8694062E 00   3 .5143311E 00   5 -.1876890E 00

** DEPENDENT UNKNOWNS **

IRES 1 .1050000E-03   IRES 3 .1323658E-01   ICUR 2 .3970973E-01
IC/B 1 .9199419E 02   IC/B 2 .1000565E 03

STATE CHANGES: TRN 1 SAT | Transistor 1 went from active to saturated. |

| The remainder of the individual worst case analysis outputs are not shown. They include WORST CASE MIN NODE 1 and MIN and MAX for NODE 3, NODE 5, IRES 1, IRES 3, ICUR 2, IC/B1, and IC/B2. |

W.C. RESULTS

NODE			MINIMUM	NOMINAL	MAXIMUM
V 1			.6325540E 00	.7519016E 00	.8694062E 00
V 3			.3973133E 00	.3441077E 01	.7187888E 01
V 5			-.4407532E 00	.2628830E 01	.6454986E 01
DEP.	ID	NO.	MINIMUM	NOMINAL	MAXIMUM
IRES	1	1	.9500000E-04	.1000000E-03	.1050000E-03
IRES	3	5	.1212649E-01	.1970443E-01	.2876694E-01
ICUR	2	10	.3637946E-01	.5911329E-01	.8630081E-01
IC/B	1	21	.2500301E 02	.5005301E 02	.1002819E 03
IC/B	2	29	.2500011E 02	.5000356E 02	.1000565E 03

PROVIDE CONTROL MESSAGES

:STOP | STOP provides an orderly way to terminate a CIRC run and log-off the computer with the BYE command (in which the BTM software types the E.) |
:BYE

Figure 4.35—*Cont.*

voltages, currents, power, etc.) versus the circuit parameters (supply voltages, resistor values, transistor betas, etc.). The results of this evaluation in the form of sensitivity rankings is the major output. Data are stored in memory to allow setup of the worst-case conditions, and CIRC outputs nominal states for diodes and transistors. Next CIRC performs and outputs the actual worst-case analysis for maximum and minimum values of all selected unknowns.

AC Nominal and Open-Loop Analysis

The circuit to be analyzed is shown in Fig. 4.10. The feedback path, R6, is a simple resistor of 2700 ohms when received for nominal analysis. The question and answer dialog to describe the circuit is shown in Fig. 4.36. Figure 4.37 concludes with the dialog requesting a nominal solution printout.

One major purpose of computer-aided design program is to study the effect of parameters on a circuit's performance. Often it is desired to do this by having a parameter of particular interest run through a whole range of values. This can be done is CIRC, and a plot of results in the form of selected variables as the parameter varies can be obtained.

Another feature of CIRC-AC is the ability to automatically perform open-loop analysis of feedback circuits. The program is set up so that a loop must be opened in the middle of a resistor. A unit voltage will be set up to

	Conversational CIRC Run	Comments
| (0) | ```
!RUN
LOAD MODULE FID:AC,CIRC7
;G
EXECUTION
``` | (0) This shows a typical procedure to access CIRC-AC under a BTM account. Various XDS Sigma computer installations may have different procedures. |
	PROVIDE CONTROL MESSAGES	
(1)	** CIRC-AC **	(1) Control message: "accept circuit description".
(2)	NUMBER OF NODES :8	(2) Circuit has 8 nodes.
(3)	NUMBER OF VOLTAGE SUPPLIES :1	(3) Circuit has 1 voltage supply.
(4)	ARE SPECIAL EQS USED :NO	(4) No special equations used.
(5)	SPECIFY THE QUANTITY OF: RES ,IND ,CAP ,TRN ,CYS ,VZS ,MUT , :9,1,2,3	(5) Circuit contains 9 resistors, 1 inductor, 2 capacitors, and 3 transistors.
	VOLTAGE SUPPLIES SPECIFY:	

Figure 4.36 Conversational CIRC-AC Run.
```

| Conversational CIRC Run | Comments |

```
         NOM VAL (MAGNITUDE,ANGLE),
         (FOR PAR.S INDICATED)

(6)      SUP   1 = PAR.S   1 AND   2
         :1
```

(6) Supply 1 is at $1 \angle 0°$ volt.

```
         JUNCTION TEMPERATURE  (DEG. CENT.)
         SPECIFY:
         NOM VAL (FOR PAR.S INDICATED)

(7)      TEMP  1 = PAR.   3
         :25
```

(7) Transistor junction temperature is $25°C$.

```
         RESISTORS

         SPECIFY:
         R TERM 1 INFO, R TERM 2 INFO,
         NOM VAL (FOR PAR.S INDICATED)

(8)      RES   1
         :S1,N1
(9)      RES   1 = PAR.   4
         :2
(10)     RES   2
         :N1,G
(11)     RES   2 = PAR.   5
         :10000
(12)     RES   3
         :N2,N8
(13)     RES   3 = PAR.   6
         :3900
(14)     RES   4
         :N3
         RES   4 = PAR.   7
         :1E4
(15)     RES   5
         :N4
(16)     RES   5 = PAR.   8
         :1E4
(17)     RES   6
         :N3,N4
(18)     RES   6 = PAR.   9
         :2700
(19)     RES   7
         :N5,N6
(20)     RES   7 = PAR.  10
         :1000
(21)     RES   8
         :N6
(22)     RES   8 = PAR.  11
         :680
(23)     RES   9
         :N7
(24)     RES   9 = PAR.  12
         :.1
```

(8) Resistor 1 is connected between supply 1 and node 1.

(9) Resistor 1 has a value of 2700Ω.

(10) Resistor 2 is connected between node 1 and ground.

(11) Resistor 2 has a value of 100Ω.

(12) Resistor 3 is connected between node 2 and node 8.

(13) Resistor 3 has a value of 3900Ω.

(14) Resistor 4 is connected between node 3 and ground.
Resistor 4 has a vlue of $10K\Omega$.

(15) Resistor 5 is connected between node 4 and ground.

(16) Resistor 5 has a value of $10K\Omega$.

(17) Resistor 6 is connected between node 3 and node 4.

(18) Registor 6 has a value of $2.7K\Omega$.

(19) Resistor 7 is connected between node 5 and node 6.

(20) Resistor 7 has a value of $1K\Omega$.

(21) Resistor 8 is connected between node 6 and ground.

(22) Resistor 8 has a value of 680Ω.

(23) Resistor 9 is connected between node 7 and ground.

(24) Resistor 9 has a value of 0.1Ω.

Figure 4.36—*Cont.*

Conversational CIRC Run   Comments

INDUCTORS

SPECIFY:
L TERM 1 INFO, L TERM 2 INFO,
TUNED CAP. REF.,
NOM VAL (FOR PAR.S INDICATED)

(25) IND   1
     :N8,G   (25) Inductor 1 is connected between node 8 and ground.
(26) (IND   1),(QL 1) = PAR.S 13 AND 14
     :2E-3   (26) Inductor 1 has a value of 2MH.

CAPACITORS

SPECIFY:
C TERM 1 INFO, C TERM 2 INFO,
NOM VAL (FOR PAR.S INDICATED)

(27) CAP   1
     :N1,N3   (27) Capacitor 1 is connected between node 1 and node 3.
(28) CAP   1 = PAR.   15
     :60E-12   (28) Capacitor 1 has a value of 60 pf.

(29) CAP   2
     :N4,N5   (29) Capacitor 2 is connected between node 4 and node 5.
(30) CAP   2 = PAR.   16
     :270E-12   (30) Capacitor 2 has a value of 270 pf.

TRANSISTORS

SPECIFY:
BASE INFO, COLL. INFO, EMIT.
INFO, COPY DATA INFO, A IF FACO
NOM VAL (FOR PAR.S INDICATED)

(31) TRN   1
     :N1,N6,N2   (31) Transistor 1 is connected to node 1, node 6, and node 2.
(32) IE   1 = PAR.   17
     :1E-3   (32) DC emitter current is 1MA.
(33) (BLF   1),(EPH 1) = PAR.S 18 AND 19
     :75   (33) Low-frequency beta is 75, excess phase is zero.
(34) FTB   1 = PAR.   20
     :250E6   (34) Beta gain-bandwidth frequency is 250 MHZ.
(35) RB   1 = PAR.   21
     :150   (35) Base bulk resistance is 150 .
(36) CTE   1 = PAR.   22
     :0   (36) Emitter transition capacitance is 0.
(37) (CC   1),(RC 1) = PAR.S 23 AND 24
     :6E-12   (37) Collector capacitance is 6 pf.
                                                        Reverse collector resistance is infinite.

(38) TRN   2
     :N3,G,N2,C1   (38) Transistor 2 is connected to node 3, ground, and node 2.
                                                        Data from transistor 1 is to be copied. (C1 means
                                                        "Copy data of transistor 1".)
(39) IE   2 = PAR.   25
     :1E-3   (39) DC emitter current is 1MA.

Figure 4.36—Cont.

Conversational CIRC Run	Comments

```
        SAME DATA AS TRN    1

        BLF   = PAR.  26
        EPH   = PAR.  27
        FTB   = PAR.  28
        RB    = PAR.  29
        CTE   = PAR.  30
        CC    = PAR.  31
        RC    = PAR.  32
```

(40) `TRN   3`
     `:N6,N4,N7`

(40) Transistor 3 is connected to node 6, node 4, and node 7.

(41) `IE    3 = PAR.  33`
     `:-.1`

(41) Negative number entered indicates value is $r_e = 0.1\Omega$.

(42) `(BLF  3),(EPH  3) = PAR.S 34 AND 35`
     `:50,12`

(42) Low frequency beta is 50; excess phase is 12° at $f_{\alpha co}$.

(43) `FTB   3 = PAR.  36`
     `:-12E6`

(43) Negative number entered indicates value is beta cutoff frequency $f_{\beta co} = 12$MHZ.

(44) `RB    3 = PAR.  37`
     `:200`

(44) Base bulk resistance = 200Ω.

(45) `CTE   3 = PAR.  38`
     `:0`

(45) Emitter transition capacitance 150.

(46) `(CC   3),(RC  3) = PAR.S 39 AND 40`
     `:2E-12`

(46) Collector capacitance is 2pf. Reverse collector resistance is infinite.

```
        FREQ DATA
```

(47) `START`
     `:1000`

(47) Solution start frequency is 1KHZ.

(48) `CONTROL(A,M,L,N)`
     `:A`

(48) Control is add increment.

(49) `INCREMENT`
     `:1`

(49) Frequency increment is 1HZ.

(50) `END`
     `:1`

(50) End frequency is 1HZ – CIRC will do only 1KHZ solution since end frequency is less than start.

Figure 4.36—*Cont.*

```
        PROVIDE CONTROL MESSAGES
```

(1) `:NODE`

   `SPECIFY:`

   `NODE NUMBERS (0 TO END ENTRIES)`
   `(LIMIT - 12 ENTRIES PER LINE,`
   `15 NODES MAXIMUM)`

(1) Control message: accept node selection.

(2) `:N4`

   `PROVIDE CONTROL MESSAGES`

(2) Node selected for output is node 4.

(3) `:DEPENDENT`

   `SPECIFY:`

   `DEP. UNK. ID'S (0 TO END ENTRIES)`
   `(LIMIT - 12 ENTRIES PER LINE,`
   `35 DEP. UNKNOWNS MAXIMUM)`

(3) Control message: accept selection of dependent unknowns.

Figure 4.37 Continuation of Conversational Run with Unknown Selection and Nominal Solutions.

(4) :IRES1,IB1,ICAP1

(4) Selected dependent unknowns are currents for resistor 1, the base of transistor 1 and capacitor 1.

PROVIDE CONTROL MESSAGES

(5) :NOMINAL

(5) Control message: Do nominal solution.

```
NOMINAL SOLUTIONS

    FREQ. =   .1000000E 04 ────────────────── 1KHZ

** NODE VOLTAGES **
                          ┌─────────────────────────┐
                          │ Solution with selected output │
                          └─────────────────────────┘
     V     AMPLITUDE
           PHASE

     4    .1013577E 01  ◄──────────────── Node 4 voltage
         -.1667586E-01  ◄──────────────── Node 4 phase


** DEPENDENT UNKNOWNS **
                                                              R1 current
                                                              Q1 base current
                                                              C1 current

    IRES  1  .7879053E-04   IB   1   .8366024E-07   ICAP  1   .2004601E-08
             .2970289E-01                .4009215E 02                -.9130010E 02


    SPECIFY:
    VARIABLE TO PLOT (0 TO END ENTRIES),TYPE,MIN,MAX,TYPE,MIN,MAX
```

(6) :0

(6) Zero entry — no plots requested.

PROVIDE CONTROL MESSAGES

(7) :STOP

(7) Control message: stop CIRC analysis.

!

Figure 4.37—*Cont.*

drive the input terminal, and the output terminal will be properly loaded. The resistor input terminal input terminal is specified by the user.

An example of these features is also presented (based on the circuit of Fig. 4.10). Since the circuit description was given in the conversational example, only the sketch of the data deck required for the analysis, Fig. 4.38, and some of the results, Fig. 4.39, are presented.

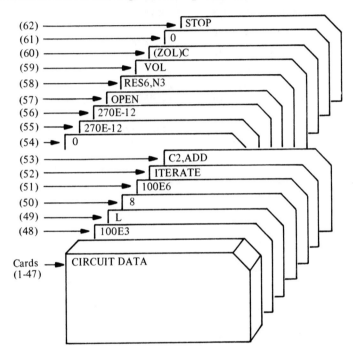

(1-47)	Circuit deck.
(48)	Start frequency.
(49)	Logarithmic control.
(50)	Eight frequency solutions per decade.
(51)	End frequency.
(52)	Control message: repeat calculations with parameter values changed as indicated.
(53)	Parameter to be changed is capacitor 2. Mode of change is "add incremental values".
(54)	First parameter value is 0.
(55)	Increment step size is 270pf.
(56)	Last parameter is 270pf.
(57)	Control message: Do open-loop analysis.
(58)	Open loop in the middle of resistor 6, drive into node 3.
(59)	Plot open-loop voltage (VOL).
(60)	Plot open-loop impedance – complex, (ZOL)C.
(61)	Zero terminates plotting mode.
(62)	Control message: STOP CIRC RUN.

Figure 4.38 Deck Set-up for Automatic Parameter Iteration and Open-loop Analysis.

PARAMETER IS CAP   2
NOMINAL VALUE IS   .0000000E 00

Open loop solutions — $C_2 = 0$

```
FREQ(OPEN LOOP)
  GAIN(MAG)         PHASE              GAIN(D B)
    ZL REAL           ZL IMAG

 .1000000E 06
 .9999748E 03      .1576580E 03        .5999978E 02
 .5165109E 04     -.3028137E 03

 .1333521E 06
 .9661389E 03      .1506783E 03        .5970079E 02
 .5134984E 04     -.3983336E 03

 .1778279E 06
 .9126904E 03      .1419186E 03        .5920647E 02
 .5083349E 04     -.5185243E 03

 .2371374E 06
 .8340207E 03      .1312997E 03        .5842354E 02
 .4997510E 04     -.6629854E 03

 .3162278E 06
 .7288112E 03      .1189913E 03        .5725230E 02
 .4862133E 04     -.8233325E 03

 .4216965E 06
 .6033686E 03      .1054141E 03        .5561165E 02
 .4665672E 04     -.9789912E 03

 .5623413E 06
 .4707489E 03      .9111495E 02        .5345579E 02
 .4412017E 04     -.1099798E 04
```

Solutions omitted for brevity

```
 .1778279E 08
 .7194359E 00     -.1088261E 02       -.2860158E 01
 .1651225E 04     -.6731969E 03

 .2371374E 08
 .3261660E 00     -.6346901E 01       -.9731227E 01
 .1561488E 04     -.5262238E 03

 .3162278E 08
 .1452872E 00      .9982539E 01       -.1675545E 02
 .1508379E 04     -.4028910E 03

 .4216965E 08
 .8368339E-01      .4050709E 02       -.2154721E 02
 .1478546E 04     -.3049831E 03

 .5623413E 08
 .6951693E-01      .6255808E 02       -.2315819E 02
 .1462312E 04     -.2298387E 03

 .7498942E 08
 .6056247E-01      .6991661E 02       -.2435593E 02
 .1453543E 04     -.1732556E 03

 .1000000E 09
 .4999801E-01      .6991135E 02       -.2602095E 02
 .1448706E 04     -.1310939E 03
```

Figure 4.39 Automatic Parameter Iteration and Open-loop Analysis Solutions.

4.6 REFERENCES

1. *CIRC-DC Reference Manual and User's Guide*, No. 901697, price $3.75, Xerox Corporation, 701 So. Aviation Blvd., El Segundo, Calif. 90245.

2. *CIRC–AC Reference Manual and User's Guide*, No. 901698, price $4.25, Xerox Corporation, 701 So. Aviation Blvd., El Segundo, Calif. 90245.

3. *CIRC–TR Reference Manual and User's Guide*, No. 901786, price approximately $4.00, Xerox Corporation, 701 So. Aviation Blvd., El Segundo, Calif. 90245.

4. LINVILL, J. G., "Lumped Models of Transistors and Diodes," *Proc. IRE*, 46, June 1958, 1141–1152.

5. MCNAIR, R. D., "Computer Aided Design of Lumped Parameter Systems," M.S. Thesis submitted to UCLA, Dept. of Engineering, Jan. 1965.

6. KLEINER, C. T., G. KINOSHITA, and E. D. JOHNSON, "Simulation and Verification of Transient Nuclear Radiation Effects on Semiconductor Electronics," *IEEE Trans. Nuclear Science*, NS 11, Nov. 1964, pp. 82–104.

7. KLEINER, C. T., "Predicting the Effect on an Aonizing Pulse of Radiation on Semiconductor Electronics," National Aeronautical Electronics Conference, Dayton, Ohio, May 12, 1965.

8. MOLL, J. L., "Large-Signal Transient Response of Junction Transistors," *Proc. IRE*, 42, Dec. 1954, pp. 1773–1784.

9. HAMILTON, D. J., F. A. LINDHOLM, and J. NARUD, *Large Signal Models for Junction Transistors*, Engineering Research Laboratories, University of Arizona, Tucson. Also see D. K. Lynn, C. S. Meyer, and D. J. Hamilton, *Analysis and Design of Integrated Circuits*, McGraw-Hill, New York, 1967, Appendix A.

10. JOHNSON, E. D., C. T. KLEINER, L. R. MCMURRAY, E. L. STEELE, and F. A. VASSALLO, "Transient Radiation Analysis by Computer Program (TRAC)," *Technical Report—Harry Diamond Laboratories*, Autonetics Division, N.A. Rockwell Corp., June 1968.

11. KLEINER, C. T., and E. D. JOHNSON, "SPARC—System of Programs for Development and Analysis of Components and Subsystems by RECOMP II," *EM8083*. Autonetics Division, North American Aviation Corp., Feb. 1962.

5

CIRCUS 2

BENJAMIN DEMBART
LOREN MILLIMAN
Boeing Company
Seattle, Washington

5.1 INTRODUCTION

The CIRCUS-2 program finds the dc steady-state and time domain transient solutions for an electronic network driven by arbitrary forcing functions [1]. This program accepts input statements which are written in a user-oriented language and produces output as tabular time histories, plots, or data recorded on magnetic tape. Networks may be built from resistors, capacitors, inductors, admittances, impedances, sources, and nonlinear elements which can be defined by the user. A library of transistors and diodes is included with the program to provide a nucleus for further development of nonlinear element modeling capability.

A typical analysis of a circuit using CIRCUS might proceed along the following lines. Appropriate information about the network is coded and the input data are prepared on punched cards which are submitted to a computer where the CIRCUS-2 program has been installed. CIRCUS reads these cards, interprets them and lists them on the printer. If there are errors in the input data, appropriate diagnostic messages are written. If the errors are such that the program cannot proceed, CIRCUS stops work on the problem and searches for another data case. When the input data seem to be error-free, CIRCUS forms the basic equations of the network using Kirchhoff's laws, Ohm's law, and similar relations for other element types. CIRCUS computes the dc steady state of the network and uses that result as the initial conditions for a transient solution. (Alternatively, the user may specify the initial conditions for a transient solution.) After the transient solution has been calculated, output that the user has requested (both listings and plots) is generated on the printer. Network parameters may be changed and the circuit rerun, so that the effect of changes can be determined immediately. Partial solutions may be saved on magnetic tape for further analysis, for input to other circuits being analyzed by CIRCUS, or for use by other programs. After completing the analysis for one data case, CIRCUS continues to read data cards so that several data cases can be analyzed in one computer run.

CIRCUS-2 was based on the earlier CIRCUS-1 program. Although the input language is much the same for both programs, CIRCUS-2 incorporates many improved programming techniques that make it a more versatile, more reliable, and more efficient means for circuit analysis. Some of the techniques used by CIRCUS-2 include

- Efficient sparse matrix methods which generate solutions directly in the form of computer instructions.
- Strongly stable numerical solution of differential equations.
- Dynamic data storage.
- User-defined models.
- Combination of separately analyzed networks by incorporation of solutions which were recorded on magnetic tape.
- Rational function or convolution representation of elements defined in the frequency domain.

CIRCUS-2 is written in the FORTRAN programming language (95%) and in assembler language (5%). It is available on the IBM/360–370, CDC 6600, CDC 7600, and Honeywell 6000 computers.

Since CIRCUS-2 uses dynamic storage extensively, there are no problem size limitations built into the program. Problem size is limited by available core storage, peripheral storage, and computer time. Problem complexity is restricted only by the ability of the user to model network components. Some topological configurations are not allowed, such as voltage sources in parallel or current sources in series. There are networks for which CIRCUS-2 may not be able to compute the dc steady state. The user may provide a set of dc initial conditions or a guess at the steady state in order to begin a transient solution.

5.2 PROGRAM STRUCTURE

The CIRCUS-2 program is a single, large computer code consisting of over 25,000 source cards. By far the bulk of the code is devoted to accepting input statements and converting this information to a system of data and matrices that are suitable for efficient mathematical analysis of the problem. The remainder of the code is devoted to the steady-state and transient analysis of the circuit and the display of the results.

Several sections of the CIRCUS-2 program are required for the analysis of all circuits: the topological matrix generator, the sparse matrix processor, steady-state and transient analysis routines, and the output processor. Two additional sections of the CIRCUS-2 program are the model compiler and the device processor, which build the model and device libraries. Each section of CIRCUS will be described in detail below, but first some program features that are common to all sections of the program will be described.

The most important feature is dynamic allocation of computer storage. In CIRCUS-2 dynamic allocation means two things. First, it means that all available storage can be utilized in the solution of a problem. Small problems can be solved in a small region, while large problems which require a larger region can be accommodated without modification of any kind to the program. Second, it means that storage is not allocated for data until it is required. It is not possible to have too many resistors while there is unused space allocated for capacitors. Storage is always fully utilized.

Another CIRCUS-2 program feature is an efficient input/output processor that allows for simultaneous input, output, and problem solution, sequential or random access I/O, and the elimination of large, storage-consuming buffers.

5.2.1 Topological Matrix Generator

In CIRCUS-2 the equations that simulate the response of a network are derived from the tableau formulation method [2]. For each branch in the network, a branch equation is written: $V = RI$ for resistors, $dV/dt = I/C$ for capacitors, and $dI/dt = V/L$ for inductors. In addition to the branch equations a maximal number of independent Kirchhoff voltage and current equations must be written. Each branch voltage is written as the difference of two node voltages, and the sum of all the branch currents entering every node is set to zero.

When these equations are written in matrix form, the resulting matrix is called the *network tableau matrix*. For large networks this matrix is very large but very sparse. Efficient sparse matrix methods are required to handle it [3].

The topological matrix generator derives the structure of the network tableau matrix from the nodal connections specified on the input cards. The nonzero positions of the matrix are identified, and wherever possible the value of each matrix entry is computed (this is always possible for the Kirchhoff equations and linear branch equations). Where this is not possible, estimates of the values are computed.

Next, all the branch voltages and currents that are required as inputs to the subroutines that model nonlinear elements are identified, and the positions in the matrix that are defined by nonlinear elements are located. A procedure is generated to move the required voltages and currents into the model subroutines and to take the computed values from the model and put them in the matrix. Once this has been accomplished, the problem has been defined as a matrix equation and the next step is to analyze the matrix.

5.2.2 Sparse Matrix Processor

The matrix equations are solved by LU decomposition [4]. The procedure that is used in CIRCUS-2 is to generate two subprograms in machine language that perform only the required nonzero arithmetic operations.

Before these subprograms can be generated, the equations are reordered in an attempt to minimize the total amount of computation required to solve the problem. This step is critical and is the key to successful sparse matrix procedures [5].

Once the reordering operation is complete, every arithmetic operation in the solution procedure is examined and classified as follows:

1. Any operation involving one of the zero coefficients in the sparse matrix is ignored.
2. If both of the nonzero coefficients are fixed in value, then the operation is classified as a *constant* operation that need be performed only once. Machine instructions are generated to perform this operation, and these instructions are added to the constant operation subprogram.
3. If at least one of the coefficients has values that change during the course of the analysis, then the operation is classified as a *variable* operation. Machine instructions are generated for this operation and added to the variable operation subprogram.

Thus, two subprograms are generated simultaneously: a constant subprogram that need be executed only once, and a variable subprogram that must be executed at every iteration of the solution procedure.

This process is repeated three times: once for the equations that represent the steady-state problem, once for the transient problem, and once to solve the problem of finding remaining branch voltages and currents given the capacitor voltages and inductor currents.

After processing the sparse matrix, the preprocessing phase of the program is complete and the analysis phase can begin. Because of the complexity of the sparse matrix processing task, CIRCUS-2 generally will spend more time in the preprocessing phase than was typical for circuit analysis programs of the previous generation (CIRCUS-1, for example). This extra cost is recovered many times over in the analysis phase on all but the simplest of circuit analysis applications.

5.2.3 Steady-State Analysis

The steady-state solution may be used as the starting point for transient analysis, or a parametric study of the circuit may be made at steady state. In either case steady-state analysis is the first type of analysis performed on the circuit.

The steady-state problem is one of finding the values of all branch voltages and currents so that $dV/dt = 0$ for each capacitor and $dI/dt = 0$ for each inductor. These equations are substituted into the tableau, and the resulting matrix equation is solved using Newton's method.

The first step is to execute the subprogram of constant operations. Then any equations that are used for modeling circuit components are evaluated. The resulting element values are inserted in the tableau matrix, and the system of linear equations required for one Newton's method step is solved using the variable operation subprogram. The modeling equations are reevaluated, and the procedure is iterated until it converges. The steady-state solution is then passed along as the starting point for transient analysis.

5.2.4 Transient Analysis

In CIRCUS-2, transient analysis is performed by a numerical integration method due to H. Rosenbrock [6]. There is some disagreement among numerical analysts as to whether this method is explicit or implicit, because in order to take a step with this method it is necessary to solve a system of equations, just as with an implicit integration method, but, unlike implicit methods, the system that must be solved is always linear, even if the network is not. Since the system of equations is linear, it can be solved directly without iteration, and some authors call the method explicit.

The particular version of Rosenbrock's method that is used in CIRCUS-2 is third-order and is numerically stable. Thus, networks with widely spaced time constants (which created difficulties for programs of the previous generation) cause no problems with CIRCUS-2.

Before transient analysis begins, the subprogram of constant operations for the transient solution is executed. Each numerical integration step consists of three passes through the variable operation subprogram (since this is a third-order method). The three results are then combined to update the branch currents and voltages and to estimate the optimum size for the next integration step. After an integration pass, it may happen that some branch current and voltage may not lie on the nonlinear characteristic curve that is specified for the element. So after each integration pass, voltages and currents are adjusted to satisfy all tableau equations with the capacitor voltages, inductor currents, and sources held fixed.

As the transient solution proceeds, the results of the analysis are recorded on a bulk storage device where they are available to the output processor at the completion of the transient analysis.

5.2.5 Output Processor

The output processor reads the results of transient analysis and processes them in any of three ways. They can be displayed in the form of tabular time histories and printer plots, or they can be recorded on magnetic tape for use by another program (or for use by CIRCUS-2 in the analysis of another network).

The data in the tabular listings are presented at the time points specified by the user. No attempt is made in the transient analysis to have integration steps end at these points. Instead, integration proceeds at its own natural rate, and linear interpolation is used to compute the values of the variables at intermediate time points.

For printer plots, points will be plotted at each time point specified by the user. Interpolation will be used to compute the value of the plotted variable. In addition, the end point of each integration step is plotted. Thus, no phenomenon can occur so fast that it is missed when the user specifies an interval that is too coarse.

The variables that are recorded are treated in exactly the same way as variables that are plotted.

5.2.6 Model Processor

In CIRCUS-2 the model for a nonlinear circuit component is defined by a topological structure of simple elements and a set of functional relationships to define the values of the elements. The model compiler reads input statements that define a model, processes the information to put it into a form that can be used by CIRCUS, and records this information on a model library where it will be available for subsequent use.

The information that is recorded on the library consists of the model name, the names of the external nodes, the topological interconnection of the elements, a dictionary of all the symbolic parameters (names that appear in the topology and equations sections of the model), and a machine language subprogram to calculate the functional relations defined in the equations section of the model.

The key ingredients of the model are the last two, the dictionary and the subprogram. It is the dictionary that provides a common language for model parameters so that communication can take place among the analyst, the model, and CIRCUS. The topological matrix generator determines the inputs required by the model and the outputs supplied by the model from this dictionary. The device processor associates device parameters with model parameters by means of the order of symbolic parameters in the dictionary. Finally, the analyst can communicate the names of model parameters he desires to see displayed or change through names in the dictionary.

The machine language subprogram contains the real power of the modeling capability. It allows the functional relations to be described in a simplified version of the FORTRAN language. This allows considerable flexibility in the definition of these relations.

The model processor has a simplified FORTRAN compiler as a major component. The compiler reads the statements that define the equations and translates them into a machine language subprogram. This allows the model complier to be run as part of the CIRCUS program (rather than a separate job or job step). Also, the equations are stored in a form that is very efficient for computer utilization (i.e., machine language), and if the model is to be used many times, repeated compilation is not necessary.

5.2.7 Device Processor

Generally a model simulates a generic type of circuit component (e.g., a transistor or a flip-flop). It is the purpose of a device to specify the numerical parameters that completely define the behavior of a particular component. The device processor reads input statements that define these numerical parameters, and then records the information on a device library in a format that is efficient for later use.

An input statement is a list of parameter names followed by the numerical value (or values) that defines the parameter. To interpret the names on an input statement in the context of a model, the model dictionary is required.

The first step in processing device data is identifying the models that correspond to the devices being processed. Then the dictionaries from those models are read and the data from the input are put into order according to the dictionary. At this point parameter names are no longer required because each value can be identified by its position in the array of values. Finally, the device parameter value array is recorded on the device library along with the device and model names for identification.

This completes the survey of the major components of the CIRCUS-2 computer code. Sections do not stand alone, but each is designed to complement the others. They are all tied together by an executive program from above and utility subprograms from below which integrate the system into one program.

5.3 NETWORK ELEMENTS

CIRCUS allows for three types of network components: lumped linear elements, frequency-dependent elements, and user-defined devices. Before describing these components, a few terms must be defined in order to minimize confusion. An *element* is a two-terminal, lumped entity which may be a resistor, capacitor, inductor, voltage source, or current source. A *model* is a connected set of elements together with equations that define the properties of the elements. Elements in a model do not need to be linear. A *device* is a model together with values for parameters which were not explicitly specified in the model. A *component* can be a linear element, a device, or a model which does not require parameter specification.

The linear elements in CIRCUS are voltage sources, current sources, resistors, capacitors, and inductors. Source elements are idealized; that is, a voltage source has zero impedance and a current source has infinite impedance. Sources may be fixed-valued or time-dependent. The time dependence can be a repeating trapezoidal pulse, a sinusoid, or a table of values. When a tabular source is used, the time points must be in nondescending order, but there is no restriction on the number of points in the table or on the voltage or current values. Reactive elements are idealized also. Series and shunt resistances may be added by the user to simulate lossy elements as needed. Of course, mutual coupling may be specified between reactive elements.

Frequency-dependent elements may be either admittances or impedances. These elements are defined by specifying the poles and zeros which characterize it. If the inverse Fourier transform is known rather than poles and zeros, then a frequency-dependent element can be modeled by convolution integrals. This method is discussed in Sec. 5.6.

Nonlinear components are built in two parts. The basic part, called a *model*, defines the structure of the component, while the other, a *device*, supplies values which distinguish one particular component. This topic is covered in detail in Sec. 5.6. There are many types of nonlinearities that can be modeled. Several examples can be found in an Ebers-Moll (large-signal) transistor model [7] such as

- A current-dependent current source to simulate transistor gain.
- Voltage-dependent capacitors to simulate depletion and diffusion effects.
- A current-dependent resistor to simulate base spreading.

The effect on a network of changing one or more element values can be determined in one computer run by using the element change capability. Linear elements and elements in a device can be modified without respecifying other CIRCUS data that do not change. Parameter variation effects can be studied in either steady-state or a transient analysis.

5.4 INPUT LANGUAGE

Communication between the user and CIRCUS is done using statements in a problem-oriented language. Each statement is a list which contains key words, names, and numbers. To do an analysis with CIRCUS, the user supplies a series of statements that define the network and a series of statements to direct the program.

5.4.1 Key Words

A *key word* is a word or phrase having special meaning to the CIRCUS-2 program. Key words will always be spelled in capital letters in the examples, and they must be spelled exactly as shown when preparing input for CIRCUS-2. A list of key words is given in Table 5.1. The remainder of this section describes how these key words are used in CIRCUS and what supporting data are required.

Table 5.1 CIRCUS-2 Key Words

The following key words may be used with circuit description data statements either before or after an EXECUTE statement:

CARD LENGTH	LIST
DIAGNOSTICS	MULTIPLE EXECUTE
END	NOLIST
END OF JOB	PLOT PAGE LIMIT
END MULTIPLE EXECUTE	PLOT INTERVALS
EXECUTE	PRINT INTERVALS
HOLD FINAL CONDITIONS	RECORD INTERVALS
INCREMENT STOP TIME	SAVE
INITIAL CONDITIONS	STEADY STATE GUESS
LIMITS	STOP TIME

The following key words may only be used prior to an EXECUTE statement:

D. C. STEADY STATE	OPTIMIZATION CRITERIA
DEVICES	PLOT
FUNCTIONS	PREFIX
GLOBAL DEVICE PARAMETERS	PRINT
MODELS	
NO DEVICE LIBRARY	RECORD
NO MODEL LIBRARY	RESTART

The following key words may only be used after an EXECUTE statement:

*	GLOBAL CHANGE
CHANGE	

5.4.2 Names

For the sake of easy reference, every element, model, device, component, node, and function is assigned a name. Symbolic parameters and tables in models are also assigned names. A name is a series of *alphanumeric* characters. The alphanumeric characters are the letters **A–Z** and the numerals **0–9**. In general, names may be entirely alphabetic, entirely numeric, or mixed.* Frequently, names will be joined together to form compound names. In this case, the names are separated by a period (.). The allowable length for various names is shown in Table 5.2.

Table 5.2 Allowable Length of Names

Name	Maximum Length of Name (Characters)
Element, component, or node	4
Symbolic parameter, table or function	6
Device	8
Model	80

*Exceptions are that the first characters of element names, symbolic parameters, and table names must be alphabetic in order to satisfy other CIRCUS conventions.

5.4.3 Numbers

Numbers may be input to CIRCUS using either standard format or a format that corresponds to scientific notation. The standard format is exactly like normal usage; integers may be written with or without decimal points. The format corresponding to scientific notation differs somewhat from normal usage: 2.63×10^{-6} is represented by **2.63E-6**, where "E" is analogous to "times 10 to the power." The decimal point of the number may be omitted, but the exponent must not have a decimal point. Values must be in the range 10^{-38} to 10^{38} and may be positive or negative.* Some examples showing correct and incorrect number representations are given in Table 5.3.

Table 5.3 *Numbers*

Valid	Not Valid
26.4	$2.63 \times 10 - 6$
−11	24.E4.4
26.3E–7	.1E39
263 E–8	26.E–40
.263E–5	

5.4.4 Scaling Factors

In many circuits, element values are specified in units other than the fundamental units. That is, resistor values are expressed in kilohms, capacitors in microfarads or picofarads, current in milliamperes, etc. Numbers can be scaled by adding a suffix that represents the multiplicative factor as shown in Table 5.4.

Table 5.4 *Scaling Factors*

Symbol	Factor	Associated Scale Name
P	10^{-12}	Pico
N	10^{-9}	Nano
U	10^{-6}	Micro (μ)
T	10^{-3}	Milli (thousandth)
K	10^{3}	Kilo
M	10^{6}	Mega
G	10^{9}	Giga

*On some computers (such as the IBM System/360), numbers may exceed this range during calculation, but the exponent field of an input number must be less than 39 in magnitude.

5.4.5 Data Card Format

The key words, names, and numbers in an input statement are separated by *separators* and are punched onto one or more data cards. The separators are comma (,), equal mark (=), and left parenthesis ((). The appropriate separator depends on the statement type. A statement can be continued onto another card simply by coding an appropriate separator after the last entry on the card and beginning the next card with the next entry. The upper limit on the number of cards for a statement is equal to the number of entries in the statement. The last entry of a statement is not followed by a separator, and the first entry of the next statement must begin on the next card.

When statements are processed, all blanks are ignored so that blanks may be used in names, numbers, and key words to improve readability. No entry in any input statement is required to begin in any particular column on the card (except when defining model equations).

5.4.6 Input Preparation

The following steps are taken in preparation for a steady-state and/or transient analysis:

1. Assign node names to each distinct node of the network.

2. Assign component labels to each distinct component (resistor, inductor, device, etc.).

3. Select models for each network device type (transistor, diode, etc.). The model may come from a permanent model library or may be defined by a set of input statements.

4. Define device parameters for every device used in the network. The parameters may be obtained from a permanent device library, or they may be specified by input statements.

5. Describe the network by writing a set of statements (one per component) giving the label, node names, and value or device name for each component.

6. Provide display specifications. A set of statements is written defining which network variables will be displayed and the time points when the output will be displayed.

7. Define execution control. A statement is required for each execution option desired; an EXECUTE statement signals that the data list is complete and initiates processing.

Suppose that a transient analysis starting from a dc steady state is required for the example network in Fig. 5.1(a). In particular, we wish to examine the rise and fall times of the output voltage in response to a trape-

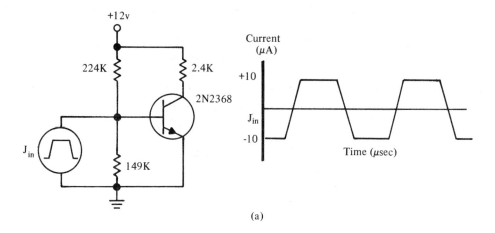

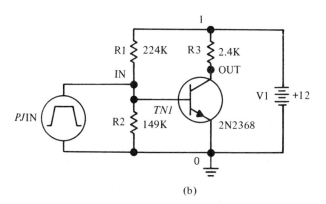

Figure 5.1 Input Preparation Example Circuit.

zoidal-wave input current. The first step is to assign labels to each circuit component and node names to each node as shown in Fig. 5.1(b). The next step is to choose a model for the transistor. An NPN charge control transistor model is specified. The statement in group A of Table 5.5 indicates that the prefix **TN** will designate this model. The model is assumed to reside on a permanent model library so that the model is not described. The specific device type is **2N2368**. The device parameters for a 2N2368 charge control transistor are assumed to reside in the permanent device library so the user need not specify them.

At this point, the input data for the circuit description can be written. These statements form group B of Table 5.5.

Group C of Table 5.5 indicates that a tabular listing of base current, emitter current, and the voltage of the output node OUT is to be printed at intervals of 2 microseconds for a duration of 100 microseconds.

Group D specifies that the transient analysis is to run to 100 microsec-

Table 5.5 *Input for Example Circuit*

```
LIMITS = 10                                          ⎫
PREFIX, TN = NPN CHARGE CONTROL TRANSISTOR           ⎬ A
                                                     ⎭
    'ONE TRANSISTOR AMPLIFIER'                       ⎫
PJIN, 0, IN, −10 UA, 10 UA, 10 USEC, 10 USEC, 30 USEC,│
10 USEC, 80 USEC                                     │
R1, IN, 1, 224K                                      │
R2, IN, 0, 149K                                      ⎬ B
R3, 1, OUT, 2.4K                                     │
TN1, 0, IN, OUT, 2N2368                              │
V1, 1, 0, 12.                                        ⎭
PRINT, I.RB.TN1, I.RE.TN1, VN.OUT                    ⎫
PRINT INTERVALS, 2E-6, 100E-6                        ⎬ C
                                                     ⎭
STOP TIME = 100 USEC                                 ⎫
EXECUTE                                              ⎬ D
END OF JOB                                           ⎭
```

onds. The EXECUTE card initiates execution, and the END OF JOB card terminates execution.

5.4.7 Circuit Description Rules

1. Every circuit must have exactly one *reference node* or ground. This node is at zero potential, and the potential at every other node (*node voltage*) is measured relative to this node. A node may be specified as a ground or reference node either by naming it 0 (zero) or by using the following statement:

    REFERENCE NODE = node name

    where *node name* is the alphanumeric name of some node in the circuit. If a REFERENCE NODE statement is used, the listed node will be the reference node even if the circuit has a node named 0.

2. Each node in a circuit must have a path through circuit elements to ground. The circuit shown in Fig. 5.2 is illegal. This circuit can be made legal by changing the name of node 5 (or 4) to 0.

3. It is illegal to put voltage sources in parallel or current sources in series.

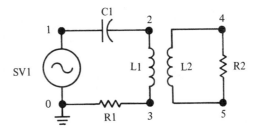

Figure 5.2 Circuit Lacking Path to Ground.

5.4.8 Circuit Description Data Statements

CIRCUS can analyze circuits which consist of the following components: resistors, capacitors, inductors, mutual coupling, fixed and time-varying voltage and current sources, rational impedances and admittances, and devices. The following sections specify the form of the descriptive data, required for each component.

5.4.8.1 Fixed-Value Element Input Statements

The format for a linear resistor, capacitor, inductor, fixed voltage source, or fixed current source input statement is

$$name, node_1, node_2, value$$

where

name = alphanumeric name of four or fewer characters beginning with the letter **R, C, L, V,** or **J**, respectively, for a resistor, capacitor, inductor, fixed voltage source, or fixed current source; this is the name of the element

$\left.\begin{array}{l}node_1\\node_2\end{array}\right\}$ = alphanumeric names of four or fewer characters denoting the nodes to which the element is connected

value = value of the element in ohms, farads, henries, volts, or amps, respectively, or in some other consistent set of units

Example.

R3, 4, 0, 15K
CIN, A3, 4, 6.4P
LOAD, OUT, IN, 7E−6
V100, P100, 0, 100
J3MA, B3, E3, 3.E−3

Positive current is assumed to flow from $node_1$ to $node_2$ (i.e., from the first listed node to the second) in an element, and it is assumed that $node_1$ is plus and $node_2$ is minus so the voltage across an element is given by (voltage at $node_1$) − (voltage at $node_2$).

5.4.8.2 Mutual Coupling

The format for specifying mutual inductance is

$$name, inductor_1, inductor_2, value$$

where

name = alphanumeric name of four or fewer characters beginning with the letter **K**; this is the name of the coupling coefficient

$\left.\begin{array}{l}inductor_1\\inductor_2\end{array}\right\}$ = names of the coupled inductors in any order

value = value of the coupling coefficient

Example.

K13, LTR4, LTR6, .7

The value must be between -1 and 1. If inductor$_1$ has the value L_1 and inductor$_2$ has the value L_2 and the value of the coupling coefficient is K, then the mutual inductance M is given by

$$M = K\sqrt{L_1 L_2}$$

If the coupling coefficient is positive, a positive voltage on inductor$_2$ induces a negative current in inductor$_1$. If the coefficient is negative, a positive voltage on inductor$_2$ induces a positive current in inductor$_1$.* For example, if the following three statements are given,

LA, 1, 2, .01
LB, 8, 6, .01
KAB, LA, LB, K

then Fig. 5.3 depicts the coupling coefficient sign convention.

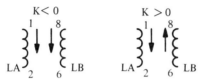

Figure 5.3 Mutual Coupling.

5.4.8.3 Pulsed Sources

The format for a pulsed source input statement is

name, node$_1$, node$_2$, v_0, v_1, t_0, t_r, t_d, t_f, t_p

where

name = alphanumeric name of four or fewer characters beginning with the letters **PV** for a pulsed voltage source, or **PJ** for a pulsed current source

$\left.\begin{array}{l}node_1\\node_2\end{array}\right\}$ = alphanumeric names of four or fewer characters denoting the nodes to which the element is connected

v_0 = initial value of pulsed source

v_1 = peak value of pulsed source

t_0 = time at which first pulse begins (delay time) in seconds

t_r = rise time in seconds

t_d = duration of peak level v_1 in seconds

t_f = fall time in seconds

t_p = period in seconds

*The form

name, capacitor$_1$, capacitor$_2$, value

will produce a similar coupling for capacitors.

Examples.

PJIN, PIN3, PIN6, 0, 3.E−5, 10.E−9, 0, 5.E−9, 20.E−9, 30.E−9
PV2, IN, 0, 2.6, −2.6, 0, 1.E−6, 12.E−6, 1.E−6, 6.E−6

Current and voltage sign conventions are the same as for fixed sources. v_0 may be less than, equal to, or greater than v_1; the signs of v_0 and v_1 should be chosen in accordance with the sign convention. The value v_0 will always be used in solving for dc initial conditions. At least one of the values t_r, t_d, or t_f must be greater than zero. A pulsed source waveform is shown in Fig. 5.4.

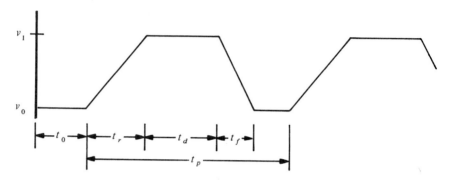

Figure 5.4 Pulsed Source.

5.4.8.4 Sine-Wave Sources

The format for a sine-wave source input statement is

$$\text{name, } node_1, node_2, v_1, v_a, t_0, t_p$$

where

- *name* = alphanumeric name of four or fewer characters beginning with the letters **SV** for a sine-wave voltage source, or **SJ** for a sine-wave current source
- $\left.\begin{array}{l}node_1\\node_2\end{array}\right\}$ = alphanumeric names of four or fewer characters denoting the nodes to which the element is connected
- v_1 = mean value of sine-wave source
- v_a = amplitude of sine-wave source
- t_0 = time at which sine wave begins (delay time) in seconds; if t_0 is less than zero, the sine wave is assumed to be running at time zero; if t_0 is greater than or equal to zero, the source value is constant at v_1 until time t_0
- t_p = period in seconds

Examples.

SV3, K6, K7, 3.5, .5, −.5E−6, .2E−6
SJUT, 6, 0, 0, 100.E−6, 5.E−9, 10.E−9

The sign of v_1 should be chosen in accordance with the sign convention. If dc initial conditions are desired, t_0 should be greater than or equal to zero for every sine-wave source. A sine-wave source waveform is shown in Fig. 5.5.

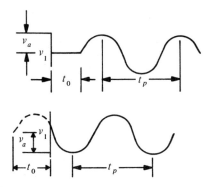

Figure 5.5 Sine Wave Source.

5.4.8.5 Tabular Sources

The format for a tabular source input statement is

$$name, node_1, node_2, (t_1, \ldots, t_n), (v_1, \ldots, v_n)$$

where

$name$ = alphanumeric name of four or fewer characters beginning with the letters **TV** for a tabular voltage source, or **TJ** for a tabular current source

$\left.\begin{array}{c}node_1\\node_2\end{array}\right\}$ = alphanumeric names of four or fewer characters denoting the nodes to which the element is connected

$t_1, \ldots, t_n$ = tabulated values of time in ascending order, two adjacent values of t may be equal to create a step in the source

$v_1, \ldots, v_n$ = tabulated values of the source

Examples.

TVB, PLUS, MINS, (50 US, 130 US, 160 US, 270 US, 270 US),
(0. , −7 , −2 , 5 , −4)
TJ4, 7, 4, (0, 0, 1N, 2N, 10N, 1),
(0, 1.E−3, 5.E−4, 2.E−4, 1.E−6, 0)

The signs of v_i should be chosen in accordance with the sign convention. The value v_1 will be used in computing initial conditions. There is no limit to

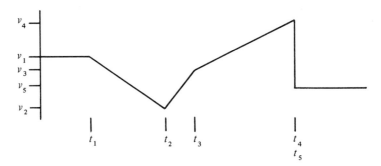

Figure 5.6 Tabled Source.

the number of points in a tabular source table. A tabled source waveform is shown in Fig. 5.6.

5.4.8.6 Rational Impedances and Admittances

If the impedance or admittance of a two-terminal element is known as a function of the Laplace parameter s ($s = 2\pi jf$, where f is the frequency), then this element can be input directly into CIRCUS-2. The impedance or admittance can be written as a rational function in factored form,

$$Z(s) = C\frac{(s - A_1)(s - A_2) \cdots (s - A_n)}{(s - B_1)(s - B_2) \cdots (s - B_m)}$$

where the constants A_i and B_i represent the system zeros and poles, respectively, and may be real or complex. The order of the denominator polynomial must exceed the order of the numerator polynomial by exactly 1.

The format for an impedance or admittance is

$$\text{name, node}_1, \text{node}_2, C, (RA_1, IA_1, RA_2, \ldots, RA_n, IA_n),$$
$$(RB_1, IB_1, RB_2, \ldots, RB_m, IB_m)$$

where

name = alphanumeric name of four or fewer characters beginning with the letter **Z** for an impedance, or **Y** for an admittance

$\left.\begin{matrix}\text{node}_1\\\text{node}_2\end{matrix}\right\}$ = alphanumeric names of four or fewer characters denoting the nodes to which the element is connected

C = the coefficient C

$RA_1, IA_1, \ldots, RA_n, IA_n$ = real and imaginary parts of the zeros; if A_i is real, then $IA_i = 0$. If A_i is complex, then its complex conjugate is assumed to be a zero and should not be listed

$RB_1, IB_1, \ldots, RB_m, IB_m$ = real and imaginary parts of the poles; the same conventions are used as for the zeros

Example.

The admittance

$$Y(s) = 3.0 \frac{(s^2 + 1)}{s^3 + 6s^2 + 11s + 6} = \frac{3.0(s + j)(s - j)}{(s + 1)(s + 2)(s + 3)}$$

is entered into CIRCUS as

Y13, 7, 4, 3.0, (0., 1.), (−1., 0., −2., 0., −3., 0.)

where the admittance was named Y13 and was assumed to be connected between nodes 7 and 4.

5.4.8.7 Devices

Before a component (other than those described in Secs. 5.4.8.1 through 5.4.8.6) can be used in a CIRCUS analysis, a model for the component type must be defined and parameters describing the particular component must be supplied. Models and devices can be used from a permanent library, or either models or devices can be defined as part of the circuit analysis. The procedure for model and device definition is given in Sec. 5.6. The following sections describe how to include a particular device as a circuit component.

5.4.8.7.1 Prefix Statement

When a model is created, it is given a name of 80 or fewer alphanumeric characters. To simplify the reference to a model name, the user assigns a distinct label prefix to each model by use of a PREFIX statement. The format for a PREFIX statement is

PREFIX, *md* = *model name*

where

md = assigned label prefix consisting of exactly two alphanumeric characters

model name = name of the model

Examples.

PREFIX, TN = NPN CHARGE CONTROL TRANSISTOR
PREFIX, QN = NPN CHARGE CONTROL TRANSISTOR
PREFIX, TX = H PARAMETER TRANSISTOR

At least one PREFIX statement is required for each model in the circuit. If there is more than one PREFIX statement, all PREFIX statements must be placed consecutively. PREFIX statements must appear before devices reference statements. In any circuit description using the example above, the prefixes TN and QN would be interchangeable.

5.4.8.7.2 Device Reference Statements

After label prefixes have been assigned to all models to be used in a network analysis, devices can be entered into the circuit description as follows:

$$name, node_1, \ldots, node_n, device$$

where

name = alphanumeric name of four or fewer characters beginning with the label prefix of the model

$\left. \begin{array}{c} node_1 \\ \cdot \\ \cdot \\ \cdot \\ node_n \end{array} \right\}$ = nodes to which the device is connected, listed in the same order as in the model

device = name assigned to the specific device

Examples.

TN6, E, B, C, 2N1613
TX7, E, B, C, 2N1613

A model can be defined in such a manner that it has no parameters (Sec. 5.6). In this case, the model is itself a device and the device reference statement takes the form

$$name, node_1, \ldots, node_n$$

where *name* and *node$_i$* are as above.

5.4.9 Basic Control Statements for Transient Analysis

Stop Time Statement

To perform a transient analysis, the user must specify a maximum circuit transient time. The transient analysis will continue until the transient time reaches the maximum time. The format for specifying a maximum time is

STOP TIME = *time*

Execute Statement

The user specifies that all the data and control information required for the analysis has been listed and that the analysis should begin by the statement

EXECUTE

When an EXECUTE statement is encountered, all the previously defined analyses will be performed, and then the succeeding statements will be processed for further instructions.

Terminating a Circuit Analysis

To specify that all statements defining a circuit analysis have been listed and that definition of a new circuit is to follow, an

END

statement should be used. The END statement erases all memory of the previous network and initiates processing of the new one. Statements should follow an END statement as if they were at the beginning of a new job.

When all statements defining the analyses of the final circuit have been listed and execution should be terminated, an

END OF JOB

statement should be used. The END OF JOB statement is always the last statement in a CIRCUS deck.

5.4.10 Naming Conventions

Up to this point, we have defined the basic statements needed to specify all components in a network. Other CIRCUS data statements described in the sections to follow need to reference properties of the components and of the network nodes, such as the voltage across an element or a particular node voltage. In order to do this, each network parameter is assigned a unique symbolic name. The naming conventions are as follows:

Component Name. A name of four or fewer characters given to the component when it is entered into the network. For devices, the first two characters are a prefix associated with the model. For example, if

DI2, 9, 4, 1N914
DI3, 6, 0, 1N914
TN6, E, B, C, 2N1613

are device reference statements, the component names are *DI2*, *DI3*, and *TN6*.

Device Class. A device class is a label prefix assigned to the model modified by the device name. For example, devices DI2 and DI3 above belong to device class *DI.1N914*.

Device Parameter. The name of a device parameter is the name of the parameter within the model modified by the name of the component. For example, if the NPN TRANSISTOR has a parameter named BN, the name of BN in component TN6 is *BN.TN6*.

Element Name. The name of an element is the name assigned to it by CIRCUS when it is entered into the network. If the element is internal to a device, the name is then the name of the element within the model modified by the name of the component. *L3* is an example of a circuit element, and *RB.TN6* is an example of an internal element.

Element Voltage. An element voltage is referred to by *V. element name*, for example, *V.R1*, *V.TJIN*, and *V.JE.TN6*.

Element Current. An element current is referred to by *I. element name*, for example, *I.C3*, *I.SV99*, and *I.RB.TN6*.

Node Name. The name of a node internal to a device is the name of the node within the model modified by the name of the component. An example is *BASE.TN6*.

Node Voltage. A node voltage is referred to by *VN. node name*, for example, *VN.6*, *VN.PIN3*, and *VN.BASE.TN6*.

5.4.11 Functions

A user may wish to compute a parameter that reflects the behavior of the network rather than the behavior of a particular element in the network. A difference of node voltages, a ratio of branch currents, or a sum of dissipated power for several elements are frequently of interest. To accomplish this, a function that computes the desired network parameter may be defined according to the rules for a FORTRAN statement function. The format is

FUNCTIONS
$function_1 \ (v_1, \ldots, v_k) = f_1(v_1, \ldots, v_k)$
$function_n \ (v_1, \ldots, v_k) = f_n(v_1, \ldots, v_k)$
END

where

$function_i$ = alphanumeric function name of six or fewer characters

$v_1, \ldots, v_k$ = distinct dummy alphanumeric names of six or fewer characters, called arguments

$f_i(v_1, \ldots, v_k)$ = a FORTRAN statement function in $v_1, \ldots, v_k$.

Examples.
  FUNCTIONS
  VDIF (VN1, VN2) = VN1 − VN2
  GAIN (I1, I2) = ABS (I2/I1)
  POWR (I1, V1, I2, V2, I3, V3) = I1∗V1 + I2∗V2 + I3∗V3
  END

The procedure for writing a statement function is the same as that for an expression in a model equation (Sec. 5.6.3.4). Any predefined function subprogram (Table 5.9) may be used in defining the statement function; however, the name of the predefined function subprogram must not be used as an argument. The arguments may be any of the following: device parameter, element name, element voltage, element current, node voltage, or a constant written as described in Sec. 5.4.3.

5.4.12 Parameter Variation

After a transient analysis, it may be desirable to modify some of the network or device parameters and repeat the transient analysis. Two types of statements are available for this purpose: an ∗ statement is used to modify values of circuit elements; a CHANGE statement is used to replace one device with another and to modify device parameters.

5.4.12.1 Circuit Element Modification

The format for an ∗ statement of an R, C, L, K, V, or J type of element is

$$\ast name = \begin{bmatrix} value \\ function\ (a_1, \ldots, a_n) \end{bmatrix}$$

The format for an ∗ statement for a PV, PJ, SV, SJ, TV, or TJ type of element is

$$\ast name = List$$

or

$$\ast name = \begin{bmatrix} \text{SHIFT} \\ \text{SCALE} \end{bmatrix} = \begin{bmatrix} value \\ function\ (a_1, \ldots, a_n) \end{bmatrix}$$

where

$name$ = name of the element to be modified

$value$ = new value of the element

$function$ = name of a function defined in a FUNCTIONS list

$a_1, \ldots, a_n$ = arguments to the function; the legal arguments are T representing the time at the end of the previous transient, ∗ representing the previous value of the parameter, and I representing a count of the number of transient solutions

$List$ = list of values necessary to define a pulsed, sine-wave, or tabled source

SHIFT = indicates that the amount specified by the value or function is to be added to each dependent element in the table

SCALE = indicates that the amount specified by the value or function multiplies each dependent element in the table

Examples.

```
FUNCTIONS
CCHANG (X) = X + 10.E−12
END
EXECUTE
*RCTS = 55.
*COUT = CCHANG (*)
*PV2R = 5., −1., 10N, 1N, 98N, 1N, 200N
*SJ16 = SCALE = 2.
*TVB = (1E−4, 2E−4, 4E−4, 5E−4), (0, 2E2, 4E2, 0)
```

5.4.12.2 Device Replacement

A device may be replaced by another device if both the original device and the replacement device were built from the same model. Replacement may be made for a component or for a device class. The format for specifying device replacement is

$$\text{CHANGE, } (old_1, new_1), \ldots, (old_n, new_n).$$

where

old_i = component name or a device class (Sec. 5.4.10)

new_i = device name

Example.

```
PREFIX, TN = NPN CHARGE CONTROL TRANSISTOR
PREFIX, DI = DIODE
DI2, 8, 4, 1N914
DI3, 6, 0, 1N914
TN6, E, B, C, 2N1613
  .
  .
  .
EXECUTE
CHANGE, (TN6, 2N1711), (DI.1N914, 1N916)
EXECUTE
```

After the CHANGE statement, component TN6 will be *NPN* charge control transistor 2N1711 and components DI2 and DI3 (and any other components belonging to device class DI.1N914) will be diodes 1N916 for all subsequent analyses until changed by other device replacement statements. Any changes made to the parameters of devices 2N1711 and 1N916 prior to the device replacement will not apply to TN6, D12, or DI3. Changes made to the parameters following device replacement do apply.

Each time that a device replacement CHANGE statement is processed, the device libraries are sequentially searched until every replacement request has been satisfied. Therefore, the most efficient means of specifying device replacement is to put all required replacements into one CHANGE statement as shown in the example. Separate CHANGE statements must be used for device replacement and for device parameter modification.

5.4.12.3 Device Parameter Modification

Any device parameter may be modified by means of a CHANGE statement. Tables may be modified by multiplying each dependent variable by a scale factor, adding to each dependent variable a shift value, or supplying a whole new table. Any number of scalar parameters and tables to be scaled or shifted can be modified with one CHANGE statement as long as they are parameters of the same device or device class. The format for a CHANGE statement is

$$\textbf{CHANGE}, device, p_1 = V(p_1), \ldots, p_n = V(p_n)$$

where

 $device =$ the name of the device or device class being modified.

 $p_i =$ the name of the parameter (scalar or tabular) being modified.

 $V(p_i) =$ may take one of several forms.

If p_i is a scalar parameter, $V(p_i)$ may take the form

 $value =$ new value for the parameter

or

 $function\ (a_1, \ldots, a_n) =$ same as for $\star$ statement

If p_i is a tabular parameter, $V(p_i)$ may take one of the forms

 SCALE $= value$

 SCALE $= function\ (a_1, \ldots, a_n)$

 SHIFT $= value$

 SHIFT $= function\ (a_1, \ldots, a_n)$

or $V(p_n)$ may take the form of a list of values defining a new table.

In the following example, assume that RANDOM is a function subprogram generating random numbers according to some distribution.

Example.
```
FUNCTIONS
ADD 6 (X) = X + 6.
RNDM (X) = RANDOM (X)
END
    .
    .
    .
EXECUTE
```

CHANGE, DI2, RS = 1.2E8, RB = ADD6 (*)
CHANGE, DI3, THETA = RNDM (I)
CHANGE, TN.2N1613, TCN = SCALE = 1.3,BN = (1.E−5, 5.E−4,
   1.E−4, 5.E−4), (37.1, 45.3, 49.6, 36.9)
EXECUTE

5.4.12.4 Global Change

If several different devices have parameters with the same name and it is desired that they have the same values, a GLOBAL CHANGE statement can be used to modify their values. The format for a GLOBAL CHANGE statement is

$$\text{GLOBAL CHANGE, } p_1, v(p_1), \ldots, p_k, v(p_k)$$

where p_i and $v(p_i)$ have the same meaning as for a CHANGE statement. If a GLOBAL CHANGE statement is used, every device in the circuit will be searched for a parameter with the name p_i, and each one found will have its value changed to the value specified by $v(p_i)$.

The CIRCUS program has no means for determining whether p_i is spelled correctly because it is not necesssary for every device to contain parameter p_i. The user must be especially careful when preparing input and when checking results to ensure that all changes were made correctly.

Example.

GLOBAL CHANGE, CTEMP = −55, GAMMAD = 1.E8,
RB = ADD6 (*)

5.4.13 Additional Steady-State Control Statements

5.4.13.1 Specifying Initial Conditions

Normally CIRCUS-2 computes a steady state for the network and uses that steady-state value as initial conditions for the transient analysis. The user can direct the program to skip the steady-state computation and use given values for initial conditions. The format is

$$\text{INITIAL CONDITIONS, } name_1 = v_1, \ldots, name_n = v_n$$

where

$name_i$ = name of an element voltage or current

v_i = initial value for that voltage or current

Voltages for every capacitor and currents for every inductor must be listed. No other parameters need be given.

Example.

INITIAL CONDITIONS, I.LXI.MAG3 = 3.E−4, V.C3 = 2.5

5.4.13.2 Steady-State Guess

CIRCUS-2 uses an iterative method to compute steady state. If a network is multistable, it will have more than one valid steady-state solution. The iterative process may converge to any one of these solutions. If the user wants one particular solution of the several steady states and he has some idea about what this solution will look like, he should supply a first guess at the solution. If a network is very unstable, a good first guess can be a significant aid in solving the steady-state problem. The user can supply an initial guess at the steady-state as follows:

$$\text{STEADY STATE GUESS, } name_1 = v_1, \ldots, name_n = v_n$$

where

$name_i$ = name of an element voltage or current

v_i = guessed value

A guess of zero will be made for each element current or voltage not listed (Sec. 5.3.2.5).

Example.

$$\text{STEADY STATE GUESS, V.CE.TN6} = .5, \text{V.CC.TN6} = -1.$$

5.4.13.3 Multiple Steady-State Analysis

A variation of parameters analysis on the dc steady state of a network can be performed. The nominal circuit (the circuit as defined by the circuit description data statements) is solved for steady state. Then circuit parameters are varied through a series of values, and a steady-state analysis is performed on each of the modified circuits. The format for specifying steady state-analysis is

> D.C. STEADY STATE
> *Parameter variation list$_1$*
> .
> .
> .
> *Parameter variation list$_n$*
> END

Each parameter variation list has a list of values for all parameters that will be varied simultaneously. More than one parameter variation list can be used to vary parameters independently. The format for a parameter variation list is

$$name_1 = (V_1, \ldots, V_k), \ldots, name_m = (V_1, \ldots, V_k)$$

where

$name_i$ = name of an element or device parameter

$(V_1, \ldots, V_k)$ = list of values for the preceding parameter

If $name_i$ is the name of a tabular device parameter, the list takes the special form

$$name_i = \textbf{SHIFT} = (V_1, \ldots, V_k)$$

or

$$name_i = \textbf{SCALE} = (V_1, \ldots, V_k)$$

Example.

    D.C. STEADY STATE
    RB.TN1 = (950, 1050), PJIN = SHIFT = (.05E−5, −.05E−5),
       R3 = (2.6E + 3, 2.2E + 3)
    IES.TN1 = (4.75E − 16, 5.E − 16, 5.25E − 16),
       R2 = (147E + 3, 149E + 3, 151E + 3)
    R1 = (220E + 3, 222E + 3, 226E + 3, 228E + 3)
    END

The first parameter variation list specifies that three parameters (RB.TN1, PJIN, and R3) are to be simultaneously varied through two values. The second parameter variation list specifies that two parameters (IES.TN1 and R2) are to be varied through three values simultaneously but independently of the parameters in the first variation list. The last parameter variation list specifies that R3 is to be varied through four values independently of the parameters in the first two variation lists. Thus, a nominal solution and 24-variant dc steady-state solutions will be found.

5.4.13.4 Selection of Optimum Parameter Values

If a series of dc steady-state solutions is to be computed, a criterion for choosing an optimal solution from the series is specified as follows:

$$\textbf{OPTIMIZATION CRITERION} = name$$

where *name* is any legal steady-state print name. The optimal solution is the solution that gives the smallest value of the named variable. The values of the parameters that generate the optimal solution replace the nominal values for the transient solution, and the initial condition for the network is the steady state associated with the optimal solution.

The optimization criterion is ordinarily used with a function. For example, if it is desired that the voltage at node 3 be as close to 5 volts as possible, then the following statements might be used:

    FUNCTIONS
    NEAR5 (X) = ABS (X−5)
    END
    OPTIMIZATION CRITERION = NEAR5 (VN.3)

where *ABS* indicates absolute value. As VN.3 approaches 5 volts, VN.3—5 approaches zero. Hence, the smallest value of the function NEAR5 is zero, and it is obtained when VN.3 equals 5.

5.4.14 Multiple Transient Analysis

If the user wishes to perform a series of transient analyses while systematically varying some circuit or device parameters, he should use the statement

MULTIPLE EXECUTE, n

where n is the number of transients in the series. A MULTIPLE EXECUTE statement must be followed by an

END MULTIPLE EXECUTE

statement. Statements specifying changes to circuit and device parameters are placed between the MULTIPLE EXECUTE and END MULTIPLE EXECUTE statements.

Example.

```
        FUNCTIONS
        ADD6 (X) = X + 6
        RNDM (X) = RANDOM (X)
        END
        .
        .
        .
        MULTIPLE EXECUTE, 10
        *RB.TN6 = ADD6 (*)
        CHANGE, TN6, BN = SCALE = RNDM(I)
        END MULTIPLE EXECUTE
```

5.4.15 Utility Statements

The following statements are available to provide additional flexibility in defining a problem and performing an analyses:

Hold Final Conditions

When a HOLD FINAL CONDITIONS statement is used, a transient solution is computed up to the time indicated in the STOP TIME statement. After parameter changes are made, the transient is continued from the point in time where it stopped.

Save-Restart

If a SAVE statement is placed in the input, data describing the state of the network will be written on a magnetic tape, upon termination of the transient analysis. A RESTART statement is all that is required to recover the network from the tape and resume analysis.

Card Length

If a user wants to utilize fewer than 80 columns of the input cards for data, a card of the form

$$\text{CARD LENGTH} = n$$

is used, where n is the last column of the card that is used for input data.

Limits

To put a limit on the amount of computer time spent on any circuit, the following statement is used:

$$\text{LIMITS} = s$$

where s is the computer time limit in seconds.

Nolist—List

A NOLIST statement in the input deck will turn off the usual listing of input statements. No input card after a NOLIST card will be listed until a LIST card is encountered.

5.4.16 Input Order

When used in a CIRCUS analysis, the following statements should always be listed in the order

> CARD LENGTH
> LIMITS
> MODELS
> DEVICES
> PREFIX

All other statements should follow the above listed statements.

Each of the statements shown in Table 5.6 is followed by some form of END statement. Only statements of the type described in the appropriate

Table 5.6 *Input Order*

Statement	Associated End Statement
MODELS	END OF INPUT
DEVICES	END OF INPUT
GLOBAL DEVICE PARAMETERS	END OF INPUT
FUNCTIONS	END
D.C. STEADY STATE	END
MULTIPLE EXECUTE	END MULTIPLE EXECUTE

section may appear between any of these statements and its associated END statement.

Any statements other than GLOBAL DEVICE PARAMETERS that modify circuit or device parameters (i.e., CHANGE and * statements) must be placed after the first EXECUTE or MULTIPLE EXECUTE statement.

Aside from these restrictions, input statements may be listed in any order.

5.5 OUTPUT SPECIFICATIONS

CIRCUS can display the results of a circuit analysis in each of three types of printed output, and it can record time history data on magnetic tape. The printed output can be tabulated time histories, time history plots, or plots of one variable against another variable. This output is generated on the computer line printer. The user has complete freedom to select which variables are displayed and to choose the time intervals at which the time history is tabulated. He also can control the format of the time history plots. Item versus item plots are put onto one printer page.

Any of the following can be displayed or recorded: device parameters, element names, element voltages, element currents, node voltages, or any functions of the preceding. Two types of control statements are required: a list of output variables, and the time values at which the output is to be displayed. The following pairs of statements are used:

PRINT, *list*
PRINT INTERVALS, *tlist*
PLOT, *list*
PLOT INTERVALS, *tlist*
RECORD, *list*
RECORD INTERVALS, *tlist*

For printed or recorded output, *list* has the form

$$v_1, \ldots, v_n$$

where v_i is the name of an output variable. For plotted output, *list* has the form

$$v_1, \ldots, v_k, (x_1, y_1), \ldots, (x_n, y_n)$$

where v_i is the name of an output variable to be plotted against time and x_i and y_i are names of output variables to be plotted against one another. Variable versus variable plot requests must follow all variable versus time plot requests. Any item in a PRINT or PLOT list that CIRCUS-2 does not recognize will be ignored. All items that are validly specified will be displayed. If there is any error in a RECORD list, the program will terminate the solution of the circuit containing the error.

For all intervals statements, *tlist* has the form

$$\Delta t_1, t_1, \ldots, \Delta t_n, t_n \qquad (t_1 < t_2 < \cdots < t_n)$$

Output is displayed at intervals of Δt_1 between 0 and t_1, at intervals of Δt_2 between t_1 and t_2, etc.

The intervals statements provide the means for time editing the output from CIRCUS. If a particular intervals statement is not supplied, the output is not edited for that display format (that is, requested variables will be displayed at every computed time point). When PRINT INTERVALS are specified, the printed output will be tabulated at the specified time points only. For plot or record output, the requested variable will be displayed at both specified time points and computed time points. The use of a RECORD INTERVALS statement is somewhat unusual because the addition of interpolated points is unnecessary for most RECORD applications.

Examples.

    PRINT, V.R3, I.RB.TN6, VDIF (VN.6, VN.IN)
    PLOT, GAIN(I.RB.TN6, I.RC.TN6), BN.TN6, (I.RB.TN6, I.RC.TN6)
    PRINT INTERVALS, 1.E−9, 100.E−9, 1.E−8, 1.E−6
    PLOT INTERVALS, 1.E−8, 1.E−6

The items that will be printed are the voltage across resistor R3, the current through resistor RB in device TN6, and the value of the function VDIF for the values of the arguments (the voltage at node 6 and the voltage at node IN). This set of variables will be printed every nanosecond for the first 100 nanoseconds, and then at intervals of 10 nanoseconds until 1 microsecond. The value of the function GAIN and the value for the parameter BN in device TN6 will be plotted against time at intervals of 10 nanoseconds for 1 microsecond (101 points). In addition, the values of GAIN and BN.TN6 at the end of each integration step will be plotted. A third plot will display the current through resistor RC in device TN6 (ordinate) against the current through resistor RB in device TN6 (abscissa).

The results from a dc steady-state analysis may be obtained for a set of variables that is different from those requested for the transient analysis. If this independence is required, a

    STEADY STATE PRINT, *list*

statement is used, where *list* is the same as that described for a PRINT statement. If a STEADY STATE PRINT statement is not input, the variables listed in the PRINT statement will be used for both steady-state and transient output. For each steady-state solution, the values of the parameters being varied and the listed output variables are displayed in a table.

The user may get a title printed on each page of output by including a title statement. When the first nonblank character of a statement is an apostrophe ('), that statement is considered to be a case title. A title statement must be punched on one card and cannot be continued onto a second card. Any number of title cards may be used as comments in a job, and they may be placed anywhere in a CIRCUS deck except in models. When more than one title card is used in a case, only the title preceding the EXECUTE or MULTIPLE EXECUTE statement will be printed on the requested output.

5.6 MODELS

The vast majority of circuits that will be analyzed on a digital computer contain nonlinear components. Many nonlinear components, such as transistors, diodes, and ferrite core transformers, are commonly used in circuit design. A less common application, but one just as important, is to determine how a circuit will behave under conditions of extreme stress. Then, components ordinarily considered linear, such as resistors and capacitors, may well behave according to some nonlinear relationship. When any nonlinear component is used in a CIRCUS analysis, its basic structure is defined by a *model*. The parameter values that distinguish one component of a given structure from another component with the same structure are defined in a *device*. Within this framework of model and device, almost any type of component can be simulated in CIRCUS.

5.6.1 Models and Devices in Circus

To present more clearly the idea of what a model is and what a device is (in the CIRCUS sense), we shall develop a model for an *NPN* transistor in common-emitter configuration using h parameters. Once the model has been defined, we can build devices using this model. A standard h-parameter representation of a transistor is shown in Fig. 5.7.

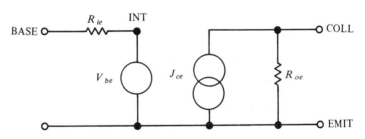

Figure 5.7 h-Parameter Model.

The elements R_{ie} and R_{oe} are fixed-valued resistors, while J_{ce} depends on the current through R_{ie} and V_{be} depends on the voltage across R_{oe}. To begin translating this information into CIRCUS format, a few arbitrary choices are made. The first is to select a *model name*. Then names are chosen for the elements and the nodes to which they are connected. It is best to choose names which have some direct association with the model or the elements in it, since this simplifies future references to the model. We shall use the names that were shown in Fig. 5.7. Also, the nodes that are accessible for external connection are identified.

The model is described to CIRCUS by the following statements:

MODEL NAME = H PARAMETER TRANSISTOR
TOPOLOGY
RIE, BASE, INT
VBE, INT, EMIT
JCE, COLL, EMIT
ROE, COLL, EMIT
EXTERNAL NODES, (BASE, EMIT, COLL)

Now we can define how the current source depends on the base current and how the voltage source depends on the collector-to-emitter voltage by the following statements:

EQUATIONS
JCE = I.RIE*HFE
VBE = V.ROE*HRE
D(JCE, I.RIE) = HFE
D(VBE, V.ROE) = HRE
RETURN
END

Consider the preceding statements one at a time. The first is a CIRCUS control card which specifies that model-defining equations will follow. This card is format-free, but the other cards must be in a prescribed format which will be given later. The next two statements specify the conditions for J_{ce} and V_{be} using the *symbolic parameters* I.RIE, V.ROE, HFE, and HRE and the *operators* * and =. I.RIE is the symbolic form for specifying "the current through resistor RIE," while V.ROE is the corresponding symbolic form for "voltage across resistor ROE." The parameters HFE and HRE are not assigned values at this time, but values will be specified later in a device. (Distinct transistor types, such as the 2N1613 and the 2N1711, will require disvices to specify different values for HFE and HRE, but both devices may use this transistor model.)

The next two statements define the derivatives of J_{ce} and V_{be} with respect to I.RIE and V.ROE, respectively. Derivatives must be defined for all variable-valued components (the derivative of a fixed-valued component is, of course, zero). The special form $D(cv, iv)$ indicates "the derivative of a component value with respect to an independent variable or time."

The statement

RETURN

indicates that we have completed a logical path through the model. For more complicated models, there may be several logical paths in the model (corresponding to different operating regions, for example). The conclusion of each such path is indicated by a RETURN statement.

The final statement

END

indicates that the equation definition section is complete for this model.

Several points should be noted about the manner in which equations are defined for a model:

1. The equation definition section begins with an EQUATIONS statement and ends with an END statement.

2. Each statement begins on a separate card. If a single statement is too lengthy to fit on 1 card, it may be continued onto as many as 19 additional cards.

3. The operators +, −, *, /, and = are used much as in ordinary algebraic notation. Parentheses may be included as required.

4. Symbolic parameters are used rather than numerical constants so that the model represents the entire class *NPN Transistor*, rather than one particular transistor type.

5. Derivatives are specified for variable-valued elements.

The previous example shows in a straightforward way how a model is developed for CIRCUS. There are two types of nonlinear elements included in this model, a current-controlled current source (JCE) and a voltage-controlled voltage source (VBE). There are also two fixed valued elements, RIE and ROE. The value of these fixed elements will not change during a CIRCUS execution, but either one may be changed between executions, as described in Sec. 5.4.12.

Values for the symbolic parameters defined in the model are supplied by devices. Devices built using the H PARAMETER TRANSISTOR model would look like the following:

DEVICE NAME = 2N1613, MODEL NAME = H PARAMETER TRANSISTOR
HFE = 90, HRE = 2.2E−4, RIE = 150, ROE = .156 MEG
DEVICE NAME = 2N1711, MODEL NAME = H PARAMETER TRANSISTOR
HFE = 160, HRE = 1.60E−4, RIE = 110, ROE = .7 MEG

It is devices such as these that are used to represent nonlinear elements in CIRCUS.

5.6.2 Basic Modeling Rules

In general, any element in a model may be variable- or fixed-valued. The choice is made by the user when he prepares input data for the model. If an element is variable, then he supplies the equations which define how it is dependent on element voltages, currents through elements, other element values, or time. The dependence may be described by any expression that has a finite derivative, and this dependence may include tabular functions of one or two independent variables. If an element is fixed-valued, then the value for the element may be specified by an equation in the model, or the value may be assigned later in a device description. The latter method is used most often.

The rules governing definition of a model to CIRCUS are few and, in general, not restrictive. Those constraints that do exist simplify processing of the model by CIRCUS, which results in lower computer costs when using the program.

1. The elements may be resistors, capacitors, inductors, voltage sources, or current sources. Any element value may be fixed or variable. When an element value is variable, the value and its derivatives are specified by the user. Mutual reactance among reactive elements is permitted and, inso far as the basic rules are concerned, is treated like another type of element.

2. Element names and node names are limited to four alphanumeric characters. The first character of an element name identifies what type of element it is, as is shown in Table 5.7. Different elements must have distinct names,

Table 5.7 *Element Label Prefixes for Models*

Element Label Prefix	Element Type
R	Resistor
L	Inductor
C	Capacitor
V	Voltage source
J	Current source
K	Coupling coefficient
M	Mutual coupling

just as different nodes must have distinct names. However, a node and an element may have the same name.

3. Symbolic parameter names are limited to six alphanumeric characters. The first character must be alphabetic.

4. The independent variables on which variable-valued elements depend must be from within the model.

5.6.3 Model Description Data Statement Formats

Several distinct statements are needed to describe a model completely. Some statements may be made on one data card, while other require multiple data cards. The following statements are required:

      MODEL NAME
      EXTERNAL NODES
      TOPOLOGY

The following statements are optional:

> EQUATIONS
> DEFAULT DEVICE PARAMETERS
> FIXED PARAMETERS
> ARRAYS

The MODEL NAME statement must be the first statement in each model. The statement

> END OF INPUT

is required following the last model and should be used there only.

5.6.3.1 Model Name Statement

Each model is given a unique name, 80 alphanumeric characters or less in length. Either of two formats may be used to supply the model name, depending on the length of the name.

Two-card format:

> MODEL NAME =
> *any name up to 80 characters long*

One-card format:

> MODEL NAME = *any name up to 70 characters long*

Example.

> MODEL NAME = NPN CHARGE CONTROL TRANSISTOR

Blanks may be used anywhere in the statement since they are ignored when CIRCUS processes the statement. When the two-card format is used, only blanks may follow the equal mark on the first card.

5.6.3.2 External Nodes Statement

The external nodes statement identifies the nodes that will be used to connect the model to an external network. Each model must have at least two external nodes. The node names are arbitrary, four alphanumeric characters or less in length, and may be in any order. However, the order that is used when referring to the model in a circuit description data statement must correspond to the order defined in the external nodes statement. An external nodes statement may use one or more data cards. The format is

> EXTERNAL NODES = $(node_1, \ldots, node_k)$

where

$node_i$ = node name which appears in the model topology

Example.

> EXTERNAL NODES = (EMIT, BASE, COLL)

5.6.3.3 Topology

The topology statement defines the elements in the model and specifies the nodes to which they are connected. Each element must be one of the basic types shown in Table 5.7. The first character of the four-character name identifies the element type.

The topology statement must contain at least two cards; the format of the first card is

TOPOLOGY

Every other card has the format

$pnam$, $node_1$, $node_2$

where

$pnam$ = alphanumeric element name of four or fewer characters.

$node_1$ = alphanumeric names of four or fewer characters

$node_2$ = denotes the nodes to which the element is connected

Examples.

TOPOLOGY
RB, BASE, TX
CE, TX, EMIT
JE, TX, EMIT

The positive direction of current flow in a two-terminal element is defined to be from the first-mentioned node to the second-mentioned node. The first-mentioned node is defined to be at a higher voltage than the second-mentioned node. There is no ambiguity when these rules are applied to resistors, inductors, or capacitors. When specifying sources, use that rule appropriate to the source type. That is, for positive voltage sources, specify the high-voltage node first, and then the low-voltage node. For current sources, specify the positive direction of current flow as from the first-mentioned node to the second-mentioned node, internal to the source.

5.6.3.4 Equations

The interrelation of model parameters is defined by statements in the equations section. An equation is used to assign a value to each variable element in the model. The partial derivative of the element value with respect to each variable on which it depends is also specified by an equation. A model may consist entirely of fixed-valued, two-terminal elements, in which case there is no functional dependence between elements. For such a model, no equations section is required in the model description.

Equation Format

Equation statements are written in a symbolic language which is a subset of the FORTRAN programming language. The format of the first card of the equations section is

EQUATIONS

The word EQUATIONS may appear anywhere on a data card since it is not column-restricted. All other statements in the equations section must conform to the standard FORTRAN format. These statements are written one to a line in columns 7–72. If a statement is too long for one line, it may be continued onto a maximum of 19 successive lines by placing a nonblank, nonzero character in column 6 of each continuation line. For the first line of a statement, column 6 must be blank or zero.

Columns 1–5 of the first line of a statement may contain a statement label. The statement label is composed of five or less alphanumeric characters (except that the character C in column 1 has a particular meaning; see below). Blanks in a statement label are ignored. Neither order nor value of statement labels is significant. Duplicate statement labels are not allowed. The purpose of statement labels will be shown in the section on control statements.

Columns 73–80 are not used by CIRCUS and may contain identification or sequencing information.

Comments that help explain the equations may be written in columns 2–80 of a line if the character C is placed in column 1. Comments may appear anywhere in the equations section, except after the END statement.

Statement Types

There are three categories of statements that may be used for model definition.

1. *Arithmetic statements* define calculations that are to be performed.

2. *Subprogram reference statements* extend the capacity of the modeling language to include predefined and user-supplied programs.

3. *Control statements* govern the sequence of execution of the arithmetic and subprogram reference statements.

Expressions

Expressions specify a relationship among symbolic variables and constants. The simplest expression consists of a single constant or variable. Constants and variables may be combined with operation symbols to indicate an algebraic calculation (Table 5.8).

Table 5.8 *Operation Symbols*

Symbol	Operation
+	Addition
−	Subtraction
*	Multiplication
/	Division
**	Exponentiation

Certain rules are required in order to avoid ambiguity in defining an expression.

Rule 1.

Any expression may be enclosed in parentheses. Parentheses may be used exactly as in algebraic expressions to specify the order in which arithmetic operations are to be performed. Within parentheses, or where parentheses are omitted, the following order of operations is followed:

1. Evaluation of function subprograms.

2. Exponentiation.

3. Multiplication and division (left to right).

4. Addition and subtraction (left to right).

Neither parentheses nor concatenation with a constant coefficient may be used to indicate multiplication; the asterisk operation symbol must be used for this purpose.

Rule 2.

No two operators may appear in sequence.

Rule 3.

The expression A**B**C is permitted and is evaluated as

$$A**(B**C)$$

Arithmetic Statements

The general format of an arithmetic statement is

$$a = b$$

where

$a = $ variable name

$b = $ valid arithmetic expression

Subprogram Reference Statements

In addition to the five basic arithmetic operations, a function reference of the form

$$f(w, x, \ldots, z)$$

may be used where

f = the function name

( ) = denotes the calling sequence

w, x, z = variables supplied to the function

These variables are called *arguments*. The model equations may reference any of the functions listed in Table 5.9 or a user-written function subprogram.

Table 5.9 *Predefined Function Subprograms*

Subprogram Name	Mathematical Function	Description
EXP (X)	Exponential	e^X
ALOG (X)	Natural logarithm	$\ln X$
ALOG10 (X)	Common logarithm	$\log_{10} X$
ARSIN (X)	Arcsine	$\sin^{-1} X$
ARCOS (X)	Arcosine	$\cos^{-1} X$
ATAN (X)	Arctangent	$\tan^{-1} X$
SIN (X)	Sine	$\sin X$
COS (X)	Cosine	$\cos X$
TAN (X)	Tangent	$\tan X$
SQRT (X)	Square root	$\sqrt{X}$
AMOD (X,Y)	Modular arithmetic	Positive remainder obtained when X is divided by Y
INT (X)	Truncation to an integer	Greatest integer in X
ABS (X)	Absolute value	$|X|$
AMAX1 (X,Y)	Largest value	Maximum of X, Y
AMINI (X,Y)	Smallest value	Minimum of X, Y
CONVLD	Convolution integral	See Sec. 5.6.3.8
CONVLM	Convolution integral	See Sec. 5.6.3.8

A function reference is equivalent to a single-valued variable when used in an arithmetic statement.

Tabular Parameter Reference Statement

Parameters in a model may be either constant-valued or tabular functions of one or two independent variables. The value of a tabular function is determined by a table reference. If the tabular function has one independent variable, the reference statement has the form

TABLE *name* (*idvn, dx, dzdx*)

where

name = symbolic name of the table

idvn = independent variable, which may be a symbolic name or an expression

dx = difference between the next larger tabulated value of the independent variable and the value of *idvn*

dzdx = slope of the function

If the tabular function has two independent variables, the reference statement has the form

TABLE *name* (*x-idvn, y-idvn, dx, dzdx, dy, dzdy*)

where

name = symbolic name of the table

x-idvn = *x* independent variable

y-idvn = *y* independent variable

dx = difference between the next large tabulated value of the *x* independent variable and the value of *x-idvn*

dzdx = slope of the function in the *x* direction

dy = difference between the next larger tabulated value of the *y* independent variable and the value of *y-idvn*

dzdy = slope of the function in the *y* direction

Examples.

BETA = TABLE BN (IBC, DX, DBDI)
JCE = TABLE ICVBE (V.JCE, I.RIE, DUMMY, D(JCE, V.JCE), DUMMY, D (JCE, I.RIE))

A table reference may be used as an operand in an expression just as though it were a symbolic variable. The arguments *name* and *idvn* are supplied by the user. Values for *dx* and *dzdx* are returned to the model. The slope of the function (*dzdx*) is used in calculating derivatives. If time is an independent variable for a tabular parameter, it is important to limit the integration step size in order to prevent skipping points in the table. A statement of the form

DTMAX = DX

will properly limit the step size. If several tables in the same model have time as an independent variable, the corresponding statement for limiting step size is

DTMAX = AMIN1 (DX1, DX2, ..., DXN)

Subroutine Call Statements

A model may CALL the subprogram listed in Table 5.10 or a user-written subprogram. Linkage to a user-written subprogram is supplied for anyone who needs to do a more complicated calculation than is either possible or practical in a model. A subprogram call from a model has the general form

$$\text{CALL } subname \ (w, x, \ldots, z)$$

where

$subname$ = unique alphanumeric name of six or fewer characters beginning with an alphabetic character

w, x, z = arguments to the subprogram; arguments may be both sent to and returned from a subroutine

Table 5.10   Subroutine Subprograms

CALL READER ($unit, t, tmax, v_1, dv_1, v_2, dv_2, \ldots, v_n, dv_n$)

Reader reads a tape with the format of a tape generated by a RECORD statement and performs linear interpolation where

$unit$ = data tape reference number between 0 and 9; different tapes used in a particular circuit must have different unit numbers

t = time at which the values are to be computed

$tmax$ = smallest time value greater than t for which there is recorded information

$v_1, v_2, \ldots, v_n$ = interpolated values of the recorded variables in the order that they were listed in the RECORD statement

$dv_1, \ldots, dv_n$ = time derivatives of $v_1, \ldots, v_n$, respectively

The arguments n, $unit$, and t must be supplied by the model. Subprogram READER returns $tmax$, v_i, and dv_i.

Adding User-Written Subprograms on the IBM/360–370

Any subprogram that is to be added to CIRCUS-2 should be coded in FORTRAN or in Assembler language using standard FORTRAN linkage conventions. All variables that enter through the calling sequence, whether scalars or arrays, are double precision. Arrays are singly dimensioned, and the length of an array is stored in position 0 of the array as a double-precision number.

To add subprograms to CIRCUS on an IBM/360–370 computer, first compile all programs to be added. Either the FORTRAN G or FORTRAN H compiler may be used. The second step is to assemble the CIRCUS subprogram SUBLIST (shown in Table 5.11). PROGRAM cards must be added

Table 5.11 *Subprogram SUBLIST*

```
//      EXEC ASMFC
//SYSIN  DD *
        MACRO
        PROGRAM &NAME
        DC      CL8'&NAME'
        DC      V(&NAME)
        MEND
SUBLIST START
        DC      A(END-SUBLIST-4)
*
* TO ADD THE USER SUBPROGRAM USERP TO THE PROGRAM LIST,
*
* CODE
*
*       PROGRAM USERP
*
* AND PLACE IT AFTER THIS CARD
        PROGRAM AMOD
        PROGRAM INT
END     EQU     *
        END
```

to SUBLIST for every user subprogram that is referenced from a CIRCUS model. PROGRAM cards contain the word PROGRAM beginning in column 10 and the subprogram name beginning in column 18.

The third step is to use the Linkage Editor to build a user load module. An INCLUDE card must be added to reference the previous user load module, as shown in Table 5.12 (*L4CVUSER* is the user load module supplied with CIRCUS).

Now the CIRCUS program can be executed. The name of newly created user load module must be supplied in the PARM field of the EXEC statement. The library containing the new load module must be the same as the library containing the CIRCUS program, or the two libraries must be concatenated as the STEPLIB library.

After steps 1, 2, and 3 have been successfully performed, the resulting user load module may be permanently saved so that these steps need not be repeated. More than one user load module may be saved. The appropriate one is selected at execution time by supplying the module name in the PARM field.

This process can be simplified considerably by writing a cataloged procedure to define all the steps.

Table 5.12 *Linkage Editor Step*

```
//     EXEC    LKED,PARM.LKED='LIST,LET,MAP'
//SYSLIB DD    DSN=SYS1.FORTLIB,DISP=SHR
//OLD    DD    DSN=CIRCUS.LOAD.MODULES,DISP=SHR
//SYSIN  DD    *
      INCLUDE   OLD(L4CVUSER)
      ENTRY    USERL
      NAME    TEMPNAME
//     EXEC    CIRCUS2,USER=TEMPNAME
//*
//*              A PROCEDURE NAMED CIRCUS2 IS SUPPLIED
//*              WITH THE IBM/360-370 VERSION OF THE
//*              CIRCUS-2 PROGRAM.
//*
//SYSIN  DD    *
```

Control Statements

1. GO TO Statement. The normal sequence of evaluation of executable statements is the sequence in which the statements are written. This sequence can be altered by using an unconditional transfer statement of the form

$$\text{GO TO } n$$

where n is one of the unique statement labels.

2. Arithmetic IF Statement. In addition to the unconditional transfer, there is a conditional transfer statement which has the form

$$\text{IF}(a)n_1, n_2, n_3$$

where

$a = $ any valid arithmetic expression

$n_1 = $ statement label of the statement to be executed when the numerical value of (a) is negative

$n_2 = $ statement label of the statement to be executed when the numerical value of (a) is exactly zero

$n_3 = $ statement label of the statement to be executed when the numerical value of (a) is positive

3. RETURN Statement. There must be at least one RETURN statement in every model equations section. A RETURN statement is placed at each logical conclusion of model equation computation.

4. END Statement. The END statement defines the end of the EQUATIONS section of the model. Physically, it must be the last statement in the EQUATIONS section.

5. STOP Statement. If a condition occurs in the model which makes further analysis of the circuit unnecessary or meaningless, execution can be terminated by a STOP statement.

5.6.3.5 Default Device Parameter Statement

Before a device can be used, every device parameter must be assigned a value. Some device parameters are normally set to a single known constant for most applications regardless of the particular device. For those device parameters, it is convenient to assign a default value in the model description rather than specifying the same number once for each device. To specify default values for device parameters, the following statements are used:

> **DEFAULT DEVICE PARAMETERS**
> **SINGLE VALUED PARAMETERS**
> $dp_1 = v_1, dp_2 = v_2, \ldots, dp_n = v_n$
> **ONE DIMENSIONAL TABLES**
> $name_1$, (idv list), (depv list)
> .
> .
> .
> $name_n$, (idv list), (depv list)
> **TWO DIMENSIONAL TABLES**
> $name_1$, (x-idv list), (y-idv list), (depv list)
> .
> .
> .
> $name_n$, (x-idv list), (y-idv list), (depv list)

where the meanings of *dp*, *v*, *name*, *idv list*, *depv list*, *x-idv list*, and *y-idv list* are the same as in Secs. 5.4.2.2, 5.4.2.3, and 5.4.2.4.

The DEFAULT DEVICE PARAMETERS statement may be used to assign nominal values to voltages and currents for the starting point of the initial conditions solution. Values assigned in a STEADY STATE GUESS override default device parameter values. For example, if the base-emitter and base-collector junction voltages of a particular transistor are set to place the transistor in an active state, every device using the transistor model will be biased in the active region at the beginning of steady-state calculation. If a particular component is not active, a STEADY STATE GUESS may be used to assign other values to the junction voltages of that component.

5.6.3.6 Fixed Parameters

Each variable appearing in a model is either a device parameter or its value is computed. Device parameters are variables whose values are input in a device definition and not modified by the model equations. In most cases, each variable can be classified by scanning the model, but when a variable appears as an argument of a subroutine subprogram, the variable cannot be classified directly. Therefore, any device parameter that is used in the calling sequence of a subroutine should be specified as a fixed parameter or be assigned a default value. The fixed parameter statement is

$$\textbf{FIXED PARAMETERS,} \; name_1, name_2, \ldots, name_n$$

where $name_i$ is a device parameter name.

The use of a FIXED PARAMETERS statement is optional. As a device is being defined, CIRCUS tests to see that every device parameter has been assigned a value or has a default value. If a FIXED PARAMETER statement was not used in the model description when it should have been used and if a value was not supplied when defining the device, unpredictable results will occur since CIRCUS cannot detect this input data error.

5.6.3.7 Arrays

Frequently a model needs an array of values to perform a calculation. For this reason CIRCUS has the capability of defining any symbolic name as an array. Arrays can be accessed directly in the equations section of a model, or they can be passed to subprograms through the calling sequence, where they may be used as any other singly dimensioned FORTRAN variable.

The values in an array and its length are determined when a device is built, but each array in a model must be identified as such. The format is

$$\textbf{ARRAYS,} \; name_1, name_2, \ldots, name_n$$

where $name_i$ is to be an array. Section 5.6.3.8 shows examples of how arrays are used.

5.6.3.8 Convolution Integrals in CIRCUS Models

Two function subprograms are available for the computation of convolution integrals. If the input and output of a linear system are related in the frequency domain by the equation

$$\hat{O}(F) = \hat{H}(F)\hat{I}(F)$$

where $\hat{I}$, $\hat{O}$, and $\hat{H}$ are the input, output, and transfer functions, respectively, then by transforming to the time domain we get the convolution integral

$$O(T) = \int_0^T H(T-t)I(t)\, dt$$

where I, O, and H are the inverse Fourier transforms of $\hat{I}$, $\hat{O}$, and $\hat{H}$.

Subprogram CONVLD computes this convolution integral. The calling sequence is

$$O = \textbf{CONVLD} \; (H, \textit{IPAST}, I, \textit{DODT}, \textit{LOC}, \textit{DELTAT}, \textbf{DTMAX})$$

where

H = array containing the convolution kernel data at equally spaced time intervals Δt; H_K corresponds to $H((K-1)\Delta t)$; $H(O) = H_1$ should be 0; if H_1 is not 0, the following formula should be used:

$$O = .5*H_1*DELTAT*I + CONVLD(\ldots)$$

$IPAST$ = array having the same length as H; $IPAST$ is used by CIRCUS for saving past values of I

I = input I

$DODT$ = derivative of the output with respect to the input

LOC = array of length 19 used by CIRCUS for scratch

$DELTAT$ = time interval Δt

DTMAX = predefined variable DTMAX (see Table 5.13).

Table 5.13 *Predefined Variable Names Available in the Equations Section of a Model Description*

DT	Time step size during transient analysis. Supplied by CIRCUS. DT is zero during steady-state analysis.
DTMAX	The largest allowable transient analysis integration time step size. The model may supply a time step limit to CIRCUS.
RJSS	Steady-state rejector flag. CIRCUS sets SSTFLG = 0 to indicate that a particular steady-state analysis is complete. If an abnormal situation occurs, the model equations may set RJSS = 1. CIRCUS then continues steady-state processing.
RJSTEP	Integration step rejector flag. The model equations can set RJSTEP = 1 to signal CIRCUS to reject the current integration step. This is usually done only if some variable attains a value that is outside the range of what is considered reasonable.
RKIND	Runge-Kutta pass indicator. CIRCUS sets RKIND = 1 on first pass when all variables are exact. Otherwise, CIRCUS sets RKIND = -1.
SSTFLG	Steady state/transient flag:

Value	Meaning
-1.0	Steady-state analysis
0.0	Steady-state analysis completed
1.0	Transient analysis

T	Current time. This value is supplied by CIRCUS to the model equations. T is zero during steady-state analysis.

In the analysis of transmission lines, there are often three or more convolutions: a self-admittance at each end of each line, and transfer admittances. The input to a self-admittance is the node-to-ground voltage, and the input to a transfer admittance is the node-to-node voltage. Since past values of the node-to-ground voltages are being saved in arrays for the self-admittances, there is no need to save past values of the node-to-node voltage as these values can be computed from differences of the two node-to-ground voltages.

Subprogram CONVLM computes the convolution integral

$$O(T) = \int_0^T H(T-t)(I_1(t) - I_2(t))\, dt$$

The calling sequence is

$O =$ **CONVLM** $(H, IPAST_1\ IPAST_2, I_1, I_2, DODT, LOC, DELTAT)$

where

$H =$ same as for CONVLD

$\left.\begin{array}{c}IPAST_1\\IPAST_2\end{array}\right\} =$ arrays that have previously been used for past values of I_1 and I_2 by means of CONVLD

$\left.\begin{array}{c}I_1\\I_2\end{array}\right\} =$ effective input is $I_1 - I_2$

$\left.\begin{array}{c}DODT\\LOC\\DELTAT\end{array}\right\} =$ same as for CONVLD

It is not necessary that CONVLM be used just because a particular admittance function happens to fit the format defined here. CONVLM does provide a convenient and efficient means for evaluating transfer admittances, and it is especially useful when solving large problems.

5.6.3.9 Predefined Variables

Certain names have been assigned predefined meanings for use in the EQUATIONS section of a model. Table 5.13 lists these names. Such names may be used for their predefined purpose only. Only *RJSS* and *RJSTEP* can be modified in the model.

5.6.4 Model Input

To input models in a CIRCUS run, the following statements are required:

MODELS

.
 . (model description data statements)
 .

END OF INPUT

Model description input must precede device description input and PREFIX statements.

5.6.5 Device Definition

Each device used in a circuit analysis will (in general) require parameters and tables to fully characterize it. These parameters may reside in a permanent device library or may be supplied by means of input statements. If device parameters are not supplied by input statements, the program will automatically search the permanent device library for them.

When device parameters are input, it may not be necessary to supply values for every parameter. Default values may have been assigned to some or all of the device parameters when the model was defined. The user may accept the default value (in which case he need not supply a value for the parameter in his input statements), or the user may override this default value by supplying another value in his input statements. Even if the user accepts the default value for every model parameter, he must still build a device.

5.6.6 Device Description Data Statements

A device description for a CIRCUS model may include

1. A device name and the name of a model to which the device applies.
2. Single-valued device parameter names and values.
3. One-dimensional tabular parameter names and table values.
4. Two-dimensional tabular parameter names and table values.

The device name/model name statement must be the first statement in each device parameter set.

5.6.6.1 Device Name/Model Name Statement

Each device must have a unique name and a specification of the model to which the parameters apply. These are specified by the statement

DEVICE NAME = *dname*, **MODEL NAME** = *mname*

where

dname = alphanumeric name of eight or fewer characters

mname = name of the model to which the parameters apply

The name *mname* must be spelled exactly as it is spelled on the model name statement. If *mname* is too long to fit in the space remaining on the device name/model name card, the device name/model name statement may be continued onto another card. The format is

DEVICE NAME = *dname*, **MODEL NAME** =   (on the first card)
mname   (on the second card)

Example.
>   DEVICE NAME = 2N1613, MODEL NAME =
>   NPN CHARGE CONTROL TRANSISTOR

5.6.6.2 Single-Valued Device Parameters

Single-valued device parameters include fixed-valued elements appearing in the topology of the model and each variable used in the equation section of the model which is not assigned a value by an equation. Predefined variables are not device parameters. If single-valued parameters specification statements are placed immediately following the device name/model name statement, the SINGLE VALUED PARAMETERS card may be omitted.

The format for the specification of single-valued device parameters is

>   SINGLE VALUED PARAMETERS
> $dp_1 = v_1, dp_2 = v_2, \ldots, dp_n = v_n$

where

> dp_i = device parameter name used in defining the model
> v_i = value to be assigned to the single-valued device parameter

Examples.
>   SINGLE VALUED PARAMETERS
>   RB = 50, RC = 3.22,
>   A1 = 26P, PHI1 = .9, N1 = .42, IES = 1.3P, THETAN = 36.3,
>   A2 = 33P, PHI2 = .9, N2 = .37, ICS = 2.2P, THETAI = 35.7

5.6.6.3 One-Dimensional Tabular Device Parameters

The format for the specification of one-dimensional tabular device parameters is

>   ONE DIMENSIONAL TABLES
> $name_1$, (idv list), (depv list)
>   .
>   .
>   .
> $name_n$, (idv list), (depv list)

where

> $name_i$ = one-dimensional tabled device parameter name
>   idv list = independent variable list whose elements are separated by commas
>   depv list = dependent variable list whose elements are separated by commas

There must be a one-to-one correspondence between the independent variable elements and the dependent variable elements. Independent variable elements must be in ascending order. For values of the independent argument outside the range specified in the table, CIRCUS uses the value of the dependent variable at the nearest end point of the table.

5.6.6.4 Two-Dimensional Tabular Device Parameters

The format for the specification of two-dimensional tabular device parameters is

TWO DIMENSIONAL TABLES

$name_1$, (x-idv list), (y-idv list), (depv list)
.
.
.
$name_n$, (x-idv list), (y-idv list), (depv list)

where

$name_1$ = two-dimensional tabled device parameter name

x-idv list = x independent variable list whose elements are separated by commas

y-idv list = y independent variable list whose elements are separated by commas

depv list = dependent variable list whose elements are input as a matrix having the form $d(x_1, y_1), d(x_2, y_1) \ldots, d(x_n, y_1), d(x_1, y_2) \ldots, d(x_n, y_m)$, where $d(x_i, y_j)$ is the dependent variable value

If there are n x-idv list values and m y-idv list values, there must be $n \times m$ depv list values.

5.6.6.5 Arrays

The format for the specification of values for an array is

ARRAYS

$name_1$, list
.
.
.
$name_n$, list

where

$name_i$ = name of an array

list = list of numbers which define the initial values in the array; the length of this list determines the length of the array; repeated values may be specified as (N, V), where N is the number of times the value is to be repeated and V is the value

Examples.

ARRAYS
Y1, 2., 3., 3., 3., 3., 4, 5.
Y2, 2., (4, 3.), 4., 5.

Arrays Y1 and Y2 in this example are initialized to the same values.

5.6.6.6 Device Parameter Input

To input device parameters in a CIRCUS run, the following statements are required:

DEVICES
.
. (device description data statements)
.
END OF INPUT

Device description input must follow model description input and precede device reference data statements.

5.6.7 Global Device Parameters

Several devices used in a circuit may share a parameter or table with the same name and physical significance. If this parameter is to be assigned the same value (or table of values) for each of these devices, the following statements may be used:

GLOBAL DEVICE PARAMETERS
SINGLE VALUED PARAMETERS
$dp_1 = v_1, dp_2 = v_2, \ldots, dp_n = v_n$
ONE DIMENSIONAL TABLES
$name_1$, (*idv list*), (*depv list*)
.
.
.
$name_n$, (*idv list*), (*depv list*)
TWO DIMENSIONAL TABLES
$name_1$ (*x-idv list*), (*y-idv list*), (*depv list*)
.
.
.
$name_n$ (*x-idv list*), (*y-idv list*), (*depv list*)
END OF INPUT

where

dp_i = device parameter name

v_i = value assigned to parameter dp_i

$table_i$ = table name

$list_i$ = list of values defining $table_i$.

When a GLOBAL DEVICE PARAMETERS statement is encountered, all the devices in the circuit are searched to determine if they have a parameter with the name dp_1. If a device has a parameter named dp_1 and this parameter was not assigned a value when the device was defined,* then the parameter dp_1 will be assigned to the value v_1. All other single-valued parameters and tables are processed in the same way. The END OF INPUT statement indicates that the list of global device parameters is complete. Each value in the list of device parameters must be followed by a comma, except for the last value. Any number of tables or no tables at all may be listed.

5.7 SAMPLE CIRCUITS

5.7.1 Digital Logic Inverter

The input for the first sample circuit (see Fig. 5.8) is shown in Table 5.14. In this example the model for the NPN TRANSISTOR model and the

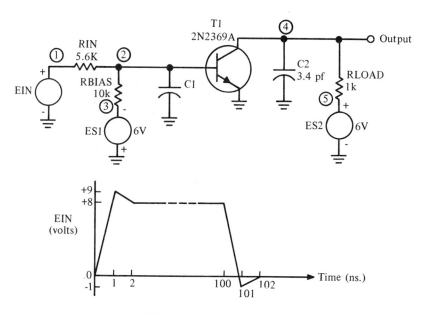

Figure 5.8 Inverter Circuit.

*When a device is defined, every device parameter that does not have a default value must be assigned a value. Thus, the global parameter feature can assign values only to parameters that have default values.

Table 5.14 *Inverter Input*

```
MODELS                                  DEVICES
MODEL NAME = NPN TRANSISTOR             DEVICE NAME = 2N2369A,
EXTERNAL NODES = ( B, C, E )            MODEL NAME = NPN TRANSISTOR
TOPOLOGY                                SINGLE VALUED PARAMETERS
   RBB, B, 4                            IES = 3.5745E-12, THETAN = 28.,
   CE, 4, E                             ICS = 7.3801E-12, THETAI = 32.,
   CC, 4, C                             CTE = 3.0E-12, TCE = 6.72E-9,
   JE, 4, E                             CTC = 2.0E-12, TCC = 185.E-9,
   JC, 4, C                             ALPHAN = .97819, ALPHAI = .47379,
EQUATIONS                               RBB = 20
C---- DIODE CHARACTERISTICS             END OF INPUT
   JBE= IES*( EXP( THETAN*V.CE ) − 1. ) ' DIGITAL LOGIC INVERTER CIRCUIT '
   JBC= ICS* ( EXP( THETAI*V.CC ) −1. ) PREFIX, TN = NPN TRANSISTOR
C---- VOLTAGE DERIVATIVES               TVIN,1,0,( ON,1N,2N,100N,101N,102N ),
   DJBE = THETAN*( JBE+IES )                 ( 0 ,9 ,8 ,8  , −1 ,0  )
   DJBC = THETAI*( JBC+ICS )            RIN, 1,2,5.6K
C---- JUNCTION CAPACITANCE              RBIS,2,3,10K
   CE = CTE + TCE*JBE                   VS1,0,3,6.
   CC = CTC + TCC*JBC                   C1, 2, 0, 3.6P
C---- CAPACITANCE DERIVATIVES           TN1,2,4,0,2N2369A
   D( CE,V.CE ) = TCE*DJBE              C2,4,0,3.4P
   D( CC,V.CC ) = TCC*DJBC              RLOD,4,5,1K
C---- JUNCTION CURRENTS                 VS2,5,0,6.
   JE = JBE − ALPHAI*JBC                PRINT, V.C1, V.C2, V.CE.TN1, V.CC.TN1,
   JC = JBC − ALPHAN*JBE                              JE.TN1, JC.TN1
C---- CURRENT DERIVATIVES               PLOT , V.C1, V.C2, ( V.C1, V.C2 )
   D( JE, V.CE ) = DJBE                 PRINT INTERVALS, 2N, 182N, 5N, 200N
   D( JE,V.CC ) = −ALPHAI*JBC           PLOT INTERVALS, 2N,200N
   D( JC,V.CC ) = DJBC                  STOP TIME = 200N
   D( JC,V.CE ) = −ALPHAN*JBE           DIAGNOSTICS
RETURN                                  EXECUTE
END                                     END
END OF INPUT
```

device parameters for a 2N2369A device are defined by input statements. The types of analysis performed here are dc steady-state and transient analysis. The dc analysis is performed with nominal values only. A transient is run up to 200 nanoseconds, and output is displayed in the form of a tabular list and plots of the two capacitor voltages.

The tabulated time domain results are shown in Table 5.15. The printer

Table 5.15 *Inverter Output*

'DIGITAL LOGIC INVERTER CIRCUIT'

TIME (NSEC)	V C1	V C2	V CE TN1	V CC TN1	JE TN1	JC TN1
0.0	−2.154	6.000	−2.154	−8.154	−7.79E−14	−3.88E−12
2.0	−1.862	6.089	−1.877	−7.966	−7.79E−14	−3.88E−12
4.0	−1.545	6.159	−1.559	−7.717	−7.79E−14	−3.88E−12
6.0	−1.259	6.194	−1.273	−7.467	−7.79E−14	−3.88E−12
8.0	−0.985	6.221	−0.998	−7.219	−7.79E−14	−3.88E−12
10.0	−0.743	6.226	−0.755	−6.981	−7.79E−14	−3.88E−12
12.0	−0.500	6.232	−0.512	−6.743	−7.79E−14	−3.88E−12
14.0	−0.296	6.225	−0.295	−6.523	1.61E−10	−1.97E−11
16.0	−9.37E−02	6.218	−0.104	−6.321	3.26E−11	−3.59E−11
18.0	0.108	6.210	9.81E−02	−6.112	5.00E−06	−4.89E−06
20.0	0.270	6.197	0.252	−5.945	1.65E−04	−1.62E−04
22.0	0.432	6.183	0.405	−5.777	3.26E−04	−3.19E−04
24.0	0.594	6.169	0.559	−5.610	4.86E−04	−4.76E−04
26.0	0.705	6.082	0.694	−5.388	9.78E−04	−9.57E−04
28.0	0.723	5.672	0.711	−4.961	1.59E−03	−1.56E−03
30.0	0.733	5.229	0.721	−4.509	2.07E−03	−2.02E−03
32.0	0.740	4.784	0.728	−4.056	2.51E−03	−2.46E−03
34.0	0.746	4.342	0.733	−3.609	2.95E−03	−2.89E−03
36.0	0.750	3.905	0.738	−3.167	3.38E−03	−3.31E−03
38.0	0.755	3.472	0.742	−2.730	3.81E−03	−3.73E−03
40.0	0.758	3.045	0.746	−2.299	4.23E−03	−4.14E−03
42.0	0.762	2.623	0.749	−1.873	4.64E−03	−4.54E−03
44.0	0.765	2.206	0.752	−1.453	5.05E−03	−4.94E−03
46.0	0.767	1.793	0.755	−1.038	5.46E−03	−5.34E−03
48.0	0.770	1.386	0.758	−0.629	5.86E−03	−5.73E−03
50.0	0.772	0.984	0.760	−0.224	6.25E−03	−6.12E−03
52.0	0.774	0.585	0.762	0.177	6.65E−03	−6.50E−03
54.0	0.774	0.270	0.762	0.492	6.59E−03	−6.42E−03
56.0	0.774	0.229	0.761	0.532	6.41E−03	−6.17E−03
58.0	0.774	0.215	0.762	0.547	6.40E−03	−6.11E−03
60.0	0.774	0.207	0.762	0.555	6.41E−03	−6.06E−03
62.0	0.774	0.201	0.762	0.561	6.41E−03	−6.02E−03
64.0	0.775	0.197	0.762	0.565	6.41E−03	−5.99E−03
66.0	0.775	0.194	0.763	0.568	6.42E−03	−5.96E−03
68.0	0.775	0.192	0.763	0.571	6.42E−03	−5.94E−03
70.0	0.775	0.190	0.763	0.573	6.42E−03	−5.92E−03

Table 5.15—*Cont.*

'DIGITAL LOGIC INVERTER CIRCUIT'

TIME (NSEC)	V C1	V C2	V CE TN1	V CC TN1	JE TN1	JC TN1
72.0	0.775	0.189	0.763	0.574	6.42E–03	–5.90E–03
74.0	0.775	0.188	0.763	0.575	6.42E–03	–5.89E–03
76.0	0.775	0.187	0.763	0.576	6.42E–03	–5.88E–03
78.0	0.775	0.186	0.763	0.577	6.42E–03	–5.87E–03
80.0	0.775	0.185	0.763	0.578	6.42E–03	–5.86E–03
82.0	0.775	0.185	0.763	0.578	6.43E–03	–5.85E–03
84.0	0.775	0.184	0.763	0.579	6.43E–03	–5.85E–03
86.0	0.775	0.184	0.763	0.579	6.43E–03	–5.84E–03
88.0	0.775	0.184	0.763	0.579	6.43E–03	–5.84E–03
90.0	0.775	0.184	0.763	0.580	6.43E–03	–5.84E–03
92.0	0.775	0.183	0.763	0.580	6.43E–03	–5.83E–03
94.0	0.775	0.183	0.763	0.580	6.43E–03	–5.83E–03
96.0	0.776	0.183	0.763	0.580	6.43E–03	–5.83E–03
98.0	0.776	0.183	0.763	0.580	6.43E–03	–5.83E–03
100.0	0.776	0.183	0.763	0.580	6.43E–03	–5.83E–03
102.0	0.738	0.185	0.754	0.569	4.97E–03	–4.54E–03
104.0	0.737	0.213	0.753	0.539	4.96E–03	–4.73E–03
106.0	0.731	0.365	0.747	0.382	4.31E–03	–4.21E–03
108.0	0.719	1.025	0.735	–0.290	3.08E–03	–3.02E–03
110.0	0.710	1.736	0.726	–1.011	2.39E–03	–2.33E–03
112.0	0.699	2.424	0.715	–1.710	1.75E–03	–1.71E–03
114.0	0.685	3.082	0.700	–2.382	1.16E–03	–1.14E–03
116.0	0.664	3.701	0.679	–3.023	6.45E–04	–6.31E–04
118.0	0.630	4.268	0.644	–3.624	2.43E–04	–2.38E–04
120.0	0.571	4.745	0.583	–4.162	4.54E–05	–4.44E–05
122.0	0.482	5.098	0.493	–4.605	4.06E–06	–3.97E–06
124.0	0.372	5.341	0.382	–4.959	1.99E–07	–1.95E–07
126.0	0.251	5.503	0.260	–5.243	1.04E–08	–1.01E–08
128.0	0.127	5.621	0.135	–5.486	1.82E–10	–1.82E–10
130.0	2.31E–03	5.693	9.88E–03	–5.683	6.77E–11	–7.02E–11
132.0	–0.119	5.751	–0.112	–5.863	4.06E–13	–4.36E–12
134.0	–0.237	5.789	–0.230	–6.019	1.22E–13	–4.08E–12
136.0	–0.351	5.821	–0.345	–6.166	–7.74E–14	–3.88E–12
138.0	–0.455	5.840	–0.449	–6.289	–7.76E–14	–3.88E–12
140.0	–0.559	5.858	–0.553	–6.412	–7.78E–14	–3.88E–12
142.0	–0.656	5.873	–0.651	–6.524	–7.79E–14	–3.88E–12

Table 5.15—*Cont.*

'   DIGITAL LOGIC INVERTER CIRCUIT   '

TIME (NSEC)	V C1	V C2	V CE TN1	V CC TN1	JE TN1	JC TN1
144.0	−0.740	5.882	−0.735	−6.617	−7.79E−14	−3.88E−12
146.0	−0.824	5.891	−0.820	−6.711	−7.79E−14	−3.88E−12
148.0	−0.908	5.900	−0.904	−6.804	−7.79E−14	−3.88E−12
150.0	−0.992	5.909	−0.989	−6.897	−7.79E−14	−3.88E−12
152.0	−1.057	5.914	−1.054	−6.968	−7.79E−14	−3.88E−12
154.0	−1.120	5.919	−1.117	−7.037	−7.79E−14	−3.88E−12
156.0	−1.184	5.925	−1.180	−7.105	−7.79E−14	−3.88E−12
158.0	−1.247	5.930	−1.244	−7.174	−7.79E−14	−3.88E−12
160.0	−1.308	5.935	−1.305	−7.240	−7.79E−14	−3.88E−12
162.0	−1.354	5.938	−1.352	−7.290	−7.79E−14	−3.88E−12
164.0	−1.401	5.942	−1.398	−7.340	−7.79E−14	−3.88E−12
166.0	−1.447	5.946	−1.445	−7.390	−7.79E−14	−3.88E−12
168.0	−1.493	5.949	−1.491	−7.440	−7.79E−14	−3.88E−12
170.0	−1.539	5.953	−1.537	−7.489	−7.79E−14	−3.88E−12
172.0	−1.572	5.955	−1.570	−7.525	−7.79E−14	−3.88E−12
174.0	−1.605	5.958	−1.603	−7.561	−7.79E−14	−3.88E−12
176.0	−1.638	5.960	−1.636	−7.596	−7.79E−14	−3.88E−12
178.0	−1.671	5.963	−1.669	−7.632	−7.79E−14	−3.88E−12
180.0	−1.703	5.965	−1.702	−7.667	−7.79E−14	−3.88E−12
182.0	−1.732	5.968	−1.731	−7.698	−7.79E−14	−3.88E−12
185.0	−1.766	5.970	−1.765	−7.735	−7.79E−14	−3.88E−12
190.0	−1.822	5.974	−1.821	−7.796	−7.79E−14	−3.88E−12
195.0	−1.876	5.979	−1.875	−7.853	−7.79E−14	−3.88E−12
200.0	−1.917	5.982	−1.916	−7.898	−7.79E−14	−3.88E−12
END PRNT	54					

plots of the voltages on C1 and C2 are shown in Figs. 5.9 and 5.10, respectively.

5.7.2 18-Volt Regulator

Figure 5.11 is a schematic diagram of a voltage regulator circuit. The voltage supply for the regulator is a full-wave-rectified, 10-kHz, 50-volt source. The regulator is being analyzed for sensitivity to changes in the load resistor RL. The input is shown in Table 5.16.

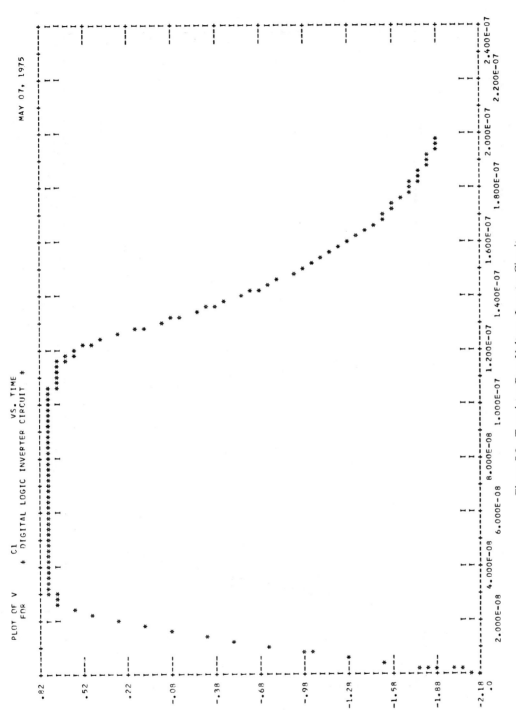

Figure 5.9 Transistor Base Voltage, Inverter Circuit.

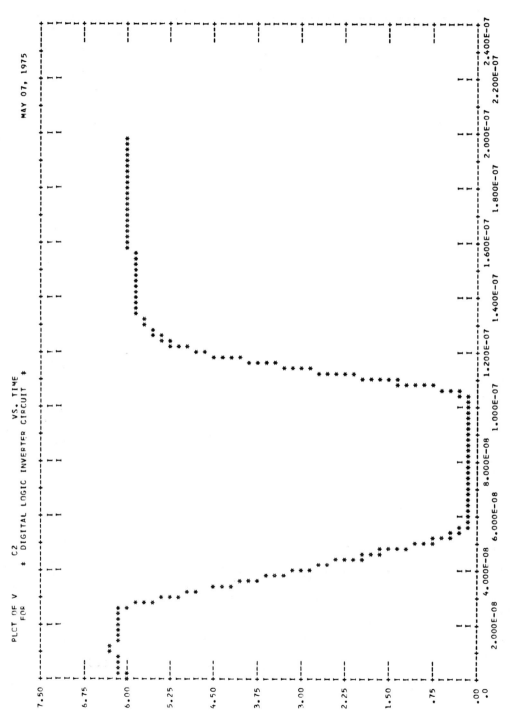

Figure 5.10 Transistor Collector Voltage, Inverter Circuit.

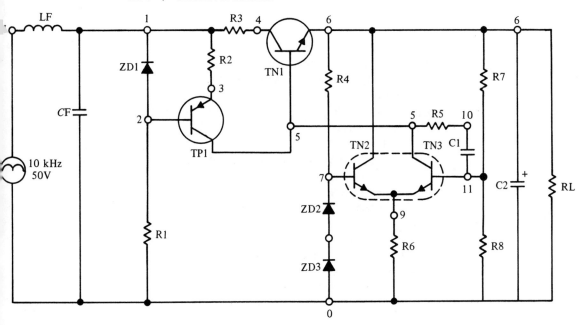

Figure 5.11 *18-Volt Regulator.*

Table 5.16 *18-Volt Regulator Input*

```
MODELS
MODEL NAME = SOURCE
EXTERNAL NODES = ( 1,2 )
TOPOLOGY
V1,1,2
EQUATIONS
        IF ( SSTFLG ) 10,10,20
10      V1 = 31.83
        D(V1,T) = 0
        RETURN
20        CONTINUE
        A = 6.28E4*T
        V1 = SIN(A)*50
        D = 6.28E4*COS(A)*50
        IF ( V1 ) 1,2,2
1       V1 = −V1
        D = −D
2       D(V1,T) = D
        RETURN
        END
```

Table 5.16—*Cont.*

```
        MODEL NAME = ZENER DIODE
        EXTERNAL NODES = ( ANOD,CATH )
        TOPOLOGY
        CD,ANOD,CATH
        JD,ANOD,CATH
        EQUATIONS
                JD = TABLE JZEN ( V.JD,DX,DJ )
                D(JD,V.JD) = DJ
                RETURN
                END
        END OF INPUT
                DEVICES
        DEVICE NAME = 1N821, MODEL NAME = ZENER DIODE
        SINGLE VALUED PARAMETERS
                CD = 5.E-9
        ONE DIMENSIONAL TABLES
        JZEN, ( −7.5, −7.0, −6.6, −6.5, −6.4, −6.3, −6.25,−6.2, −6.0,
                0., 0.5, 0.6, 0.7 , 0.8 ),
                (−26.52,−1.14,−.091,−.049,−.026,−.014,−.010,−74−4,−21−4,
                0.,49−4,0.99,1.978,39.734 )
        DEVICE NAME = 1N746, MODEL NAME = ZENER DIODE
        SINGLE VALUED PARAMETERS
                CD = 6.E-9
        ONE DIMENSIONAL TABLES
        JZEN, ( −4.0, −3.5, −3.4, −3.3, −3.25,−3.2, −3.1,
                0., 0.5, 0.6, 0.7 , 0.8 ),
                (−11.23,−.215,−.061,−.018,−94−4,−50−4,−14−4,
                0.,49−4,0.99,1.978,39.734 )
                END OF INPUT
                ' 18 VOLT REGULATOR  '
        PREFIX, QV, SOURCE
        PREFIX,    TN = NPN CHARGE CONTROL TRANSISTOR
        PREFIX,    TP = PNP CHARGE CONTROL TRANSISTOR
        PREFIX,    ZD = ZENER DIODE
        R1,    2,    0,    1.21E3
        R2,    1,    3,    3.90E2
        R3,    1,    4,    3.92E1
        R4,    6,    7,    5.60E2
        R5,    5,    10,   1.00E2
        R6,    9,    0,    9.10E2
        R7,    6,    11,   4.53E2
        R8,    11,   0,    1.00E3
```

Table 5.16—*Cont.*

```
RL,    6,   0,   1.80E2
C1,   10,  11,   1.00E-7
C2,    6,   0,   4.50E-4
TN1,   5,   4,   6,    2N3551
TN2,   7,   6,   9,    2N2219
TN3,  11,   5,   9,    2N2219
TP1,   2,   5,   3,    2N2904
ZD1,   2,   1,   1N746
ZD2,   8,   7,   1N821
ZD3,   0,   8,   1N821
CF,    1,   0,   .1E-6
LF,   1N,   1,   5.E-3
QV1,  1N,   0
STEADY STATE GUESS,
      V.C1 = 6.278 ,   V.C2 =     18.16,
      V.CC.TN1 = -8.136 , V.CE.TN1 =  0.606,
      V.CC.TN2 = -5.648 , V.CE.TN2 =  0.666,
      V.CC.TN3 = -6.271 , V.CE.TN3 =  0.659,
      V.CC.TP1 = -9.896 , V.CE.TP1 =  0.698,
      V.CD.ZD1 =    -3.3 , V.CD.ZD2 =  -6.2,
      V.CD.ZD3 =    -6.2 , V.CF     =  32. ,
      I.LF     =    .017
PRINT,
      VN.4, VN.5, VN.6, VN.7, VN.9, VN.11
PRINT INTERVALS, 1.E-5, 1.E-3,, 5.E-5, 2.E-2
PLOT,
      VN.5, VN.6, I.RE.TN1, I.RB.TN1, I.RC.TP1
PLOT INTERVALS = 1.E-5, 2.E-2
STOP TIME = .5E-3
DIAGNOSTICS
HOLD FINAL CONDITIONS
EXECUTE
*RL = 100.
STOP TIME = 1.E-3
HOLD FINAL CONDITIONS
EXECUTE
*RL = 50.
STOP TIME = 3.E-3
EXECUTE
END OF JOB
```

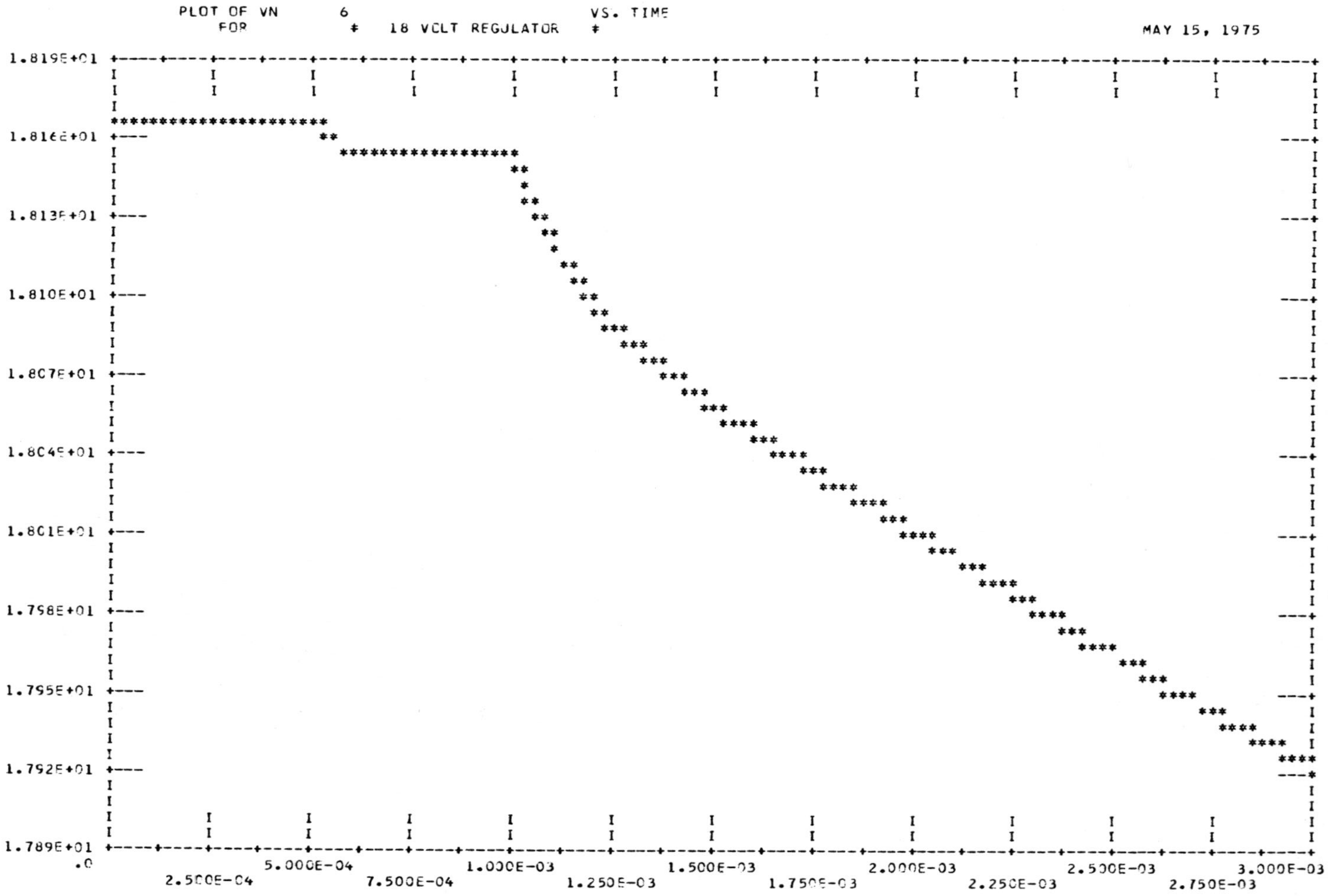

Figure 5.12 18-Volt Regulator Output.

The NPN CHARGE CONTROL TRANSISTOR and PNP CHARGE CONTROL TRANSISTOR models reside on a permanent model library. Models for the 10-kHz source and zener diode are supplied by input statements. All the parameters necessary to define the source model are contained within the model, so no device is specified for this model. The zener diode model is simply a capacitor and a voltage-controlled current source defined by a table.

The device parameters for the 2N3551, 2N2219, and 2N2904 transistors reside on the permanent device library. The parameters for the 1N746 and 1N821 zener diodes are supplied by input statements.

A guess is made at the steady state of the network to assist the program in the steady-state analysis. A transient is run to 0.5 milliseconds and stopped, RL is changed to 100 ohms, and the transient is continued to 1.0 milliseconds. RL is changed again to 50 ohms, and the transient is run to 3.0 milliseconds. The fact that this circuit, containing semiconductors with time constants in the nanosecond range, can be run to milliseconds is due to the numerically stable integration routine.

Output from the analysis is in the form of tabulated lists of several network parameters, and printer plots. Figure 5.12 is the plot of the output voltage (voltage at node 6).

5.8 PROGRAM LIMITATIONS

Because of the dynamic allocation used in CIRCUS, it is difficult to place upper limits on the size of the problems that can be run. Circuits having more than 600 nodes and 1500 circuit elements have been solved by CIRCUS-2. Problems of this size will require about 2 seconds of CPU time for each integration step on an IBM/360 Model 65 computer. Larger networks could be solved more efficiently on larger computers.

The most common problems in using CIRCUS-2 are failure to specify derivatives properly (in the equations section of a model), failure of the steady-state solution procedure to converge, and failure to write all model equations as functions of element voltages and currents or time.

The first of these problems is actually an input error on the part of the user. However, the requirement that function derivatives be specified is not standard among circuit analysis programs. This problem is frequently encountered by users who attempt to develop complex models. The symptoms of incorrect derivative specification are slow-running transients and sometimes integration failures.

When the steady-state solution cannot be found, incorrect derivatives are frequently the cause. However, convergence failures may occur even when all derivatives are correct. A good steady-state guess will usually eliminate the problem, but sometimes INITIAL CONDITIONS must be specified, bypassing the steady-state computations entirely. Research on improved steady-state techniques is currently under way to reduce the impact of this problem.

The CIRCUS-2 modeling language requires that all functional relationships be defined in terms of element voltages, element currents, or time. Modelers who attempt to use other independent variables sometimes have difficulties. For example, using the integral of a current or voltage which was computed numerically (i.e., by the trapezoidal rule) or using the derivative obtained by numerical calculation of the tangent may cause unusually slow-running speeds. The solution is to use only the allowed independent variables and to specify integrals or derivatives by explicit equations.

5.9 ERROR DIAGNOSTICS

Although it is easy to prepare input for CIRCUS, it seems that errors inevitably occur. CIRCUS detects more than 110 errors that stem from mistakes in the input problem definition. Many of these error messages are printed immediately below the data card which caused the error. In such cases, the error message is almost always self-explanatory, and the user can detect the source of error immediately. In other cases, however, an error cannot be diagnosed until all data cards have been scanned. For example, if there are no elements connected to the reference node for a particular circuit, the error cannot be detected until all input data cards have been interpreted. The associated error message does not apply to a particular data card but to the circuit definition as a whole. Error messages in cases such as this are printed after all input cards have been listed.

The CIRCUS error diagnostics are grouped according to the section of the program in which an error was detected. A listing of the error diagnostics and a discussion of possible causes for each error (where the causes are not immediately obvious) is given below.

1. *Error messages that occur while processing device description data statements are*

DEV001   NO MATCH FOR *name* (DEVICE *dname*) IN MODEL *mname*

        name is a misspelled device parameter name or device key word.

        dname is the name of the device being processed (DEFAULT* while processing default device parameters in a model).

        mname is the model name.

DEV002   PARAMETER *name* (DEVICE *dname*) IS NOT A TABLE IN MODEL *mname*

name is a parameter that was incorrectly specified as a one-dimensional or two-dimensional table in device *dname* but not in model *mname*.

DEV004   NO KEYWORD MATCH FOR *name*

name is not a valid device description key word.

DEV005   LIBRARY DOES NOT CONTAIN MODEL *mname*

The model specified on the DEVICE NAME/MODEL NAME statement does not exist on a model library.

DEV006   THE FOLLOWING PARAMETERS ARE MISSING FROM DEVICE *dname* FOR MODEL *mname* list

Not all values were specified in device *dname* as required for model *mname*. Values may be specified either in the device or a default device parameters in the model.

2. *Error messages that occur while processing input statements* (*other than device description or model description data statements*) *are*

INP001   PREFIX STATEMENTS NOT SPECIFIED CONSECUTIVELY
INP002   ABOVE ELEMENT IMPROPERLY SPECIFIED

A circuit description data statement has an incorrect number of nodes or values.

INP003   ABOVE PREFIX IMPROPERLY SPECIFIED

A prefix statement must have exactly two separators.

INP004   STATEMENT CANNOT BE INTERPRETED AS A NETWORK MODIFICATION STATEMENT

A misspelled key word, misused key word, or circuit description data statement appeared after an EXECUTE statement.

INP005   *name* IS NOT AN ELEMENT NAME OR IS IMPROPERLY SPECIFIED

A circuit element modification statement beginning with ∗*name* is incorrect. Either *name* is misspelled or the wrong number of values was specified.

INP006   TABLE CHANGE REQUEST IMPROPERLY SPECIFIED
INP007   *name* IS NOT IN A DEVICE LIBRARY
INP008   *name* IS NOT A DEVICE NAME

A device replacement change statement is incorrectly written.

INP009   *name* CANNOT BE FOUND IN A MODEL LIBRARY

name appeared in a PREFIX statement and referenced an underfined model.

INP010 *text* IS AN IMPROPER NUMBER REPRESENTATION

text is a number with more than one decimal point, a decimal point in the exponent field, or an exponent greater than 38 in magnitude.

INP011 END OF FILE ENCOUNTERED WHILE READING FROM UNIT *n*

The input deck was not terminated with an END OF JOB card. This error does not necessarily invalidate the job.

INP012 ILLEGAL DATA STATEMENT ABOVE

A CARD LENGTH statement was improperly used.

INP013 ITEM AGAINST ITEM PLOTS IMPROPERLY SPECIFIED PLOTS SUPPRESSED

Item versus item plot requests must follow all item versus time plot requests.

INP014 UNRECOGNIZED PARAMETER VARIATION STATEMENT ABOVE

Parameter variation for multiple transient analyses is specified by functions. See Sec. 5.4.14.

INP015 NO PARAMETER VARIATION STATEMENTS FOLLOW MULTIPLE EXECUTE STATEMENT
MULTIPLE EXECUTION WILL BE SUPPRESSED

INP016 *name* HAS A NON-POSITIVE PERIOD

A pulsed source or sine-wave source is improperly specified.

INP019 ABOVE TABLE IS IMPROPERLY SPECIFIED

Table are specified in the form

name, (*idv list*), (*depv list*)

or

name, (*x-idv list*), (*y-idv list*), (*depv list*)

The parentheses must be included, and the number of dependent variables must correspond to the number of independent variables.

INP020 ELEMENT UPDATE LIST FOR *name* IMPROPERLY SPECIFIED

Incorrect number of values for a sine-wave or pulsed source update. The node labels are not respecified when updating.

INP022 INDEPENDENT ARGUMENTS OF TABLE *name* NOT IN ASCENDING ORDER

```
       NNNNNNNNFFFFFFFF      0.0           1.000E 00
                        X.XXXE XX     X.XXXE XX     X.XXXE XX
                        X.XXXE XX     X.XXXE XX     X.XXXE XX
                              .             .             .
                              .             .             .
```

NNNNNNNN is the number of independent arguments (in binary) and the list of numbers on the subsequent lines is the table—independent arguments followed by dependent arguments. (If the table has more than 250 independent arguments, only the first 250 values will be printed.)

INP023   *name* CANNOT BE FOUND IN THE DEVICE TABLE

This error is usually caused by using a component name that is more than four characters long.

INP024   THE ABOVE STEADY STATE PARAMETER VARIATION LIST IS IMPROPERLY SPECIFIED

The list of values for each parameter being varied must be enclosed in parentheses. See Sec. 5.4.13.3.

INP025   *name* IS NOT A FIXED VALUE AND CAN BE VERIED ONLY BY A SHIFT OR SCALE FACTOR

INP026   *name* IS A FIXED VALUE AND CANNOT BE VARIED BY A *text* FACTOR

INP027   *name* CANNOT BE MATCHED AGAINST ANY ELEMENT OR PARAMETER NAME

Only elements or device parameters may be varied. This error may be caused by failure to end a previous data card with a separator or by concatenating separators.

INP028   *name* IS AN INVALID NAME FOR INITIAL CONDITIONS

name is misspelled, a separator is missing, or the network data deck has been modified without correcting the INITIAL CONDITIONS statement.

3. *Error messages that occur while processing model definition data statements are*

MOD101   DUPLICATE STATEMENT NUMBER
MOD102   IMPROPER STATEMENT NUMBER

Statement label field contains nonalphanumeric characters.

MOD103   UNDETERMINED ELEMENT TYPE

An element name in the topology list begins with a character other than C, J, K, L, R, or V.

MOD104   DEPENDENT VARIABLE *name* NOT IN MODEL TOPOLOGY

A dependent variable *name* specified in a derivative must be in the topology list.

MOD105   MORE THAN 19 CONTINUATION CARDS
MOD106   NO END CARD

MOD107   IMPROPER IF STATEMENT

An IF (expression) must be followed by three statement labels separated by commas.

MOD108   NO TOPOLOGY STATEMENTS
MOD109   OPERATOR *character* ILLEGAL IN CONTEXT

Operators may not be concatenated except that ** indicates exponentiation.

MOD110   MISSING STATEMENT NUMBER *label*
MOD111   PARAMETER *name* IN DEVICE *dname* IS AN INDEPENDENT SOURCE AND NOT A LEGAL INPUT

To specify the voltage across a voltage source or current through a current source in a model, use the component name. That is, the voltage across voltage source V1 is *V1*, instead of *V.V1*.

MOD112   SYNTAX ERROR

Statement processing error which does not fit into any other category.

MOD113   INSUFFICIENT SPACE

More core storage required to process a model. See UTL002.

MOD114   INCORRECT NUMBER OF ARGUMENTS

A reference to a predefined function subprogram (Table 5.9) must contain the correct number of arguments.

4. *Error messages associated with invalid network topology are*

TOP002   THE REFERENCE NODE *name* DOES NOT CORRESPOND TO ANY NODE IN THE NETWORK

TOP003   THE NETWORK CONTAINS A LOOP OF VOLTAGE SOURCES OR A CUTSET OF CURRENT SOURCES

Neither configuration is allowed.

TOP004   NODE *name* IS NOT CONNECTED TO GROUND

There must be a path to ground through elements from every node in a circuit. A path through mutual coupling of reactive elements is not sufficient.

TOP005   UNABLE TO CHOOSE PIVOT AT STAGE k OF THE LU DECOMPOSITION, STEP n

Before CIRCUS solves the network equations, they are set up in matrix form and then reordered to reduce computation time. During this process, a singular matrix was found.

k is the stage in the reordering pivot selection process.

n indicates which matrix is singular:

n	Matrix
1	Full tableau (transient)
2	Nonstate variables
3	Steady state

If the error occurs during steady-state matrix pivot selection, the problem is usually due to a capacitor-current source cutset or an inductor-voltage source loop in the network topology. This problem can be solved by eliminating the offending topology or circumvented by specifying initial conditions.

If the error occurs during full tableau or nonstate variable matrix pivot selection, the cause is probably overspecification of a variable in a model. For example, consider the following model (Fig. 5.13):

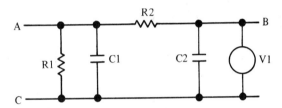

Figure 5.13 Overspecified Variable.

MODEL NAME = OVERSPECIFIED VARIABLE SOURCE
EXTERNAL NODES = (A, B, C)
TOPOLOGY
    R1, A, C
    R2, A, B
    C1, A, C
    C2, B, C
    V1, B, C
EQUATIONS
    V1 = K*V.C1
    D(V1, V.C1) = K

Since $V.C2$ is a state variable and $V1$ is in parallel with $C2$, the voltage across $V1$ must equal the voltage across $C2$. Since $V.C1$ is also a state variable, a conflict exists between the model equations and Kirchhoff's laws. A singular matrix will result.

TOP006   POOR PIVOT CHOICE AT STEP k v_1 v_2

A matrix can be reordered for computational efficiency if the magnitudes of the new diagonal elements are consistent with criteria that ensure a negligible amount of error in the solution process. A poor pivot choice does not necessarily mean that the solution is invalid. There is no direct way to determine what change in the network or in the parameter values will guarantee satisfactory pivots.

k is defined under TOP005.

v_1 is the value of the pivot.

v_2 is the magnitude of the largest element in the row containing v_1.

If the ratio of v_1 to v_2 is less than 0.0001, the pivot choice is probably unsatisfactory and a change to the input data should be made. Potential sources of trouble are extreme variation in the values of resistors or specifying unrealistic nominal values for model parameters.

5. *Error messages that occur during solution of the network equations are*

TRN001 INSUFFICIENT DATA TO PERFORM RESTART

Either the tape designated as a CIRCUS restart tape is in fact not a CIRCUS SAVE tape, or else the run in which the SAVE tape should have been created failed.

TRN002 EXECUTION SUPPRESSED DUE TO PREVIOUS ERRORS
ERROR SCAN CONTINUES

TRN003 THE FOLLOWING INDEPENDENT INDUCTOR CURRENTS OR CAPACITOR VOLTAGES WERE NOT SPECIFIED IN THE INITIAL CONDITIONS LIST

$name_1$ $name_2$ ...

Initial conditions must be specified for each state variable. When specifying currents or voltages for the omitted parameters, be certain to include *I.* or *V.* as required.

TRN004 UNABLE TO COMPUTE INITIAL CONDITIONS
TRANSIENT ANALYSIS SUPPRESSED

TRN005 INTEGRATION FAILURE AT TIME *time*
NON-STATE VARIABLES FAIL TO CONVERGE

The transient analysis was terminated at the specified time because CIRCUS was unable to compute values for the voltages and currents of the controlled sources that were consistent with the model equations and Kirchhoff's laws. This problem is usually caused by a missing or incorrect derivative or a jump discontinuity in a source equation.

TRN006 INTEGRATION FAILURE AT TIME *time*
TIME STEP TOO SMALL

The transient analysis was terminated at the specified time because CIRCUS was finding unacceptable integration error at all reasonable time steps. This problem can be caused by large fluctuations in a controlled source or an unusually large STOP TIME.

6. *Other messages associated with errors in setting up problems for CIRCUS are*

UTL001 *pgm* FOUND NO MATCH FOR *name*

name is a misspelled key word, model name, device name, or component name.

pgm is the name of the CIRCUS subprogram that detected the error.

UTL002 NOT ENOUGH CORE AVAILABLE TO PREPROCESS INPUT

CIRCUS-2 uses a storage allocation method which first takes all core storage allocated for the job and then assigns it as storage space is required. If the amount of storage available is not sufficient, processing is terminated. To solve the problem, rerun it requesting more core storage.

UTL003 Z MATRIX TOO LARGE

See UTL002.

UTL004 NOT ENOUGH CORE IS AVAILABLE FOR PLOTS
PLOT REQUEST WILL BE IGNORED

The problem has been solved; printed output and record output will be generated although plotted output will not.

UTL005 NOT ENOUGH CORE AVAILABLE TO RESTART
n WORDS OF ADDITIONAL CORE IS NEEDED

See UTL002.

UTL006 n WORDS OF ADDITIONAL CORE REQUIRED FOR PRINTING TRANSIENT RESULTS

The estimate of additional core required is exact since allocation of a print buffer is the last step in the dynamic storage allocation. Rerun the job requesting at least n additional words.

UTL007 AT LEAST n WORDS OF ADDITIONAL CORE REQUIRED

This estimate of additional core required may not be accurate since it occurred at an early stage of dynamic storage allocation. See UTL002.

UTL010 SAVE TAPE REQUESTED BUT NOT WRITTEN

Due to earlier errors, no valid data were available to be saved. The input data must be corrected and the problem rerun.

UTL012 *text* IS AN INVALID QUALIFIER FOR *name*

The qualifier should have been a *V.* or an *I.* in a model equation or in an output display request.

UTL013   *text* FAILS TO MATCH NETWORK PARAMETER NAMES

> An output display request contained *V.text*, *I.text*, or *VN.text*. Either *text* is misspelled or the network data deck has been modified without correcting the output display request statements.

> 7. The following error messages indicate either an error within CIRCUS or a computer hardware error. If it is apparent that CIRCUS is at fault, please contact the authors of the CIRCUS program.

***001   *text* CANNOT BE FOUND IN THE DEVICE PARAMETER DICTIONARY
***002   ERROR OR END OF FILE ON UNIT *n*

> Abnormal end of data. Allowable values for *n* are

n	Unit
1	Permanent model library
2	Temporary model library
3	Permanent device library
4	Temporary device library
7	Function
8	CIRCUS utility unit
9	Transient solution temporary output
10	Plotted output temporary unit
11	Record output
12	Save
13	Restart
14	CIRCUS utility unit

***003   OUTPUT NUMBER *n* IN DEVICE *dname* IS IMPROPERLY SPECIFIED
***004   BAD MATRIX ELEMENT
           i j type

> *i* and *j* are the row and column indices. *Type* indicates whether the element is a constant or a variable.

***005   ELEMENT *i j* ELIMINATED
***006   ILLEGAL PSEUDO OP. *text*
***007   NO SPACE FOR PRINT BUFFER

5.10 REFERENCES

1. Dembart, B., and L. Milliman, "CIRCUS-2, a Digital Computer Program for Transient Analysis of Electronic Circuits (Revision 2)," prepared under Harry Diamond Laboratories Contract #DAAG39-67-0070, Washington, D.C., 1973.

2. Hachtel, G. D., R. K. Brayton, and F. G. Gustavson, "The Sparse Tableau Approach to Network Analysis and Design," *IEEE Trans.*, CT-18, No. 1, Jan. 1971, pp. 101–113.

3. ROSE, D. J., and R. A. WILLOUGHBY, *Sparse Matrices and Their Applications*, Plenum, New York, 1972.

4. FORSYTHE, G., and C. B. MOLER, *Computer Solutions of Linear Algebraic Systems*, Prentice-Hall, Englewood Cliffs, N.J., pp. 27-33, 1967.

5. NORIN, R. S., and C. POTTLE, "Effective Ordering of Sparse Matrices Arising from Nonlinear Electrical Networks," *IEEE Trans.*, CT-18, No. 1, Jan. 1971, pp. 139-145.

6. ROSENBROCK, H. H., "Some General Implicit Processes for the Numerical Solution of Differential Equations," *Computer J.*, 5, No. 4, Jan. 1963, pp. 329-330.

7. DANIELS, M. E., "Device Modeling for Circuit Analysis Programs," in *Computer Oriented Circuit Design*, Franklin F. Kuo and Waldo G. Magnuson, Jr., eds., Prentice-Hall, Englewood Cliffs, N.J., 1969, pp. 317-328.

6

ECAP II

GERALD R. HOGSETT

IBM Corporation
San Jose, California

FOREWORD

An overview and synopsis of the main features and language formats of the IBM Program Products ECAP II for the IBM 1130 Computing System and for the IBM Operating System for System/360 and System/370 is presented in this chapter. Readers wishing to find additional information, including any updates to the program products, should consult the ECAP II documentation which is listed in the References at the end of this chapter.

The author would like to express his appreciation for the collaborative efforts in the development of these program products to Anita Ford, Roger Boe, Dr. F. H. Branin, Martin Goldberg, Dave Goodwin, L. E. Kugel, R. L. Lunde, Almerin O'Hara, Indru Jhangiani, R. A. Payne, and E. F. Sarkany. Thanks are also due to Dr. K. O. Wang of the University of Cincinnati, Cincinnati, Ohio; D. A. R. Zein of IBM, East Fishkill, New York; and G. Martin of the Université Libre de Bruxelles, Belgium, for their contributions to the programs.

6.1 INTRODUCTION*

The IBM Electronic Circuit Analysis Program II (ECAP II) is a general-purpose electronic circuit analysis program that performs both direct current (dc) and transient analyses of linear and nonlinear circuits. The program

---

*Portions hereof are reprinted with the permission of International Business Machines Corporation from its manual "Electronic Circuit Analysis Program II (ECAP II) for the IBM Operating System and 1130 Computing System—Program Description Manual" (*SH20–1015*), © 1971 by International Business Machines Corporation.

makes use of improved numerical integration and nonlinear equation solution algorithms. It also embodies many advanced programming techniques. ECAP II is a completely new program which is a state-of-the-art improvement over the corresponding features in the IBM Type II program, entitled 1620 Electronic Circuit Analysis Program (known as ECAP), which has been in use for several years.

Two program products are provided, one for the IBM Operating System on System/360 and System/370 and one for the 1130 Computing System. The input and output formats, program features, and analysis techniques are the same for the two program products. For this reason, the terms ECAP II, program, or system apply to either or both program products and therefore are used in the singular in this chapter.

The emphasis in the design of ECAP II has been directed toward providing a basic dc and transient analysis program which is easy to use, handles large problems, and provides accurate, reliable operation in minimum computation time. These properties, plus the open-ended and modular construction, permit the program to serve as a base for further development, such as the later insertion of alternative algorithms or additional analysis facilities.

ECAP II provides a problem-oriented language that allows the engineer to describe a circuit in familiar electrical terminology. The language is also used to request output, to select the type of analysis desired, to select run controls, and to perform utility operations. The user does not need extensive knowledge of programming or the mathematics of circuit theory. The system supplies a set of functions (DIODE, MULT, PULSE, and so forth) which can be used in specifying certain element dependencies. Occasionally the user may wish to add his own special functions to the system which can then be used in specifying the values for linear or nonlinear elements. To do this, a knowledge of FORTRAN or the assistance of an application programmer is required.

Although ECAP II is oriented primarily toward the analysis of electronic circuits, it can simulate any physical system which can be represented by a network model. This is facilitated through the ECAP II language and its model library facility, which permit storage of electrical analogs for non-electrical components. The program might be used to simulate a variety of physical phenomena, such as heat flow, fluid flow, and mechanical vibrations.

The program can be operated in the batch mode under the IBM Operating System or the 1130 Computing System. Additionally, on the 1130 system, an interactive mode of operation is provided, in which language statements may be entered and diagnostic messages may be displayed on the 1130 console typewriter. In this way, a circuit description may be modified (both topology and parameter values), input errors may be corrected, and analyses may be requested directly at the console. In addition, a concurrent output option permits observation of the output as the analysis progresses, so that execution may be halted when expected results are not obtained. The use of the program in this interactive mode is discussed in detail in the ECAP II Operations Manual for the 1130 system. (See item 11 in the References at the end of this chapter.)

Major Features

The following paragraphs summarize the 11 major facilities of the program:

1. *Transient analysis.* Provides transient response of linear and nonlinear circuits corresponding to user-specified driving functions (time-varying or constant). A variable-step, variable-order, implicit integration algorithm (see Reference 6) is used. Accuracy is the controlling factor on the size of time step rather than stability. This permits the user to trade off accuracy for running time even in circuits with small time constants and still obtain a stable solution. Several run controls which control the transient solution may be set by the user, or automatic defaults may be used. Transient initial conditions may be user-supplied or computed by a prior dc analysis.

2. *dc Analysis.* Provides a dc steady-state solution of linear and nonlinear circuits. dc parameter studies may be carried out using a one-at-a-time parameter variation feature which automatically varies one parameter through a range of values, providing corresponding dc solutions while holding other parameters at nominal values.

3. *Single circuit description.* Used for both dc and transient analyses.

4. *Advanced iteration technique.* Employed in both dc and transient analyses for solving nonlinear (or linear) network equations. A modified Newton-Raphson algorithm (see Reference 8) provides rapid convergence.

5. *Input language.* Statements can be entered in any order. Statements can be modified or replaced at any time, without reentering all statements, for easy circuit modification and return. For convenience, standard data items, such as component names, nodal connections, and values, are entered in a fixed order, while most optional data items are entered in any order. Data items are entered in free format.

6. *Model library facility.* Permits storage of general device models for later inclusion in a circuit. A model library entry (usually called a model) may represent a device model, a subcircuit, or an entire circuit. Both the network topology and parameter values are stored. There are no built-in models, so the user can define and store models for new devices having arbitrary topology and element types. A substitutable parameter feature

permits modification of model parameters for each use in a circuit, without changing the stored model. Nested models are permitted, so that models may reference other lower-level models. A maximum of 20 levels of nesting are permitted.

7. *Element dependencies.* Any ECAP II element (for example, a resistor, capacitor, or voltage source) may be specified as dependent on circuit response currents and voltages in order to model nonconstant (linear and nonlinear) circuit components. The dependency may be described by a table of values, by reference to one of the system-supplied functions (for example, DIODE), or through reference to user-written FORTRAN subprograms. Through the subprogram capability, the user may add his own standard functions to the system to be used in describing the values of nonconstant elements.

8. *User-defined parameters.* May be used to define data values (for example, PTEMP=25, BETA=60). Network elements may then be parameterized by making them dependent on one or more parameters. Parameters themselves may be defined as functionally dependent on other parameters or network responses, through use of the subprogram capability. In this way, auxiliary computations or operations may be carried out during an analysis, for example, to cause conditional termination or to generate nonstandard output such as overstresses and device states.

9. *Diagnostic messages.* Issued during all phases of system operation to help the user detect input errors. Errors are assigned appropriate severity levels, depending on the probability of obtaining useful results. Minor errors are signaled with warning messages, while other errors may be severe enough to inhibit an analysis.

10. *Sparse matrix methods and dynamic storage allocation techniques.* Permit the handling of much larger network problems in a given size main storage than would otherwise be possible. The sparse matrix methods store only the nonzero elements in the network equations. In addition to a reduction in storage requirements, the sparse matrix methods provide a reduction in running times and increased accuracy due to the fact that fewer computations need be performed.

11. *Output.* May be requested in the form of tabular listings or printer plots. On the 1130 Computing System, plots may also be obtained on the 1627 Plotter. A wide variety of standard output variables is permitted, including element voltages and currents, node voltages, element values, and parameter values. Output may be requested selectively (for example, specific element currents) or by class (all element currents). Output may be obtained concurrently with an analysis (for example, at each step of a transient analysis) and/or at the end of the analysis.

6.2 ANALYSIS FACILITIES

dc Analysis

A dc analysis is requested with the following command:
ANALYZE,DC

The dc analysis option finds the equilibrium state or dc operating point of a circuit.

For a circuit which contains storage elements (prefix symbols C and L) and so would exhibit a transient response to time-varying sources (driving functions), the dc solution corresponds to the equilibrium state which exists after all transients have decayed, that is, when the driving functions are removed or become constant and the response voltages and currents have settled to constant values with respect to time.

Such an equilibrium solution could be obtained by performing a transient analysis and letting the solution go to steady state. However, much more computer time would be required. Accordingly, a direct solution for the dc response is provided.

A dc analysis is obtained by performing a transient analysis with a single, giant time step. This, in effect, models any capacitors in a circuit as virtual open circuits (small Gs) and inductors to be modeled as virtual short circuits (small Rs), which corresponds to their behavior under steady-state conditions. The equivalent G and R values are directly proportional to the original capacitor and inductor values and inversely proportional to the giant time step. Thus, as mentioned before, the proportional relationships between response variables in special circuitry, such as a capacitive voltage divider, are preserved. Time-dependent sources are replaced by constant sources set to the value specified for TIME = TSTART (usually TIME = 0.0).

dc Parameter Studies

In addition to providing a dc steady-state solution, corresponding to circuit components set at their nominal values, the dc analysis option permits a single circuit parameter to be automatically varied through a prespecified range of values. The other parameters are held at their nominal values.

At each successive value of the varied parameter, a corresponding dc solution is automatically produced. This option is commonly referred to as one-at-a-time parameter variation. For example, resistor RLOAD could be varied from 50 to 150 ohms, in steps of 10 ohms, obtaining a dc solution at each step, by inserting the following statement before an ANALYZE, DC command:
VARY, RLOAD = 50, 150, 10

Initial Response Estimates

For nonlinear circuits, initial estimates of the response variables may be specified to assist the nonlinear solution algorithm in converging on a solution. The quantities that can be specified are conductance voltages, resistance currents, currents through dependent voltage sources, and voltages across dependent current sources. Quantities not specified are assumed, by default, to have initial values of zero or values consistent with the specified initial values of the other response variables.

For example, initial estimates for the voltage across diodes CR1 and CR2, modeled with nonlinear current sources JCR1 and JCR2, could be supplied by an initial conditions statement as follows:

$$\text{INIT, VJCR1} = .6\text{MV, VJCR2} = -.6\text{MV}$$

For circuits which have more than one equilibrium state, the initial estimates can be used to bias the solution algorithm so that it converges to a preferred solution.

dc Analysis Run Controls

Special run controls are provided for controlling the nonlinear solution algorithm during a dc analysis. Default values are supplied for these controls such that for many problems there is no need to change their values. However, the more experienced user may use them to trade off solution accuracy for computational speed.

The controls provided are

1. DCCONV. This is the maximum allowable relative convergence error. When the actual relative convergence error becomes less than this value, the iterative solution process is ended and the requested response variables are output.

2. DCITR. This control sets the maximum number of iterations to be carried out during the nonlinear solution procedure. If convergence is not obtained within this number of iterations, the values of the response variables obtained during the last iteration are supplied as output and an error message is printed.

Transient Analysis

The transient analysis option produces a time domain solution, providing the transient response of a network to driving functions specified by the user. The driving functions may be time-varying or constant.

Transient Analysis Initial Conditions

The initial conditions for the transient analysis may be either user-specified or computed automatically. The transient initial conditions consist of the initial currents through inductors and the initial voltages across capacitors as well as conductance voltages, resistance currents, currents through dependent voltage sources, and voltages across dependent current sources. These initial conditions correspond to the state of the circuit at the start of a transient analysis (at TIME=TSTART). If desired, the initial conditions may be computed automatically through use of the following command:

ANALYZE, DC, TR

In this case, a dc analysis if performed first, with any user-supplied initial values serving as initial estimates for the dc iteration processes. The resultant values of the response variables are then used as the initial conditions for the transient analysis, which follows automatically.

For some circuits, such as oscillators, although a dc solution can be obtained, it may not be useful in providing initial conditions for a transient analysis. For other circuits as well, the dc equilibrium solution may produce responses quite different from the transient initial conditions actually desired by the user.

In these cases, a transient analysis can be performed alone using the following command:

ANALYZE, TR

The transient initial conditions may be supplied by the user, via an initial conditions statement (INIT) as in the example below. Here, the initial voltage across C1 and the initial current through LCHOKE are given:

INIT, VC1 = 20, ILCHOKE = −6MA

Another use for manual specification of initial conditions is to continue a solution from the point at which a previous solution ended. In addition, the specification of initial conditions may be used to bypass the beginning portion of a transient response, which may not be of interest to the user. The computation time which would otherwise be required to produce the beginning portion of the response is avoided.

Run Controls for Transient Analysis

The transient analysis program employs a variable-order, variable-step, implicit integration algorithm. Although several run controls are provided for controlling the algorithm, the user is generally concerned with only four controls.

The purpose of providing additional controls, beyond the basic ones, is to allow the more experienced user to exert finer control over the integration algorithm. Generally, however, the default values of these auxiliary run controls provide efficient operation for the majority of circuits. The controls are described briefly below, with the four most commonly used controls listed first:

TSTART. Start time (default=0) for a transient analysis.

TSTOP. Stop time for the analysis.

TRUNC. Maximum allowable truncation error per unit time. This controls the integration accuracy (and the computation time required).

OUTINT. Minimum output interval. Allows the user to adjust the time interval between output values (default provides output at each time step).

DTMIN. Minimum allowable step size. Applies to automatic reduction of the step size to maintain accuracy during rapid changes in responses.

DTMAX. Maximum step size. Applies to automatic increase in step size during flat portions of a response.

DTSTART. Starting step size at the beginning of the integration.

MAXORD. Maximum order of the integration. Order is adjusted automatically to optimize step size.

TRCONV. Same usage as DCCONV but applies to nonlinear solution algorithm during transient analysis.

TRITR. Same usage as DCITR but applies to nonlinear solution algorithm during transient analysis.

6.3 PROCESSING DESCRIPTION

The ECAP II language is the basic interface between the user and the program. Through the language the user describes a circuit as well as the processing to be performed on the circuit data. The language itself is discussed in detail in Sec. 6.5. In this section we shall concentrate only on introducing examples of the language as it is used to carry out various operations on a circuit.

Some of the ways in which the circuit data may be processed have already been discussed. In the remainder of this section we shall present a more comprehensive discussion of the circuit processing facilities provided by the language.

First, the overall structure of the program and the processing performed by the major program phases is discussed. Then the individual processing operations associated with specific input language commands are described.

Finally some typical ECAP II runs for processing the data describing a circuit are illustrated.

Program Structure and Processing Phases

The overall structure of ECAP II is shown in Fig. 6.1. ECAP II is divided into three major processing phases: input, analysis, and output. Input statements are read by the input processing program (input phase). Each statement is checked for validity, saved in an area of storage called the circuit work space, converted into internal storage format, and stored in the data base.

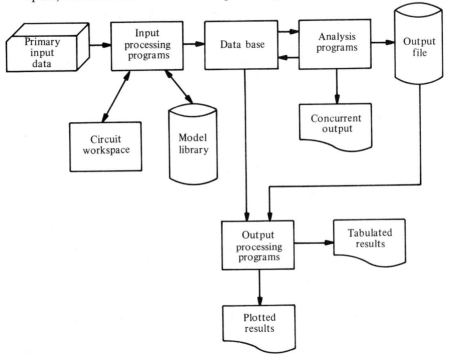

Figure 6.1 Overall Program Organization.

If the statement is a model reference (that is, if it represents a subnetwork previously defined and stored in the model library), the corresponding statements are read from the model library and included in the data base. If the statement is a command (that is, it requests that some particular processing be performed), it is checked for validity and converted to internal data format, and then the requested processing is performed.

When the command requesting an analysis is read, control is passed to the analysis programs (analysis phase). These programs use the data base both as input (the circuit description) and as intermediate storage (for matrices and tables). At each step of the analysis, output data are generated and either placed in the output file, printed concurrently, or both, as requested by the user. Concurrent output is requested by a special output description statement called a MONITOR statement. This statement is used most commonly on the 1130 Computing System where output will be printed at each step of the analysis. When executing under the operating system, the concurrent output is queued by the operating system and thus is not actually obtained until the analysis is complete.

When the analysis is complete, control is passed to the output processing programs (output phase). These programs, as directed by user input (that is, the output description statements), produce various tabular listings or plots of the solution data.

Control passes back to the input phase when all output has been processed. Additional input can now be processed. At this point, the user has a variety of options, which include

1. Changing the circuit description.

2. Saving the circuit in the model library.

3. Loading a previously saved circuit or subcircuit.

4. Reanalyzing the circuit after changes have been made.

5. Requesting additional output for further study of the results saved in the output file.

These program functions are requested by use of the command statements. Three categories of commands are provided: (1) to control entry and modification of a circuit, (2) to maintain the model library, and (3) to control program execution.

Interaction with the Circuit Description

As stated above, the primary input data are automatically stored in the circuit work space. The circuit work space represents one of the most important design features of the ECAP II language. It permits language statements to be accepted in any order, with each statement being processed as it is encountered in the input. This interactive property of the language facilitates operation of the program in a terminal-oriented mode, as may be done from the typewriter console on the 1130 Computing System or at a remote terminal of the operating system.

The circuit work space is an area of storage into which language statements are placed as they are encountered in the primary input. Also, statements included in a stored circuit or subcircuit (model) may be retrieved from the model library and placed in the work space using the LOAD command.

Generally, the statements in the work space describe a circuit, which is then referred to as the active circuit. In this sense, the circuits or subcircuits in the model library may be thought of as inactive circuits, which can be made active by loading them into the circuit workspace. Only the active circuit is affected by ANALYZE and STORE commands or by the other commands (described below) provided for interacting with the circuit work space.

At any given time, the statements in the work space may consist of any one or all of the following types: circuit description, run description, and output description. (Commands are not saved in the work space—they initiate processing on the work space.) For example, a stored circuit might contain only circuit description statements (describing only the equivalent circuit). These may be loaded into the circuit work space, with the run description and output description statements for a particular run being supplied through the primary input device (card reader, console typewriter, and so forth).

In summary, the user may think of the circuit work space and the active circuit as being analogous to a workbench containing a circuit breadboard. Elements may be replaced, removed, and added to the circuit, which may then be analyzed, modified, and reanalyzed as many times as necessary. The circuit may then be placed on the shelf (stored in the model library) for later inclusion in a larger network or later retrieval and analysis. Finally, the work space may be cleared to accept a new circuit. A variety of commands is supplied which provides the user with the facility to interact with the work space. These are listed below with their function.

Command	Function
CIRCUIT	Clears the work space and data base in preparation for reading a new circuit description. The active circuit, if one is in the work space, is lost.
LOAD	Clears the work space and loads it with a previously stored circuit or subcircuit. The function of LOAD is similar to that of CIRCUIT, except that the work space and data base are initially loaded with a circuit from the model library.
REMOVE	Removes 1 or more statements from the work space and data base.
LIST	Lists the contents of the work space. This is particularly useful for checking the active circuit description for validity after a series of changes.
STORE	Copies the active circuit into the model library as a new entry. This entry can then be used during the current or later runs as a model in larger circuits, or, using the LOAD command, loaded as the active circuit. (The STORE-LOAD combination can be used to save the current run, thereby eliminating the need of reentering the original description and any changes that have been made).

Automatic Replace Feature

An additional means of interaction with the work space (active circuit) is through the use of the automatic replace feature. Any statement previously entered can be replaced by entering a complete replacement which has the same name. For example, if R17,4–GND=20K had been entered as one of the statements describing the active circuit, the later entry R17,4–24=4.3* PAR4 will replace it. Note that both the topology and the means of value specification can be changed.

Only when the name of a newly entered statement does not match the names of any existing statements will the new statement be added to the work space.

Interaction with the Model Library

The following set of commands is provided for entry into and maintenance of the model library:

Command	Function
STORE	Creates a new entry in the model library by copying the contents of the work space (active circuit).
DELETE	Deletes a named entry from the model library. All record of it is lost.
LIST	Provides three options: 1. A list of the named entry (a particular model). 2. A list of the names of all entries in the file table of contents. 3. A list of the file in its entirely, that is, all statements of all models.

Program Control Commands

Three commands are provided for control of program execution:

Command	Function
ANALYZE	Analyzes the active circuit with either the dc or transient programs (or both when initial conditions are requested).
OUTPUT	Reexecutes the output processing programs in order to produce additional tabular or plotted output.
EXIT	Transfers control of ECAP II back to the computing system at the end of an analysis session or executes one or more user programs with control returning to ECAP II when the execution is completed.

Examples of Typical ECAP II Runs

The following examples illustrate some typical ECAP II runs, using various combinations of commands. Three runs are shown; however, the separation of data into separate runs is somewhat artificial in order to illustrate the overall functions of each run. All the statements from each run could be combined into one input deck and submitted together in a batch mode of operation.

On the 1130 Computing System, each statement might be separately entered from the console typewriter in an interactive mode of operation.

Run 1.

Specifying Models.

```
          CIRCUIT
                 .
                 .   circuit description statements
                 .
          STORE, 2N321, EXT = B−C−E
          CIRCUIT
                 .
                 .   circuit description statements
                 .
          STORE, 2N445, EXT = B−C−E
          LIST, 2N941
          LOAD, 2N941
          RB, B − BP = 27
          REMOVE, CBE
          DELETE, 2N941
          STORE, 2N941, EXT = B−C−E
          LIST
```

Explanation:

Two new models (2N321 and 2N445) are specified and stored in the model library. Note that the second CIRCUIT command is used to clear the work space to accept the second model, after the first model has been stored. Otherwise, the second model would include the statements of the first model.

After the second model is stored, a listing of the statements describing the 2N941 model is obtained to provide a record of the model before it is changed. The LOAD command clears the work space and loads model 2N941. The existing RB element statement within the model is replaced by the new RB statement, and capacitor CBE is removed. The current 2N941 model in the library is first deleted, and then its replacement is stored (copied from the work space into the model library). Finally, the active circuit (contents of the work space) is listed as a permanent record of the changes made to the 2N941 model.

Run 2.

Specifying a Circuit.
    CIRCUIT
        .
        .

    Q7, B7 — C7 — E7 = 2N321
    Q11, B11 — C11 — E11 = 2N445   circuit description statements
    Q13, B13 — C13 — E13 = 2N941
        .
        .

    PRINT7, IRB.Q13
        .   output description statements
        .
    STORE, MAINCKT

Explanation:

A main circuit is specified, referencing the models stored in the previous run. The circuit description and output description statements are stored in the model library. Several analysis runs with varying initial conditions and run controls are anticipated, so the run description statements are not stored. They are supplied, along with the ANALYZE command, on the next run. Note that no external nodes are supplied on the STORE command, since the circuit will not be referenced within a larger circuit.

Run 3.

Specifying Analysis Runs.
    LOAD, MAINCKT
    INIT1, IL4 = .5MA, VCIN = 6
    TSTOP = 200   run description
    TRUNC = .1   statements
    MONITOR1, IRLOAD
    ANALYZE, TR
    INIT1, IL4 = 1MA, VCIN = 10, IL3 = 0.2MA
    REMOVE, MONITOR1
    ANALYZE, TR
        .   successive
        .   modifications
        .   and analyses

```
PLOT3, IRLOAD, IL4
PRINT8, VCN, TIME
OUTPUT, PRINT7, PLOT3, PRINT8
EXIT
```

Explanation:

The circuit stored in the previous run is loaded into the work space from the model library. The run description statements, an additional output statement (MONITOR), and the ANALYZE command are supplied from the primary input device. On the first analysis run, a concurrent print out of IRLOAD is requested via the MONITOR statement. (On the 1130 system, the analysis could be terminated through use of one of the console switches, should the IRLOAD response not match expected results.)

After the first analysis run, the initial conditions are modified (the original INIT1 statement is replaced with the modified INIT1 statement), the MONITOR1 statement is removed, and a second analysis is requested. This may be followed by successive modification-analysis combinations.

After the last analysis, the additional output description statements, PLOT3 and PRINT8, are entered. The OUTPUT command requests additional output of the data previously saved in the output file from the last analysis run. Only the output requested via the specific statements referenced on the OUTPUT command is obtained. In the example, the output requested on the PLOT3, PRINT8, and PRINT7 statements is obtained, where the PRINT7 statement was included in the stored circuit of run 2.

Note:

The LOAD command would not be required if run 3 immediately followed run 2, since MAINCKT from run 2 would then still be in the work space.

6.4 NETWORK ELEMENTS

Basic Elements

A network may contain a wide variety of devices, for example, diodes, transistors, tunnel diodes, and field-effect transistors. If equivalent circuit models for these devices have already been developed and stored in the model library, they may be included in the network's equivalent circuit description by use of a special language statement called a model reference statement.

Each equivalent circuit, representing either an entire circuit or a device model, must be composed of a set of allowed circuit components which are recognized by the program. A separate language statement is provided to describe the value and topological connections for each of the allowed components. The language statement for a given component is characterized by its statement name, which begins with a prefix designating the type of the component. Table 6.1 lists the allowed ECAP II components.

Table 6.1 Allowed ECAP II Circuit Components

Category	Component Types	Prefix Symbol
Elements	Resistor	R
	Conductor	G
	Inductor	L
	Capacitor	C
	Voltage source	E
	Current source	J
Element couplings	Transference	TF
	Mutual coupling	M
Model references	Model reference	Q, D, CR, T, TX, X, MOD (all equivalent)
Auxiliary components	Ammeter	IM
	Voltmeter	VM

Because of the flexibility permitted by ECAP II in specifying component dependencies, the effort required by the user to describe equivalent circuit models is kept to a minimum. The components shown in Table 6.1 and the functional dependencies permitted by ECAP II for describing them are discussed below.

Elements

The most basic ECAP II components are categorized as elements (R, G, L, C, E, J) in Table 6.1. The voltage and current sign conventions and nomenclature for the variables associated with an element are shown in Fig. 6.2.

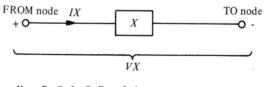

X = R, G, L, C, E or J element
IX = Element current
VX = Element voltage

Figure 6.2 ECAP II Element.

A uniform sign convention which is easy for the user to remember has been chosen for all basic elements, namely, positive current is assumed to flow through the element from the positive terminal (the FROM node) to the negative terminal (the TO node) of each element.

Note that this convention for positive current is opposite to the normal flow of current through sources, which is from the negative to the positive terminal. However, it has the advantage of being easy to remember, since it consistently applies to passive elements as well as sources. In addition, it leads to the correct sign for power in any element, whether sink or source. For example, the current in a voltage source is negative so that the power dissipation, the product of voltage and current, is also negative.

The basic response variables computed by ECAP II are element voltages (designated as VR, VC, VL, and so forth) and element currents (IR, IC, IL, and so forth). Node voltages (VN) relative to ground can be obtained as well.

Elements are categorized by ECAP II as constant or nonconstant. The values of constant elements are given as numeric constants. Nonconstant elements may be functions of circuit responses, time, numeric constants, and user-defined parameters. User-defined parameters are arbitrarily named (except for certain restrictions on the first letter) and may be assigned any desired meaning, such as temperature, base width, and conductivity. In this way the functional dependencies associated with both linear and nonlinear elements can be described.

For example, a diode may be modeled as a J element dependent on its own response voltage, VJ. An element may also be described as dependent on the response current or voltage of other elements in the circuit. In this way, the linear and nonlinear dependent or controlled sources typically used in device models can be represented.

Functional dependencies associated with nonconstant elements may be described either by tables or mathematical functions, depending on which is most convenient or appropriate for describing a given element. A standard set of mathematical functions and a table interpolation function, called system-supplied functions, are supplied with ECAP II. However, a user who has some knowledge of FORTRAN may define additional functions, if necessary, and store them in the program library. These are called user-written functions or subprograms.

Figure 6.3 contains some examples showing how elements are used in modeling circuit components, including the use of function references for describing element dependencies. For each component, a typical equivalent circuit representation is given along with the ECAP II language statements used to describe the elements. These examples illustrate the problem-oriented nature of the language and the ease with which functional dependencies can be described. Some ECAP II features illustrated in Fig. 6.3 are discussed in the following paragraphs.

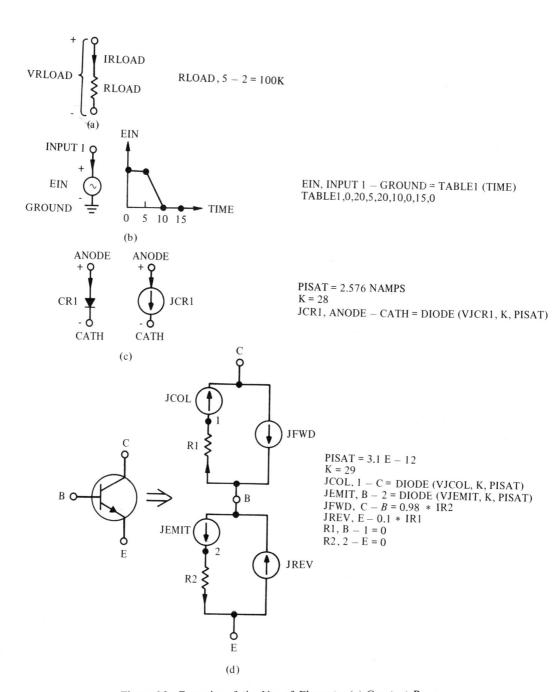

Figure 6.3 Examples of the Use of Elements: (a) Constant Resistor; (b) Time-Dependent Voltage Source; (c) Diode Model; (d) Transistor Model.

In comparing Fig. 6.3(a) with Fig. 6.2, note how the resistor representation corresponds to the general representation for an element. The resistor name, RLOAD, is arbitrary except that it must begin with the proper prefix, R, as given in Table 6.1. The name appears as the statement name for the language statement. The element's voltage and current are then referred to as VRLOAD and IROAD, respectively. In this example, numbers have been used to label the nodes, and they are entered in the sequence FROM node-TO node in the corresponding language statement. In general, any combination of letters and/or numbers (alphameric labels) may be used for node names. The resistor value of 100 kilohms is inserted after the equals sign, using the standard electrical notation 100K. The language allows other standard electrical units to be similarly expressed.

Figure 6.3(b) illustrates a time-dependent voltage source. The dependency is described by a TABLE statement containing number pairs which give the coordinates of points along the desired curve. The key letters TAB, appearing after the equal sign in the FIN statement, invoke the system-supplied table interpolation function. EIN is evaluated using the current value of TIME as the independent variable to interpolate along the curve described by the TABLE1 statement. In this example, alphameric node names have been chosen for the nodes to which element EIN is connected. The name GND or the number 0 (zero) is reserved for the ground node.

Figure 6.3(c) illustrates the use of a nonlinear current source (J element) to model a diode. The user-defined parameters PISAT and K have been used to supply parameter values in the diode characteristic equation

$$J = PISAT(e^{KV} - 1)$$

where

J = diode current
V = diode voltage
PISAT = reverse saturation current
K = a constant

This nonlinear current-voltage dependency is specified in the language statement for the current source JCR1 by reference to the DIODE function, one of the ECAP II system-supplied functions. The independent variable in this use of the DIODE function is VJCR1, the voltage across the JCR1 element, which appears as the first of the three arguments required by the DIODE function.

Figure 6.3(d) contains a simplified Ebers-Moll model for a transistor, consisting primarily of current sources. The nonlinear characteristic of the collector-base junction diode is modeled with the current source JCOL, and the base-emitter junction diode is modeled with the JEMIT current source.

The effects of the forward and reverse current gains are represented by JFWD and JREV, respectively. JFWD is a controlled current source which is dependent on the current through the base-emitter diode, following the linear relationship JFWD = 0.98∗IR2, where 0.98 is the forward current gain. Similarly, JREV = 0.1∗IR1, where 0.1 is the inverse current gain. R1 and R2 are dummy resistors (of zero resistance) inserted in the model to monitor the currents IR1 and IR2. IR1 and IR2 represent the diode currents (instead of IJCOL and IJEMIT) because they are primary response variables. The significance of primary response variables is discussed later in the section entitled "Primary and Secondary Response Variables" in Sec. 6.5.

Element Couplings

The element couplings listed in Table 6.1 provide a convenient means for modeling interaction between elements, such as current gain, transconductance, and mutual inductance.

These interelement couplings induce a current or voltage in one element which is proportional to the current, current rate, voltage, or voltage rate in another element. Such induced quantities are commonly called dependent or controlled sources. Unidirectional couplings are called transferences (TF) and equal bidirectional couplings are called mutual couplings (M). Only linear interelement couplings (that is, linear controlled sources) may be modeled by use of the TF or M type of element couplings. The effect of nonlinear couplings may be modeled by nonconstant E or J source elements.

The primary advantage of couplings (TF and M) lies in the variety of linear dependencies which can be modeled with them. For example, they can be used to model current and voltage differentation in addition to the more common dependencies mentioned above. The various coupling types are listed in Table 6.2.

Model References

Once a given model has been stored in the model library, it may be used in describing other models or entire circuits. A given instance (or usage) of a model within a circuit description is called a model reference. In this sense, then, a model reference can be considered an individual circuit component and so is included in Table 6.1. A special language statement, called a model reference statement, is used for inserting models into a circuit description.

For example, suppose that the diode model of Fig. 6.3(c) were stored in the model library. This could be accomplished by inserting the following language statement after the three statements shown in Fig. 6.3(c):

STORE,1N221,EXT=ANODE−CATH

This statement stores the diode model in the model library under the name 1N221. The entry, EXT=ANODE−CATH specifies the names of the

Ch. 6 / Hogsett

Table 6.2 Transference Types

FROM Controlling Variable	Transference Type	TO Controlled Variable	Effect Induced in Controlled Variable
VC	TC DMU	IC or IG or IJ VL or VR or VE	$I' = TC*VC$ $V' = DMU*VC$
IL	DBETA TL	IC or IG or IJ VL or VR or VE	$I' = DBETA*IL$ $V' = TL*IL$
VG or VJ	TG MU	IC or IG or IJ VL or VR or VE	$I' = TG*(VG\ or\ VJ)$ $V' = MU*(VG\ or\ VJ)$
IR or IE	BETA TR	IC or IG or IJ VL or VR or VE	$I' = BETA*(IR\ or\ IE)$ $V' = TR*(IR\ or\ IE)$

external nodes of the model. Later, when this diode model is included in a circuit, these external node names are replaced in the circuit description (but not in the model library) by the node names specified in a model reference statement.

For example, the 1N221 diode model can be used in describing the transistor model in Fig. 6.3(d) by first removing the PISAT and K statements and then inserting the model reference statements

$$\text{DCOL},1-\text{C}=1\text{N}221$$
$$\text{DEMIT},\text{B}-2=1\text{N}221$$

in place of the statements JCOL and JEMIT, to obtain the following set of statements describing the transistor:

$$\text{DCOL},1-\text{C}=1\text{N}221$$
$$\text{DEMIT},\text{B}-2=1\text{N}221$$
$$\text{JFWD},\text{C}-\text{B}=.98*\text{IR}2$$
$$\text{JREV},\text{E}-\text{B}=.1*\text{IR}1$$
$$\text{R}1,\text{B}-1=0$$
$$\text{R}2,2-\text{E}=0$$

DCOL and DEMIT are model reference names. These may begin with any of the prefixes listed in Table 6.1 which identify the statement type as a model reference. When the program encounters the DCOL and DEMIT statements, the 1N221 diode model is automatically inserted into the transistor equivalent circuit from nodes 1 to C and from nodes B to 2. These nodes replace the dummy ANODE and CATH nodes used in defining the diode model of Fig. 6.3(c).

Substitutable Parameters. This feature of ECAP II permits modification of the values of elements or parameters contained within a library model for a given use in a circuit. The modifications or changes apply only to a given instance of the use of a model (a model reference) and do not affect the stored model. For example, a given transistor model can be used to model transistors of different types (with varying parameters).

This feature can be illustrated by again considering the transistor model of Fig. 6.3(d), with the model reference statements inserted to include the 1N221 diode model.

The 1N221 diode model is now being used as a general model for the transistor junction diodes. For this use of the diode model, the values of the PISAT and K parameters stored as part of the diode model definition may not properly represent the parameters for the transistor junctions. The appropriate values of PISAT and K for each junction in the transistor model can be included in the DCOL and DEMIT model references as follows:

$$\text{DCOL},1-\text{C}=1\text{N}221,\text{PISAT}=7.1\text{E}-12,\text{K}=32$$
$$\text{DEMIT},\text{B}-2=1\text{N}221,\text{PISAT}=3.1\text{E}-12,\text{K}=29$$

The original values of PISAT = 2.576 NAMPS and K = 28 remain unchanged in the description of the 1N221 diode model still in the library.

To summarize, the statements for defining the transistor model of Fig. 6.3(d) and storing it in the model library under the name XISTOR would appear as follows:

```
DCOL,1−C=1N221,PISAT=7. 1E−12,K=32
DEMIT,B−2=1N221,PISAT=3. 1E−12,K=29
JFWD,C−B=.98*IR2
JREV,E−B=.1*IR1
R1,B−1=0
R2,2−E=0
STORE,XISTOR,EXT=B−C−E
```

The XISTOR model could then be included in a main circuit description with a model reference statement which might typically appear as

$$Q7,3-7-9=\text{XISTOR}$$

where the base, collector, and emitter of Q7 are connected to main circuit nodes 3, 7, and 9, respectively.

Nested Models. Model references are the means by which the user takes advantage of the nested model feature of the program. Model nesting occurs when a model references (or calls) one or more lower-level models. In the above example, only one level of nesting is represented; that is, the transistor includes two diodes, both of which are at the same level. However, the program permits definition of models with up to 20 levels of nesting. As a result, a given model reference statement inserted in a circuit description includes the referenced model together with any models which it calls, and any models called by those models, and so on until all references to models are resolved.

An interesting, but perhaps only academic, characteristic of this feature is that the definition of recursive models is possible. That is, a model, if properly constructed, can refer to itself within the body of its definition. The expansion of a recursive model will result in a hierarchy of nesting up to the twentieth level, at which time the recursion will cease and a nonterminal warning/message will occur. This technique could be used to model repetitions structures such as transmission lines or other distributed components.

Circuit Description Independent of Analysis Type. Except for the considerations of accuracy and/or fidelity when specifying a circuit or model, the user does not need to be concerned with the type of analysis to be performed. That is, any model may be used for a dc analysis, a transient analysis, or both. When a dc analysis is performed, any storage elements, Ls and Cs, are automatically modeled by the program with equivalent dissipative elements, small Rs ("short circuits") and small Gs ("open circuits"), respectively. The resistance and conductance values in the equivalent models are proportional to the original inductance and capacitance values. Because of this, the proportional relationships between response variables, in such circuit configurations as capacitive voltage dividers, are preserved.

Auxiliary Components

The auxiliary components listed in Table 6.1 include ammeters (IM) and voltmeters (VM). These may be connected anywhere in a circuit to measure voltages and currents. Generally, they are used when the response variable of interest cannot be obtained conveniently by a standard output variable, such as an element voltage, element current, or node voltage. Figure 6.4 illustrates some situations in which meters can be advantageous.

A typical output statement associated with the circuit of Fig. 6.4 is

PRINT,IR1,IR2,IR3,IM1,IM2,VM1

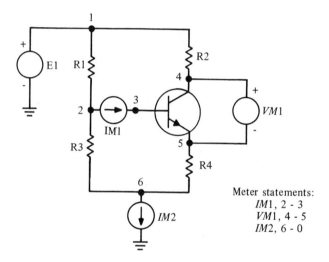

Figure 6.4   Examples of Meters.

6.5 THE ECAP II LANGUAGE

Statement Conventions

Format

Statements are entered on cards for the operating system. On the 1130, statements may be entered on cards or on the 1130 console typewriter. Typewriter input is similar to card input. The typewriter equivalent of a single card is a 1-to-72-character line, followed by the EOF character, which is the equivalent of the physical end of a card. The following discussion on conventions, while about card formats, is also applicable to typewriter input.

Statements may be entered in columns 1–72 of a card. A statement may be entered in free format, that is, anywhere within columns 1–72. Data within columns 73–80 are ignored by the program; these columns may be used for identifying information and sequence numbers.

Figure 6.5 illustrates the card formats for statements. A statement may be continued on one or more cards by inserting a plus sign (+) in column 1 of the continuation cards. The ampersand (&) is treated the same as the plus sign and may also be used. The number of continuation cards permitted is limited only by the size of the circuit work space.

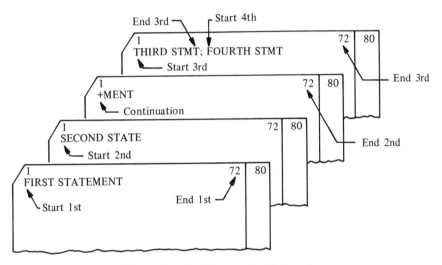

Figure 6.5 Card Formats for Statements.

To facilitate making later corrections, it is generally good practice to enter only one statement per card. In this case the physical end of the card delimits the end of the statement, provided there is no plus sign in column 1 of the following card. Multiple statements may be entered on a card, by separating the statements with a semicolon (;). The appearance of a semicolon anywhere within columns 1–72 signals the end of the preceding statement, whether or not another statement follows on the same card.

Statements may be punched using any characters contained in the extended binary coded decimal interchange code (EBCDIC) character set, as provided by the IBM 029 Card Punch or the 1130 Console Typewriter. The IBM 026 or 024 Card Punch containing the binary coded decimal (BCD) character set may be used, but the characters #, %, <, @, and & are treated by the program as =, (,), ', and + respectively.

Blank cards in the input card deck are ignored by the program, except between a card and its continuations (which would be an error). If a blank card appears between a card and its continuation, the first continuation card is assumed to begin a new statement and, hence, would evoke an error message.

Blank characters may be inserted within a statement to improve readability. With certain exceptions, blanks are removed by the program to conserve space in the work space. Leading and trailing blanks are always removed. With the exception of literals and comments, blanks within the text of a statement are also removed. That is, regardless of the number of blanks inserted within the text of a statement, the program prints the statement in a standard format. The standard format contains a blank character before and after major delimiters such as ,, —, and =. For example, a statement punched as

```
        CC1
         ↓
         R    LOAD, OUTPUT  —  GND  =     100K
```

appears in a program printout as

```
        CC1
         ↓
         RLOAD , OUTPUT  —  GND  =  100K
```

Literals

A literal is a string of alphameric characters (A–Z and 0–9) and special characters ($, <, (,), :, /, and so forth) enclosed in single quotation marks. The general form for a literal is

'literal'

Literals are used for entering descriptive data to appear in printed and plotted output. They may also be used in comments, although this use is not frequent. The data between the quotes are stored and output by the program exactly as it is entered. For example, a literal would be used to specify a title as follows:

TITLE5,'RUN5—TEST OF MODEL 2N990'

- Blanks within literals are not removed by the program.
- The single quote character cannot be used within a literal. If desired, the double quotation character (") may be used. Note that the double quote is an unique EBCDIC character and not a pair of single quotes.
- The semicolon cannot be used within a literal, since this would signal the end of the statement.

Basic Statement Format

In the above section entitled "Statement Conventions" we presented the overall physical properties of a statement, where the statement is considered an entity. In this section we shall discuss the logical makeup of a statement and presents rules for constructing the logical portions of a statement.

A statement may consist of three sections: a statement name, data, and a comment. The two general formats and required orders for entering these are

     statement name , data : comment
     statement name = data : comment

The statement name is separated from the data by a delimiter which is either an equals sign (=) or comma (,). A comment is delimited from the rest of the statement by a colon (:).

Statements do not require all three sections. The data section may not be required, depending on the requirements of a specific statement. A comment is always optional and may appear by itself, with a name, or with both a name and data. A comment must always be preceded by a colon (:). The parts of the basic statement are discussed following the examples.

Examples.

1. Statement containing a statement name, data, and comment:

   Q13, B13−C13−E13 = 2N522 : MODEL REFERENCE

2. Statements containing a statement name and data:

   a. RLOAD , N1−N2 = 100K
   b. PAR10 = 5

3. Statement containing a statement name only:

     CIRCUIT

4. Statement containing a comment only:

    : A COMMENT STATEMENT

Statement Name

One function of the statement name is to identify the statement type, so that the program can perform the associated action (as in the case of a command) and properly interpret the data following the statement name, if any.

Statement names (except for parameter statements) must begin with a prescribed prefix which uniquely specifies the statement type. The lack of a prescribed prefix on a statement name distinguishes it as a parameter name.

Another function of the statement name is to differentiate between statements of the same type, that is, to provide a unique statement label. A given statement may then be referenced on other statements by its statement name. The unique statement names (labels) are created by the user, by adding alphameric characters (A–Z and 0–9) following the prescribed prefix. Even for statement names that are not referenced or for run control statements, the user may want to add characters to enhance the meaning of a statement name.

The following examples of statement names illustrate both the prescribed prefixes and optional user additions:

Prescribed Prefix	Optional User Addition	
R	LOAD	
R	10	
C	BE	Addition creates unique names
C	5	(statement labels) and/or meaning enhancement
PRINT	7	
DTMIN	IMUM	
MAXORD	ER	Addition provides only meaning enhancement
TRUNC	ATION ERROR	

The prescribed prefix for each type of statement is described later under the individual statement descriptions.

For statements used to specify circuit components (those with the prefix R, C, L, and so forth), the statement name is synonymous with the name of the component. Thus, the element, coupling, model reference, and meter names referred to under the individual statement descriptions follow the rules given here for statement names.

To illustrate the referencing of statements by their statement names, the RLOAD and PRINT7 statements could be removed from a circuit description through use of the REMOVE statement (a command):

REMOVE, RLOAD, PRINT7

The statement name is also synonymous with the value of a component. For example, the values of R10 and CBE are printed by the output description statement

PRINT, R10, CBE

Note: The user normally assigns unique names to circuit entities. (The term *entity* is used to refer to any named item in circuit description statements and run controls.) This permits reference to those names later in the circuit description, for example, if those entities might be removed or their values printed. However, assigning unique names is necessary only for referencing an entity by its name. That is, a circuit or model can be described using only the prescribed prefixes as entity names (or statement names) so long as no references are made to these entities. For example, statements such as

REMOVE, R, PRINT   (invalid)
PUNCH, L, C   (invalid)

are invalid because R, PRINT, L, and C are not unique names.

Data Section

The data section of a statement consists of data entries separated by special characters used as delimiters.

The specific form of a data section (order of the entries, required and optional entries, delimiters used, and so forth) varies from statement to statement and is described later.

Comment Section

A comment is used for entering descriptive information. It is not interpreted by the program; however, it is stored with the circuit in the model library and appears in any listings of the circuit input statements.

Alphameric and special characters, except for the semicolon (;), may be used in a comment. Groups of successive blanks within a comment are reduced to one blank character. For example, a comment punched as

    : MULTIPLE   BLANKS REDUCED   TO   ONE

is stored and listed by the program as

    : MULTIPLE BLANKS REDUCED TO ONE

Comments may be continued on continuation cards and may be of any length. However, when listed by the program, two blanks are inserted after every 99 characters.

A literal may be inserted anywhere within a comment, or a comment may consist entirely of a literal. As discussed in the section entitled "Literals" earlier in this section, blanks in literals are not suppressed.

Network Nomenclature

In this section we shall present the conventions used for naming (identifying) network entities.

In general, entity names may contain from 1 to 99 alphameric characters. Blank characters are acceptable and are removed by the program. The blanks are not included in the character count. However, the character count does include the qualifiers and periods used in qualified names, which are described later in this section.

The next six subsections contain further rules for naming specific types of entities.

Element, Element Coupling, and Model Reference Names

Elements, element couplings, and model references are the basic entities (circuit building blocks) recognized by the program. All circuits and models must be reduced to equivalent circuits composed only of these basic components.

The names of these components are constructed following the general rules given previously for statement names. The prescribed prefixes, which identify each component type, were listed earlier in Table 6.1. To form unique names for each entity in a given circuit or model the user would add alphabetic and/or numeric characters to the prescribed prefixes. Some examples of unique component names are

Element names	RIN, C10, LCHOKE, EREF, J2
Coupling names	M10, ML5L1, TF12, TFBETA
Model reference names	Q2, MOD1, DIODE1, XTOR4

Element Voltage and Current Names

The voltage across an element or the current through an element may be requested as output by including the voltage or current names on an output description statement. Element voltage or current names may also be used as arguments of function references used to define the nonconstant elements.

An element voltage or current name is formed by prefixing V or I, respectively, to the element name.

General form.

```
V element name
I element name
```

Examples.

VR7, VRLOAD, IRLOAD, ILCHOKE, IC10

Node Names

A node is defined as the junction of two or more elements. Since models are also referenced as single entities, nodes are also formed when models are connected to elements or to other models.

The node names GND and 0 (zero) are recognized by the program as the ground or reference node. It is not necessary to specify a ground node unless node voltages are requested as output. However, if no ground node is specified, the last node name defined in the input stream is assumed to be the ground node.

Examples of node names are

1, 4, INPUT, V1, NODE5, VOUT, BASE.Q1   (qualified name)

Node Voltage Names

The voltage from any circuit node to the ground or datum node is called a node voltage. Node voltages may be requested as program outputs, or they may be used as variables (arguments) in function references that define auxiliary parameters.

A node voltage name is formed by prefixing VN to the node name.

General Form.

> VN node name

Examples.
> VN12, VNOUTPUT, VNBASE.Q1

Parameter Names

Parameters are used to describe auxiliary network quantities. They may also supply a value which is common to many entities and which may be changed from run to run. Changing the parameter changes all the entities defined on the parameter.

Parameter names have no prescribed prefix and are selected arbitrarily by the user. However, a parameter name cannot conflict with the names of other entities. Further details on naming parameters are given under the description of the parameter statement.

Examples of parameters names are
> PAR1, PTEMP, BETA, K, SWITCH

Qualified Names

Qualified names are used in a main circuit description only when it is necessary to refer to entities inside a model which is referenced (or called) by the main circuit. Since models can call other models (nested models), this convention applies as well when a statement in a model contains an entry referring to an entity within a lower-level model. The term entity is used in a general sense to refer to any named item in a circuit description.

An example of a qualified name is
> VJ1.CR1.Q2

which refers to the voltage across element J1, contained in the CR1 model reference, which is contained in the Q2 model reference.

General Rule. A qualified name is formed by starting with the entity name and attaching, in order, the names of the model references in which it is contained, proceeding from the lowest- to the highest-level model reference. A period (or decimal point) is used to separate the individual names (qualifiers) making up the qualified name.

Entity names which may require qualification are element names, coupling names, node names, node voltage names, element current and voltage names, and parameter names. In addition, any main circuit references to output or run description statements within model references must be qualified, for example,

REMOVE, PRINT7.MOD1, TSTOP1.MOD1

The need for qualifying entity names when referencing entities within models is illustrated in Fig. 6.6, which shows a main circuit description containing two model references, Q1 and Q2, both of which refer to the same model, XISTR, stored in the model library. The systematic naming procedure automatically applied by the program to assign unique internal labels to each entity is illustrated.

The user must assign unique model reference names if entities within the model are to be referenced. Such a procedure was followed in choosing the unique model reference names, Q1 and Q2, in the main circuit description of Fig. 6.6(a).

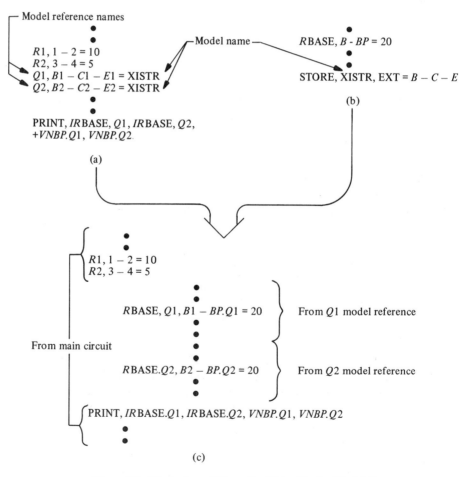

Figure 6.6 Illustration of Name Qualifying During Model Expansion.

Each reference to a model in the main circuit description is expanded by the program to include the entities contained in the model. While expanding each model reference, the program preserves uniqueness by automatically qualifying the names of entities in the model. It does this by adding the model reference name to the names of the entities within the model, as illustrated in Fig. 6.6(c).

The user follows the same procedure when it is necessary to refer to entities within a model called by another model or by the main circuit. This is illustrated by the PRINT statement in the main circuit description, which references element currents and node voltages within the Q1 and Q2 model references. Note that the names of nodes which are internal to the model must be qualified, such as node BP in the example. External nodes of the model (nodes B, C, and E) are replaced by the nodes of the model references (B1, C1, E1, and B2, C2, E2); therefore, they do not need to be qualified.

Notes:

1. Because the program must internally qualify names to uniquely identify circuit entities, it is good practice for the user to keep the names of entities as short as possible. This reduces storage requirements and shortens processing times. The maximum length (99 characters) for names includes the extra characters required to form the fully qualified name of an entity. Another reason for keeping names short is that only the first 16 characters of the names are used to label entities in the printed or plotted output.
2. Statements entered by the user may not have qualified statement names. For example, the following statement is invalid:

    RBB.Q1,1−2=20   (invalid)

External References

Statements within a model may reference entities which are external to the model. These external entities may be either in a higher-level model which contains the model in question or in the main circuit itself.

The ECAP II input processing program attempts first to resolve all entity references within a model itself. Any symbols left unresolved are then processed in an attempt to resolve them in successively higher-level models. This process is continued until either the symbol is resolved in one of the higher-level models or in the main circuit itself.

The example in Fig. 6.7 illustrates the ECAP II symbol resolution process. There are three levels of circuitry illustrated: model DIODE, which is used in model XISTOR, which itself is used in the main circuit. The diode JD references parameters K and PISAT, which are not supplied in the model itself. The transistor model uses DIODE and supplies only the K factor. Then, when XISTOR is used in the main circuit, PISAT is supplied.

The transistor model uses a table TABJ1 to describe the value of J1. When the model is used as Q1, TABJ1 is deleted by the TABJ1/ entry. A replacement table (of the same exact name) is supplied in the main circuit.

```
CIRCUIT : DIODE MODEL (REQUIRES EXTERNAL SPEC. OF K AND PISAT)
JD , A — K = DIODE(VJD,K,PISAT)
STORE , DIODE , EXT = A — K
DGM233 *I*          0 FOLLOWING MODEL HAS BEEN STORED
DGM233              DIODE
CIRCUIT : TRANSISTOR MODEL (REQUIRES EXTERNAL SPEC. OF PISAT FOR DIODE D1)
J1 , 4 — 7 = TABJ1(VJ1)
D1 , 7 — E = DIODE
    JD , 7 — E = DIODE(VJD,K,PISAT)
K = 28 : PROVIDE Q/KT FACTOR FOR JD IN DIODE D1
TABJ1 , −3 , 4 , 7 , 5
STORE , XISTOR , EXT = E — B — C
DGM233 *I*          0 FOLLOWING MODEL HAS BEEN STORED
DGM233              XISTOR
CIRCUIT : MAIN CIRCUIT
R3 , IN — 0 = 100K
J4 , IN — 8 = DIODE(VJ4,K,4.3E−12)
Q1 , 4 — 9 — GND = XISTOR , TABJ1 /
    J1 , 4.Q1 — 7.Q1 = TABJ1(VJ1)
    D1 , 7.Q1 — 4 = DIODE
        JD , 7.Q1 — 4 = DIODE(VJD,K,PISAT)
    K = 28 : PROVIDE Q/KT FACTOR FOR JD IN DIODE D1
TABJ1 , −4 , 3.5 , 0 , 3.5 , 1 , 9 : REPLACEMENT TABLE
PISAT = 7.3NANOAMPS : PROVIDE REV. SAT. CURRENT FOR JD IN DIODE D1 IN XISTOR Q1
K = 32 : Q/KT FACTOR FOR J4 (WILL NOT BE CONFUSED FOR K=28 IN XISTOR)
ANALYZE , DC
```

Figure 6.7 Example of External Reference.

Internal References

Statements outside of a model may reference entities within a model. This was illustrated above in Fig. 6.6, where statements in the main circuit requested printout of quantities inside the models. This concept can be exploited further by referencing, in one model, entities in either lower-level models contained in this model or entities in an entirely separate model. All that is necessary to take advantage of this feature is to properly qualify the entity names which are referenced.

Value Specification

In addition to assigning names to the various network entities following the conventions given above, values must be given for element, couplings, and parameters. Values are also supplied for auxiliary quantities described in more detail later, such as initial conditions, tables, model changes, and run controls.

Except as noted under the individual statement descriptions presented later, values may be expressed as numeric constants or as function references (used to describe nonconstant entities).

Numeric Constants

Numeric constants may be entered with optional signs, decimal points, exponents, and scale factors. Blanks embedded within a numeric constant are ignored.

Examples of valid numeric constants include

<pre>
4
4.
.3
3.14159
5.E3   (means 5 × 10³)
−6.15E−12   (means −6.15 × 10⁻¹²)
7 .   (interpreted as 7.)
</pre>

If the leftmost sign (for the mantissa) is missing, the number is assumed to be positive.

Exponents are expressed in the form $E \pm dd$, where dd represents a one- or two-digit exponent. The exponent is assumed to be positive if the sign is missing. The form $E \pm dd$ indicates that the preceding decimal number is to be multiplied by 10 raised to the power $\pm dd$.

On the 1130 Computing System, at least nine significant digits are carried. The magnitude of a numeric constant may not be greater than 2^{127} or less than 2^{-128} (approximately 10^{38} and 10^{-39}). A value of zero is permitted.

Under the operating system, at least 15 significant digits are carried. The magnitude of a numeric constant may not be greater than 16^{63} or less than 16^{-63} (approximately 10^{75} and 10^{-75}). A value of zero is permitted.

Examples of invalid numeric constants include

<pre>
3,100   (embedded comma)
1.E   (missing a one- or two-digit exponent)
1.2E+113   (three-digit exponent)
23.5E+97   (value exceeds magnitude permitted)
</pre>

Scale Factors

A scale factor following a numeric constant multiplies the numeric constant by a corresponding multiplier. Its use is optional.

Scale Factor	Unit Prefix	Multiplier
T	Tera	10^{12}
G	Giga	10^{9}
MEG	Mega	10^{6}
K	Kilo	10^{3}
M	Milli	10^{-3}
U	Micro	10^{-6}
N	Nano	10^{-9}
P	Pico	10^{-12}

Any additional characters can be appended to a scale factor to create mnemonics for common electrical units. However, only the scale factor is interpreted by the program. For example, the scale factor P might be expressed as PF or PICOFARADS.

Examples of the use of scale factors are

$$\left.\begin{array}{l}\text{10MEG}\\ \text{10MEGOHMS}\end{array}\right\} \text{means } 10 \times 10^6$$

$$\text{100K} \quad\} \text{means } 100 \times 10^3$$

$$\left.\begin{array}{l}\text{.005USEC}\\ \text{5 NSEC}\\ \text{5E}-\text{3USEC}\end{array}\right\} \text{all equivalent to } 5 \times 10^{-9}$$

Units

When specifying element values, the user must make sure that all element values form a consistent set of electrical units. This applies to element values given as function references (mathematical or tabular) as well as to values given as numeric constants.

To assist the user in maintaining a consistent set of units, the program assumes that values have been expressed in the following set of units. Units assumed by the program:

Quantity	Units
Resistance	Ohms
Conductance	Mhos
Inductance	Henries
Capacitance	Farads
Voltage	Volts
Current	Amperes
Time	Seconds
Frequency	Hertz
Phase	Degrees

The scale factors, discussed previously, allow element values to be conveniently expressed in the set of units assumed by the program. For example, a given resistor, R2, with a value of 5000 ohms might be specified in any of the following ways, depending on which is most convenient:

$$\text{R2, } 1-2=5000$$
$$\text{R2, } 1-2=5\text{K}$$
$$\text{R2, } 1-2=.005 \text{ MEG}$$

Note: It is recommended that scale factors be used with some caution. They can be applied only to values given as numeric constants and not to scale the resulting value from function references. Inconsistent unit specifications can result if the user is not aware of the units used in defining the function.

In certain circuits, selection of a set of units which result in numerical values close to unity can improve numerical accuracy.

If no scale factors at all are used (within the main circuit or within any models referenced by it), the user may interpret the numerical values of the circuit components and responses in terms of any desired self-consistent set of units. Examples of such sets are

Quantity	Audio Units	VHF Units	UHF Units
Voltage	Volts	Volts	Volts
Current	Milliamperes	Microamperes	Milliamperes
Resistance	Kilohms	Megohms	Kilohms
Conductance	Millimhos	Micromhos	Millimhos
Inductance	Henries	Henries	Microhenries
Capacitance	Microfarads	Picofarads	Picofarads
Time	Milliseconds	Microseconds	Nanoseconds
Frequency	Kilohertz	Megahertz	Gigahertz

Values Expressed as Function References

The values of nonconstant (functionally dependent) circuit entities or auxiliary quantities are expressed as function references. The general form for a function reference is

function name (arg, arg, ... , arg)

The function name in the reference is the name under which the FORTRAN subprogram (which defines the function) is specified to the ECAP II program. For built-in functions the function name is specified in the dictionary file.* For load-on-call functions, the function name is the name of the FORTRAN subprogram in the program library.

The function name is optionally followed by arguments separated by commas and enclosed in parentheses. The arguments are the independent quantities of the function. They generally correspond to response variables and/or parameters which are operated on during the execution of the function to produce a result value for a nonconstant entity.

Before a function can be referenced, a corresponding FORTRAN subprogram which defines the function must be added to the program library. Certain functions are supplied with the program; these are listed in Table 6.3. The user may add additional functions, called user-written functions, which may then be referenced by function references. Further details

---

*The dictionary file is an ECAP II file which contains system- and user-specified information needed during ECAP II execution. For more details on this and other ECAP II files, the reader is referred to the References at the end of this chapter.

Table 6.3 ECAP II Supplied Functions

Reference Form	Description or Remarks
MULT (x,z) or x∗z	$y = xz$, e.g., R1, N1–N2 = $\begin{cases} \text{MULT (PAR1, IR2)} \\ \text{or} \\ \text{PAR1}*\text{IR2} \end{cases}$
EQU (x) or x	$y = x$, e.g., R1, N1–N2 = $\begin{cases} \text{EQU (PAR1)} \\ \text{or} \\ \text{PAR1} \end{cases}$
TAB (x)	Invokes system supplied table interpolation routine x = independent table variable e.g., y = TAB1 (TIME)
SIN (M, f, θ, x)	M = magnitude (zero-to-peak) θ = phase (degrees) f = frequency (hertz) t_p = period (sec) x = independent variable (TIME) $y = \begin{cases} \text{SIN}(2\pi fx - \theta) \\ \text{or} \\ \text{COS}(2\pi fx - \theta) \end{cases}$
COS (M, f, θ, x)	
PULSE (x, T0, TR, TU, TF, TD, DL, UL)	x = independent variable (TIME) T0 = pulse delay TR = total rise time TU = time at up level TF = total fall time TD = time at down level DL = down level UL = up level TP = period
DIODE (VJ, θ, IS)	$J = IS[\exp(\theta VJ) - 1]$ e.g. J1, N1–N2 = DIODE (VJ1, 36, 1E–8)

Note: y represents the value of an element coupling or parameter

of function references as well as their specification (definition) are available in the ECAP II manuals (References 10 and 11).

The following example illustrating the use of a function reference should help clarify the general form discussed above:

JEMIT, BASE=EMIT=DIODE(VJEMIT,PAR1,PAR2)

The current source JEMIT is a function of its own voltage and two parameters. The particular functional dependency is defined by the DIODE function which supplies a value for JEMIT. The DIODE function solves the conventional diode equation. The arguments in the reference replace dummy arguments used in defining the DIODE function, so that effectively the diode equation solved is

$$\text{JEMIT} = \text{PAR } 2(e^{\text{PAR}1*\text{VJEMIT}} - 1)$$

Primary and Secondary Response Variables

The network equations compiled by ECAP II are based on a hybrid (or state variable) formulation involving primary and secondary response variables. The details of the formulation and solution techniques may be found in the ECAP II Program Description Manual (Reference 10 in the References).

The primary response variables are

Currents	Voltages
IR	VG
IL	VC
IE (for dependent E)	VJ (for dependent J)

IE and VJ are defined as primary response variables only when the corresponding sources E and J themselves are dependent sources, that is, sources which dependent on one or more primary response variables. (Independent sources are not response variables since their values are independently specified—giving rise to the term *independent source*.)

In the example given in the previous section, the voltage VJEMIT is a primary response variable, since JEMIT is dependent on a primary response variable. However, the current IJEMIT is a secondary response variable (see below).

Once obtained, the primary response variables are used in auxiliary equations to compute secondary response variables. The computational procedure is shown in Fig. 6.8. Parameters, elements, primary responses, and then secondary responses are computed in that order.

The secondary response variables are

Currents	Voltages
IG	VR
IC	VL
IJ	VE
IE (for independent E)	VJ (for independent J)
IM (ammeter)	VM (voltmeter)
	VN (node voltage)

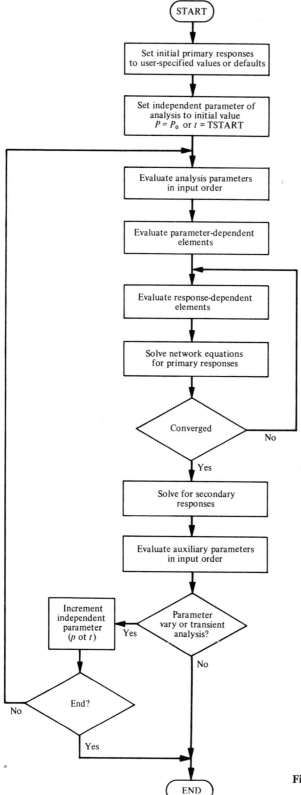

Figure 6.8 ECAP II Computation Procedure.

For example, in Fig 6.3, the dummy resistors R1 and R2 were inserted in order to obtain the primary response variables IR1 and IR2 used in describing the nonconstant elements JFWD and JREV. IJEMIT and IJCOL could not have been used for this purpose because they are secondary response variables.

Since the primary response variables are computed before the secondary response variables, only the primary response variables may be used as arguments of function references used in describing nonconstant elements, whether linear or nonlinear. The ECAP II variables which may be used as arguments of functions describing elements, couplings, and parameters are shown in Table 6.4.

Table 6.4 *Argument Restrictions for ECAP II Entities*

Entities \ Permitted arguments	Constants	Constant parameters	Constant elements	Constant couplings	TIME	Analysis parameters*	Parameter-dependent elements	Parameter-dependent couplings	Response-dependent elements	Primary responses	Secondary responses	Auxiliary parameters
Analysis Parameters	Y	Y	Y	Y	Y	P						
Parameter-Dependent Elements	Y	Y	Y	Y	R1	Y						
Parameter-Dependent Couplings	Y	Y	Y	Y	R1	Y						
Response-Dependent Elements	Y	Y	Y	Y	R1	Y	Y	Y		Y		
Auxiliary Parameters	Y	Y	Y	Y	Y	Y	Y	Y	Y	Y	Y	P

Y   means indicated argument is permissible for indicated entity.

P   means that argument must be defined in input deck before entity upon which it is used.

R1 capacitance and inductance elements cannot be functions of TIME.
    Element couplings which induce a voltage- or current-rate cannot be functions of TIME.

*The use of a parameter value as an argument of an analysis parameter, an element, or a coupling, causes the parameter to be classified as an analysis parameter.

6.6 STATEMENT DESCRIPTIONS

In the following sections, the individual statements are defined in detail. The sections are organized into five general categories:

1. Circuit description statements.

2. Run description statements.

3. Output description statements.
4. Commands.
5. Miscellaneous control statements.

The following notation is used to define the entries on each of the statements. All entries in uppercase letters (e.g., TAB) must be entered exactly as shown. All entries in lowercase must be replaced by a user-selected entry according to the rules given in the "Entries" and "Comments" sections. All punctuation must be entered as shown. This includes only the following eight characters: , = ( ) — * / and '. Two notational symbols are introduced: brackets, [ ], and ellipses, ... Brackets indicate optional entries, and ellipses indicate optional repetition of the previously bracketed item.

In the following sections, the general form of a statement is given first. Then the items used in the general form are defined in tabular format. Any new terms introduced in the tabular format are then similarly defined.

Circuit Description Statements

element [name], node — node = value

Item	Entries	Examples	Comments
element [name]	1. R [name]	R or RLOAD	Resistor
	2. G [name]	G or G7	Conductor
	3. L [name]	L or LCHOKE	Inductor
	4. C [name]	C or CBYPASS	Capacitor
	5. E [name]	E or EIN	Voltage source
	6. J [name]	J or JFWD	Current source
node	1. Any sequence of alphabetic and/or numeric characters	IN, 7, BASE.Q1 (Node names, including any qualification, are limited to 99 characters.)	Circuit node name
	2. 0 or GND		Reserved node names for ground, or datum, node
	3. NC		Used only on model reference statements to indicate no connection to a model external node
value	1. constant	Defines a constant circuit element	3.75 KILOHMS 4.92E7 24
	2. function ref	Defines a nonconstant circuit element	DIODE (VJ1, 28, 32P) MULT (7, PAP3) 7.5*VR9

Item	Entries	Examples	Comments
name	Any sequence of alphabetic and/or numeric characters	LOAD, 77, CHOKE	Used to create a unique statement name (Statement names, which include the prefix and any qualification, are limited to 99 characters.)
constant	Numeric part [scale factor] (See the section entitled "Value Specification" for a definition of constants.)		
function ref	1. function (argument [,argument] . . .)	DIODE (VJ1, 33, 9P)	Explicit reference to a named system or user-supplied FORTRAN subprogram
	2. argument	R33	Implicit reference to the system-supplied EQU (argument) function
	3. argument∗argument	PAR1∗7.5	Implicit reference to the system-supplied function MULT (argument, argument)
	4. TAB name (argument)	TAB1(TIME)	Implicit reference to the system-supplied table interpolation routine, passing as arguments the table "TABname" and the independent variable "argument"
function	OS: 1 to 8 alphabetic or numeric characters 1130: 1 to 5 alphabetic or numeric characters	DIODE, PULSE, SIN, SUM, PWRCK	Name of the system- or user-supplied FORTRAN subprogram (Table 6.3 lists the system-supplied functions and gives their definitions.)
argument	1. constant	3.72E-3	Supplies a constant value to a function reference
	2. parameter name	PAR3	Supplies the most current value of a parameter as an "analysis" parameter

Item	Entries	Examples	Comments
	3. element name	R33	Supplies the most current value of an element
	4. coupling name	TF74	Supplies the most current value of a coupling
	5. TIME		Supplies the value of time in a transient analysis (Time cannot be used in a dc analysis.)
	6. primary response		(See definitions below.)
	7. secondary response		
primary response	1. V element name	VCBYPASS	Where element begins with G, C, J (J must be dependent J)
	2. I element name	IRLOAD	Where element begins with R, L, E (E must be a dependent E)
secondary response	1. I element name	IG7	Where element name begins with G, C, J, E (E must be an independent E.)
	2. IM name	IM33	Ammeter
	3. V element name	VRLOAD	Where element name begins with R, L, E, J (J must be an independent J.)
	4. VM name	VMOUT	Voltmeter
	5. VN node	VNBASE	Node-to-datum voltage

> coupling [name], element name — element name = value

Item	Entries	Examples	Comments
coupling [name]	1. TF [name]	TFGR	Transferance
	2. M [name]	ML1L2	Mutual coupling
element name	Any element name, including qualition if any	J12.Q1 L33	The first element name is the "from" branch of a transferance and the second is the "to" branch.

Item	Entries	Examples	Comments
value	See definition of this entry for element statements.		Note the restrictions for arguments of a function reference for a coupling, as listed in Table 6.4.
name	As defined for element statements		

$$\boxed{\text{parameter name} \;=\; \text{value}}$$

Item	Entries	Examples	Comments
parameter name	1. A [name] 2. B [name] 3. F [name] 4. H [name] 5. K [name] 6. O [name] 7. P [name] 8. S [name] 9. U [name] 10. W [name] 11. Y [name] 12. Z [name]	A1 B FLUX H K3 OUT PAR7 STOP UPLEV WIDTH Y7 Z	Any name beginning with 1 of the 12 letters at the left indicates that the statement is a parameter. The "parameter name" cannot be one of the following: ANALYZE, FILES, OUTPUT, OUTINT, PRINT, PLOT, PRINTPLOT, PUNCH, STORE, SAVE, SCALE.
value	See the definition of this entry for element statements.		Note the restrictions for arguments of function reference for parameters, as listed in Table 6-4.
name	As defined for element statements		

$$\boxed{\text{modref [name], node } - \text{ node [}-\text{node]} \cdots = \text{ model name [,change]} \cdots}$$

Item	Entries	Examples	Comments
modref [name]	1. Q [name] 2. X [name] 3. T [name] 4. TX [name] 5. CR [name] 6. D [name] 7. MOD [name]	Q17 X33 TA TX7 CR14 DA MODEL1	Any name beginning with 1 of the 7 prefixes shown at the left indicates that the statement is a model reference. Note that Q1 is distinct from X1, T1, MOD1, etc.

Item	Entries	Examples	Comments
node	See the definition of this entry for element statements.	IN1 0 NC	Note the use of the special node name "NC" indicating "no external connection" to this model node.
model name	Any sequence of 1 to 8 alphabetic and/or numeric characters	2N706 SUBCKT 1N93A	This is the name under which the referenced model was stored in the model library.
change	1. element name = value 2. coupling name = value 3. parameter name = value 4. modref name = model name 5. run control = constant R3=7.2K Q7=2N706 DCCONV=1E−3 6. statement name/ PRINT1/ TAB4/		Any of the 5 entries at the left can be used to override (for this specific use of the model only) the values which were stored in the model library. The slash character (/) following a statement name causes the named statement to be omitted from the model (for this specific use of the model only). The named statement can be any circuit description, run description, or output description statement which was originally stored with the model.

TAB name, constant [,constant] · · · [INC=constant[,MIN=constant]]

Item	Entries	Examples	Comments
TAB name	TAB name	TABLE1 TABA	Table statement name
constant	See the section entitled "Value Specification" for a definition of constants.	7.5MV 9E–3 42USEC	Successive constants are interpreted as $x_0, y_0, x_1, y_1, \ldots, x_n, y_n$, pairs unless the entry "INC=constant" is present. In the latter case, the constants are interpreted as $y_0, y_1, y_2, \ldots, y_n$.

Item	Entries	Examples	Comments
INC=constant	INC=constant	INC=3MSEC	Indicates an evenly spaced incremental table
MIN=constant	MIN=constant	MIN=5.4USEC	Specifies a nonzero value of x_0 corresponding to y_0

$$\boxed{\text{meter name, node} - \text{node}}$$

Item	Entries	Examples	Comments
meter name	1. VM name 2. IM name	VMBE IMCOL	Voltmeter Ammeter
node	As defined above under element statements		Can be any circuit node including nodes inside models

Run Specification Statements

$$\boxed{\text{INIT [name], primary response} = \text{constant [, primary response} = \text{constant]} \cdots]}$$

Item	Entries	Examples	Comments
INIT [name]	INIT [name]	INIT1 INITIAL	If more than one statement is used, each must have either no [name] or unique [name] fields.
primary response	Defined above under "Circuit Description Statement"		
constant	Defined above under "Circuit Description Statements"		
[name]	Defined above under "Circuit Description Statements"		

Examples.

INIT1, VC3=−7.5MV, ILCHOKE = 3.4 UA
INITIAL CONDITIONS, VG9=0.0

In the second example, the voltage across G9 is specified to be initially 0.0. Since the default value is 0.0 anyway, one might ask whether this statement would be necessary. However, ECAP II gives priority in the state vector to those primary responses for which the user has specified an initial value. In the absence of the above statement, VG9 may not have ended up in the state vector, and therefore its default initial value would have been derived from combinations of state vector values (and therefore not necessarily constrained to 0.0). There is no guarantee that VG9 will end up in the state vector in any case. It only has priority over those branches for which no initial value has been specified. For example, if all the initial condition values were specified for a loop of branches containing G9, one or more of those branches would not be selected as a state variable.

$$\boxed{\text{run control [name]} = \text{constant}}$$

Item	*Entries*	*Examples*	*Comments*
run control [name]	TSTART [name] TSTOP [name] TRUNC [name] OUTINT [name] DTMIN [name] DTMAX [name] DTSTART [name] MAXORD [name] DCCONV [name] TRCONV [name] DCITR [name] TRITR [name]	TSTART TSTOP TRUNCERROR OUTINTERVAL DTMINIMUM DTMAXIMUM DTSTART MAXORDER DCCONVERGENCE TRCONV DCITR TRITRLIMIT	See Table 6.5 for a definition of the run controls, including their default values. Only one of each type of run control may be entered in any analysis task. The [name] field can be used to make the statement more readable.
[name]	Defined above under "Circuit Description Statements"		
constant	Defined above under "Circuit Description Statements"		

Table 6.5 *Default Values for Run Controls*

Run Control Name	*Default Value*	*Comment*
TSTART	0	Start time for transient analysis
TSTOP	10^{-6}	Stop time for transient analysis
TRUNC	10^5	Truncation error per unit time
OUTINT	0 (every step)	Minimum output interval
DTMIN	10^{-20}	Minimum step size
DTMAX	(TSTOP-TSTART)/20	Maximum step size

Table 6.5—Cont.

Run Control Name	Default Value	Comment
DTSTART	10^{-6} (TSTOP-TSTART) or DTMIN	Starting step size (set to largest of the two values shown at the left)
MAXORD	6	Maximum integration order
DCCONV	10^{-6}	dc convergence criterion
TRCONV	10^{-6}	Transient convergence criterion
DCITR	25	Maximum iterations for convergence during dc analysis
TRITR	8	Maximum iterations for convergence during transient analysis

VARY [name], independent parameter = constant, constant, [*] constant

Item	Entries	Examples	Comments
VARY [name]	VARY [name]	VARY1	Only one vary statement will be used in one analysis task. If more than one is entered, only the last will be used.
independent parameter	element name coupling name parameter name	R17 TF3 P9	The element, coupling, or parameter will be classified as a parameter-dependent entity. (Further, a parameter will be classified as an analysis parameter.)
constant	Defined above under "Circuit Description Statements"		
[*]	[*]	*	Specifies a multiplicative variation. Absence of the asterisk indicates additive variation.

Output Description Statements

```
PRINT[name], list [,TITLE=title]
PUNCH[name], list [,TITLE=title]
PLOT[name], list [,VS=item]
PRINTPLOT[name], list [,VS=item][,TITLE=title][,INT=constant]
MONITOR[name], list
SAVE[name], list
```

Item	Entries	Examples	Comments
PRINT [name]	PRINT [name]	PRINT7	
PUNCH [name]	PUNCH [name]	PUNCHA	
PLOT [name]	PLOT [name]	PLOT4	
PRINTPLOT [name]	PRINTPLOT [name]	PRINTPLOTX	
MONITOR [name]	MONITOR [name]	MONITOR1	
SAVE [name]	SAVE [name]	SAVE8	
List	item [,item] . . .	VR3, VG9, SAVE3	
TITLE=title	1. TITLE='literal'	TITLE='RESPONSE'	One-line titles can be entered as a literal.
	2. TITLE=TITLEname	TITLE=TITLE4	This entry references a TITLE statement, upon which the TITLE can be found.
VS=item	VS=item	VS=TIME	Default abscissa is the independent parameter of the analysis (i.e., time for transient and object of VARY for dc).
INT=constant	INT=constant	INT=32USEC	Default PRINTPLOT interval is the range of the independent parameter of the analysis divided by 100.
item	1. variable [([newname] [,SCALEname])]	IJ1 VR3 (OUTPUT) VG9 (,SCALE3)	Specifies a program variable and, optionally, a new name and/or scale (for plotting)
	2. SAVEname	SAVEA	Specifies that the list of variables on the SAVE statement are to be merged with the other items in the list. (Note that SAVEname cannot be used as a list item on SAVE and MONITOR statements.)
	3. variable class	V I VN	Specifies that all variables of a given class are to be output

Item	Entries	Examples	Comments
variable	1. element name 2. coupling name 3. parameter name 4. primary response 5. secondary response 6. TIME	R33 TF7 PAR1 VG9 VR3 TIME	All 6 entries are defined above under "Circuit Description Statements."
newname	1. string 2. literal	OUTPUT 'BASE VOLTS'	New names are limited to 16 characters.
variable class	1. V 2. I 3. VN	V I VN	Element voltages Element currents Node voltages
string	A sequence of alphabetic and/or numeric characters	OUTPUT RESPONSE	Blanks are eliminated in strings.
literal	'string'	'BASE VOLTS'	Blanks are not eliminated in literals.

SCALEname, constant, constant

Item	Entries	Examples	Comments
SCALEname	SCALEname	SCALE3	
constant	Constants are defined in the section entitled "Value Specification."		The constants define the minimum and maximum values of the axis upon which the variable(s) are to be plotted.

Example.

PLOT3, VR3(, SCALE10), VG9(, SCALE4)
SCALE1, −7.5, 7.5
SCALE4, 0, 10

TITLEname, literal[,literal] ...

Item	Entries	Examples	Comments
literal	Character string enclosed in quotes	'TITLE LINE 1' 'TITLE LINE 2'	

Commands

```
CIRCUIT[,model name][,NEW]
LOAD[,model name][,NEW]
STORE[,model name][,EXT=node−node[−node]···][,DES=literal]
DELETE, model name
```

Item	Entries	Examples	Comments
model name	See definition above under "Circuit Description Statements."		
node	See definition above under "Circuit Description Statements."		
literal	See definition above under "Output Description Statements."		

```
REMOVE, statement name[,statement name] ...
LIST[,library reference]
ANALYZE, type
OUTPUT, output statement [,output statement] ...
EXIT [,TO=exist list]
```

Item	Entries	Examples	Comments
statement name	Any circuit description, run description, or output description statements which is present in the circuit work space	R33 DCITR PUNCH7 J4.Q1	Statements within models can be removed by entering the fully qualified name.
library reference	1. model name	2N706	Causes the named model to be listed
	2. LIBRARY	LIBRARY	Causes the entire model library to be listed
	3. INDEX	INDEX	Causes the model library catalog to be listed
type	1. DC	DC	Requests a dc analysis (includes a parameter VARY if a VARY statement was previously entered)
	2. TR	TR	Requests a transient analysis without an initial condition solution

Item	Entries	Examples	Comments
	3. TR, DC or DC, TR	TR, DC DC, TR	Requests an initial condition solution followed by a transient analysis
output statement	1. PRINT name 2. PUNCH name 3. PLOT name 4. PRINTPLOT name	PRINT3 PUNCHA PLOT5 PRINTPLOT1	The named output statements are executed against a preexisting output file.
exit list	1. PLAN	PLAN	Transfers control back to the PLAN system; succeeding input is read by PLAN
	2. pgmname[,pgm name] . . .	PGMA, PGMB	Transfers execution control to the named program modules in turn, followed by return to the ECAP II input processor
pgm name	1. OS System: 1- to 8-character program name 2. 1130 System: 1- to 5-character program name	MYMODULE MYMOD	The name of a user-written program which resides in the program library available to the PLAN loader

Miscellaneous Control Statements

```
RUNTITLE, literal
FILES[,MODEL=[name][,[drive][,[nomod][,ndict]]]]
  [,OUTPUT=[name][,drive]]
CARD
TYPE
```

Item	Entries	Examples	Comments
literal	Defined above under "Output Description Statements"	'J.DOE–9/9/99'	Run title literal is limited to 76 characters.
name	1. 1130 System: 1- to 5-character name of file 2. OS system: 1- to 3-digit integer 'nnn'	MDLIB 007	Name as defined in LEFT/FLET. 'nnn' as entered on ddname PLFSynnn

Item	Entries	Examples	Comments
drive	1. 1130 system: 1-digit drive no.	3	Physical drive on which file can be found
	2. OS system: 1-digit drive code 'y'	4	'y' as entered on ddname PLFSynnn
nomod	1- or 2-digit integer	33	Number of models which can be accommodated per dictionary in model library
ndict	1- or 2-digit integer	40	Number of dictionaries which can be accommodated in model library

6.7 SAMPLE PROBLEMS

The following sample problems illustrate many of the features of the program which were presented earlier, presenting them as they would be typically used in performing an analysis.

Digital Logic Inverter

Figure 6.9(a) is the circuit schematic of a digital logic inverter. The transient response of the circuit for the given input signal of Fig. 6.9(b) is desired.

The language statements used to specify the equivalent circuit are shown in Fig. 6.9(c). The input signal EIN is described by referencing the table statement TAB1, which supplies the data of Fig. 6.9(b) in the form of time-voltage pairs of values. The model reference statement Q1 references the 2N2369A transistor in the model library. The value of the base spreading resistance, named RBB in the model, is changed to 20 ohms, which is the value measured for the actual transistor used in the circuit. The nominal value of RBB for the 2N2369A-type transistor, as stored in the model library, is 30 ohms. The change applies only to the use of the model in the inverter circuit, illustrating the substitutable parameter feature of the program.

The Ebers-Moll equivalent circuit of Fig. 6.10(a) represents the 2N2369A transistor. The dummy resistors R1 and R2, both of value zero, are used to monitor the collector and emitter diode currents, J_{bc} and J_{be}, respectively. The currents IR1 and IR2 are primary response variables, whereas the cur-

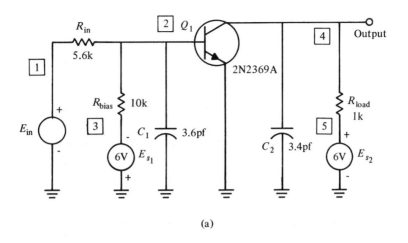

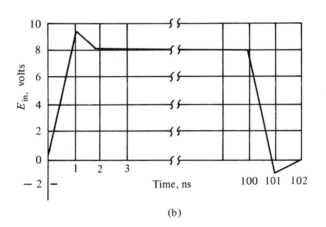

ES1,0−3=6
ES2,5−0=6
EIN,1−0=TAB1(TIME)
TAB1, 0,0, 1NS,9, 2NS,8, 100NS,8, 101NS,−1, 102NS,0, 110NS,0
C1, 2 − 0 = 3.6PF
C2, 4 − 0 = 3.4PF
RIN,1−2=5.6K
RBIAS,2−3=10K
RLOAD,5−4=1K
Q1,C−2−4=2N2369A,RBB=20

(c)

Figure 6.9 (a) Inverter Circuits; (b) Input Signal; (c) Input Statement.

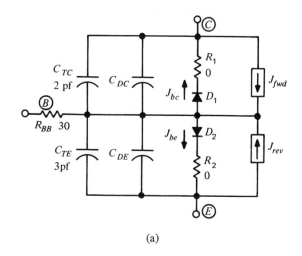

(a)

$$J_{bc} = 3.5745 \times 10^{-9}(e^{28V_{be}} - 1) \text{ mA}$$
$$J_{be} = 7.3801 \times 10^{-9}(e^{32V_{bc}} - 1) \text{ mA}$$

$$C_{DC} = 185 \times 10^{-9} J_{bc} \text{ F}$$
$$C_{DE} = 6.72 \times 10^{-9} J_{be} \text{ F}$$

$$J_{fwd} = \alpha_N J_{be} = \alpha_N I_{R2}$$
$$J_{rev} = \alpha_I J_{bc} = \alpha_I I_{R1}$$

$$\alpha_N = 0.97819$$
$$\alpha_I = 0.47379$$

```
CIRCUIT, 2N2369A  : TRANSISTOR MODEL
RBB,B-4=30       : BASE RESISTANCE
R1,5-C = 0       : DUMMY BRANCH TO MEASURE JBC
R2,6-E = 0       : DUMMY BRANCH TO MEASURE JBE
CTE,4-E = 3.0PF
CDE,4-E = 6.72E-9 * IR2
CTC,4-C = 2.0PF
CDC,4-C = 185.E-9 * IR1
D1,4-5 = JCNDIODE,K=32,PISAT=7.3801E-12 : CHANGE ITS CONSTANTS
D2,4-6 = JCNDIODE,K=28,PISAT=3.5745E-12 : CHANGE ITS CONSTANTS
JREV,E-4 = 0.47379*IR1 : REVERSE ALPHA CONTROLLED SOURCE
JFWC,C-4 = 0.97819*IR2 : FORWARD ALPHA CONTROLLED SOURCE
STORE, EXT = E - B - C, DES =' TRANSISTOR MODEL FOR HIGH SPEED SWITCHING
+ APPLICATIONS. '
```

(b)

Figure 6.10 Transistor Equivalent Circuit (2N2369A).

rents in the diodes are secondary responses. The ECAP II input statements for the transistor model are given in Fig. 6.10(b).

The charge storage effects in the transistor are modeled with the transition capacitances CTE and CTC and the diffusion capacitances CDE and CDC. The diffusion capacitances are nonlinear functions of the junction diode currents, as specified by the CDE and CDC statements in Fig. 6.9(b).

The model reference statements D1 and D2 specify the junction diodes of the transistor by referencing a previously stored diode model named JCNDIODE. The JCNDIODE model contains a nonlinear current source (J element) which references the system-supplied DIODE function. The DIODE function solves the diode characteristic equation to obtain the junction diode currents. The model parameters K and PISAT are adjusted to the appropriate values for the transistor by the model changes on the D1 and D2 statements. Statements RBB through JFWD are circuit description statements.

The CIRCUIT and STORE statements are examples of another type of language statement, called commands. Commands initiate a specific action or response from the program. The CIRCUIT command clears the circuit work space, into which a new circuit description (which follows) is placed. In this case, the CIRCUIT command also supplies the model name, 2N2369A, under which the model is later stored in the model library. The STORE command stores the transistor model in the model library under the model name, 2N2369A, which was supplied on the CIRCUIT command. The model name can be supplied on either the STORE or the CIRCUIT command.

Figure 6.11 shows that portion of the input data deck which contains the language statements needed to perform the inverter circuit analysis. A

```
                        ╱─ Command
                       ╱
                    CIRCUIT
                  ┌ ES1,0-3=6
                  │ ES2,5-0=6
                  │ EIN,1-0=TAB1(TIME)
                  │ TAB1, 0,0, 1NS,9, 2NS,8, 100NS,8, 101NS,-1, 102NS,0, 110NS,0
      Circuit     │ C1,  2  -  0  =  3.6PF
    Description   │ C2,  4  -  0  =  3.4PF
                  │ RIN,1-2=5.6K
                  │ RBIAS,2-3=10K
                  │ RLOAD,5-4=1K
                  └ Q1,0-2-4=2N2369A,RBB=20
                  ┌ SAVE1, VC1('BASE VOLTS'), VC2('OUT VOLTS')
                  │ TITLE1,'DIGITAL LOGIC INVERTER','VC1=BASE VOLTAGE','VC2=OUTPUT VOLTAGE'
       Output     │ PRINT1, VC1, VC2, CDE.Q1, CDC.Q1, TIME
    Description   │ PRINTPLOT1 , SAVE1, TITLE = TITLE1
                  └ PLOT1, SAVE1, VS = TIME
        Run       ┌ TRUNC = 25E5
    Description   └ TSTOP =  25NSEC
                    ANALYZE, TR, DC
                       ╲
                        ╲─ Command
```

Figure 6.11 Inverter Circuit Input Data.

complete input deck would also contain system control and job control statements associated with the computing system being used. The input data deck begins with a CIRCUIT command, followed by the language statements describing the inverter circuit as presented in Fig. 6.11.

These are followed by the output description statements. The SAVE1 statement specifies an output list containing the names of the output variables, VC1 and VC2, to be saved in an output file, VC1 and VC2 are the voltages across C1 and C2 of Fig. 6.9(a). They are renamed BASE VOLTS and OUT VOLTS, respectively, for output identification proposes. The output list specified on the SAVE1 statement is referenced in the PRINTPLOT and PLOT statements, via the statement name, SAVE1. The SAVE1 output list then serves as the output list for both the PRINTPLOT and the PLOT statements. In this sense, the name SAVE1 may be thought of as the name of an output list. This conveniently allows an output list to be specified only once and then used in as many other statements as desired.

The TITLE1 statement specifies a title and two subtitles to be printed on output associated with any PRINT or PRINTPLOT statements which reference the TITLE1 statement.

The PRINT1 statement requests a tabular listing of VC1 and VC2, the diffusion capacitances CDE and CDC in transistor Q1, and TIME. The variable names CDE.Q1 and CDC.Q1 are examples of qualified names, used to refer to elements within model reference.

Both the PRINTPLOT1 and PLOT1 statements request that the variables on the SAVE1 statement be plotted versus time. The PRINTPLOT1 statement requests a printer plot of the variables, with a title given by the TITLE1 statement. The PLOT1 statement requests a plot on an IBM 1627 Plotter, which is supported only on the 1130 Computing System.

The TRUNC and TSTOP statements are examples of run description statements, or, more specifically, run controls. TRUNC specifies the maximum allowable truncation error per unit time, which controls the accuracy of the numerical integration, and hence the execution time required for the analysis. TSTOP specifies that the analysis is to end at 200 nanoseconds of simulated time.

The ANALYZE command requests a transient (TR) analysis, preceded by a dc analysis, to obtain the initial conditions. The order of the keywords TR and DC is immaterial. The dc analysis is always performed first.

Running the Analysis

To run the analysis, the statements of Fig. 6.11 preceded by appropriate system and job control statements, are input to the program. The system normally lists all input statements, including the statements within referenced models. Either or both of these listing options can be inhibited. These options are described in the ECAP II operations manuals. The generated listing is shown in Fig. 6.12.

```
                            ECAP EXECUTION                              RUN NO.    24
          CIRCUIT
          ES1 , 0 - 3 = 6
          ES2 , 5 - 0 = 6
          EIN , 1 - 0 = TAB1(TIME)
          TAB1 , 0 , 0 , 1NS , 9 , 2NS , 8 , 100NS , 8 , 101NS , -1 , 102NS , 0 , 110NS , 0
          C1 , 2 - 0 = 3.6PF
          C2 , 4 - 0 = 3.4PF
          RIN , 1 - 2 = 5.6K
          RBIAS , 2 - 3 = 10K
          RLOAD , 5 - 4 = 1K
          Q1 , 0 - 2 - 4 = 2N2369A , RBB = 20

               RBB , 2 - 4.Q1 = 20  : BASE RESISTANCE
               R1 , 5.Q1 - 4 = 0   : DUMMY BRANCH TO MEASURE JBC
               R2 , 6.Q1 - 0 = 0   : DUMMY BRANCH TO MEASURE JBE
               CTE , 4.Q1 - 0 = 3.0PF
               CDE , 4.Q1 - 0 = 6.72E-9*IR2
               CTC , 4.Q1 - 4 = 2.0PF
               CDC , 4.Q1 - 4 = 185.E-9*IR1
               D1 , 4.Q1 - 5.Q1 = JCNDIODE , K = 32 , PISAT = 7.3801E-12 : CHANGE ITS CONSTANTS

                    K = 32
                    PISAT = 7.3801E-12
                    J1 , 4.Q1 - 5.Q1 = DIODE(VJ1,K,PISAT)

               D2 , 4.Q1 - 6.Q1 = JCNDIODE , K = 28 , PISAT = 3.5745E-12 : CHANGE ITS CONSTANTS

                    K = 28
                    PISAT = 3.5745E-12
                    J1 , 4.Q1 - 6.Q1 = DIODE(VJ1,K,PISAT)

               JREV , 0 - 4.Q1 = 0.47379*IR1 : REVERSE ALPHA CONTROLLED SOURCE
               JFWD , 4 - 4.Q1 = 0.97819*IR2 : FORWARD ALPHA CONTROLLED SOURCE

          SAVE1 , VC1('BASE VOLTS') , VC2('OUT VOLTS')
          TITLE1 , 'DIGITAL LOGIC INVERTER' , 'VC1=BASE VOLTAGE' , 'VC2=OUTPUT VOLTAGE'
          PRINT1 , VC1 , VC2 , CDE.Q1 , CDC.Q1 , TIME
          PRINTPLOT1 , SAVE1 , TITLE = TITLE1
          PLOT1 , SAVE1 , VS = TIME
          TRUNC    = 25E5
          TSTCP    = 200NSEC
          ANALYZE , TR , DC
```

Figure 6.12 ECAP II Input Listing.

In the listing, the model references Q1 (in the main circuit) and D1 and D2 (in the transistor model) are expanded to include the elements within each model reference. The resulting composite circuit, as depicted in the listing, is actually analyzed.

The listing is indented successively at each corresponding level of nested model references. The model changes which were inserted on the model reference statements have been applied to the corresponding elements and parameters within the model references (RBB in Q1 and K and PISAT in D1 and D2).

Note also that the external nodes of the models (as entered in a STORE command) are replaced by the circuit nodes to which the model reference is connected (as supplied in a model reference statement). Thus, nodes E, B, and C of the transistor model are replaced by circuit nodes 0, 2 and 4, respectively. Internal nodes of models are qualified with the model reference name. Thus, internal node 4 in the transistor model is changed to 4.Q1.

The output from the analysis is shown in Figs. 6.13, 6.14, and 6.15, which illustrate all the output forms provided by ECAP II, except for the punched card output. In this example, output was obtained at each time step

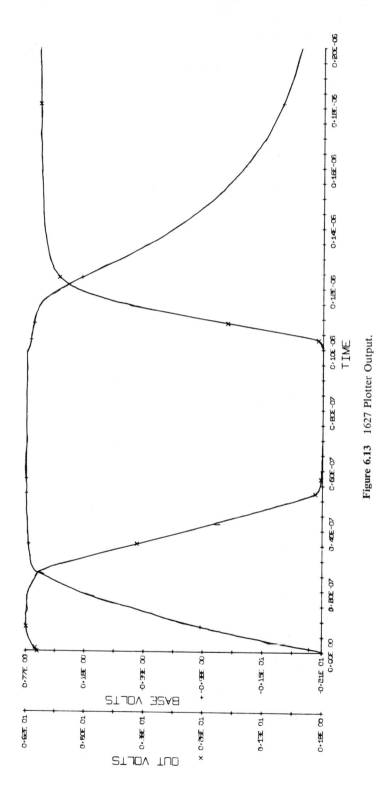

Figure 6.13 1627 Plotter Output.

```
DIGITAL LOGIC INVERTER
VC1=BASE VOLTAGE
VC2=OUTPUT VOLTAGE
1 BASE VOLTS    -0.21538E 01   -0.15679E 01   -0.98208E 00   -0.39619E 00    0.18968E 00    0.77556E 00
2 OUT  VOLTS     0.18325E 00    0.13930E 01    0.26028E 01    0.38125E 01    0.50223E 01    0.62321E 01

TIME                --- INDEPENDENT PARAMETER

0.00000E 00    1........................................................................2...+
0.20000E-08    .       1                  .                  .                  .           2  .
0.40000E-08    .                .1         .                  .                  .           2.
0.60000E-08    .                .          1                  .                  .           2
0.80000E-08    .                .          .          1       .                  .           2
0.10000E-07    .                .          .                  1                  .           2
0.12000E-07    .                .          .                  .          1       .           2
0.14000E-07    .                .          .                  .       1          .           2
0.16000E-07    .                .          .                  .                  1           2.
0.18000E-07    .                .          .                  .                  .          2+
0.20000E-07    +................+..........+..................+..................+.........1+
0.22000E-07    .                .          .                  .                  1     2     .
0.24000E-07    .                .          .                  .                  .   1  2    .
0.26000E-07    .                .          .                  .                  .     2     1.
0.28000E-07    .                .          .                  .                  .  2        1.
0.30000E-07    .                .          .                  .          2       .           1.
0.32000E-07    .                .          .                  .             2    .           1.
0.34000E-07    .                .          .                  .        2         .           1.
0.36000E-07    .                .          .             2    .                  .           1.
0.38000E-07    .                .          .                  .                  .           1+
0.40000E-07    +................+..........+......2...........+..................+...........1
0.42000E-07    .                .          .   2              .                  .           1
0.44000E-07    .                .          2                  .                  .           1
0.46000E-07    .                .     2    .                  .                  .           1
0.48000E-07    .          2     .          .                  .                  .           1
0.50000E-07    .        2       .          .                  .                  .           1
0.52000E-07    .  2             .          .                  .                  .           1
0.54000E-07    .2               .          .                  .                  .           1
0.56000E-07    .2               .          .                  .                  .           1
0.58000E-07    2                .          .                  .                  .           1
0.60000E-07    2................+..........+..................+..................+...........1
0.62000E-07    2                .          .                  .                  .           1
0.64000E-07    2                .          .                  .                  .           1
0.66000E-07    2                .          .                  .                  .           1
0.68000E-07    2                .          .                  .                  .           1
0.70000E-07    2                .          .                  .                  .           1
0.72000E-07    2                .          .                  .                  .           1
0.74000E-07    2                .          .                  .                  .           1
0.76000E-07    2                .          .                  .                  .           1
0.78000E-07    2                .          .                  .                  .           1
0.80000E-07    2................+..........+..................+..................+...........1
0.82000E-07    2                .          .                  .                  .           1
0.84000E-07    2                .          .                  .                  .           1
0.86000E-07    2                .          .                  .                  .           1
0.88000E-07    2                .          .                  .                  .           1
0.90000E-07    2                .          .                  .                  .           1
0.92000E-07    2                .          .                  .                  .           1
0.94000E-07    2                .          .                  .                  .           1
0.96000E-07    2                .          .                  .                  .           1
0.98000E-07    2                .          .                  .                  .           1.
0.10000E-06    2................+..........+..................+..................+...........1.
0.10200E-06    2                .          .                  .                  .           1.
0.10400E-06    .  2             .          .                  .                  .           1.
0.10600E-06    .                .    2     .                  .                  .           1.
0.10800E-06    .                .          .        2         .                  .           1
0.11000E-06    .                .          .                  .      2           .           1
0.11200E-06    .                .          .                  .            .2    .           1.
0.11400E-06    .                .          .                  .                  .  2     1  .
0.11600E-06    .                .          .                  .                  .      2. 1 .
0.11800E-06    +................+..........+..................+..................+....2..1...+
0.12000E-06    .                .          .                  .                  .        1 2
0.12200E-06    .                .          .                  .                  .       .1  2
0.12400E-06    .                .          .                  .                  .   1       2
0.12600E-06    .                .          .                  .              1   .           2
0.12800E-06    .                .          .                  .         1        .           2
0.13000E-06    .                .          .                  .    1             .           2
0.13200E-06    .                .          .               1  .                  .           2
0.13400E-06    .                .          .         1        .                  .           2
0.13600E-06    .                .          .    1             .                  .           2
0.13800E-06    .                .          . 1                .                  .           2
0.14000E-06    +................+..........1..................+..................+...........2
0.14200E-06    .                .       1  .                  .                  .           2
0.14400E-06    ..               .     1    .                  .                  .           2
0.14600E-06    .                .   1      .                  .                  .           2
0.14800E-06    .                .1.        .                  .                  .           2
0.15000E-06    .                1          .                  .                  .           2
0.15200E-06    .             1  .          .                  .                  .           2
0.15400E-06    .           1    .          .                  .                  .           2
0.15600E-06    .         1      .          .                  .                  .           2.
0.15800E-06    .       1        .          .                  .                  .           2.
0.16000E-06    +......1.........+..........+..................+..................+.........2.+
0.16200E-06    .    1           .          .                  .                  .           2
0.16400E-06    .   1            .          .                  .                  .           2
0.16600E-06    .  1             .          .                  .                  .           2
0.16800E-06    .  1             .          .                  .                  .           2
0.17000E-06    . 1              .          .                  .                  .           2
0.17200E-06    . 1              .          .                  .                  .           2
0.17400E-06    . 1              .          .                  .                  .           2
0.17600E-06    . 1              .          .                  .                  .           2
0.17800E-06    . 1              .          .                  .                  .           2.
0.18000E-06    +.1..............+..........+..................+..................+...........2
0.18200E-06    . 1              .          .                  .                  .           2
0.18400E-06    .  1             .          .                  .                  .           2
0.18600E-06    .  1             .          .                  .                  .           2
0.18800E-06    .   1            .          .                  .                  .           2
0.19000E-06    .    1           .          .                  .                  .           2
0.19200E-06    .    1           .          .                  .                  .           2
0.19400E-06    .     1          .          .                  .                  .           2
0.19600E-06    .      1         .          .                  .                  .           2
0.19800E-06    .       1        .          .                  .                  .           2
0.20000E-06    +........1.......+..........+..................+..................+..........2+
END OF DATA
```

Figure 6.14 Printerplot Output.

RUN NO 24

VC1	VC2	CDE.Q1	CDC.Q1	TIME
-0.2153842E 01	0.5999998E 01	-0.1448731E-17	-0.3562729E-15	0.0000000E 00
-0.2153792E 01	0.6000000E 01	-0.2403212E-19	-0.1368710E-17	0.1129347E-10
-0.2153698E 01	0.6000006E 01	-0.2403212E-19	-0.1368710E-17	0.2258694E-10
-0.2153563E 01	0.6000017E 01	-0.2403212E-19	-0.1368710E-17	0.3388041E-10
-0.2153303E 01	0.6000045E 01	-0.2403211E-19	-0.1368710E-17	0.4967320E-10
-0.2152293E 01	0.6000191E 01	-0.2403211E-19	-0.1368709E-17	0.9389302E-10
-0.2150855E 01	0.6000458E 01	-0.2403211E-19	-0.1368709E-17	0.1381128E-09
-0.2146754E 01	0.6001404E 01	-0.2403209E-19	-0.1368709E-17	0.2265525E-09
-0.2135386E 01	0.6004500E 01	-0.2403203E-19	-0.1368706E-17	0.3857438E-09
-0.2092655E 01	0.6017345E 01	-0.2403178E-19	-0.1368696E-17	0.7359646E-09
-0.2030111E 01	0.6037026E 01	-0.2403140E-19	-0.1368679E-17	0.1086185E-08
-0.1977027E 01	0.6054318E 01	-0.2403104E-19	-0.1368663E-17	0.1366362E-08
-0.1946004E 01	0.6064185E 01	-0.2403083E-19	-0.1368654E-17	0.1534468E-08
-0.1915905E 01	0.6073368E 01	-0.2403062E-19	-0.1368645E-17	0.1702574E-08
-0.1881050E 01	0.6083510E 01	-0.2403038E-19	-0.1368634E-17	0.1904302E-08
-0.1813272E 01	0.6101850E 01	-0.2402992E-19	-0.1368612E-17	0.2307756E-08
-0.1760311E 01	0.6115047E 01	-0.2402957E-19	-0.1368594E-17	0.2630520E-08
-0.1677394E 01	0.6133953E 01	-0.2402901E-19	-0.1368566E-17	0.3146942E-08
-0.1533035E 01	0.6162191E 01	-0.2402804E-19	-0.1368515E-17	0.4076501E-08
-0.1260905E 01	0.6201357E 01	-0.2402620E-19	-0.1368411E-17	0.5935620E-08
-0.5607056E 00	0.6224839E 01	-0.2402418E-19	-0.1368292E-17	0.8166563E-08
-0.6327996E 00	0.6232119E 01	-0.2402196E-19	-0.1368152E-17	0.1084369E-07
0.2303729E 00	0.6193703E 01	0.3695803E-17	-0.1367756E-17	0.1941051E-07
0.6790183E 00	0.5965244E 01	0.3272153E-11	-0.1367461E-17	0.2626396E-07
0.7139513E 00	0.5766845E 01	0.8372965E-11	-0.1367360E-17	0.2763466E-07
0.7270974E 00	0.5543263E 01	0.1189146E-10	-0.1367257E-17	0.2873121E-07
0.7348607E 00	0.5128055E 01	0.1474345E-10	-0.1367072E-17	0.3048569E-07
0.7396997E 00	0.4794755E 01	0.1689757E-10	-0.1366924E-17	0.3188928E-07
0.7440866E 00	0.4531528E 01	0.1906441E-10	-0.1366807E-17	0.3301215E-07
0.7482390E 00	0.4209560E 01	0.2138560E-10	-0.1366665E-17	0.3435959E-07
0.7505805E 00	0.3950028E 01	0.2286633E-10	-0.1366550E-17	0.3543755E-07
0.7529044E 00	0.3696625E 01	0.2436554E-10	-0.1366439E-17	0.3651551E-07
0.7562339E 00	0.3349735E 01	0.2679285E-10	-0.1366286E-17	0.3802464E-07
0.7586249E 00	0.3073967E 01	0.2864535E-10	-0.1366164E-17	0.3923195E-07
0.7607446E 00	0.2799347E 01	0.3040560E-10	-0.1366043E-17	0.4043926E-07
0.7630970E 00	0.2473822E 01	0.3248017E-10	-0.1365900E-17	0.4188803E-07
0.7665734E 00	0.1963428E 01	0.3580599E-10	-0.1365675E-17	0.4420607E-07
0.7702852E 00	0.1365633E 01	0.3973308E-10	-0.1365413E-17	0.4698771E-07
0.7747466E 00	0.5536537E 00	0.4503599E-10	0.1088540E-14	0.5088201E-07
0.7755632E 00	0.3276204E 00	0.4604083E-10	0.4654487E-12	0.5213201E-07
0.7755479E 00	0.3188540E 00	0.4594196E-10	0.1807607E-11	0.5219451E-07
0.7754046E 00	0.3022648E 00	0.4563188E-10	0.3456963E-11	0.5233201E-07
0.7748465E 00	0.2790157E 00	0.4486118E-10	0.7134809E-11	0.5260701E-07
0.7742723E 00	0.2654057E 00	0.4421545E-10	0.1200734E-10	0.5288201E-07
0.7738827E 00	0.2553419E 00	0.4381253E-10	0.1481275E-10	0.5321201E-07
0.7737656E 00	0.2477434E 00	0.4368979E-10	0.1883026E-10	0.5354201E-07
0.7737307E 00	0.2425332E 00	0.4365327E-10	0.2222589E-10	0.5380601E-07
0.7736845E 00	0.2340388E 00	0.4362412E-10	0.2914464E-10	0.5433401E-07
0.7737188E 00	0.2279193E 00	0.4367943E-10	0.3549672E-10	0.5486201E-07
0.7738602E 00	0.2196952E 00	0.4386808E-10	0.4641633E-10	0.5591801E-07

RUN NO 24

VC1	VC2	CDE.Q1	CDC.Q1	TIME
0.7740444E 00	0.2131848E 00	0.4407583E-10	0.5702742E-10	0.5718521E-07
0.7742282E 00	0.2069767E 00	0.4430214E-10	0.7051638E-10	0.5870585E-07
0.7744988E 00	0.1999357E 00	0.4464409E-10	0.8910902E-10	0.6113888E-07
0.7752789E 00	0.1858827E 00	0.4562937E-10	0.1426254E-09	0.7087097E-07
0.7754887E 00	0.1834752E 00	0.4589204E-10	0.1557563E-09	0.8087097E-07
0.7754648E 00	0.1837262E 00	0.4586291E-10	0.1543905E-09	0.9087097E-07
0.7754767E 00	0.1834722E 00	0.4587798E-10	0.1554102E-09	0.9587097E-07
0.7754961E 00	0.1832596E 00	0.4590339E-10	0.1568736E-09	0.9837097E-07
0.7415241E 00	0.1911619E 00	0.3706861E-10	0.9538547E-10	0.1008709E-06
0.7364835E 00	0.2072979E 00	0.3427218E-10	0.5204996E-10	0.1021209E-06
0.7341757E 00	0.2717918E 00	0.3176282E-10	0.6059022E-11	0.1031209E-06
0.7305803E 00	0.3893035E 00	0.2835883E-10	0.1237015E-12	0.1037209E-06
0.7268987E 00	0.5268265E 00	0.2559856E-10	0.1350660E-14	0.1042009E-06
0.7242119E 00	0.6552825E 00	0.2382327E-10	0.1866150E-16	0.1045849E-06
0.7216321E 00	0.8195749E 00	0.2211993E-10	-0.1275593E-17	0.1050457E-06
0.7194727E 00	0.9870546E 00	0.2089187E-10	-0.1364889E-17	0.1055065E-06
0.7159918E 00	0.1254899E 01	0.1894812E-10	-0.1365376E-17	0.1062438E-06
0.7130567E 00	0.1470107E 01	0.1745100E-10	-0.1365471E-17	0.1068336E-06
0.7107248E 00	0.1643001E 01	0.1632453E-10	-0.1365546E-17	0.1073055E-06
0.7074065E 00	0.1883630E 01	0.1485729E-10	-0.1365654E-17	0.1079661E-06
0.7038083E 00	0.2120049E 01	0.1341626E-10	-0.1365759E-17	0.1086267E-06
0.6989677E 00	0.2398472E 01	0.1169069E-10	-0.1365865E-17	0.1094194E-06
0.6922968E 00	0.2726676E 01	0.9667784E-11	-0.1366029E-17	0.1103707E-06
0.6845115E 00	0.3048555E 01	0.7743013E-11	-0.1366173E-17	0.1113220E-06
0.6731354E 00	0.3423882E 01	0.5594503E-11	-0.1366342E-17	0.1124635E-06
0.6584262E 00	0.3782923E 01	0.3671034E-11	-0.1366506E-17	0.1136050E-06
0.6177709E 00	0.4368665E 01	0.1140872E-11	-0.1366780E-17	0.1156597E-06
0.5157167E 00	0.4993405E 01	0.6162074E-13	-0.1367098E-17	0.1185366E-06
0.4137062E 00	0.5280569E 01	0.3429475E-14	-0.1367269E-17	0.1205500E-06
0.2960202E 00	0.5475441E 01	0.1198612E-15	-0.1367406E-17	0.1225637E-06
0.2212295E 00	0.5558489E 01	0.1492641E-16	-0.1367475E-17	0.1237718E-06
0.2237535E-01	0.5696934E 01	0.3117975E-19	-0.1367623E-17	0.1269131E-06
-0.2834652E 00	0.5796401E 01	-0.2394223E-19	-0.1367801E-17	0.1319392E-06
-0.4500515E 00	0.5830006E 01	-0.2402209E-19	-0.1367888E-17	0.1349548E-06
-0.6021168E 00	0.5858265E 01	-0.2402158E-19	-0.1367967E-17	0.1379705E-06
-0.7933022E 00	0.5888894E 01	-0.2402297E-19	-0.1368065E-17	0.1421923E-06
-0.1024475E 01	0.5914288E 01	-0.2402449E-19	-0.1368177E-17	0.1481030E-06
-0.1182163E 01	0.5925683E 01	-0.2402556E-19	-0.1368251E-17	0.1528315E-06
-0.1412652E 01	0.5941072E 01	-0.2402711E-19	-0.1368359E-17	0.1613428E-06
-0.1613553E 01	0.5956662E 01	-0.2402847E-19	-0.1368454E-17	0.1713428E-06
-0.1759187E 01	0.5969109E 01	-0.2402945E-19	-0.1368523E-17	0.1813428E-06
-0.1865348E 01	0.5977917E 01	-0.2403017E-19	-0.1368573E-17	0.1913428E-06
-0.1934019E 01	0.5983222E 01	-0.2403063E-19	-0.1368606E-17	0.2000000E-06

Figure 6.15 Tabular Output.

444

of the analysis to illustrate a property of the variable step integration algorithm. Note that the step size is relatively small in regions in which the response variables change rapidly and larger when the response is flat.

Application to Nonelectrical System Analysis

As stated in the beginning of this chapter, ECAP II can be of use in the solution of problems in areas other than the purely electrical. For many years, programs have been available for the simulation of algebraic and differential equations (e.g., Reference 14). These programs provide the user the facility for entering his equations either directly or in block diagram form. Difficulty has been encountered, however, when the system being modeled contained both differential equations and electrical circuitry, as in the case, for example, of an electromechanical actuator or transducer.

The model library facility and the nonlinear element capability of ECAP II provide a simple, straightforward method of modeling the analog function blocks required in such systems. Figure 6.16 shows several well-known transfer functions along with associated electrical analogs and their definitions as models in the ECAP II language. These circuit analogs provide the given functions with complete isolation between input and output, as required by the given transfer function.

These models, and others, once stored in the users model library, can then be used to model a variety of analog systems. One such system is shown in Fig. 6.17. This particular system is a speed control regulator with a nonlinear amplifier, A, in the forward loop. Feedback control is picked up by a transducer at the output (node C). This signal is compared to the input control R at summer block M21.

Using the previously described model blocks LOPASS, HIPASS, and SUM, this system would be described to ECAP II as shown in Fig. 6.18. The gain and time-constant factors K1 and PTAU, both set to 1.0 in the models, are overridden by their actual values when each model is used.

The analysis of this system with a step change in control voltage of 88 volts is displayed in Fig. 6.19.

It should be obvious that, at any point in the system where circuitry is more appropriate, it could be used in the normal manner of ECAP II input. The advantages to be gained by use of ECAP II are that

1. Mixed systems can be handled.

2. Large systems can be accommodated.

3. The implicit integration technique used in ECAP II is inherently more stable and less time-constant-limited than are the explicit integration techniques.

Definition	Network	Transfer Characteristic
CIRCUIT L1, IN−X=PTAU R1, X−0=1 E1, CUT−0=IR1*K1 K1=1 PTAU=1 STORE, LOPASS, EXT = IN−OUT	L1, R1, E1	$V_{OUT} = \dfrac{K1}{PTAU \cdot s + 1} \cdot V_{IN}$
CIRCUIT C1, IN−X=PTAU R1, X−0=1 E1, CUT−0=IR1*K1 PTAU=1 K1=1 STORE, HIPASS, EXT = IN−OUT	C1, R1, E2	$V_{OUT} = \dfrac{K1 \cdot PTAU \cdot s}{PTAU \cdot s + 1} \cdot V_{IN}$
CIRCUIT E1, OUT−0=IR1 R1, X−0=1 JA, 0−X=K1*IRA JB, X−0=K2*IRB RA, IN1−0=1 RB, IN2−0=1 K1=1 K2=1 STORE, SUM, EXT=IN1−IN2−OUT	JA, JB, RA, RB, R1, E1	$V_{OUT} = K1 \cdot V_{IN1} - K2 \cdot V_{IN}$

Figure 6.16 Transfer Function Models.

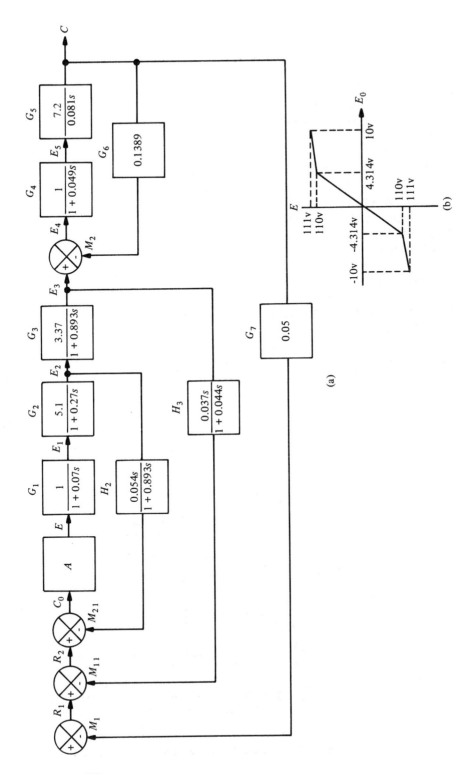

Figure 6.17 Nonlinear Speed Control System Block Diagram: (a) Block Diagram; (b) Function Block A Transfer Characteristic.

447

NON-LINEAR SPEED CONTROL SYSTEM ANALYSIS
ERROR,R−0=88
X1,R−C−R1=SUM,K2=.05
X2,R1−M11−R2=SUM,K2=.841
X3,R2−M21−E0=SUM,K2=.060047
R1,E0−0=1
E1,E−0=TAB1(IR1)
TAB1,−10,−111, −4.314,−110, 0,0, 4.314,110, 10,111
X4,E−E1=LOPASS,PTAU=.07
X5,E1−E2=LOPASS,PTAU=.027,K1=5.1
X6,E2−E3=LOPASS,PTAU=.893,K1=3.37
X7,E3−C−E4=SUM,K2=.1389
X8,E4−E5=LOPASS,PTAU=.044
X9,E5−C=LOPASS,R1=0,PTAU=.081,K1=7.2
X10,E2−M21=HIPASS,PTAU=.893
X11,E3−M11=HIPASS,PTAU=.044
SAVE1,VR1('ERROR VOLTAGE'),VNC('OUTPUT RPM')
PRINT1,SAVE1,TIME,TITLE=TITLE1
PRINTERPLOT1,SAVE1,VS=TIME,TITLE=TITLE1
PLOTS   1,SAVE1,VS=TIME,TITLE=TITLE1
TITLE1, 'NONLINEAR SPEED CONTROL SYSTEM','VR1=ERROR','VC2=OUTPUT'
TSTOP=1
TRUNC=1
ANALYZE,TR

Figure 6.18 Nonlinear Speed Control System ECAP II Input.

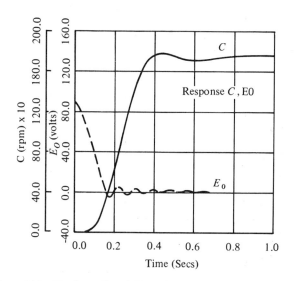

Figure 6.19 Nonlinear Speed Control System Analysis Results.

6.8 ERROR MESSAGES

ECAP II incorporates an extensive set of validity checks on the user input data and on the progress of an analysis run. As a result there are too many individual messages to enumerate here. Only the general philosophy and format of the messages are given.

There are five levels of messages which ECAP II produces:

I = information
W = warning
C = conditional
S = serious
T = terminal

Information messages are used to indicate successful completion of certain ECAP II operations, such as the successful storing of a model in the library. Warning messages are used to indicate possible user error. Execution will continue, but the results should be checked for validity. An example of a warning message would be the case of attempting to store a model in the library when one of the same name already exists.

Conditional messages are treated as either warning or serious according to a user-selected option. Certain errors are not serious enough to inhibit execution, but the results are in doubt. One would normally treat conditional errors as serious, unless one were attempting to isolate a problem area in ECAP II and did not want conditional errors to inhibit further execution.

Serious errors are those for which further execution would be meaningless. An example would be the case of a singular coefficient matrix in the analysis phase of the program.

Terminal errors are of such serious nature that continued execution of ECAP II is impossible. These are normally due to errors in the program.

Message Formats

DGMnnn *1* icode   text

    DGMnnn = ECAP II component code followed by three-digit error number
    1 = error level (I, W, C, S, or T)
    icode = integer used on some messages to clarify meaning of text
    text = text of error message

DGMnnn   Text

This format is used in conjunction with the previous format for some messages to provide a character string text for clarification.

Example.

```
DGM233   *I*   0   FOLLOWING MODEL HAS BEEN STORED
DGM233             2N2369A
```

PLAN Messages

During execution of ECAP II certain messages may be produced by the PLAN subsystem under which ECAP II operates. These messages are in a similar format:

DFJnnn s ecode SEQ= seq ID=id PGM=pgm text

The meanings of these messages can be found in the PLAN documentation (Reference 12).

The most common PLAN messages are those which are concerned with files and with input processing. For example, if ECAP II discovers a terminal-class error, it will terminate execution immediately. At this point, PLAN will begin reading the remaining ECAP II input. Since ECAP II input was not intended to be read by PLAN, the following PLAN error message could result:

```
DFJ222 R  00530 SEQ=004 ID=  PGM=DFJPSCAN    STATEMENT OVER 450
                                             CHARACTERS
```

6.9 REFERENCES

1. BRANIN, F. H., G. R. HOGSETT, R. L. LUNDE, and L. E. KUGEL, "ECAP II—A New Electronic Circuit Analysis Program," *IEEE J. Solid State Circuits*, SC-6, No. 4, Aug. 1971, pp. 146–166.

2. BRANIN, F. H., G. R. HOGSETT, R. L. LUNDE, and L. E. KUGEL, "ECAP II—A New Electronic Circuit Analysis Program," *IEEE Spectrum*, 8, June 1971, pp. 14–25.

3. TINNEY, W. F., and J. W. WALKER, "Direct Solutions of Sparse Network Equations by Optimally Ordered Triangular Factorization," *Proc. IEEE*, 55, Nov. 1967, pp. 1801–1809.

4. *Proc. Sparse Matrix Symp.*, IBM Report RA-1, March 1969.

5. GEAR, C. W., "The Automatic Integration of Stiff Ordinary Differential Equations," in *Information Processing 68*, A. J. H. Morell, ed., North-Holland, Amsterdam, 1969, pp. 186–193.

6. BRAYTON, R. K., F. G. GUSTAVSON, and G. D. HACHTEL, "The Use of Variable-Order Variable-Step Backward Differentiation Methods for Nonlinear Electrical Networks," in *Mem. Mexico 1971 Internal. IEEE Conf. Systems, Networks, and Computers, Oaxtepec*, Mexico. pp. 102–106.

7. BRANIN, F. H., JR., and L. E. KUGEL, "The Hybrid Method of Network Analysis," presented at IEEE Symp. Circuit Theory, San Francisco, Calif., Dec. 1969.

8. BRANIN, F. H., JR., "A Controlled-Step Newton Method for Solving Nonlinear Equations," *Digest Rec. ACM-SIAM-IEEE Conf. Mathematical and Computer Aids to Design*, Anaheim, Calif., Oct. 1969, 378.

9. *ECAP II Application Description Manual (GH20–0983)*, IBM Corp., Mechanicsburg, Pa., 1971.

10. *ECAP II Program Description Manual (SH20–1015)*, IBM Corp., Mechanicsburg, Pa., 1971.

11. ECAP II Operations Manual (SH20–1025), IBM Corp., Mechanicsburg, Pa., 1971.

12. *PLAN Application Description Manual (GH20–0490)*, IBM Corp., Mechanicsburg, Pa., 1969.

13. JENSEN, R. W., and M. D. LIEBERMAN, *IBM Electronic Circuit Analysis Program—Techniques and Applications*, Prentice-Hall, Englewood Cliffs, N.J., 1968.

14. *CSMP App. Des. Manual (GE19-0036)*, IBM Corp., Mechanicsburg, Pa.

7

LISA
Linear Systems Analysis

EDWARD T. JOHNSON

IBM Corporation
San Jose, California

7.1 INTRODUCTION

The use of s-plane or Laplace transform techniques for analysis and design have formed the basis for much linear circuit theory in recent years. The insight offered from knowledge of the pole and zero locations about circuit stability, frequency response, and transient response can be of great aid to the circuit designer. Without the help of the computer, however, the practical use of these methods is limited.

LISA, written for the IBM System/360 and System 370, is an integrated package of FORTRAN IV routines for analyzing linear systems using Laplace transform techniques. It analyzes electrical networks, transfer functions, two-block control systems, and any system of linear equations whose coefficient matrix has polynomial elements.

Inputs may be a topological circuit description, transfer function, matrix equation, or block diagram. Outputs may be poles and zeros, frequency and transient response, root locus, or sensitivity. Both listings and plots may be obtained.

The flexibility of LISA can be demonstrated through applications that start with an equivalent circuit diagram. The engineer enters the circuit description as values of R, L, C, mutual inductance, independent current sources, and voltage- or current-dependent current sources. The dependent sources are used for active device models as the π or T transistor circuit models. Initial conditions can also be given for transient analysis.

From this information, the program automatically sets up the nodal matrix equation in the complex frequency variable $s = \sigma + j\omega$. In addition, the engineer defines the functions required by the problem. These may be voltages, voltage ratios, currents, current ratios, impedances, admittances, determinants, cofactors, or any of the common three-terminal grounded two-port parameters.

The program can then compute one or more of the following:

- Poles and zeros and scale factor.
- Variations of the poles and zeros to component variations.
- Locus of roots over a range of values for any component.
- Time response to various driver functions.
- Bode response over a range of frequencies.
- Sensitivity of a function to component variations at a given frequency.

Answers appear as numerical listings or, if applicable, in the form of plots.

In the case of linear networks, the nodal equations under the Laplace transform give rise to a system of equations

$$Y(s)V(s) = I(s) \qquad (7.1)$$

where $Y(s)$ is the nodal admittance matrix, $V(s)$ is the unknown node-voltage vector, $I(s)$ is the vector of current sources, and $s = \sigma + j\omega$ is the Laplace transform variable.

POLY and TRFN are two methods available in LISA to compute transfer function poles and zeros from matrix equation (7.1). POLY solves for the coefficients of the polynomials which define the various components of $V(s)$ by a modified Crout reduction technique. It then finds the roots of the polynomials. TRFN expresses the components of $V(s)$ as ratios of determinants by Cramer's rule and then finds the roots of the voltages of interest directly from these determinants by using Muller's method [1].

In addition, LISA has a method called ACCA for computing the real frequency response directly by substituting $j\omega$ for s in Eq. (7.1) and then solving the resultant set of linear equations for the voltages of interest. This is done over a specified range of ω values.

Another option, called BEV4, is used for analyzing control systems. It works on the basic two-block diagram shown in Fig. 7.1. Given the transfer functions $H(s)$ and $F(s)$ and the gain factor K, BEV4 computes the following

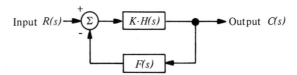

Figure 7.1 Basic Closed-Loop Feedback System.

three transfer functions:

$$K \cdot H(s) \qquad \text{forward response (called FORWARD)} \qquad (7.2)$$

$$K \cdot H(s) \cdot F(s) \qquad \text{open-loop response (called OPEN)} \qquad (7.3)$$

$$\frac{K \cdot H(s)}{1 + K \cdot H(s) \cdot F(s)} \qquad \text{closed-loop response (called CLOSED)} \qquad (7.4)$$

For root locus computation, the characteristic equation

$$1 + K \cdot H(s) \cdot F(s) = 0 \qquad (7.5)$$

is solved for specific values of the gain K.

7.2 PROGRAM STRUCTURE

The LISA input language provides compute, definition, data, and output commands which allow the user to control the operational flow shown in Fig. 7.2. Data and transfer function computations are saved when commands are executed so the results can be used with later commands. This permits a user to loop on the program structure or enter it at any appropriate point.

7.3 NETWORK ELEMENTS

A user enters descriptive data on the network elements of his circuit following a READ, CIRCUIT command. The program stores the circuit table and sets up matrix equation (7.1). If an equation contains a $1/s$ factor (due to inductors, mutuals, initial inductor currents, and step functions), the program multiplies that equation by s.

Components

Component input cards can be entered in any order. Allowed are $\pm R$, $\pm C$, $\pm L$, mutual coupling between any two inductors, initial conditions, independent current generators (I or E/R, impulse or step) and three- or four-node voltage and current-dependent current generators. The dependent current generators are used for π or T transistors and other active device models.

Component Names

Up to four alphanumeric characters may be used,* with the following restrictions:

*Names with more than four characters are allowed but will be truncated.

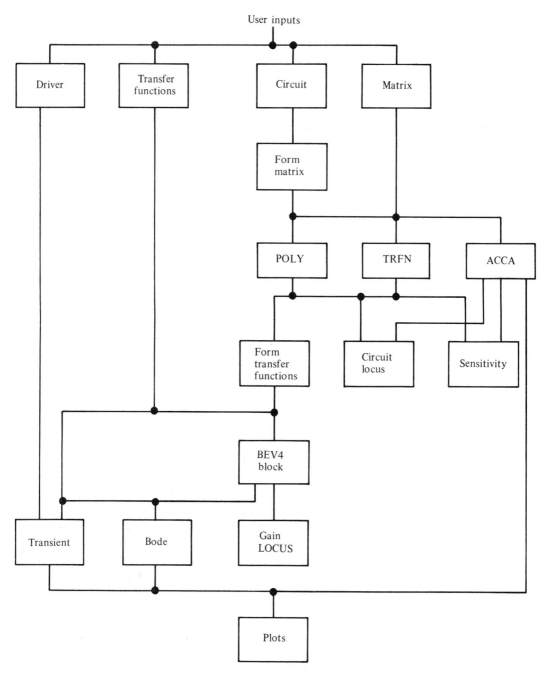

Figure 7.2 LISA Operational Flow.

1. A LISA command word is not allowed, as it ends the circuit list.
2. Component names must begin with the letter

        R for resistors

        C for capacitors

        L for inductors

        K for coefficient of coupling $\Big\}$ either may be used for mutuals
        M for mutual inductance

     I or J for independent current sources

   E or V for independent voltage sources in a resistor branch

        T for three-node dependent current sources

        D for four-node dependent current sources

Initial conditions have no letter restriction.

Component Values

Usually, circuit values are scaled to a consistent set of units. The E convention of FORTRAN is used for powers of 10. A decimal point is advisable but not necessary.

Node Labeling

Circuit nodes are either numbered sequentially from 0 (for ground) to n or given unique symbolic names using GND or GROUND for the datum node. This must be kept consistent with the function definitions. For symbolic names, any alphanumeric names are allowed, provided the first character is a letter, such as GND, V12, and A.

Component Card Format

Each component is described on a separate card using a free format with only a comma to separate fields. For readability, the equal sign may replace the comma.*

Component descriptions take the following forms:

1. *R, L, C, and initial conditions.* The general format is

      name, node i, node j, value, tolerance, initial-condition name, value.

If a field is not applicable, it is simply left out. The tolerance field may be

---

*A dash is also allowed as a delimiter between nodes for compatability with certain other circuit analysis programs, for example, $R, 1 - 2 =$ value.

either just the value or given as TOL = value for readability.* For the tolerance value, use 0.05 for 5%, 0.10 for 10%, etc. Usually, the initial capacitor voltages and initial inductor currents are ignored. The initial condition name and value field can then be omitted. If initial conditions are included, the sign convention is

Node i o—−|+—o Node j   Node i o—⁀⁀⁀—o Node j

For example,

    RCB, 1, 2 = 220., .05
    C15, 0, 3, 2E−6, TOL = .1, ICC15 = .7
    L1, 3, 2, .001
    LA, 3, 4, .002, ICL2, .55

2. *Independent current sources.* The general format is

    name, node i, node j, value, tolerance, resistor name, option.

The resistor name field is omitted from an I or J current source. The sign and the first letter convention is

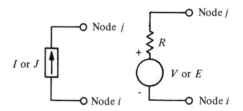

The voltage-resistor branch is converted to its Norton's equivalent current source by the program. The resistor must also be given in the resistor list as a separate component from node i to node j. The option field may be IMPULSE (assumed if left out) or STEP. For example,

    RGEN, 0, 1, 1000, TOL = .05
    V1, 0, 1, 1.,RGEN
    IDRIVE, 3, 4, 1., .05, STEP
    JSOR, 0, 2, 5

3. *Mutual inductance.* The general format is

    name, inductor name, inductor name, value, tolerance.

Either K or M may be used as the first letter, allowing the user to enter either the coefficient of coupling (K ≠ 1) or the mutual value. If K is

---

*The tolerance is not actually used in LISA.

given, M will be computed and printed for user convenience and vice versa. The two inductor names which are coupled are given. These are also entered in the inductor list as separate components. The sign convention is that the dots are on the first or node i side of the inductor. Also, initial condition currents are assumed from node i to node j. Conflicts are resolved by using $-\rho$ (rho). For example,

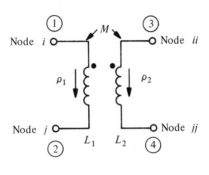

L1, 1, 2, 1.,RHO1 = .005
L2, 3, 4, 3.,RHO2 = .001
K, L1, L2, .75

4. *Dependent current sources.* Either a voltage or current-dependent current source is allowed. The three-node model is specifically used for the transistor tee and π models, while the four-node model is general. In either case, if a resistor name is given, it must also have been entered in the resistor list as a separate component.

The format for the three-node model is

   name, collector node, base prime node, emitter node, value, tolerance, resistor name.

The first letter is T. The resistor name is used for the tee model only, and is the emitter-current resistor. For example,

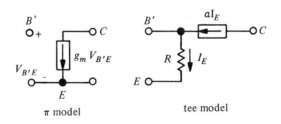

TGM, 1, 2, 3, .022   (for π model)
TALPHA 4, 5, 6, .98, RE   (for tee model)
TAL2, 7, 8, 0, .90, TOL = .52,RE2

The four-node model, a general case of the three-node model, has a format of

name, node i, node j, node k, node l, value, tolerance, resistor name.

The first letter is D. For example,

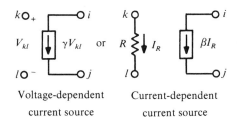

Voltage-dependent current source   Current-dependent current source

DX, 1, 2, 3, 4, .013
DBETA, 5, 6, 7, 8, 20., RBASE
DALPHA, 4, 5, 5, 6, .98, RE   (same as tee model)
DGM, 3, 0, 1, 0, .03   (same as grounded emitter π)

Many active devices may be modeled using the dependent or controlled-current source. Passive elements of the model are treated as normal passive components.

7.4 INPUT LANGUAGE AND OUTPUT SPECIFICATIONS

The LISA input language has a basic set of command or action words. Option words tailor a command word to the action the user desires at that moment. Free format is used throughout. That is, words and data can be placed anywhere on the card separated by commas or equal signs. In general, it is best to omit extra commas when deleting options rather than to leave blank fields or dangling commas.

General user feeling should be that his command is immediately executed. The main requirement on the order between commands is that enough information be given ahead of the command to be executed at the moment. For example, the commands

READ, CIRCUIT
DEFINE, function name

must come before the command

COMPUTE, POLY, function names

so that the circuit and function to be computed are known at compute time. This requirement is relaxed in the case of the DATA command, in that it may be used anywhere before or just after a COMPUTE command.

Once a function is defined or read in, it need only be referred to later by its user-assigned name. This not only allows control but helps the user to jump from one computing mode to another. Similarly, once data are entered, they need only be referred to by name or implication, remaining in their current state until read over.

The basic LISA command set is

READ, CIRCUIT
    name, i, j, value, tol, ic name, ic value   (R, L, C)
    name, i, j, value, tol, R name, option   (I, J or E, V)
    name, L_1 name, L_2 name, value, tolerance   (K, M)
    name, iC, jBP, kE, value, tol, R name   (T pi or T tee)
    name, i, j, k, l, value, tol, R name   (D voltage-dependent or
                                                            D current-dependent
                                                            current source)

    [all current directions assumed i ⟶ j or k ⟶ l]

READ, MATRIX
    row i, col j, degree, coeff in descending order
    [one card per nonzero element]
    [insert S field after degree field if symmetric]

READ, DRIVER
    (IMPULSE, STEP, or RAMP), t_0, a   }
    (SINE, COSINE, SINH, or COSH), t_0, a, ω}   assumes $\alpha = 0$
                PERIOD, tau

READ, function name
    COEFF, scale, degree, coeff in descending order
    COEFF, value   [for a constant]
    ROOTS, scale, # roots, name = real, imag, mult, ...
    [card 1 is numerator, card 2 is denominator]

DEFINE, function name = K∗($\pm$V(i)$\pm$V(j))∗S∗Z/denominator
DEFINE, PORTS = input node, output node
COMPUTE, POLY, function names
COMPUTE, TRFN, function names
COMPUTE, ACCA, function names
COMPUTE, BEV4, value for K, forward function name, feedback function name

---

COMPUTE, BODE, function names
COMPUTE, TRANSIENT, function names and DRIVER
COMPUTE, LOCUS, function names
COMPUTE, SENSITIVITY, function names
COMPUTE, VARIATION, function names
DATA, FREQUENCY = # points per decade, # decades, start freq,
DATA, FREQUENCY = start freq   , end freq , # points , $\begin{Bmatrix} RPS \\ CPS \\ Hz \end{Bmatrix}$, LOG
DATA, FREQUENCY = start freq   , end freq , freq step , $\begin{Bmatrix} RPS \\ CPS \\ Hz \end{Bmatrix}$, LIN
DATA, GAIN   = # points per decade, # decades, start gain

DATA, GAIN = start gain , end gain , # points , LOG
DATA, GAIN = start gain , end gain , gain step, LIN
DATA, TIME = start time, end time, time step
DATA, LOCUS = circuit component name, value 1, value 2, . . .
DATA, SENSITIVITY = freq,* $\genfrac{}{}{0pt}{}{\text{RPS}}{\text{CPS}}_{\text{Hz}}$,* % variation, circuit component names or ALL
CHANGE, CIRCUIT, VALUE
    component-name = value, name = value, . . .
CHANGE, CIRCUIT, TOPOLOGY, ADD
    [same format as READ, CIRCUIT with one card per component]
CHANGE, CIRCUIT, TOPOLOGY, REMOVE
    component-name, name, . . .
CHANGE, CIRCUIT, TOPOLOGY, SHORT
    node-name = node-name, node-name = node-name, . . .
CHANGE, CIRCUIT, SCALE
    R∗value, C∗value, L∗value, T∗value, . . .
CHANGE, MATRIX, ADD
    [same format as READ, MATRIX with one card per element]
CHANGE, MATRIX, REMOVE
    $i_1, j_1, i_2, j_2, \ldots$
TITLE, alphanumeric data   printed at top of a page
NOTE, alphanumeric data   printed once when used
LABEL, alphanumeric data   label for plots
PLOT, function names and DRIVER
PPLOT, function names and DRIVER
PRINT, NORMAL
PRINT, USER, NOCARD
PRINT, NOTHING
PRINT, function names and DRIVER
PRINT, CIRCUIT, SYMBOL
PRINT, MATRIX
PRINT, RESIDUE, function names
EXIT

Each command is now covered in detail.

READ

This command tells the program and the user that descriptive cards follow. These descriptions are given in lists that are terminated by the next LISA command.

The CIRCUIT option is used to begin the nodal description of a circuit.

*Only needed for COMPUTE, ACCA mode—user may delete if not applicable.

The program will automatically set up the describing nodal admittance matrix at this time.

The MATRIX option is used to read the nonzero terms of a matrix, where each element can be an nth degree polynomial.

The DRIVER option is used to read the transient driver function, which is described by using a combination of the following seven basic functions and PERIOD:

Format	$r(t)$	$R(s) = L[r(t)]$
IMPULSE, t_0, a	$a\,\delta(t)$	a
STEP, t_0, a, α	a or $ae^{\alpha t}$	$\dfrac{a}{s}$ or $\dfrac{a}{s-\alpha}$
RAMP, t_0, a, α	at or $ate^{\alpha t}$	$\dfrac{a}{s^2}$ or $\dfrac{a}{(s-\alpha)^2}$
SINE, t_0, a, ω α	$a \sin \omega t$ or $ae^{\alpha t} \sin \omega t$	$\dfrac{a\omega}{s^2+\omega^2}$ or $\dfrac{a\omega}{(s-\alpha)^2+\omega^2}$
COSINE, t_0, a, ω, α	$a \cos \omega t$ or $ae^{\alpha t} \cos \omega t$	$\dfrac{as}{s^2+\omega^2}$ or $\dfrac{as}{(s-\alpha)^2+\omega^2}$
SINH, t_0, a, ω	$a \sinh \omega t$	$\dfrac{a\omega}{s^2-\omega^2}$
COSH, t_0, a, ω	$a \cosh \omega t$	$\dfrac{as}{s^2-\omega^2}$
PERIOD, tau		

In each format above, t_0 is the beginning time of that basic function.

A composite or superposition of a number of these functions will occur if more than one is listed, allowing the formation of a great variety of drivers. In addition, such a function may be made periodic by including the card PERIOD, *tau* after the list

Example.

Form a periodic square wave for the driver of the form

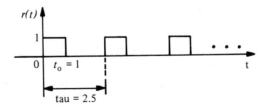

The data cards are

READ, DRIVER
STEP, 0, 1
STEP, 1, −1
PERIOD, 2.5

If the READ option is neither MATRIX, CIRCUIT, nor DRIVER, then it is assumed to be a transfer function name. The function is computed at this time and is saved in a table called TFEX under the assigned name for later use

The numerator is given first, and then the denominator, in either coefficient or root form. Two cards are required, having the format

1. *Coefficient format*

    COEFF, scale factor, degree, coefficients in descending order

    A constant can be given as just: COEFF, value.

2. *Root format*

    ROOTS, scale factor, # of roots given, root name = real part, imaginary part, multiplicity, ...

The conjugate of a complex root will be assumed. A multiplicity of 1 is assumed if this field is omitted. Only the real part of simple real root need be given.

For either COEFF or ROOTS, a card is extended by using ... after a comma. For example,

$$\text{GAIN} = \frac{2s^3 + 3s^2 + 4}{5s^2(s + 1)(s + 2 + j6)}$$

is read in as

    READ, GAIN
        COEFF, 1, 3, 2, 3, 0, 4
        ROOTS, 5, 3, PA = 0, 0, 2, PB = -1, PC = -2, 6

DEFINE

This command is used to define a function that will later be used in computations If the option is PORTS, then the three-terminal grounded two-port will be defined by the input and output nodes given. These nodes will be automatically related to the current circuit at compute time.

If the modifier is not PORTS, then it is assumed to be a function name. This differs from the read-in function in that it is not yet known but will be computed. The form is general enough so that voltages, voltage ratios, currents, current ratios, impedances, admittances, determinants, and cofactors may be defined and later computed. The general form is a FORTRAN-like expression:

    DEFINE, name = K∗($\pm$V(i_1)$\pm$V(i_2))∗S∗Z/denominator

where name is a user-assigned function name; K is a numerical constant;

Z is the name of a circuit resistor, capacitor, or inductor; S is the complex frequency variable; and $V(i_1)$ and $V(i_2)$ are circuit voltages at nodes i_1 and i_2. The denominator is identical in form to the numerator, and the two parts are separated by a slash. The * means multiplication. Thus, the named function is defined in terms of the still unknown voltages, which are themselves functions of s. Any degenerate, or simplified, form is allowed.

For example, assume that an inductor named L9 is connected from node 6 to node 7 in a circuit:

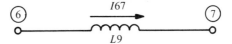

The function of interest might be

DEFINE, V7 = V(7)

or

DEFINE, I67 = V(6) − V(7)/S * L9

In addition, the letter X may be used in place of the letter V. If X is used, it means the numerator solution polynominal, as in a Cramer's rule solution by determinants. The system determinant is given as X(0). Any matrix cofactor may be found by using the proper X notation and matrix driver.

COMPUTE

This command is used to cause computation to start in the selected mode, operating on the functions listed. The program assumes that all necessary information has been previously set up using the commands READ, DEFINE, DATA, etc.

Three methods of solving the matrix equation for the unknown functions are possible with LISA. Each has its advantages and limitations. The engineer's choice will depend on the relative importance of these factors and the characteristics of his design.

The first method, POLY, is used to compute polynomial coefficients and find the roots. Calculating speed is an advantage of this method, but coefficient roundoff error and overflow problems limit its use. Resulting transfer functions are saved in the TFEX table.

TRFN, another method, finds the roots directly from the proper determinants by iteration without forming the coefficients. Roundoff error and overflow are considerably reduced at the cost of increasing the computing time in comparison with POLY speeds. Resulting transfer functions are saved in the TFEX table.

The third method, called ACCA, works only in the $s = j\omega$ or real-frequency domain, and roots are not found. However, the method is fast and accurate. Results are magnitude and phase.

These three alternative methods make it possible for the engineer to check the accuracy of one set of computations against another.

The functions listed for the POLY, TRFN, and ACCA options may also be any of the two-port parameters: Z11, Z12, Z21, Z22, Y_{ij}, H_{ij}, G_{ij}, A_{ij} (transmission), and B_{ij} (inverse transmission), for $i, j = 1, 2$. These are the grounded three-terminal two-port parameters, the input/output terminals (ports) having been previously given by the DEFINE, PORTS command.

Option BEV4 causes computation in the s domain on the two-block feedback system of Fig. 7.1, using as the forward and feedback functions the named functions. These functions may themselves have resulted from a COMPUTE, POLY or COMPUTE, TRFN command given previously or a read-in function.

BEV4 expects to find these functions in the TFEX table. It will then compute the three functions given in Eqs. (7.2), (7.3) and (7.4) and will assign the function names CLOSED, OPEN, and FORWARD, storing the results back in TFEX under these program assigned names.

Default Options. The program assumes that $K = 1.0$ and $F(s) = 1.0$ if either or both are not given.

Option BODE causes computation of the listed functions for $s = j\omega$ over a given range of ω. It is assumed that the named functions of s are known, either from being read in or computed previously, and are found in the TFEX table.

Similarly, the TRANSIENT option causes computation of the time response over a given range to the previously read DRIVER function. Since the computed function is normally the impulse response, this results in a multiplication in the s domain and convolution in the time domain. Again, the named functions are assumed to be in the TFEX table. If the user also desires the driver function to be computed as a distinct quantity, the program-assigned name DRIVER must be given.

The LOCUS, SENSITIVITY, and VARIATION options can only be given and used within the context of a previous COMPUTE command having the option POLY, TRFN, ACCA, or BEV4. For each case, all components with the same name are changed at the same time in the same manner. The original circuit and matrix are restored at the end.

The LOCUS option causes computation of the root locus. If the context is the BEV4 mode, the locus of the block-diagram characteristic equation [Eq. (7.5): $1 + K \cdot H(s) \cdot F(s) = 0$] over the range of the gain K is found. If the context is the POLY or TRFN mode, the poles and/or zeros of the listed functions are found over the range of the specified circuit component. If the context is ACCA, the listed functions are computed over the range of the specified circuit component at the frequency given on the DATA, SENSITIVITY card.

The SENSITIVITY option causes computation of the variation in root locations of a listed function to specified variations of circuit components. The context may be POLY or TRFN. In the ACCA mode, the magnitude and phase variation at a single frequency is found.

The SENSITIVITY option is based on computing an approximation to the true partial derivative by making a very small change in the component. A linear extrapolation to the specified variation is then computed from this.

The VARIATION option is similar to the SENSITIVITY option except that the component is given its specified variation immediately and the computation is not based on the partial derivative. The results will usually differ if both options are compared since the relation between the function and any component is usually nonlinear.

The SENSITIVITY and VARIATION options are not allowed in the BEV4 context.

Before any COMPUTE command is executed, the program will look ahead for DATA commands immediately following the COMPUTE command. The look-ahead stops when the first command that is not DATA is read, and the computation is executed.

DATA

This command is used to enter data other than CIRCUIT, MATRIX, DRIVER, and function descriptions which are done with the READ command. It differs from using the READ command in two ways. First, all the data are given on the same card in a string (use three periods after a comma to extend the card). Second, the DATA command may appear immediately after (or anywhere before) a COMPUTE command and still be used for that particular computation.

Once entered, data remains available for use by any COMPUTE command that needs the data.

The FREQUENCY option specifies the range of frequencies for COMPUTE, ACCA or COMPUTE, BODE. The range usually is taken on a logarithmic scale, using the number of points per decade given over the number of decades desired and starting at the given frequency (stated in RPS or HZ with the default option in HZ). The RPS, CPS, or HZ specification given here will be used for the horizontal axis and units label on later Bode plots. This will not affect any frequency scaling that the user has introduced in scaling his component units but is used merely to specify the 2π factor in $\omega = 2\pi f$.

Two additional options for specifying frequency points are available. They are selected by specifying LOG or LIN in the last field of the DATA, FREQUENCY card. If the three number fields are F1, F2, and F3, then

1. *LOG case*. (F3 + 1) points are generated starting with F1, using the recursion $f_{k+1} = R \cdot f_k$, where ($k = 1, \ldots$, F3) and

$$R = \text{ABS } (F_2/F_1) ** (1/F_3)$$

2. *LIN case* (linear). Points are generated starting with F1, by adding F3 until F2 is reached.

The GAIN option specifies the range of gain for K in the block diagram of Fig. 7.1 and is used for root locus via the COMPUTE, LOCUS command when in the COMPUTE, BEV4 context. The range for gain is handled exactly as the range for frequency.

The TIME option specifies the range and step size for the COMPUTE, TRANSIENT command. The starting time specifies when computation will start; this is not necessarily the same time that a driver is applied.

The LOCUS option is used for the COMPUTE, LOCUS command when dealing with circuits in the POLY, TRFN, or ACCA context. The name of the circuit component, followed by the desired values, is given here as data.

The SENSITIVITY option gives data for the COMPUTE, SENSITIVITY or COMPUTE, VARIATION command when dealing with circuits in the POLY, TRFN, or ACCA context. The frequency given will be used in the ACCA context only. Variation is given as ± 0.05 for 5% and is the amount each listed circuit component will be varied. The word ALL can be substituted for the list to check every circuit component.

CHANGE

This command is used with the CIRCUIT option to change the current circuit state. It allows the user to study the effects of using different component values and limited topological variations of the given circuit without having to read in the entire description again. The following options to CHANGE, CIRCUIT will be covered here. Descriptive cards follow each command. A new nodal admittance matrix is generated for each CHANGE command.

The VALUE option changes the component given to the value given. This may be given in a string separated by commas, i.e.,

$$R1 = \text{value}, \text{CFB} = \text{value}, \ldots$$

The TOPOLOGY option itself has three options: ADD, REMOVE, and SHORT. In each case, it is possible and allowable to add or delete circuit nodes. The program will give a message describing the condition for user checking if nodes are deleted.

The user may or may not desire to redefine his functions or ports after this command is used. A general rule is that if nodes are deleted, the node labels used in the current circuit description, and function and port definitions, must have been given as symbols rather than numbers. No ambiguity will then occur. Note, however, that node deletions apply only to the circuit list and that the symbolic node names used in function or port definitions are not changed even if deleted.

Descriptive cards following the ADD option are given exactly as in READ, CIRCUIT. These components are then added to the circuit component table.

The REMOVE option will remove the components listed in a string on the following card, i.e.,

$$R1, CFB, \ldots$$

To delete a node, all connected components are removed.

The SHORT option is used to connect floating nodes or to actually short-out components (thus removing them).

In general, only adjacent nodes or branches may be shorted. As this operation always combines two nodes, a redundant node exists and is deleted. The node to be deleted is taken to be the first-mentioned node, and the format is

$$A = B, V1 = D$$

This can be interpreted as "set node A equal to node B," or "replace node V1 by node D." This may be given in a string separated by commas.

The CHANGE, CIRCUIT, SCALE command allows the user to multiply all components of a given type, such as all resistors, by a scale factor. Any or all of the component types may be given, using the first letter as a component type identifier.

TITLE, NOTE, and LABEL

These commands are used to enter one line of alphanumeric data. Once entered, they will be used until read over, which may be done at any time. TITLE appears at the top of the printout and is usually changed for each new problem. NOTE is printed just once at the time of occurrence and is used for comments. LABEL is used for plot labels and may appear immediately after (or anywhere before) a PLOT or PPLOT command.

PLOT, PPLOT

This command causes graphical output and is used after a COMPUTE command. If no function names are given, all the functions listed in the previous COMPUTE statement will be plotted; otherwise, selection occurs.

PLOT is used for offline plotting on a mechanical line plotter. The command PPLOT is used for printer plots. PPLOT will always be available, while the PLOT command will depend on availability at the user's particular location.

If the COMPUTE context is BODE or ACCA, one graphical picture is generated for each function named or implied by default.

In the COMPUTE, TRANSIENT context, this rule holds only for the default option. That is, if no function names are given on the PLOT card, one graphical picture is generated per implied function, with time on the horizontal or x axis.

If, however, functions are specified, the interpretation is

$$\text{PLOT, } x, y_1, y_2$$

That is, the name in the x field is taken to be the horizontal or x axis, and the name or names in the y field are the vertical or y axis. The names TIME and DRIVER are recognized by LISA. Each plot card generates one graphical picture.

Example.

PLOT, TIME, V1	(V1 versus t)
PLOT, TIME, V2, DRIVER	(V2 and DRIVER versus t)
PLOT, V1, V2	(V2 versus V1, phase plane)

PRINT

This command gives the user selection and control of the off-line printout if desired. The three basic options of NORMAL, USER, and NOTHING set the mode for the entire program. The NORMAL mode is assumed; if this is desired throughout the program, the PRINT command need never be used.

The NOTHING option suppresses all printout. The user may change the PRINT mode at any point in his command list.

The USER option suppresses the majority, but not all, of the NORMAL printout. The experienced user, who knows what to expect, can be highly selective when he desires but can at any time return to PRINT, NORMAL for convenience. Usually, the user's card images are printed in a group for verification; this can also be suppressed with the PRINT, USER option of NOCARD.

Under the PRINT, USER mode, the user can then apply the following PRINT options at points of interest. After a COMPUTE operation, the functions whose names modify the PRINT command will be printed. If no function names are given, all the functions listed in the previous COMPUTE statement will be printed.

The RESIDUE option causes printout of the residues of the named functions. The program expects to find these functions in the TFEX table.

EXIT

This command must be given as the last statement. It is used only once for program termination.

7.5 MODELS

LISA does not have a model macro expansion facility or specific built-in models other than the passive and dependent current source circuit elements described earlier. These can be used to build such idealized circuit models as a gyrator, ideal transformer, current negative impedance converter, and voltage amplifier, which are included as part of the total circuit description.

7.6 EXAMPLES

Example 1: An Emitter-Follower Circuit

A design problem for an emitter-follower circuit illustrates how LISA operates in practice. The engineer starts with a circuit diagram, selects the hybrid-π as the transistor model for this circuit, and completes the equivalent circuit (Fig. 7.3). For convenience, the base resistance and generator resistance are combined. Nodes on the equivalent circuit are numbered in sequence for reference, starting with 0 for ground. Symbolic names, such as V1, A, or GND, can be used as an alternative method of referring to the nodes.

Figure 7.4 shows the LISA input cards for the emitter-follower example. The NOTE card following the TITLE is a comment which gives the units of scaling for the example. The circuit is entered with the READ, CIRCUIT command. These description cards are terminated by the next LISA command, DEFINE, identifying the transfer function to be computed.

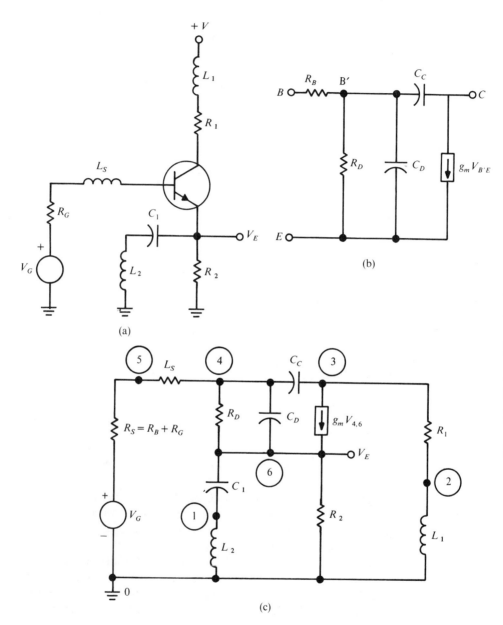

Figure 7.3 Emitter-follower Design for Example 1.

```
       THE FOLLOWING DATA CARD IMAGES WERE READ BY THE        L I S A        SYSTEM INPUT PHASE (LISAX-V2)

TITLE,EMITTER FOLLOWER ANALYSIS                                              00000010
NOTE,SCALE IS K-OHM,UH,PF,NSEC,GIGA-RAD/SEC,VOLT,MA                          00000020
READ,CIRCUIT                                                                 00000030
R1,2,3=.15                                                                   00000040
L1,2,0,.1                                                                    00000050
R2,6,0,.2                                                                    00000060
C1,0,1,15                                                                    00000070
L2,0,1,.06                                                                   00000080
RD,4,6,.49                                                                   00000090
C0,4,6,60                                                                    00000100
CC,3,4,2.68                                                                  00000110
DGM,3,6,4,6,155.1                                                            00000120
LS,4,5,.001                                                                  00000130
VG,0,5,1,RS                                                                  00000140
RS,0,5,.05                                                                   00000150
DEFINE,MS=V(6)                                                               00000160
COMPUTE,PCLY,MS                                                              00000170
COMPUTE,BODE,MS                                                              00000180
DATA,FREQ=50,3,.001,RPS                                                      00000190
PLOT                                                                         00000230
LABEL,EMITTER FOLLOWER MS (W=G-RPS, T=NSEC)                                  00000240
READ,DRIVER                                                                  00000250
STEP,0,1                                                                     00000260
DATA,TIME=0,80,.25                                                           00000270
COMPUTE,TRANSIENT,MS                                                         00000280
PLOT                                                                         00000290
NOTE,CHECK POLY POLES AND ZEROS                                              00000300
COMPUTE,TRFN,MS                                                              00000310
COMPUTE,BODE,MS                                                              00000320
COMPUTE,ACCA,MS                                                              00000330
NOTE,TRY PARAMETER VARIATIONS ON LS AND RS                                   00000340
CHANGE,CIRCUIT,VALUE                                                         00000345
LS=.05636                                                                    00000346
COMPUTE,PCLY,MS                                                              00000347
CHANGE,CIRCUIT,VALUE                                                         00000350
LS=.2                                                                        00000360
COMPUTE,PCLY,MS                                                              00000370
CHANGE,CIRCUIT,VALUE                                                         00000380
RS=.1                                                                        00000390
COMPUTE,PCLY,MS                                                              00000400
NOTE,FIND ROOT LOCUS OF MS TO RS                                             00000410
COMPUTE,LOCUS,MS                                                             00000420
DATA,LOCUS=RS,.075,.1,.125,.15                                               00000430
EXIT                                                                         00000510
```

Figure 7.4 LISA Input Cards for Example 1.

For the emitter-follower problem, the decision to compute the function MS has many advantages. Because $V_G(s)$ is set to the impulse function, MS becomes

$$MS = V_E(s)/V_G(s) = V_E(s)/1 = V(6) = N(s)/D(s).$$

This function, therefore, represents a voltage ratio for the Bode response or a voltage for the transient response; its denominator is the characteristic function for stability studies.

With the command, COMPUTE, POLY, MS, the computer calculates the poles and zeros of the function MS. Output from the computation (Fig. 7.5) lists the pole and zero coefficients in descending order (i.e., $1.608 \times 10^6 s^7 + 4.172 \times 10^8 s^6 + \cdots$) and shows the roots. The results, after cancellation, are

$$MS = N(s)/D(s)$$
$$= 200(s-z_1)(s-z_2)(s-z_3)/(s-p_1)(s-p_2)(s-p_3)(s-p_4),$$

where

$p_1 = -0.1108 \pm j0.8346 \qquad z_1 = -0.75 \pm j1.780$

$p_2 = -1.177 \pm j1.822 \qquad z_2 = \pm j1.054$

$p_3 = -2.648 \qquad z_3 = -2.619$

$p_4 = -254.2$

By computing MS in the s plane, the engineer has immediate information on stability at his disposal. In Fig. 7.5, for example, the poles are all in the left half-plane, indicating that the circuit is stable.

The statement COMPUTE, BODE, MS now directs LISA to provide the magnitude and phase of MS over the frequency range given on the DATA card. This card specifies a range of 50 points per decade over three decades, starting at $\omega = 0.001$ Grad/s $= 1$ Mrad/s. Figure 7.6 shows a partial listing of the results. The PLOT and LABEL commands produce a Bode plot (Fig. 7.7).

The engineer enters the transient driver with the READ, DRIVER command. He asks for just the step response, beginning with $t = 0$ and with unit height (1.0). LISA actually allows seven basic Laplace transform drivers: impulse, step, ramp, sine, cosine, sinh, and cosh. These drivers can be superimposed to obtain a variety of pulse shapes, which can be periodic.

The DATA, TIME statement supplies the starting time (0 nanosecond), the ending time (80 nanoseconds), and the time step (0.25 nanosecond) for the transient response computation. COMPUTE, TRANSIENT, MS, and PLOT commands produce a listing of results (Fig. 7.8) and a plot of MS versus time (Fig. 7.9) for the step driver.

EMITTER FOLLOWER ANALYSIS

NOTE   SCALE IS K-OHM,UH,PF,NSEC,GIGA-RAD/SEC,VOLT,MA

CIRCUIT TABLE   NUMBER OF NODES = 6

		PRIMARY COMPONENTS		RELATED COMPONENTS		
NAME	NODE NUMBERS	VALUE	TOLERANCE	NAME	VALUE	COMMENTS
NK= 4						
R1	2- 3	0.15000	0.0			
R2	6- 0	0.20000	0.0			
RD	4- 6	0.49000	0.0			
RS	0- 5	5.00000E-02	0.0			
NC= 3						
C1	6- 1	15.000	0.0			
CD	4- 6	60.000	0.0			
CC	3- 4	2.6800	0.0			
NL= 3						
L1	2- 0	1.00000E-01	0.0			
L2	0- 1	6.00000E-02	0.0			
LS	4- 5	1.00000E-03	0.0			
NGEN= 1						
VG	0- 5	1.0030	0.0	RS	5.00000E-02	IMPULSE
NTX= 0						
NDEP= 1						
DGM	3- 6- 4- 6	155.10	0.0			VOLT DEP CURR SOURCE
MUTUAL= 0						

Figure 7.5 Emitter-follower Poles and Zeros for MS for Example 1.

NON-ZERO ELEMENTS OF AUGMENTED MATRIX...COEFFICIENTS LISTED IN ORDER OF DESCENDING DEGREE

MATRIX SIZE N = 6

```
Y( 1, 1)    DEGREE = 2       15.00000        0.0         16.66667
Y( 1, 6)    DEGREE = 2      -15.00000        0.0          0.0
Y( 2, 2)    DEGREE = 1        6.666668      10.00000
Y( 2, 3)    DEGREE = 1       -6.666668       0.0
Y( 3, 2)    DEGREE = 0       -6.666668
Y( 3, 3)    DEGREE = 1        2.679999       6.666668
Y( 3, 4)    DEGREE = 1       -2.679999     155.1000
Y( 3, 6)    DEGREE = 0       -155.1000
Y( 4, 3)    DEGREE = 1       -2.679999
Y( 4, 4)    DEGREE = 2       62.68000        0.0         1000.000
Y( 4, 5)    DEGREE = 0      -1000.000        2.040817
Y( 4, 6)    DEGREE = 2      -60.00000       -2.040817    0.0
Y( 5, 4)    DEGREE = 0      -1000.000
Y( 5, 5)    DEGREE = 1       20.00000      1000.000
Y( 5, 7)    DEGREE = 1       20.00000        0.0
Y( 6, 1)    DEGREE = 1      -15.00000        0.0
Y( 6, 4)    DEGREE = 1      -60.00000     -157.1408
Y( 6, 6)    DEGREE = 1       75.00000      162.1408
```

EMITTER FOLLOWER ANALYSIS   POLY

```
TRANSFER FUNCTION   MS  =   ZNUM*N(S)*KN / ZDEN*D(S)*KD
ZNUM =  1.0000    (      )  ,  N(S) = +V( 6)                ,  KN =  1.0000
ZDEN =  1.0000    (      )  ,  D(S) = NOT REQUESTED - SET =1.0  ,  KD =  1.0000
```

POLES AND ZEROS FOR MS

NUMERATOR COEFFICIENTS, DEGREE = 6. DESCENDING ORDER

```
3.2160000E 08    1.3246748E 09    2.8207465E 09    4.6146765E 09    2.7371256E 09
3.4920202E 09    0.0
```

Figure 7.5—*Cont.*

```
       REAL              IMAG
       0.0               0.0
       0.0               1.054092E 00
       0.0              -1.054092E 00
      -7.500001E-01      1.780125E 00
      -7.500001E-01     -1.780125E 00
      -2.619013E 00      0.0

DENOMINATOR COEFFICIENTS, DEGREE = 7. DESCENDING ORDER

   1.6080000E 06    4.1719936E 08    2.1563159E 09    5.2469842E 09    7.5568251E 09
   4.3164672E 09    3.6144691E 09    0.0

       REAL              IMAG
       0.0               0.0
      -1.108415E-01      8.346870E-01
      -1.108415E-01     -8.346870E-01
      -1.177269E 00      1.822899E 00
      -1.177269E 00     -1.822899E 00
      -2.648344E 00      0.0
      -2.542278E 02      0.0

TOTAL MULTIPLIER = 2.00000E 02

PROCESSED ROOTS FOR MS   MULTIPLIER = ( 2.0000000E 02)*10**  0

                    NUMERATOR
       0.0              1.0540924E 00
      -7.5000012E-01    1.7801247E 00
      -2.6190128E 00    0.0

                    DENOMINATOR
      -1.1084145E-01    8.3468699E-01
      -1.1772690E 00    1.8228989E 00
      -2.648344CE 00    0.0
      -2.5422778E 02    0.0
```

Figure 7.5—*Cont.*

EMITTER FOLLOWER ANALYSIS

BODE DIAGRAM FOR MS

BODE

FREQ(RAD)	FREQ(CPS)	SIGMA	OMEGA	DEGREES	MAGNITUDE	MAG(DB)	DELAY
1.0000E-03	1.5915E-04	9.66122E-01	-3.96492E-04	-0.024	9.66122E-01	-0.299	4.104E-01
1.0471E-03	1.6666E-04	9.66120E-01	-4.15178E-04	-0.025	9.66120E-01	-0.299	4.104E-01
1.0965E-03	1.7451E-04	9.66120E-01	-4.34745E-04	-0.026	9.66120E-01	-0.299	4.104E-01
1.1482E-03	1.8273E-04	9.66120E-01	-4.55233E-04	-0.027	9.66120E-01	-0.299	4.104E-01
1.2023E-03	1.9135E-04	9.66122E-01	-4.76688E-04	-0.028	9.66122E-01	-0.299	4.104E-01
1.2589E-03	2.0036E-04	9.66122E-01	-4.99155E-04	-0.030	9.66122E-01	-0.299	4.104E-01
1.3183E-03	2.0981E-04	9.66122E-01	-5.22679E-04	-0.031	9.66122E-01	-0.299	4.104E-01
1.3804E-03	2.1969E-04	9.66121E-01	-5.47313E-04	-0.032	9.66121E-01	-0.299	4.104E-01
1.4454E-03	2.3005E-04	9.66121E-01	-5.73107E-04	-0.034	9.66121E-01	-0.299	4.104E-01
1.5136E-03	2.4089E-04	9.66121E-01	-6.00117E-04	-0.036	9.66121E-01	-0.299	4.104E-01
1.5849E-03	2.5224E-04	9.66123E-01	-6.28400E-04	-0.037	9.66123E-01	-0.299	4.104E-01
1.6596E-03	2.6413E-04	9.66123E-01	-6.58016E-04	-0.039	9.66123E-01	-0.299	4.104E-01
1.7378E-03	2.7658E-04	9.66122E-01	-6.89027E-04	-0.041	9.66122E-01	-0.299	4.104E-01
1.8197E-03	2.8961E-04	9.66123E-01	-7.21501E-04	-0.043	9.66123E-01	-0.299	4.104E-01
1.9055E-03	3.0326E-04	9.66121E-01	-7.55505E-04	-0.045	9.66124E-01	-0.299	4.104E-01
1.9953E-03	3.1756E-04	9.66121E-01	-7.91110E-04	-0.047	9.66122E-01	-0.299	4.104E-01
2.0893E-03	3.3252E-04	9.66122E-01	-8.28394E-04	-0.049	9.66122E-01	-0.299	4.104E-01
2.1878E-03	3.4819E-04	9.66122E-01	-8.67438E-04	-0.051	9.66122E-01	-0.299	4.104E-01
2.2909E-03	3.6460E-04	9.66122E-01	-9.08318E-04	-0.054	9.66122E-01	-0.299	4.104E-01
2.3988E-03	3.8179E-04	9.66123E-01	-9.51129E-04	-0.056	9.66124E-01	-0.299	4.104E-01
2.5119E-03	3.9978E-04	9.66123E-01	-9.95954E-04	-0.059	9.66123E-01	-0.299	4.104E-01
2.6303E-03	4.1862E-04	9.66122E-01	-1.04289E-03	-0.062	9.66123E-01	-0.299	4.104E-01
2.7542E-03	4.3835E-04	9.66122E-01	-1.09204E-03	-0.065	9.66123E-01	-0.299	4.104E-01
2.8840E-03	4.5901E-04	9.66122E-01	-1.14351E-03	-0.068	9.66124E-01	-0.299	4.104E-01
3.0200E-03	4.8064E-04	9.66122E-01	-1.19740E-03	-0.071	9.66123E-01	-0.299	4.104E-01
3.1623E-03	5.0329E-04	9.66122E-01	-1.25384E-03	-0.074	9.66123E-01	-0.299	4.104E-01
3.3113E-03	5.2701E-04	9.66122E-01	-1.31293E-03	-0.078	9.66123E-01	-0.299	4.104E-01
3.4674E-03	5.5185E-04	9.66123E-01	-1.37481E-03	-0.082	9.66124E-01	-0.299	4.104E-01
3.6308E-03	5.7786E-04	9.66123E-01	-1.43961E-03	-0.085	9.66125E-01	-0.299	4.104E-01
3.8019E-03	6.0509E-04	9.66123E-01	-1.50746E-03	-0.089	9.66125E-01	-0.299	4.104E-01
3.9811E-03	6.3361E-04	9.66123E-01	-1.57850E-03	-0.094	9.66125E-01	-0.299	4.104E-01
4.1687E-03	6.6347E-04	9.66123E-01	-1.65289E-03	-0.098	9.66126E-01	-0.299	4.104E-01
4.3652E-03	6.9474E-04	9.66127E-01	-1.73080E-03	-0.103	9.66129E-01	-0.299	4.104E-01
4.5709E-03	7.2748E-04	9.66125E-01	-1.81237E-03	-0.107	9.66127E-01	-0.299	4.104E-01

Figure 7.6 Frequency Response for Example 1.

FREQ(RAD)	FREQ(CPS)	SIGMA	OMEGA	DEGREES	MAGNITUDE	MAG(DB)	DELAY
4.7863E-03	7.6176E-04	9.66126E-01	-1.89779E-03	-0.113	9.66128E-01	-0.299	4.104E-01
5.0119E-03	7.9766E-04	9.66126E-01	-1.98724E-03	-0.118	9.66128E-01	-0.299	4.104E-01
5.2481E-03	8.3526E-04	9.66128E-01	-2.08090E-03	-0.123	9.66130E-01	-0.299	4.104E-01
5.4954E-03	8.7462E-04	9.66128E-01	-2.17898E-03	-0.129	9.66130E-01	-0.299	4.104E-01
5.7544E-03	9.1584E-04	9.66129E-01	-2.28168E-03	-0.135	9.66132E-01	-0.299	4.104E-01
6.0256E-03	9.5900E-04	9.66129E-01	-2.38922E-03	-0.142	9.66132E-01	-0.299	4.104E-01
6.3096E-03	1.0042E-03	9.66129E-01	-2.50183E-03	-0.148	9.66133E-01	-0.299	4.104E-01
6.6069E-03	1.0515E-03	9.66131E-01	-2.61975E-03	-0.155	9.66134E-01	-0.299	4.104E-01
6.9183E-03	1.1011E-03	9.66132E-01	-2.74323E-03	-0.163	9.66136E-01	-0.299	4.105E-01
7.2443E-03	1.1530E-03	9.66135E-01	-2.87253E-03	-0.170	9.66139E-01	-0.299	4.105E-01
7.5858E-03	1.2073E-03	9.66135E-01	-3.00793E-03	-0.178	9.66140E-01	-0.299	4.105E-01
7.9433E-03	1.2642E-03	9.66136E-01	-3.14970E-03	-0.187	9.66141E-01	-0.299	4.105E-01
8.3176E-03	1.3238E-03	9.66138E-01	-3.29816E-03	-0.196	9.66143E-01	-0.299	4.105E-01
8.7096E-03	1.3862E-03	9.66141E-01	-3.45364E-03	-0.205	9.66147E-01	-0.299	4.105E-01
9.1201E-03	1.4515E-03	9.66141E-01	-3.61643E-03	-0.214	9.66148E-01	-0.299	4.105E-01
9.5499E-03	1.5199E-03	9.66144E-01	-3.78691E-03	-0.225	9.66152E-01	-0.299	4.105E-01
1.0000E-02	1.5915E-03	9.66146E-01	-3.96544E-03	-0.235	9.66154E-01	-0.299	4.105E-01
1.0471E-02	1.6666E-03	9.66150E-01	-4.15238E-03	-0.246	9.66159E-01	-0.299	4.105E-01
1.0965E-02	1.7451E-03	9.66153E-01	-4.34814E-03	-0.258	9.66162E-01	-0.299	4.105E-01
1.1482E-02	1.8273E-03	9.66156E-01	-4.55313E-03	-0.270	9.66166E-01	-0.299	4.106E-01
1.2023E-02	1.9135E-03	9.66161E-01	-4.76780E-03	-0.283	9.66173E-01	-0.299	4.106E-01
1.2589E-02	2.0036E-03	9.66162E-01	-4.99258E-03	-0.296	9.66175E-01	-0.299	4.106E-01
1.3183E-02	2.0981E-03	9.66172E-01	-5.22798E-03	-0.310	9.66182E-01	-0.299	4.106E-01
1.3804E-02	2.1969E-03	9.66172E-01	-5.47450E-03	-0.325	9.66187E-01	-0.299	4.107E-01
1.4454E-02	2.3005E-03	9.66176E-01	-5.73264E-03	-0.340	9.66193E-01	-0.299	4.107E-01
1.5136E-02	2.4089E-03	9.66181E-01	-6.00297E-03	-0.356	9.66200E-01	-0.299	4.107E-01
1.5849E-02	2.5224E-03	9.66188E-01	-6.28607E-03	-0.373	9.66209E-01	-0.299	4.107E-01
1.6596E-02	2.6413E-03	9.66194E-01	-6.58253E-03	-0.390	9.66217E-01	-0.299	4.108E-01
1.7378E-02	2.7658E-03	9.66200E-01	-6.89301E-03	-0.409	9.66225E-01	-0.298	4.108E-01
1.8197E-02	2.8961E-03	9.66211E-01	-7.21815E-03	-0.428	9.66238E-01	-0.298	4.108E-01
1.9055E-02	3.0326E-03	9.66217E-01	-7.55864E-03	-0.448	9.66247E-01	-0.298	4.109E-01
1.9953E-02	3.1756E-03	9.66227E-01	-7.91524E-03	-0.469	9.66259E-01	-0.298	4.109E-01
2.0893E-02	3.3252E-03	9.66238E-01	-8.28872E-03	-0.491	9.66274E-01	-0.298	4.110E-01
2.1878E-02	3.4819E-03	9.66249E-01	-8.67983E-03	-0.515	9.66288E-01	-0.298	4.110E-01
2.2909E-02	3.6460E-03	9.66261E-01	-9.08946E-03	-0.539	9.66303E-01	-0.298	4.111E-01
2.3988E-02	3.8179E-03	9.66274E-01	-9.51847E-03	-0.564	9.66321E-01	-0.298	4.111E-01
2.5119E-02	3.9978E-03	9.66290E-01	-9.96782E-03	-0.591	9.66342E-01	-0.297	4.112E-01
2.6303E-02	4.1862E-03	9.66306E-01	-1.04384E-02	-0.619	9.66362E-01	-0.297	4.112E-01

Figure 7.6—*Cont.*

FREQ(RAD)	FREQ(CPS)	SIGMA	OMEGA	DEGREES	MAGNITUDE	MAG(DB)	DELAY
2.7542E-02	4.3835E-03	9.66326E-01	-1.09314E-02	-0.648	9.66388E-01	-0.297	4.113E-01
2.8840E-02	4.5901E-03	9.66343E-01	-1.14476E-02	-0.679	9.66411E-01	-0.297	4.114E-01
3.0200E-02	4.8064E-03	9.66366E-01	-1.19884E-02	-0.711	9.66441E-01	-0.296	4.115E-01
3.1623E-02	5.0329E-03	9.66389E-01	-1.25549E-02	-0.744	9.66471E-01	-0.296	4.116E-01
3.3113E-02	5.2701E-03	9.66417E-01	-1.31483E-02	-0.779	9.66506E-01	-0.296	4.117E-01
3.4674E-02	5.5185E-03	9.66445E-01	-1.37700E-02	-0.816	9.66543E-01	-0.296	4.119E-01
3.6308E-02	5.7786E-03	9.66477E-01	-1.44211E-02	-0.855	9.66584E-01	-0.295	4.120E-01
3.8019E-02	6.0509E-03	9.66510E-01	-1.51033E-02	-0.895	9.66628E-01	-0.295	4.122E-01
3.9811E-02	6.3361E-03	9.66549E-01	-1.58181E-02	-0.938	9.66678E-01	-0.294	4.124E-01
4.1687E-02	6.6347E-03	9.66590E-01	-1.65670E-02	-0.982	9.66732E-01	-0.294	4.125E-01
4.3652E-02	6.9474E-03	9.66636E-01	-1.73517E-02	-1.028	9.66792E-01	-0.294	4.127E-01
4.5709E-02	7.2748E-03	9.66683E-01	-1.81739E-02	-1.077	9.66854E-01	-0.293	4.130E-01
4.7863E-02	7.6176E-03	9.66739E-01	-1.90355E-02	-1.128	9.66926E-01	-0.292	4.132E-01
5.0119E-02	7.9766E-03	9.66799E-01	-1.99386E-02	-1.181	9.67005E-01	-0.291	4.135E-01
5.2481E-02	8.3526E-03	9.66866E-01	-2.08851E-02	-1.237	9.67091E-01	-0.291	4.138E-01
5.4954E-02	8.7462E-03	9.66936E-01	-2.18771E-02	-1.296	9.67184E-01	-0.290	4.141E-01
5.7544E-02	9.1584E-03	9.67015E-01	-2.29171E-02	-1.358	9.67287E-01	-0.289	4.145E-01
6.0256E-02	9.5900E-03	9.67102E-01	-2.40074E-02	-1.422	9.67400E-01	-0.288	4.149E-01
6.3096E-02	1.0042E-02	9.67198E-01	-2.51507E-02	-1.490	9.67525E-01	-0.287	4.153E-01
6.6069E-02	1.0515E-02	9.67302E-01	-2.63496E-02	-1.560	9.67661E-01	-0.286	4.158E-01
6.9183E-02	1.1011E-02	9.67417E-01	-2.76071E-02	-1.635	9.67811E-01	-0.284	4.163E-01
7.2444E-02	1.1530E-02	9.67545E-01	-2.89262E-02	-1.712	9.67977E-01	-0.283	4.169E-01
7.5858E-02	1.2073E-02	9.67683E-01	-3.03101E-02	-1.794	9.68157E-01	-0.281	4.176E-01
7.9433E-02	1.2642E-02	9.67833E-01	-3.17623E-02	-1.880	9.68354E-01	-0.279	4.183E-01
8.3176E-02	1.3238E-02	9.68000E-01	-3.32866E-02	-1.969	9.68572E-01	-0.277	4.190E-01
8.7096E-02	1.3862E-02	9.68183E-01	-3.48868E-02	-2.064	9.68811E-01	-0.275	4.199E-01
9.1201E-02	1.4515E-02	9.68382E-01	-3.65672E-02	-2.163	9.69073E-01	-0.273	4.208E-01
9.5499E-02	1.5199E-02	9.68602E-01	-3.83322E-02	-2.266	9.69360E-01	-0.270	4.219E-01
1.0000E-01	1.5915E-02	9.68847E-01	-4.01869E-02	-2.375	9.69681E-01	-0.267	4.230E-01
1.0471E-01	1.6666E-02	9.69111E-01	-4.21360E-02	-2.490	9.70027E-01	-0.264	4.242E-01
1.0965E-01	1.7451E-02	9.69405E-01	-4.41856E-02	-2.610	9.70411E-01	-0.261	4.256E-01
1.1482E-01	1.8273E-02	9.69727E-01	-4.63414E-02	-2.736	9.70834E-01	-0.257	4.271E-01
1.2023E-01	1.9135E-02	9.70080E-01	-4.86099E-02	-2.869	9.71298E-01	-0.253	4.288E-01
1.2589E-01	2.0036E-02	9.70470E-01	-5.09982E-02	-3.008	9.71809E-01	-0.248	4.306E-01
1.3183E-01	2.0981E-02	9.70896E-01	-5.35142E-02	-3.155	9.72370E-01	-0.243	4.327E-01
1.3804E-01	2.1969E-02	9.71367E-01	-5.61662E-02	-3.309	9.72990E-01	-0.238	4.349E-01
1.4454E-01	2.3005E-02	9.71886E-01	-5.89634E-02	-3.472	9.73673E-01	-0.232	4.374E-01
1.5136E-01	2.4089E-02	9.72458E-01	-6.19157E-02	-3.643	9.74427E-01	-0.225	4.401E-01
1.5849E-01	2.5224E-02	9.73086E-01	-6.50342E-02	-3.824	9.75257E-01	-0.218	4.432E-01
1.6596E-01	2.6413E-02	9.73774E-01	-6.83308E-02	-4.014	9.76169E-01	-0.209	4.465E-01

Figure 7.6—*Cont.*

FREQ(RAD)	FREQ(CPS)	SIGMA	OMEGA	DEGREES	MAGNITUDE	MAG(DB)	DELAY
1.7378E-01	2.7658E-02	9.74540E-01	-7.18200E-02	-4.215	9.77183E-01	-0.200	4.503E-01
1.8197E-01	2.8961E-02	9.75382E-01	-7.55161E-02	-4.427	9.78301E-01	-0.191	4.544E-01
1.9055E-01	3.0326E-02	9.76311E-01	-7.94361E-02	-4.652	9.79537E-01	-0.180	4.591E-01
1.9953E-01	3.1756E-02	9.77318E-01	-8.35993E-02	-4.889	9.80907E-01	-0.167	4.642E-01
2.0893E-01	3.3252E-02	9.78471E-01	-8.80268E-02	-5.141	9.82423E-01	-0.154	4.700E-01
2.1878E-01	3.4819E-02	9.79727E-01	-9.27429E-02	-5.408	9.84107E-01	-0.139	4.764E-01
2.2909E-01	3.6460E-02	9.81114E-01	-9.77750E-02	-5.691	9.85974E-01	-0.123	4.836E-01
2.3988E-01	3.8179E-02	9.82653E-01	-1.03155E-01	-5.993	9.98052E-01	-0.104	4.917E-01
2.5119E-01	3.9978E-02	9.84357E-01	-1.08920E-01	-6.314	9.90365E-01	-0.084	5.008E-01
2.6303E-01	4.1862E-02	9.86249E-01	-1.15112E-01	-6.657	9.92944E-01	-0.062	5.110E-01
2.7542E-01	4.3835E-02	9.88352E-01	-1.21780E-01	-7.024	9.95827E-01	-0.036	5.227E-01
2.8840E-01	4.5901E-02	9.90692E-01	-1.28982E-01	-7.418	9.99053E-01	-0.008	5.359E-01
3.0199E-01	4.8064E-02	9.93296E-01	-1.36787E-01	-7.841	1.00674E 00	0.023	5.509E-01
3.1623E-01	5.0329E-02	9.96199E-01	-1.45277E-01	-8.297	1.00674E 00	0.058	5.681E-01
3.3113E-01	5.2701E-02	9.99442E-01	-1.54551E-01	-8.790	1.01132E 00	0.098	5.879E-01
3.4674E-01	5.5185E-02	1.00307E 00	-1.64728E-01	-9.326	1.01650E 00	0.142	6.107E-01
3.6308E-01	5.7786E-02	1.00712E 00	-1.75955E-01	-9.910	1.02238E 00	0.192	6.373E-01
3.8019E-01	6.0509E-02	1.01166E 00	-1.88413E-01	-10.550	1.02906E 00	0.249	6.683E-01
3.9811E-01	6.3361E-02	1.01676E 00	-2.02328E-01	-11.254	1.03670E 00	0.313	7.049E-01
4.1687E-01	6.6347E-02	1.02248E 00	-2.17989E-01	-12.035	1.04545E 00	0.386	7.484E-01
4.3652E-01	6.9474E-02	1.02889E 00	-2.35764E-01	-12.906	1.05555E 00	0.470	8.005E-01
4.5709E-01	7.2748E-02	1.03607E 00	-2.56133E-01	-13.886	1.06726E 00	0.565	8.636E-01
4.7863E-01	7.6176E-02	1.04408E 00	-2.79731E-01	-14.998	1.08091E 00	0.676	9.410E-01
5.0119E-01	7.9766E-02	1.05296E 00	-3.07409E-01	-16.275	1.09692E 00	0.803	1.037E 00
5.2481E-01	8.3526E-02	1.06267E 00	-3.40333E-01	-17.758	1.11584E 00	0.952	1.159E 00
5.4954E-01	8.7462E-02	1.07301E 00	-3.80123E-01	-19.507	1.13835E 00	1.126	1.315E 00
5.7544E-01	9.1584E-02	1.08348E 00	-4.29076E-01	-21.604	1.16535E 00	1.329	1.520E 00
6.0256E-01	9.5900E-02	1.09298E 00	-4.90491E-01	-24.171	1.19791E 00	1.568	1.795E 00
6.3096E-01	1.0042E-01	1.09863E 00	-5.69125E-01	-27.385	1.23729E 00	1.849	2.175E 00
6.6069E-01	1.0515E-01	1.09507E 00	-6.71706E-01	-31.524	1.28467E 00	2.176	2.712E 00
6.9183E-01	1.1011E-01	1.07002E 00	-8.06879E-01	-37.019	1.34015E 00	2.543	3.491E 00
7.2444E-01	1.1530E-01	9.97605E-01	-9.81875E-01	-44.545	1.39975E 00	2.921	4.628E 00
7.5858E-01	1.2070E-01	8.28171E-01	-1.18666E 00	-55.089	1.44708E 00	3.210	6.215E 00
7.9433E-01	1.2642E-01	4.98180E-01	-1.34809E 00	-69.718	1.43720E 00	3.150	8.040E 00
8.3176E-01	1.3238E-01	3.70649E-02	-1.29577E 00	-88.361	1.29630E 00	2.254	9.078E 00
8.7096E-01	1.3862E-01	-3.12534E-01	-9.54996E-01	-103.121	1.00484E 00	0.042	8.199E 00
9.1201E-01	1.4515E-01	-3.81663E-01	-5.45085E-01	-124.999	6.65420E-01	-3.538	6.104E 00
9.5499E-01	1.5199E-01	-2.79424E-01	-2.55946E-01	-137.511	3.78928E-01	-8.429	4.161E 00
1.0000E 00	1.5915E-01	-1.39713E-01	-9.30057E-02	-146.348	1.67838E-01	-15.502	2.798E 00

Figure 7.6—*Cont.*

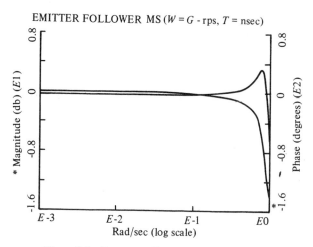

Figure 7.7 Frequency Response Plot for Example 1.

Since residue methods are used, the partial fraction expansion can also be obtained.

The next three COMPUTE cards check numerical results. Because pole/zero computation is subject to error, checking should be done for each new circuit. The TRFN method is used to compute the poles and zeros of MS for comparison with the previous POLY method roots. TRFN is slower but is usually more accurate than POLY. The COMPUTE, BODE command then gives the frequency response using the transfer function just computed by TRFN. Finally the COMPUTE, ACCA command gives the frequency response again. The ACCA frequency response results are the most accurate. If the BODE and ACCA frequency results compare over the bandwidth of the circuit, the poles and zeros computed probably are correct.

The next 10 cards are used to study the effect of RS and LS on circuit stability.

These statements produce computations similar to those seen in Fig. 7.5. The results are illustrated in a sketch (Fig. 7.10) showing the movement of the dominant poles. The critical pole moves to the $j\omega$ axis, and then into the right half-plane as L_S increases. Adding more base resistance stabilizes the circuit.

The next three cards compute the root locus of the transfer function MS to resistor R_S. As the previous command was COMPUTE, POLY, MS, the context is POLY, and the computations are done using the POLY method. The locus calculations can also be done in the TRFN or ACCA context. For ACCA, they are done at a single frequency point

EMITTER FOLLOWER ANALYSIS

TRANSIENT RESPONSE FOR MS   TRAN

T	F(T)	T	F(T)	T	F(T)
0.0	-8.34465E-07	1.2500E 01	1.05445E 00	2.5000E 01	9.61830E-01
2.5000E-01	6.15665E-01	1.2750E 01	1.04305E 00	2.5250E 01	9.66756E-01
5.0000E-01	5.25911E-01	1.3000E 01	1.02896E 00	2.5500E 01	9.71389E-01
7.5000E-01	5.05753E-01	1.3250E 01	1.01293E 00	2.5750E 01	9.75546E-01
1.0000E 00	5.34727E-01	1.3500E 01	9.95758E-01	2.6000E 01	9.79074E-01
1.2500E 00	5.93237E-01	1.3750E 01	9.78236E-01	2.6250E 01	9.81856E-01
1.5000E 00	6.65832E-01	1.4000E 01	9.61138E-01	2.6500E 01	9.83812E-01
1.7500E 00	7.42122E-01	1.4250E 01	9.45175E-01	2.6750E 01	9.84902E-01
2.0000E 00	8.16192E-01	1.4500E 01	9.30974E-01	2.7000E 01	9.85128E-01
2.2500E 00	8.85316E-01	1.4750E 01	9.19047E-01	2.7250E 01	9.84525E-01
2.5000E 00	9.48590E-01	1.5000E 01	9.09785E-01	2.7500E 01	9.83164E-01
2.7500E 00	1.00583E 00	1.5250E 01	9.03442E-01	2.7750E 01	9.81145E-01
3.0000E 00	1.05686E 00	1.5500E 01	9.00133E-01	2.8000E 01	9.78590E-01
3.2500E 00	1.10124E 00	1.5750E 01	8.99836E-01	2.8250E 01	9.75636E-01
3.5000E 00	1.13818E 00	1.6000E 01	9.02403E-01	2.8500E 01	9.72434E-01
3.7500E 00	1.16677E 00	1.6250E 01	9.07567E-01	2.8750E 01	9.69133E-01
4.0000E 00	1.18617E 00	1.6500E 01	9.14968E-01	2.9000E 01	9.65880E-01
4.2500E 00	1.19577E 00	1.6750E 01	9.24167E-01	2.9250E 01	9.62814E-01
4.5000E 00	1.19541E 00	1.7000E 01	9.34671E-01	2.9500E 01	9.60055E-01
4.7500E 00	1.18538E 00	1.7250E 01	9.45960E-01	2.9750E 01	9.57705E-01
5.0000E 00	1.16647E 00	1.7500E 01	9.57507E-01	3.0000E 01	9.55843E-01
5.2500E 00	1.13986E 00	1.7750E 01	9.68801E-01	3.0250E 01	9.54524E-01
5.5000E 00	1.10709E 00	1.8000E 01	9.79371E-01	3.0500E 01	9.53773E-01
5.7500E 00	1.06991E 00	1.8250E 01	9.88802E-01	3.0750E 01	9.53593E-01
6.0000E 00	1.03019E 00	1.8500E 01	9.96750E-01	3.1000E 01	9.53961E-01
6.2500E 00	9.89803E-01	1.8750E 01	1.00295E 00	3.1250E 01	9.54832E-01
6.5000E 00	9.50562E-01	1.9000E 01	1.00724E 00	3.1500E 01	9.56141E-01
6.7500E 00	9.14108E-01	1.9250E 01	1.00953E 00	3.1750E 01	9.57807E-01
7.0000E 00	8.81868E-01	1.9500E 01	1.00984E 00	3.2000E 01	9.59741E-01
7.2500E 00	8.55003E-01	1.9750E 01	1.00825E 00	3.2250E 01	9.61845E-01
7.5000E 00	8.34374E-01	2.0000E 01	1.00493E 00	3.2500E 01	9.64019E-01
7.7500E 00	8.20527E-01	2.0250E 01	1.00013E 00	3.2750E 01	9.66166E-01
8.0000E 00	8.13685E-01	2.0500E 01	9.94127E-01	3.3000E 01	9.68195E-01
8.2500E 00	8.13761E-01	2.0750E 01	9.87244E-01	3.3250E 01	9.70027E-01
8.5000E 00	8.20376E-01	2.1000E 01	9.79825E-01	3.3500E 01	9.71591E-01
8.7500E 00	8.32892E-01	2.1250E 01	9.72218E-01	3.3750E 01	9.72837E-01
9.0000E 00	8.50452E-01	2.1500E 01	9.64759E-01	3.4000E 01	9.73727E-01
9.2500E 00	8.72031E-01	2.1750E 01	9.57760E-01	3.4250E 01	9.74242E-01
9.5000E 00	8.96487E-01	2.2000E 01	9.51499E-01	3.4500E 01	9.74381E-01
9.7500E 00	9.22613E-01	2.2250E 01	9.46202E-01	3.4750E 01	9.74158E-01
1.0000E 01	9.49200E-01	2.2500E 01	9.42047E-01	3.5000E 01	9.73601E-01
				3.7500E 01	9.61738E-01
				3.7750E 01	9.61138E-01
				3.8000E 01	9.60785E-01
				3.8250E 01	9.60680E-01
				3.8500E 01	9.60814E-01
				3.8750E 01	9.61169E-01
				3.9000E 01	9.61718E-01
				3.9250E 01	9.62426E-01
				3.9500E 01	9.63255E-01
				3.9750E 01	9.64162E-01
				4.0000E 01	9.65105E-01
				4.0250E 01	9.66040E-01
				4.0500E 01	9.66929E-01
				4.0750E 01	9.67735E-01
				4.1000E 01	9.68428E-01
				4.1250E 01	9.68985E-01
				4.1500E 01	9.69389E-01
				4.1750E 01	9.69631E-01
				4.2000E 01	9.69709E-01
				4.2250E 01	9.69629E-01
				4.2500E 01	9.69403E-01
				4.2750E 01	9.69048E-01
				4.3000E 01	9.68586E-01
				4.3250E 01	9.68044E-01
				4.3500E 01	9.67448E-01
				4.3750E 01	9.66827E-01
				4.4000E 01	9.66210E-01
				4.4250E 01	9.65622E-01
				4.4500E 01	9.65088E-01
				4.4750E 01	9.64626E-01
				4.5000E 01	9.64254E-01
				4.5250E 01	9.63982E-01
				4.5500E 01	9.63817E-01
				4.5750E 01	9.63759E-01
				4.6000E 01	9.63807E-01
				4.6250E 01	9.63951E-01
				4.6500E 01	9.64180E-01
				4.6750E 01	9.64481E-01
				4.7000E 01	9.64836E-01
				4.7250E 01	9.65227E-01
				4.7500E 01	9.65635E-01

Figure 7.8 Transient Response for Example 1.

TRANSIENT RESPONSE (CONTINUED)

T	F(T)	T	F(T)	T	F(T)	T	F(T)
1.0250E 01	9.75080E-01	2.2750E 01	9.39150E-01	3.5250E 01	9.72754E-01	4.7750E 01	9.66043E-01
1.0500E 01	9.99181E-01	2.3000E 01	9.37567E-01	3.5500E 01	9.71667E-01	4.8000E 01	9.66431E-01
1.0750E 01	1.02056E 00	2.3250E 01	9.37296E-01	3.5750E 01	9.70401E-01	4.8250E 01	9.66786E-01
1.1000E 01	1.03845E 00	2.3500E 01	9.38278E-01	3.6000E 01	9.69019E-01	4.8500E 01	9.67093E-01
1.1250E 01	1.05227E 00	2.3750E 01	9.40403E-01	3.6250E 01	9.67588E-01	4.8750E 01	9.67342E-01
1.1500E 01	1.06165E 00	2.4000E 01	9.43517E-01	3.6500E 01	9.66170E-01	4.9000E 01	9.67525E-01
1.1750E 01	1.06642E 00	2.4250E 01	9.47435E-01	3.6750E 01	9.64827E-01	4.9250E 01	9.67637E-01
1.2000E 01	1.06662E 00	2.4500E 01	9.51944E-01	3.7000E 01	9.63612E-01	4.9500E 01	9.67679E-01
1.2250E 01	1.06249E 00	2.4750E 01	9.56819E-01	3.7250E 01	9.62571E-01	4.9750E 01	9.67652E-01
5.0000E 01	9.67560E-01	5.7750E 01	9.66690E-01	6.5500E 01	9.66337E-01	7.3250E 01	9.66199E-01
5.0250E 01	9.67412E-01	5.8000E 01	9.66607E-01	6.5750E 01	9.66295E-01	7.3500E 01	9.66179E-01
5.0500E 01	9.67216E-01	5.8250E 01	9.66508E-01	6.6000E 01	9.66247E-01	7.3750E 01	9.66157E-01
5.0750E 01	9.66984E-01	5.8500E 01	9.66398E-01	6.6250E 01	9.66197E-01	7.4000E 01	9.66135E-01
5.1000E 01	9.66727E-01	5.8750E 01	9.66281E-01	6.6500E 01	9.66146E-01	7.4250E 01	9.66114E-01
5.1250E 01	9.66458E-01	5.9000E 01	9.66164E-01	6.6750E 01	9.66097E-01	7.4500E 01	9.66094E-01
5.1500E 01	9.66190E-01	5.9250E 01	9.66052E-01	6.7000E 01	9.66052E-01	7.4750E 01	9.66076E-01
5.1750E 01	9.65932E-01	5.9500E 01	9.65949E-01	6.7250E 01	9.66012E-01	7.5000E 01	9.66061E-01
5.2000E 01	9.65697E-01	5.9750E 01	9.65858E-01	6.7500E 01	9.65979E-01	7.5250E 01	9.66050E-01
5.2250E 01	9.65493E-01	6.0000E 01	9.65784E-01	6.7750E 01	9.65954E-01	7.5500E 01	9.66043E-01
5.2500E 01	9.65327E-01	6.0250E 01	9.65729E-01	6.8000E 01	9.65937E-01	7.5750E 01	9.66039E-01
5.2750E 01	9.65204E-01	6.0500E 01	9.65693E-01	6.8250E 01	9.65930E-01	7.6000E 01	9.66039E-01
5.3000E 01	9.65127E-01	6.0750E 01	9.65678E-01	6.8500E 01	9.65931E-01	7.6250E 01	9.66043E-01
5.3250E 01	9.65097E-01	6.1000E 01	9.65682E-01	6.8750E 01	9.65940E-01	7.6500E 01	9.66049E-01
5.3500E 01	9.65113E-01	6.1250E 01	9.65706E-01	6.9000E 01	9.65957E-01	7.6750E 01	9.66059E-01
5.3750E 01	9.65171E-01	6.1500E 01	9.65746E-01	6.9250E 01	9.65979E-01	7.7000E 01	9.66071E-01
5.4000E 01	9.65267E-01	6.1750E 01	9.65800E-01	6.9500E 01	9.66008E-01	7.7250E 01	9.66084E-01
5.4250E 01	9.55394E-01	6.2000E 01	9.65864E-01	6.9750E 01	9.66038E-01	7.7500E 01	9.66099E-01
5.4500E 01	9.65546E-01	6.2250E 01	9.65937E-01	7.0000E 01	9.66071E-01	7.7750E 01	9.66113E-01
5.4750E 01	9.65714E-01	6.2500E 01	9.66013E-01	7.0250E 01	9.66105E-01	7.8000E 01	9.66127E-01
5.5000E 01	9.65891E-01	6.2750E 01	9.66091E-01	7.0500E 01	9.66137E-01	7.8250E 01	9.66141E-01
5.5250E 01	9.66069E-01	6.3000E 01	9.66165E-01	7.0750E 01	9.66168E-01	7.8500E 01	9.66152E-01
5.5500E 01	9.66239E-01	6.3250E 01	9.66233E-01	7.1000E 01	9.66194E-01	7.8750E 01	9.66162E-01
5.5750E 01	9.66394E-01	6.3500E 01	9.66293E-01	7.1250E 01	9.66216E-01	7.9000E 01	9.66170E-01
5.6000E 01	9.66530E-01	6.3750E 01	9.66343E-01	7.1500E 01	9.66233E-01	7.9250E 01	9.66175E-01
5.6250E 01	9.66641E-01	6.4000E 01	9.66380E-01	7.1750E 01	9.66244E-01	7.9500E 01	9.66177E-01
5.6500E 01	9.65724E-01	6.4250E 01	9.66404E-01	7.2000E 01	9.66249E-01	7.9750E 01	9.66177E-01
5.6750E 01	9.66776E-01	6.4500E 01	9.66415E-01	7.2250E 01	9.66249E-01	8.0000E 01	9.66175E-01
5.7000E 01	9.66798E-01	6.4750E 01	9.66413E-01	7.2500E 01	9.66243E-01		
5.7250E 01	9.66789E-01	6.5000E 01	9.66398E-01	7.2750E 01	9.66232E-01		
5.7500E 01	9.66752E-01	6.5250E 01	9.66372E-01	7.3000E 01	9.66218E-01		

NOTE CHECK POLY PULES AND ZEROS

Figure 7.8—*Cont.*

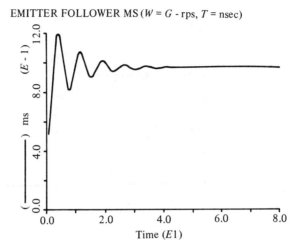

Figure 7.9 Transient Response to Unit-step for Example 1.

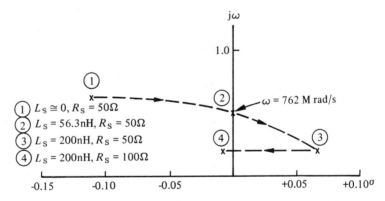

Figure 7.10 Effects of Parameter Variations on Circuit Stability for Example 1.

given on the DATA, SENSITIVITY card. The EXIT card ends the problem.

Example 2: ac Voltage Gain Problem

The sample circuit and its small-signal equivalent circuit are shown in Figs. 7.11 and 7.12. Since LISA does not directly handle an independent voltage source, we could add a small source resistance to VIN and use the LISA voltage source/resistance branch. However, since the transfer function of interest is GAIN = V(N2)/V(NG), we can also use

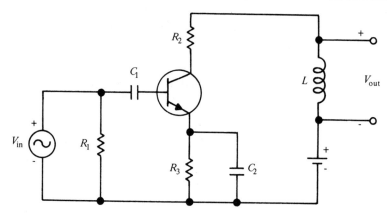

Figure 7.11 Circuit for Example 2.

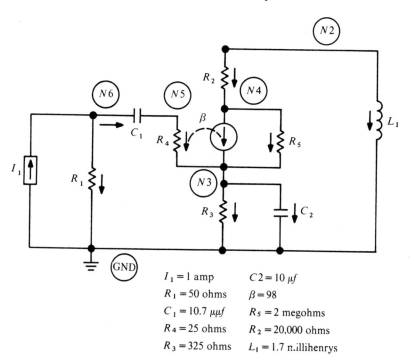

$I_1 = 1$ amp $C_2 = 10\ \mu f$
$R_1 = 50$ ohms $\beta = 98$
$C_1 = 10.7\ \mu\mu f$ $R_5 = 2$ megohms
$R_4 = 25$ ohms $R_2 = 20{,}000$ ohms
$R_3 = 325$ ohms $L_1 = 1.7$ millihenrys

Figure 7.12 Small Signal Equivalent Circuit for Example 2.

an arbitrary independent current source as a driver. For this example, symbolic node labels are used.

The LISA input cards for the problem are shown in Fig. 7.13. Besides GAIN, the input and output nodes for the grounded two-port are defined, permitting the computation of the two-port parameter

$$G21 = v_2/v_1,\ (i_2=0) = V(N2)/V(N6) = \text{GAIN}$$

```
THE FOLLOWING DATA CARD IMAGES WERE READ BY THE  L I S A  SYSTEM INPUT PHASE (LISAX-V2)

TITLE,SAMPLE CIRCUIT                                   00000010
READ,CIRCUIT                                           00000020
I1,GND,N6,1                                            00000030
R1,N6,GND,50                                           00000050
C1,N6,N5,1C.7E-12                                      00000060
R4,N5,N3,25                                            00000070
R3,N3,GND,325                                          00000080
C2,N3,GND,10E-6                                        00000090
DI1,N4,N3,N5,N3,98,R4                                  00000100
R5,N4,N3,2E6                                           00000110
R2,N2,N4,20E3                                          00000120
L1,N2,GND,1.7E-3                                       00000130
DEFINE,GAIN=V(N2)/V(N6)                                00000140
DEFINE,PORTS=N6,N2                                     00000160
DATA,FREQ=5,9,1E3,HZ                                   00000180
DATA,SENSITIVITY=.1,ALL                                00000200
COMPUTE,TRFN,GAIN,G21                                  00000230
COMPUTE,BODE,GAIN                                      00000240
COMPUTE,ACCA,GAIN                                      00000250
NOTE,SCALE TO GHZ,NS,KOHM,PF,AND UH                    00000264
CHANGE,CIRCUIT,SCALE                                   00000265
R*1E-3,C*1E12,L*1E6,I*1E3                              00000266
COMPUTE,POLY,GAIN,G21                                  00000270
COMPUTE,SENSITIVITY,GAIN                               00000272
COMPUTE,BODE,GAIN                                      00000275
DATA,FREQ=5,9,1E-6,HZ                                  00000276
COMPUTE,ACCA,GAIN                                      00000277
EXIT                                                   00000280
```

Figure 7.13 LISA Input Cards for Example 2.

Figure 7.14 shows the transfer function GAIN results from TRFN. The TRFN results are checked using BODE and ACCA.

Next, the circuit is scaled to avoid coefficient overflow in the COMPUTE, POLY command. The GAIN results from POLY appear in

```
POLES AND ZEROS FOR GAIN

    NUMERATOR ROOTS, LEADING COEFF. = -2.09727E -20 DEGREE =  3

              REAL           IMAG
              0.0            0.0
              0.0            0.0
              0.0            0.0

    DENOMINATOR ROOTS, LEADING COEFF. =  2.67508E -27 DEGREE =  3

              REAL           IMAG
              0.0            0.0
             -1.188423E 09   0.0
             -3.738143E 09   0.0

TOTAL MULTIPLIER = -7.84003E 06
```

Figure 7.14   GAIN from TRFN (Unscaled).

```
POLES AND ZEROS FOR GAIN

    NUMERATOR COEFFICIENTS, DEGREE =  3.   DESCENDING ORDER

      -2.0971990E 13     -6.4529160E 06      0.0                 0.0

              REAL           IMAG
              0.0            0.0
              0.0            0.0
             -3.076920E-07   0.0

DENOMINATOR COEFFICIENTS, DEGREE =  3.   DESCENDING ORDER

       2.6749990E 06      1.3178545E 07      1.1883611E 07       3.6573248E 00

              REAL           IMAG
             -3.077622E-07   0.0
             -1.188416E 00   0.0
             -3.738143E 00   0.0

TOTAL MULTIPLIER = -7.84000E 06
```

Figure 7.15   GAIN from POLY (Unscaled).

Fig. 7.15. Note that TRFN could not distinguish the very small pole and zero near the origin from 0.0 but that POLY did. The pole/zero sensitivity of GAIN to all circuit components is computed (using the POLY method as the context is POLY). In this case, the sensitivity results allow the following first-order interpretation to be made:

$$\text{GAIN} = \frac{Ks^2(s+z_1)}{(s+p_1)(s+p_2)(s+p_3)}$$

where

$$z_1 = 1/R_3C_2 = p_1, \qquad p_2 = R_5/L_1, \qquad p_3 = 1/R_4C_1$$

The POLY results are checked using BODE and ACCA with scaled frequency data.

7.7 LIMITATIONS

Rules

All names used must begin with a character A–Z. The numbers 0–9 are allowed to complete a name. Blanks are ignored and removed. Only the first 4 characters are retained. In general, the comma (,) is used as a delimiter or separator, with the equals (=) also allowed for readability. It is best to avoid dangling commas and blank fields between delimiters. Generally, only columns 1–72 on a card are used.

Number fields can be given with or without a decimal point, and with or without a sign. The E convention of FORTRAN is used for powers of 10.

The function names DRIVER, PORT, PORTS, OPEN, OPENLO, CLOSED, FORWARD, and Z11, Z12, Z21, Z22, Y_{ij}, H_{ij}, G_{ij}, A_{ij}, B_{ij} for $i,j = 1, 2$ have special meaning to LISA and should not be used except when intended.

Restrictions

When using the following LISA commands or features, the present version has these restrictions:

1. *READ, CIRCUIT*. A maximum of 125 components, a maximum total of 600 nonzero matrix coefficients, and a maximum of 50 independent nodes are allowed. Inductors should be connected to ground if possible to avoid extra roots at $s = 0$.

2. *READ, DRIVER*. A maximum of 50 basic driver functions is allowed for any one driver description.

3. *READ, function name.* A maximum of 25 coefficients (degree $= 24$) or a maximum of 25 distinct roots (multiple roots are counted; conjugate roots are not counted) for any one numerator or denominator.

4. *DEFINE.* The last four function definitions are saved. The fifth definition causes the first definition to be lost, etc. Only the last PORT definition is saved.

5. *COMPUTE, PRINT, PLOT, PPLOT.* A maximum of four function names may be specified on any card. Where appropriate, DRIVER may be given in addition.

6. *TFEX table.* The last 12 computed and read-in functions are saved in the TFEX table. The thirteenth function causes the first function to be lost, etc. A pushdown method is used, with the leftmost function on a COMPUTE card being treated as the first in. If two functions in the TFEX table have the same name, the last one computed is used.

7. *DATA, LOCUS.* A maximum of 50 component values is allowed.

8. *DATA, SENSITIVITY.* A maximum of 50 component names can be specified. If ALL is given, this restriction does not apply.

9. *COMPUTE, POLY.* Due to the nature of the present POLY algorithm, the $Y(1, 1)$ term should not be zero. Further, if a diagonal term goes to zero during processing, the solution will be in error. This can occur in certain highly idealized problems, and if it does, an error message is printed. The TRFN method does not contain this restriction.

10. *COMPUTE, TRFN.* If the spread of the roots is very large (that is, the smallest and largest roots or largest and smallest time constants are far apart), the program may encounter numerical difficulties. This can occur in root finding or in root processing. It is sometimes necessary to eliminate those components which contribute to the numerical problems but have little or no consequence on the results in the range of interest. The frequency response computed using ACCA, however, will generally be accurate in these cases because its calculation does not need a root solution.

11. *COMPUTE, TRANSIENT.* As the single-sided or unilateral Laplace transform is used, time begins at $t = 0$ (or $t = 0+$). This means that the time response to a periodic or aperiodic driver that is applied at $t = 0$ will result in a transient or decay period until steady state is reached. However, if only the steady state is desired, the starting time on the DATA, TIME card can be large, and just a snapshot of the steady state will be taken.

The normal program methods imply transfer functions that result from a single driver with no circuit initial conditions. However, initial condition problems and multiple driver problems are readily handled by LISA when set up properly.

7.8 ERROR DIAGNOSTICS

Input Phase

The program will print out the following message:

ERROR MESSAGE i (INPUT PHASE)

where i is the error number. In some cases, an additional message will further explain the error. The program will continue processing and will usually recover. However, computation involved with the error may be skipped, may give more error messages, or may produce bad results.

An explanation of the error numbers, along with the subroutines, is shown below:

i	Explanation	Subroutine
1	Option for COMPUTE command not recognized	(INPUT)
2	Error condition in data transfer during read	(PONOV)
3	Command not recognized	(INPUT)
4	End of input data set—EXIT card simulated	(PONOV)
5	Option COEFF, COEFFI, ROOT, or ROOTS not found for read-in function	(RTFREE)
6	Read-in function name not found	(WRTFRD)
7	No read-in function found for output	(PRCNTL)
8	Function definition type not recognized—only V and X are allowed in this version	(TRANVX)
9	Function definition V(0), V(GND), or V(GROUND) has no meaning to LISA	(DTFSTO)
10	Defined function name not found	(DTFSET)
11	Resistor, capacitor, or inductor name given in function definition not found in circuit component list	(DTFSET)
12	No acceptable functions recognized for COMPUTE	(WEXMAT)
13	Defined function name field not symbolic	(RDEFTF)
14	Defined function free-format error	(RDEFTF)
15	Too many options on command card	(INPUT)
16	More than 125 circuit components	(RCRDLF)
17	Circuit component count check failed	(RCRDLF)
18	Resistor related to V, T, or D not found in circuit component list	(RESDEP)

i	Explanation	Subroutine
19	Circuit component type not recognized—ignored	(LDAUP)
20	Free-format error in a number field	(FLTNUM)
21	Circuit component type not recognized in READ, CIRCUIT	(RCRDLF)
22	A READ, MATRIX term has too high a degree for the number of coefficients given	(RMFREE)
23	Matrix has N > 50	(INPUT)
24	CHANGE command option not recognized	(INPUT)
25	Only ADD, REMOVE, or SHORT recognized as options for CHANGE, CIRCUIT, TOPOLOGY command	(CHCRKT)
26	Component name for change not in circuit list	(CHCRKT)
27	Circuit component count check failed during CHANGE	(RCPACK)
28	Circuit mutual inductance factor $1 - k^2 = 0$	(RCMUT)
29	Warning—mutual inductance factor $1 - k^2$ is negative	(RCMUT)
30	Circuit node count failed during CHANGE	(REDUCE)
31	Missing node number during CHANGE	(REDUCE)
32	Node reduction check failed during CHANGE	(REDUCE)
33	Node reduction check failed during CHANGE	(REDUCE)
34	Symbolic node name given in command CHANGE, CIRCUIT, TOPOLOGY, SHORT not found in symbolic node name list	(SHORT)
35	Symbolic node name in function definition not found in symbolic node list	(DTFSET)
36	Circuit mutual inductance count check failed	(RCRDLF)
37	Circuit component type not recognized during CHANGE	(RCNCNT)
38	Circuit component count check failed during CHANGE	(RCNCNT)
39	Circuit component count for R not correct	(WRCRKT)
40	Circuit component count for C not correct	(WRCRKT)
41	Circuit component count for L not correct	(WRCRKT)
42	Circuit component count for I or V not correct	(WRCRKT)
43	Circuit component count for T not correct	(WRCRKT)
44	Circuit component count for D not correct	(WRCRKT)
45	Circuit component count for M not correct	(WRCRKT)
46	One or both of the inductors related to the circuit mutual inductance not found	(RCMUT)
47	Function definition type not X or V	(SETFAC)
48	Resistor name for E- or V-type source not given	(RCRDLF)
49	Both nodes of a component are 0—component ignored	(LDAUP)
50	Option for SET or RESET command not recognized	(RSET)
51	Symbolic node name for two-port definition not found in symbolic node name list	(DTFSET)
52	Function setup check failed	(WEXMAT)
53	More than 4 excess columns in matrix setup	(WEXMAT)
54	Option for DATA command not recognized	(RDATA)
55	Transient-driver type name not recognized	(RDRIVE)
56	More than 50 transient driver types—the rest are ignored	(RDRIVE)
57	COMPUTE option not recognized in ACCA context	(PREP)
58	Statement or function out of COMPUTE context—could be due to previous error	(INPUT, PREP)

Ch. 7 / Johnson

i	Explanation	Subroutine
59	—Not used by LISA—	
60	PRINT or TYPE command out of context	(PRCNTL)
61	Output command ignored due to previous error	(PRCNTL)
62	Function name not recognized	(WXVAR)
63	Matrix setup in error	(WXVAR)
64	A two-port parameter name was not recognized	(SETFAC)
65	Data for COMPUTE, LOCUS not given	(INPUT)
66	Rows > 50 or columns > 54 in CHANGE, MATRIX	(CHMAT)
67	Check for null function failed—function definition probably in error	(TRANVX)
68	Free-format error on circuit component card	(ARRANG)
69	Free-format error on function definition card	(FUNDEF)
70	More than 600 coefficients in matrix derived from circuit	(LDAUP)
71	Matrix derived from circuit may be wrong due to insufficient initialization in program	(LDAUP)
72	Matrix derived from circuit in error	(COMPA)
73	Matrix derived from circuit in error	(COMPA)
74	Function name on PLOT or SCOPE card not recognized—default option of blank used	(XPLOT)
75	Mode or context error	(XPLOT)
76	PLOT or SCOPE command out of context or mode	(XPLOT)
77	PLOT or SCOPE command ignored due to previous error	(XPLOT)
78	No acceptable functions recognized for COMPUTE	(PREP)
79	Degree > 24 or number of distinct roots > 25 for read-in function	(RTFREE)
80	Format error in data for CHANGE, CIRCUIT, SCALE	(CHCRKT)
81	Missing node number—see message	(NODCHK)
82	No datum or ground node given	(NODCHK)
83	COMPUTE command LOCUS, SENSITIVITY, or VARIATION out of context	(INPUT)
84	Warning—component value just read = 0.0	(RCRDLF)
85	Warning—changed component value = 0.0	(CHCRKT)
86	Warning—changed scale value = 0.0	(CHCRKT)
87	Number of circuit components = 0	(NODCHK)

Compute Phase

The following are errors that occurred during computation in the execute phase:

No.	Explanation	Mode	Action
1	Matrices exceed available storage	ACCA, POLY, TRFN	Job discontinued
2	Insufficient storage for scratch	POLY, TRFN	Job discontinued
3	Insufficient storage to find roots	POLY	Ignore root calculation, continue job

No.	Explanation	Mode	Action
4	Double cofactor for system of 2 nodes or illegal solution component requested	POLY, TRFN	Ignore and continue
5	Insufficient storage for frequency points	BODE, ACCA	Job discontinued
6	Insufficient storage for output	ACCA	Job discontinued
7	Function not in TFEX table or was not computed due to error	ALL	Ignore and continue
8	Insufficient storage for driver	TRANSIENT	Job discontinued
9	More than 2000 time points requested	TRANSIENT	Truncate to 2000 and continue
10	Insufficient storage for scaled roots	BODE, TRANSIENT	Job discontinued
11	Gain factor for function out of range	BODE, TRANSIENT	Ignore function
12	Error in calculation of solution component	TRFN	Continue job
13	Insufficient storage for output	BODE, TRANSIENT	Ignore output, request and continue job
14	Insufficient storage for time points	TRANSIENT	Job disconitnued
15	Matrix is singular	ACCA, POLY, TRFN	Continue job
16	Error in root finder	BEV4	Continue job
17	Degree of numerator greater than or equal to degree of denominator results in impulsive-type operators; residues of poles are correct; values of impulse operators not found	RESIDUE, TRANSIENT	Continue job

7.9 REFERENCES

1. MULLER, D. E., "A Method for Solving Algebraic Equations Using an Automated Computer," MTAC, 1956, pp. 208–215.

2. SYN, W. M., "BEV4—A Transfer Function Linear Analysis Program," *IBM Technical Report TR 02.339*, San Jose, Calif., Dec. 1964.

3. BECK, J. W., "A Semiautomatic Network Analysis Program," *IBM Technical Report TR 02.315*, San Jose, Calif., July 15, 1964.

4. DECKERT, K. L., and E. T. JOHNSON, "LISA—A Program for Linear Systems Analysis," paper presented at WESCON, August 23, 1966.

5. JOHNSON, E. T., "Here's A Powerful Design Program," *Electronic Design* 23, Oct. 11, 1966, pp. 56–61.

6. DECKERT, K. L., and E. T. JOHNSON, "Users Guide for LISA" (7090/94 version), *IBM Technical Report TR 02.409*, San Jose, Calif., Aug. 2, 1967.

7. Deckert, K. L., and E. T. Johnson, "User's Guide for LISA/360, A Program for Linear System Analysis," *IBM Technical Report TR 02.432*, San Jose, Calif., July 31, 1968. Available as Type III program PID No. 360D–16.4.009 from IBM Corporation, Program Information Dept. (PID), 40 Saw Mill River Road, Hawthorne, New York 10532.

8. Johnson, E. T., "GSPAN: A Program for Graphic S-Plane Analysis," *IBM Technical Report TR 02.433*, San Jose, Calif., Aug. 1, 1968.

9. Johnson, E. T., "LISAX—Extended Features for LISA/360," *IBM Technical Report TR 02.504*, San Jose, Calif., May 14, 1971.

8

MARTHA

PAUL PENFIELD, Jr.
Massachusetts Institute of Technology

8.1 INTRODUCTION

MARTHA is a notation for denoting electrical networks, and it is also a computer program which uses this notation.

General Description

In MARTHA—the notation—every network is either an element (resistor, capacitor, transmission line, etc.) or one or two previously defined networks wired together. The basic idea is illustrated in Fig. 8.1. In this example, S and P are *wiring functions*, and R, L, and C are *element functions*. The functions used in Fig. 8.1 are sufficient to describe any series-parallel network containing linear resistors, capacitors, and inductors. For more complicated networks, other, similar, functions in MARTHA are used.

As a notation, MARTHA is an alternative to the widely used schematic diagram. It can be used for communication from one person to another (for example, for documentation purposes), or from a person to a computer, or from a computer to a person. It is useful both in analysis programs (where the user writes it and the computer reads it) and in synthesis programs (where the computer writes it and either the computer or the user reads it).

MARTHA—the computer program—is a network-analysis program embedded in the interactive language APL, and the syntax resembles that of APL in many ways. The simplicity and versatility of the input language make MARTHA relatively easy to use. It differs from most of the other programs in this book in that it is naturally oriented toward ports rather than toward nodes. In MARTHA node matrices are never calculated or used, so that time-consuming matrix inversion is avoided altogether. There are no difficulties associated with capacitor or inductor loops or tie sets, and it is never

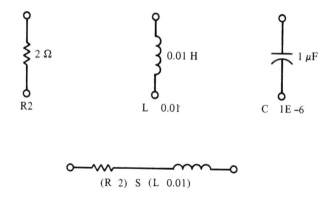

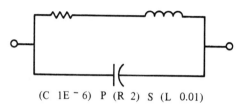

Figure 8.1 Illustration of MARTHA Notation for a Model of a Parallel Tuned Circuit. The functions R, L, and C define elements, and the functions S and P perform the wiring. The functions R, L, and C are *monadic* (having one argument, located on their right), and S and P are *dyadic* (having two arguments, one on each side). These five functions are sufficient to denote any linear series-parallel RLC network.

necessary to find the eigenvalues of any large matrix. Analysis is carried out at all frequencies simultaneously, using APL's fast array handling.

Capabilities

The program MARTHA performs frequency-domain analysis of one-port or two-port networks which are made up of one-port or two-port linear elements wired together so that at every stage in the construction only one-ports and two-ports are used. Examples of networks of this type are most amplifiers, filters, and microwave systems. Networks may be active or passive, may be reciprocal or nonreciprocal, and may be lumped or distributed (or a combination of both).

Engineers may use MARTHA for circuit analysis without knowledge of APL. The simple and uniform notation in MARTHA is advantageous to beginners, as is the interactive nature of MARTHA. However, users who know APL can write their own functions to control MARTHA. For example, network definitions, including both parameter values and topology, can be varied under program control. Advanced users can write synthesis algorithms using MARTHA notation for the resulting network. In other words, MARTHA can be used as a programming language as well as simply a program.

Although MARTHA is intended for general-purpose network analysis, different users can make use of rather specialized portions of MARTHA to make up their own special-purpose analysis and synthesis systems.

Microwave engineers can use the distributed elements in MARTHA, and wave variables (with complex, frequency-dependent normalization if desired), and Smith-chart plotting. Aspects of interest to amplifier designers include various transistor models, calculation of several measures of gain and stability, and MARTHA's ability to plot Nichols charts and the U/A *gain plane*. Filter designers can make use of MARTHA's ability to scale frequency and impedance and to perform high-pass and band-pass transformations. For active filters, there are four operational-amplifier models of differing complexity. Of interest to experimentalists is MARTHA's ability to work with tables of measured performance and use them for calculation and interpretation as elements. Engineers concerned with model making will find an extensive repertoire of basic elements including 16 controlled sources, negative-impedance convertors, gyrators, and elements whose response goes with complex frequency to an integral power.

The user can define many different networks and wire them together or analyze them when desired. Two or more networks can be analyzed at the same time, and the results compared. There are more than 100 different response functions that can be requested. The results of an analysis can be printed, or plotted versus frequency, or versus any network parameter (on linear or log scale), or versus another response. The response of one network can be plotted against the response of another network or against a numerical table of values. Alternatively, the results can be stored for later display or calculation, possibly using other results.

MARTHA incorporates an extensive set of tools for defining, editing, manipulating, and interpreting tables of numerical values. Such a numerical function of frequency (FOF) can be interpreted as a quantity to be printed or plotted (alongside the results of a normal MARTHA analysis) or as the impedance, admittance, or scattering coefficient of a numerically defined element.

Documentation

MARTHA is described in the book *MARTHA User's Manual*, by Paul Penfield, Jr., The M.I.T. Press, Cambridge, Mass., 1971. This book completely covers versions of MARTHA dated 71. For versions of MARTHA dated

73 (and this includes all versions on commercial time-sharing computers) several improvements are described in the pamphlet "*MARTHA* User's Manual, 1973 Addendum," by Paul Penfield, Jr., M.I.T. Department of Electrical Engineering, Cambridge, Mass., 1973. Users of MARTHA can determine their version date by referring to the line which starts *CIRCUIT ANALYSIS BY MARTHA* at the top of each print or plot: either 71∘ or 73∘ appears in that line.

Several other publications describing MARTHA or the ideas behind it appear in the References at the end of this chapter. See References 1–4.

Besides these publications, there is extensive on-line documentation in MARTHA; for information on how to access this, type

```
)LOAD 100 HOWMARTHA
DESCRIBE
```

Availability

MARTHA is available both from commercial time-sharing computer companies and for use on separate machines that run APL. Inquiries should be directed to the Manager of Software Services, The M.I.T. Press, 28 Carleton Street, Cambridge, Mass. 02142, or to Steinbrecher Corporation, 73 Pine Street, Woburn, Mass. 01801, phone (617) 935–8460.

8.2 PROGRAM STRUCTURE

In APL, programs are stored in *work spaces*. MARTHA consists of nine work spaces of which one is purely documentation and seven constitute the *MARTHA library*. The basic work space,* 100 MARTHA, contains about 70 cooperating *foreground* APL functions that the user may call directly, several *background* utility functions called by the foreground functions, and a few global variables. The foreground functions fall into six categories. First are functions which create elements, such as the functions R, L, and C. Second are functions which wire networks together, for example, S and P. Third are functions that calculate the response of the network, for example, impedance, admittance, reflection coefficient, or VSWR of one-port networks, or any of the two-port parameters or various gain or stability measures of two-port networks. Fourth are functions which can modify the response functions by taking the real part, imaginary part, magnitude, etc. Fifth are functions that help specify the format of the output. Finally, there are some miscellaneous functions to aid in defining networks.

*On some computers, the number in the work-space names is different from 100.

The MARTHA library contains additional, more specialized functions in these same categories. It is a standard feature of the APL work-space–storage system that individual functions, or groups of functions, can be copied into the user's active work space; by so doing the user can select those functions from the library that he needs and leave the rest behind, thereby creating his own personalized version of MARTHA.

8.3 NETWORK ELEMENTS

Table 8.1 and Fig. 8.2 show the network elements defined in MARTHA. The two-port networks WR and WTHRU are constants, and all the others are monadic functions. The simple functions R, L, and C (resistors, inductors, and capacitors) have been illustrated earlier. The function L, when used with an argument of length 3 (that is, a vector with three numbers in it), produces a mutual inductor; thus, the function L produces different elements according to the length of its argument. The same is true of the functions TEM, WG, and OPAMP. The wave-guide function WG produces a length of wave guide if its argument has three numbers in it, but if only two are present (the length is absent), the result is the frequency-dependent characteristic impedance of the wave guide. The wave-guide analysis is valid both above the cutoff frequency, where the characteristic impedance is real, and below the cutoff frequency, where it is imaginary.

Examples.

R 40	(resistor, 40 ohms)
L .015	(15-millihenry inductor)
L .015 .02 .01	(two-port mutual inductor)
IT 3	(ideal transformer, turns ratio 3 : 1)
WG 1E9 377	(matched load of wave guide with cutoff frequency 1 GHz, 377-ohm characteristic impedance at $f = \infty$)
WG 1E9 377 .6	(60-centimeter length of wave guide)
WG 1E9 377 90 DEGREESAT 2E9	(quarter-wave section of guide)
WG 1E9 377 .4 FORDIEL 2.5	(dielectric-filled guide)

Note the two auxiliary functions DEGREESAT and FORDIEL. The first is used to specify an electrical length in degrees at a reference frequency (in this case 2 GHz) and the second is used for dielectrically loaded transmission lines or wave guides.

Table 8.1 Elements Defined in MARTHA*

ELEMENT	TYPE	NAME	ARGUMENT VECTOR	REQUIRED	EQUATIONS
RESISTOR	1-PORT	R	RESISTANCE RES IN OHMS		$V=RES \times I$
CAPACITOR	1-PORT	C	CAPACITANCE CAP IN FARADS		$I=S \times CAP \times V$
INDUCTOR	1-PORT	L	INDUCTANCE IND IN HENRIES	$IND \neq 0$	$V=S \times IND \times I$
STRAIGHT-THROUGH CONNECTION	2-PORT	WTHRU	(NONE)		$V1=V2; I1=-I2$
POLARITY REVERSE	2-PORT	WR	(NONE)		$V1=-V2; I1=I2$
MUTUAL INDUCTOR	2-PORT	L	INPUT SELF-INDUCTANCE L1 IN HENRIES OUTPUT SELF-INDUCTANCE L2 IN HENRIES MUTUAL INDUCTANCE M IN HENRIES	$M \neq 0$	$V1=S \times (L1 \times I1)+M \times I2$ $V2=S \times (M \times I1)+L2 \times I2$
IDEAL TRANSFORMER	2-PORT	IT	TURNS RATIO N	$N \neq 0$	$V1=N \times V2; I1=-I2 \div N$
OPERATIONAL AMPLIFIER	2-PORT	OPAMP	OPEN-CIRCUIT VOLTAGE GAIN A OUTPUT IMPEDANCE ROUT IN OHMS INPUT IMPEDANCE RIN IN OHMS	$A \neq 0$ $RIN \neq 0$	$V1=RIN \times I1$ $V2=(A \times V1)+ROUT \times I2$
OPERATIONAL AMPLIFIER	2-PORT	OPAMP	OPEN-CIRCUIT VOLTAGE GAIN A OUTPUT IMPEDANCE ROUT IN OHMS	$A \neq 0$	$I1=0$ $V2=(A \times V1)+ROUT \times I2$
OPER. AMPLIFIER	2-PORT	OPAMP	VOLTAGE GAIN A	$A \neq 0$	$I1=0; V2=A \times V1$
FIELD-EFFECT TRANSISTOR MODEL, GROUNDED-SOURCE	2-PORT	FET	GATE-SOURCE CAPACITANCE CGS IN FARADS GATE-DRAIN CAPACITANCE CGD IN FARADS TRANSCONDUCTANCE GM IN MHOS	$GM \neq 0$	$I1=S \times (CGS \times V1)+CGD \times V1-V2$ $I2=(GM \times V1)+S \times CGD \times V2-V1$
BIPOLAR-TRANSISTOR MODEL, GROUNDED-EMITTER	2-PORT	HYBRIDPI	RESISTANCE RX IN OHMS RESISTANCE RPI IN OHMS CAPACITANCE CPI IN FARADS CAPACITANCE CMU IN FARADS TRANSCONDUCTANCE GM IN MHOS	$RPI \neq 0$ $GM \neq 0$	$V1=VPI+RX \times I1$ $I1=(VPI \times (S \times CPI)+RPI)+$ $S \times CMU \times VPI-V2$ $I2=(GM \times VPI)+S \times CMU \times V2-VPI$
LOSSLESS TRANSMISSION LINE	2-PORT	TEM	CHARACTERISTIC IMPEDANCE Z0 IN OHMS LENGTH LEN IN METERS	$Z0 \neq 0$	$A=*-J \times O2 \times LEN \times F \div 3E8 \div DIEL*.5$ $(V2-Z0 \times I2)=A \times V1+Z0 \times I1$ $(V1-Z0 \times I1)=A \times V2+Z0 \times I2$
TRANSMISSION LINE CHARACTERISTIC IMPEDANCE	1-PORT	TEM	CHARACTERISTIC IMPEDANCE Z0 IN OHMS	$Z0 \neq 0$	$V=Z0 \times I$
LOSSLESS WAVEGUIDE DOMINANT MODE	2-PORT	WG	CUTOFF FREQUENCY FC IN HERTZ INFINITE-FREQUENCY CHARACTERISTIC IMPEDANCE ZINF IN OHMS LENGTH LEN IN METERS	$ZINF \neq 0$ $\sim FC \in F$	$Z0=ZINF \div (1-(FC \div F) \times 2) \times .5$ $A=J \times LEN \times ZINF \times F \div 20 \times 3E8 \div DIEL \times .5$ $(V2-Z0 \times I2)=(*-O2 \times A) \times V1+Z0 \times I1$ $(V1-Z0 \times I1)=(*-O2 \times A) \times V2+Z0 \times I2$
WAVEGUIDE CHARACTERISTIC IMPEDANCE	1-PORT	WG	CUTOFF FREQUENCY FC IN HERTZ INF.-FREQ. CHAR. IMP. ZINF IN OHMS	$ZINF \neq 0$ $\sim FC \in F$	$V=I \times ZINF \div (1-(FC \div F) \times 2) \times .5$

*In the equations, S is $j2\pi f$. The library work space 100 MARTHAE contains many additional elements.

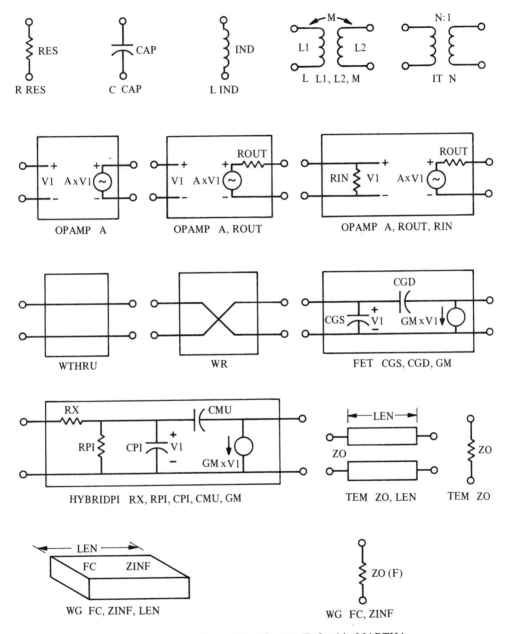

Figure 8.2 Elements Defined in MARTHA.

The library work space 100 MARTHAE contains about 50 additional elements and models. One of them is the voltage-controlled voltage source, VCVS. That is a two-port element, with the controlling branch on the input and the controlled branch on the output. Current can be used as well as voltage for either the controlling or controlled branch, so there are three additional elements, named VCCS, CCVS, and CCCS. In addition, if flux linkage or charge is possible as the controlling or controlled variable, there are 12 additional controlled sources, including, for example, a charge-controlled current source, QCCS. These are useful for modeling.

Another function useful for modeling is ZPDE. If its argument is of length 2, the first an integer and the second a coefficient, the result is a one-port network whose impedance is equal to the coefficient times complex frequency s raised to the integer power. If the integer is 0, the result is a resistor; if it is 1, the result is an inductor. Values of the integer from -5 through 5 are possible. ZPDE also produces, if its argument is of length 3, 4, or 5, a two-port similarly defined element where the numbers in the argument are from the two-port impedance matrix. Other functions named YPDE, HPDE, and ABCDPDE operate similarly. These are useful for modeling using a power series expansion.

Other elements in the MARTHA library include gyrators, nullors, negative-impedance convertors, and attenuators and isolators, both for TEM lines and wave guides. Included also are functions for converting numerically defined functions of frequency (FOF's) into elements. The FOF can be interpreted as impedance, admittance, or reflection coefficient of a one-port network, or as impedance, admittance, hybrid, ABCD, or scattering matrix of a two-port network.

The work space 100 MARTHAE contains more elements than can be described here. Complete documentation appears in References 1 and 2.

The library work space 100 MARTHAX contains several auxiliary functions for working with MARTHA, many of which are useful models. Examples are functions to calculate characteristic impedance of coaxial or microstrip transmission lines, or to calculate coaxial discontinuity capacitances, or to calculate cutoff frequency and characteristic impedance of common wave guides.

8.4 INPUT LANGUAGE

MARTHA is interactive. The user sits at a computer terminal and types his input line and gets an immediate reply.

Some of the things a user must type have nothing to do with network analysis. After the user dials the telephone number of the computer which carries MARTHA (which may, of course, be many miles distant) he must log in by identifying himself. He must also load MARTHA from the computer's public library and perhaps copy some of the MARTHA library and perhaps copy some of his previous results, including previously defined networks or models or previously calculated results. At any time the user can save his work up to that point, or start over, or log off the computer.

The rest of the user's time can be spent analyzing networks. A network can be defined at any time, and an analysis can be requested at any time. The analysis might be of a previously defined network or of a newly defined network, possibly composed in part of previously defined networks.

The general form of an analysis request in MARTHA is

$$\begin{Bmatrix} \text{PRINT} \\ \text{PLOT} \\ \text{PLOG} \\ \text{SMITH} \\ \text{STORE} \end{Bmatrix} \quad \langle \text{output list} \rangle \text{ OF } \langle \text{network} \rangle$$

The first word used (PRINT, PLOT, PLOG, SMITH, or STORE) fixes the basic format of the output; PLOG is used for a plot with a logarithmic scale for the independent variable. The form of the *output list* is discussed in Sec. 8.5. The word OF is required to separate the output list from the network description.

The form of the network description is unique to MARTHA. Two types of APL functions are used to define networks. The first creates elements, for example, resistors, capacitors, etc., and the second creates networks out of other networks by wiring them together. Element-definition functions were described in Sec. 8.3. An example of a wiring function is S, which is dyadic and, like all dyadic APL functions, has an argument on each side. Thus, if A and B are one-port (two-terminal) networks, then A S B is the new one-port network formed by putting the two networks A and B in series. Similarly, the function P wires two networks in parallel. Figure 8.1 shows a schematic diagram and a description in MARTHA notation of a simple series-parallel network. The functions S and P are sufficient to wire together all linear series-parallel networks. For more complicated topology, other wiring functions, designed to work on two-port networks, are defined in MARTHA.

There are 14 wiring functions in MARTHA, including the functions S and P. These are shown in Fig. 8.3. To create two-port networks out of one-port networks, the functions WS and WP are used. These are monadic; that is, they have one argument instead of two. The argument appears on the right of the function, as is true of all monadic APL functions. Two two-port networks can be wired together several ways. The functions WPP, WPS, WSP, and WSS connect the two inputs and the two outputs, using either parallel or series connections at each port. These are useful in denoting feedback amplifiers; for example, the function WSS might be used for

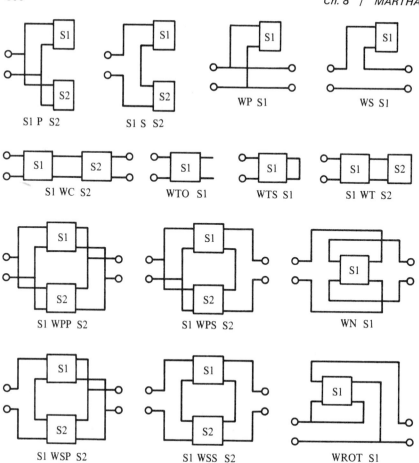

Figure 8.3 Wiring Functions Defined in MARTHA.

emitter degeneration. The cascade function WC is very common. The monadic functions WN and WROT are useful in converting grounded-emitter transistors to grounded-base or grounded-collector configurations. Symmetrical filters can be denoted easily with WN. If the left-hand side of the filter is called A, then the overall filter is A WC WN A. Finally, three techniques for converting a two-port network to a one-port network by terminating the output port are shown. The output port can be open-circuited (WTO) or short-circuited (WTS) or terminated in another network (WT). Note that WTO and WTS are monadic but that WT is dyadic, expecting a two-port network as its left argument and a one-port network as its right argument.

A precedence rule for the wiring functions must be established. By that is meant a rule for determining which of the functions are to be considered executed before others. For example, in ordinary algebraic notation, the expression $A \times B + 3$, written without parentheses, indicates that the multiplication is to be performed before the addition. This is an example of the common rule that exponentiation is performed before multiplication and division and that those are performed before additions and subtractions. By way of contrast, the language APL has a simpler precedence rule. The rule is that all functions have equal precedence and are executed strictly in the order indicated, from right to left, unless parentheses are used to delimit expressions which are to be evaluated first. Thus, in APL, $A \times B + 3$ would be $A \times (B+3)$ rather than $(A \times B) + 3$. The precedence rule for the wiring functions in MARTHA is the same as that of APL. As a consequence, every monadic wiring function takes as its argument the entire expression to its right, and every dyadic function takes as its left argument the one object immediately to its left and as a right argument the entire expression to the right. Parentheses can, of course, be used in the usual way to surround expressions which are to be evaluated first. Parentheses are often required for left arguments of dyadic functions but are never required (though they are permitted) for right arguments of functions. As an example, the expression A1 S A2 P A3 refers to the network of Fig. 8.4(a) rather than the network of Fig. 8.4(b), which would be written (A1 S A2) P A3.

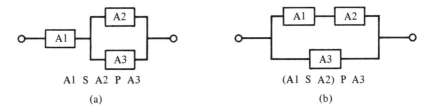

Figure 8.4 Illustration of the Precedence Convention for Wiring Functions in MARTHA.

Some of the wiring functions expect one-port networks and others two-port networks as arguments. What happens if the wrong kind of network is used? For example, what happens in A S B if B is a two-port network? To allow all wiring functions to operate on both one-port and two-port networks, two *automatic conversion conventions* are adopted in MARTHA:

1. If a wiring function expects a one-port network and encounters a two-port network, then the output is open-circuited and the input is used. This is equivalent to using the function WTO.

2. If a wiring function expects a two-port network and encounters a one-port network, then the wiring function WP is automatically invoked to convert to a two-port network.

Several examples of networks are shown both in MARTHA notation and in schematic diagrams in Fig. 8.5. These include the widely used Darlington transistor connection, a half-lattice and a Wheatstone bridge.

MARTHA incorporates several other functions that are not, strictly speaking, wiring functions, although they have as an argument a network and return as a result a network based on the argument. All these operate on both one-port and two-port networks and return a network with the same number of ports. The monadic function WAD converts a network to its adjoint [6–8], which is the network whose impedance and admittance matrices are the transposes of the corresponding matrices for the original network. The function ZSCALE is dyadic; its left argument is a number, and the result is

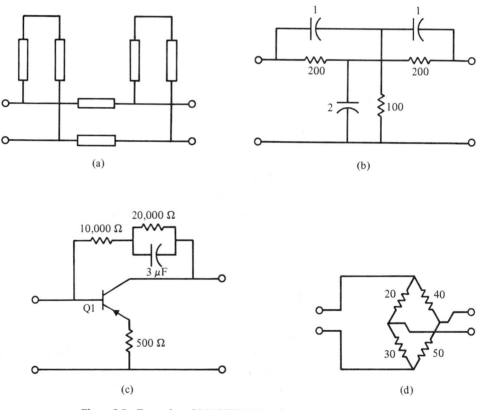

Figure 8.5 Examples of MARTHA Notation:
(a) Double-stub tuner with 50-ohm lines and adjustable line lengths L1 and L2, and fixed separation 5 inches:
(*WTS TEM* 50,*L1*) *WC* (*TEM* 50,5×0.0254) *WC WTS TEM* 50,*L2*
(b) Twin-tee filter, values in ohms and microfarads:
((*WS R* 200) *WC* (*C* 2E¯6) *WC WS R* 200) *WPP* (*WS C* 1E¯6) *WC* (*R* 100) *WC WS C* 1E¯6
(c) Feedback amplifier:
(*Q1 WSS R* 500) *WPP WS* (*R* 10E3) *S* (*R* 20E3) *P C* 3E¯6
(d) Wheatstone bridge:
((*WS R* 40) *WC R* 50) *WPS WR WC* (*WS R* 20) *WC R* 30

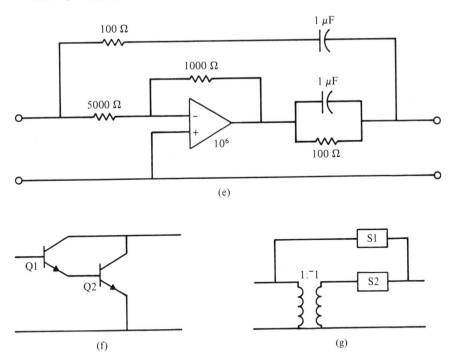

Figure 8.5—*Cont.*
(e) Active all-pass filter [5]:
`(WS (R 100) S C 1E`$^-$`6) WPP (WS R 5000) WC ((OPAMP `$^-$`1E6) WPP WS R 1000) WC WS (R 100) P C 1E`$^-$`6`
(f) Darlington transfer connection:
`WROT WROT (WROT Q2) WC WROT Q1`
(g) Half-lattice network:
`(WS S1) WPP (IT `$^-$`1) WC WS S2`

a network with the same topology as the network of its right argument, but with all elements scaled in impedance. Similarly, FSCALE performs a frequency scaling. These functions are useful in filter designs, where perhaps a 1-ohm, 1-Hz prototype is known. In a similar way, FINVERT performs a low-pass to high-pass transformation, and FBP a low-pass to band-pass transformation. The latter two may be used in succession to define a band-elimination filter. The function WCC returns a network with every element replaced by its *complex conjugate*; it is useful in defining conjugate-match loads for optimum power transfer. Finally, the function WDUAL takes the dual of a network, changing topology, and element values, and element types in the process.

This last set of functions is useful when MARTHA is used for design work. The MARTHA library contains a function named WHATIS which prints network descriptions of any network. Thus, for example, if PROTO is a 1-ohm, 1-Hz prototype filter, then

    DESIGN←1000 ZSCALE 1E6 FSCALE PROTO

defines a 1000-ohm, 1-MHz filter. Then

    WHATIS DESIGN

will print its definition, and

    PLOT (DB IG OF DESIGN), DB IG OF DESIGN WC WS R 800

will analyze it (in one case after it is wired with another resistor).

8.5 OUTPUT SPECIFICATIONS

The general form of an analysis request in MARTHA is

$$\begin{Bmatrix} \text{PRINT} \\ \text{PLOT} \\ \text{PLOG} \\ \text{SMITH} \\ \text{STORE} \end{Bmatrix} \quad \langle \text{output list} \rangle \text{ OF } \langle \text{network} \rangle$$

In this section the output list is described.

The output list is, basically, a list of response functions of the network to be calculated. It also can include modifiers on the response functions and some format requests.

Response Functions

MARTHA can calculate over 100 different response functions, of which the 30 in Table 8.2 are thought to be of wide interest and the remainder of specialized interest. The response functions of Table 8.2 are in MARTHA and the rest in the library work space 100 MARTHAR.

For one-port networks, the response functions are the impedance Z, admittance Y, reflection coefficient SC, and voltage standing-wave ratio VSWR. Of these, VSWR is real and the others are complex. If the network in question is a two-port network, then (in accordance with the automatic conversion convention for wiring functions), its output port is open-circuited and the input port is used. In calculating SC and VSWR, a normalization impedance is required; in MARTHA the normalization impedance for one-port networks is known as ZN and must be set by the user before analysis. It may be either an impedance (a real number) or a real or complex numerical function of frequency (FOF) or any one-port network defined in MARTHA notation (in which case the impedance of the network will be used).

Table 8.2 Response Functions in MARTHA*

NAME	C/R	MEANING	DEPENDS ON
Z	C	IMPEDANCE OF A 1-PORT NETWORK $V = Z \times I$	NETWORK
Y	C	ADMITTANCE OF A 1-PORT NETWORK $I = Y \times V$	NETWORK
SC	C	REFLECTION COEFFICIENT OF 1-PORT $B = SC \times A$	NETWORK, ZN
Z11 Z12 Z21 Z22	C C C C	IMPEDANCE MATRIX $V1 = (Z11 \times I1) + (Z12 \times I2)$ $V2 = (Z21 \times I1) + (Z22 \times I2)$	NETWORK
Y11 Y12 Y21 Y22	C C C C	ADMITTANCE MATRIX $I1 = (Y11 \times V1) + (Y12 \times V2)$ $I2 = (Y21 \times V1) + (Y22 \times V2)$	NETWORK
H11 H12 H21 H22	C C C C	HYBRID MATRIX $V1 = (H11 \times I1) + (H12 \times V2)$ $I2 = (H21 \times I1) + (H22 \times V2)$	NETWORK
S11 S12 S21 S22	C C C C	SCATTERING MATRIX $B1 = (S11 \times A1) + (S12 \times A2)$ $B2 = (S21 \times A1) + (S22 \times A2)$	NETWORK, ZNIN, ZNOUT
ZIN	C	INPUT IMPEDANCE $V1 \div I1$	NETWORK, ZL
YIN	C	INPUT ADMITTANCE $I1 \div V1$	NETWORK, ZL
SIN	C	INPUT REFLECTION COEFFICIENT $B1 \div A1$	NETWORK, ZL, ZNIN
ZOUT	C	OUTPUT IMPEDANCE $V2 \div I2$ WHEN OUTPUT EXCITED	NETWORK, ZG
YOUT	C	OUTPUT ADMITTANCE $I2 \div V2$ WHEN OUTPUT EXCITED	NETWORK, ZG
SOUT	C	OUTPUT REFLECTION COEFFICIENT $B2 \div A2$ WHEN OUTPUT EXCITED	NETWORK, ZG, ZNOUT
VG	C	VOLTAGE GAIN $V2 \div EG$	NETWORK, ZG, ZL
AG	R	AVAILABLE GAIN $POUT,AV \div PIN,AV$	NETWORK, ZG
IG	R	INSERTION GAIN $POUT(NETWORK) \div POUT(WTHRU)$	NETWORK, ZG, ZL
PG	R	POWER GAIN $POUT \div PIN$	NETWORK, ZL
TG	R	TRANSDUCER GAIN $POUT \div PIN,AV$	NETWORK, ZG, ZL

*Of these 30, 26 are complex and 4 real. Over 70 additional response functions are in the library work space 100 MARTHAR.

For two-port networks, there is a wide variety of response functions. If a two-port response function is requested of a one-port network, then (in accordance with the automatic conversion convention for wiring functions) the wiring function WP is assumed to be invoked. The response functions available include all the common two-port parameters (impedance, admittance, hybrid, *ABCD* matrix, and others) and the corresponding matrices for wave variables (including the scattering matrix and scattering transmission matrix). For calculations of the wave-variable responses, normalization impedances at both the input and the output are required; these are known as ZNIN and ZNOUT and may be different both from each other and from ZN. Each of these, like ZN, may be a real constant, real or complex numerically defined function, or any MARTHA one-port network.

Many of the two-port response functions also depend on the generator and/or load impedance. A two-port network is assumed to be terminated by generator and load as shown in Fig. 8.6. The variables ZG, ZL, and EG

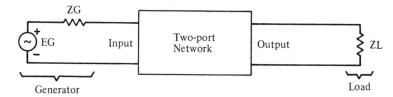

Figure 8.6 Termination of Two-port Networks Assumed When Some of the Response Functions are Calculated. The generator and load impedances ZG and ZL are not part of the network definition.

may, like ZN, be specified in several different ways, but they must be specified before analysis. Examples of response functions include the input impedance, admittance, reflection coefficient, and VSWR (these all depend on ZL) and corresponding output quantities (these all depend on ZG). Other examples are various measures of gain, including power gain, voltage gain, open-circuit voltage gain, voltage ratio, transducer gain, insertion gain, insertion voltage gain, unilateral gain, conjugate-match gain, and available gain. Also available are various characteristic impedances of the network, including image impedance, iterative impedance, and conjugate-match impedance, along with stability factors for amplifier, and the input and output voltages, currents, and wave variables.

Modifiers

The complex response functions appear normally in the form of real and imaginary parts. This may be changed by the use of modifiers. Table 8.3 lists the most important modifiers in MARTHA. For complex responses, the real part, imaginary part, magnitude, magnitude in decibels, angle, and phase delay are all useful. The appropriate modifiers are placed before the responses in question.

Table 8.3 Modifiers in MARTHA*

Modifier	Meaning
RE	Real part
IM	Imaginary part
MAG	Magnitude
RAD	Phase in radians
DEG	Phase in degrees
DB	Magnitude in decibels
PD	Phase delay
REC	Reciprocal

*Of these six, RE, IM, RAD, PD, and DEG are ignored if applied to a real response, and DB is $20 \times 10*$MAGNITUDE for complex responses and $10 \times 10*$ABSOLUTE VALUE for real responses. If no modifier is used, the real and imaginary parts of complex responses will result. Other modifiers are in the library work space 100 MARTHAM.

Example.   `PRINT MAG Z, DEG Z OF (R 1) P L 0.02`

Each of the modifiers acts only on the one response immediately to its right, contrary to the normal APL convention that functions have as their right argument the entire expression to their right. Thus, parentheses are not ordinarily used in the output list.

Formats

Various format requests can also be inserted in the output lists, generally at any point in the list. For prints, the number of significant figures printed can be changed from its normal value of 5 by the function PLACES.

Example.   `PRINT 7 PLACES DB IG OF CRYSTALFILTER`

For plots, several additional options are available. Plots are normally made against frequency as the independent variable, but if desired any one of the responses will act as the independent variable if it is preceded in the output list by VS. Alternatively, if PAIRS appears in the output list, the first response function will be plotted against the second, and the third against the fourth, etc. Since complex outputs are considered, for this purpose, as two real outputs, PAIRS easily produces loci in the complex plane. Plots are normally 50 spaces wide and 50 lines high, to fit conveniently on one page (except those made by the function SMITH, which are designed to fit a standard-size Smith chart). Other widths and heights can be specified in the output list. Plotting characters other than the standard ones can be specified by the function SYMBOLS. Normally MARTHA plots all dependent variables with different scales, selecting each so that the scales consist of round numbers but significant detail is still shown in each plot. The dependent variables will all have the same scale (with usually some loss of detail) if SS appears in the output list. The horizontal and vertical scales can be set to arbitrary values by the functions HSCALE and VSCALE. This is useful in magnifying certain critical regions of the plot; points falling outside the specified scales are simply ignored.

These plotting format functions are illustrated by some examples:

```
PLOG 'MD' SYMBOLS MAG Z, DEG Z OF NETWORK

PLOT SS Z11, Z12, Z22 OF FILTER

PLOT ⁻20 20 HSCALE ⁻180 180 VSCALE DB RR VS DEG RR OF AMPLIFIER

SMITH S11, SIN, S12 OF FILTER

PLOT PAIRS (Z OF AMP1), ZIN, Z11 OF AMP2
```

The third request prints the Nichols chart for the return ratio RR of the amplifier, and the fourth request produces a standard-size Smith chart.

8.6 MODELS

A computation using MARTHA (or any other program) will be accurate only if reasonable care is used in modeling the network. Devising suitable models for a network is usually the most challenging part of any analysis. MARTHA cannot, of course, do the modeling job for the user, but it does offer him a selection of elements useful for models, the possibility of numerically defined elements, and the option of creating other elements in the form of *user-defined elements*.

Types of Models Allowed

Models in MARTHA may be any network with topology definable in the MARTHA notation, using any of the built-in MARTHA elements, along with any linear one-port or two-port element not already in MARTHA, for which the user is able to supply an APL algorithm for computing its impedance (for one-port elements) or its *ABCD* matrix (for two-port elements).

Note that MARTHA does not distinguish models from elements or from networks containing several elements wired together. If any user finds that he uses a given configuration frequently, then he can write a simple APL function that returns that particular network upon demand. For users who wish a model that is not representable by a network made up from built-in MARTHA elements, MARTHA allows user-defined elements without restriction as to complexity of the required algorithm.

Several elements of special interest in modeling were discussed in Sec. 8.3. MARTHA can also handle numerically defined elements. To create one of these, the user types in a table of values and then calls one of the functions ZFOF, YFOF, SFOF, and ABCDFOF to interpret the table as numerical values of impedance, admittance, or reflection coefficient for one-port networks or Z, Y, S, H, or $ABCD$ matrices for two-port networks. The resulting numerical models are then treated by MARTHA like other elements. Arbitrary frequency dependence is available this way.

Built-in Models

MARTHA contains several built-in models, of which the most important are probably the hybrid-pi and FET transistor models and four simple models of operational amplifiers. These are considered as elements in MARTHA and were discussed in Sec. 8.3.

Input Techniques and Format

For user-defined models, a distinction should be made between those models that are networks of MARTHA elements and those that are not. For the former, a simple APL function with an argument will usually suffice. As an example, consider the Marcuvitz model [9] for a window formed from a pair of semicircular obstacles along the sides of a wave guide (Fig. 8.7). The

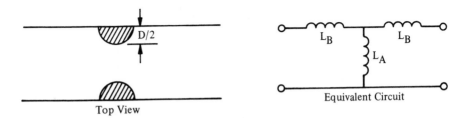

Figure 8.7 Symmetrical Window Between Two Semicircular Inductive Posts in a Waveguide and the Model Given by Marcuvitz [9].

equivalent circuit shown has inductors with inductances

$$L_A = \frac{Z_\infty}{2\pi f_c}\left(\frac{c}{2\pi f_c D}\right)^2$$

$$L_B = -\frac{Z_\infty}{16\pi f_c}\left(\frac{2\pi f_c D}{c}\right)^4$$

Note that L_B is negative. Z_∞ is the characteristic impedance of the wave guide at infinite frequency, f_c is the cutoff frequency, and c is the speed of light. A function that returns this two-port networks is

```
    ∇ B←WINDOW A
[1]   X←300000000÷(6.28×A[1]×A[3])
[2]   LA←A[2]×(X*2)÷(6.28×A[1])
[3]   LB←-A[2]÷((X*4)×16×3.14×A[1])
[4]   B←(WS L LB) WC(L LA) WC WS L LB
    ∇
```

where the argument A is a vector of length 3 containing the cutoff frequency, the impedance, and the diameter. A relatively small amount of APL programming ability is required to read or write this model. The normal APL function-definition scheme is used, and all APL primitive functions (and MARTHA functions) may be used.

Next, consider the case where the model is so complicated that a network with MARTHA elements is not sufficient to describe it. The technique now is to use MARTHA's user-defined-element capability. A user-defined element is *defined* by the MARTHA function UDE. The argument of this function may be any vector of parameter values which describe the element. However, the user must also write a function which instructs the computer in calculating the impedance or *ABCD* matrix of the element. This second function is more difficult. It must be a monadic APL function entitled NEWELEMENT. It is not called directly by the user but instead will be called by one of the MARTHA functions when analysis is done. At that time the frequency is known (it need not be known during the definition of the network) and NEWELEMENT may refer to it. The function NEWELEMENT must return either a two-dimensional matrix according to Table 8.4 or a MARTHA network, possibly incorporating such a matrix as an element. The first dimension of the matrix is in each case the number of frequencies, and the second dimension is 2, 6, or 8, depending on whether the network is a one-port network, a reciprocal two-port network, or a nonreciprocal two-port network.

As an example, consider a length of coaxial transmission line with skin-effect loss. The TEM lines in MARTHA are lossless, and the frequency dependence of the skin effect is not given exactly by any MARTHA element. For this example, the necessary parameters to specify the line are the inner

Table 8.4 *Required Shape and Interpretation of Matrices Returned by the User-Written function NEWELEMENT*

	Number of Columns		
Column	2	6	8
1	Re Z	Re A	Re A
2	Im Z	Im A	Im A
3		Re B	Re B
4		Im B	Im B
5		Re C	Re C
6		Im C	Im C
7			Re D
8			Im D
Assumed		$D = \dfrac{(1 + BC)}{A}$	
Properties of result	One-port	Two-port reciprocal	Two-port

and outer radii and the length. A function that creates the user-defined element is very simple:

```
    ∇ B←COPPERLINE A
[1]   B←UDE 2.61E⁻7,A
    ∇
```

The argument A is assumed to consist of the inner and outer radii and the length, all in meters. This function merely defines the element; it does not do any calculations.

During MARTHA analysis, the function NEWELEMENT is called when this element is encountered. The real calculations are performed by that function. The *ABCD* matrix for this line is given by

$$A = D = \cosh \gamma l$$

$$B = Z_0 \sinh \gamma l$$

$$C = \frac{\sinh \gamma l}{Z_0}$$

where l is the length and

$$\gamma = \sqrt{(R + j\omega L)(G + j\omega C)}$$

$$Z_0 = \sqrt{\frac{R + j\omega L}{G + j\omega C}}$$

and where the per-unit-length quantities L, C, R, and G are given by [10]

$$L = \frac{\mu_0}{2\pi} \ln \frac{r_0}{r_i}$$

$$C = \frac{2\pi\epsilon_0}{\ln r_0/r_i}$$

$$G = 0$$

$$R = 2.61 \times 10^{-7} \sqrt{f}\, \frac{1}{2\pi}\left(\frac{1}{r_0} + \frac{1}{r_i}\right)$$

A function NEWELEMENT that incorporates these formulas is

```
       ∇ E←NEWELEMENT A;RS;RO;RI;LEN;CC;LL;RR;RO;XO;ALPHAL;BETAL
[1]    RS←A[1]
[2]    RO←A[2]
[3]    RI←A[3]
[4]    LEN←A[4]
[5]    CC←6.28×8.854E¯12÷⍟RO÷RI
[6]    LL←2E¯7×⍟RO÷RI
[7]    RR←RS×(,F*0.5)×((÷RO)+÷RI)÷6.28
[8]    RO←(((LL÷CC)+(((LL÷CC)*2)+(RR÷CC×6.28×,F)*2)*0.5)÷2)*0.5
[9]    XO←-RR÷2×RO×CC×6.28×,F
[10]   ALPHAL←LEN×RR÷2×RO
[11]   BETAL←LEN×RO×CC×6.28×,F
[12]   E←((ρ,F),6)ρ0
[13]   E[; 1 2]←⌽(2 1 ∘.○BETAL)× 6 5 ∘.○ALPHAL
[14]   E[; 3 4]←⌽(2 1 ∘.○BETAL)× 5 6 ∘.○ALPHAL
[15]   E[; 5 6]←(E[; 3 4]×RO,[1.5] RO)+E[; 4 3]×XO,[1.5]-XO
[16]   E[;5]←E[;5]÷(XO*2)+RO*2
[17]   E[;6]←E[;6]÷(XO*2)+RO*2
[18]   E[; 3 4]←(E[; 3 4]×RO,[1.5] RO)-E[; 4 3]×XO,[1.5]-XO
       ∇
```

This function is not trivial, and some familiarity with APL is necessary either to read or write it.

The number 2.61E-7 in the function COPPERLINE is a measure of the resistivity of copper and is used in line [7] of NEWELEMENT. By including it as part of the element definition, other element-definition functions can use the same function NEWELEMENT for other materials, for example,

```
       ∇ B←BRASSLINE A
[1]    B←UDE 5.01E¯7,A
       ∇
```

An important aspect of every model is its range of validity. This includes permissible values for both parameters associated with the model as well as allowed frequency ranges. Users are advised to have their model-making functions check the parameter values to see that they are reasonable. To check on frequency range, MARTHA can be made to print a warning whenever

Ch. 8 / Penfield

one or more frequencies is outside the permissible range. To set the permissible range, the network is operated on by the function FLIMITS, which is a dyadic APL function with the left argument a pair of frequency values limiting the allowed range. For example, the model for the wave-guide window given above is valid (according to Marcuvitz [9]) only between f_c and $3f_c$. If the function WINDOW had the following fifth line

[5]   B←(A[1],3×A[1]) FLIMITS B

then whenever that network is analyzed, if any of the frequencies is outside that range, that fact will be reported, but analysis will continue.

8.7 EXAMPLES

Three fully worked examples are given in this section. The first is a transistor amplifier, which is also treated in other chapters of this book. The second is a crystal filter, and the third is a coaxial low-pass filter. The lines typed by the user are identified by being indented six spaces. The computer response generally is not indented.

Transistor Amplifier

Figure 8.8 shows the circuit diagram for the amplifier, together with the model to be used for the transistor. The bias source has been omitted. Because of the versatility of the technique of defining networks in MARTHA, the amplifier may be defined in several ways. One way which is probably appropriate for this network is suggested by Fig. 8.9, where the input and output coupling networks, and the emitter circuit, are shown as separate two-port networks. Those circuits are straightforward and are defined first (see EXAMPLE 1). The hybrid-pi transistor model used in MARTHA is generally regarded as the best linear model at high frequencies. This model (see Fig. 8.2) can be made to coincide with the transistor model to be used if RX, CPI, and CMU are all set to zero and GM is BETA ÷ RPI. Then the output resistor R5 must be added separately. The transistor is defined using this approach, and note that it is not necessary to separately calculate BETA ÷ RPI, since the formula can be typed in and the calculated result used as part of the argument for the function HYBRIDPI.

Refer now to EXAMPLE 1 (Fig. 8.10). First the work space containing MARTHA is requested from the APL public library. Then the four subnetworks are defined and then wired together and given the name AMP1. This completes the network definition.

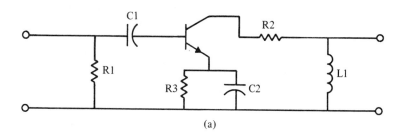

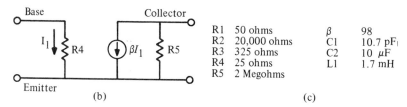

R1	50 ohms	β	98	
R2	20,000 ohms	C1	10.7 pF	
R3	325 ohms	C2	10 μF	
R4	25 ohms	L1	1.7 mH	
R5	2 Megohms			

(c)

Figure 8.8 (a) Transistor Amplifier; (b) Model for Transistor; (c) Component Values.

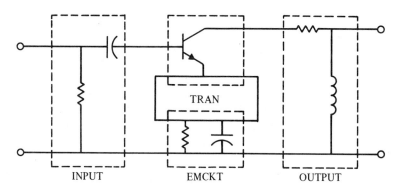

Figure 8.9 Transistor Amplifier Shown as Four Subnetworks Wired Together.

Before analysis, the frequency vector F must be specified. A study of Fig. 8.8 reveals no mechanism for limiting high-frequency response (this is perhaps an unrealistic circuit). To view as wide a frequency range as is necessary, a logarithmic frequency sweep is used. The frequency vector F is set to 10 raised to the power (the asterisk is used in APL for power) of a vector ranging from 0 to 11; this covers all interesting ranges.

Next, a choice of response function must be made. In this example, the voltage gain VG is most logical, and the magnitude (in decibels) and phase (in degrees) are requested. The output request begins with the word PRINT to set the basic format, and this is followed by the output list, then the word

OF, and finally the network definition. As expected, the voltage gain saturates at high frequencies at the rather high value of 138 decibels, and the phase approaches -180 degrees.

These results must be taken with a grain of salt. Like all other network-analysis programs, MARTHA can give reasonable answers only if reasonable models are used, and in the present case the circuit is probably not well modeled above 1 MHz. This fact does not destroy the usefulness of the circuit as an example of how to use MARTHA, but it does imply the need for better models.

As an illustration of the use of MARTHA in judging different models,

```
A BEGINNING OF EXAMPLE 1.

      )LOAD 100 MARTHA
SAVED  11.08.26 06/28/74

      INPUT←(R 50) WC WS C 10.7E-12
      EMCKT←(R 325) P C 10E-6
      OUTPUT←(WS R 20E3) WC L 1.7E-3
      TRAN←(HYBRIDPI 0, 25, 0, 0, 98÷25) WC R 2E6

      AMP1←INPUT WC (TRAN WSS EMCKT) WC OUTPUT

      F←10* 0 1 2 3 4 5 6 7 8 9 10 11

      PRINT DB VG, DEG VG OF AMP1

CIRCUIT ANALYSIS BY MARTHA.     73∘D    6/30/74   0:37
MARTHA COPYRIGHT (C̄) 1973 MASSACHUSETTS INSTITUTE OF TECHNOLOGY

          F            DB VG          DEG VG
      -----------   -----------    -----------
       1.0000E0      -2.0314E2       6.7017E-5
       1.0000E1      -1.6314E2       6.4350E-4
       1.0000E2      -1.2314E2       1.2648E-3
       1.0000E3      -8.3139E1      -2.3785E-4
       1.0000E4      -4.3139E1      -3.9761E-3

       1.0000E5      -3.1390E0      -3.9921E-2
       1.0000E6       3.6861E1      -3.9923E-1
       1.0000E7       7.6848E1      -3.9894E0
       1.0000E8       1.1567E2      -3.7407E1
       1.0000E9       1.3642E2      -1.3854E2

       1.0000E10      1.3787E2      -1.7551E2
       1.0000E11      1.3789E2      -1.7955E2

      AMP2←AMP1 WC WP C 10E-12

      TRAN3←(HYBRIDPI 5, 25, 1000E-12, 50E-12, 98÷25) WC R 2E6
      AMP3←INPUT WC (TRAN3 WSS EMCKT) WC OUTPUT
```

Figure 8.10 Example 1.

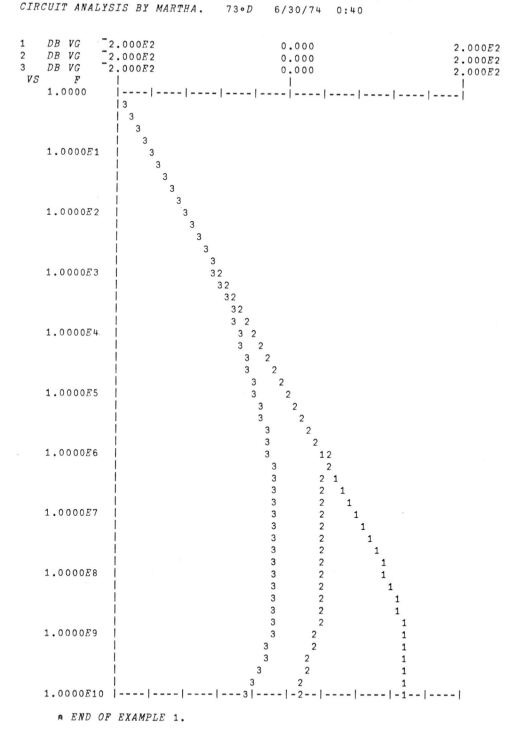

Figure 8.10—Cont.

some rather simple improvements are made. First, a stray capacitance is placed across the inductor L1, and the resulting network is named AMP2. This is, of course, still a crude model. Second, the transistor is redefined with what might be typical values of RX, CMU, and CPI. The new network is called AMP3. In each case, the network definition is easy because the previously defined subnetworks are used.

The voltage gains in the three cases are compared by simultaneously analyzing and plotting the gains of the three networks. To cover such a wide frequency range, a logarithmic plot is necessary, so the function PLOG is used (the function PLOT produces a linear plot). Note that MARTHA normally automatically selects the scales but is here requested to use the same scale for the three different responses.

Crystal Filter

This crystal filter (Fig. 8.11) is adapted from a filter shown to me privately by Mr. William B. Lurie, who ascribed the design to Prof. G. Szentirmai. There are two stages, each with a half-lattice construction which is modeled with the aid of an ideal transformer. The four crystals all have the same equivalent circuit, with different element values. The filter has a passband about 4000 Hz wide, between 8 and 8.004 MHz. It is designed to operate between

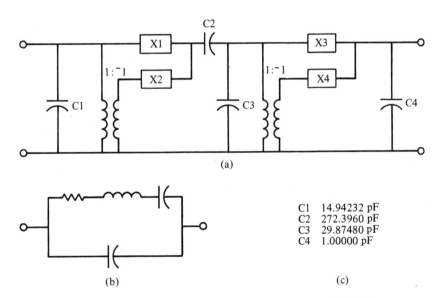

C1 14.94232 pF
C2 272.3960 pF
C3 29.87480 pF
C4 1.00000 pF

Figure 8.11 (a) Crystal Filter. (b) Model for the Crystals X1, X2, X3, and X4. (c) Component Values. The element values in the crystal models are not listed since they appear in Example 2 (Fig. 8.12). The resistances in the crystal models are calculated from assumed values of Q.

⍝ BEGINNING OF EXAMPLE 2.

)LOAD 100 MARTHA
SAVED   11.08.26 06/28/74

      X1←(C 8.60776E⁻12) P (C 0.0116013E⁻12) S (L 0.0341232) S R 0.1752
      X2←(C 17.1140E⁻12) P (C 0.0177387E⁻12) S (L 0.0223008) S R 0.1121
      X3←(C 18.7184E⁻12) P (C 0.0138144E⁻12) S (L 0.0286568) S R 0.144
      X4←(C 20.0690E⁻12) P (C 0.0147852E⁻12) S (L 0.0267554) S R 0.1345

      STAGE1←(C 14.94232E⁻12) WC (WS X1) WPP (IT ⁻1) WC WS X2
      STAGE2←(C 29.87480E⁻12) WC (WS X3) WPP (IT ⁻1) WC WS X4

      CRYSTALFILTER←STAGE1 WC (WS C 272.396E⁻12) WC STAGE2 WC C 1E⁻12

      ZG←500
      ZL←500
      F←7.999E6 + 0, 200×⍳30

      PLOT 30 HIGH DB IG, DEG VG OF CRYSTALFILTER

CIRCUIT ANALYSIS BY MARTHA.   73∘D   6/30/74   0:54
MARTHA COPYRIGHT (C̄) 1973 MASSACHUSETTS INSTITUTE OF TECHNOLOGY

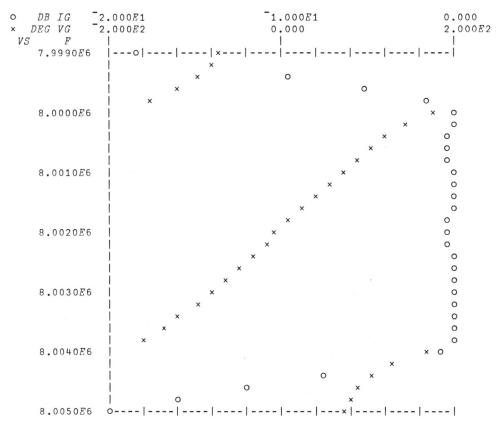

Figure 8.12   Example 2.

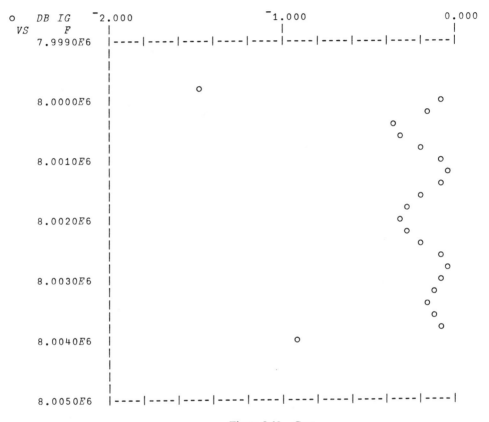

Figure 8.12—*Cont.*

source and load of 500 ohms. Refer to EXAMPLE 2 (Fig. 8.12). First MARTHA is loaded, and then the four crystals are defined. Next, the two stages and the overall filter are defined. Next, the generator and load impedances ZG and ZL are set to 500 ohms, and the frequency vector is set so as to encompass the passband. In specifying the frequency vector, the index generator was used. A plot 30 lines high of the insertion gain IG expressed in decibels and the phase of the voltage gain in degrees is requested. Next, to see the ripple in the passband a little more clearly, the horizontal scale is deliberately set to the range from -2 to 0 decibels, with the aid of the function HSCALE from the MARTHA library. This request overrides the normal MARTHA practice of automatically setting the scale. Note that when a

```
F←8.002E6 + 200×ι30
         PRINT DB IG, DEG VG, MAG ZIN OF CRYSTALFILTER
CIRCUIT ANALYSIS BY MARTHA.    73∘D    6/30/74   1:1
```

F	DB IG	DEG VG	MAG ZIN
8.0022E6	⁻2.7428E⁻1	⁻1.8240E1	5.0138E2
8.0024E6	⁻1.8555E⁻1	⁻3.2660E1	4.7038E2
8.0026E6	⁻8.9823E⁻2	⁻4.7854E1	4.7521E2
8.0028E6	⁻3.9720E⁻2	⁻6.3898E1	5.2096E2
8.0030E6	⁻6.3555E⁻2	⁻8.0654E1	5.9390E2
8.0032E6	⁻1.3452E⁻1	⁻9.7992E1	6.4698E2
8.0034E6	⁻1.7197E⁻1	⁻1.1624E2	6.1842E2
8.0036E6	⁻1.0630E⁻1	⁻1.3673E2	5.0307E2
8.0038E6	⁻8.4973E⁻2	⁻1.6221E2	3.8952E2
8.0040E6	⁻9.1322E⁻1	1.6514E2	5.0864E2
8.0042E6	⁻3.6485E0	1.3134E2	1.0891E3
8.0044E6	⁻7.7195E0	1.0583E2	2.8661E3
8.0046E6	⁻1.1993E1	8.8834E1	2.7605E4
8.0048E6	⁻1.6059E1	7.7184E1	4.9553E3
8.0050E6	⁻1.9882E1	6.8697E1	2.7793E3
8.0052E6	⁻2.3519E1	6.2179E1	2.0690E3
8.0054E6	⁻2.7039E1	5.6972E1	1.7124E3
8.0056E6	⁻3.0519E1	5.2684E1	1.4960E3
8.0058E6	⁻3.4040E1	4.9075E1	1.3495E3
8.0060E6	⁻3.7707E1	4.5983E1	1.2432E3
8.0062E6	⁻4.1680E1	4.3297E1	1.1621E3
8.0064E6	⁻4.6255E1	4.0941E1	1.0981E3
8.0066E6	⁻5.2139E1	3.8866E1	1.0461E3
8.0068E6	⁻6.2280E1	3.7103E1	1.0029E3
8.0070E6	⁻6.7171E1	⁻1.4507E2	9.6644E2
8.0072E6	⁻5.7815E1	⁻1.4640E2	9.3517E2
8.0074E6	⁻5.4557E1	⁻1.4777E2	9.0803E2
8.0076E6	⁻5.2904E1	⁻1.4905E2	8.8424E2
8.0078E6	⁻5.1997E1	⁻1.5022E2	8.6319E2
8.0080E6	⁻5.1516E1	⁻1.5131E2	8.4443E2

Figure 8.12—*Cont.*

network is to be analyzed a second time for the same frequency vector, the previous results, saved under the name SAME, can be used.

Next, it is desired to inspect the transition region and the stop band, so the frequency vector is redefined to cover the upper half of the passband and 4000 Hz above. The insertion gain, phase, and magnitude of the input impedance are printed.

An interesting phenomenon is observed at approximately 8.007 MHz. The insertion gain passes through a (negative) peak and the phase changes by 180 degrees. Apparently there is a transmission zero somewhere in the vicinity, and to view this in more detail, the frequency vector is redefined so as to expand the range between 8.0068 and 8.0070 MHz. The depth of the

```
F←8.0068E6 + 10×ι20
```

PRINT DB IG, DEG VG OF CRYSTALFILTER

CIRCUIT ANALYSIS BY MARTHA.   73∘D   6/30/74   1:4

F	DB IG	DEG VG
8.0068E6	⁻6.3118E1	3.7031E1
8.0068E6	⁻6.4030E1	3.6964E1
8.0068E6	⁻6.5030E1	3.6902E1
8.0068E6	⁻6.6140E1	3.6846E1
8.0068E6	⁻6.7390E1	3.6800E1
8.0069E6	⁻6.8823E1	3.6769E1
8.0069E6	⁻7.0507E1	3.6761E1
8.0069E6	⁻7.2558E1	3.6793E1
8.0069E6	⁻7.5194E1	3.6907E1
8.0069E6	⁻7.8915E1	3.7229E1
8.0069E6	⁻8.5389E1	3.8437E1
8.0069E6	⁻1.0546E2	⁻1.6972E2
8.0069E6	⁻8.4100E1	⁻1.4631E2
8.0069E6	⁻7.8498E1	⁻1.4539E2
8.0069E6	⁻7.5168E1	⁻1.4511E2
8.0070E6	⁻7.2803E1	⁻1.4502E2
8.0070E6	⁻7.0976E1	⁻1.4499E2
8.0070E6	⁻6.9491E1	⁻1.4500E2
8.0070E6	⁻6.8243E1	⁻1.4503E2
8.0070E6	⁻6.7171E1	⁻1.4507E2

```
)COPY 100 MARTHAF PLACES
SAVED  12.45.42  06/28/74

     F←8.00691E6 + ι10
```

PRINT 7 PLACES DB IG, DEG VG OF CRYSTALFILTER

CIRCUIT ANALYSIS BY MARTHA.   73∘D   6/30/74   1:6

F	DB IG	DEG VG
8.006911E6	⁻8.639767E1	3.873101E1
8.006912E6	⁻8.753608E1	3.910974E1
8.006913E6	⁻8.884346E1	3.961350E1
8.006914E6	⁻9.037903E1	4.031436E1
8.006915E6	⁻9.223957E1	4.135279E1
8.006916E6	⁻9.459954E1	4.304337E1
8.006917E6	⁻9.782233E1	4.626150E1
8.006918E6	⁻1.028618E2	5.461697E1
8.006919E6	⁻1.119975E2	⁻1.025601E2
8.006920E6	⁻1.054552E2	⁻1.697215E2

A END OF EXAMPLE 2.

Figure 8.12—*Cont.*

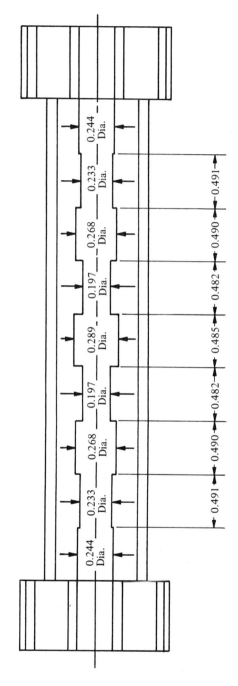

Figure 8.13 Seven-section Coaxial Low-pass Filter. Dimensions are in inches. The diameter of the outer conductor is 0.561 inches, and the nominal characteristic impedance of the terminating lines is 50 ohms.

notch in insertion gain (105 decibels) and still rather sharp transition in phase suggest that the zero is very close to the $j\omega$ axis, and it is interesting to expand the passband again, between 8.00691 and 8.00692 MHz. Note that the printing will not resolve these small changes in frequency and so a larger number of significant figures is requested with the aid of the function PLACES from the MARTHA library. The frequency sweep is now in 1-Hz steps, and the high resolution available in this analysis is due to the fact that MARTHA (like APL) uses double-precision arithmetic.

Coaxial Low-Pass Filter

The third example (Fig. 8.13) illustrates MARTHA's ability to work with distributed as well as lumped networks. The original design of this 7-section low-pass filter was done by Levy and Rozzi [11], who stated that this filter is not particularly good except as an example of their design method, which they then used on a more practical, 23-section filter. Only the 7-section filter is analyzed here.

The coaxial discontinuity capacitances of the filter are important to its design. These capacitances are models which should be used at every junction between conductors of different size, as shown in Fig. 8.14. MARTHA has

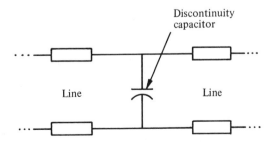

Figure 8.14 Discontinuity-Capacitor Model for the Junction Between Coaxial Lines of Different Size.

a function named COAXDISCAP in the MARTHA library which calculates the values of these capacitors when supplied with the appropriate radii of the line. Like all lengths in MARTHA, the radii must be given in meters, and so a conversion from the dimensions in inches in Fig. 8.13 is necessary. In EXAMPLE 3 (Fig. 8.15), the first four discontinuity capacitors C1 through C4 are defined (the other four are equal because the filter is symmetric), and the conversions from inches to meters and from diameters to radii are done during the definitions (multiplication by 0.0254 meters per inch and 0.5).

Next, the individual lengths of transmission lines L1 through L4 are defined. In MARTHA a length of line is specified by its characteristic impedance and physical length. In our case we need to calculate the characteristic impedance from the dimensions. The function COAX in the MARTHA library is copied and then used to do this. It expects as an argument a vector consisting of the inner and outer radii and length, all in meters.

Since the filter is symmetric, it is possible to define the left side (consisting of the first three lines with the discontinuity capacitors placed at all junctions) and then use this definition twice in the final definition of the filter. Making the definition in this way eliminates the need to enter the same numerical data twice and therefore tends to reduce keyboard errors.

```
            )LOAD 100 MARTHA
    SAVED   11.08.26 06/28/74
            )COPY 100 MARTHAX COAXDISCAP
    SAVED   12.41.11 06/28/74

            C1←C COAXDISCAP 0.5×0.0254×0.561 0.244 0.233
            C2←C COAXDISCAP 0.5×0.0254×0.561 0.233 0.268
            C3←C COAXDISCAP 0.5×0.0254×0.561 0.268 0.197
            C4←C COAXDISCAP 0.5×0.0254×0.561 0.197 0.289

            )COPY 100 MARTHAX COAX
    SAVED   12.41.11 06/28/74

            L1←TEM COAX 0.0254×0.2805 0.1165 0.491
            L2←TEM COAX 0.0254×0.2805 0.1340 0.490
            L3←TEM COAX 0.0254×0.2805 0.0985 0.482
            L4←TEM COAX 0.0254×0.2805 0.1445 0.485

            LEFTSIDE←C1 WC L1 WC C2 WC L2 WC C3 WC L3 WC C4
            FILTER←LEFTSIDE WC L4 WC WN LEFTSIDE

            ZG←50
            ZL←50

            )COPY 100 MARTHAR VSWRIN
    SAVED   12.46.22 06/28/74

            F←1E8×0,ı32
```

Figure 8.15 Example 3.

Ch. 8 / Penfield

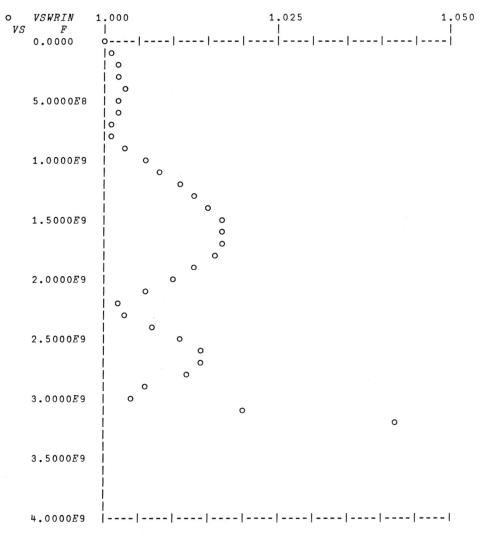

Figure 8.15—*Cont.*

The filter is designed to operate from a 50-ohm generator and into a 50-ohm load, so the generator and load impedances ZG and ZL are set to 50. The VSWR at the input is wanted, and this response function VSWRIN is copied from the MARTHA library. The frequency vector is set, and a plot is requested. The result, of course, is in agreement with the analysis by Levy and Rozzi [11].

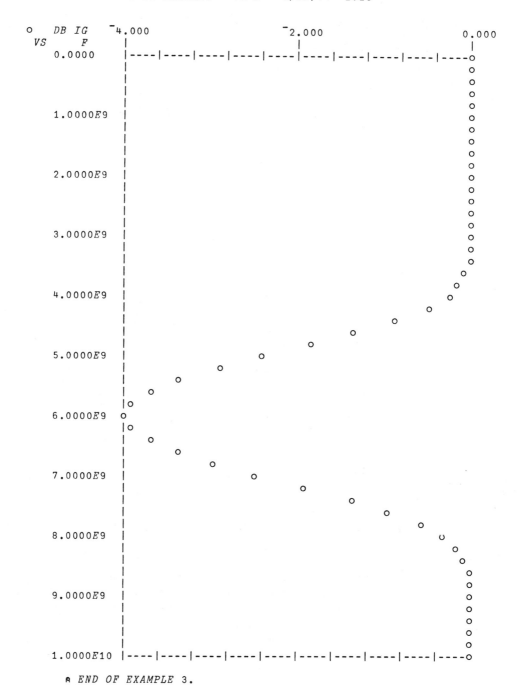

Figure 8.15—*Cont.*

Filters of this sort are periodic and have other passbands. In this case there is another passband centered around 12 GHz, so the stop band extends from 3 to 9 GHz. Levy and Rozzi stated that this seven-section filter had poor stop-band attenuation, and the plot of the insertion gain up to 10 GHz reveals just how true this is. The maximum attenuation is only about 4 decibels.

8.8 LIMITATIONS

Every network-analysis program has two types of limitations not discussed so far. One has to do with computer resource limits and the other with problems that are ill-conditioned.

Resource Limitations

Many programs have limitations on the number of nodes, or number of elements, or number of network definitions allowed. These limitations arise from the amount of space set aside for certain variables. MARTHA has, for practical purposes, no such precise limitations since it is embedded in APL, which uses dynamic storage for all variables. The resource limitations of MARTHA are somewhat more difficult to describe.

The primary limitation is caused by the finite size of APL work spaces (usually 32K bytes, each byte being 8 bits). The MARTHA functions require about 18K, leaving about 14K for the user's network definitions and temporary storage of results. If possible, MARTHA automatically analyzes with all frequencies at once; if this is not possible, it automatically selects a smaller number of frequencies and repeats the analysis as many times as necessary until all frequencies are accounted for. However, even this can be insufficient in some cases with lengthy output lists. Experience has shown no trouble for analysis with up to 50 frequencies, provided not too much of the available space is taken by unrelated functions and data. The APL error message WS FULL indicates there is a problem of this sort. Some of the computers that carry MARTHA have larger APL work spaces, in which case this ceases to be a real problem. If it is necessary to analyze with a large number of frequencies and available space cannot be created by erasing unnecessary objects, then MARTHA has a function entitled ATATIME which can be used to repeat the analysis for a small number of frequencies.

Ill-Conditioned Networks

Every network-analysis scheme has its own set of networks for which it performs poorly, or perhaps not at all. During analysis, MARTHA represents one-port networks by their impedances and two-port networks by their

ABCD matrices. Thus, MARTHA is unable to handle networks for which these representations do not exist. For one-port networks, this is only open circuits (or because of overflow or underflow problems, networks with impedance magnitudes greater than about 10^{37} or less than 10^{-37} ohms). For two-port networks, this is any network for which the output voltage and current are related by one equation not containing the input voltage or current. This class of networks includes those which produce no effect at the output when the input is excited, that is, either those with disconnected inputs and outputs or those that are backwards unilateral. Generally when such a situation exists, it is easy to redefine the network so as to avoid the ill-conditioned case.

Certain element values, particularly zero, can lead to ill-conditioned networks, and Table 8.1 lists several such cases.

8.9 ERROR DIAGNOSTICS

MARTHA recognizes two classes of errors. One class is errors so serious that MARTHA cannot reasonably proceed, and in the other case only some of the calculations may be in error and MARTHA continues.

For nonfatal errors, MARTHA prints a warning but continues. Examples include use of frequencies outside the range specified by the function FLIMITS and frequencies at which by coincidence the network is ill-conditioned. Generally in the latter case trouble is encountered when a denominator vanishes; the warning message is ATTEMPT TO DIVIDE BY ZERO. This is also printed when a user asks for a response function that is actually infinite, such as the admittance of a 0-ohm resistor. After this message, some of the results, but not all, may be in error.

Fatal errors are caught either by MARTHA or by APL. Among those caught by MARTHA are the wrong length for an argument for an element-definition function. Errors caught by MARTHA induce an explanatory message, usually indicating what was expected, followed by the message MARTHA ERROR and a diagnostic arrow pointing to the place where MARTHA got into trouble. These errors may often be regarded as a useful prompting, because by deliberately committing them, a user can remind himself what is expected. A complete list of these error messages appears in Reference 2.

Errors not caught by MARTHA, but rather by APL, are reported by the message NAME ERROR where NAME is the name of one of the functions in MARTHA. This message may or may not be informative, but it is followed by an indication of the place where MARTHA got into trouble.

To recover from an error a user should type a single right arrow → followed by a carriage return and then proceed after correcting the source of the error.

8.10 REFERENCES

This chapter has not covered all aspects of using MARTHA; in particular, no attempt has been made to describe the MARTHA library fully, because it periodically gets additions. Complete instructions for using MARTHA appear in References 1 and 2 below. Each work space in the MARTHA library has a variable entitled DESCRIBE which gives an up-to-date description of the contents. A succinct summary of MARTHA usage is available on-line in the work space 100 HOWMARTHA.

1. PENFIELD, JR., P., *MARTHA User's Manual*, The M.I.T. Press, Cambridge, Mass., 1971.
2. PENFIELD, JR., P., *MARTHA User's Manual, 1973 Addendum*, M.I.T. Department of Electrical Engineering, Cambridge, Mass., 1973.
3. PENFIELD, JR., P., "Description of Electrical Networks Using Wiring Operators," *Proc. IEEE*, 60, No. 1, Jan. 1972, pp. 49–53.
4. GREENSPAN, M., K. I. THOMASSEN, and P. PENFIELD, JR., "General-Purpose Microwave Circuit Analysis Incorporating Waveguide Discontinuity Models," *Digest of Technical Papers*, 1972 IEEE-GMTT International Microwave Symposium, Arlington Heights, Ill., May 22–24, 1972, pp. 104–106.
5. BHATTACHARYYA, B. B., "Realization of an All-Pass Transfer Function," *Proc. IEEE*, 57, No. 11, Nov. 1969, pp. 2092–2093.
6. BORDEWIJK, J. L., "Inter-reciprocity Applied to Electrical Networks," *Applied Scientific Research*, B6, Nos. 1–2, 1956, pp. 1–74.
7. DIRECTOR, S. W., and R. A. ROHRER, "The Generalized Adjoint Network and Network Sensitivities," *IEEE Trans. Circuit Theory*, CT-16, No. 3, Aug. 1969, pp. 318–323.
8. PENFIELD, JR., P., R. SPENCE, and S. DUINKER, *Tellegen's Theorem and Electrical Networks*, The M.I.T. Press, Cambridge, Mass., 1970, Appendix D.
9. MARCUVITZ, N., *Waveguide Handbook*, McGraw-Hill, New York, 1951, Sec. 5.10.
10. RAMO, S., J. R. WHINNERY, and T. VAN DUZER, *Fields and Waves in Communication Electronics*, Wiley, New York, 1965, Tables 5.14 and 8.09.
11. LEVY, R., and T. E. ROZZI, "Precise Design of Coaxial Low-Pass Filters," *IEEE Trans. Microwave Theory and Techniques*, MTT-16, No. 3, March 1968, pp. 142–147.

9
SCEPTRE

A. BRENT WHITE
MICHAEL A. CHIPMAN

Air Force Weapons Laboratory
Kirtland AFB
Albuquerque, New Mexico

9.1 INTRODUCTION

SCEPTRE (System for Circuit Evaluation and Prediction of Transient Radiation Effects) is a FORTRAN IV circuit analysis program which employs a free-form input language and state variable methods to simulate problems of interest to electrical engineers. While the code may be considered to be a general-purpose network simulation program, the input language and analysis options available have been directed at the analysis of electrical circuitry, generally, and highly nonlinear radiation effects on circuitry, specifically. Analysis options available include nonlinear transient, dc, and ac calculations. Optimization, worst-case, Monte Carlo, and sensitivity calculations are available in the dc sense, and convolution kernels may be used in conjunction with transient analyses. User-defined FORTRAN IV subroutines may be included in circuit simulations allowing the ultimate in modeling flexibility and user control of program execution. The ease of input and multiplicity of analyses available give electrical engineers the capability to simulate a wide range of circuit problems.

Brief History

SCEPTRE was originally developed by IBM Federal Systems Division, Owego, New York, for the transient radiation effects group of the Air Force Weapons Laboratory Kirtland AFB, New Mexico, in 1966. The code was a dc and nonlinear transient program only which employed state variable methods, explicit integration for transient analysis, and Newton-Raphson techniques for dc analysis. The uniqueness of the program was provided by the implementation of algorithms, i.e., the transformation of the network description into a FORTRAN IV subroutine containing the equations of the network. The original code suffered a bit from the exorbitant execution times required by the explicit integration routines on "real-world" problems. dc solutions were troubled by the lack of convergence on circuitry with highly nonlinear characteristics or high gain feedback. Further, the size of the matrix employed in the dc algorithm and the amount of computer time required for its conversion grew drastically as the circuits of interest became larger. IBM, with AFWL support, modified the code through the years to alleviate these problems and to provide additional user utility in a variety of smaller areas.

In 1969 sparse matrix techniques were included in the dc solution algorithm in order to reduce core storage and running time. 1970 saw the inclusion of the Gear implicit integration method in the code with the result that real-world, i.e., "stiff", problems could be simulated cost effectively. Efforts in 1971 resulted in an additional method for dc calculations, i.e., implicit integration of transient equations to steady state at TIME=0, to be employed where the Newton-Raphson method failed. In 1972–1973 GTE Sylvania, Waltham, Mass., assumed the development effort for the AFWL and produced a version of the code which addionally offered analysis, convolution kernels, optimization, sensitivity, worst-case, and Monte Carlo calculations.

Basic Features of SCEPTRE

The following features of SCEPTRE make the use of the code attractive:

1. Free-form input. Descriptions of networks to the code are topology-oriented and free of formatting problems foreign to the engineer.
2. Flexible nonlinear input. Circuit elements may be described as constants, tabular data, or arbitrary functions of other network quantities.
3. Flexible modeling capability. The arbitrary functional capability allows any device to be modeled to the degree of accuracy required for an analysis.
4. State-of-the-art numerical methods. The code simulates networks in a cost-effective manner.
5. Variety of run controls. The user may control simulations to a fine degree rather than "giving" the network to the code and losing control.
6. User-defined subroutines. The user may perform analyses not provided for in the code, describe nonlinearities which may only be represented via FORTRAN, and control code execution in any fashion desired.
7. Error messages. SCEPTRE provides an error message for virtually every possibility of user error.

8. Minimization of computational delays. The sequencing of instructions by SCEPTRE ensures that computational delays in transient analyses will be few even when complex functional dependencies are present.
9. Multiple analyses available. Allow the analyst to use one code to run various analyses on circuitry.

The features of SCEPTRE which are not particularly attractive are the following:

1. Overhead. The amount of computer time used in setting up problems can sometimes get out of hand, though this has been alleviated in recent versions.
2. Problem capacity. SCEPTRE is *hard-wired* to handle up to 300 circuit elements.

Program Availability

SCEPTRE may be obtained free of charge from the Air Force Weapons Laboratory. The latest version is available for IBM 360/370 and CDC 6000/7000 computers. Older versions (transient and dc only) are available for IBM 7094, GE 635/645, Honeywell 6000, and Univac 1108 (Exec II) computers.

Users desiring a copy of the program should send a $\frac{1}{2}$-inch, 2400-foot tape to

AFWL/ELP

Attn: SCEPTRE Project Officer

Kirtland AFB, NM 87117

The tape should be accompanied by a covering letter specifying which version of the code is desired and the return address for the tape.

The SCEPTRE documentation referenced at the end of this chapter can be obtained from the National Technical Information Service (NTIS), 5285 Port Royal Road, Springfield, VA 22151. Document AD (accession numbers) may be obtained from the SCEPTRE Project Officer.

9.2 PROGRAM OPERATION

SCEPTRE is a two-phase program. The first phase processes the input description and formulates equations to be used by the second phase, which generates the analysis answers requested for the given network/equation description. The operation of the program from a control card point of view will be discussed in program operation context. The mathematical formulation, a subject for an extensive chapter in itself, will be discussed specifically for the formulation of the transient equations used by SCEPTRE and in general terms for the dc and ac solutions. The reader will hopefully gain a "feel" for what SCEPTRE does with an input description and how solutions are generated.

Transient Formulation

Starting with the knowledge that a network tree may be chosen such that all circuit elements are either members of the tree (called branches) or not (called links) and that, if the branches are properly chosen, a matrix **B** may be written such that

$$\mathbf{I}_B = \mathbf{B}^T \mathbf{I}_L \quad \text{and} \quad \mathbf{V}_L = -\mathbf{B}\mathbf{V}_B$$

or all branch element currents equal **B** transpose times the vector of link currents, and all link voltages equal minus **B** times the branch voltages. If the link currents and branch voltages can be determined, then the remaining network quantities will be known, given that **B** is known. Let the members of the tree be elements of the network chosen such that no loops are closed and are chosen in a preferred order, E (voltage source), C (capacitor), R (resistor), L (inductor), and J (current source). Elements which close a loop will be links.

Let the following classes of elements be defined:

Class 1   Link capacitors
Class 2   Link resistors
Class 3   Link inductors
Class 4   Branch capacitors
Class 5   Branch resistors
Class 6   Branch inductors
Class 7   Voltage sources
Class 8   Current sources

Voltage sources will be excluded as links, and current sources will be excluded as branches.

In the following discussion **I** refers to a vector of currents, **V** to a vector of voltages, single subscript to element class, double subscript to a two-dimensional matrix, and overdot ($\cdot$) to time derivative, e.g.,

$\mathbf{V}_5$ Vector of class 5 (branch resistor) voltages

$\mathbf{I}_3$ Vector of class 3 (link inductor) currents

$\mathbf{R}_{22}$ Matrix with class 2 (link resistor) elements on diagonal

$\mathbf{B}_{36}$ Portion of **B** matrix applicable between classes 3 and 6

$\dot{\mathbf{V}}_4$ Vector of branch capacitor time derivatives

Tree Branches

	Class 4	Class 5	Class 6	Class 7
Class 1	B_{14}	0	0	B_{17}
Class 2	B_{24}	B_{25}	0	B_{27}
Class 3	B_{34}	B_{35}	B_{36}	B_{37}
Class 8	B_{84}	B_{85}	B_{86}	B_{87}

Links (rows: Class 1, Class 2, Class 3, Class 8)

Figure 9.1 Transient-Solution **B** Matrix in SCEPTRE.

The matrix **B** as shown in Fig. 9.1 will be a matrix of 1s, 0s, and −1s with branch elements horizontal and link elements vertical.

Note that some elements of the **B** matrix are zero-valued. The selection order for the tree (E, C, R, L, J) precludes, as an example, the possibility of a resistor link closing a loop containing an inductor branch ($\mathbf{B}_{26}=0$) since the resistor would have been chosen over the inductor, a branch element. Armed with this knowledge the following equations may be written in matrix form:

$$\mathbf{V}_L = -\mathbf{B}\mathbf{V}_B$$

Therefore,

$$\mathbf{V}_1 = -\mathbf{B}_{14}\mathbf{V}_4 - \mathbf{B}_{17}\mathbf{E}_7 \tag{9.1}$$

$$\mathbf{V}_2 = -\mathbf{B}_{24}\mathbf{V}_4 - \mathbf{B}_{25}\mathbf{V}_5 - \mathbf{B}_{27}\mathbf{E}_7 \tag{9.2}$$

$$\mathbf{V}_3 = -\mathbf{B}_{34}\mathbf{V}_4 - \mathbf{B}_{35}\mathbf{V}_5 - \mathbf{B}_{36}\mathbf{V}_6 - \mathbf{B}_{37}\mathbf{E}_7 \tag{9.3}$$

$$\mathbf{V}_8 = -\mathbf{B}_{84}\mathbf{V}_4 - \mathbf{B}_{85}\mathbf{V}_5 - \mathbf{B}_{86}\mathbf{V}_6 - \mathbf{B}_{87}\mathbf{E}_7 \tag{9.4}$$

$$\mathbf{I}_B = \mathbf{B}^T \mathbf{I}_L$$

Therefore,

$$\mathbf{I}_4 = \mathbf{B}^T_{14}\mathbf{I}_1 + \mathbf{B}^T_{24}\mathbf{I}_2 + \mathbf{B}^T_{34}\mathbf{I}_3 + \mathbf{B}^T_{84}\mathbf{J}_8 \tag{9.5}$$

$$\mathbf{I}_5 = \mathbf{B}^T_{25}\mathbf{I}_2 + \mathbf{B}^T_{35}\mathbf{I}_3 + \mathbf{B}^T_{85}\mathbf{J}_8 \tag{9.6}$$

$$\mathbf{I}_6 = \mathbf{B}^T_{36}\mathbf{I}_3 + \mathbf{B}^T_{86}\mathbf{J}_8 \tag{9.7}$$

$$\mathbf{I}_7 = \mathbf{B}^T_{17}\mathbf{I}_1 + \mathbf{B}^T_{27}\mathbf{I}_2 + \mathbf{B}^T_{37}\mathbf{I}_3 + \mathbf{B}^T_{87}\mathbf{J}_8 \tag{9.8}$$

Differentiating Eqs. (9.1) and (9.7) gives

$$\dot{\mathbf{V}}_1 = -\mathbf{B}_{14}\dot{\mathbf{V}}_4 - \mathbf{B}_{17}\dot{\mathbf{E}}_7 \tag{9.9}$$

$$\dot{\mathbf{I}}_6 = \mathbf{B}^T_{36}\mathbf{I}_3 + \mathbf{B}^T_{86}\dot{\mathbf{J}}_8 \tag{9.10}$$

At this point we may say we know the values of V_4 and I_3 since these values will be predicted by the integration routine at each time step. E_7 and J_8 are known since these are voltage and current sources, respectively. Derivation of the remaining network quantities for the transient solution follows. The resistive quantities are determined by writing

$$V_2 = R_{22}I_2 \tag{9.11}$$

$$V_5 = R_{55}I_5 \tag{9.12}$$

The current I_2 may be derived by manipulating Eqs. (9.2), (9.6), (9.11), and (9.12) to get

$$I_2 = M_R^{-1}\{-B_{24}V_4 - B_{27}E_7 - B_{25}R_{55}[B_{35}^T I_3 + B_{85}^T J_8]\} \tag{9.13}$$

where

$$M_R = R_{22} + B_{25}R_{55}B_{25}^T \tag{9.14}$$

The I_2 vector is now known in terms of V_4, I_3, E_7, and J_8. I_5, V_2, and V_5 may now be determined from Eqs. (9.6), (9.11), and (9.12). SCEPTRE may alternatively solve resistive quantities by dropping out V_5 in a similar fashion if that method is more advantageous in terms of the matrix inverse involved (i.e., M_R^{-1}).

The capacitance values to be solved for the I_1, I_4, V_1, and $\dot{V}_4$, V_4, again, is predicted by the integration routine. The fundamental capacitance equation gives

$$\dot{V}_4 = S_{44}I_4 \tag{9.15}$$

$$\dot{V}_1 = S_{11}I_1 \tag{9.16}$$

Manipulating Eqs. (9.5), (9.9), (9.15), and (9.16) and solving for I_1 gives

$$I_1 = M_S^{-1}\{-B_{14}S_{44}[B_{24}^T I_2 + B_{34}^T I_3 + B_{84}^T J_8] - B_{17}\dot{E}_7\} \tag{9.17}$$

where

$$M_S = S_{11} + B_{14}S_{44}B_{14}^T$$

I_1 has been determined from known quantities. Equations (9.5), (9.1), and (9.15) together with I_1 will give the equations for I_4, V_1, and $\dot{V}_4$.

There remains at this point only the inductor quantities $\dot{I}_3$, V_3, V_6, and I_6. Vector I_3 is known. The relationship for inductors allow

$$V_3 = L_{33}\dot{I}_3 + L_{36}\dot{I}_6 \tag{9.18}$$

$$V_6 = L_{63}\dot{I}_3 + L_{66}\dot{I}_6 \tag{9.19}$$

Combining Eqs. (9.3), (9.10), (9.18), and (9.19) and solving for $\dot{I}_3$ gives

$$\dot{I}_3 = M_L^{-1}\{-B_{34}V_4 - B_{35}V_5 - B_{37}E_7 - [B_{36}E_{66}B_{86}^T + L_{36}B_{86}^T]\dot{J}_8\} \tag{9.20}$$

where

$$M_L = L_{33} + L_{36}B_{36}^T + B_{36}L_{63} + B_{36}L_{66}B_{36}^T$$

V_3, V_6, and I_6 may now be obtained from Eqs. (9.10), (9.18), (9.19), and (9.7), respectively.

The only remaining network quantities to be determined are I_7 and V_8, which may be obtained directly from Eqs. (9.8) and (9.4) since all other quantities are now known.

The only other items to discuss as regards the transient equations are the handling of the source derivatives in Eqs. (9.17) and (9.20) and the handling of linearly dependent sources. These topics are not necessary for a general understanding of the solution process used by SCEPTRE and are not presented here. The reader should see Reference 4 for further information.

dc, Monte Carlo, Optimization, Sensitivity, Worst-Case and ac Solution

The following paragraphs present in gross terms the mathematical algorithms employed for SCEPTRE solutions of dc and ac problems.

dc Solution

The dc solution employed by SCEPTRE is a Newton-Raphson iterative scheme or a Jacobian derived via a **B** matrix and equations resulting from choosing the network tree according to the preference order E, L, R, C, J. Equations are written in a fashion similar to the transient formulation, certain assumptions on network elements made which simplify the solution process (see section on pathological problems), certain network elements differentiated with respect to each other, and a Jacobian suitable for iteration to dc conditions set up. The iteration is performed in Phase II and continues until the convergence criteria for each quantity are met.

Monte Carlo

Monte Carlo solutions are available with the dc algorithm. Very simply, each network quantity given bounds or mean and standard deviation is varied for successive dc solutions by randomly picking values for these quantities, iterating to a dc solution, and collecting the requested output quantities in terms of minimum and maximum values observed over all iterations, the mean, and the standard deviation.

Sensitivity, Worst Case, and Optimization

Sensitivity, worst-case, and optimization solutions are generated via the use of the adjoint network as put forth by Director and Rohrer [13, 14] and with the dc iteration already discussed. The sensitivity calculation produced is not normalized and is simply $\partial \mathbf{H}/\partial x$, where $\mathbf{H}$ is a network function and x is a network element. Optimization is performed by the minimization of user-defined objective functions using the Davidon method, which is essentially a steepest descent start at the optimization process with a gradual change to Newton-Raphson iteration in the area of a solution. Worst case is performed by extrapolating the gradient vector for the objective function linearly from the nominal operating point to the limits specified for the independent variables given in a particular run. Worst-case values are calculated for both the high and low ends of the objective function gradient. Optimization may be used for calculation of the worst-case value for a network function.

ac Calculation

ac analysis of circuits is performed by complex variable operations with state variable equations similar to the transient equations. Nonlinear network quantities are linearized about "time zero" values prior to the ac calculations.

SCEPTRE Execution

The following discussion describes the operation of the coding employed in SCEPTRE in terms of the sequential operations which one would translate into control cards for a particular computer. The tasks performed by the program at each step are discussed and the files used by the code are mentioned. Hopefully the two-phase nature of SCEPTRE will become clear to the reader.

1. Attach the two SCEPTRE files to the system. Typically this is done by requesting tape to be mounted and copying the two phases onto disk files. Permanent files of one form or another are certainly available.
2. Mount the permanent model library tape if required. Logical unit 10 is the file used by Phase I to search for permanent model library references.
3. Load and execute Phase I of SCEPTRE. Phase I processes the SCEPTRE input descriptions (MODEL DESCRIPTION, CIRCUIT DESCRIPTION, RUN CONTROLS, ELEMENTS, RERUN DESCRIPTION, etc.), formulates the network equations, and sets up solution controls for Phase II, which will actually generate the requested solutions. Phase I creates a FORTRAN subprogram which contains the dc and/or transient equations, as required for a requested analysis. ac equations are passed to Phase II via tables of element array indices. The FORTRAN subprogram written is called SIMUL8 and may be split into a controlling routine, the dc equation portion (called SIMIC), the transient equations (called SIMTR) and ac equations (called SIMAC). Additionally, Phase I sets up tables

containing the network values described in a form compatible with the FORTRAN READ statements, generated as part of the SIMUL8 program. Logical unit 8 is the file which contains all this information at the end of Phase I execution.

4. Compile the Phase I generated SIMUL8 program. Some versions of SCEPTRE process a user-defined subprogram and include it on the logical unit 8 file, and others require the analyst to insert an additional compile step for the subprogram(s). This compile should generally be accomplished after the compilation of SIMUL8.

5. Load and execute the Phase II utility routines with the compiled SIMUL8 program and execute. Phase II will read the network values and execution controls from logical unit 8 and perform the analyses requested. Phase II operation is divided into simulation and output presentation. Reruns are automatically performed by determining whether additional network data are present on logical unit 8 at the end of output presentation and, if so, the read/simulation/output cycle is reentered.

9.3 NETWORK ELEMENTS AND UNITS

The list of network elements available in SCEPTRE has been kept as simple as possible without limiting the flexibility or potential of the code. The basic set of network elements available are resistors, capacitors, inductors, voltage sources, current sources, mutual inductance, stored network models, and the convolution kernel.

Flexibility and potential with a limited elements list are achieved with the techniques used in SCEPTRE to resolve and treat complex dependencies without numerical instability or computational delays. It is therefore possible to create voltage-controlled voltage sources, current-controlled voltage sources, etc., limited only by the ingenuity and experience of the user. Functional dependencies may be specified by several methods including user-supplied FORTRAN function subprograms.

While the versatility of this approach is apparent, responsibility is placed on the user to ensure that the dependencies specified and that decision-making processes in subroutines do not result in inconsistencies.

Responsibility is also placed on the user to select and maintain an appropriate and consistant set of units when describing circuit parameter values to SCEPTRE. For example if high-speed transient behavior of networks is of primary interest, then the units in column A of Table 9.1 may be the most practical. On the other hand, a study of the behavior of a control system containing electromechanical elements may dictate the choice of units in column B of Table 9.1.

Table 9.1 *Examples of Consistent Sets of Units*

Parameter	Unit (set A)	Unit (set B)
Resistance	Kilohms	Ohms
Capacitance	Picofarads	Farads
Inductance	Microhenries	Henries
Current	Milliamperes	Amperes
Voltage	Volts	Volts
Frequency	Gigahertz	Hertz
Time	Nanoseconds	Seconds

If another system of units is desired, the most effective procedure is to select units of voltage, current, and time that correspond to the magnitudes of those values that are anticipated and to determine units of R, L, C, frequency, etc., from fundamental relationships. Regardless of the unit system ultimately selected, the user must ensure that all parameters are expressed in terms of the consistent set of units being used.

9.4 INPUT LANGUAGE

The SCEPTRE circuit description language is a user-oriented free-format language that offers wide flexibility and potential to the user. The use of delimiters results in a visual display of the input which is intuitive and easily interpreted by the engineer. Input information may be punched on cards anywhere on the input card in columns 1–72 with any desired spacing or intervening blanks. In general, several complete statements can be punched on a card, separated by commas, and continuation from one card to another can be made immediately after a delimiter, e.g., $+$, $-$, $*$, $/$, $,$, $.$, $($, $)$, and $=$, with the delimiter appearing as the last nonblank character on the card.

Networks are described in SCEPTRE language under major headings (or modes) and subheadings (or groups). In general, the major headings allow the user to activate general features and options available e.g., storing of models, description of circuits, repetition of simulations with changed parameters, continuation of a simulation or additional output generation, while the subheadings provide the data necessary for the execution of that option or feature.

When preparing input for SCEPTRE, the first elements of the language which must be considered are the mode and group headings. Table 9.2 gives a listing of the modes and groups allowed under each mode.

Table 9.2 *Modes (Major Headings) and Allowable Groups (Subheadings)*

Groups (Subheadings)	MODEL DESCRIPTION	CIRCUIT DESCRIPTION	RERUN DESCRIPTION(n)	CONTINUE	REOUTPUT	END
MODEL	X					
ELEMENTS	X	X	X			
DEFINED PARAMETERS	X	X	X			
OUTPUTS	X	X			X	
FUNCTIONS	X	X	X			
RUN CONTROLS		X	X	X	X	
INITIAL CONDITIONS	X	X	X			
SENSITIVITY		X				
MONTE CARLO		X				
WORST CASE		X				
OPTIMIZATION		X				

Modes or Major Headings

The **MODEL DESCRIPTION** heading introduces one of the important features of SCEPTRE, that of stored models with arbitrary topology. One or more models may be stored either temporarily or permanently under a single **MODEL DESCRIPTION** heading. Each model begins with a **MODEL name** card which gives the type of model (permanent or temporary) and nodal interconnection data. The name may be any group of alphanumeric characters up to a total of 18. The nodal interconnection data may contain a maximum of 25 nodes. Comments pertaining to a particular model, up to a total of 11, may appear between the **MODEL name** card and the first group card within the model. Any one or all 6 groups may appear after the optional comments.

The section on modeling will further discuss the specific groups used in the most common models.

The **CIRCUIT DESCRIPTION** heading is used whenever any network (whether that network is electrical, an electrical analogy, or a set of differential equations) is to be solved. Any of the 10 groups may be used. Note that defined parameters and their derivatives allow the solution of any set of first-order, nonlinear, simultaneous differential equations without any circuitry involved.

The **RERUN DESCRIPTION(n)** heading is used whenever the same network, with different parameters, is to be simulated. This allows a considerable savings in overhead since the equations remain constant, with only the values changing. Again, any one or all of the allowable groups may be used.

The **CONTINUE** heading is intended for use only when a previously started run is to be restarted for the purpose of continuing the simulation. This feature can be used to check the progress of a long simulation or as a data save feature in case of computer failure. The subheading allowed in this mode is **RUN CONTROLS**. Before attempting a **CONTINUE** run, all data generated in Phase II and the simulation program generated by Phase I must be saved.

The **RE-OUTPUT** heading is used to repeat the output of a run without repeating the calculations. Again it is necessary to save some of the data generated by Phase II if the **RE-OUTPUT** feature is to be used.

The **END** card is unique in our discussion of mode headings since it is the only one which *must* appear. Every SCEPTRE input data deck *must* be terminated with an END card. No group or mode cards are allowed after the END card.

Groups or Subheadings

The groups are subheadings under each mode specification. Between a mode card and the first group card within that mode there may appear up to 11 comment cards. These comments need not follow any specific format.

The data supplied within each of the groups consists of descriptive statements which are constructed as properly punctuated sequences of symbols. The group subheading and subsequent statements may be punched on a card with arbitrary spacing or location from columns 1–72. During a single computer run a user may not use all six modes, although all of the groups under **CIRCUIT DESCRIPTION** could well be used. Sequences of symbols and punctuation are used to convey information in each of the groups. Definitions for each of the symbols used follow:

ELEMENT name. This denotes the name given to each component of a circuit. No more than five alphanumeric characters may be used to name an element, including the type prefix R, L, C, J, E, or M (see Table 9.3). Model circuit element designations are limited to no more than four alphanumeric characters, e.g., R105, CC12A, TA10.

Node. This denotes the designation assigned to each node of a circuit. No more than six alphanumeric characters may be used to name a node, e.g., GND, A1, 105.

Number. A numerical constant that may be written as a signed quantity in either integer or decimal form with or without an exponent. Up to 13 characters may be used to represent a number, e.g., 528, 52.8, -5, $-5.28E-2$.

Constant. Same as number except that a decimal point must be included in the specification of the numerical constant.

Value. The term *value* will be used to denote any of the following: number, defined parameter, table, equation, expression, or external function, e.g., 52.8, PX1, TABLE 10(VCC), EQUATION5 (PX1, VL3), X10(5.28*VCC*(3.3/VR3)), X2(FCAP(VCE)).

Special value. This will be used for any of the following: value, constant*resistor current, constant*resistor voltage, value*current source, **DIODE TABLE**, or **DIODE EQUATION (X1, X2)**.

Variable. A variable can mean any of the following:

1. The voltage or current associated with any element, e.g., VR1, IE2.
2. Any source or source derivative, e.g., JE, E1, DJE, DE1.
3. Any defined parameter, e.g., PX1.
4. Any element value, e.g., RA, C528, E10.
5. Time as TIME.
6. Any internal SCEPTRE parameter, e.g., XSTPSZ, XTISSS (see Reference 4).

Velement name or **Ielement name.** These special prefixes denote the element voltage or current of an element name. For example, the voltage across a capacitor C15 would be referred to as VC15. Similarly, the current through that capacitor would be referred to as IC15.

TABLE name (independent variable). Tables are used when a variable circuit quantity is given in a tabular form. The table used must be given a unique name prefixed by the word **TABLE** or simply the letter **T** and followed by a single independent variable. The name may consist of up to five alphanumeric characters. The independent variable can be any of the quantities defined under variable. If an independent variable, including the enclosed parenthesis, is not supplied, **TIME** will automatically be chosen, e.g., TX1(VCC), TABLE V. TABLE V is an implicit function of **TIME**.

Table 9.3 Entries Under Elements

Name	Nodes		Value Specification[9]
R ,	⎫		⎧ number
C ,	⎪		⎪ TABLE name (independent
L ,	⎬ Node—node	= Value	⎪ variable)
E ,	⎪		⎨
J ,	⎭		⎪ Defined parameter
			⎪ EQUATION name (argument list)
M ,	L name—L name		⎪ EXPRESSION name (mathematical
			⎪ definition)
See notes			⎪ External function
1 and 2			⎩ (Argument list)

Linearly Dependent Sources[13]

E	, Node—node	= Constant∗VR
J	, Node—node	= Constant∗IR

Primary Dependent Current Sources[3,4,13]

J	, Node—node	= DIODE TABLE name
		DIODE EQUATION (x1, x2)

Secondary Dependent Current Sources[4,5,13]

J	, Node—node	= Value∗J(J is a primary dependent current source)

Voltage and Current Source Derivatives[6,8]

DE	= Value
DJ	= Value

Model Calls

Circuit Designation[7]	Nodes	Model Name
	⎧ Node—node	= MODEL 1NXXX(PERM)
Name ,	⎨ Node—node	= MODEL 2NXXX(TEMP)
	⎩ Node—node … node	= MODEL XYZ(TEMP)

Elements with Bounds

R ⎫		
E ⎬	, Node—node	= ⎧ Number (number,number)
J ⎭		⎩ Number (number)

ac Sources

E ⎫		⎧ (entry,entry)
J ⎭	, Node—node	⎨ (entry,entry), DEGREES
		⎪ (entry,entry), RADIANS
		⎩ (entry,entry), COMPLEX

where each entry is one of the
following   Constant
            Defined parameter
            TABLE name
            FREQ

Table 9.3—Cont.

Convolution Model Calls[10,12]

Circuit	Nodes		Convolution Kernel[11]
Kname	, Node—node	=	{FCONVE (constant) {FCONVJ (constant)

1. All C, R, L, and M entries will be treated as constants in the ac calculations. Before making the ac calculations, SCEPTRE will evaluate any entries which are given as functions. For example, a capacitor may be a function of voltage and a resistor may be a function of time. SCEPTRE will evaluate these functions at TIME=0, using supplied or calculated initial conditions. The appropriate values so obtained will be used in the ac calculations. *Elements as a function of frequency are not permitted in ac calculations.*

2. All voltage and current sources which are given as constants (dc sources) or as functions of TIME (transient sources) will be given a value of zero in ac calculations, and a WARNING message will be printed out whenever the argument TIME is used.

3. To obtain an ac analysis around a circuit's dc operating point, the user must either supply initial conditions or must enter RUN INITIAL CONDITIONS under the RUN CONTROLS subheading of CIRCUIT DESCRIPTION. If initial conditions are supplied, then SCEPTRE determines all element values dependent on these dc voltages and currents.

4. Primary and secondary dependent current source specifications can be used to represent certain semiconductor junctions when requesting initial conditions solutions.

5. By definition this class of current source can appear only if the appropriate diode source has previously been included. The secondary source must be specified only as a value times the primary source.

6. Although this entry is permitted in ac calculations, a time-source, and hence its derivatives, is not meaningful. The source will be treated as in note 2 above and the derivatives ignored.

7. Model circuit designation names can be any combination of no more than four alphanumeric characters. Names unique from other element names are recommended.

8. If the topology of the circuit dictates that a time derivative is required for a transient analysis (e.g., capacitor tie set and independent voltage source, or inductor cut set and independent current source), then the same is true for an ac analysis. However, for ac no card entry is required. A time derivative of an ac source means simply a multiplication by $j\omega$. SCEPTRE will detect this situation and will automatically handle it.

9. A complex defined parameter, W, cannot be used as a value specification.

10. Convolution models are either series combinations of voltage source and resistor (FCONVE) or parallel combinations of current source and resistor (FCONVJ).

11. The constants in the convolution model call are arbitrarily assigned integers identifying the impedance or admittance functions stored on disk.

12. A discussion of the convolution feature of SCEPTRE is beyond the scope of this text. For details on the use of convolution, the reader is referred to Reference 4 at the end of this chapter.

13. These special sources can be used to specify dependencies for an ac analysis. They will be treated as nonzero complex-valued sources and will be automatically linearized.

EQUATION name (argument list). When a circuit element quantity may be described by a closed-form equation, this identifier should be used. The equation must be given a name prefixed by the word **EQUATION** or alternatively the letter **Q** followed by one or more arguments separated by commas and enclosed in parentheses. The equation name may consist of up to five alphanumeric characters. The argument list may consist of any variable, constant, or table, e.g., EQUATION 9 (PX1, TIME, 52.8), Q105 (VCC, 1.5, TABLE A).

EXPRESSION name (mathematical definition). This is an alternative to the equation entry that may be used to describe a variable quantity. It is particularly useful when an expression is to be used only once, as it requires no further description under the function group heading. The expression must be given a name prefixed by the word **EXPRESSION** or the letter **X** followed by a FORTRAN mathematical definition. The expression name may consist of up to five alphanumeric characters. To avoid any possible confusion with internal parameters one should use numerical designations, such as **X1**, **X25**, etc. The mathematical definition is enclosed in parentheses, e.g., X5 (VCC*TABLE 3(IR4)).

To assemble the input data deck, each section is punched and put into its respective group. That group in turn is placed under the mode of interest. Each of the specifications is uniquely named by the code and no further bookkeeping on the part of the user need be accomplished. Take an example an element **R1** which might appear under the **ELEMENTS** heading of the **MODEL DESCRIPTION** mode. A different **R1** might appear under **ELEMENTS** of the **CIRCUIT DESCRIPTION** mode and SCEPTRE will not confuse the elements.

The input data deck describing a network is formed by punching the heading card and the associated data sequence for each of the defined data groups. The sequences for each of the groups in terms of the symbols defined will now be discussed.

Element Definition

All elements, i.e., resistances, capacitances, inductances (including mutuals), voltage and current sources, source derivatives, and model circuit designations, which are to be component parts of the network under analysis, must be introduced in the **ELEMENTS** group. Each element name must begin with the appropriate prefix, i.e., **R**, **C**, **L**, **E**, **J**, or **M** for resistors, capacitors, inductors, voltage sources, current sources, or mutual inductance, respectively. Model circuit designation names can be any combination of up to four alphanumeric characters. No more than two characters are recommended, however, and the user may find it convenient to avoid using prefixes reserved for other elements to avold confusion.

Entries allowed under this group are summarized in Table 9.3. The general form for entries under the **ELEMENTS** group heading is

$$\text{ELEMENT } name, node - node \ldots = value$$

Each network element must be defined by stating first the element *name* and then the nodes between which the branch is connected and finally the component value. The interconnecting nodes are specified in from-to order, corresponding to the assumed direction of positive current flow. More than one element can be described on a card if the element descriptions are separated by commas.

Several often-used special values have been included in the formation of SCEPTRE. When analyzing small-signal transistor equivalent circuits, it is necessary to have linearly dependent sources. Additionally, when using the large-signal Ebers-Moll transistor equivalent circuit, it is necessary to have a class of secondary dependent sources. Special processing algorithms for these sources have been included in the program, and they should always be entered directly in the elements subheading in the proper format. Primary dependent current sources which are used to represent diodes or transistor junctions must also be specified in a particular format. **DIODE EQUATION (X1, X2)** should be used when any diode or transistor junction has been entered and the user wishes to employ the conventional closed-form representation of the diode current generator:

$$I = I_s(e^{\theta V} - 1)$$

where

$$\theta = \frac{q}{\eta k T}$$

The value of I_s in the diode equation is represented by the value of **X1** and the value of **X2** corresponds to θ. SCEPTRE will automatically assume that the independent variable to be used is the voltage across that particular current generator. For this reason the voltage need not be specified. It should be noted here that both arguments to the diode equation must be described as constants, that is, they must have decimal points included.

An alternative method of describing a diode or transistor junction is the use of a tabular function by using the key words **DIODE TABLE name**. A table look-up call is generated using the voltage across the current generator as the independent variable.

Mutual inductances are to be entered according to the general format

$$Mname, Lname - Lname = value$$

A physical limitation of the principle of mutual inductance must be observed in order to have a physically realizable circuit; i.e., the mutual inductance between any two inductors must be less than the square root of the product of the self-inductances of the components between which the mutual inductance exists. Further, the matrix of coupling for a network must be positive-definite.

Topologies are sometimes encountered where a variable voltage source or current source requires its appropriate derivative. Source derivitives will be required whenever dependent voltage source-capacitor loops or dependent current source-inductor cut sets exist in the network. The general form for source derivative entry is

$$DEname = value$$
$$DJname = value$$

where the *name* is that of the appropriate E or J source. A SCEPTRE diagnostic message is printed if source derivitives are required and have not been supplied. The need for source derivitives can be avoided altogether by placing appropriate valued resistors in the network to eliminate **E-C** loops or **J-L** cut sets.

Special additional information is required when running either the ac or the three dc options, Monte Carlo, worst case, and optimization. For the dc options it is necessary to specify parameters for distribution of the variable elements for the Monte Carlo solution, and for worst-case and optimization calculations minimum and maximum values of independent variables must be specified. In all cases the element information is provided by bounds added in parentheses after the element values. With the exception of the means and standard deviations for the Monte Carlo solution, the bounds are provided by statements under **ELEMENTS** of the form

Element name, node — node = value (value, value)

or

Element name, node — node = value (value)

The first of these forms gives two values in parentheses. SCEPTRE reads the smaller number as the lower bound and the larger as the upper bound. The second form has one number in parentheses. This format assumes that the number is a percentage variation allowed in the nominal value of the **ELEMENTS**. For worst case and optimization the lower and upper bounds are taken to be the limiting values for the elements. For these calculations the nominal value must lie within the stated limits. For Monte Carlo calculations, the element distribution mean and standard deviation are computed from the lower and upper bounds as

$$\text{mean} = \frac{\text{upper bound} + \text{lower bound}}{2}$$

$$\text{standard deviation} = \frac{\text{upper bound} - \text{lower bound}}{6}$$

This exception to the use of numbers exclusively in specifying elements within bounds is as follows: Elements with bounds may be specified as values or diode equations if the values or diode equations are expressed in terms of defined parameters with bounds under **DEFINED PARAMETERS**.

Source voltage and currents for ac calculations are complex numbers and therefore require both real and imaginary parts (or magnitude and phase) for their definition. The format for entering an ac source is

$$E name, node - node = (entry, entry)\ type$$

The word entered for type identifies the meaning of the two entries in parentheses. Type may be either **DEGREES, RADIANS**, or **COMPLEX**. If **DEGREES** or **RADIANS** is entered, the first entry in parentheses is the magnitude in the polar coordinate expression of the voltage or current and the second entry is phase angle. If type is specified as **COMPLEX**, the first entry in parentheses is the real portion of the complex expression in cartesian coordinates and the second entry is the imaginary portion. If type is not specified, it has the default value of **DEGREES**.

An *entry*, as described above, may be one of the following: a constant, the problem frequency denoted by **FREQ**, a **TABLE name** (where the independent variable must be stated in the **TABLE name** because the default value of the independent variable must be supplied and must not be **TIME** as **TIME** is set to zero for ac calculations), or a defined parameter.

The maximum allowed number of independent ac sources is 50. The maximum number allowed for a linearly dependent ac source plus secondary dependent ac current sources is also 50.

Defined Parameter Definition

The group of inputs called **DEFINED PARAMETERS** give SCEPTRE a great deal of versatility. They may be used to combine circuit variable for the purposes of output, they may define network variables, they may be used as an argument of an equation or table, or they may specify a system of differential equations. The input format for defined parameters required that the first letter be **P** followed by no more than five alphanumeric characters. The general form for entries under **DEFINED PARAMETERS** is

$$P name = value$$

and the format for the derivative of a defined parameter is

$$DP name = value$$

For the special case in which the derivative of a defined parameter quantity is supplied, the first two characters must be DP followed by no more than four alphanumeric characters.

Use of the defined parameter is illustrated in several of the example problems presented in Sec. 9.7 of this chapter.

Special types of defined parameters are required in some situations when using the ac option and the dc options with other than initial conditions. Real-valued defined parameters with bounds may be used as independent variables in dc calculations. When defined parameters are used in this manner, bounds are specified under **DEFINED PARAMETERS** in a manner similar to the bounds placed on elements in the **ELEMENTS** group heading. The format is

$$P name = value\ (value, value)$$

or

$$P name = value\ (value)$$

Again, the first form gives two values in parentheses which are read as the lower and upper bounds. The second has one value in parentheses, which is interpreted as the percentage variation allowed in the nominal value of the dependent variable. For worst-case and optimization calculations the nominal value must lie within the region defined by the upper and lower bounds.

When the user is exercising the optimization, sensitivity, or worst-case features of the dc program, adjoint calculations are made. In these instances the user must specify the closed-form differential for each parameter that is used as a dependent variable. The defined parameter must be a function of one or more adjoint dependent variables and may be a function of an independent variable. Valid dependent variables and independent variables for these calculations may be found in Table 9.4.

Table 9.4 **Dependent and Independent Variables in dc Calculations**

Name	*Dependent Variables* Description (see Table 9.3)
VC	Voltage across a capacitor
IE	Current through an independent voltage source
J	Value of a primary dependent current source
VJ	Voltage across an independent current source
IL	Current through an inductor
IR	Current through a resistor
VR	Voltage across a resistor
P	Defined parameter*

Name	*Independent Variables* Description (see Table 9.3)
R	Resistor
E	Independent voltage source
J	Independent current source
P	Defined parameter†

*For a sensitivity, optimization, or worst-case dependent variable Pname, the differential GPname must also be supplied.

†A defined parameter independent variable can only be the factor in a secondary dependent current source definition, e.g., J0=Pname∗J9.

The user should enter differentials of defined parameters under the **DEFINED PARAMETERS** subheading of **CIRCUIT DESCRIPTION**. For a defined parameter, **Pname**, the total differential is given by

$$G Pname = list$$

List is a sum of products of the form **PX∗DY**, where **PX** is a defined parameter and **Y** is either an independent or dependent variable name.

For example, if the following defined parameter is given,

$$PWR = X1 \ (IRLD**2 + ILX**2)$$

then if an option requiring adjoint calculations is made, the following definition must be made:

$$GPWR = P2*DIRLD + P3*DILX$$

and

$$P2 = X94 \ (2.*IRLD)$$
$$P3 = X81 \ (2.*ILX)$$

Another class of defined parameters which have complex values are included to enable complex outputs to be calculated within the ac analysis portion of SCEPTRE. These are analogous in principle to the real-valued defined parameters which are prefixed with a **P**. The appropriate prefix for the complex-valued defined parameter is **W** followed by no more than five alphanumeric characters. Unlike the real-valued defined parameter, *a complex-valued defined parameter cannot be used to define ELEMENTS;* i.e., it cannot appear in an equation, expression, table, or function. W parameters can be used only for the calculation of output values.

The acceptable entry for W parameters under the DEFINE PARAMETERS subheading is

$$Wname = value$$

Here the term *value* may be applied to a real-valued defined parameter, a table, an equation, an expression, or a user-supplied function. It is the user's responsibility when writing FORTRAN programs to ensure that the correct declaration and usage of complex-valued quantities are maintained. An alternative format is allowed for specifying the **COMPLEX**-valued defined parameters which is analogous to that used for specifying ac sources. This format is

$$Wname = (value, value) \ type$$

where the details are as described in the paragraph in ac sources.

Optimization, Worst Case, Sensitivity, and Monte Carlo

Group headings under the **CIRCUIT DESCRIPTION** mode heading are provided for the user to identify the variables he wishes treated in a sensitivity, worst-case, optimization, or Monte Carlo calculation. The group headings are **OPTIMIZATION, WORST CASE, MONTE CARLO** and

SENSITIVITY followed by the parameteric data appropriate to the heading used. Each set of parameters is enclosed in parentheses. The objective function or dependent variables are listed first, followed by a slash and a list of independent variables. Individual variables in the list are separated by commas. Each set must name at least one dependent variable. The form is

(*dependent variable list/independent variable list*)

The allowable dependent and independent variables are listed in Table 9.4.

An example of valid sensitivity input data cards is

SENSITIVITY
(IL3, PX/JX,EY)
(VR1, IR1, VCX/P1,E1)

In this example the following partial derivatives would be calculated:

$$\frac{\partial IL3}{\partial JX}, \frac{\partial IL3}{\partial EY}, \frac{\partial PX}{\partial JX}, \frac{\partial PX}{\partial EY}, \frac{\partial VR1}{\partial P1}, \frac{\partial VR1}{\partial E1}, \frac{\partial IR1}{\partial P1}, \frac{\partial IR1}{\partial E1},$$

$$\frac{\partial VCX}{\partial P1}, \frac{\partial VCX}{\partial E1}$$

An example of optimization input data is

OPTIMIZATION
(IR1, P1/R1,P2,E1)
(VC1/J1,R2)

In this example

IR1 is optimized with respect to **R1, P2, E1**.

P1 is optimized with respect to **R1, P2, E1**.

VC1 is optimized with respect to **J1, R2**.

An example for a worst-case calculation would take a form like those above under a

WORST CASE

subheading.

SCEPTRE has no specific limit on the number of dependent variables or independent variables. Rather, the limit is on sets of variables. A set is a list of dependent variables together with its list of independent variables. The total number of all sets entered for worst case plus sensitivity and optimization must not exceed 100. Additionally, four times the total number of dependent variables plus the total number of independent variables must not exceed 400.

Initial Conditions Specifications

There are two methods of specifying the initial or steady-state values for network inductor currents and capacitor voltages. The first method is to make use of the automatic initial conditions solution capability provided as a feature of SCEPTRE. The second is for the user to input manually the required initial conditions. The set of all capacitor voltage and inductor current values existing at the start of a problem are sufficient for this purpose. The probability that the Newton-Raphson iteration for automatic initial conditions will converge is greatly enhanced if the user supplies an initial guess by manually inserting approximate initial conditions at the start of the run. Otherwise, the Newton-Raphson iteration begins from zero values. Initial capacitor voltages and inductor currents may be supplied by listing the desired values in the following format:

$$\text{VC}name = value, \ldots, \text{IL}name = value$$

Any initial condition not specified will be taken as zero. If all initial conditions are zero, neither the heading card nor the data are required.

Care must be taken to establish the proper polarities for initial conditions. Initial inductor currents are positive if they flow in the same direction as the assumed direction of positive current flow for the inductor. The initial capacitor voltages are positive when they are consistent with the assumed voltage polarity. In addition to the state variables, i.e., capacitor voltages and inductor currents, one other class of initial conditions may be legitimately entered. This class is intended for the case in which the user wishes to supply the starting values for diode current generators when using the dc portion of the program. In this situation the initial voltage across diodes or transistor junctions may be specified. The format is

$$\text{VJ}name = number$$

If the dc portion of the program is used, the iterative procedure will begin with any initial conditions entries that are supplied.

Functions Specification

In the previous discussion of elements and defined parameters it was noted that one of the *values* which could be specified was an equation and another was a table. Specific definition sequences are defined under the **FUNCTIONS** group card. If no such references have been made, neither the group card nor the data need be supplied.

Each different equation used to specify the variation of an element or defined parameter is defined by giving the equation a name, a dummy variable list, and a mathematical definition. The general format is

$$\text{EQUATION } name \, (list) = (mathematical\ definition)$$

The dummy variable list must contain the same number of entries as does the argument list in the original equation reference. Each dummy variable may contain up to six alphanumeric characters, the first of which must not be a number or the letters **I** through **N** inclusive.

The mathematical definition must be included in parentheses and must be written in terms of the dummy variables along with any constants and allowable subprogram functions which apply. It is important to mention that there would be no need for the user to reserve the dummy variables for a single equation. A single set of dummy variables may be freely used in other equations to represent other circuit quantities. The equation is most powerful when it can be used as a single mathematical sequence for several purposes. An example might be in the large-scale transistor model presented by Ebers and Moll. The two junction capacitances are computed from the same mathematical relationship. A different set of constants and circuit variables is used when computing the value of each capacitance, but the **form** of the equation is the same for the emitter-base and the collector-base capacitance. For example, the two junction capacitors in an Ebers-Moll transistor model (see Sec. 9.6) may be entered as follows:

```
ELEMENTS
CE, 1-E = EQUATION 1 (5.,70.,J1)
CC, 1-2 = EQUATION 1 (8.,370.,J2)
  .
  .
  .

FUNCTIONS
EQUATION 1 (A,B,C) = (A+B*C)
```

The user will find the **EXPRESSION** considerably more efficient when handling an equation which is used by only a single reference.

The mathematical definition may be any FORTRAN-compatible combination of the allowable operations, functions, or variables. The following mathematical operations and corresponding symbols are included in SCEPTRE:

Operation	Symbol
Exponentiation	**
Multiplication	*
Division	/
Addition	+
Subtraction	−

The order in which the operations are performed is indicated by the order in which they are listed. To avoid confusion the liberal use of parentheses is encouraged.

Any subprogram available in the FORTRAN IV subprogram library may be used in an equation or expression. Additionally, any FORTRAN function subprogram written by the user may be referenced by an equation. However, when one of these functions is referenced, all variables appearing as the arguments of operational functions must be given in terms of dummy variables. Mathematical expressions are not limited to 72 characters and may be continued on subsequent cards provided that the last character on each card is a delimiter. The 72-character limit for any card must be honored.

Every *TABLE name* which has been referenced under **ELEMENTS** or **DEFINED PARAMETERS** must be explicitly defined under the **FUNCTIONS** group heading. The tabular data themselves are represented by a series of numbers in pairs separated by commas in which the first number represents the independent variable, the dependent variable data point immediately following. Any number of point pairs can be supplied per card. The data point pairs must be supplied such that the independent variable is in increasing or monotonic order. No specific limit to the number of point pairs which may constitute any table is defined. Only single-valued functions are allowed, but it is permissible to supply two consecutive independent variable values which are equal but which have different dependent variable values. This may be done to produce step functions. When the table values are updated at each solution time step, linear interpolation is used between the points supplied. For independent variable values falling outside the range of values supplied in the table, linear extrapolation is performed to determine the correct value from the table. Therefore, proper termination may be necessary. When the independent variable takes on a value exactly at a step point, the ordinate used depends on the direction from which the step was approached. In situations where the same table will serve to define more than one quantity, the user need explicitly define the table only once. A common situation occurs when two current generators in the network are defined by the same tabular data, differing only in the independent variable.

The general format for tabular data definition is

> TABLE *name, number, number,* . . .

or

> DIODE TABLE *name, number, number,* . . .

or

> TABLE *name*
> *value, value*
> *value, value*
> .
> .
> .

Run Control Specification

The **RUN CONTROL** group heading precedes the auxiliary information needed to control the run. This information does not directly affect the network. Most of these quantities have automatically present entries which hold unless they are specifically changed by the user. The present entries are given in Table 9.5. The most frequently used entries under the **RUN CONTROLS** group are discussed in detail hereafter.

Table 9.5 Default Run Control Quantities in SCEPTRE

Quantity	Default Information
STOP TIME	None
COMPUTER TIME LIMIT	Depends on version
MAXIMUM INTEGRATION PASSES	20,000
START TIME	0
INTEGRATION ROUTINE	XPO
MINIMUM STEP SIZE	1×10^{-5} (STOP TIME) for XPO, TRAP, RUK 1×10^{-14} (STOP TIME) for IMPLICIT
MAXIMUM STEP SIZE	2×10^{-2} (STOP TIME)
STARTING STEP SIZE	1×10^{-3} (STOP TIME) for XPO, TRAP, RUK 1×10^{-8} (STOP TIME) for IMPLICIT

	XPO	TRAP	RUK	IMPLICIT
MINIMUM ABSOLUTE ERROR	.0001	.00005	.00005	.0001
MAXIMUM ABSOLUTE ERROR	.005	.001	.005	
MINIMUM RELATIVE ERROR	.002	.0005	.00005	
MAXIMUM RELATIVE ERROR	.005	.01	.005	

Quantity	Default Information
NEWTON-RAPHSON PASS LIMIT	100
RELATIVE CONVERGENCE	.001
ABSOLUTE CONVERGENCE	.0001
MAXIMUM PRINT POINTS	1000
COMPUTER SAVE INTERVAL	15
VECTOR EQUATIONS	100 (transient) 70 (dc)
XPLOT DIMENSION	10
YPLOT DIMENSION	5
TYPE FREQUENCY RUN	LINEAR
NUMBER FREQUENCY STEPS	10
COMPRESSION COUNT	.3*(Input Function Buffer)
RUN MONTE CARLO*	10
INITIAL RANDOM NUMBER	127263527
DISTRIBUTION	GAUSSIAN
RUN OPTIMIZATION*	30
MINIMUM FUNCTION ESTIMATE	0
OPTIMIZATION CRITERION	10^{-7}
OPTIMIZATION RANDOM STEPS	0
RANDOM STEP SIZE CONTROL	.2
INITIAL H MATRIX FACTOR	1
RUN WORST CASE*	NOMINAL

*Default value applies when run control is entered without "=number." If the run control itself is omitted, the indicated calculation will not occur.

Run control entries may be made in any order and as many may be placed on a card as will fit if they are separated by commas.

Run Controls to Specify Mode of Analysis

SCEPTRE can compute dc, ac, or transient solutions by themselves or specific combinations of these three types of solutions. The allowable combinations are dc with ac and dc with transient. Runs combining ac with transient are not allowed. Unless otherwise specified by the user, only the transient solution will be computed.

If the user desires only the dc solution, the following run control must be entered:

> RUN INITIAL CONDITIONS ONLY

The user then has the option of specifying any or all of the following:

> RUN IC VIA IMPLICIT
> RUN SENSITIVITY
> RUN MONTE CARLO
> RUN WORST CASE
> RUN OPTIMIZATION

A dc plus transient analysis is requested by entering the run control

> RUN INITIAL CONDITIONS

or any of

> RUN SENSITIVITY
> RUN MONTE CARLO
> RUN WORST CASE
> RUN OPTIMIZATION

and

> STOP TIME = number

The option is available to use either the Newton-Raphson algorithm or the initial conditions via implicit algorithm for dc calculations. The implicit algorithm is selected by also entering the run control

> RUN IC VIA IMPLICIT

Run control cards to perform a dc plus ac analysis are as for dc plus transient solutions with the **STOP TIME = number** entry replaced by the run control

> RUN AC

Transient-only solutions require only the run control

> STOP TIME = number

ac-only solutions require only the run control

> RUN AC

These and other run controls will now be discussed in greater detail, along with a discussion of the options which they control.

All transient runs must have a problem duration in the time units consistent with those used to describe the circuit. The form is

$$\text{STOP TIME} = number$$

Since the user will never know in advance the computer time required for the solution of a given run, it may sometimes be desirable to enter a limit that will automatically terminate the run if that limit is exceeded. The limit in minutes may be entered as

$$\text{COMPUTER TIME LIMIT} = number$$

When this computer time limit is exceeded, the results calculated to that point will be presented.

A default limit of 20,000 passes is imposed upon the integration routines. If in the event the user wishes to increase or decrease this limit, he may do so by entering

$$\text{MAXIMUM INTEGRATION PASSES} = number$$

Generally transient problems start at **TIME** $= 0$, and a **START TIME** entry is not required. If you need to start a run at some other time, the appropriate entry is

$$\text{START TIME} = number$$

Four integration routines are available for use with SCEPTRE. Three are explicit (**XPO**, **TRAP**, and **RUK**) and one is implicit. The default option is **XPO** and therefore this method will be selected unless the user specifically requests another. When one of the other integration routines is desired, the format is

$$\text{INTEGRATION ROUTINE} = routine\ name$$

where *routine name* refers to **TRAP**, **RUK**, or **IMPLICIT**.

When using the **IMPLICIT** integration method, there are two algorithms used to compute the Jacobian. An automatic feature has been incorporated to decide whether the symbolic Jacobian or the differenced Jacobian would best fit the circumstances of the problem, but the user may exercise control. The two possible statements are

$$\text{USE SYMBOLIC JACOBIAN}$$
$$\text{USE DIFFERENCED JACOBIAN}$$

Any transient run will automatically terminate whenever a solution time step size is required that is smaller than a minimum specified step size. If this quantity is not supplied, the program will automatically compute a minimum limit equal to 1×10^{-5} times the specified **STOP TIME** if an explicit integration routine is selected, or 1×10^{-4} times the **STOP TIME** if implicit is selected as the integration option. If a minimum step size independent of the problem stop time is desired, the format is

MINIMUM STEP SIZE = number

Similarly, an upper bound to the time solution increment to be used prevents a complete transient solution from being composed of only a few solution points. If this quantity is not supplied, the program will automatically compute a maximum time limit equal to 2×10^{-2} times the problem **STOP TIME**. If a specific maximum step size independent of the problem duration is desired, the format is

MAXIMUM STEP SIZE = *number*

An initial solution time increment is automatically computed by the program and will be equal to 1×10^{-3} times the problem **STOP TIME** for explicit integration and 1×10^{-8} times the **STOP TIME** for the implicit routine. A specific initial step size independent of the problem duration may be supplied using the following format:

STARTING STEP SIZE = *number*

Unless otherwise specified the integration routines will operate with the preset error criterion. Only the most experienced users and those completely familiar with numerical methods should exercise control over the error criterion. For this reason the user is referred to Reference 4 for methods of use and formats for changing these preset values.

SCEPTRE offers five dc solution modes: **INITIAL CONDITIONS, SENSITIVITY, MONTE CARLO, WORST CASE,** and **OPTIMIZATION**. Any of the five may be used alone or as a source of initial conditions data for a transient or ac run. The initial conditions solution takes part in all dc calculations. It may be called separately, and if any other dc option is used, that option utilizes two or more passes of the initial conditions calculation to produce results. The initial conditions solution may use either the Newton-Raphson iteration scheme or may be calculated using a transient method built into the code. To compute steady-state values the user must make one of two entries. The language for these run controls is

RUN INITIAL CONDITIONS

and

RUN INITIAL CONDITIONS ONLY

Use of the former will cause the dc steady-state calculations to be made for use in conjunction with either an ac or transient calculation. The latter form will exclude any other solution modes. Use of any of the dc options and their associated run controls will force calculation of the initial conditions. A complete description of the procedure for computing initial conditions using the implicit integration scheme is discussed in Reference 4. If this scheme is desired due to failure of the Newton-Raphson method to converge or the necessity of computing the initial conditions for a circuit which has a topology not allowed in the Newton-Raphson formulation, the extra entry below must be included:

> RUN IC VIA IMPLICIT

The user should note that the method chosen—either the Newton-Raphson or implicit—for determining the initial conditions will be utilized in all dc solutions he requests on the same run. Therefore, the choice of the method for a particular problem can have a considerable impact on the running time when several dc calculations are involved. The user should thus exercise caution in selection of the implicit initial conditions calculation method.

The order of calculation of the dc options is sensitivity, Monte Carlo, worst case, and then optimization—regardless of the order in which the cards are entered. Thus, if the user desires to employ the results of a particular dc computation as the initial conditions for a transient or dc solution to follow, he must not request any of the dc options which appear after it in the above-mentioned order. Under normal circumstances the initial conditions for any rerun are either inserted by the user or computed by the dc portion of the program. However, there may be situations in which it is desirable to have the results of the master run, either dc or transient, used as the starting values for one or more associated reruns. If it is intended that all associated reruns use the final values of the master run as initial conditions, the appropriate entry under **RUN CONTROLS** is

> IC FOR RERUNS = MASTER RESULTS

If it is intended that each rerun use the final result of the preceding rerun in the sequence, the entry is

> IC FOR RERUNS = PRECEDING RESULTS

The Newton-Raphson iteration method should produce convergence, and hence dc solutions occur in less than 30 passes for most networks. If convergence is slow or does not occur, some realistic limit should be placed on the number of passes that are made. Unless otherwise specified a limit of 100 passes is built in. To change this limit, the proper entry under **RUN CONTROLS** is

> NEWTON-RAPHSON PASS LIMIT = *number*

The user may often wish to monitor certain voltages or currents in a network to determine their relation with some predetermined quantity. If these relations are satisfied, there may be no further interest in continuing the run. The run may be terminated at that point by the entry

> TERMINATE IF (*variable . relational operator . variable*)

where the variables refer to any network variable or constant. The user frequently supplies some element voltage or current for this purpose. The following relational operators may be used:

.LE. less than or equal to
.LT. less than
.GT. greater than
.GE. greater than or equal to
.EQ. equal to
.NE. not equal to
.AND. logical "and"
.OR. logical "or"

If it is desired that all remaining reruns be canceled if a particular termination condition is met, the word **STOP** is used instead of **TERMINATE**. Example card sequences are

    RUN CONTROLS
    TERMINATE IF(VCC.GT. 10.1)
        .
        .
    STOP IF((IRL.LE.5.).OR.(VL3.GE. 1.53E$-$2))
        .
        .

Special Run Controls for dc Options

To fully utilize the dc options recently included in SCEPTRE, one must understand several new run controls. Sensitivity calculations are requested by making the entry

        RUN SENSITIVITY

No further additional run controls are required for the sensitivity calculations since all computations are made automatically.

The Monte Carlo calculation is initiated by the use of

        RUN MONTE CARLO

or

        RUN MONTE CARLO = *number*

where number specifies the count of Monte Carlo iterations to be performed. If the count is unspecified, it is taken by default to be 10. The distribution to be used for the random variables in a Monte Carlo run is selected by the use of one of the controls

        DISTRIBUTION = GAUSSIAN

or

        DISTRIBUTION = UNIFORM

If neither is specified, the Gaussian distribution is used. The initial random number used in calculating the values to be used may be specified by the use of the control

INITIAL RANDOM NUMBER = *number*

The number used should be a nine-digit positive odd integer. If no initial number is specified, a suitable number is provided by SCEPTRE. If it is important to have all reruns start with the same random number, the statement

INITIAL RANDOM NUMBER = DEFAULT

may be used. Without this statement, a different random number will be selected for each rerun.

The worst-case calculation will be executed if one of the following run controls is provided:

RUN WORST CASE = LOW
RUN WORST CASE = NOMINAL
RUN WORST CASE = HIGH

The word following the equal sign determines the set of values to be left as element voltages, currents, and defined parameters at the conclusion of the worst-case calculation. These values are used as the initial conditions for a subsequent transient calculation. **LOW** and **HIGH** refer, respectively, to the table values which produce the smallest and largest value of the last objective function requested in the list entered under the **WORST CASE** heading of the **CIRCUIT DESCRIPTION**. If the equal sign and information to the right of it is not supplied on this run control or if the word **NOMINAL** is used, the nominal values are restored to the tables at the conclusion of the worst-case calculation.

The optimization process will be run if the run control

RUN OPTIMIZATION = *integer*

appears and if the appropriate entries are made under the optimization subheading. The integer which appears after the equal sign indicates the maximum number of optimization iterations which will be allowed for any objective function. If it is omitted, 30 iterations will be allowed. Since optimization can be a time-consuming process and may in fact fail if the network has pathological behavior, additional **RUN CONTROLS** have been provided to assist the user. If the run control

LIST OPTIMIZATION DETAILS

is used, auxiliary intermediate results will be provided to the user. If he supplies the run control

PUNCH OPTIMIZATION RESULTS

the most recently updated values of the approximate inverse Hessian matrix of the gradient vector and other parameters will be punched out. Thus, if the user wishes to limit the number of optimization iterations until he examines the progress of the process, he can save the intermediate results for use by reinitializing a later run. To utilize the optimization restart feature, these punched values must be provided as data cards in the SCEPTRE deck and the card:

INITIAL H MATRIX FACTOR = 0

must be included in the **RUN CONTROLS**. If this run control is used with a nonzero number, no attempt will be made to read previously punched values. Instead the number will be used to scale the initial approximation to the inverse Hessian. Careful use of any preknowledge about the scale of the inverse matrix can improve optimization performance. If this run control is not entered, the initial approximation is the unit matrix.

The tolerance within which the value of the objective function should be located may be controlled by the user. To do this one enters the run control

OPTIMIZATION CRITERION = *number*

If this card does not appear, the iterations will stop whenever the estimated improvement will be less than 10^{-7}. Note that in the presence of a broad shallow minimum the objective functions will be relatively insensitive to the values of the independent variable. In this case even the small objective function tolerance will permit the program wide latitude in its choice of final values of the independent variables. Under some conditions, particularly when a shallow minimum as described above is encountered, the algorithm may not locate a true minimum because the error criterion permits it to stop at a neighboring point in the space of independent variables. To protect against this possibility, two additional run controls have been implemented. If the run control

OPTIMIZATION RANDOM STEPS = *integer*

appears, displacements in random directions from the assumed minimum will be made the indicated number of times and the function reminimized each time. The smallest value resulting from this random motion and reoptimization should be accepted as a true minimum. If the run control

RANDOM STEP SIZE CONTROL = *number*

is used, the independent variable steps an amount estimated to increase the value of the function by $(number)^2/2$ above its minimum at each random displacement. If this run control is not used, the value of 0.2 is used.

The formulation of the Davidon method employed by SCEPTRE requires the objective function minimum value to be positive. Since this is not always the case, provision has been made within SCEPTRE to effectively maintain the objective function positive by addition of a suitable positive number. However, if the user has some estimate of the true minimum value, he may enter his own reference value and thereby improve the performance of the optimization. The correct form of this run control is

MINIMUM FUNCTION ESTIMATE = *number*

However, that minimum value entered will apply to all objective functions specified under the **OPTIMIZATION** heading.

Special Run Controls for ac Analysis

The ac analysis is performed only when the run control

> RUN AC

is included. This entry causes a different path to be taken through both the setup portion of SCEPTRE and the solution section of the program.

Frequencies at which the analysis is to be carried out are specified in two different ways. If the response at a single frequency is desired, a single entry under **RUN CONTROLS** may be made. The correct format for that run control is

> FREQUENCY = *number*

Number in the above entry is the desired single frequency. If, however, the response is desired over a range of frequencies, the following sequence of three entries is used:

> INITIAL FREQUENCY = *number*
> FINAL FREQUENCY = *number*

where the *number* in each case denotes a frequency, and

> NUMBER FREQUENCY STEPS = *number*

where *number* is the number of intervals desired. If the last entry is omitted, the default value of frequency steps is 10, which will give 11 calculations.

The user is allowed to control the type of frequency spacing to be used. The form of the run control is

> TYPE FREQUENCY RUN = *type*

where *type* may be either **LINEAR** or **LOG**. These run controls specify the frequency spacing during the run only and do not control the type of plots generated.

If circuit elements (other than source values) are not varied in reruns, the computation time for the eigen solution portion of the analysis may be saved by the inclusion of the run control

> USE FIXED AC MATRIX IN RERUNS

Special Run Controls for Diagnostics

Numerous run controls exist in SCEPTRE for diagnostic or other purposes. A few of the more often used controls are listed below. For a complete discussion of diagnostic controls, the reader is referred to Reference 4 at the end of this chapter.

The user may request SCEPTRE to print the **SIMUL8, SIMIC, SIMTR,** and **SIMAC** programs with a list of data read by these programs by entering the run control

> WRITE SIMUL8 DATA

Nodal interconnection data may be requested from SCEPTRE as an aid to checking for input errors by entering the run control

LIST NODE MAP

Information about the matrices created from the network description by SCEPTRE can be requested by entering the run controls

PRINT A MATRIX
PRINT B MATRIX
PRINT EIGENVALUES
PRINT EIGENVECTORS

Program Data Limits

There are certain program data limits which must be observed. These limits are given in Table 9.6. The limits shown in this table are checked by SCEPTRE as the data are read into storage. If any limits are exceeded, a message stating the limit is produced and the computer run is terminated.

Table 9.6 *Program Data Limits*

Description of Data	Maximum Number	
Heading cards	11	
Elements	300	
Nodes	301	
Source derivatives	50	
Defined parameters, P	100	
Defined parameters, W	50	
Defined parameter differential equations	100	
Mutual inductances	50	
Arguments in equation value specification	50	
Model table changes	15	
Model equation changes	15	
Model output suppressions	10	
Output requests	100	
Supplied initial conditions	100	
Equation functions (1 equation per card)	80	(approximately)
Cards per equation function	20	
Table functions	80	(approximately)
Optional termination conditions	10	
Models on library tape (combined)	250	
Characters in model name	18	
Model terminals (external nodes)	25	
Model internal nodes	301	
Linearly dependent ac sources plus secondary dependent ac current sources	50	
Independent ac voltage and current sources	50	
Maximum convolution kernels	50	

Table 9.7 shows the limits placed on alphanumeric characters used in names. The circuit designation used for models is always appended to any name used in a model description, and the combined total characters must not exceed the **CIRCUIT DESCRIPTION** limits. For example, an element **RBB** in a model designated as **T7** in the **CIRCUIT DESCRIPTION** becomes **RBBT7**.

Table 9.7 *Maximum Alphanumeric Character Lengths Allowed*

Item	Circuit Description Maximum Number	Model Description Maximum Number*
Nodes names	6	3
Element names	5	2
Defined parameter names	6	3
Table names	5	2
Equation names	5	2
Model names	18	18
Output labels	6	3
Circuit designation (for calling models)	3*	—

*Recommended.

Vectorized Notation

Compilers on most modern computers begin having problems when a routine with more than 300–400 uniquely named variables is encountered. Since each element introduces three unique variable names (the element name, the voltage across the element, and the current through the element), it is easy to see that a circuit containing 70 elements will begin to tax the capacity of the compiler. For this reason, circuits containing more than 70 elements are converted to a vectorized notation. When using the vector notation, one refers to a member of a vector rather than the unique name originally used for the circuit variable. In other words, **R15** may become **X(48)**, **IR15** would become **IX(48)**, and the voltage across **R15** would become **VX(48)**.

Model Description and Storage of Models

All active devices must be represented by combinations of **R, L, C, M, E,** and **J**. Most users will find that repeated use is made of certain models of active devices or of standard combinations of passive elements, such as filter sections, biasing networks, connective elements, etc. These situations can always be handled by inserting the components one by one in a **CIRCUIT**

DESCRIPTION under **ELEMENTS**. A more convenient approach is to describe or network once and store it for future use. The stored model feature is not free. Its use requires extra tape manipulation and hence more computer time. However, the time saved in circuit preparation and the added flexibility usually compensates for the additional computer cost.

The stored model feature of SCEPTRE allows a model to be stored either permanently or temporarily. Permanent model storage should be reserved for proven models which will see reasonably frequent use. Temporary storage stores a model for only one run. This type of storage will probably prove most useful for experimental models or for models which are seldom used.

All stored models are transferred from storage to the main circuit where they are used by reference to their external nodes or terminals. The user must ensure that corresponding nodes in both the main circuit and the stored model match in sequence. Internal nodes of any stored model are of no particular significance in that node names in a model will not be confused with nodes having the same name in the circuit. For permanent retention the following format should be used after the **MODEL DESCRIPTION** card:

MODEL *name* (*node*- . . . -*node*) (PERM)

The *name* may contain up to 18 alphanumeric characters. There may be a maximum of 25 nodes included in the nodal interconnection data. Following the **MODEL name** would be group headings **ELEMENTS, DEFINED PARAMETERS, OUTPUTS,** and **FUNCTIONS** as specified earlier. The group heading **RUN CONTROLS** will never appear in a stored model.

For the special case of the first model permanently stored on any individual tape, the **MODEL DESCRIPTION** card must have an additional parameter. The word **INITIAL** must appear in parentheses after any other data on the **MODEL DESCRIPTION** card. One other optional field is *allowed* on the **MODEL DESCRIPTION** card. The word **PRINT** may appear separately in parentheses after other data. This will cause a printed listing of *all* models stored on the permanent library tape. The word **PRINT** is omitted if no listing is desired. After the model is stored permanently, it may be called upon as often as required.

The user should use as few alphanumeric characters as possible to represent individual elements in stored models because the program must combine the element name with the circuit designation given under **ELEMENTS** of the **CIRCUIT DESCRIPTION**. If the stored model containing an element **C3** is used in a network and the model is designated **T11**, the program must refer to the capacitance **C3** as **C3T11**. In this case the maximum number of five alphanumeric characters has been used for the given element. If either the circuit designation or the element name contained more characters, the run would be aborted. This consideration is independent of the model name.

Each model intended for temporary storage must be described element by element under **MODEL DESCRIPTION** in the run in which it is to be used. The words **INITIAL** and **PRINT** are never used on the **MODEL DESCRIPTION** card for temporary storage. The rest of the entry format is the same as for permanent storage except for the word TEMP:

MODEL *name* (*node* — *node* — ... — *node*) (TEMP)

If neither **TEMP** nor **PERM** is used, the program will automatically assume temporary storage.

When frequent use is made of the stored model feature of SCEPTRE, one will often encounter the situation in which the topology of his stored model is satisfactory but the size of a few of the model elements must be changed. Changes can be effected easily for any individual run, but no permanent changes to the stored model are possible. All change must be included along with the statement that locates the stored model in the main circuit. This will always be under the *main circuit* **ELEMENTS** heading. The method used is to insert the word **CHANGE** followed by any changes necessary after the model identification data on the element card which locates the stored model. For example, consider that a stored model has been called out as

T1, 7−8−12 = MODEL 2N1734B

In this case, the model will be inserted into the main circuit as it was originally stored. If a change is desired in one or more internal elements or defined parameters of the stored model, the proper entry could be

T1, 7−8−12 = MODEL 2N1734B   (CHANGE CC = 50)

In this case, **CC** of the original model has been changed to 50 units of capacitance, regardless of its original form or size. All other elements in the stored model would remain as originally stored.

Another practical situation could be reflected by the entry

T1, 7−8−12 = MODEL 2N1734B (CHANGE CC = TABLE 7 (VCCT1), JA = .2∗J2T1, JE = DIODE TABLE 4)

In this case, **CC** of the original stored model has been changed to a tabular function which must be represented (**TABLE 7**) under the **FUNCTIONS** subheading of the main circuit. Also, **JA** has been changed to a different mathematical definition and **JE** has been changed to another tabular function which must appear under **FUNCTIONS**. The other elements of the stored model will remain in their original forms.

As a final example, assume that it is desired to change **CC** in a stored model to a different equation. Then

T1, 7−8−12 = MODEL 2N1734B (CHANGE CC = EQUATION 5 (VCCT1, VCX))

In this case, **CC** of the stored model has been changed to a closed-form function which must be represented by **EQUATION 5** under **FUNCTIONS** of the main circuit description. All elements or element voltages or currents (within a model) that are referenced on the right-hand side of the changed expression or in the main circuit description must include the model designation as a suffix (**VCCT1, J2T1**). **VCX** is the voltage across a capacitor, **CX**, that is not in a model and therefore uses no suffix.

Output requests in the stored model may be inhibited as well. If any quantity is not desired as an output for a given run, that output may be suppressed in the statement that locates the stored model in the main circuit. An example of the proper format is

T12, 7−9−GND = MODEL 2N918 (TEMP, SUPPRESS JE)

The normal output routine will produce a listing of the entire circuit without a detailed printout of any stored models. The proper language for a detailed printout of any permanently stored model is shown in the following example:

T9, 16−14−18 = MODEL 2N2222 (PERM, PRINT)

A table in a stored model may be changed by the entry

T1, 7−8−12 = MODEL 2N1741B (PERM, CHANGE TABLE 1 = TABLE 7)

which replaces **TABLE 1** of the stored model with **TABLE 7**. **TABLE 7** must be supplied under **FUNCTIONS** of the main circuit. No correlation is required between the number of point pairs contained in the original table and the point pairs in the new table. It is never possible to directly change **DIODE TABLE X** = **DIODE TABLE Y**. The same effect can be achieved by changing the element. For example, suppose that the element **JE** in the permanently stored model **2N1741B** had been equated to **DIODE TABLE X**. Then **DIODE TABLE X** could be changed to **DIODE TABLE Y** by making the entry

T1, 7−8−12 = MODEL 2N1741B (PERM, CHANGE JE = DIODE TABLE Y)

An equation in a stored model may be changed by the entry

T1, 7−8−12 = MODEL 2N1741B (PERM, CHANGE Q1 = Q2)

which replaces the original **EQUATION 1** of the stored model with **EQUATION 2**. **EQUATION 2** must be supplied under **FUNCTIONS** of the main circuit. The only special stipulation is that the same independent variables that held for the original equation must also apply to the new equation. It is never possible to directly change an equation in a stored model to a table or a table in a stored model to an equation; (**CHANGE EQUATION 1** = **TABLE 1**) *is not allowed*. However, an element or a defined parameter in a stored model can be changed to a constant, table, or equation.

When several different types of changes are desired, the changed statements are separated by commas. The overall change statement is enclosed in one set of parentheses, for example,

T1, 7−8−12 = MODEL 2N1741B (PERM, CHANGE CC = 50., CE = 30., SUPPRESS ALL, PRINT)

An entire model may be removed from the user's permanent library tape by making an entry under **MODEL DESCRIPTION** of the following form:

MODEL *name* (DELETE)

Reruns

The rerun feature will permit the user to run multiple versions of a master run which differ from the master in one or more ways. The user need supply only the quantities which have been changed from the master run. An unlimited number of reruns may be made from a single master run, but each rerun will require approximately as much computer *solution* time as did the master. A savings will be realized in the setup overhead, however.

The **RERUN DESCRIPTION** card is followed by any of the five possible group headings given in Table 9.2. The format of the card is

RERUN DESCRIPTION (n)

where **n** is the number of reruns desired. The (n) designation may be omitted if just one rerun is desired. The user may not describe the values of more than one variable on the same card anywhere in the rerun description section. The subheading **ELEMENTS** is used if any constant-valued element is to be changed for any of the reruns. The general form of the card under ELEMENTS is

Element name = *number*, . . .

where there are n numbers corresponding to the number of reruns desired. **DEFINED PARAMETERS** may be changed in a similar manner. The format for these changes is

P*name* = *number*, . . .

Any elements or defined parameters which are unchanged under a **RERUN DESCRIPTION** will retain any bounds previously given. However, if the user specifies an element or defined parameter value, he may at the same time change, delete, or add bounds to the element value whether or not bounds had been previously specified. Each value, as used in the above paragraph, must be complete with bounds if they are so desired.

INITIAL CONDITIONS may be changed for any number of transient runs which start at different operating points. The format is consistent with the **ELEMENTS** and **DEFINED PARAMETERS** and is as follows:

VC*name* = *number*, . . .

or

IL*name* = *number*, . . .

FUNCTIONS may be changed in only two special ways. First, it is permissible to change any part of a table which describes an element or defined parameter of the original run. Both the independent and dependent variables may be changed but are limited to the exact number of entries used in the master run. The only change which may be made to equations under **RERUN DESCRIPTION** is to defined parameter values. Any defined parameter used in the mathematical definition or in the calling sequence for the equation may be changed.

Any of the solution control values, except automatic termination messages, which were defined under the **RUN CONTROLS** of a master run, may be changed in any rerun. The automatic termination messages of the master run apply to all reruns. If four reruns are to be made in which the problem duration differs for each rerun, an appropriate sequence would be

      RUN CONTROLS
      STOP TIME = 100, 125, 150, 175

There are certain features of the master run which cannot be changed in any of the corresponding reruns. These are in the areas of output, element form, and operational mode.

There can be no difference between the quantities which are outputs in the master run and those that are outputs of the reruns. For this reason the **OUTPUTS** group heading is not allowed under rerun.

The form in which an element is described may not be changed between the master run and any subsequent reruns. If an element originally appears as a constant, its rerun version must also be a constant. A master run may consist of one of the following combinations of solutions: dc only, transient only, ac with dc, or transient with dc. A general rule is that no rerun may contain more than its master run did. If the master run was an ac or transient solution only, all associated reruns must be the same. If the master run was **INITIAL CONDITIONS ONLY,** all associated reruns must also be initial conditions solutions. However, if the master run was an initial conditions solution plus a transient solution, there is some leeway. The user should determine whether the changes in the network will affect the initial conditions. If it is decided that the initial conditions will not change, the dc run which was made for the master will suffice for all reruns. In this case no special entry is made for the reruns. If, however, it is decided that the new initial conditions run is required for each rerun, the user must request this with the run control **RUN INITIAL CONDITIONS**. This single entry will allow for recomputation of initial conditions for all reruns in that group. The user may ask for new initial conditions solutions for each rerun even if the master run conditions are valid. However, this is wasteful of computer time and should be avoided. Two special run controls have been provided to reduce the number of iterations required to come to an initial conditions solution:

      IC FOR RERUNS = MASTER RESULTS
      IC FOR RERUNS = PRECEDING RESULTS

The first causes the iterative scheme to begin with the initial conditions results from the master run, and the second will cause the scheme to use the results of a preceding rerun as the initial estimate for beginning the iterative procedures.

User-Supplied FORTRAN Subprograms

If complex functions or decision-making algorithms are required to describe an element or parameter in SCEPTRE, it is possible to supply these as FORTRAN IV function subprograms. There is, however, one restrictive condition that must be placed on the programmer. This is that the function name begin with the letter **F**. The general format of the use of a function is shown below:

$$\text{Network Variable} = \text{F}name\,(list)$$

The function name and its calling sequence of variable may be used anywhere a **value** is required.

An alternative method avoiding the requirement that the function name begin with **F** is as follows:

$$\text{Network Variable} = \text{Q}name\,(list)$$
$$\text{Q}name\,(list) = (function\ name\,(list))$$

the latter card follows a **FUNCTIONS** group card.

Inclusion of the FORTRAN functional subroutines is machine-dependent and therefore will not be given in this discussion. Users are referred to Reference 4 for additional information.

Use of CONTINUE Heading

The purpose of this feature is to allow the user to continue transient runs which have been run previously and have been properly terminated. Without this feature, the user would be forced to begin the problem again. This would waste computation. Most continue runs will consist only of an extended problem duration time, but the user has more flexibility with the continue feature.

The changes permitted under this heading are most of the changes allowed under **RUN CONTROLS**. No changes are allowed to the network under this heading, so the subheadings **ELEMENTS DEFINED PARAMETERS, OUTPUTS, INITIAL CONDITIONS,** and **FUNCTIONS** can never appear under **CONTINUE**.

The following representative sequence is appropriate for the case where an original run is carried to its specified duration and the user subsequently wishes to continue the run to a new duration of 2000 units of time:

```
CONTINUE
RUN CONTROLS
STOP TIME = 2000
END
```

This feature is intended for transient runs only and has no control over dc or ac computations (including initial conditions computations). When the original run contains a number of reruns, the continue option will apply only to the last run that was processed, which will usually be the last rerun.

RE-OUTPUT Heading

The user may often desire additional copies of output from previous runs without the necessity of repeating the runs. This can be done in SCEPTRE through the use of the re-output feature if the original run has been preserved on tape. In addition to repeating the original output, the re-output feature permits some degree of modification. Any quantity that was originally output in printed form may be obtained in both printed and plotted form by the use of this feature. For example (see Sec. 9.5 for a description of the output for a master run),

```
RE-OUTPUT
OUTPUTS
VR1, VR2
VC10, PLOT
END
```

RE-OUTPUT also permits the subheading **RUN CONTROLS. MAXIMUM PRINT POINTS** may be used under run controls. For example,

```
RE-OUTPUT
VR1, VR2
VC10, PLOT
RUN CONTROLS
MAXIMUM PRINT POINTS = 2000
END
```

On ac runs it is also possible to obtain a plot via the **RE-OUTPUT** feature if it was not requested on the original run. As is the case for other runs, a plot cannot be obtained of an output which was not printed on the original run. In the ac case this restriction extends to type of plot. If the original run requested a print of magnitude and phase in degrees versus frequency, a plot versus radians cannot be obtained. A Nyquist plot can be obtained only if a complex printout (with or without a Nyquist plot) had been requested on the initial run.

9.5 OUTPUT DEFINITION

To be useful, a circuit analysis code must be able to present the output variables in some coherent manner. SCEPTRE output consists of printed tabular listings or printer plots of requested variables as functions of time or of some other independent variable or combinations of both.

General Output Specifications

The following general types of quantities may serve as either dependent or independent variables: the voltage or current associated with any passive element, the voltage or current associated with any source, any element value, any state variable derivative, or any defined parameter or its derivative. To request outputs in printed tabular form, one uses the general format under the subheading **OUTPUTS**:

variable, variable, ..., variable

To request that your outputs be presented in plotted form, it is necessary only to place a comma, and then the work **PLOT** is used as the last entry on each output request card for which plots are desired:

variable, ..., variable, PLOT

Note that more than one output quantity can be requested on a single card. They need only be separated by commas. Also, if more than one variable appears on a card followed by the word **PLOT**, all the requests on that card will be plotted. To make the plotted outputs more useful and more general, the user is allowed to rename any dependent or independent variable except **TIME**. Also, other network quantities or defined parameters may be used as the independent variable. The plot rename and choice of independent variable options can be represented in terms of the general format

variable (rename), PLOT (variable (rename))

At times it is necessary to limit the printed output of SCEPTRE. The default of 1000 print points which will produce vast volumes of paper if many variables are requested as output. This number may be changed for any particular run by entering

MAXIMUM PRINT POINTS = *number*

under the RUN CONTROLS subheading. The indentical entry with a number greater than 1000 will cause the opposite effect. If the particular number 0 (zero) is entered, all printed output will be suppressed, but the normal printer plots will appear.

An additional means by which the number of solution points may be precisely controlled and chosen more selectively is the use of the print interval. Here the user supplies the desired time interval and only those solution points which fall on or immediately after integer multiples of this specified interval will be present as output. The required RUN CONTROLS entry is

PRINT INTERVAL = *number*

A specialized plot format is available in which up to nine quantities may be plotted against a common abscissa. The ordinate for each variable is separately scaled, and unique graphic characters are used to represent each quantity. Use of this feature requires that all the quantities which are to be plotted together be requested on the same card or sequence of cards under OUTPUTS followed by a specific plot name. For example,

>   OUTPUTS
>   VC1, VC2, P13, PLOT
>   VR11, VCET1, JET6, IE4, PLOT 1

In this case, quantities VC1, VC2, and P13 would be plotted singly as usual. Quantities VR11, VCET1, JETS, and IE4 would be plotted together since they have been requested together with a plot that has been "named" simply as 1. Any plot name may include up to six alphanumeric characters, and more than one name may be used to plot different combinations of quantities.

The composite plot feature also requires a PLOT INTERVAL entry under RUN CONTROLS. While this is a RUN CONTROL entry, it is discussed here because it relates only to composite plots, and its omission will cause the requested plots to appear in their usual separate formats.

The physical length of any composite plot may be controlled. The number of pages encompassed by any composite plot is determined by the problem duration (STOP TIME) and a user-supplied entry called PLOT INTERVAL. The former divided by the latter will determine the number of lines required, which will in turn determine the number of pages required. For the system S/360, 66 lines will fill 1 page. Therefore, a problem duration of 1000 and a PLOT INTERVAL of 5 will require 200 lines and 4 pages. This entry always appears under RUN CONTROLS and the format is simply

>   PLOT INTERVAL = number

Only one PLOT INTERVAL will be recognized regardless of the number of composite plots that are requested.

ac Outputs

The results of ac calculations can be obtained as outputs in either tabular or plotted form. The general rules for output requests apply equally to ac outputs, except that ac outputs are not given as functions of time. All ac tabular outputs and all but one type of plot are given as functions of frequency. The Nyquist plot gives the imaginary part of a complex function versus the real part.

The standard forms of requests, under **OUTPUTS**, are

>   *variable, variable, . . . , variable*

and/or

>   *variable*
>   *variable*
>   *variable*

This form will produce a tabular printout of magnitude versus frequency and phase in degrees versus frequency for each variable. This is the default entry.

The word **DEGREES** is optional. If the output is desired in radians, the proper entry is

variable, variable, RADIANS

The entry

variable, COMPLEX

produces a printout of the real part of each variable versus frequency and the the imaginary part versus frequency.

If the word **PLOT** is added to the entry

variable, COMPLEX, PLOT

a plot of the same data is also obtained.

The Nyquist plot shows the imaginary part versus the real part for each variable. The proper entry for the Nyquist plot is

variable, variable, NYQUIST, PLOT

In this entry only, the word **PLOT** is optional.

9.6 MODELS FOR USE IN SCEPTRE

The overall philosophy behind the SCEPTRE code is to provide as much potential and user flexibility as possible. Because of this approach, there are no models with fixed topologies or fixed functional relationships built into SCEPTRE. As described earlier in this chapter, however, models may be defined and easily included as subelements with the network description under a MODEL DESCRIPTION heading or called from permanent storage. If called from either permanent or temporary storage, functional dependencies, element values, and parameter lists may be altered as needed, thus providing the potential ease of use offered by built-in models.

Modeling for SCEPTRE (or for that matter any circuit analysis code) is an art involving the definition and application of equivalent circuits to represent, in acceptable detail, features of active or passive networks or devices. Complex equivalent circuits, in general, result in long computer execution times, while simplified equivalent circuits tend to misrepresent the device because of inaccuracies. Modeling is therefore a compromise that requires careful engineering judgments. As a general modeling guide, to prevent numerical instabilities or long execution times, dependencies resulting in network derivatives considerably in excess of absolute network values should be avoided, along with dependencies which allow sudden or instantaneous changes in derivatives of network values to occur.

To permit users as much latitude as possible in the choice of equivalent circuits, almost any combination of the allowable elements—resistors, capacitors, inductors, mutual inductances, voltage or current sources—may be used to represent an active device. This is true whether the user relies on the stored model feature or elements to enter each component of the equivalent circuit individually under the CIRCUIT DESCRIPTION. A complete discussion of equivalent circuits is an endless task. A few points will be given in this section that may prove useful to the user.

Diode Models

A general diode model based on the charge control model is shown in Fig. 9.2 in which C_d represents the sum of the junction and diffusion capaci-

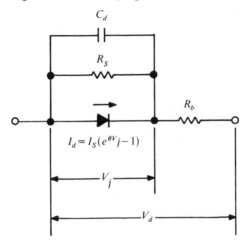

Figure 9.2 General Diode Model.

tances of the diode and R_b and R_s refer to the bulk and shunt resistances, respectively. The heart of the model is the diode junction model, the current of which is given by the diode equation, which in turn is dependent on the diode junction voltage. The diode junction itself is just a voltage-dependent current generator and is entered in SCEPTRE as such. The closed-form representation of the diode junction requires that the user possess accurate values of I_s and theta (see Sec. 9.4) for use in the descriptive equation.

A sequence of cards that could be used to describe a diode under MODELS or CIRCUIT DESCRIPTION is

```
ELEMENTS
JD, A−B = DIODE EQUATION (1.6E−6, 37.2)
CD, A−B = 20
RS, A−B = 2000
RB, B−C = 0.05
```

The SCEPTRE representation of this diode is shown in Fig. 9.3. The description for the current generator is based on the explanation given for the current generator in Sec. 9.4.

An alternative procedure would be to obtain the characteristics of the diode by measuring the diode current as a function of the voltage across it. The current generator could then be described in tabular form. The effect

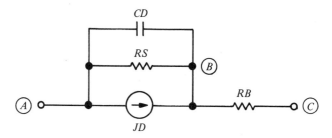

Figure 9.3 SCEPTRE Diode Representation.

of the bulk resistance is included by this method, and the user may omit the shunt resistance R_s. However, the omission of the equivalent shunt capacitance is another matter. The difficulty with omitting that capacitance is that the current generator is voltage-dependent and the voltage in question must be taken across some element. If the current generator is dependent on a capacitor voltage, the current source will be updated at the start of each solution step based on known internal state variables. If, however, the equivalent capacitor is removed and the dependence is placed on the voltage across the current source itself or some shunt resistance, the current source may be updated based on the information from the previous solution increment. Thus, a computational delay will have been created, and significant errors can result. For this reason the user is cautioned against removing the shunt capacitance associated with diodes for all transient applications.

When a large-signal diode model is used, the user will find it most convenient to assume the reference direction for the capacitor in the same direction as the current source. The reason for this is that a positive capacitor voltage will correspond to forward bias on the diode. While this convention is not mandatory, it simplifies interpretation of the solution in most cases.

Transistor Models

A conventional Ebers-Moll equivalent circuit for *npn* and *pnp* transistors are shown in Figs. 9.4 and 9.5. This model is in wide use because it can accommodate all regions of operation with a minimum amount of complexity and because it operates with conventional electrical quantities. The currents through the diode junction models are represented by voltage-dependent nonlinear current generators (DIODE EQUATION). The expressions for the open-circuit and short-circuit currents are as follows:

$$I_{re} = I_{es}(e^{\theta_e V_{b'e}} - 1) = \frac{I_{e0}}{1 - \alpha_I \alpha_N}(e^{\theta_e V_{b'e}} - 1)$$

$$I_{rc} = I_{cs}(e^{\theta_c V_{b'c'}} - 1) = \frac{I_{c0}}{1 - \alpha_I \alpha_N}(e^{\theta_c V_{b'c'}} - 1)$$

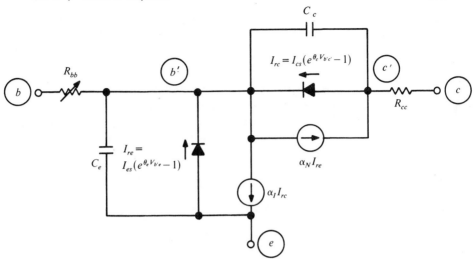

Figure 9.4 Basic Ebers-Moll *npn* Transistor Equivalent Circuit.

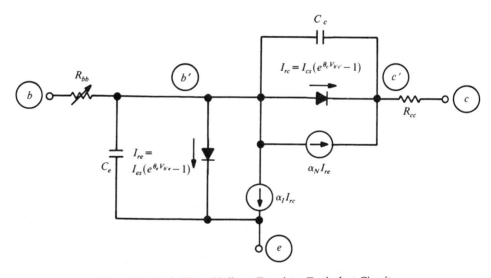

Figure 9.5 Basic Ebers-Moll *pnp* Transistor Equivalent Circuit.

Thus, by inspection the short-circuit junction current is greater than the open-circuit junction current by a factor of $1/(1 - \alpha_I \alpha_N)$. The current sources I_{re} and I_{rc} are defined in SCEPTRE as *primary dependent sources*, and to achieve numerical convergence for the initial condition computation they must be entered in the general form

$$J name, node - node = \text{DIODE EQUATION } (X_1, X_2)$$

or if tabular data are used,

$$J name, node - node = \text{DIODE TABLE } name$$

where X_1 equals I_{es} or I_{cs} and X_2 equals θ_e or θ_c. The current sources $\alpha_N I_{re}$ and $\alpha_I I_{rc}$ are defined as *secondary dependent sources* and should be entered as

$$Jname, node - node = value * primary\ dependent\ current\ source$$

where *value* represents α_N or α_I and can be a number, table, defined parameter, equation, or mathematical expression.

The total capacitance associated with a junction is usually expressed as a sum of the junction capacitance and the diffusion capacitance. Mathematically this is

$$C_e = \frac{C_{0e}}{(V_{0e} - V_{b'e})^{n_e}} + \theta_e T_e (I_{re} + I_{es}) \quad \text{for the emitter junction}$$

and

$$C_c = \frac{C_{0c}}{(V_{0c} - V_{b'c'})^{n_c}} + \theta_c T_s (I_{rc} + I_{es}) \quad \text{for the collector junction}$$

(See Reference 11 for development of relationship and definition of terms.) In the SCEPTRE input language the total emitter capacitance element may be entered as

$$CE,\ B-I = EQUATION\ 1\ (20,\ .5,\ VCE,\ .5,\ 38,\ JE,\ 1E-6)$$

where the value 38 is equal to the product of θ_e and T_e. The collector capacitance element has a similar form.

The mathematical relationship implied by EQUATION 1 in the element value specification for CE is supplied under a FUNCTIONS card, and in its most basic form would be entered as

$$EQUATION\ 1(A,B,C,D,E,F,G) = (A/(B-C)**D + E*(F+G))$$

Numerical problems are frequently encountered when this basic form is used however, particularly when implicit integration is employed. This is because the large step size attempted by the integration routine frequently requires reduction when nonlinearities are encountered during solution. When the step size requires reduction, an erroneous state vector has temporarily been computed and the functions, e.g., EQUATION 1, have been evaluated with unrealistic network values. In EQUATION 1, an unrealistically large value for VCE would result in a negative capacitance being computed for CE. This erroneous result will almost always prevent further solution progress. To prevent this numerical problem, it is convenient to write the previous equation for total capacitance in the form

$$C_e = \frac{C_{0e}}{\left[V_{0e}\left(1 - \frac{V_{b'e}}{V_{0e}}\right)\right]^{n_e}} + \theta_e T_e (I_{re} + I_{es})$$

It is now possible to use a minimum function to select the smallest of the ratio $V_{b'e}/V_{0e}$ or an arbitrary constant, say 0.9, to limit the $1 - V_{b'e}/V_{0e}$ term to a small positive value (0.1 for this example). The value of the junction capacitance term is normally negligible with respect to the diffusion term when $V_{b'e}/V_{0e}$ reaches 0.9 and can be limited at this point.

The equation for total capacitance can therefore be written

$$C_e = \frac{C_{0e}}{\left\{V_{0e}\left[1 - \min\left(\frac{V_{b'e}}{V_{0e}}, 0.9\right)\right]\right\}^{n_e}} + \theta_e T_e(I_{re} + I_{es})$$

It is also sometimes advantageous to limit the diffusion term to an arbitrary maximum, say 1000 if the units are picofarads, by writing the diffusion term

$$\min\left[\theta_e T_e(I_{re} + I_{es}), 1000.0\right]$$

The final improved relationship to be entered under the FUNCTIONS card for EQUATION 1 becomes

EQUATION 1(A,B,C,D,E,F,G)=((A/((B*(1.0−AMIN1((C/B),0.9)))**D)) + AMIN1((E*(F+G)),1000.0))

where AMINI is a FORTRAN callable library function which performs the minimum function mentioned earlier. This relationship insures that the value computed for CE will be positive, non-zero, and not excessively large regardless of the network values supplied to the equation throughout the simulation.

In general, when supplying any function, expression, or FORTRAN function subprogram, the user must keep in mind the possibility exists that the function will be evaluated with unrealistic arguments and should provide for the return of reasonable values under these circumstances.

A sequence of cards that could be used to describe an *npn* or *pnp* transistor under the subheading ELEMENTS in either a MODEL DESCRIPTION or CIRCUIT DESCRIPTION mode is

1. For the *npn*,

    ELEMENTS
    CE, B−E = EQUATION 1 (57., 1., VCE, .43, 189., JE, 2.9E−9)
    CC, B−C = EQUATION 1 (10., .8, VCC, .32, 7500., JC, 1.46E−8)
    JE, B−E = DIODE EQUATION (2.9E−9, 36.7)
    JC, B−C = DIODE EQUATION (1.46E−8, 28.2)
    JI, E−B = .534*JC
    JN, C−B = .99*JE
    RB, B1−B = 1.
    RC, C1−C = .06
    OUTPUTS
    VCE, VCC, VRB, PLOT
    ICC, ICE, IRC
    FUNCTIONS
    EQUATION 1 (A, B, C, D, E, F, G) = (A/(B−C)**D+E*(F+G))

2. For the *pnp*,

ELEMENTS
CE, E−B = EQUATION 1 (9., .9, VCE, .55, 7.05, JE, 2.67E−5)
CC, C−B = EQUATION 1 (11., .6, VCC, .34, 1620., JC, 2.12E−4)
JE, E−B = DIODE EQUATION (2.67E−5, 36.7)
JC, C−B = DIODE EQUATION (2.12E−4, 34)
JI, B−E = .2598∗JC
JN, B−C = .985∗JE
RB, B−B1 = .1
RC, C−C1 = .005
OUTPUTS
VCE, VCC, VRB, PLOT
ICC, ICE, IRC
FUNCTIONS
EQUATION 1 (A, B, C, D, E, F, G) = (A/(B−C)∗∗D+E∗(F+G))

The SCEPTRE representation of these transistors is shown in Fig. 9.6.

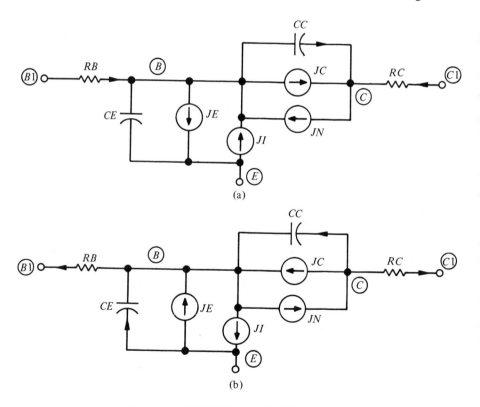

Figure 9.6 SCEPTRE Ebers-Moll Representations.

Another approximation that is often made is to enter the bulk resistance as a constant where, in fact, it really is a nonlinear function of emitter current. If the user needs a greater degree of sophistication in his model, any of the variables passed through the equation call list may be entered as variable quantities in equation or tabular data form by linking through defined parameters.

An alternative form of the Ebers-Moll equivalent circuit may be used. Here the effects of the bulk resistors have been incorporated into the representation of the junction diodes. The current generators themselves are represented by tabular data obtained from laboratory measurements. Both versions of the basic equivalent circuit contain approximations of one form or another, and others may be made for the sake of simplicity. One example is as follows: When the inverse α of a transistor is negligible, that current generator may be omitted. In any instance it is always true that diodes or transistor junctions must always be represented as current generators by the designation DIODE EQUATION (X_1, X_2) for closed-form representation and the designation DIODE TABLE for tabular functions. Current generators that are functions of other current generators must always be represented as a *special value*, as defined earlier.

The H-, Y-, Z-, and R-parameter equivalent circuits are known as small-signal equivalent circuits. The feature that all these models have in common is that they are intended for use at some particular operating point in the linear region. Saturation and cutoff cannot be accommodated, and large-signal swings are subject to considerable error. Due to these factors, small-signal equivalent circuits are not nearly so versatile as the general nonlinear equivalent circuit presented earlier, but for some applications the user can legitimately utilize their inherent simplicity. Voltage sources that are linearly dependent on resistor voltages and current sources that are linearly dependent on resistor currents are handled directly. When voltage sources depend on resistor current or current sources depend on resistor voltages, the user must make the appropriate conversion prior to entry into SCEPTRE.

Several references are given at the end of this chapter to direct attention to state-of-the-art techniques which have been developed in the last few years in the modeling of integrated circuits and the simplification of large nonlinear circuits.

9.7 EXAMPLES

As an aid to understanding the general use and application of SCEPTRE and to illustrate some of the salient features of the code, the process by which an example problem is submitted to the code for analysis is described in the following example. Subsequent examples are designed for checkout purposes or as sample entry forms.

Example 9.1. Transient Analysis of a Control System

Let us assume that the transient behavior of the system of Fig. 9.7 is to be investigated. Whether or not the behavior of this particular system could be derived without the aid of SCEPTRE is not germane to this exercise. This system has as components an integrated-circuit linear operational amplifier, transistors, resistors, capacitors, inductors, voltage sources, and current sources. The system also contains an electromechanical actuator which is represented by its second-order transfer function in the frequency domain. A logical approach to the analysis is to begin by developing models and SCEPTRE input card sequences for the linear amplifier, transistors, and actuator. Once these models are available, we can generate the card sequences for input under CIRCUIT DESCRIPTION that describe the control system to SCEPTRE.

An equivalent circuit for an operational amplifier adequate for this investigation is shown in SCEPTRE representation in Fig. 9.8. Each node has been given an alphanumeric identifier according to the rules described in Sec. 9.4. Methods for obtaining parameters, element values, and dependencies are described in the References at the end of this chapter and while representing a significant part of the problem for the user are not discussed here.

The card sequence generated to input this model to SCEPTRE is as follows:

```
   MODEL HA 2700 (N-I-OUT-GND)
*      MODEL C1(GAMMA,NEUTRON)
*      BASIC UNITS   OHMS AMPS VOLTS SECS FARADS
*      USER SUPPLIED RADIATION DATA
*      *    PGR - GAMMA RATE IN RAD(SI)/SEC
*      *    PNF - NEUTRON FLUENCE IN N/CM**2
*      *    PGD - GAMMA PULSE DELAY FROM SIMULATION TIME ZERO
   ELEMENTS
EA1,GND-3=Q2(PA,VCIN)
EA3, GND-12 = Q1(PSAT,P1,P3)
EX,5-XI=QX(PGR,TABLEEX(PTG),5.E-8)
CIN,N-I=1.E-12
CA1,4-6=2280.E-6
CA2,5-GND=45.E-12
RIN,N-I=1.E6
RI,11-GND=1.E-3
RN,NN-GND=1.E-3
RA1,3-4=100.
RA2,4-5=100.E3
RA6,12-OUT=100.
RA7,6-GND=.028
JI,  I-11 = 0.
JN,  N-NN = 0.
```

```
JO, OUT-GND = 0.
JX,XI-GND=QX(PGR,TABLEJX(PTG),1.875E-11)
   DEFINED PARAMETERS
PGR=0.
PGD=0.
PNF=0.
P1=0.
P3=0.
PSAT=TABLE 1(VJX)
PTG=QTG(TIME,PGD)
PAO=2.E5
PAF=QAF(PAO,PNF)
PA=QA(PAO,PAF,TABLEN(PTG))
   FUNCTIONS
Q1(A,B,C) = (A+B*C)
Q2(A,B)=(A*B)
QX(A,B,C)=(A*B*C)
QTG(A,B)=(A-B)
QAF(A,B)=(A/(1.+2.71E-18*A*B))
QA(A,B,C)=(A-(A-B)*C)
TABLE 1
-20,-13,-13,-13,13,13,20,13
TABLEEX
-1.E-6,0,0,0,.01E-6,1,.3E-6,1,.31E-6,0,1.E-6,0
TABLEJX
-1E-6,0,.1E-6,0,.11E-6,1,.3E-6,1,.31E-6,.05,2E-6,.05,
2.01E-6,0,1E-6,0
TABLEN
-1E-6,0,10E-6,0,11E-6,1,20E-6,1
```

The card sequence begins with a MODEL *name* card (intended to follow a MODEL DESCRIPTION card), and a series of 8 comment cards to further identify the model. The next card to appear is the ELEMENTS card. The 17 elements which follow are entered between the proper nodes in the arbitrary node sequence shown in Fig. 9.8 and given values, some of which are specified as tables or equations with a list of actual parameters to be used in computation. The particular node sequence chosen becomes important in determining polarities of element voltages or currents for functions or output. In like manner, there are 10 defined parameters, some of which also have tables or equations specified as values. The tables and mathematical operations implied by the values of the elements and defined parameters are defined under the FUNCTIONS subheading. The number of actual and formal parameters appearing as arguments must agree in number and must appear in proper sequence. For example, the value of the formal parameter B in equation Q1 under the FUNCTIONS subheading has the value P1, a defined parameter, which is defined equal to zero under the DEFINED PARAMETERS subheading. At first glance this may seem contradictory or unnecessary, but the intent is to provide a means of activating certain radiation options available in the model. These options are activated by changing the value of parameter P1 when the model is called as an element from the CIRCUIT DESCRIPTION, using the model change feature. No outputs have been requested

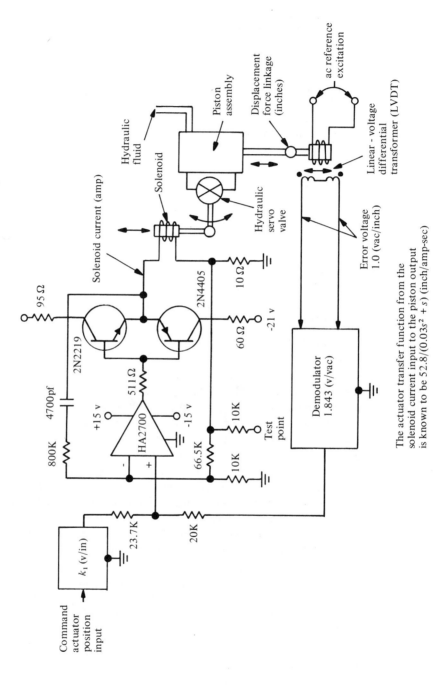

Figure 9.7 Schematic Diagram of Example 9.1 System.

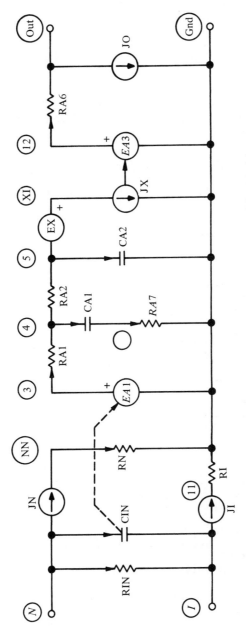

Figure 9.8 SCEPTRE Representation of Linear Amplifier Model for Transient Analysis.

from this model. The elements EX and JX are radiation response elements and account for most of the dependencies under FUNCTIONS and many of the DEFINED PARAMETERS entries.

Transistor Models

The equivalent circuit chosen for the transistor models is similar to the one described in Sec. 9.6 and is shown in SCEPTRE form in Fig. 9.9. The appropriate card sequences for these two transistor models, one an *npn* and the other a *pnp*, are as follows:

```
     MODEL 2N2219(B-E-C)
*       USER SUPPLIED RADIATION DATA
*         *     PGR - GAMMA RATE IN RAD(SI)/SEC
*         *     PGD - GAMMA PULSE DELAY FROM SIMULATION TIME ZERO
*         *     PNF - NEUTRON FLUENCE IN N/CM**2
*         BASIC UNITS   OHMS VOLTS FARADS AMPS SECONDS
     ELEMENTS
CC,B1-C1=13.E-12
CE,B1-E=22.E-12
RB,B-B1=50.
RC,C-C1=2.8
JE,B1-E=DIODE EQUATION(3.02E-14,40.)
JC,B1-C1=DIODE EQUATION(1.19E-13,38.)
JN,C1-B1=PAN*JE
JI,E-B1=PAI*JC
JR,C1-B1=QR(PGR,TABLER(PTG),8.5E-12)
     DEFINED PARAMETERS
PGR=0.
PGD=0.
PNF=0.
PAO=.9927
PAF=QAF(PAO,PNF)
PAN=QAN(PAO,PAF,TABLEN(PTG))
PAI=QAI(PAN)
PTG=QTG(TIME,PGD)
     FUNCTIONS
QR(A,B,C)=(A*B*C)
QAF(A,B)=(A/(1.+A*2.051E-16*B))
QAN(A,B,C)=(A-(A-B)*C)
QAI(A)=(.702*A)
QTG(A,B)=(A-B)
TABLER
-1E-6,0,0,0,.04E-6,1,.1E-6,0,1E-6,0
TABLEN
-1E-6,0,10E-6,0,11E-6,1,20E-6,1


     MODEL 2N4405(B-E-C)
*       USER SUPPLIED RADIATION DATA
*         *     PGR - GAMMA RATE IN RAD(SI)/SEC
*         *     PGD - GAMMA PULSE DELAY FROM SIMULATION TIME ZERO
*         *     PNF - NEUTRON FLUENCE IN N/CM**2
*         BASIC UNITS   OHMS VOLTS FARADS AMPS SECONDS
     ELEMENTS
CE,E-B1=55.E-12
```

```
CC,C1-B1=43.E-12
RB,B1-B=20.
RC,C1-C=1.4
JE,E-B1=DIODE EQUATION(1.02E-13,39.)
JC,C1-B1=DIODE EQUATION(1.71E-13,38.)
JN,B1-C1=PAN*JE
JI,B1-E=PAI*JC
JR,B1-C1=QR(PGR,TABLER(PTG),14.E-12)
   DEFINED PARAMETERS
PGR=0.
PGD=0.
PNF=0.
PAO=.9914
PAF=QAF(PAO,PNF)
PAN=QAN(PAO,PAF,TABLEN(PTG))
PAI=QAI(PAN)
PTG=QTG(TIME,PGD)
   FUNCTIONS
QR(A,B,C)=(A*B*C)
QAF(A,B)=(A/(1.+A*4.566E-16*B))
QAN(A,B,C)=(A-(A-B)*C)
QAI(A)=(.6829*A)
QTG(A,B)=(A-B)
TABLER
-1E-6,0,0,0,.04E-6,1.,1E-6,.3,.3E-6,0,1E-6,0
TABLEN
-1E-6,0,10E-6,0,11E-6,1,20E-6,1
```

As with the linear amplifier, the sequence begins with a MODEL *name* card followed by entries under ELEMENTS, DEFINED PARAMETERS, and FUNCTIONS. Here again, the element JR in both models is a radiation response element and accounts for many of the FUNCTIONS and DEFINED PARAMETERS entries for these models.

Actuator Transfer Function

The next problem to be solved is the representation of the electromechanical actuator transfer function in the time domain within the SCEPTRE input syntax. It is at this point that we need to apply either the general modeling technique of a subsequent example problem, Example 9.12, dealing specifically with this problem, or apply techniques for deriving state-space equations directly (see References 26a and 27b). For this exercise, we shall apply the techniques of state equation derivation.

Transfer Function Implementation Using Defined Parameters

It is known that the transfer function of the actuator corresponds to some differential equation. Let us take the transfer function of the actuator

$$\frac{X(s)}{U(s)} = \frac{52.8}{0.03s^2 + s} \qquad (9.21)$$

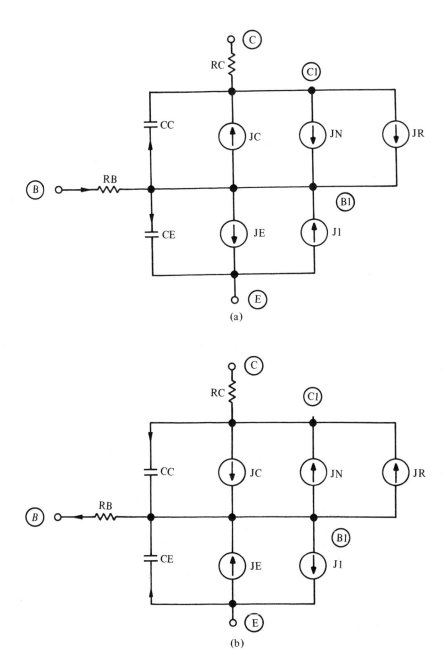

Figure 9.9 SCEPTRE Representation of Transistor Models: (a) *npn* Model; (b) *pnp* Model.

and rearrange it as follows:

$$0.03s^2 X(s) + sX(s) = 52.8 U(s) \tag{9.22}$$

By taking the inverse Laplace transform, we obtain

$$x''(t) + \frac{1}{0.03}x'(t) = \frac{52.8}{0.03}u(t) \tag{9.23}$$

In terms of the operator $p = d/dt$, this equation can be written as

$$p^2 x + \frac{1}{0.03}px = 1760u \tag{9.24}$$

or

$$p^2 x = 1760u - \frac{1}{0.03}px \tag{9.25}$$

We can now draw an analog computer diagram to synthesize x as shown in Fig. 9.10(a). If we assign the integrator outputs as state variables x_1 and x_2, then the diagram becomes that of Fig. 9.10(b). It is now clear from Fig. 9.10(b) that the following relationship holds:

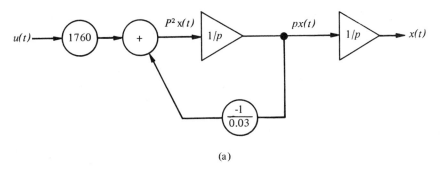

(a)

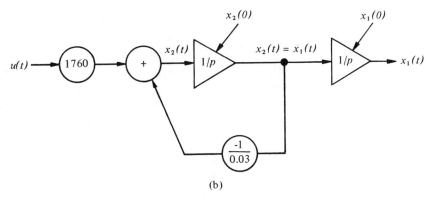

(b)

Figure 9.10 Analog Computer Diagrams Representing Actuator Transfer-Function.

$$x_1' = x_2$$
$$x_2' = -\frac{1}{0.03}x_2 + 1760u(t) \qquad (9.26)$$

This relationship is the state equation with the auxiliary equations

$$x = x_1 \quad \text{and} \quad x_1(0), x_2(0) \quad \text{given}$$

We can implement the state equation directly in SCEPTRE using the defined parameter derivative feature of the code. Let us first define under the DEFINED PARAMETERS subheading the parameters PX1 and PX2 to be equal to the states x_1 and x_2 at time equals zero. The time zero state values can be computed, if desired, knowing that at time zero the input voltage to the operational amplifier must be zero to produce zero solenoid current, or steady-state conditions. Since no solenoid current is flowing, node 70 (see Fig. 9.13) is at ground potential and node 57 must also be at ground potential. With the values of the source EC, resistor R124, and resistor R125, it is a simple calculation to show that the demodulator output, represented by the source EA, must have a value of $+1.662$ volts. Dividing by the demodulator gain gives $+0.902$ volts ac from the linear-voltage differential transformer or 0.902 inches initial actuator displacement for steady state. Since x_1 is equal to the displacement of the actuator, $x_1(0)$ is equal to 0.902. By taking the derivative $d/dt(x_1(0)) = x_2(0)$, the time zero value of the state x_2 is found to be zero.

If the initial states are not known a priori, they can also be set as zero and solved for by using the RUN IC VIA IMPLICIT option in SCEPTRE after the entire system has been described and is ready for a transient run. The RUN IC VIA IMPLICIT option is required because the Newton-Raphson formulation does not recognize defined parameter differential equations, illustrating an important use of the former dc option in SCEPTRE.

The state equation tells us that the derivative of the parameter PX1, i.e., DPX1, is equal to the parameter PX2 and that the derivative of PX2, i.e., DPX2, is equal to a function of PX2 and the input forcing function $u(t)$. The input to the actuator transfer function in this circuit is the solenoid current and will be seen later to be the current in an inductor labeled LTORK, i.e., ILTORK, in the SCEPTRE form schematic of Fig. 9.13. The card sequence to input the actuator transfer function using defined parameter derivatives to compute the displacement of the actuator (PX1) as a function of solenoid current (ILTORK) is as follows:

```
DEFINED PARAMETERS
PX1=.902
PX2=0.0
DPX1=PX2
DPX2=QXX(PX2,ILTORK)
FUNCTIONS
QXX(A,B)=((-1./.03)*A+1760.*B)
```

Transfer Function Implementation Using Network Synthesis

An alternative possible approach to transfer function implementation is to factor the transfer function in both the numerator and denominator and partition the function as products of easily synthesized RLC networks in cascade, with proper isolation between partitions. While this procedure may not always be practical in the synthesis of physical systems, it is certainly possible for a SCEPTRE model, where simple dependent voltage and current sources can be used to drive and isolate the partitioned networks to eliminate the loading effects of one partition on another.

We can partition the actuator transfer function as follows:

$$\frac{X(s)}{U(s)} = 52.8 * \frac{1}{s} * \frac{1}{0.03s + 1} \tag{9.27}$$

The RLC model corresponding to this partition is shown in Fig. 9.11. An appropriate card sequence for this model is

```
MODEL ACTUATOR(1-5-0)
*     TRANSFER FUNCTION IMPLIMENTATION FOR TRANSIENT ANALYSIS
*     THE FUNCTION IMPLEMENTED HAS THE FORM -
*         X(S)/U(S) = 52.8/(.03*S*S + S)
   ELEMENTS
ELEMENTS
E1,1-0=0.
E2,0-3=Q1(VC1)
E3,0-5=Q2(VC2)
C1,2-0=1.
C2,4-0=.03
R1,3-4=1.
J1,0-2=Q1(IE1)
   FUNCTIONS
Q1(A)=(A)
Q2(A)=(52.8*A)
```

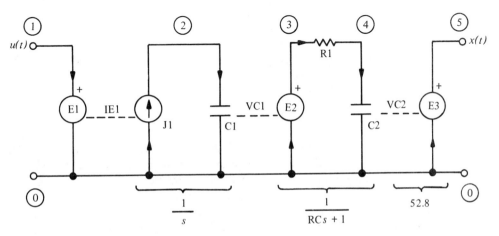

Figure 9.11 SCEPTRE RLC Actuator Model for Transient Analysis.

Initial conditions for the model can be computed using the steady-state actuator displacement calculated earlier, recognizing from Figs. 9.7 and 9.11 that the following relationship holds:

$$\text{VC1}(0) = \text{VC2}(0) = \frac{\text{initial displacement}}{52.8} = \frac{E3(0)}{52.8} = \frac{0.902}{52.8} = 0.0171 \quad (9.28)$$

These initial conditions can either be inserted directly in the model under an INITIAL CONDITIONS subheading or entered in the CIRCUIT DESCRIPTION under an INITIAL CONDITIONS subheading (if appended by the element name given the model), as illustrated later in the development of this example (see the CIRCUIT DESCRIPTION implementing actuator model).

If the user is intending to investigate the behavior of the system in the frequency domain as well as the time domain, the latter approach to the actuator transfer function implementation would be advantageous as the ac SCEPTRE syntax will accept this form provided that the partitioned function is implemented using linearly dependent sources, recognized by SCEPTRE as complex-valued sources, which will not be set equal to zero during an ac analysis. This requires the addition of resistors to the topology of Fig. 9.11 to provide the values required by the linearly dependent source format. The modified topology compatible with ac analysis is shown in Fig. 9.12. The resistor values should be chosen such that the desired response is not affected within the desired accuracy and such that the problem is not made too stiff. An ac analysis compatible card sequence for the actuator model would appear as follows:

```
      MODEL ACTUATOR(1-5-0)
  *      AC COMPATIBLE TRANSFER FUNCTION IMPLEMENTATION
  *      THE FUNCTION IMPLEMENTED HAS THE FORM -
  *         X(S)/U(S) = 52.8/(.03*S*S + S)
      ELEMENTS
  E1,0-3=1.0*VR2
  E2,0-5=52.8*VR4
  C1,2-0=1.
  C2,4-0=.03
  R1,1-0=1.E-2
  R2,2-0=1.E3
  R3,3-4=1.
  R4,4-0=1.E2
  J1,0-2=1.0*IR1
```

It should be noted that similar changes must be made in the linear amplifier model if an ac analysis is required. The transistor models, however, are of a form which will be automatically linearized by SCEPTRE and are compatible with ac analysis syntax.

The defined parameter derivative approach developed earlier for implementing transfer functions is excluded for ac analysis because the SCEPTRE

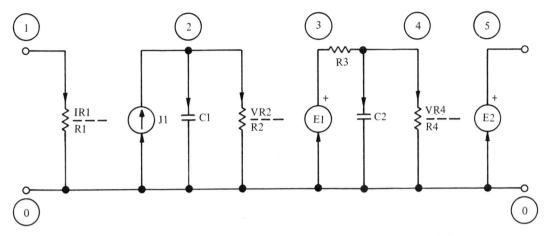

Figure 9.12 SCEPTRE RLC Actuator Model for ac Analysis.

ac formulation does not include provision for defined parameter differential equations. The user is also cautioned not to attempt to input transfer functions for ac analysis under the FUNCTIONS subheading or using function subprograms, because in most circuits a computational delay will result where the program will be forced to use past values of some variables for present calculations. It is expected that computational delays in ac analysis will cause considerably larger errors than computational delays in the transient mode of analysis since the frequency step size is not a function of the behavior of network values or any internal error criteria.

Control System SCEPTRE Circuit Description

The task remaining is to describe the system network to SCEPTRE under the heading CIRCUIT DESCRIPTION and supply appropriate INITIAL CONDITIONS and RUN CONTROLS entries for the transient run. The nodes of the schematic diagram of Fig. 9.7 are given alphanumeric labels, and element polarity conventions are assigned as shown in Fig. 9.13. The zero-valued current sources are placed in the circuit to function as infinite impedance voltmeters. In like manner, zero-valued voltage sources may be used as zero-impedance ammeters as was done in the actuator model using the element E1 (Fig. 9.11). The solenoid coil has been modeled using the elements RTORK and CTORK to represent the parasitic effects in the coil and LTORK representing the bulk property of the coil. These values are easily obtained from data sheet specifications or measurements on the physical device.

An appropriate sequence of cards, implementing the actuator transfer function using defined parameter derivatives, is as follows:

```
CIRCUIT DESCRIPTION
*     TORQUER CLOSED-LOOP STEP RESPONSE TEST CIRCUIT
*     PARAMETERS FOR A 2 PERCENT STEP INCREASE FROM 88 PERCENT POINT
*      IN DISPLACEMENT SCHEDULE
*     BASIC UNITS  OHMS VOLTS FARADS HENRIES AMPS SECONDS
*     ACTUATOR IMPLEMENTED USING DEFINED PARAMETER DERIVITIVES
 ELEMENTS
R125,57-11=20000.
J57,57-0=0.0
C105,73-81=4700.E-12
E13,0-80=21.
E14,78-0=21.
JR1,73-82=0.0
JR2,83-0=0.0
JT,73-0=0.0
JDR,72-0=0.0
JOT,71-0=0.0
O3,57-70-71-0=MODEL HA 2700
R126,70-0=10.E3
R127,81-70=800.E3
R128,71-72=511.
R129T,80-79=95.
R131,70-82=66.5E3
R132,82-0=10.
R133,83-82=10.E3
RT,74-78=60.
RTORK,73-TI=50.
LTORK,82-TI=.2
CTORK,73-82=.127E-6
T4,72-73-74=MODEL 2N4405
T3,72-73-79=MODEL 2N2219
R124,56-57=23.7E3
EC,0-56=TABLE C(TIME)
EA,0-11=QA(PDIS)
 DEFINED PARAMETERS
PX1=.902
PX2=0.
DPX1=PX2
DPX2=QXX(PX2,ILTORK)
PDIS=PX1
INITIAL CONDITIONS
ILTORK=0.
VCCT3=-2.1E+01
VCCT4=-2.09162E+01
VCTORK=0.
 OUTPUTS
VJ57,EA,EC,VJOT,VJDR,VJT,ILTORK,PDIS,PLOT
 FUNCTIONS
TABLE C
0.,-1.97,.01,-1.97,.01,-2.07,1.,-2.07
QA(A)=(A*1.843)
QXX(A,B)=((-1./.03)*A+1760.*B)
 RUN CONTROLS
INTEGRATION ROUTINE=IMPLICIT
MAXIMUM PRINT POINTS=0
STOP TIME=.45
MINIMUM STEP SIZE=1.E-50
MAX INTEGRATION PASSES=1.E6
WRITE SIMUL8 DATA
 END
```

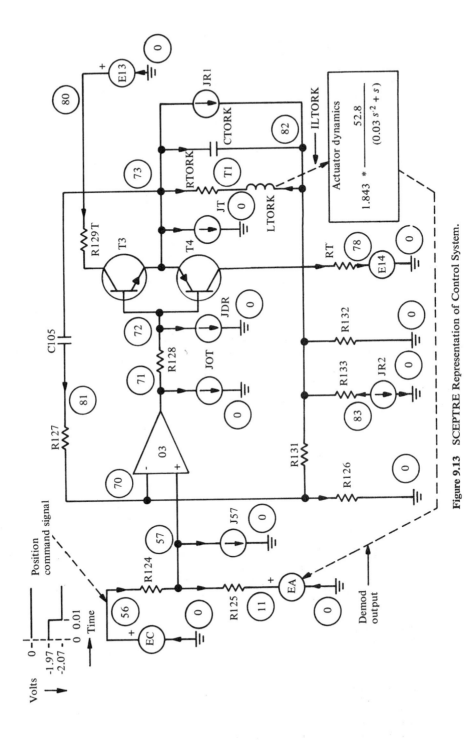

Figure 9.13 SCEPTRE Representation of Control System.

A card sequence using either actuator model is shown below and results in the same solution as the sequence implementing the transfer function using defined parameter derivatives. Changes required in the SCEPTRE CIRCUIT DESCRIPTION topology to implement the actuator model are shown in Fig. 9.14.

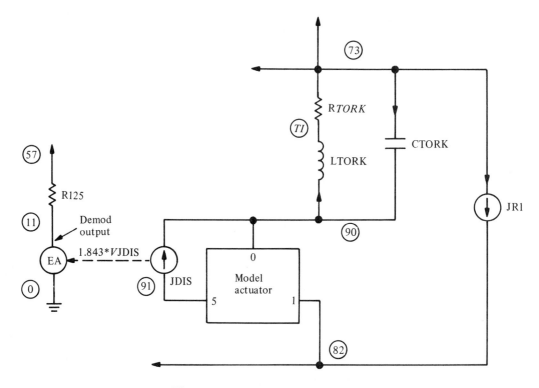

Figure 9.14 Topology Changes to Use Actuator Model.

```
CIRCUIT DESCRIPTION
*      TORQUER CLOSED-LOOP STEP RESPONSE TEST CIRCUIT
*      PARAMETERS FOR A 2 PERCENT STEP INCREASE FROM 88 PERCENT POINT
*      IN DISPLACEMENT SCHEDULE
*      BASIC UNITS   OHMS VOLTS FARADS HENRIES AMPS SECONDS
*      ACTUATOR IMPLEMENTED USING RLC MODEL
ELEMENTS
R125,57-11=20000.
J57,57-0=0.0
C105,73-81=4700.E-12
E13,0-80=21.
E14,78-0=21.
JR1,73-82=0.0
JR2,83-0=0.0
JT,73-0=0.0
JDR,72-0=0.0
JOT,71-0=0.0
O3,57-70-71-0=MODEL HA 2700
```

```
R126,70-0=10.E3
R127,81-70=800.E3
R128,71-72=511.
R129T,80-79=95.
R131,70-82=66.5E3
R132,82-0=10.
R133,83-82=10.E3
RT,74-78=60.
RTORK,73-TI=50.
LTORK,90-TI=.2
CTORK,73-90=.127E-6
T4,72-73-74=MODEL 2N4405
T3,72-73-79=MODEL 2N2219
R124,56-57=23.7E3
EC,0-56=TABLE C(TIME)
EA,0-11=QA(PDIS)
TX,82-91-90=MODEL ACTUATOR
JDIS,91-90=0.0
 DEFINED PARAMETERS
PDIS=QX(VJDIS)
INITIAL CONDITIONS
ILTORK=0.
VCCT3=-2.1E+01
VCCT4=-2.09162E+01
VCTORK=0.
VC2TX=.0171
VC1TX=.0171
 OUTPUTS
VJ57,EA,EC,VJOT,VJDR,VJT,ILTORK,PDIS,PLOT
 FUNCTIONS
TABLE C
0.,-1.97,.01,-1.97,.01,-2.07,1.,-2.07
QA(A)=(A*1.843)
QX(A)=(A)
 RUN CONTROLS
INTEGRATION ROUTINE=IMPLICIT
MAXIMUM PRINT POINTS=0
STOP TIME=.45
MINIMUM STEP SIZE=1.E-50
MAX INTEGRATION PASSES=1.E6
WRITE SIMUL8 DATA
 END
```

The system is being exercised for this example by a step change in commanded actuator position by changing element EC as a function of time (Fig. 9.15). The resulting actuator displacement, PDIS, is plotted as a function of time in Fig. 9.16. The commanded step in position is small enough so that the entire system operates within the "linear" region, but this is not a constraint and nonlinear behavior may be investigated as desired. The effect of crossover distortion in the power amplifier driving the solenoid is seen in the plot of VJDR versus time in Fig. 9.17. The error signal at the junction of R124 and R125 driving the HA2700 operational amplifier is plotted as a function of time in Fig. 9.18.

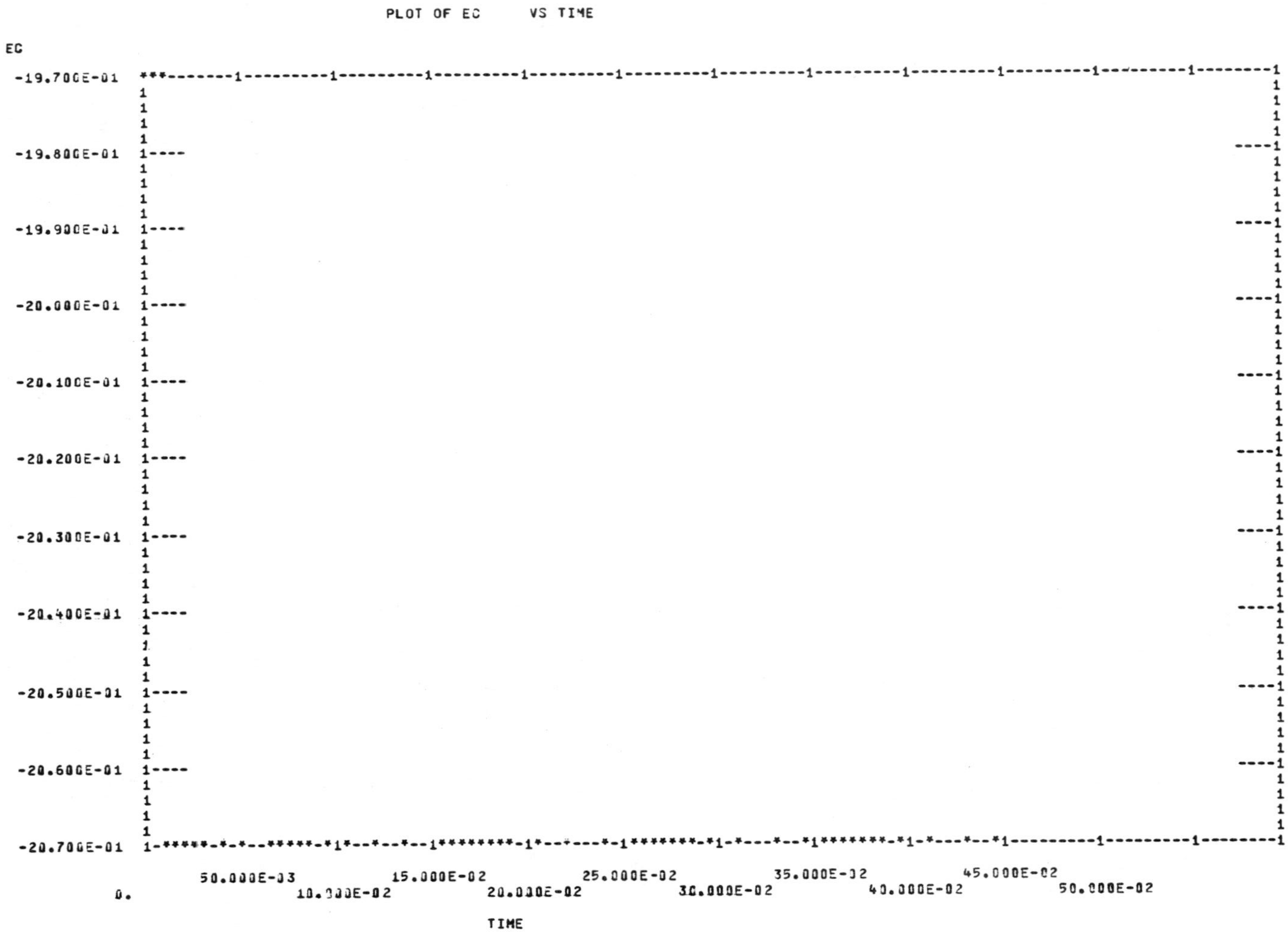

Figure 9.15 Plot of EC vs. Time.

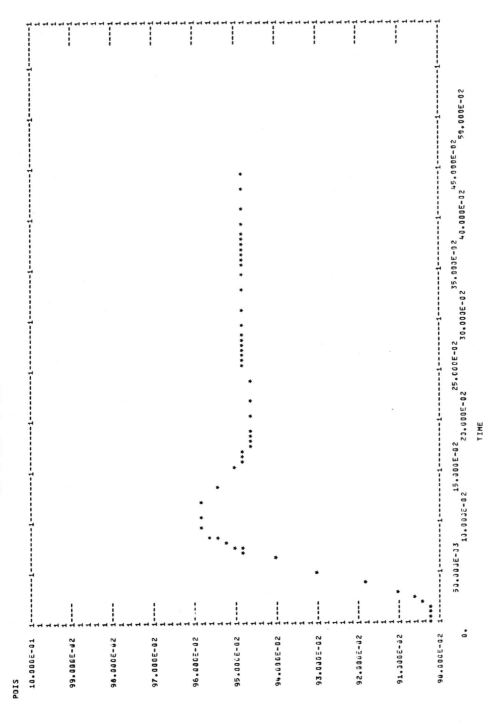

Figure 9.16 Plot of PDIS vs. Time.

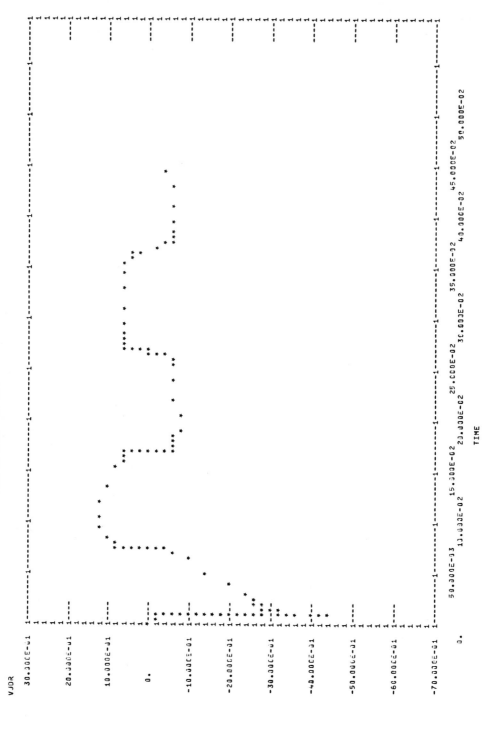

Figure 9.17 Plot of VJOR vs. Time.

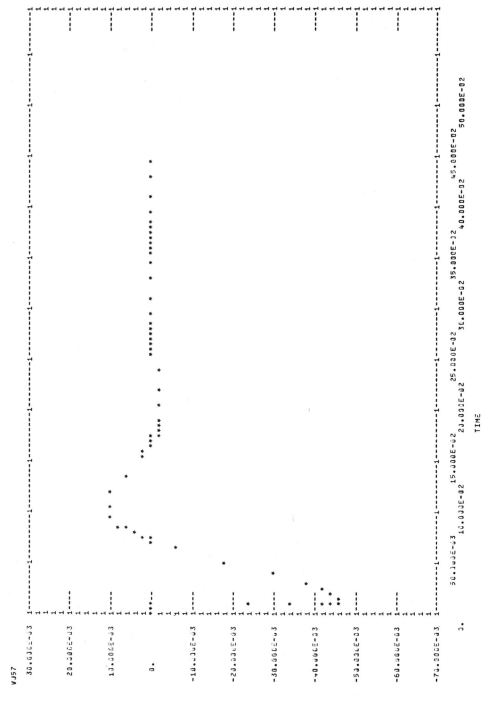

Figure 9.18 Plot of VJ57 vs. Time.

Computer system time (CDC 6600 central processor time) for runs using either defined parameter derivatives or the transient analysis compatible actuator model was approximately 45 seconds including Phase 1 setup overhead. When using the ac analysis compatible actuator model for transient analysis, computer system time increased materially. The increased system time is probably due to a poor choice of resistor values in the model, creating an extremely stiff problem.

Example 9.2. Inverter Circuit Loaded with *RC* Network

Figure 9.19 shows a schematic of an inverter circuit loaded with an *RC* network It is desired to analyze the effects of a transient radiation environment on this circuit. The forcing function will be the primary photocurrent

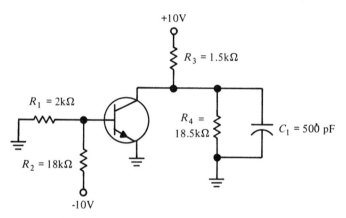

Figure 9.19 Example 9.2 Schematic Diagram.

appropriate to the environment. Two reruns are to be made, and these differ from the master run only in the magnitude of the photocurrent (effectively applying an upper and lower tolerance to the I_{pp}). The stored model feature will not be used, and the initial conditions will be supplied as known quantities; therefore, the mode of analysis will be transient only.

The schematic of Fig. 9.19 is redrawn in SCEPTRE form in Fig. 9.20. Here J1 and J2 represent the transistor junctions, JA and JB represent the conventional current-controlled current generators of the Ebers-Moll equivalent circuit, and JX represents the primary photocurrent caused by the effects of the radiation on the transistor in the inverter. A valid sequence of cards would be

```
CIRCUIT DESCRIPTION
*    INVERTER CIRCUIT LOADED WITH RC NETWORK
*    UNITS - KOHMS, UHENRIES, PFARADS, VOLTS, MAMPS, NSECONDS, GHZ
 ELEMENTS
E1,7-1=10.
E2,1-6=10.
CE,3-1=EQUATION 1(5.,70.,J1)
CC,3-4=EQUATION 1(8.,370.,J2)
C1,5-1=500
R1,1-2=2
R2,2-7=18
R3,6-5=1.5
R4,5-1=18.5
RB,2-3=.3
RC,5-4=.015
J1,3-1=DIODE EQUATION(1.E-7,35.)
J2,3-4=DIODE EQUATION(5.E-7,37.)
JA,1-3=.1*J2
JB,4-3=.98*J1
JX,4-3=TABLE 1(TIME)
 OUTPUTS
VCE,VCC,VC1,IR3,J1,PLOT
JX,PLOT
XPASNO
XSTPSZ
 INITIAL CONDITIONS
VC1=9.25
VCE=-1
VCC=-10.25
 FUNCTIONS
TABLE 1
0,0
40,.8
100,.5
200,.25
500,0
600,0
EQUATION 1(A,B,C)=(A+B*C)
 RUN CONTROLS
STOP TIME=800
INTEGRATION ROUTINE=TRAP
 RERUN DESCRIPTION(2)
 FUNCTIONS
TABLE 1=0,0,0
40,1.2,.4
100,.75,.25
200,.375,.125
500,0,0
600,0,0
 END
```

The results of the master run indicate that the inverter turned on at about 48 nanoseconds (base-emitter junction forward-biased) and returned to the OFF condition at about 135 nanoseconds. Since the degree of turn-on

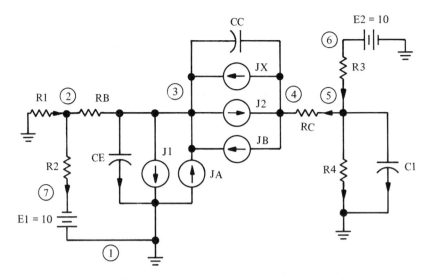

Figure 9.20 Example 9.2 SCEPTRE Form.

was small (maximum positive VBE ≈ 0.26 volt), the transistor never approached saturation in this environment. The voltage excursion seen by the RC network was about 0.18 volt.

The first 35 and the last 7 print intervals are enclosed in this subsection as Fig. 9.21. In addition, a reproduction of the plotter output for the voltage across capacitor CE (or VBE) is also enclosed, Fig. 9.22.

The first rerun effectively included the effects of a 50% increase in I_{pp} as reflected in the modified TABLE 1. The increased effects on the base-emitter junction voltage are shown in Fig. 9.23. The second rerun effectively included the effects of a 50% decrease in I_{pp}, and the corresponding reduced circuit reaction can be seen in Fig. 9.24.

Example 9.3. Transformer-Coupled Amplifier

The schematic of an emitter follower-common emitter combination driving an output transformer with a resistive load is shown in Figs. 9.25 and 9.26. The circuit will be driven by a ramp voltage input coupled through a capacitor. For this example, it will be assumed that the user wishes to temporarily store a transistor model and to use this model for both circuit transistors. The initial conditions will be computed along with the transient solution. No reruns will be made. The SCEPTRE input data is shown on page 619.

```
          MINI-SCEPTRE TRANSIENT RESULTS
*   INVERTER CIRCUIT LOADED WITH RC NETWORK
*   UNITS - KOHMS, UHENRIES, PFARADS, VOLTS, MAMPS, NSECONDS, GHZ

TIME     0.           8.32150E+00  1.63746E+01  3.22123E+01  4.02665E+01  4.80499E+01  5.63714E+01
VCE     -1.00000E+00 -9.55561E-01 -8.37482E-01 -6.62790E-01 -4.56912E-01 -2.11074E-01  1.39912E-01
VCC     -1.02500E+01 -1.02036E+01 -1.00826E+01 -9.90251E+00 -9.68967E+00 -9.43455E+00 -9.06480E+00
VC1      9.25000E+00  9.24946E+00  9.24754E+00  9.24397E+00  9.23867E+00  9.23126E+00  9.21331E+00
IR3     -1.00000E-01  5.00363E-01  5.01643E-01  5.04023E-01  5.07553E-01  5.12491E-01  5.24458E-01
J1      -1.00000E-07 -1.00000E-07 -1.00000E-07 -1.00000E-07 -1.00000E-07 -9.99381E-08  1.32876E-05
JX       0.           1.66430E-01  3.27491E-01  4.88553E-01  6.44245E-01  7.98673E-01  7.18143E-01
XPASNO   1.00000E+00  3.78000E+02  4.24000E+02  4.70000E+02  5.03000E+02  5.50000E+02  5.93000E+02  6.41000E+02
XSTPSZ   8.00000E-09  5.36871E-01  5.36871E-01  5.36871E-01  5.36871E-01  5.36871E-01  2.68435E-01  5.36871E-01

TIME     6.41561E+01  7.24776E+01  8.05306E+01  8.83153E+01  9.60999E+01  1.04153E+02  1.12474E+02  1.21064E+02
VCE      2.17000E-01  2.55525E-01  2.64843E-01  2.50544E-01  2.17648E-01  1.75267E-01  1.26359E-01  7.68265E-02
VCC     -8.97841E+00 -8.91204E+00 -8.91264E+00 -8.91760E+00 -8.94287E+00 -8.97766E+00 -9.01948E+00 -9.06186E+00
VC1      9.20425E+00  9.19463E+00  9.18554E+00  9.17760E+00  9.16906E+00  9.16105E+00  9.15348E+00  9.14628E+00
IR3      5.30499E-01  5.36913E-01  5.36913E-01  5.42971E-01  5.48628E-01  5.53961E-01  5.59303E-01  5.64344E-01  5.69145E-01
J1       1.98722E-04  7.65606E-04  1.06086E-03  6.43101E-04  2.03283E-04  2.03283E-04  4.60429E-05  8.23104E-06  1.37468E-06
JX       6.79220E-01  6.37612E-01  5.97367E-01  5.58424E-01  5.19501E-01  4.60429E-01  4.89618E-01  4.68814E-01  4.47339E-01
XPASNO   6.88000E+02  7.33000E+02  7.65000E+02  7.96000E+02  8.33000E+02  8.56000E+02  8.86000E+02  9.04000E+02
XSTPSZ   2.68435E-01  5.36871E-01  5.36871E-01  5.36871E-01  5.36871E-01  2.68435E-01  1.07374E+00  1.07374E+00  1.07374E+00

TIME     1.28044E+02  1.36634E+02  1.44687E+02  1.52203E+02  1.60256E+02  1.68309E+02  1.76689E+02  1.84415E+02
VCE      3.81983E-02 -8.74221E-03 -5.16503E-02 -9.13030E-02 -1.33504E-01 -1.75064E-01 -2.19689E-01 -2.58405E-01
VCC     -9.09547E+00 -9.13608E+00 -9.17384E+00 -9.20876E+00 -9.24621E+00 -9.28410E+00 -9.32445E+00 -9.36013E+00
VC1      9.14078E+00  9.13430E+00  9.12877E+00  9.12391E+00  9.11898E+00  9.11454E+00  9.11014E+00  9.10648E+00
IR3      5.72814E-01  5.77135E-01  5.80818E-01  5.84059E-01  5.87348E-01  5.90307E-01  5.93239E-01  5.95682E-01
J1       2.80738E-07 -2.63597E-08 -8.35979E-08 -9.59058E-08 -9.90652E-08 -9.97817E-08 -9.99542E-08 -9.99882E-08
JX       4.29891E-01  4.08416E-01  3.88283E-01  3.69493E-01  3.49360E-01  3.29227E-01  3.07753E-01  2.89962E-01
XPASNO   9.25000E+02  9.40000E+02  9.63000E+02  9.79000E+02  1.02500E+03  1.04000E+03  1.04000E+03  1.05900E+03
XSTPSZ   1.07374E+00  1.07374E+00  1.07374E+00  1.07374E+00  1.07374E+00  1.07374E+00  1.07374E+00  5.36871E-01

TIME     1.92468E+02  2.00521E+02  2.08037E+02  2.16627E+02  2.24144E+02  2.32734E+02  2.41323E+02  2.48840E+02
VCE     -3.00438E-01 -3.41708E-01 -3.79666E-01 -4.12460E-01 -4.38419E-01 -4.62654E-01 -4.85268E-01 -5.04765E-01
VCC     -9.39835E+00 -9.43737E+00 -9.47288E+00 -9.50335E+00 -9.52735E+00 -9.54891E+00 -9.56982E+00 -9.58763E+00
VC1      9.10299E+00  9.10005E+00  9.09736E+00  9.09462E+00  9.09237E+00  9.09008E+00  9.08800E+00  9.08636E+00
IR3      5.98004E-01  6.01760E-01  6.01760E-01  6.03587E-01  6.05085E-01  6.06612E-01  6.07997E-01  6.09464E+00
J1      -9.99973E-08 -9.99968E-08 -9.99998E-08 -6.01769E-01 -1.00000E-07 -1.00000E-07 -1.00000E-07 -1.00000E-07
JX       2.68829E-01  2.49566E-01  2.43302E-01  2.36144E-01  2.29880E-01  2.22722E-01  2.15564E-01  2.09300E-01
XPASNO   1.07700E+03  1.10200E+03  1.12000E+03  1.14900E+03  1.16800E+03  1.19500E+03  1.20900E+03  1.21800E+03
XSTPSZ   1.07374E+00  1.07374E+00  5.36871E-01  1.07374E+00  5.36871E-01  1.07374E+00  1.07374E+00  1.07374E+00
```

Figure 9.21 Example 9.2 Output Listing.

```
*      INVERTER CIRCUIT LOADED WITH RC NETWORK
*            MINI-SCEPTRE TRANSIENT RESULTS
*      UNITS - KOHMS, UHENRIES, PFARADS, VOLTS, MAMPS, NSECONDS, GHZ

TIME     2.56356E+02  2.64946E+02  2.72462E+02  2.80515E+02  2.88031E+02  2.96084E+02  3.04674E+02  3.12190E+02
VCE     -5.20301E-01 -5.37335E-01 -5.50977E-01 -5.66257E-01 -5.80274E-01 -5.95302E-01 -6.11383E-01 -6.24747E-01
VCC     -9.60159E+00 -9.61745E+00 -9.62994E+00 -9.64372E+00 -9.65707E+00 -9.67121E+00 -9.68582E+00 -9.69875E+00
VC1      9.08486E+00  9.08323E+00  9.08196E+00  9.08056E+00  9.07960E+00  9.07846E+00  9.07743E+00  9.07660E+00
IR3      6.10091E-01  6.11180E-01  6.12028E-01  6.12962E-01  6.13597E-01  6.14359E-01  6.15046E-01  6.15598E-01
J1      -1.00000E-07 -1.00000E-07 -1.00000E-07 -1.00000E-07 -1.00000E-07 -1.00000E-07 -1.00000E-07 -1.00000E-07
JX       2.03037E-01  1.95879E-01  1.89615E-01  1.82904E-01  1.76641E-01  1.69930E-01  1.62771E-01  1.56504E-01
XPASNO   1.24200E+03  1.26500E+03  1.27800E+03  1.29800E+03  1.31600E+03  1.33200E+03  1.35500E+03  1.37000E+03
XSTPSZ   5.36871E-01  1.07374E+00  1.07374E+00  1.07374E+00  1.07374E+00  5.36871E-01  1.07374E+00  1.07374E+00

TIME     6.40219E+02  6.48272E+02  6.56325E+02  6.64378E+02  6.72968E+02  6.80484E+02  6.88000E+02  6.96053E+02
VCE     -9.97134E-01 -9.96819E-01 -9.96827E-01 -9.96795E-01 -9.96726E-01 -9.96918E-01 -9.96967E-01 -9.97064E-01
VCC     -1.01153E+01 -1.01168E+01 -1.01193E+01 -1.01198E+01 -1.01215E+01 -1.01228E+01 -1.01243E+01 -1.01257E+01
VC1      9.11797E+00  9.11979E+00  9.12154E+00  9.12319E+00  9.12472E+00  9.12601E+00  9.12739E+00  9.12873E+00
IR3      5.88023E-01  5.86809E-01  5.85640E-01  5.84539E-01  5.83522E-01  5.82661E-01  5.81737E-01  5.80847E-01
J1      -1.00000E-07 -1.00000E-07 -1.00000E-07 -1.00000E-07 -1.00000E-07 -1.00000E-07 -1.00000E-07 -1.00000E-07
JX       0.           0.           0.           0.           0.           0.           0.           0.
XPASNO   1.97200E+03  1.99600E+03  2.01700E+03  2.03800E+03  2.05800E+03  2.07300E+03  2.09200E+03  2.11200E+03
XSTPSZ   1.07374E+00  1.07374E+00  5.36871E-01  5.36871E-01  1.07374E+00  1.07374E+00  5.36871E-01  1.07374E+00

TIME     7.04106E+02  7.12159E+02  7.21823E+02  7.28265E+02  7.36855E+02  7.44372E+02  7.52961E+02  7.61015E+02
VCE     -9.97126E-01 -9.96737E-01 -9.96758E-01 -9.96840E-01 -9.96717E-01 -9.96941E-01 -9.96882E-01 -9.96902E-01
VCC     -1.01271E+01 -1.01285E+01 -1.01302E+01 -1.01313E+01 -1.01329E+01 -1.01341E+01 -1.01356E+01 -1.01370E+01
VC1      9.13036E+00  9.13178E+00  9.13341E+00  9.13446E+00  9.13593E+00  9.13710E+00  9.13843E+00  9.13978E+00
IR3      5.79760E-01  5.78812E-01  5.77724E-01  5.77025E-01  5.76044E-01  5.75265E-01  5.74378E-01  5.73478E-01
J1      -1.00000E-07 -1.00000E-07 -1.00000E-07 -1.00000E-07 -1.00000E-07 -1.00000E-07 -1.00000E-07 -1.00000E-07
JX       0.           0.           0.           0.           0.           0.           0.           0.
XPASNO   2.12800E+03  2.15000E+03  2.16200E+03  2.16800E+03  2.17700E+03  2.18900E+03  2.20700E+03  2.23000E+03
XSTPSZ   5.36871E-01  1.07374E+00  2.14748E+00  2.14748E+00  2.14748E+00  1.07374E+00  1.07374E+00  1.07374E+00

TIME     7.68531E+02  7.76047E+02  7.84637E+02  7.92690E+02  8.00206E+02
VCE     -9.97207E-01 -9.96799E-01 -9.96869E-01 -9.97244E-01 -9.96978E-01
VCC     -1.01381E+01 -1.01397E+01 -1.01412E+01 -1.01424E+01 -1.01438E+01
VC1      9.14123E+00  9.14266E+00  9.14418E+00  9.14515E+00  9.14660E+00
IR3      5.72511E-01  5.71558E-01  5.70548E-01  5.69903E-01  5.68933E-01
J1      -1.00000E-07 -1.00000E-07 -1.00000E-07 -1.00000E-07 -1.00000E-07
JX       0.           0.           0.           0.           0.
XPASNO   2.25000E+03  2.26900E+03  2.29100E+03  2.31600E+03  2.33100E+03
XSTPSZ   5.36871E-01  5.36871E-01  1.07374E+00  1.07374E+00  1.07374E+00
```

Figure 9.21—*Cont.*

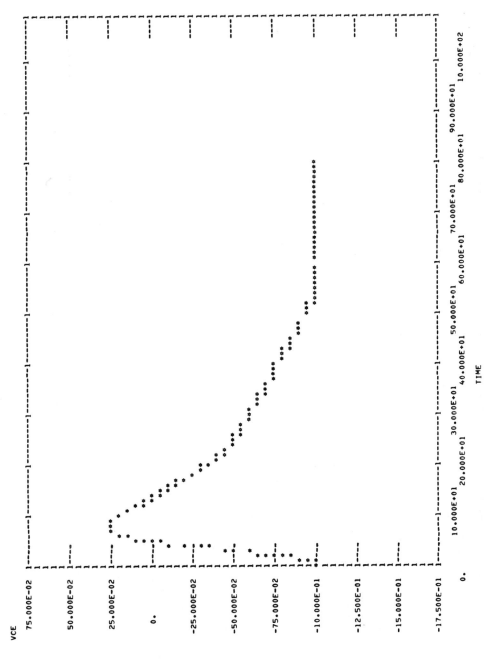

Figure 9.22 Example 9.2 Printer Plot Output.

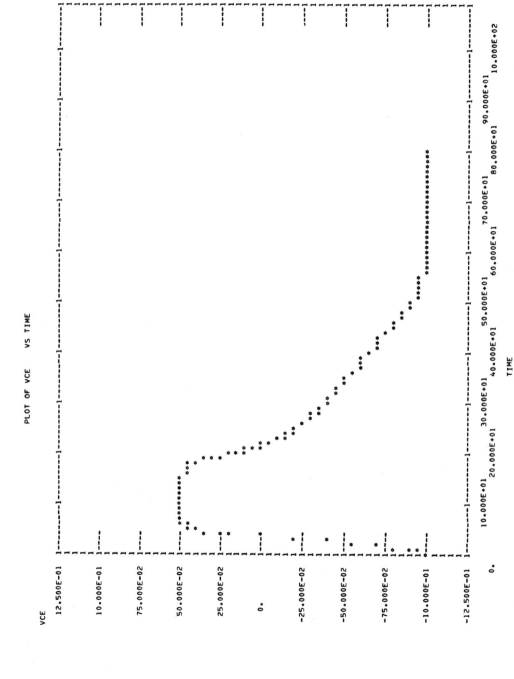

Figure 9.23 Example 9.2 Rerun 1.

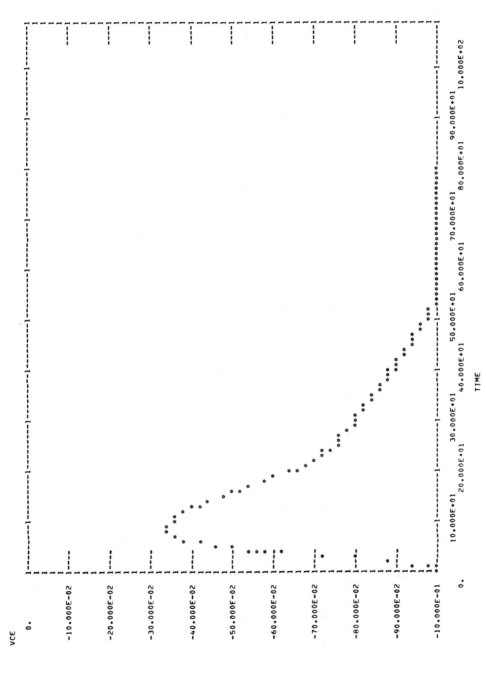

Figure 9.24 Example 9.2 Rerun 2.

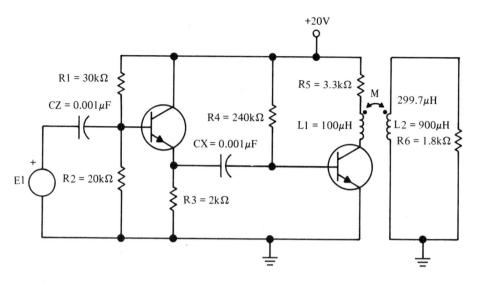

Figure 9.25 Example 9.3 Schematic.

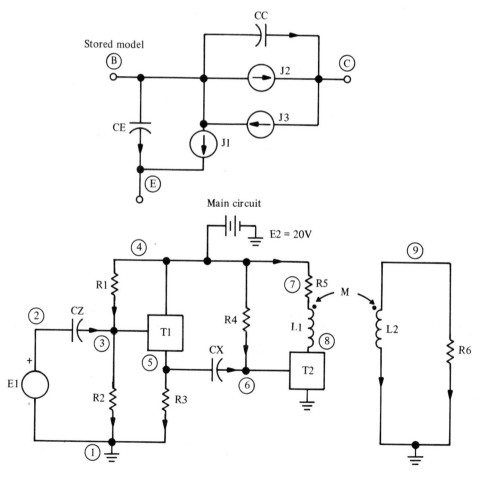

Figure 9.26 Example 9.3 SCEPTRE Form.

```
 MODEL DESCRIPTION
MODEL 2N9999AA (TEMP) (B-E-C)
*    UNITS - KOHMS, UHENRIES, PFARADS, VOLTS, MAMPS, NSECONDS, GHZ
 ELEMENTS
CE,B-E=EQUATION 1(5.,40.,TABLE 1(VCE))
CC,B-C=EQUATION 1(10.,400.,TABLE 2(VCC))
J1,B-E=DIODE TABLE 1
J2,B-C=DIODE TABLE 2
J3,C-B=P1*J1
 DEFINED PARAMETERS
P1=.98
 OUTPUTS
VCE,VCC,J1,PLOT
 FUNCTIONS
EQUATION 1(A,B,C)=(A+B*C)
DIODE TABLE 1
0,0,.3,0,.65,.05,.7,.6,.72,1.4,.73,2,.74,3.4,.77,10,.8,22
DIODE TABLE 2
0,0,.58,0,.62,.4,.64,1,.66,2,.67,3,.69,7,.7,12
 CIRCUIT DESCRIPTION
*    TRANSFORMER COUPLED AMPLIFIER
*    UNITS - KOHMS, UHENRIES, PFARADS, VOLTS, MAMPS, NSECONDS, GHZ
 ELEMENTS
E1,1-2=TABLE 1(TIME)
DERIVATIVE E1=TABLE DE1
E2,1-4=20
CZ,2-3=1E3
CX,5-6=1E3
R1,4-3=30
R2,3-1=20
R3,5-1=2
R4,4-6=240
R5,4-7=3.3
R6,9-1=1.8
T1,3-5-4=MODEL 2N9999AA (TEMP)
T2,6-1-8=MODEL 2N9999AA (TEMP,CHANGE P1=.975)
L1,7-8=100
L2,9-1=900
M,L1-L2=299.7
 OUTPUTS
VR6,VL1,VL2,PLOT
ICZ,ICX,IR3,IR2,IR1,J3T1,J2T1,CCT1,ICCT1
 FUNCTIONS
TABLE 1
0,0,50,.5,100,.5
TABLE DE1
0,.01,50,.01,50,0,100,0
 RUN CONTROLS
STOP TIME=500,COMPUTER TIME LIMIT=5,RUN INITIAL CONDITIONS
INTEGRATION ROUTINE=IMPLICIT
MAXIMUM PRINT POINTS=3000
 END
```

```
TRANSIENT ANALYSIS RESULTS

C    **********************
C     EXAMPLE 3 NETWORK
C    **********************

TIME    0.0            5.00000E-06    1.00000E-05    5.00100E-02    1.00010E-01    2.38750E-01    3.46427E-01
VCET1   7.38868E-01    7.38868E-01    7.38868E-01    7.39085E-01    7.39301E-01    7.39893E-01    7.40341E-01
VCCT1  -1.27780E 01   -1.27780E 01   -1.27780E 01   -1.27775E 01   -1.27770E 01   -1.27758E 01   -1.27748E 01
JIT1    3.24158E 00    3.24158E 00    3.24158E 00    3.27190E 00    3.30210E 00    3.38499E 00    3.47506E 00
VCET2   7.38644E-01    7.38644E-01    7.38644E-01    7.38644E-01    7.39080E-01    7.39676E-01    7.40140E-01
VCCT2  -8.93245E 00   -8.93245E 00   -8.93245E 00   -8.93230E 00   -8.93201E 00   -8.93006E 00   -8.92775E 00
JIT2    3.21023E 00    3.21023E 00    3.21023E 00    3.24081E 00    3.27124E 00    3.35467E 00    3.43077E 00
VR6     0.0            1.06528E-19    1.06528E-19   -6.57824E-13   -6.57824E-06   -2.09285E-06    4.16207E-04
VL1     7.10543E-15    8.93223E-04    1.06528E-19   -4.38771E-08   -5.54053E-06   -2.09285E-06    4.16207E-04
VL2     0.0            1.06528E-19    1.06528E-19   -6.57824E-13   -3.88040E-05   -2.29777E-05    9.17065E-04
ICZ     6.78663E-01    6.78663E-01    6.78663E-01    6.57824E-06   -5.54053E-06   -2.09285E-06    4.16207E-04
ICX     5.85449E-01    5.85452E-01    5.85456E-01    6.82298E-01    6.85862E-01    6.95277E-01    6.97627E-01
IR3     3.24158E 00    3.24158E 00    3.24158E 00    6.18674E 00    6.51706E-01    7.41993E-01    8.32277E-01
IR2     3.61101E-01    3.61101E-01    3.61101E-01    3.24170E 00    3.24183E 00    3.24218E 00    3.24245E 00
IR1     4.25933E-01    4.25933E-01    4.25933E-01    3.61124E-01    3.61148E-01    3.61212E-01    3.61262E-01
J3T1    3.17674E 00    3.17675E 00    3.17675E 00    4.25917E-01    4.25902E-01    4.25858E-01    4.25825E-01
J2T1    0.0            0.0            0.0            3.20646E 00    3.23606E 00    3.31729E 00    3.40556E 00
CCT1    1.00000E 01    1.00000E 01    1.00000E 01    1.00000E 01    1.00000E 01    1.00000E 01    1.00000E 01
ICCT1   9.32133E-02    9.32133E-02    9.32133E-02    9.31770E-02    9.31414E-02    9.30472E-02    9.30237E-02

TIME    4.54104E-01    6.02007E-01    7.49909E-01    8.97811E-01    1.38099E 00    1.86416E 00    2.34734E 00
VCET1   7.40785E-01    7.41388E-01    7.41983E-01    7.42566E-01    7.44402E-01    7.46131E-01    7.47759E-01
VCCT1  -1.27738E 01   -1.27724E 01   -1.27710E 01   -1.27696E 01   -1.27652E 01   -1.27607E 01   -1.27563E 01
JIT1    3.57260E 00    3.70548E 00    3.83616E 00    3.96458E 00    4.36839E 00    4.74882E 00    5.10698E 00
VCET2   7.40596E-01    7.41216E-01    7.41822E-01    7.42417E-01    7.44287E-01    7.46055E-01    7.47733E-01
VCCT2  -8.92439E 00   -8.91882E 00   -8.91332E 00   -8.90196E 00   -8.85905E 00   -8.79805E 00   -8.72010E 00
JIT2    3.53109E 00    3.66742E 00    3.80089E 00    3.93171E 00    4.34323E 00    4.73218E 00    5.10128E 00
VR6     8.27442E-04    1.55348E-03    2.60448E-03    3.98058E-03    1.06131E-02    2.03838E-02    3.30861E-02
VL1     1.54978E-03    2.50733E-03    3.60526E-03    4.79768E-03    1.06131E-02    1.46284E-02    2.08187E-02
VL2     8.27442E-04    1.55348E-03    2.60448E-03    3.98058E-03    9.27210E-03    2.03838E-02    3.30861E-02
ICZ     7.08853E-01    7.23201E-01    7.36068E-01    7.47627E-01    7.78261E-01    7.99716E-01    8.14177E-01
ICX     9.38836E-01    1.08305E 00    1.22361E 00    1.36063E 00    1.78562E 00    2.17838E 00    2.54221E 00
IR3     3.24272E 00    3.24312E 00    3.24350E 00    3.24390E 00    3.24521E 00    3.24657E 00    3.24798E 00
IR2     3.61312E-01    3.61381E-01    3.61449E-01    3.61518E-01    3.61741E-01    3.61964E-01    3.62186E-01
IR1     4.25792E-01    4.25746E-01    4.25700E-01    4.25655E-01    4.25506E-01    4.25357E-01    4.25209E-01
J3T1    3.50115E 00    3.63137E 00    3.75943E 00    3.88529E 00    4.28102E 00    4.65384E 00    5.00484E 00
J2T1    0.0            0.0            0.0            0.0            0.0            0.0            0.0
CCT1    1.00000E 01    1.00000E 01    1.00000E 01    1.00000E 01    1.00000E 01    1.00000E 01    1.00000E 01
ICCT1   9.29114E-02    9.27680E-02    9.26393E-02    9.25237E-02    9.22174E-02    9.20028E-02    9.18582E-02
```

Figure 9.27 Transient Analysis Results Example 9.3.

TRANSIENT ANALYSIS RESULTS

```
C     ************************
C         EXAMPLE 3 NETWORK
C     ************************
```

TIME	4.73191E 02	4.83191E 02	4.93191E 02	5.03191E 02
VCET1	7.40188E-01	7.40187E-01	7.40186E-01	7.40185E-01
VCCT1	-1.23121E 01	-1.23125E 01	-1.23129E 01	-1.23133E 01
J1T1	3.44137E 00	3.44115E 00	3.44093E 00	3.44071E 00
VCET2	7.47624E-01	7.47527E-01	7.47430E-01	7.47334E-01
VCCT2	-2.56264E 00	-2.63958E 00	-2.71602E 00	-2.79195E 00
J1T2	5.07739E 00	5.05595E 00	5.03462E 00	5.01340E 00
VR6	2.97004E-01	2.78417E-01	2.60262E-01	2.42532E-01
VL1	9.84732E-02	9.22858E-02	8.62426E-02	8.06199E-02
VL2	2.97004E-01	2.78417E-01	2.60262E-01	2.42532E-01
ICZ	4.23839E-02	4.23446E-02	4.23053E-02	4.22660E-02
ICX	-3.25127E-02	-3.25232E-02	-3.25337E-02	-3.25443E-02
IR3	3.47387E 00	3.47366E 00	3.47345E 00	3.47324E 00
IR2	3.84397E-01	3.84376E-01	3.84354E-01	3.84333E-01
IR1	4.10402E-01	4.10416E-01	4.10430E-01	4.10444E-01
J3T1	3.37255E 00	3.37233E 00	3.37211E 00	3.37190E 00
J2T1	0.0	0.0	0.0	0.0
CCT1	1.00000E 01	1.00000E 01	1.00000E 01	1.00000E 01
ICCT1	-4.23839E-04	-4.23446E-04	-4.23053E-04	-4.22660E-04

Figure 9.27—*Cont.*

A few remarks about this run are in order. Only three quantities are requested for plotting (VR6, VL1, VL2) under the main program, but in addition to these, the user will get six more from the stored models (VCE, VCC, J1 for each transistor). Note that a change has been made in the stored model to use a lower current gain for the second transistor.

The input voltage caused a conduction pulse in both transistors. The conduction of the second stage caused a voltage pulse in the primary of the transformer which was reflected into the secondary as a 1.6-volt swing. The transient in the secondary had almost abated by the problem duration time of 500 nanoseconds, even though significant current levels remained in the two transistors. The first 14 and the last 7 print intervals and the plots of the transformer primary and secondary voltages are given in Figs. 9.27, 9.28, and 9.29.

Example 9.4. Darlington Pair

The schematic of a Darlington pair appears in Fig. 9.30. The problem is to determine the dc output voltage and power requirements of this circuit under nominal conditions and after the first-stage transistor alpha has been degraded to various levels due to the effects of a steady-state radiation environment. The rerun feature will be used to accommodate the additional runs that are required for the degraded alpha versions. The problem can be prepared from Fig. 9.31, which is the SCEPTRE form of Fig. 9.32.

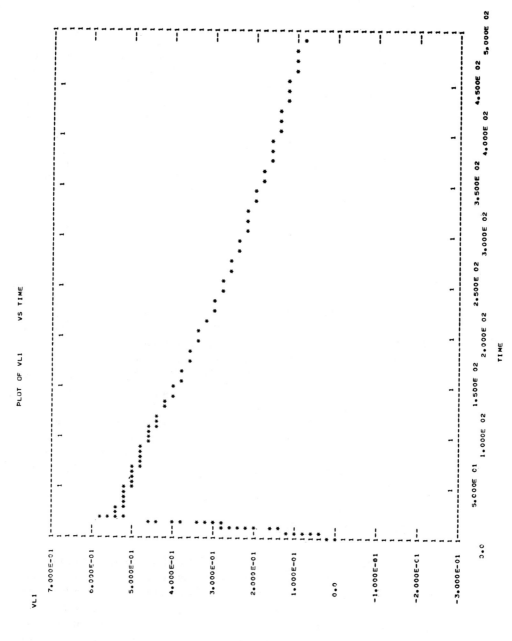

Figure 9.28 Plot of VL1 vs. Time.

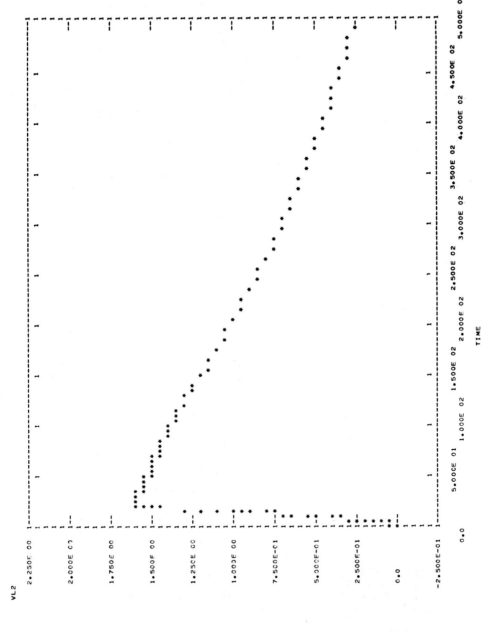

Figure 9.29 Plot of VL2 vs. Time.

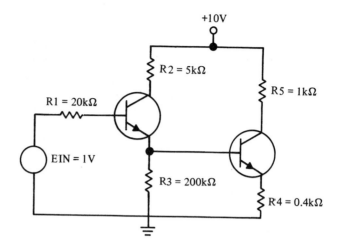

Figure 9.30 Example 9.4 Schematic Diagram.

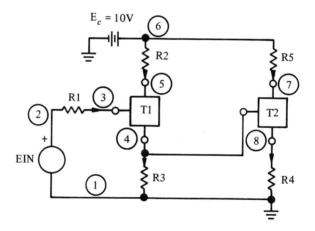

Figure 9.31 Example 9.4 SCEPTRE Form.

A valid sequence of cards for Example 9.4 would be as follows:

```
MODEL DESCRIPTION
MODEL 2N706A (TEMP) (B-E-C)
*    UNITS - KOHMS, UHENRIES, PFARADS, VOLTS, MAMPS, NSECONDS, GHZ
     ELEMENTS
CE,1-E=EQUATION1(5.,70.,J1)
CC,1-2=EQUATION1(8.,370.,J2)
RB,B-1=.3
RC,C-2=.015
J1,1-E=DIODE EQUATION(1.E-7,35.)
```

RESULTS OF INITIAL CONDITION COMI

```
*************************
EXAMPLE 4 - DARLINGTON PAIR
*************************

VCET1          3.3818865D-01

VCCT1         -8.9376695D 00

J1T1           1.3821911D-02

VCET2          4.4221916D-01

VCCT2         -8.8226781D 00

J1T2           5.2706833D-01

PEC            5.3007344D 00

PEIN           2.7598821D-04

VR3            6.5620879D-01

VR4            2.1082735D-01
```

0.0	0.0	6.0000000D 02	4.0000000D 00	0.0	0.0	0.0	1.0000000D-04
5.0000000D-03	1.0000000D 00	1.0000000D 00	2.0000000D 04	2.0000000D-04	5.0000000D-03	0.0	1.0000000D 02
1.0000000D-03	1.0000000D-04	1.0000000D 03	1.0000000D 01	5.0000000D 00	0.0	1.0000000D 01	1.1000000D 01
3.0000000D 00	4.0000000D 00	0.0	1.0000000D 00	0.0	0.0	0.0	2.0000000D 00
0.0	1.0000000D 00	1.5000000D 01	9.7242000D 04	6.5999966D 00	0.0	0.0	6.0000000D 02
0.0	0.C	0.0	1.0000000D 00	0.0	0.0	0.0	0.0
0.0	0.0	0.0	0.0	1.0000000D 00	1.0000000D 00	0.0	1.0000000D-08
0.0	1.0000000D 74	0.0	0.0	1.0000000D 00	0.0	0.0	0.0
0.0	0.0	0.0	0.0	0.0	0.0	0.0	0.0
0.0	1.0000000D 00	1.0000000D 01	0.0	0.0	0.0	0.0	0.0
0.0	0.0	0.0	0.0	0.0	0.0	0.0	0.0
0.0	0.0	0.0	2.0000000D 00	4.0000000D 00	5.0000000D 00	0.0	0.0
4.0000000D 00	0.0	C.0	2.0000000D 00	4.0000000D 00	4.0000000D 00	0.0	0.0
0.0	0.0	0.0	4.0000000D 00	0.0	2.5000000D 01	0.0	0.0
0.0	0.0	0.0	0.0	0.0	0.0	0.0	0.0
0.0	0.0	0.0	0.0	0.0	0.0	0.0	0.0
0.0	0.0	0.0	0.0	0.0	1.0000000D-07	0.0	2.0000000D-01
1.0000000D 00	0.0	0.0	0.0	0.0	0.0	0.0	0.0
0.C	0.0	0.0	0.0	0.0	0.0	0.0	0.0
0.0	0.0	0.0	0.0	0.0	0.0	0.0	0.0

Figure 9.32 Example 9.4 Output Listings.

```
RESULTS OF INITIAL CONDITION COMPUTATIONS

FEFUN      1

******************
EXAMPLE 4 - DARLINGTON PAIR
******************

VCET1    3.3769191D-01
VCCT1   -8.9456210D 00
JIT1     1.3583679D-02
VCET2    4.4163421D-01
VCCT2   -8.8381601D 00
JIT2     5.1638715D-01
PEC      5.1910074D 00
PEIN     5.4289718D-04
VR3      6.5128728D-01
VR4      2.0655488D-01

0.0         0.0         6.0000000D 02  4.0000000D 00  0.0         0.0                        1.0000000D-04
5.0000000D-03  1.0000000D 00  2.0000000D 00  2.0000000D 04  2.0000000D-04  5.0000000D-03                   1.0000000D 02
1.0000000D-03  1.0000000D-04  1.0000000D 03  1.0000000D 01  5.0000000D 00  0.0                            1.1000000D 01
3.0000000D 00  4.0000000D 00  0.0         1.0000000D 00  0.0         1.0000000D 01   3.0000000D 00
0.0         1.0000000D 00  1.5000000D 01  9.7251000D 04  6.5999966D 00  0.0                               6.0000000D 02
0.0         0.0         0.0         1.0000000D 00  0.0         0.0
0.0         1.0000000D 74  0.0         0.0         1.0000000D 00  1.0000000D 00                            1.0000000D-08
0.0         1.0000000D 00  1.0000000D 01  0.0         2.0000000D 00  4.0000000D 00  0.0
0.0         0.0         0.0         0.0         2.0000000D 00  4.0000000D 00  0.0
4.0000000D 00  0.0         0.0         0.0         4.0000000D 00  2.5000000D 01  0.0
0.0         0.0         0.0         0.0         0.0         0.0         0.0
0.0         0.0         0.0         0.0         1.0000000D-07  0.0                                       2.0000000D-01
1.0000000D 00  0.0         0.0         0.0         0.0         0.0
0.0         0.0         0.0         0.0         0.0         0.0
```

Figure 9.32—*Cont.*

RESULTS OF INITIAL CONDITION COMPUTATIONS

RERUN   2

****
EXAMPLE 4 - DARLINGTON PAIR
****

VCET1   3.369657lD-01
VCCT1   -8.9570427D 00
J1T1   1.3242272D-02
VCET2   8.4077602D-01
VCCT2   -8.8603292D 00
J1T2   5.0110724D-01
PEC   5.0340188D 00
PEIN   3.2654405D-04
VR3   6.4422545D-01
VR4   2.0044292D-01

```
0.0          0.0          6.0000000D 02   4.0000000D 00   0.0             1.0000000D-04
5.0000000D-03 1.0000000D 00 3.0000000D 00  2.0000000D 04   5.0000000D-03   1.0000000D 02
1.0000000D-03 1.0000000D-04 1.0000000D 03  1.0000000D 00   5.0000000D 00   1.0000000D 01
3.0000000D 00 4.0000000D 00 1.5000000D 00  9.7262000D 04   0.0             4.0000000D 00
0.0          1.0000000D 00 1.0000000D 00  6.5999966D 00   1.0000000D 01   6.0000000D 02
0.0          0.0          0.0             1.0000000D 00   0.0             1.0000000D-08
0.0          1.0000000D 74 0.0             1.0000000D 00   0.0             0.0
0.0          0.0          0.0             0.0             0.0             0.0
0.0          1.0000000D 00 1.0000000D 00  4.0000000D 00   5.0000000D 00   0.0
0.0          0.0          0.0             4.0000000D 00   4.0000000D 00   0.0
4.0000000D 00 0.0          2.0000000D 00  4.0000000D 00   0.0             0.0
0.0          0.0          2.0000000D 00  2.5000000D 00   0.0             0.0
0.0          0.0          4.0000000D 00  0.0             1.0000000D-07   2.0000000D-01
1.0000000D 00 0.0          0.0            0.0             0.0             0.0
0.0          0.0          0.0            0.0             0.0             0.0
0.0          0.0          0.0            0.0             0.0             0.0
```

Figure 9.32—*Cont.*

```
RESULTS OF INITIAL CONDITION COMPUTATIONS

RERUN         3


************************
EXAMPLE 4 - DARLINGTON PAIR
************************

VCET1         3.3626101D-01

VCCT1        -8.9679012D 00

J1T1          1.2920139D-02

VCET2         4.3993970D-01

VCCT2        -8.8813264D 00

J1T2          4.8665186D-01

PEC           4.8885479D 00

PEIN          1.2915639D-03

VR3           6.3752024D-01

VR4           1.9466077D-01
```

Figure 9.32—*Cont.*

```
J2,1-2=DIODE EQUATION(5.E-7,37.)
JA,E-1=.1*J2
JB,2-1=P1*J1
JX,2-1=0
 DEFINED PARAMETERS
P1=.98
 OUTPUTS
VCE,VCC,J1,PLOT
 FUNCTIONS
EQUATION1(A,B,C)=(A+B*C)
 CIRCUIT DESCRIPTION
*    DARLINGTON PAIR
*    UNITS - KOHMS, UHENRIES, PFARADS, VOLTS, MAMPS, NSECONDS, GHZ
 ELEMENTS
EC,1-6=10
EIN,1-2=1
R1,2-3=20
R2,6-5=5
R3,4-1=200
R4,8-1=.4
R5,6-7=1
T1,3-4-5=MODEL 2N706A (TEMP)
T2,4-8-7=MODEL 2N706A (TEMP)
 DEFINED PARAMETERS
PEC=X1(EC*IEC)
PEIN=X2(EIN*IEIN)
 OUTPUTS
PEC,PEIN,VR3,VR4
 RUN CONTROLS
RUN INITIAL CONDITIONS ONLY
 RERUN DESCRIPTION(3)
 DEFINED PARAMETERS
P1T1=.98,.93,.9
 END
```

The RUN INITIAL CONDITIONS ONLY ensures that no transient computations will be made. The Newton-Raphson process will iterate to the final dc solution for the master run; then do the same for each of the reruns in turn.

Figure 9.32 shows the results of the master run and the three reruns in tabular form. Even though only 4 quantities were specifically requested under CIRCUIT DESCRIPTION, 20 appear in the output listing because each transistor uses the stored model, which, in itself, has three output requests. There is little change between the results of the master run and the reruns because of the inherent stability of the circuit. The power supplied from the input source EIN increases successively in the reruns because the reduced alpha of the first stage permits more input current.

Example 9.5. Use of Small-Signal Equivalent Circuit

This example is intended to illustrate the proper use of one of a class of small-signal equivalent circuits with SCEPTRE. Figure 9.33 shows the schematic of a two-stage linear RC coupled amplifier. Let it be desired to determine the response of this circuit to a low-amplitude 100 kHz sinusoidal input.

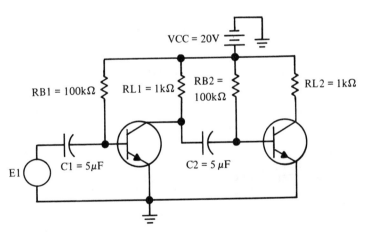

Figure 9.33 Example 9.5 Schematic Diagram.

The low-frequency small-signal h-parameter equivalent circuit shown in Fig. 9.34 will be temporarily stored. As is customary in the use of this type of equivalent circuit, all dc power supplies will be grounded since only the ac excursion around each of the individual operating points is of interest. The resulting SCEPTRE form appears in Fig. 9.35. The automatic termina-

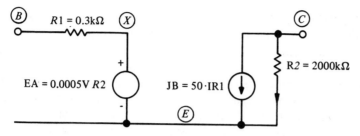

Figure 9.34 Low-Frequency h-Parameter Equivalent Circuit.

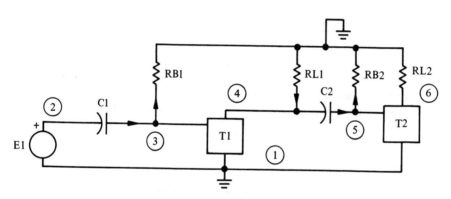

Figure 9.35 Example 9.5 SCEPTRE Form.

tion feature is invoked to halt the run if the voltage across the load resistor of the second stage reaches 20 volts.

```
MODEL DESCRIPTION
MODEL SS1(TEMP)(B-E-C)
*   H - PARAMETER EQUIVALENT CIRCUIT
*   UNITS - OHMS, HENRIES, FARADS, VOLTS, AMPS, SECONDS, HZ
  ELEMENTS
EA,E-X=.0005*VR2
R2,C-E=2000
R1,B-X=.3
JB,C-E=50.*IR1
  CIRCUIT DESCRIPTION
*   SMALL SIGNAL AMPLIFIER
*   UNITS - OHMS, HENRIES, FARADS, VOLTS, AMPS, SECONDS, HZ
  ELEMENTS
E1,1-2=X1(.001*DSIN(.000628*TIME)
C1,2-3=5E6
C2,4-5=5E6
```

```
RB1,3-1=100
RL1,1-4=1
RB2,5-1=100
RL2,1-6=1
T1,3-1-4=MODEL SS1
T2,5-1-6=MODEL SS1
 OUTPUTS
VRL1,VRL2,VC1,VC2,PLOT
 RUN CONTROLS
STOP TIME=30000
INTEGRATION ROUTINE=TRAP
TERMINATE IF(VRL2.GE.20.)
 END
```

Since the frequency of the input sinusoid is 100 kHz, its period is 10,000 nanoseconds. The problem duration has been set to 30,000 nanoseconds to accommodate three cycles of the input wave. Output plots for the voltages across the two load resistors are shown in Figs. 9.36 and 9.37. The latter waveshape peaks at about 6.6 volts, which indicates that this circuit has an overall voltage gain of 6600 at this frequency. The automatic termination condition was never activated since VRL2 remained below 20 volts throughout the run. It is worth noting that the input sinusoid, E1, was entered directly under ELEMENTS as a direct math expression and its value is completely enclosed in parentheses. An equation could have been referenced for E1 and then subsequently defined under FUNCTIONS, but this was not done for this run.

Example 9.6. Solution of Simultaneous Differential Equations

This example is intended to illustrate the flexibility of SCEPTRE through the special use of the DEFINED PARAMETERS section. Assume that the user has the problem of solving the following set of first-order, simultaneous differential equations that may be entirely independent of any electrical network:

$$\dot{X} = -6X + 5Y + 10$$

$$\dot{Y} = 5X - 7Y + 2Z$$

$$\dot{Z} = 0.2Y - 0.2Z - 0.5$$

with $X(0) = 6$, $Y(0) = 5$, and $Z(0) = 4$ as initial conditions. Note, since the derivatives of PX, PY, and PZ (DPX, DPY, DPZ) are entered, PX, PY, and PZ will be updated at each integration step. Only capacitor voltages and inductor currents are entered under the INITIAL CONDITIONS subheading.

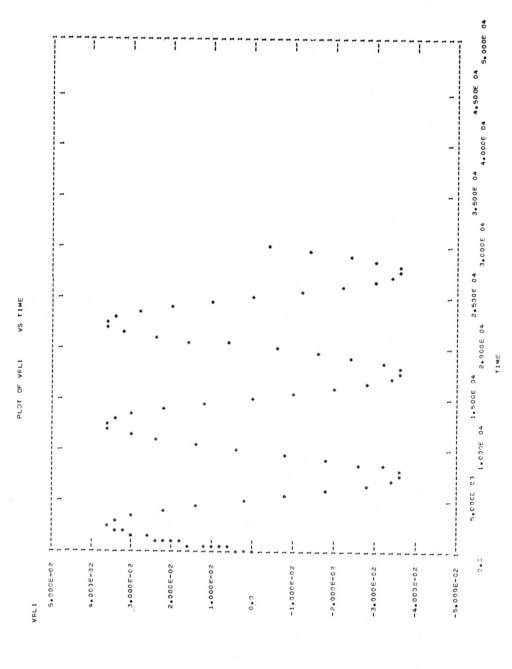

Figure 9.36 Plot of VRL1 Versus Time.

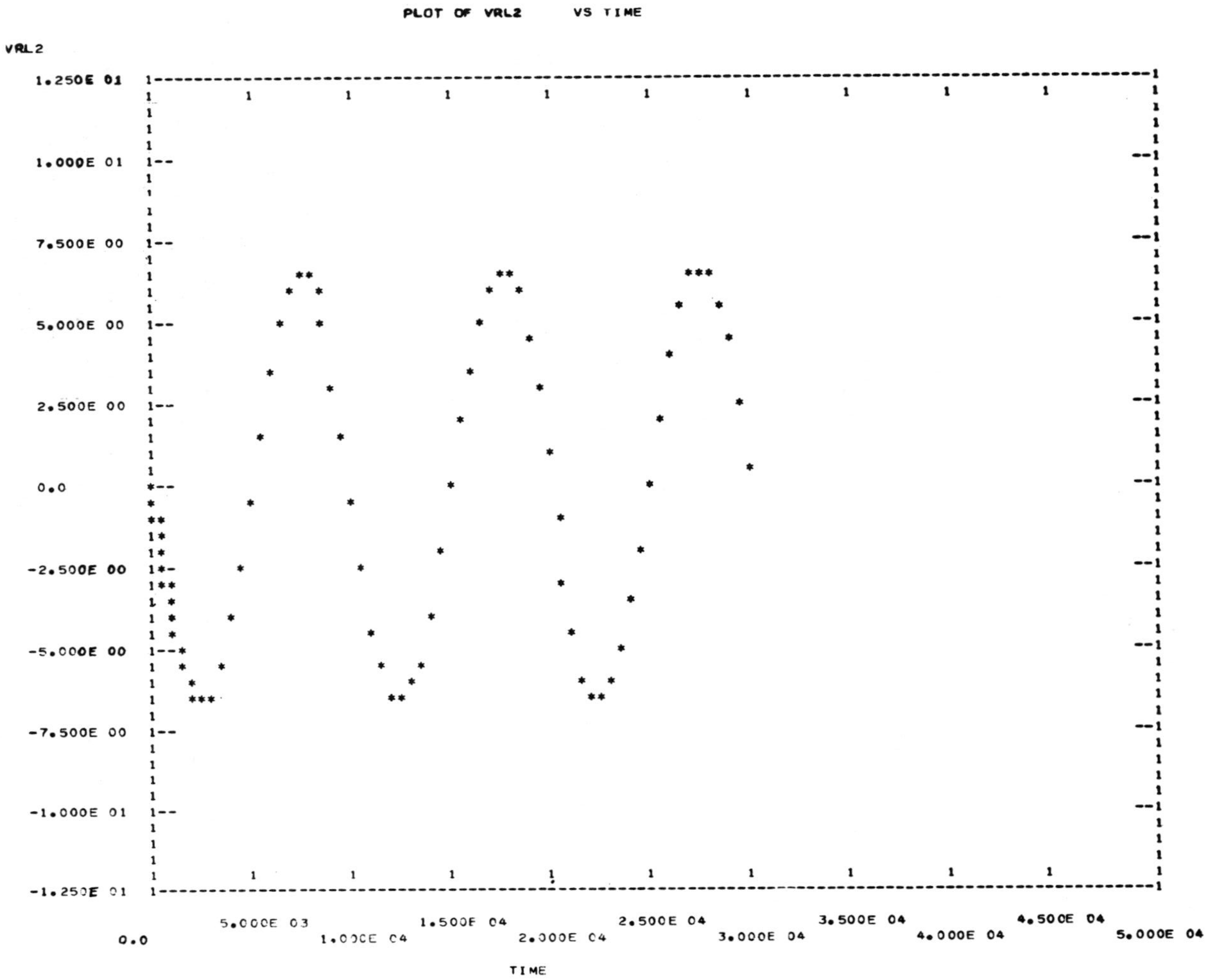

Figure 9.37 Plot of VRL2 versus Time.

The user may enter each of the derivatives under DEFINED PARAMETERS in explicit form. A proper sequence would be

```
CIRCUIT DESCRIPTION
SOLUTION OF SIMULTANEOUS DIFFERENTIAL EQUATIONS
DEFINED PARAMETERS
DPX=EQUATION 1(PX,PY)
DPY=EQUATION 2(PX,PY,PZ)
DPZ=EQUATION 3(PY,PZ)
PX=6
PY=5
PZ=4
OUTPUTS
PX(X),PY(Y),PZ(Z),XSTPSZ,PLOT
FUNCTIONS
EQUATION 1(A,B)=(-6.*A+5.*B+10.)
EQUATION 2(A,B,C)=(5.*A-7.*B+2.*C)
EQUATION 3(A,B)=(.2*A-.2*B-.5)
RUN CONTROLS
INTEGRATION ROUTINE=TRAP
STOP TIME=100
END
```

Note that the initial values of each of the variables X, Y, and Z have been entered as PX = 6, PY = 5, and PZ = 4, respectively. The differential equations themselves are entered under DEFINED PARAMETERS. The quantities X, Y, and Z will be treated in the same manner as would the state variables of the general transient problem, and, therefore, these equations will be subjected to the same step size limitations in whatever integration routine is used. These quantities are explicitly labeled as X, Y, and Z because of the format used under OUTPUTS. The step size is output through the use of the internal name XSTPSZ. The first 56 and the last 19 print intervals are shown in Fig. 9.38.

Example 9.7. Use of Monte Carlo

The Darlington Pair used for Example 9.4 and shown in Fig. 9.30 is used also to demonstrate the Monte Carlo option. While all four of the dc options (Monte Carlo, sensitivity, worst case, and optimization) can be requested in a single run, they are shown separately in this and the next three examples for clarity. In each case, the complete SCEPTRE input deck is listed, and the output due to execution of the requested feature is shown.

In the Monte Carlo example, Gaussian distribution was used (since none was requested and Gaussian is the default mode), and 10 iterations were requested. The independent variables are the defined parameter P1 as used in both model T1 and model T2. Since no initial random number was specified, SCEPTRE supplied its own. Details of each iteration were not requested. If they had been, the random element and defined parameter values and the requested output values would have been listed for each run. The outputs provided, shown in Fig. 9.39, are the statistical parameters of the independent

```
TRANSIENT ANALYSIS RESULTS

*********************
EXAMPLE  6
*********************

TIME    0.0           5.00000E-02   1.00000E-01   1.50000E-01   2.00000E-01   2.50000E-01   3.00000E-01
X       6.00000E 00   5.95875E 00   5.95973E 00   5.97912E 00   6.00566E 00   6.03409E 00   6.06204E 00
Y       5.00000E 00   5.13892E 00   5.22294E 00   5.27913E 00   5.32055E 00   5.35358E 00   5.38135E 00
Z       4.00000E 00   3.96527E 00   3.97198E 00   3.95961E 00   3.94790E 00   3.93670E 00   3.92593E 00
XSTPSZ  1.00000E-01   5.00000E-02   5.00000E-02   5.00000E-02   5.00000E-02   5.00000E-02   5.00000E-02

TIME    3.50000E-01   4.50000E-01   5.50000E-01   6.50000E-01   7.50000E-01   8.50000E-01   9.50000E-01
X       6.08845E 00   6.13678E 00   6.17710E 00   6.21000E 00   6.23661E 00   6.25791E 00   6.27473E 00
Y       5.40548E 00   5.44737E 00   5.48054E 00   5.50709E 00   5.52811E 00   5.54447E 00   5.55690E 00
Z       3.91555E 00   3.89556E 00   3.87677E 00   3.85900E 00   3.84209E 00   3.82592E 00   3.81039E 00
XSTPSZ  1.00000E-01   1.00000E-01   1.00000E-01   1.00000E-01   1.00000E-01   1.00000E-01   1.00000E-01

TIME    1.05000E 00   1.25000E 00   1.45000E 00   1.65000E 00   1.85000E 00   2.05000E 00   2.25000E 00
X       6.28778E 00   6.30658E 00   6.31526E 00   6.31694E 00   6.31378E 00   6.30731E 00   6.29856E 00
Y       5.56602E 00   5.57788E 00   5.58089E 00   5.57779E 00   5.57049E 00   5.56031E 00   5.54819E 00
Z       3.79540E 00   3.76651E 00   3.73918E 00   3.71301E 00   3.68772E 00   3.66313E 00   3.63910E 00
XSTPSZ  2.00000E-01   2.00000E-01   2.00000E-01   2.00000E-01   2.00000E-01   2.00000E-01   2.00000E-01

TIME    2.45000E 00   2.85000E 00   3.25000E 00   3.65000E 00   4.05000E 00   4.45000E 00   4.85000E 00
X       6.28328E 00   6.26542E 00   6.24025E 00   6.21441E 00   6.18861E 00   6.16316E 00   6.13815E 00
Y       5.53476E 00   5.50600E 00   5.47538E 00   5.44437E 00   5.41357E 00   5.38325E 00   5.35349E 00
Z       3.61553E 00   3.56931E 00   3.52444E 00   3.48069E 00   3.43794E 00   3.39615E 00   3.35526E 00
XSTPSZ  4.00000E-01   4.00000E-01   4.00000E-01   4.00000E-01   4.00000E-01   4.00000E-01   4.00000E-01

TIME    5.25000E 00   6.05000E 00   6.85000E 00   7.65000E 00   8.45000E 00   9.25000E 00   1.00500E 01
X       6.11365E 00   6.06583E 00   6.02005E 00   5.97623E 00   5.93429E 00   5.89416E 00   5.85575E 00
Y       5.32434E 00   5.26747E 00   5.21302E 00   5.16092E 00   5.11105E 00   5.06332E 00   5.01764E 00
Z       3.31527E 00   3.23729E 00   3.16266E 00   3.09124E 00   3.02288E 00   2.95746E 00   2.89485E 00
XSTPSZ  8.00000E-01   8.00000E-01   8.00000E-01   8.00000E-01   8.00000E-01   8.00000E-01   8.00000E-01

TIME    1.08500E 01   1.24500E 01   1.40500E 01   1.56500E 01   1.72500E 01   1.88500E 01   1.90500E 01
X       5.81899E 00   5.74914E 00   5.68524E 00   5.62679E 00   5.57315E 00   5.51712E 00   5.52185E 00
Y       4.97393E 00   4.89086E 00   4.81488E 00   4.74538E 00   4.68201E 00   4.63177E 00   4.61351E 00
Z       2.83494E 00   2.72108E 00   2.61693E 00   2.52167E 00   2.43453E 00   2.35468E 00   2.34566E 00
XSTPSZ  1.60000E 00   1.60000E 00   1.60000E 00   1.60000E 00   1.60000E 00   1.60000E 00   2.00000E-01
```

Figure 9.38 Example 9.6 Output Listing.

```
TRANSIENT ANALYSIS RESULTS

*************************
EXAMPLE 6
*************************

TIME    1.92500E 01   1.94500E 01   1.96500E 01   1.98500E 01   2.00500E 01   2.02500E 01   2.06500E 01
X       5.51188E 00   5.50818E 00   5.50192E 00   5.49683E 00   5.49133E 00   5.48608E 00   5.47558E 00
Y       4.61169E 00   4.60307E 00   4.59740E 00   4.59060E 00   4.58437E 00   4.57800E 00   4.56556E 00
Z       2.33644E 00   2.32744E 00   2.31849E 00   2.30966E 00   2.30092E 00   2.29227E 00   2.27519E 00
XSTPSZ  2.00000E-01   2.00000E-01   2.00000E-01   2.00000E-01   2.00000E-01   4.00000E-01   4.00000E-01

TIME    2.10500E 01   2.14500E 01   2.18500E 01   2.22500E 01   2.26500E 01   2.30500E 01   2.38500E 01
X       5.46533E 00   5.45530E 00   5.44549E 00   5.43588E 00   5.42649E 00   5.41729E 00   5.39937E 00
Y       4.55336E 00   4.54144E 00   4.52976E 00   4.51834E 00   4.50717E 00   4.49624E 00   4.47492E 00
Z       2.25848E 00   2.24213E 00   2.22613E 00   2.21048E 00   2.19516E 00   2.18018E 00   2.15096E 00
XSTPSZ  4.00000E-01   4.00000E-01   4.00000E-01   4.00000E-01   4.00000E-01   8.00000E-01   8.00000E-01

TIME    9.64500E 01   9.66500E 01   9.68500E 01   9.70500E 01   9.72500E 01   9.76500E 01   9.80500E 01
X       5.00789E 00   5.00703E 00   5.00728E 00   5.00706E 00   5.00705E 00   5.00688E 00   5.00673E 00
Y       4.00812E 00   4.00889E 00   4.00844E 00   4.00850E 00   4.00834E 00   4.00818E 00   4.00800E 00
Z       1.51198E 00   1.51133E 00   1.51171E 00   1.51158E 00   1.51146E 00   1.51121E 00   1.51097E 00
XSTPSZ  2.00000E-01   2.00000E-01   2.00000E-01   2.00000E-01   4.00000E-01   4.00000E-01   4.00000E-01

TIME    9.84500E 01   9.88500E 01   9.92500E 01   9.96500E 01   1.00050E 02
X       5.00658E 00   5.00644E 00   5.00630E 00   5.00617E 00   5.00603E 00
Y       4.00783E 00   4.00766E 00   4.00750E 00   4.00733E 00   4.00718E 00
Z       1.51073E 00   1.51050E 00   1.51027E 00   1.51005E 00   1.50984E 00
XSTPSZ  4.00000E-01   4.00000E-01   4.00000E-01   4.00000E-01   8.00000E-01
```

Figure 9.38—*Cont.*

```
INPUTS TO MONTE CARLO

NAME        NOMINAL         LBOUND          UBOUND          MEAN            SIGMA

PIT1        0.9600000       0.9000000       0.9800000       0.9400000       0.1333333D-01
PIT2        0.9600000       0.9000000       0.9800000       0.9400000       0.1333333D-01

INITIAL RANDOM NUMBER =    127263527

DISTRIBUTION IS GAUSSIAN

NORMAL MONTE CARLO TERMINATION AFTER    10 ITERATIONS

FINAL RANDOM NUMBER =   1007121511

OBSERVED STATISTICS

INDEPENDENT     NOMINAL         MINIMUM         MAXIMUM         SAMPLE          SAMPLE
VARIABLE        VALUE           VALUE           VALUE           MEAN            SIGMA

PIT1            0.9600000       0.9308245       0.9790631       0.9538829       0.1648086D-01
PIT2            0.9600000       0.9192627       0.9590029       0.9425582       0.1132093D-01

DEPENDENT       NOMINAL         MINIMUM         MAXIMUM         SAMPLE          SAMPLE
VARIABLE        VALUE           VALUE           VALUE           MEAN            SIGMA

VCET1           0.3512397       0.3510979       0.3651036       0.3578234       0.4015770D-02
VCCT1          -8.912637       -8.921013       -8.859640       -8.892786        0.2201369D-01
J1T1            0.2182475D-01   0.2171672D-01   0.3545560D-01   0.2772682D-01   0.3930754D-02
VCET2           0.4387467       0.4344037       0.4379516       0.4364949       0.1485083D-02
VCCT2          -8.919751       -9.030061       -8.939226       -8.977382        0.3535040D-01
J1T2            0.4667503       0.4009303       0.4539405       0.4318985       0.2225250D-01
PEC             4.690331        4.021073        4.561632        4.337242        0.2267543
PEIN            0.8725401D-03   0.6325778D-03   0.1836840D-02   0.1257592D-02   0.4307638D-03
VR3             0.6310477       0.6027551       0.6266059       0.6166474       0.9779361D-02
VR4             0.1867001       0.1603722       0.1815762       0.1727594       0.8901000D-02
```

Figure 9.39 Monte Carlo Example Outputs.

variables, plus a summary of the maximum, minimum, and mean and standard deviation of the requested output parameters.

```
 MODEL DESCRIPTION
 MODEL 2N706A (TEMP) (B-E-C)
*    UNITS - KOHMS, UHENRIES, PFARADS, VOLTS, MAMPS, NSECONDS, GHZ
 ELEMENTS
CE,1-E=EQUATION1(5.,70.,J1)
CC,1-2=EQUATION1(8.,370.,J2)
RB,B-1=.3
RC,C-2=.015
J1,1-E=DIODE EQUATION(1.E-7,35.)
J2,1-2=DIODE EQUATION(5.E-7,37.)
JA,E-1=.1*J2
JB,2-1=P1*J1
JX,2-1=0
 DEFINED PARAMETERS
P1=0.96(0.98,0.9)
 OUTPUTS
VCE,VCC,J1
 FUNCTIONS
EQUATION1(A,B,C)=(A+B*C)
 CIRCUIT DESCRIPTION
*    MONTE CARLO EXAMPLE
*    UNITS - KOHMS, UHENRIES, PFARADS, VOLTS, MAMPS, NSECONDS, GHZ
 ELEMENTS
EC,1-6=10
EIN,1-2=1
R1,2-3=20
R2,6-5=5
R3,4-1=200
R4,8-1=.4
R5,6-7=1
T1,3-4-5=MODEL 2N706A (TEMP)
T2,4-8-7=MODEL 2N706A (TEMP)
 MONTE CARLO
(VCET1,VCCT1,J1T1,VCET2,VCCT2,J1T2,PEC,PEIN,VR3,VR4/P1T1,P1T2)
 DEFINED PARAMETERS
PEC=XEC(EC*IEC)
PEIN=X2(EIN*IEIN)
 OUTPUTS
PEC,PEIN,VR3,VR4
 RUN CONTROLS
RUN INITIAL CONDITIONS ONLY
RUN MONTE CARLO=10
 END
```

Example 9.8. Use of Sensitivity

The Darlington Pair used in Example 9.4 (shown in Fig. 9.30) is used to demonstrate the sensitivity option. For this case, two sets of partial derivatives were requested; the first with five dependent and two independent variables and the second with four dependent and three independent variables. The input data are shown below. The 22 partial derivatives calculated are tabulated with dependent and independent variables identified. The output appears in Fig. 9.40.

SENSITIVITY CALCULATION(S)

SET NUMBER	DEPENDENT VARIABLE	INDEPENDENT VARIABLE	PARTIAL DERIVATIVE	NORMALIZED SENSITIVITY
1.	VCET1	P1T1	0.43639145D-01	0.11927347
		P1T2	-0.50049458	-1.3679399
	VCCT1	P1T1	0.68591209	-0.73881120D-01
		P1T2	-1.5301683	0.16481784
	J1T1	P1T1	0.33334625D-01	1.4662820
		P1T2	-0.38231269	-16.816695
	PEC	P1T1	8.0917673	1.6561938
		P1T2	20.289212	4.1527229
	VR3	P1T1	0.37233559	0.56642652
		P1T2	0.81093249	1.2336550
2.	VCET2	P1T1	0.48164235D-01	0.10538578
		P1T2	0.12301307	0.26915882
		R3	0.36378729D-05	0.16583020D-02
	VCCT2	P1T1	1.1295747	-0.12157197
		P1T2	3.3587232	-0.36148703
		R3	0.85317440D-04	-0.19130005D-02
	PEIN	P1T1	-0.20491366D-01	-22.545339
		P1T2	-0.15292508D-01	-16.825368
		R3	-0.53033638D-06	-0.12156150
	VR4	P1T1	0.31472947	1.6183185
		P1T2	0.80382962	4.1332396
		R3	0.23771702D-04	0.25465113D-01

Figure 9.40 Sensitivity Example Output.

The problem shows the use of defined parameters as both independent and dependent variables. Only those defined parameters used to relate primary and secondary current sources may be used for independent variables. The defined parameters used for dependent variables must have their total differentials provided. Thus,

$$PEC = EC*IEC$$

and its total differential is

$$EC \cdot d(IEC) + IEC \cdot d(EC)$$

Since EC is constant and is not one of the independent variables, it is not necessary to include its differential contribution to the total differential, GPEC.

```
 MODEL DESCRIPTION
 MODEL 2N706A (TEMP) (B-E-C)
 *   UNITS - KOHMS, UHENRIES, PFARADS, VOLTS, MAMPS, NSECONDS, GHZ
 ELEMENTS
 CE,1-E=EQUATION1(5.,70.,J1)
 CC,1-2=EQUATION1(8.,370.,J2)
 RB,B-1=.3
 RC,C-2=.015
 J1,1-E=DIODE EQUATION(1.E-7,35.)
 J2,1-2=DIODE EQUATION(5.E-7,37.)
 JA,E-1=.1*J2
 JB,2-1=P1*J1
 JX,2-1=0
 DEFINED PARAMETERS
 P1=0.96
 OUTPUTS
 VCE,VCC,J1
 FUNCTIONS
 EQUATION1(A,B,C)=(A+B*C)
 CIRCUIT DESCRIPTION
 *   SENSETIVITY EXAMPLE
 *   UNITS - KOHMS, UHENRIES, PFARADS, VOLTS, MAMPS, NSECONDS, GHZ
 ELEMENTS
 EC,1-6=10
 EIN,1-2=1
 R1,2-3=20
 R2,6-5=5
 R3,4-1=200
 R4,8-1=.4
 R5,6-7=1
 T1,3-4-5=MODEL 2N706A (TEMP)
 T2,4-8-7=MODEL 2N706A (TEMP)
 SENSITIVITY
 (VCET1,VCCT1,J1T1,PEC,VR3/P1T1,P1T2)
 (VCET2,VCCT2,PEIN,VR4/P1T1,P1T2,R3)
 DEFINED PARAMETERS
 PEC=XEC(EC*IEC)
 GPEC=P3*DIEC
 P3=X3(EC)
 PEIN=X2(EIN*IEIN)
 GPEIN=P4*DIEIN
 P4=X4(EIN)
 OUTPUTS
 PEC,PEIN,VR3,VR4
 RUN CONTROLS
 RUN INITIAL CONDITIONS ONLY
 RUN SENSITIVITY
 END
```

Example 9.9. Use of Worst Case

The Darlington Pair used for Example 9.4 (shown in Fig. 9.30) is used to demonstrate the worst-case option. The worst-case request specifies one set of 10 dependent and 2 independent variables. The output from the worst-case calculation on one of the requested dependent variables, VCCT1, is presented in Fig. 9.41 as typical of the entire set.

```
WORST CASE COMPUTATION - NOMINAL VALUE

OBJECTIVE FUNCTION VCCT1    =  -8.9126370

INDEPENDENT VARIABLE      VALUE          GRADIENT COMPONENT      LOWER BOUND        UPPER BOUND

        P1T1           0.96000000            0.68591209           0.90000000         0.98000000
        P1T2           0.96000000           -1.5301683            0.90000000         0.98000000

LOW  VALUE LOCATED AT DISTANCE -2.19174482D-02 ALONG GRADIENT = UPPER BOUND  0.98000000  OF INDEPENDENT VARIABLE P1T2
HIGH VALUE LOCATED AT DISTANCE  4.88945825D-02 ALONG GRADIENT = UPPER BOUND  0.98000000  OF INDEPENDENT VARIABLE P1T1


WORST CASE COMPUTATION - HIGH VALUE

OBJECTIVE FUNCTION VCCT1    =  -8.8243137

INDEPENDENT VARIABLE      VALUE          GRADIENT COMPONENT

        P1T1           0.98000000            1.5109581
        P1T2           0.91538296           -1.5095120


WORST CASE COMPUTATION - LOW VALUE

OBJECTIVE FUNCTION VCCT1    =  -8.9490961

INDEPENDENT VARIABLE      VALUE          GRADIENT COMPONENT

        P1T1           0.95103482            0.38462176
        P1T2           0.98000000           -1.6576036
```

Figure 9.41 Worst-case Example Outputs.

The function value of −8.9126 for the nominal independent variable values (P1T1, P1T2) is printed, followed by the gradient components of VCCT1, with respect to each independent variable. A projection along the positive direction of the gradient vector then intersects the upper bound of P1T2 as its closest boundary suface. Along the negative direction of the gradient, the closest boundary surface was found to be the upper bound of P1T1.

Using the values at the boundary along the positive direction, a calculation was performed which yielded a value of −8.8149 for VCCT1. At the boundary in the negative direction, the value was found to be −8.9487.

```
 MODEL DESCRIPTION
 MODEL 2N706A (TEMP) (B-E-C)
 *    UNITS - KOHMS, UHENRIES, PFARADS, VOLTS, MAMPS, NSECONDS, GHZ
   ELEMENTS
 CE,1-E=EQUATION1(5.,70.,J1)
 CC,1-2=EQUATION1(8.,370.,J2)
 RB,B-1=.3
 RC,C-2=.015
 J1,1-E=DIODE EQUATION(1.E-7,35.)
 J2,1-2=DIODE EQUATION(5.E-7,37.)
 JA,E-1=.1*J2
 JB,2-1=P1*J1
 JX,2-1=0
  DEFINED PARAMETERS
 P1=0.96(0.98,0.9)
  OUTPUTS
 VCE,VCC,J1,PLOT
  FUNCTIONS
 EQUATION1(A,B,C)=(A+B*C)
  CIRCUIT DESCRIPTION
 *    WORST-CASE EXAMPLE
 *    UNITS - KOHMS, UHENRIES, PFARADS, VOLTS, MAMPS, NSECONDS, GHZ
   ELEMENTS
 EC,1-6=10
 EIN,1-2=1
 R1,2-3=20
 R2,6-5=5
 R3,4-1=200
 R4,8-1=.4
 R5,6-7=1
 T1,3-4-5=MODEL 2N706A (TEMP)
 T2,4-8-7=MODEL 2N706A (TEMP)
  WORST CASE
 (VCET1,VCCT1,J1T1,VCET2,VCCT2,J1T2,PEC,PEIN,VR3,VR4/P1T1,P1T2)
  DEFINED PARAMETERS
 PEC=XEC(EC*IEC)
 GPEC=P3*DIEC
 P3=X3(EC)
 PEIN=X2(EIN*IEIN)
 GPEIN=P4*DIEIN
 P4=X4(EIN)
  OUTPUTS
 PEC,PEIN,VR3,VR4
```

```
RUN CONTROLS
RUN INITIAL CONDITIONS ONLY
RUN WORST CASE
END
```

Example 9.10. Use of Optimization

The Darlington pair used for Example 9.4 (shown in Fig. 9.30) is used to demonstrate the optimization option. For this case, one set of optimization parameters was specified. The set contained two objective functions and four independent variables. The input data are shown below. The two objective functions are minimized with respect to the four independent variables specified. The output appears in Fig. 9.42.

The problem shows the use of defined parameters as both objective functions and independent variables. Only those defined parameters used to relate primary and secondary current sources may be used as independent variables. The defined parameters used for objective functions must have their total differentials provided. Thus,

$$PEC = EC*IEC$$

and its total differential is

$$EC \cdot d(IEC) + IEC \cdot d(EC)$$

Since EC is constant and not one of the independent variables, it is not necessary to include its differential contribution to the total differential, GPEC.

```
 MODEL DESCRIPTION
 MODEL 2N706A (TEMP) (B-E-C)
 *   UNITS - KOHMS, UHENRIES, PFARADS, VOLTS, MAMPS, NSECONDS, GHZ
     ELEMENTS
 CE,1-E=EQUATION1(5.,70.,J1)
 CC,1-2=EQUATION1(8.,370.,J2)
 RB,B-1=.3
 RC,C-2=.015
 J1,1-E=DIODE EQUATION(1.E-7,35.)
 J2,1-2=DIODE EQUATION(5.E-7,37.)
 JA,E-1=.1*J2
 JB,2-1=P1*J1
 JX,2-1=0
  DEFINED PARAMETERS
 P1=0.96(0.98,0.9)
  OUTPUTS
 VCE,VCC,J1
  FUNCTIONS
 EQUATION1(A,B,C)=(A+B*C)
  CIRCUIT DESCRIPTION
 *   OPTIMIZATION EXAMPLE
 *   UNITS - KOHMS, UHENRIES, PFARADS, VOLTS, MAMPS, NSECONDS, GHZ
     ELEMENTS
```

```
EC,1-6=10.(9.5,10.5)
EIN,1-2=1
R1,2-3=20
R2,6-5=5
R3,4-1=200.(195.,205.)
R4,8-1=.4
R5,6-7=1
T1,3-4-5=MODEL 2N706A (TEMP)
T2,4-8-7=MODEL 2N706A (TEMP)
 OPTIMIZATION
(VR3,PEC/P1T1,P1T2,R3,EC)
 DEFINED PARAMETERS
PEC=XEC(EC*IEC)
GPEC=P3*DIEC
P3=X3(EC)
PEIN=X2(EIN*IEIN)
GPEIN=P4*DIEIN
P4=X4(EIN)
 OUTPUTS
PEC,PEIN,VR3,VR4
 RUN CONTROLS
RUN INITIAL CONDITIONS ONLY
RUN OPTIMIZATION=50
OPTIMIZATION CRITERION=1.E-5
 END
```

OPTIMIZATION RESULTS

TERMINATION CONDITION	OPTIMIZATION RUN COMPLETED
OBJECTIVE FUNCTION	PEC
VALUE OF OBJECTIVE FUNCTION AT TERMINATION	3.18310922D 00
OPTIMIZATION PASS COUNTER	4
MAXIMUM PASSES SPECIFIED	50

INDEPENDENT VARIABLE VALUES AT TERMINATION

VARIABLE	VALUE	GRADIENT COMPONENT	TRANSFORMED VALUE (RADIANS)	TRANSFORMED GRADIENT
P1T1	9.00000195D-01	8.95806035D 00	3.12180972D-03	1.11861258D-03
P1T2	9.00000005D-01	1.26168244D 01	5.21984985D-04	2.63431705D-04
R3	1.99929305D 02	3.65441281D-04	1.55665686D 00	1.82702376D-03
EC	1.00000000D 01	0.0	1.57079633D 00	0.0

INITIAL VALUES OF OPTIMIZATION PARAMETERS

OBJECTIVE FUNCTION	PEC
NUMBER OF INDEPENDENT VARIABLES	4
NUMBER OF RANDOM STEPS	0

Figure 9.42 Optimization Example Outputs.

```
INITIAL H MATRIX FACTOR           1.00000000D 00
CONVERGENCE CRITERION             1.00000000D-05
RANDOM STEP SIZE CONTROL          2.00000000D-01
MINIMUM FUNCTION ESTIMATE         0.0
H MATRIX DETERMINANT              1.00000000D 00

INITIAL VALUES OF INDEPENDENT VARIABLES

               NOMINAL            LOWER              UPPER
VARIABLE       VALUE              BOUND              BOUND

P1T1           9.60000000D-01     9.00000000D-01     9.80000000D-01
P1T2           9.60000000D-01     9.00000000D-01     9.80000000D-01
R3             2.00000000D 02     1.95000000D 02     2.05000000D 02
EC             1.00000000D 01     9.50000000D 00     1.05000000D 01

INITIAL APPROXIMATION TO H MATRIX

    1.0000000D 00      0.0              0.0              0.0
    0.0                1.0000000D 00    0.0              0.0
    0.0                0.0              1.0000000D 00    0.0
    0.0                0.0              0.0              1.0000000D 00


                                            OPTIMIZATION RESULTS

TERMINATION CONDITION                       OPTIMIZATION RUN COMPLETED
OBJECTIVE FUNCTION                          VR3
VALUE OF OBJECTIVE FUNCTION AT TERMINATION  5.65018752D-01
OPTIMIZATION PASS COUNTER                   13
MAXIMUM PASSES SPECIFIED                    50

INDEPENDENT VARIABLE VALUES AT TERMINATION
                              GRADIENT         TRANSFORMED        TRANSFORMED
VARIABLE       VALUE          COMPONENT        VALUE (RADIANS)    GRADIENT

P1T1       9.0000003D-01    4.50857415D-01    -3.56057675D-04    -6.42124959D-06
P1T2       9.0000004D-01    5.48437991D-01    -4.45162598D-04    -9.76576291D-06
R3         1.99900222D 02   2.58026021D-05     1.55083931D 00     1.28987320D-04
EC         1.00000000D 01   0.0                1.57079633D 00     0.0
```

Figure 9.42—*Cont.*

Example 9.11. Use of ac Analysis

This example is intended to illustrate the proper use of the ac program which is designed to conduct a small-signal ac analysis around a circuit's dc operating points. Figures 9.43 and 9.44 show a circuit containing an active device in both schematic and SCEPTRE form.

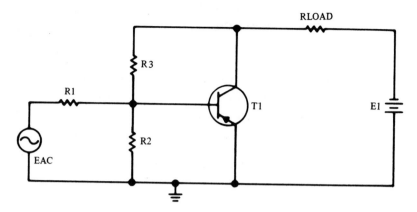

Figure 9.43 Example 9.11 Schematic.

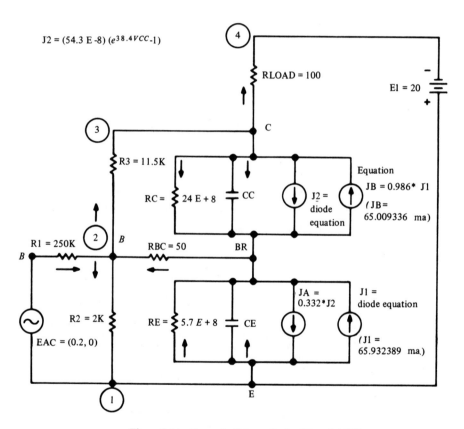

Figure 9.44 Example Schematic for IC and AC Run.

The dc operating point setup is either supplied by the user or is done automatically by the program. The only response required by the user, in addition to requesting the ac run by supplying the appropriate run controls, is the inclusion of the cards for the dc mode he desires. Run controls must

646

be supplied if initial conditions are to be calculated, or the subheading card INITIAL CONDITIONS must be included (followed by the IC data) if the user is supplying the initial conditions. A valid sequence of cards would be

```
 MODEL DESCRIPTION
 MODEL 101M (B-C-E)
 *   UNITS - OHMS, HENRIES, FARADS, VOLTS, AMPS, SECONDS, HZ
  ELEMENTS
 RBC,BR-B=50.
 RC,C-BR=2.4E9
 RE,E-BR=5.7E8
 CC,C-BR=EQUATION 1(5.5E-12,.8,VCC,.3,2.88E-7,J2,3.12E-12)
 CE,E-BR=EQUATION 2(3.5E-12,17.E-9,J1)
 JA,BR-E=.332*J2
 JB,BR-C=.986*J1
 J1,E-BR=DIODE EQUATION (60.E-9,38.4)
 J2,C-BR=DIODE EQUATION (54.3E-8,38.4)
  FUNCTIONS
 EQUATION 1(A,B,C,D,E,F,G)=(A/(B-C)**D*E*(F+G))
 EQUATION 2(A,B,C)=(A+B*C)
  OUTPUTS
 VCC,VCE,NYQUIST,PLOT
 CC,CE
 JA,JB,DEGREES
 J1,J2,RADIANS,PLOT
  CIRCUIT DESCRIPTION
 *   AC ANALYSIS EXAMPLE
 *   UNITS - OHMS, HENRIES, FARADS, VOLTS, AMPS, SECONDS, HZ
  ELEMENTS
 EAC,1-B=(0.2,0.)
 R1,B-2=25.E+4
 R2,2-1=2000.
 R3,2-3=11500.
 RLOAD,3-4=100.
 E1,1-4=-20.
 T1,2-3-1=MODEL 101M
  OUTPUTS
 VR2,VR3,COMPLEX,PLOT
  RUN CONTROLS
 RUN. INITIAL CONDITIONS
 RUN AC
 INITIAL FREQUENCY=1.E5
 FINAL FREQUENCY=1.E6
 NUMBER FREQUENCY STEPS=20.
 TYPE FREQUENCY RUN=LINEAR
  END
```

The automatic dc operating point setup procedure results in the initial condition program, SIMIC, being written and executed. This is followed by one pass through the transient program, SIMTR, in order to determine the values of any elements, or outputs, that may be dependent on the initial conditions. Then the ac analysis proceeds using fixed resistor values, representing the dc operating points, instead of the original CLASS J9 diode sources. The user will receive a listing of these various dc results as shown in

```
                                        IC OUTPUTS BEFORE AC RUN

     VCCT1          -1.3023927D 01

     VCET1           3.6223429D-01

     CCT1            5.4649591D-12

     CET1            1.1243506D-09

     JAT1           -1.8027600D-07

     JBT1            6.5009336D-02

     J1T1            6.5932389D-02

     J2T1           -5.4300000D-07

     VR2            -4.0836859D-01

     VR3             1.2977793D 01


                                     I/C TRANSIENT VALUES AT TIME EQUALS ZERO

 VCCT1 =-1.3023927D 01

 VCET1 = 3.6223429D-01


 J9 SOURCE *J1T1       * IS REPLACED BY RESISTOR= 0.3949750

 J9 SOURCE *J2T1       * IS REPLACED BY RESISTOR= 0.2400000D 13
 4D000000014000000440000033C00000030000000C000000010000004000000030000000C000000
    CCDDDDRRCC
           1              8       0.332000
           2              7       0.986000
```

Figure 9.45 Example 9.11 Initial Conditions Output.

Fig. 9.45. In particular, in part (c), he will be given the values of the replacement resistors used by the ac program.

The ac sources may be connected between any two nodes. However, since all ac sources are set to zero in a dc run, they should be decoupled from the circuit by a dc-blocking capacitor or a sufficiently large resistor value, as R1 in Fig. 9.43. Similarly, all dc sources are set to zero in an ac run; i.e., batteries are short-circuited, and current sources are open-circuited.

Figure 9.46 is a sample of a typical listing produced by the ac run. Figure 9.47 is a sample of a typical ac Nyquist plot.

Example 9.12. Transfer Function Simulation

An extremely useful method of applying SCEPTRE to true system problems has evolved that adds even more flexibility to the program. This method depends on the use of transfer functions that define the output/input relationship of systems or subsystems. Typically, these transfer functions appear in the form of a ratio of polynomials as

$$F(s) = \frac{E_0(s)}{E_i(s)} = \frac{\sum_{i=0}^{m} a_i s^i}{\sum_{j=0}^{n} b_j s^j} = \frac{a_0 + a_i s + \cdots + a_{m-1} s^{m-1} + a_m s^m}{b_0 + b_i s + \cdots + b_{n-1} s^{n-1} + b_n s^n} \quad (9.29)$$

AC ANALYSIS RESULTS

FREQ		1.00000E 05	1.45000E 05	1.90000E 05	2.35000E 05
VCCT1	(COMPLEX)	-3.38927E-03 9.35998E-05	-3.38644E-03 1.35607E-04	-3.38258E-03 1.77491E-04	-3.37769E-03 2.19212E-04
VCET1	(COMPLEX)	-1.36412E-05 3.72039E-07	-1.36300E-05 5.39009E-07	-1.36146E-05 7.05487E-07	-1.35952E-05 8.71322E-07
CCT1	(DEGREES)	5.46496E-12 0.0	5.46496E-12 0.0	5.46496E-12 0.0	5.46496E-12 0.0
CET1	(DEGREES)	1.12435E-09 0.0	1.12435E-09 0.0	1.12435E-09 0.0	1.12435E-09 0.0
JAT1	(DEGREES)	4.69027E-16 1.78418E 02	4.68833E-16 1.77707E 02	4.68567E-16 1.76996E 02	4.68230E-16 1.76287E 02
JBT1	(DEGREES)	3.40661E-05 1.78438E 02	3.40519E-05 1.77735E 02	3.40326E-05 1.77034E 02	3.40082E-05 1.76333E 02
J1T1	(RADIANS)	3.45498E-05 3.11433E 00	3.45354E-05 3.10207E 00	3.45158E-05 3.08982E 00	3.44910E-05 3.07759E 00
J2T1	(RADIANS)	1.41273E-15 3.13398E 00	1.41215E-15 3.10157E 00	1.41135E-15 3.08917E 00	1.41033E-15 3.07678E 00
VR2	(COMPLEX)	3.78463E-05 3.23542E-08	3.78486E-05 4.68590E-08	3.78486E-05 6.13317E-08	3.78503E-05 7.57486E-08
VR3	(COMPLEX)	3.41347E-03 -9.31954E-05	3.41066E-03 -1.35021E-04	3.40681E-03 -1.76724E-04	3.40195E-03 -2.18265E-04

FREQ		2.80000E 05	3.25000E 05	3.70000E 05	4.15000E 05
VCCT1	(COMPLEX)	-3.37179E-03 2.60735E-04	-3.36488E-03 3.02023E-04	-3.35698E-03 3.43039E-04	-3.34810E-03 3.83748E-04
VCET1	(COMPLEX)	-1.35718E-05 1.03637E-06	-1.35443E-05 1.20048E-06	-1.35129E-05 1.36351E-06	-1.34776E-05 1.52532E-06
CCT1	(DEGREES)	5.46496E-12 0.0	5.46496E-12 0.0	5.46496E-12 0.0	5.46496E-12 0.0
CET1	(DEGREES)	1.12435E-09 0.0	1.12435E-09 0.0	1.12435E-09 0.0	1.12435E-09 0.0
JAT1	(DEGREES)	4.67823E-16 1.75578E 02	4.67346E-16 1.74871E 02	4.66801E-16 1.74165E 02	4.66186E-16 1.73461E 02
JBT1	(DEGREES)	3.39786E-05 1.75633E 02	3.39440E-05 1.74935E 02	3.39044E-05 1.74238E 02	3.38598E-05 1.73543E 02
J1T1	(RADIANS)	3.44611E-05 3.06538E 00	3.44260E-05 3.05319E 00	3.43858E-05 3.04103E 00	3.43405E-05 3.02890E 00
J2T1	(RADIANS)	1.40911E-15 3.06442E 00	1.40767E-15 3.05207E 00	1.40603E-15 3.03976E 00	1.40418E-15 3.02747E 00
VR2	(COMPLEX)	3.78523E-05 9.00968E-08	3.78547E-05 1.04363E-07	3.78575E-05 1.18536E-07	3.78605E-05 1.32603E-07
VR3	(COMPLEX)	3.39607E-03 -2.59609E-04	3.38919E-03 -3.00718E-04	3.38133E-03 -3.41557E-04	3.37249E-03 -3.82090E-04

FREQ		4.60000E 05	5.05000E 05	5.50000E 05	5.95000E 05
VCCT1	(COMPLEX)	-3.38826E-03 4.24116E-04	-3.32747E-03 4.64109E-04	-3.31576E-03 5.03696E-04	-3.30314E-03 5.42845E-04
VCET1	(COMPLEX)	-1.34385E-05 1.68577E-06	-1.33956E-05 1.84474E-06	-1.33491E-05 2.00209E-06	-1.32989E-05 2.15770E-06
CCT1	(DEGREES)	5.46496E-12 0.0	5.46496E-12 0.0	5.46496E-12 0.0	5.46496E-12 0.0
CET1	(DEGREES)	1.12435E-09 0.0	1.12435E-09 0.0	1.12435E-09 0.0	1.12435E-09 0.0
JAT1	(DEGREES)	4.64756E-16 1.72760E 02	4.64756E-16 1.72060E 02	4.63942E-16 1.71362E 02	4.63063E-16 1.70667E 02
JBT1	(DEGREES)	3.38102E-05 1.72850E 02	3.37559E-05 1.72159E 02	3.36968E-05 1.71470E 02	3.36330E-05 1.70784E 02
J1T1	(RADIANS)	3.42903E-05 3.01680E 00	3.42352E-05 3.00474E 00	3.41752E-05 2.99272E 00	3.41105E-05 2.98075E 00
J2T1	(RADIANS)	1.40212E-15 3.01522E 00	1.39987E-15 3.00301E 00	1.39742E-15 2.99084E 00	1.39477E-15 2.97871E 00
VR2	(COMPLEX)	3.78639E-05 1.46551E-07	3.78677E-05 1.60371E-07	3.78717E-05 1.74049E-07	3.78761E-05 1.87576E-07
VR3	(COMPLEX)	3.36269E-03 -4.22283E-04	3.35195E-03 -4.62104E-04	3.34028E-03 -5.01520E-04	3.32771E-03 -5.40500E-04

FREQ		6.40000E 05	6.85000E 05	7.30000E 05	7.75000E 05
VCCT1	(COMPLEX)	-3.28963E-03 5.81527E-04	-3.27526E-03 6.19713E-04	-3.26005E-03 6.57375E-04	-3.24403E-03 6.94488E-04
VCET1	(COMPLEX)	-1.32452E-05 2.31145E-06	-1.31881E-05 2.46323E-06	-1.31276E-05 2.61293E-06	-1.30640E-05 2.76045E-06
CCT1	(DEGREES)	5.46496E-12 0.0	5.46496E-12 0.0	5.46496E-12 0.0	5.46496E-12 0.0
CET1	(DEGREES)	1.12435E-09 0.0	1.12435E-09 0.0	1.12435E-09 0.0	1.12435E-09 0.0
JAT1	(DEGREES)	4.62121E-16 1.69975E 02	4.61116E-16 1.69286E 02	4.60051E-16 1.68599E 02	4.59925E-16 1.67916E 02
JBT1	(DEGREES)	3.35645E-05 1.70101E 02	3.34916E-05 1.69420E 02	3.34142E-05 1.68743E 02	3.33325E-05 1.68069E 02
J1T1	(RADIANS)	3.40411E-05 2.96882E 00	3.39671E-05 2.95694E 00	3.38886E-05 2.94512E 00	3.38057E-05 2.93335E 00
J2T1	(RADIANS)	1.39193E-15 2.66662E 00	1.38891E-15 2.95459E 00	1.38570E-15 2.94261E 00	1.38231E-15 2.93069E 00
VR2	(COMPLEX)	3.78808E-05 2.00942E-07	3.78857E-05 2.14136E-07	3.79105E-05 2.27149E-07	3.78965E-05 2.39972E-07
VR3	(COMPLEX)	3.31427E-03 -5.79015E-04	3.29996E-03 -6.17035E-04	3.28481E-03 -6.54535E-04	3.26886E-03 -6.91487E-04

Figure 9.46 Example 9.11 Tabular Output.

AC ANALYSIS RESULTS

FREQ		8.2000E 05		8.65000E 05		9.10000E 05		9.55000E 05	
VCCT1	(COMPLEX)	-3.22722E-03	7.31027E-04	-3.20965E-03	7.66971E-04	-3.19134E-03	8.02296E-04	-3.17233E-03	8.35984E-04
VCET1	(COMPLEX)	-1.29972E-05	2.90569E-06	-1.29273E-05	3.04855E-06	-1.28546E-05	3.18897E-06	-1.27790E-05	3.32685E-06
CCT1	(DEGREES)	5.46496E-12	0.0	5.46496E-12	0.0	5.46496E-12	0.0	5.46496E-12	0.0
CET1	(DEGREES)	1.12435E-09	0.0	1.12435E-09	0.0	1.12435E-09	0.0	1.12435E-09	0.0
JAT1	(DEGREES)	4.57742E-16	1.67237E 02	4.56501E-16	1.66561E 02	4.55205E-16	1.65888E 02	4.53855E-16	1.65220E 02
JBT1	(DEGREES)	3.32465E-05	1.67398E 02	3.31564E-05	1.66731E 02	3.30623E-05	1.66067E 02	3.29643E-05	1.65408E 02
J1T1	(RADIANS)	3.37186E-05	2.92165E 00	3.36272E-05	2.91000E 00	3.35318E-05	2.89842E 00	3.34323E-05	2.88691E 00
J2T1	(RADIANS)	1.37874E-15	2.91883E 00	1.37500E-15	2.90703E 00	1.37110E-15	2.89530E 00	1.36703E-15	2.88363E 00
VR2	(COMPLEX)	3.79023E-05	2.52597E-07	3.79084E-05	2.65015E-07	3.79148E-05	2.77220E-07	3.79213E-05	2.89204E-07
VR3	(COMPLEX)	3.25212E-03	-7.27869E-04	3.23463E-03	-7.63657E-04	3.21640E-03	-7.98830E-04	3.19747E-03	-8.33368E-04

FREQ		1.00000E 06							
VCCT1	(COMPLEX)	-3.15264E-03	8.71016E-04						
VCET1	(COMPLEX)	-1.27007E-05	3.46212E-06						
CCT1	(DEGREES)	5.46496E-12	0.0						
CET1	(DEGREES)	1.12435E-09	0.0						
JAT1	(DEGREES)	4.52453E-16	1.64555E 02						
JBT1	(DEGREES)	3.28625E-05	1.64752E 02						
J1T1	(RADIANS)	3.33291E-05	2.87547E 00						
J2T1	(RADIANS)	1.36281E-15	2.87204E 00						
VR2	(COMPLEX)	3.79281E-05	3.00962E-07						
VR3	(COMPLEX)	3.17786E-03	-8.67253E-04						

Figure 9.46—*Cont.*

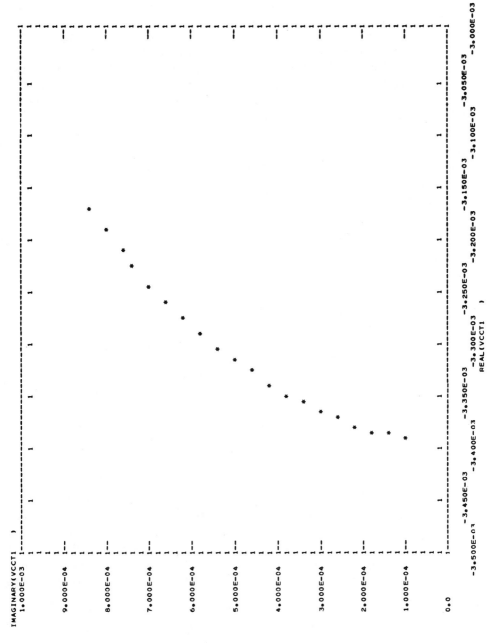

Figure 9.47 Example 9.11 Nyquist Plot.

As a practical matter, it is almost always true that the order of the numerator is less than that of the denominator ($m < n$). A procedure will be given by which any transfer function with $m < n$ may be readily and accurately simulated on SCEPTRE. The case $m = n$ may also be accommodated,* while the rather impractical case $m > n$ is not discussed.

The first task must be to devise an automatic method of converting the general polynomial function of the complex variables given in Eq. (9.29) into a form that is compatible with the mathematical formulation of SCEPTRE. If, from Eq. (9.29), we define

$$\frac{E(s)}{E_i(s)} = \frac{1}{\sum_{j=0}^{n} b_j s^j} \longrightarrow E_i(s) = E(s) \sum_{j=0}^{n} b_j s^j \qquad (9.30)$$

and

$$\frac{E_0(s)}{E(s)} = \sum_{i=0}^{m} a_m s^m \longrightarrow E_0(s) = E(s) \sum_{i=0}^{m} a_i s^i \qquad (9.31)$$

then the inverse transforms of Eqs. (9.30) and (9.31) yield,† respectively,

$$e_i(t) = b_n e(t)^{(n)} + b_{n-1} e(t)^{(n-1)} + \cdots + b_1 \dot{e}(t) + b_0 e(t) \qquad (9.32)$$

and

$$e_0(t) = a_m e(t)^{(m)} + a_{m-1} e(t)^{(m-1)} + \cdots + a_1 \dot{e}(t) + a_0 e(t) \qquad (9.33)$$

where the notation $e(t)^{(n)}$ is used to represent the nth derivative of $e(t)$, the general voltage variable.

The next step must be to properly simulate the operations given in Eqs. (9.32) and (9.33) within the framework of the SCEPTRE input language. Simulation of Eq. (9.33) implies an output voltage source which is equal to a combination of derivatives. The highest of these derivatives is obtained by a transposition of Eq. (9.32), which yields

$$e(t)_{(n)} = \frac{e_i(t) - b_{n-1} e(t)^{(n-1)} - \cdots - b_1 \dot{e}(t) - b_0 e(t)}{b_n} \qquad (9.34)$$

Once the highest derivative is known, all others may be obtained by successive integration.‡ Conversion of the mathematical operations inherent in Eqs. (9.33) and (9.34) to SCEPTRE language requires recourse to the defined parameter feature of the program. If the nodes of the four-terminal system of Fig. 9.48(a) are as shown,§ the user must define one current source and one voltage source under ELEMENTS, as implied in Fig. 9.48(b). The former is simply set equal to zero; its function is to serve as an infinite input impedance across which will appear the system input voltage $e_i(t)$. The voltage

---

*This case may sometimes lead to computational delay.

†Provided that the initial values of $e(t)$ and all its derivatives are zero.

‡This practice is identical in principle to that used in analog computer programming where the highest derivative is fed into a series of integrators.

§Nodes 2 and 8 may be common without loss of generality.

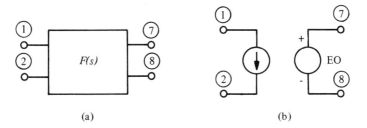

Figure 9.48 A Transfer Function Block and the Equivalent SCEPTRE Representation.

source, given earlier by Eq. (9.33), may be equivalently written as

$$\text{EO} = a_m P_{m+1} + a_{m-1} P_m + \cdots + a_1 P_2 + a_0 P_1^* \qquad (9.33a)$$

where the $a_0, \ldots, a_m$ coefficients are as defined from Eq. (9.29) and the defined parameters $P_1, \ldots, P_{m+1}$ represent the appropriate derivatives from Eq. (9.33). In addition, the user must define two types of expressions under DEFINED PARAMETERS. The first will simulate the highest-order derivative of the system which was given by Eq. (9.34) and is here written equivalently as

$$\text{DP}_n = \frac{\text{VJI} - b_{n-1} P_n - \cdots - b_1 P_2 - b_0 P_1}{b_n} \qquad (9.34a)$$

where the $b_0, \ldots, b_n$ coefficients are as defined in the denominator of Eq. (9.29). Finally, a series of $n - 1$ expressions must also be entered in the general form

$$\begin{aligned} \text{DP}_{n-1} &= P_n \\ &\vdots \\ \text{DP2} &= P_3 \\ \text{DP1} &= P_2 \end{aligned} \qquad (9.35)$$

Note that n differential equations must be supplied—just what one would expect in order to simulate an nth order system.

Now that the format has been described, a specific example will be considered to see what is actually involved. Let it be desired to simulate the transfer function

$$F(s) = \frac{s^2 + 7s - 10}{s^4 + 2s^3 + 10s^2 + 200s + 1000}$$

with the same node designation used in Fig. 9.48. Note that here $m = 2$, $n = 4$. The entire SCEPTRE input will be given in uppercase, followed by appropriate commentary in lowercase type.

*For the special case $m = n$, this relation must be revised to $\text{EO} = a_m DP_m + a_{m-1} P_m + \cdots + a_1 P_2 + a_0 P_1$.

ELEMENTS
JI, 1−2 = 0
EO, 8−7 = X1 (DP2 + 7. *P2 − 10. *P1)   from Eq. (9.33a)
DEFINED PARAMETERS
DP4 = X2 ((VJI − 2. *P4 − 10. *P3 − 200. *P2 − 1000. *P1)/1)

from Eq. (9.34a)

DP3 = X3 (P4)
DP2 = X4 (P3)   from Eq. (9.35)
DP1 = X4 (P2)
P1 = 0
P2 = 0
P3 = 0   These establish initial values; they may be omitted, and the only
P4 = 0   consequence will be a series of warning messages

It is clear that if a particular system requires two or more transfer functions of the above type, the user must repeat the procedure for each one. If there are very many transfer functions involved or if those that are involved are of higher order, the input task becomes at bit tedious and error-prone. The remedy for this, of course, is the stored model of SCEPTRE. It is suggested that* the user store on his permanent tape one model for each degree transfer function of interest. The process is simple and need be done only once. An example of the storage procedure for a general second-order transfer function will be given. This model will then be called for use as a specific second-order function. Consider permanent storage of the following with $m = 2, n = 2$:

$$F2(s) = \frac{a_2 s^2 + a_1 s + a_0}{b_2 s^2 + b_1 s + b_0}$$

The required cards are

MODEL DESCRIPTION
MODEL 2ORDER (PERM) (A−B−C−D)
ELEMENTS
JI, A−B = 0
EO, D−C = X1 (PA*DP2 + PA1*P2 + PAO*P1)
DEFINED PARAMETERS
DP2 = X2 ((VJI−PB1*P2−PBO*P1)/PB2)
DP1 = X3 (P2)
PAO = 0, PA1 = 0, PA2 = 0
PBO = 0, PB1 = 0, PB2 = 0
P1 = 0, P2 = 0

Here the defined parameters are named as PAO, ..., PBO, ... to correspond with coefficients $a_0, \ldots, b_0, \ldots$. P1 and P2 serve as the dependent variables of the two differential equations. The model is general in that any of the

---

*A higher-order model could be stored and degenerated to one of the desired order by proper manipulation of defined parameters. The method suggested here uses less computer solution time, uses fewer defined parameters, and is less complicated.

coefficients can be changed (when the model is called) from the originally assigned zero values. It should be noted that even though this model is stored with provision for the case $m = n$ this will not be the case actually used unless parameter PA2 is made nonzero in the model call.

Now that the model has been stored in general second-order form, the desired coefficients needed to represent a specific second-order transfer function can be supplied when the model is called out for use under CIRCUIT DESCRIPTION.

If it is desired to simulate

$$\frac{E_o(s)}{E_i(s)} = \frac{10s + 4}{s^2 + 0.7s + 5}$$

with the nodes as labeled in Fig. 9.48, an appropriate entry sequence would be

CIRCUIT DESCRIPTION
ELEMENTS
.
.
.
.

G2, 1 − 2 − 7 − 8 = MODEL 2ORDER (PERM, CHANGE PA0 = 4)
PA1 = 10, PB0 = 5, PB1 = 0.7, PB2 = 1)

The same model could be called repeatedly to represent many second-order transfer functions with arbitrary coefficients. The procedure is even simpler to simulate ideal amplifiers, summing junctions, and limit functions (References 2 and 3). The ability of SCEPTRE to perform these operations now qualifies it as a tool for systems analysis. When one combines this with its unmatched capability for nonlinear circuit analysis, its potential for the solution of large systems that can be broken down into nonlinear circuitry and linear subsystems becomes clear.

9.8 PATHOLOGICAL PROBLEMS

Most computer codes have particular "quirks" which each code user must be aware of, and SCEPTRE is no exception. A number of quirks, or limitations, are presented below and are generally a result of the network formulation or particular programming algorithms employed by SCEPTRE. Each "problem" is discussed in terms of what the problem is, why it exists, what its effects are, and how one may work around the problem.

Topological Restrictions

Voltage source loops and current source cut sets are not allowed for either transient or dc analyses in SCEPTRE. Figure 9.49 illustrates these restrictions.

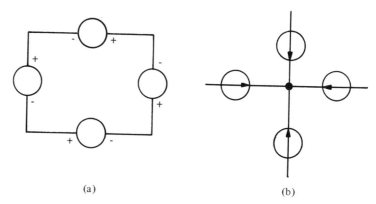

Figure 9.49 Topological Restrictions: (a) Voltage Source Loop; (b) Current Source Loop.

The restrictions arise from the assumption that all voltage sources in the network will be branches and that current sources will be links. The user would generally want to avoid these situations anyway, due to the possible violation of Kirchoff's voltage and current laws.

If the situation does arise where these topologies are required for an analysis, a resistor may be added to the voltage source loop or connected to the cut set node to "absorb" any excess voltage or current generated. Usually the problem is caused by typographical errors in the element node specifications.

The formulation of the iterative dc algorithm introduces a number of topological restrictions of which a user should be aware. It should be noted that these initial conditions restrictions do not apply to the transient solution.

1. No circuit containing a loop consisting of only voltage sources and inductors will be accomodated.

2. No circuit may contain a cut set composed entirely of current sources and capacitors.

3. Certain circuit values may not be chosen as independent variables for functional element value specifications. No resistor or inductor current may be used, and capacitor voltages may be used only if the capacitors are in parallel with a resistor or current source.

4. Circuits with inductor loops may not be solved unless the members of the loop are constant.

5. Capacitor sets are allowed only if the capacitors involved are constant.

The above restrictions are in general introduced by the desire to reduce the computation and storage requirements for dc iterations. If the iterative scheme is required by the user for networks containing the topologies mentioned, a judicious selection of resistors to be inserted into the network will generally alleviate the problem.

A small resistor added to the E-L loop or a large resistor connected from the V-C cut set to ground will allow the iterative scheme to work. Addition of a large resistor or a zero-valued valued current source in parallel with the capacitor of interest will alleviate this problem. Iterative solutions may be obtained for capacitor cut sets and inductor loops if the elements of interest are set constant. The application of any of the above-mentioned remedies may cause errors in subsequent transient calculations, in which case the user would want to calculate initial conditions in one run, input the appropriate results under the INITIAL CONDITIONS heading, and revert to his original description of the network for a transient analysis. Of course, the RUN IC VIA IMPLICIT method for integration of transient equations to steady state is always available when topological problems with the iteration algorithm occur.

The ac analysis algorithm employs network values generated by the initial conditions solution so that the restrictions applicable to dc calculations are applicable to the ac calculations. Additionally, the ac solution algorithm contains the restriction on circuits containing linearly dependent sources with the requirement for their time derivatives. The same methods presented above apply to avoidance and alleviation of these problems.

Computational Delays

The computational delay is a problem which is usually no problem but nevertheless one which requires some attention. Recall the discussion of Sec. 9.2 regarding the derivation of the transient equations. The strategy was to systematically derive the values of network voltages and currents based on already computed quantities at each time step. Where all network elements are constant, this scheme works very nicely. The problem may arise, however, when functional dependencies are introduced which refer to values not yet computed in the given solution sequence. For example, the voltage source in the circuit shown in Fig. 9.50 is a function of the voltage across R1. It is obvious that the value of the resistor voltage is dependent on the value of the voltage source, which is dependent on the value of the resistor voltage, which

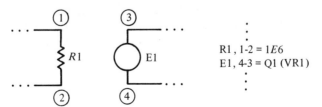

Figure 9.50 Computational Delay.

is dependent on. . . . Simply stated, one of the equations must be written first (it will usually be VR1 = E1), thereby resulting in the use of the value at the last pass instead of the desired value at the current pass—thus, the computational delay. Typically the delay presents no problem since the response of the network is usually slow in relation to the integration time step being taken. Some highly nonlinear simulations may produce erroneous results if this one time step delay exists in a "fast" portion of the network. It is generally in the analyst's best interest to at least inspect the time steps taken versus the circuit response time to ensure that reasonably correct results are obtained.

In the event that a computational delay does induce erroneous results, all is not lost. The best strategy is to search for values to base the dependency which have been computed in the solution sequence. In the previous example, the simulation program SIMTR (obtained by entering the special run control WRITE SIMUL8 DATA) may be examined for values equivalent to or from which VR1 can be calculated which have already been defined or computed in SIMTR. This may be a capacitor voltage or a series of voltage values. The VR1 dependency would then be modified to the new value for subsequent runs. If suitable quantities are not available (as would probably be the case here, since resistive quantities are computed first), the user can create a suitable state variable by inserting the appropriate capacitor or inductor into the network. Incidentally, the predictor-corrector scheme employed by the implicit integration routine tends to minimize the effects of computational delay in all cases.

Program Limits

Table 9.6 (see Sec. 9.4) illustrates the maximum capacities for various input elements which are built into SCEPTRE. Generally speaking, the limits allow the analysis of discretely described circuitry on the order of integrated circuit operational amplifiers and fairly large control systems with the use of functional *system blocks*.

Convergence Problems on Initial Conditions

The iterative Newton-Raphson algorithm for solution of initial conditions may sometimes fail to converge to a solution. Situations where this happens usually fall into the following categories:

1. Highly nonlinear networks.
2. High gain feedback networks.
3. Symmetrical networks.

Generally, the iterative scheme will approach the area of the solution and then oscillate around the solution point "forever." For the case of symmetrical networks, the best strategy for obtaining convergence is to provide asymmetrical initial conditions for the convergence to start from. For high gain feedback situations, the feedback loop(s) may be broken, approximate initial conditions calculated, and these calculated values used as a starting point for subsequent analyses with feedback restored. Highly nonlinear networks are a bit more difficult to treat. Attempts to "soften" the nonlinearities present may be made, but this can end up in destroying the original network. The best technique to try in this situation is the alternative initial conditions algorithm, i.e., the integration of the transient equations to steady state called with the run control RUN IC VIA IMPLICIT. The integration technique also works well with high gain feedback and symmetrical networks.

The fault with integration, however, is that it may be costly in terms of computer time. To alleviate this problem and to cut computer costs in general, it is always best to provide initial conditions where they are known.

9.9 ERROR DIAGNOSTICS

The free-form input data format of SCEPTRE is intended to minimize formatting errors. Other types of errors such as those of omission, ambiguity, inconsistency, and violation (of syntax, program limits, etc.) must be detected and diagnosed and the user alerted. For this purpose, the program possesses a comprehensive input data diagnostic capability. A listing of the principal error messages in the input processor program are shown in Table 9.8.

Table 9.8

1	INITIAL CONDITION LACKS A 'V' OR 'I' PREFIX
2	EQUAL SIGN MISSING
3	VARIABLE NAME EXCEED CHARACTER LIMIT
4	REDUNDANT VOLTAGE OR CURRENT SPECIFICATION
5	NO VALUE SPECIFICATION FOLLOWS EQUALS
6	INCORRECT VALUE SPECIFICATION
7	MORE THAN 100 INITIAL CONDITIONS HAVE BEEN SPECIFIED
8	THE MAXIMUM OF 100 DEFINED PARAMETERS HAS BEEN EXCEEDED
9	DEFINED PARAMETER LACKS A 'D' or 'P' PREFIX
10	DEFINED PARAMETER DERIVATIVE CANNOT BE A CONSTANT
11	IMPROPER DELIMITER SEQUENCE FOUND IN NODE STRING
12	NODE DESIGNATION MISSING
13	LIMIT OF 25 MODEL NODES HAS BEEN EXCEEDED
14	NODE STRING IMPROPERLY TERMINATED
15	PRECEDING MODEL IS NOT IN THE MODEL LIBRARY

Table 9.8—*Cont.*

16	TABLE NAME OR INDEPENDENT VARIABLE EXCEEDS 6 CHARACTERS
17	TABLE VALUE OR ARGUMENT EXCEEDS VALUE LIMIT
18	TABLE INDEPENDENT VARIABLE IMPROPERLY SPECIFIED
19	RIGHT PAREN FOLLOWING TABLE INDEPENDENT VARIABLE MISSING
20	EQUATION ARGUMENT STRING MISSING
21	NO OUTPUT REQUESTS SUPPLIED, RE-OUTPUT HAS BEEN DELETED
22	INCORRECT DELIMITER OR INSUFFICIENT DATA ON CARD
23	THE LIMIT OF ONE HUNDRED OUTPUTS HAS BEEN EXCEEDED
24	MISSING COMMA SUPPLIED BY PROCESSOR
25	OUTPUT VARIABLE OR LABEL EXCEEDS SIX CHARACTERS
26	PLOT NAME OR LABEL IS MISSING
27	PRECEDING UNRECOGNIZABLE CARD IGNORED
28	THE MAX OF 100 DEFINED PARAMETER DERIVATIVES IS EXCEEDED
29	REDUNDANT DEFINED PARAMETER OR DEFINED PARAM DERIVATIVE
30	MAX OF TEN TERMINATE IF RUN CONTROLS HAS BEEN EXCEEDED
31	IMPROPER RUN CONTROL FORMAT
32	INCOMPLETE SPECIFICATION PRIOR TO GROUP OR MODE CARD
33	LEFT PAREN MISSING FOR TERMINATION CONTROL
34	EXTRANEOUS DELIMITER IGNORED
35	SECONDARY DEPENDENT SOURCE IMPROPERLY SPECIFIED
36	INVALID NUMBER OR INTEGRATION ROUTINE
37	UNRECOGNIZABLE DATA FOLLOWS MODEL DESCRIPTION
38	MODEL NAME MISSING
39	REDUNDANT MODEL HAS BEEN SPECIFIED
40	THE MAXIMUM OF 11 HEADING CARDS HAS BEEN EXCEEDED
41	TABLE OR EQUATION SPECIFICATION EXPECTED
42	INCOMPLETE EQUATION OR UNEVEN TABLE PRIOR TO NEW FUNCTION
43	TABLE OR EQUATION IMPROPERLY SPECIFIED
44	TABLE OR EQUATION IS REDUNDANT
45	TABLE OR EQUATION NAME EXCEEDS SIX CHARACTERS
46	INVALID CONSTANT SPECIFIED
47	TABLE DATA EXCEEDS CAPACITY OF THE INPUT PROCESSOR
48	THE MAXIMUM OF 300 ELEMENTS HAS BEEN EXCEEDED
49	THE MAXIMUM OF 50 MUTUAL INDUCTANCES HAS BEEN EXCEEDED
50	THE MAXIMUM OF 50 SOURCE DERIVATIVES HAS BEEN EXCEEDED
51	COMMA OR = SIGN MISSING OR WRONG DERIVATIVE SPECIFIED
52	SOURCE ELEMENT MISSING FOR TABLE OR DIODE EQUATION
53	INCORRECT MODEL CARD — ELEMENT NAME INCOMPLETE
54	IMPROPER MODEL CHANGE SPECIFIED
55	INCONSISTENT MODEL CHANGE, EQUATION AND TABLE ARE EQUATED
56	LIMIT OF 15 TABLE OR EQUATION MODEL CHANGES EXCEEDED
57	INSUFFICIENT MODEL NODES SPECIFIED
58	MODEL NAME EXCEEDS EIGHTEEN CHARACTERS
59	MODEL CIRCUIT DESIGNATION EXCEEDS FOUR CHARACTERS
60	MODEL CHANGE VALUE IS VARIABLE
61	THE LIMIT OF TEN SUPPRESSED OUTPUTS HAS BEEN EXCEEDED
62	UNRECOGNIZABLE MODEL MODIFICATION DATA
63	MORE THAN 250 TEMPORARY OR PERMANENT MODELS SPECIFIED

Table 9.8—*Cont.*

64	AN EQUATION IS NOT A PERMISSIBLE ARGUMENT
65	IMPROPER ARGUMENT FORMAT
66	EQUATION, ARGUMENT OR VALUE EXCEEDS 1200 CHARACTERS
67	MODEL NOT IN SPECIFIED LIBRARY, ALTERNATE LIB SCAN STARTS
68	MUTUAL INDUCTANCE REFERENCES NON-EXISTENT INDUCTOR
69	SOURCE DERIVATIVE REFERENCES NON-EXISTENT SOURCE
70	DEFINED PARAM DERIV. REFERENCES NON-EXISTENT DEF. PARAM.
71	A MODEL ELEMENT CANNOT BE A MODEL
72	ELEMENT, MUTUAL INDUCTANCE OR SOURCE DERIV IS REDUNDANT
73	MODEL TABLE OR EQUATION CHANGE IS REDUNDANT
74	ELEMENT IMPROPERLY SPECIFIED
75	REFERENCED TABLE, EQUATION, DEF PARAM OR ELEMENT MISSING
76	THE FOLLOWING OUTPUT REQUEST IS NOT VALID
77	NO OUTPUT REQUESTS HAVE BEEN SUPPLIED
78	IMPROPER EXPRESSION FORMAT —COMPUTATION DELAYS MAY OCCUR
79	INVALID SYNTAX
80	TABLE XXX FORMAT ALTERED TO TXXX
81	TIME SUPPLIED AS ARGUMENT IN XTABLE FUNCTION
82	THE MAX. OF 100 PARTIAL DEPRAM. DERIV. HAS BEEN EXCEEDED
83	PART. DFPRAM DERIV REFERENCES NON-EXISTENT DFPRAM DERIV.
84	INVALID OR MISSING DELIMITER IN A DC OPTION CARD
85	INVALID VARIABLE IN A DC OPTION CARD
86	MONTE CARLO DEFAULT, 10 PASSES SUPPLIED
87	OPTIMIZATION DEFAULT, 30 PASSES SUPPLIES
88	NOMINAL VALUE DOES NOT LIE BETWEEN BOUNDS
89	AC RUN REQUESTED WITH INITIAL CONDITIONS ONLY RUN
90	COMPLEX DEFINED PARAMETER GIVEN AS REAL CONSTANT
91	LIMIT OF 50 COMPLEX DEFINED PARAMETERS EXCEEDED
92	AC SOURCES PRESENT IN NON-AC RUN
93	SYMBOLIC JACOBIAN WITH ADJOINT RUN MAY LEAD TO ERROR
94	INVALID OR MISSING DELIMITER IN A DIFFERENTIAL DATA CARD
95	INVALID VARIABLE IN A DIFERENTIAL DATA CARD

As the input data cards are read in by the input processor, each card is printed out and then scanned for errors. If an error is found, an error message stating the trouble is printed immediately following the detection of the error. The severity of errors detected in this manner is also indicated. There are three levels of severity:

Level 1 (warning only). Errors of this type are not critical and will be repaired by the input processor. The error scan is continued and the analysis phase executed providing that errors of a higher level are not detected elsewhere in the input data.

Level 2 (simulation deleted). Errors of this type are critical and thus prohibit execution of the analysis phase. However, the error scan is continued in order to detect, diagnose, and alert the user of any remaining errors in the input data.

Level 3 (execution terminated). This type of error will not permit the proper execution of the input processor or successful continuation of either the error scan or execution of the analysis phase.

Most input data errors fall into level 2, as illustrated by the following example of an initial conditions specification which should contain an I prefix designating the initial current through inductor L3:

INITIAL CONDITIONS
VC1 = 100
L3 = 0.01
***Error Message 1 — level 2 — simulation deleted — ERROR SCAN CONTINUES
INITIAL CONDITION LACKS A 'V' OR 'I' PREFIX

The program generator portion of the program also contains a diagnostic capability to inform the user of errors that can be detected only after the network topology is analyzed. Such errors may or may not cause the execution of the analysis phase to be aborted as indicated by the error messages which occur. A list of diagnostic messages that may originate in the program generator follows:

AN ILLEGAL PROCESSING SEQUENCE HAS BEEN REQUESTED

PROGRAM CAPACITY HAS BEEN EXCEEDED — XXX LIST OVER FLOWED IN FILING XXX

'WTSPTP' HAS BEEN ASKED TO WRITE XXX SOURCE STATEMENT CARDS. THE LIMIT IS XXX

THE IIIII ELEMENTS TO BE PROCESSED EXCEEDS THE LIMIT OF XXX.

XXX RESULTS IN A CURRENT SOURCE CUT SET.

XXX CAUSES J—C CUT SET PREVENTING I/C SOLUTION.

XXX RESULTS IN A VOLTAGE SOURCE LOOP.

XXX CAUSES E—L LOOP PREVENTING I/C SOLUTION.

THE ELEMENT XXX FROM XXX TO XXX IS SHORTED OUT OF THE CIRCUIT.
EXECUTION WILL BE TERMINATED AFTER COMPLETED ERROR SCAN.

WARNING ONLY THE XXX NODE OF XXX IS NOT CONNECTED TO ANY OTHER
ELEMENT IN THE CIRCUIT — EXECUTION CONTINUES.

AN ERRONEOUS TREE BRANCH EXISTS IN THE PATH OF XXX.

ERROR IN BCD CONVERSION OF XXX ENTRIES IS CLASS II

THE REFERENCE INDUCTOR XXX FOR XXX CANNOT BE FOUND IN THE ELEMENTS TABLE

PROGRAM ERROR IN 'WTEQTN'. ILLEGAL ARCUMENT TYPE CODE FOUND IN DEPENDENCY TABLE.

THE VARIABLE XXX COULD NOT BE FOUND IN THE VALUE LIST BY 'WTEQTN' — PROGRAM ERROR.

THE VARIABLE XXX IS DEPENDENT UPON ITSELF. THE RUN CANNOT BE CONTINUED.

THE DEPENDENCY TABLE HAS OVERFLOWED IN 'WTEQTN' — PROGRAM ERROR.

THE TERM XXX WILL CAUSE A COMPUTATIONAL DELAY.

THE TERM XXX COULD NOT BE FOUND IN THE VALUE LIST BY 'EQFORM' — PROGRAM ERROR.

VALUE EXPRESSION FOR TYPE 6 IS NOT CODED YET. THE RUN CANNOT BE CONTINUED.

THE VARIABLE XXX CANNOT BE FOUND IN EITHER THE LIST OF CIRCUIT ELEMENTS OR DEFINED PARAMETERS BY 'COMPCK'.

THE FOLLOWING INVALID PREFIX HAS BEEN SUPPLIED TO 'COPYDP', 'X'

'COPYDP' IS BEING REQUESTED TO GENERATE THE FOLLOWING (INVALID) RESISTOR DERIVATIVE TERM IN AN EQUATION XXX

THE FOLLOWING TERM WILL CAUSE THE DEPENDENCY TABLE TO OVERFLOW XXX

'SCANEQ' COULD NOT FIND THE TERM XXX IN THE ELEMENTS TABLE — PROGRAM ERROR

THE XXX ELEMENT COMPLETES A LOOP CONTAINING XXX WHICH WILL PROHIBIT AN INITIAL CONDITIONS SOLUTION.

THE VARIABLE ELEMENT (XXX) COMPLETES A LOOP CONTAINING AN ELEMENT OF THE SAME TYPE WHICH PROHIBITS AN IC SOLUTION

CAPACITOR XXX REFERENCED IN TABLE OR EQUATION BUT NOT SUPPLIED AS ELEMENT

ELEMENT XXX IS FUNCTION OF CAPACITOR XXX VOLTAGE FOR WHICH VOLTAGE SUBSTITUTION CANNOT BE PERFORMED (INITIAL CONDITIONS)

RESISTOR OR INDUCTOR CURRENT SUPPLIED AS TABLE INDEPENDENT VARIABLE OR EQUATION ARGUMENT FOR XXX IS NOT ALLOWED FOR INITIAL CONDITIONS PROBLEM

XXX UPON WHICH XXX IS DEPENDENT CANNOT BE FOUND IN THE CLASS 9 SECTION OF THE ELEMENTS TABLE.

AN INTEGER—TO—BCD CONVERSION ERROR OCCURRED IN A DIMENSION STATEMENT GENERATION

XXX WHICH XXX IS DEPENDENT UPON IS NOT IN THE CLASS 2 OR 5 SECTION OF ELEMENTS—PROGRAM ERROR

THE VMUTUAL SUBROUTINE SUPPLIED XXX FOR TYPE OF MUTUAL TERM TO BE ADDED TO V3P(N) or V6(N) — MUST BE IN RANGE OF 1 — 5.

THE DERIVATIVE OF THE VARIABLE SOURCE XXX HAS NOT BEEN SUPPLIED.

'GTCODE' CAN'T ENCODE THE ILLEGAL TERM XXX — PROGRAM ERROR.

NUMBER OF HEADING CARDS EXCEEDS MAXIMUM — THIS CARD IS IGNORED

SIMUL8 DATA GENERATION HAS BEEN DELETED FOR THIS RERUN

NO STOP TIME HAS BEEN SPECIFIED

ILLEGAL INITIAL CONDITION VARIABLE NAME — XXX

IMPROPER VALUE SPECIFICATION ON CARD PRIOR TO GROUP CARD

ILLEGAL VALUE SPECIFICATION

ILLEGAL TABLE SPECIFICATION

EXTRANEOUS X IGNORED

FOLLOWING VARIABLE NAME TRUNCATED TO SIX CHARACTERS XXX ———

INCORRECT DELIMITER FOLLOWS VARIABLE NAME XXXXXX AN = HAS BEEN SUPPLIED

INCORRECT DELIMITER FOLLOWS VARIABLE VALUE EEEEEE. EEEEE X, HAS BEEN SUPPLIED

ILLEGAL RUN CONTROL SPECIFICATION

RUN CONTROL DATA MISSING

INITIAL CONDITIONS WERE NOT SPECIFIED FOR THE ORIGINAL RUN

EXTRANEOUS X IGNORED

INCORRECT DELIMITER FOR FOLLOWING RUN CONTROL SPECIFICATION — XXX ———

AN EQUAL SIGN HAS BEEN SUPPLIED

INVALID NUMBER OR INTEGRATION ROUTINE SPECIFIED

NO ENTRIES IN RERUN DATA LIST

ILLEGAL VARIABLE NAME — XXX

ILLEGAL TABLE NAME SPECIFICATION — XXX

TABLE XXX — RERUN TABLE EXCEEDS LENGTH OF ORIGINAL TABLE

ILLEGAL CONTINUE SPECIFICATION

ILLEGAL RUN CONTROL SPECIFICATION

RUN CONTROL DATA MISSING

INCOMPLETE RUN CONTROL SPECIFICATION ON CARD PRIOR TO END CARD

ILLEGAL DEBUG CARD IGNORED

THE NEWTON — RAPHSON PASS LIMIT HAS BEEN EXCEEDED WITHOUT ATTAINING CONVERGENCE

MATRIX XXX IS SINGULAR

UNABLE TO LOCATE DEPENDENT PLOT VARIABLE FOR INDEPENDENT VARIABLE XXX

NO. PLOT PTS. HAVE EXCEEDED PLOT STORAGE BLOCK SIZE

DERIVATIVE XXX FOUND IN THE TERMINATION CONDITION(S) IS NOT RESOLVED

THE LENGTH OF THE FIRST SOLUTION BUFFER IS ZERO.

MORE THAN 9 OUTPUTS WERE REQUESTED ON PLOT NUMBER XXX. ONLY THE FIRST 9 REQUESTS WILL BE HONORED.

PLOT REQUEST XXX WAS NOT SPECIFIED AS AN OUTPUT IN THE ORIGINAL (CIRCUIT DESCRIPTION) RUN FOR THIS SYSTEM.

MAXIMUM PLOT LENGTH (2000 LINES) HAS BEEN EXCEEDED.

A TRACE ON ONE OF THE PLOTS IS DISPLACED FROM ZERO BY MORE THAN 10**6 TIMES ITS TOTAL RANGE. LOSS OF SIGNIFICANCE IS POSSIBLE IN THESE RESULTS.

TABLE CONTAINING IMPLICIT INTEGRATION DATA HAS OVERFLOWED. SIMULATION HAS BEEN DELETED.

TABLE CONTAINING SPARSE MATRIX DATA HAS OVERFLOWED. SIMULATION HAS BEEN DELETED.

TABLES CONTAINING SPARSE MATRIX DATA HAVE OVERFLOWED. PIVOTING OF THE MATRIX WILL BE DELETED.

9.10 REFERENCES

This section refers the interested reader to the particular works which the authors of the code used in formulating SCEPTRE and to some articles of interest in the area of modeling.

1. ASHAR, K. G., H. N. GHOSH, A. W. ALDRIDGE, and J. L. PATTERSON, "Transient Analysis and Device Characterization of ACP Circuits," *IBM J. Research Development*, 7, 1963, p. 218.

2. BASHKOW, T. R., "The A Matrix, New Network Description," *IRE Trans. Circuit Theory*, CT-4, Sep. 1957, pp. 117–120.

3. BEAUFOY, R., and J. J. SPARKES, "The Junction Transistor as a Charge-Controlled Device," *ATE J.*, 13, No. 4, Oct. 1957, pp. 310–324.

4. BECKER, DAVID, et al., "Extended SCEPTRE", *USAF Technical Report AFWL-TR-73-75*, Air Force Weapons Laboratory, Albuquerque, N. M.

5. BEEZHOLD, W., et al., "Generalized Model for Micro-Circuit Transient Radiation Response Prediction," *USAF Technical Report RADC-TR-68-115*, Rome Air Development Center, 1968. Rome, N. Y.

6. BOWERS, J., and S. SEDORE, *SCEPTRE: A Computer Program for Circuit and Systems Analysis*, Prentice-Hall, Englewood Cliffs, N. J., 1971.

7. BOX, M. J., "A Comparison of Several Current Optimization Methods, and the Use of Transformation in Constrained Problems," *Computer J.*, 9, 1966, pp. 67–77.

8. BOX, M. J., D. DAVIES, and W. G. SWANN, *Non-Linear Optimization Techniques*, Monograph No. 5, Imperial Chemical Industries, Ltd., 1969, Edinburgh, Scotland.

9. CHASE, P. E., "Stability Properties of Predictor-Corrector Methods for Ordinary Differential Equations," *J. ACM*, 9, Oct. 1962, pp. 457–468.

10. CORDWELL, W. A., "Radiation Device Modeling for the SCEPTRE Program," *Report 69-825-2371*, IBM, Owego, N.Y., June 1969.

11. CORDWELL, W. A., "Transistor and Diode Model Handbook," *USAF Technical Report AFWL-TR-69-44*, Air Force Weapons Laboratory, Oct. 1969, Albuquerque, N. M.

12. DAVIDON, W. C., "Variable Metric Method for Minimization," *AEC Research and Development Report ANL-5990*, rev. ed., 1959.

13. DIRECTOR, S. W., and R. A. ROHRER, "The Generalized Adjoint Network and Network Sensitivities," *IEEE Trans. Circuit Theory*, CT-16, Aug. 1969, pp. 318–323.

14. DIRECTOR, S. W., and R. A. ROHRER, "Automated Network Design—The Frequency-Domain Case," *IEEE Trans. Circuit Theory*, CT-16, Aug. 1969., pp. 330–336.

15. EBERS, J. J., and J. L. MOLL, "Large Signal Behavior of Function Transistors," *Proc. IRE*, 42, No. 12, Dec. 1954, pp. 1761–1772.

16. FOWLER, M., and R. WARTEN, "A Numerical Integration Technique for Ordinary Differential Equations with Widely Separated Eigenvalues," *IBM J. Research Development*, 11, Sept. 1967, pp. 537–543.

17. GEAR, C. W., "The Numerical Integration of Ordinary Differential Equations," *Math. Comp.*, 21, April 1967, pp. 146–156.

18. GEAR, C. W., "The Automatic Integration of Ordinary Differential Equations," *Comm. ACM*, 14, March 1971, pp. 176–179.

19. GUMMEL, H. K., and H. C. POON, "A Integral Charge Control Model of Bipolar Transistors," *Bell System Tech. J.*, 49, No. 5, May 1970, pp. 827–852.

20. GUMMEL, H. K., and H. C. POON, "Modeling of Emitter Capacitance," *Proc. IEEE*, 57, No. 12, Dec. 1969, pp. 2181–2182.

21. HILDEBRAND, F. B., *Introduction to Numerical Analysis*, McGraw-Hill, New York, 1956, p. 237.

22. KAPLAN, W., *Ordinary Differential Equations*, Addison-Wesley, Reading, Mass., 1958.

23. KREYSZIG, E., *Advanced Engineering Mathematics*, Wiley, New York, 1962.

24. KUH, E., and R. ROHRER, "The State-Variable Approach to Network Analysis," *Proc. IEEE*, 53, No. 7, July 1965, pp. 672–686.

25. LEEDS, J. V., and G. I. UGRON, "Simplified Multiple Parameter Sensitivity Calculation and Continuously Equivalent Network," *IEEE Trans. Circuit Theory*, CT-14, June 1967, pp. 188–191.

26. NARATI, R., *An Introduction to Semiconductor Electronics*, McGraw-Hill, New York, 1963.

26a. OGATA, K., *State Space Analysis of Control Systems*, Prentice-Hall, Englewood Cliffs, N.J., 1967.

27. PHILLIPS, A., *Transistor Engineering*, McGraw-Hill, New York, 1962.

27a. POCOCK, O. N., M. G. KREBS, and C. W. PERKINS, "Simplified Microcircuit Modeling," *AFWL Technical Report AFWL-TR-73-272*, 1973, Albuquerque, N. M.

27b. POLAK, E., and E. WONG, *Notes for a First Course on Linear Systems*, Van Nostrand Reinhold, New York, 1970.

28. PORTASIK, J., and C. GLASS, "Systems Analysis Using the SCEPTRE Computer Program," *AFWL Technical Report 68-98*, Sept. 1968, Albuquerque, N. M.

29. RAYMOND, JAMES P., and ROBERT E. JOHNSON, "Study of Generalized, Lumped Transistor Models for Use with SCEPTRE," *USAF Technical Report AFWC-TR-68-86*, Air Force Weapons Laboratory, March 1969, Albuquerque, N. M.

30. REED, M. B., and S. SESHU, *Linear Graphs and Electrical Networks*, Addison-Wesley, Reading, Mass., 1961.

31. SEDORE, S. R., "SCEPTRE: A Program for Automatic Network Analysis," *IBM J. Research Development*, 11, Nov. 1967, pp. 627–637.

32. SEDORE, S. R., "More Efficient Use of the F Matrix in Practical Circuit Analysis Programs," *IEEE Trans. Computers*, C-17, May 1968, pp. 503–506.

33. SESHU, S., and N. BALABANIAN, *Linear Network Analysis*, Wiley, New York, 1959.

34. TINNEY, W., and J. WALKER, "Direct Solutions of Sparse Network Equations by Optimally Ordered Triangular Factorization," *Proc. IEEE*, 55, No. 11 Nov. 1967, pp. 1801–1809.

35. "Variable Metric Minimization," *SHARE Routine No. 1117, AN 2013, A3*, SHARE, Inc., Suite 750, 25 Broadway, New York, 1961.

36. WARTEN, R. M., "Automatic Step Size Control for Runge-Kutta Integration," *IBM J. Research Development*, 7, Oct. 1963, pp. 340–341.

37. ZADEH, L., and DESOER, *Linear System Theory*, McGraw-Hill, New York, 1960.

38. ZOBRIST, GEORGE W., et al., *Network Computer Analysis*, Boston Technical Publishers, Inc., Cambridge, Mass., 1969.

10
SYSCAP

JOHN D. BASTIAN
ELLMAR D. JOHNSON
Rockwell International
Anaheim, California

10.1 INTRODUCTION

SYSCAP (System of Circuit Analysis Programs) is a system of electronic circuit analysis programs (dc, ac, and transient) designed to handle a wide class of electronic hardware design and evaluation problems. The program has proved to be a "most cost-effective approach to analyzing the Transient Radiation Effects of Electronic circuits (TREE)" [1]. Dependable operation and efficiency of the program have been demonstrated [2, 3, 4] on thousands of circuits. The types of problems handled include transient response and stress determination, radiation effects analysis, Fourier analysis, dc variations, failure mode and effects, power supply sequencing, parameter sensitivities, worst-case effects, statistical determination, and ac linear gain and phase relationships.

SYSCAP consists of a component data bank and three integrated programs: DICAP (Direct Current Circuit Analysis Program), ALCAP (AC Linear Circuit Analysis Program), and TRACAP (Transient Circuit Analysis Program). DICAP is a dc or static analysis program. ALCAP is a linear ac or frequency domain analysis program. TRACAP is a time history or time domain analysis program. All three programs have individual dc solving capability and a common input language and can access the component data bank (Fig. 10.1).

The objective of SYSCAP is to provide computerized analysis results which can be easily evaluated to answer design, reliability, logistics, and environmental effects questions. The program is oriented toward the design engineer, requiring an absolute minimum of computer programming knowledge. Device models are built in and data are supplied from a common data bank to maintain uniformity and reduce input coding errors. The built-in models also allow the designer to concentrate his effort on the overall circuit

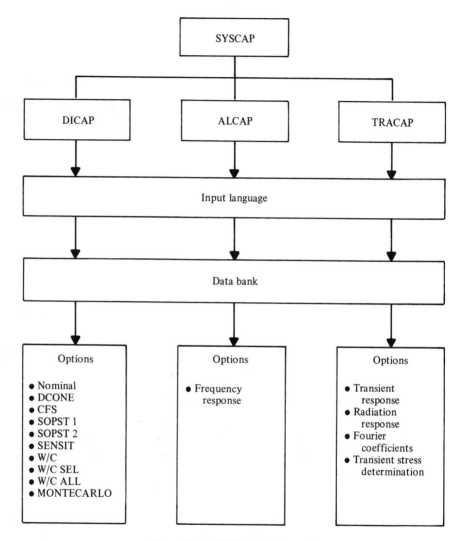

Figure 10.1 SYSCAP Configuration.

design and evaluation, rather than concerning himself with the degree of device modeling. The output data are structured in report format, thereby reducing or eliminating typing.

Historical Background

In 1964, the Autonetics Division of North American Rockwell (name changed to Rockwell International Corporation on February 1, 1973) developed TRAC [5, 6] (Transient Radiation Analysis by Computer Program). The TRAC program was the first transient circuit analysis program to utilize implicit integration and a modified Newton-Raphson iteration technique. TRAC thereby achieved a computational efficiency which was an order of magnitude greater than other programs available at that time.

The TRAC program provided automatic generation and solution of the linear and nonlinear simultaneous equations which characterize the mathematical models used to predict electronic circuit response. It was primarily used to predict the radiation response of electronic circuits. The TRAC program, or derivatives thereof, is one of the most widely used transient analysis programs in the United States, primarily because of its low running cost and availability from the Harry Diamond Laboratories.

The successful utilization of TRAC and the necessity to perform other than radiation evaluations on production or large-scale bases [7, 8] resulted in the company-funded development of SYSCAP. SYSCAP was initially developed in 1968–1969 at Rockwell International. Over 20 engineers, programmers, and management personnel (Appendix 10.1) contributed to the overall design and development. The program was licensed to Control Data Corporation in 1969 for domestic data service use and in 1971 for international data service use. In 1975 a major update was made to the program and it was renamed SYSCAP II.

Through 1974, Rockwell International Corporation alone has analyzed on TRAC and SYSCAP over 2000 circuits on such contracts as Minuteman, B-1, Space Shuttle, MARK II, SINS, and SRAM and on thin-film and MOS technology projects.

Mathematical Approach

The efficient transient analysis of modern electronic circuits which contain large numbers of semiconductor devices and where parasitic effects lead to time constants in the subnanosecond range requires the use of numerical methods which give stable and accurate solutions. To achieve economy in terms of computer solution time, the ability to adjust the time increment in the integration routine to the dominant solutions and to use a minimum amount of core storage are highly desirable.

A nodal analysis with an implicit integration approach, coupled with efficient sparse matrix and dependable nonlinear iteration techniques, enables SYSCAP to effectively handle these increasingly common problem cases. The unconditional asymptotic stability of the integration method allows the user to specify integration time increments consistent with the accuracy he requires. It also allows the program to calculate initial or steady-state conditions by using a relatively large time increment. Both the implicit character of the integration method and the nonlinearity of the circuit elements necessitate solving the nonlinear nodal admittance equations iteratively at each step. A suitably tailored Newton-Raphson method is used, and because the admittance matrix is sparse, only nonzero elements are stored in a compact, one-dimensional array. The structure is particularly well suited to handling circuits containing large numbers of semiconductor devices as only their external nodes and pseudonodes contribute terms in the admittance matrix, irrespective of the number of internal nodes or branches in their models.

General Capability

DICAP

The DICAP program provides the capability of performing any one of seven distinct types of dc analysis on a given circuit with only minor changes in the data deck. The seven types of analyses that can be performed are as follows:

Nominal. The nominal option sets all parameters at their typical values and solves the circuit matrix. The nominal option is generally used to eliminate user input data errors and to verify circuit operation.

One-at-a-Time Parameter Variation (DCONE). The DCONE analysis option is used to vary single parameters while maintaining the remainder of the parameters at their typical values. All parameters except for those which provide the shape of the transistor gain curve may be varied, such as

Transistors. Leakage and bulk resistance, breakdown and gain.

Diodes. Leakage and bulk resistance, and breakdown.

Independent voltage sources. Magnitude and series resistance.

Independent current sources. Magnitude and parallel resistance.

Resistors. Resistance value.

Dependent sources. Multiplying factor and source resistance.

The parameter being varied may be increased or decreased in value using either a constant step size or a set of discrete values.

Circuit Failure Simulation (CFS). The CFS analysis option simulates single catastrophic failures of circuit components and records the resultant overstresses (an overstress being a calculated stress that is greater than the user-entered stress limit). The following types of catastrophic part failure modes are simulated:

Transistors. Base-emitter (B-E) open and base-collector (B-C) open; B-E short and B-C short; B-E short and B-C open; B-E open and B-C short.*

Diodes. Open and short.

Voltage sources. Open and short.

Resistors. Open and short.

Nodes. Shorted to another node or ground.

*If QMIN is entered as a CFS Special Control Entry, then only two failures will be simulated, i.e., all junctions opened or shorted.

FET's. Gate open ($K_1 = 0$); gate-drain-source-shorted.

Operational amplifiers. Zero gain; inverted input open; non-inverted input open.

The overstresses that occur as a result of single catastrophic part failures are grouped into three classes. The classes are arbitrary and are defined as follows:

Class I. 0–50% overstress (0–150% stress). No permanent effect should arise from this level of overstress.

Class II. 50–400% overstress (150–500% stress). The overstressed component will suffer degraded reliability but will probably still perform its function.

Class III. Greater than 400% overstress (greater than 500% stress). The overstressed component will probably cease functioning in a normal manner.

Class III secondary failures (i.e., changes in parameter values) are not simulated by the program. The CFS analysis option produces tables and graphs that display the overstress information.

Simulation of Power Supply Turn-On (SOPSTO). The SOPSTO analysis option simulates all the possible worst-case power supply turn-on configurations. The program accomplishes this by taking all possible combinations of zero volts and nominal power supply voltage for all the power supplies (limited to a maximum of seven supplies). The overstresses that occur as the different configurations are simulated are tabulated at the end of the analysis. There are two distinct types of SOPSTO simulation:

SOPSTO 1. Simulates the off supplies with zero volts and their nominal supply impedance.

SOPSTO 2. Simulates the off supplies with zero volts and a 10-megohm supply impedance.

Sensitivity (SENSIT). The SENSIT analysis option computes the sensitivity of all circuit output solutions (node voltages and auxiliary equations), with respect to each circuit parameter. Sensitivity is defined as equal to the percentage change of any computed output solution divided by the percentage change of a circuit parameter:

$$\text{sensitivity} = \frac{\Delta V / VN}{\Delta P / PN}$$

where

ΔV = change in some computed output solution

VN = nominal value of the computed output solution

ΔP = change in the parameter

PN = nominal parameter value

A parameter change of 0.1% is used in the sensitivity computation. The parameters are changed one at a time, and the resulting changes in circuit output solutions are noted and used to calculate the sensitivities. Any sensitivity less than SENTOL on the control card (default = 0.05) is not recorded. It should be noted that the small parameter change allows an assumption of linearity and for this reason the normal method of iterating on the nonlinear side conditions is not employed. The SENSIT analysis option prints out a table for each circuit function, listing those parameters that resulted in sensitivities greater than SENTOL for the output solution being tabulated.

Worst Case (W/C). The W/C analysis option performs a worst-case maximum and minimum computation for those nodes and auxiliary equations indicated by the user. This is accomplished by setting each parameter, having a sensitivity-tolerance product greater than SENTOL on the control card (default = 0.05%), to its appropriate maximum or minimum value. The sensitivity-tolerance product is the percentage change in the output solution for a full tolerance change in the parameter being varied. This approach assumes linearity of the solution-parameter relationship. The sign of the sensitivity-tolerance product determines whether the parameter is set to its positive or negative tolerance extreme. To obtain a worst-case maximum output solution value, if the sensitivity is positive, the positive tolerance extreme is used; if the sensitivity is negative, the negative tolerance extreme is used. The circuit matrix is then solved, including iteration on the nonlinearities, to determine the maximum and minimum values for the node or auxiliary equation being examined.

The W/C analysis option has a parameter-matching feature that overrides the normal W/C parameter setting. Parameter matching admits two parameters, specified on a parameter-matching card, to be directly or inversely matched. *For matching to be initiated, at least one of the matched parameters must be set to its worst-case setting*. Matching of more than two parameters can be accomplished by having each additional parameter matched to a previously matched parameter. The parameter with the highest sensitivity-tolerance product is the primary parameter whose value is determined by the W/C program. The values of the matched parameters are obtained from the parameter-matching data cards.

The worst-case program also tests to see if any of the semiconductor devices changed state as a result of a given worst-case solution. The states of the semiconductor devices are compared to the nominal states, and any differences are printed. A change of state for some worst-case condition may indicate a design problem.

At the completion of the worst-case analysis, three summary tables are printed. The first of these tables shows the maximum and minimum values that occurred during the analysis, for each node. The second table tabulates similar information for each auxiliary equation. The third table summarizes the overstresses that occurred during the worst-case analysis.

In some instances a new maximum or minimum for solution X will occur when solution Y is being maximized or minimized. This may occur because only those sensitivity-tolerance products that exceed SENTOL are used in determining worst case. If the parameters used to obtain a worst-case value for solution X are a subset of those used for solution Y, a new maximum or minimum for solution X may occur. The new worst-case value is usually very close to the old worst-case summary tables. Additionally, a message is printed by the program when a new maximum or minimum occurs.

There are three options in performing a worst-case analysis. The first option is exercised if W/C is entered on the control card. This option will find the worst-case maximum and minimum for each node and the maximum only for each auxiliary equation. The second option is triggered by entering the key word W/C ALL on the DICAP control card. This option forces a maximum and minimum computation for every node and every auxiliary answer. The third option, selected worst case, is called by entering the key word W/C SEL on the DICAP control card. This option provides a maximum and/or minimum computation on nodes and/or auxiliary equations entered by the user.

Statistical. The statistical option provides a method to simulate the construction and testing of a large number of copies of a circuit, with components selected at random from a representative sample of actual available parts. An empirical study would involve the actual construction and measurements of many copies of the circuit and would be very time-consuming and expensive.

The statistical option chooses each parameter value randomly according to its frequency distribution and correlations with other parameters, solves the circuit matrix, and stores the values of the output solutions. The process is repeated as many times as desired to simulate any sample size. Finally, a statistical study is made on each set of values stored for each output solution and the results are tabulated.

To reduce memory storage and execution time, only one type of four possible frequency distributions may be specified:

UNIFORM. Uniform distribution between minimum and maximum tolerance limits.

NORMAL. Normal or skew-normal distribution between minimum and maximum tolerance limits with average as the mean value.

MIN-MAX. Random choice of either minimum or maximum tolerance limit.

TABLES. User-created, cumulative frequency distribution table.

In addition to a choice of parameter frequency distributions, the user may elect to have up to 50 pairs of correlated parameters. The program will then choose the value for the second parameter according to the value chosen randomly for the first parameter (Fig. 10.2).

The output results consists of the maximum value, minimum value, average value, standard deviation, and three-sigma limits for each output solution over the entire sample set. After the statistical output, the program checks for any possible overstresses in the statistical results. Histograms are available for up to 25 output solutions.

The transistor parameters are correlated within four groups: leakage, gain, bulk, and junction voltages. The typical data are the mean, and the minimum and maximum values are the -3σ and $+3\sigma$ values, respectively. The leakage group includes REL and RCL. A random value is first calculated

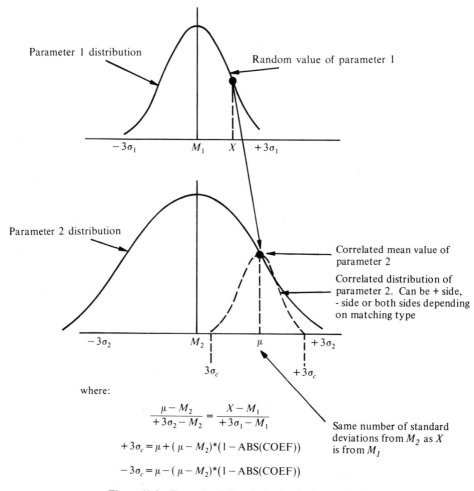

Figure 10.2 Example of Correlation in the Monte Carlo Program.

for REL. The random value for RCL is constrained to lie on the same side of the mean as REL. Correspondingly, beta normal is randomly determined, and beta inverse is constrained to lie on the same side of the mean. For the bulk group, RB is randomly determined, and RC and RE are constrained to lie on the same side of the mean as RB. Within the junction voltage group, the correlation becomes more complex. A random number generator is used to provide the number of standard deviations that the log of IES is from the log of the mean. That same random number is also used to calculate MES and MCS. Based on the nominal operating point and min-max transistor data, VCE minimum and VCE maximum are calculated. A random VCE is selected between these extremes, and the corresponding ICS is calculated (since MCS is already available). If ICS exceeds the min-max limits, then the limit value is entered. Breakdown parameters are not included in the correlated groups for the Monte Carlo option as they were in the worst-case option.

ALCAP

The ALCAP program is used to simulate the frequency response (amplitude, phase, and Nichols) of a linearized electronic circuit. A dc solution is first performed in ALCAP to determine the dc operating points for all active circuit elements. All nonlinear models, such as transistors, diodes, zeners, operational amplifiers, and FETs, are automatically linearized. Transistor current gain is selected from the variable beta curve at the nominal dc operating point. The circuit is then solved using the values for current gain previously selected in conjunction with the ac parameters, such as diffusion and depletion capacitance for a transistor. Frequency is iterated, and the circuit is solved for each frequency to obtain the frequency response. Complex arithmetic is used.

TRACAP

The TRACAP program is used to simulate the time response of electronic circuits to various forcing stimuli. These forcing functions can be those encountered in a normal environment, such as the circuit input signal and power supply voltages or additionally may include forcing functions due to a radiation environment.

The TRACAP program will compute initial conditions or allow the user to enter a set of initial conditions for the circuit.

For a radiation environment, TRACAP models the primary photocurrents induced in all *PN* junctions (transistors, diodes, zeners, and FETs) as a function of the incident ionizing radiation. Included in this effect are the generated contributions in both the depletion and diffusion regions of the junction. The depletion region contributions are voltage-dependent and occur instantaneously in relation to the ionizing radiation. The diffusion region contribution is time-dependent. Secondary effects are automatically included through beta multiplication.

Engineering-Oriented Features

The following is a partial list of engineering oriented features in order to provide the user an indication of the ease and versatility of SYSCAP:

Input Convenience. Data entry is free form and can be entered anywhere on the input program card (columns 1–80) with any desired spacing. Component topology cards and solution control cards are not ordered within their individual sections. Controls on the initial control card may be defaulted and are not ordered. Symbolic designators such as B for base, C for collector, etc., are used for entering polarized devices so that specific order does not have to be memorized. Finally, the ordering of auxiliary equations, when requested, is not critical.

Engineering Suffixes. The following common engineering suffixes in addition to scientific notation can be used for any data entries:

T:	10^{12}	K:	10^3	N:	10^{-9}
G:	10^9	M:	10^{-3}	P:	10^{-12}
MEG:	10^6	U:	10^{-6}	E:	NOTATION

Data Bank. A data bank is available for transistors and diodes. Typical minimum and maximum values are included for dc, ac, and transient simulations. Generic device types are also available. Any part of the data called from the data bank may be modified by the user for his individual run. The data bank is updated periodically.

Built-in Model. A complete range of device models are available. They include

Resistor, capacitors, and inductors

Independent voltage and current sources

Dependent voltage and current sources

Bipolar transistors

Diodes

Zeners

Field effect transistors

Transformers (linear and saturable)

Operational amplifiers

Default Options. Default options are available when the user omits an entry in order to avoid a potentially unnecessary program abort. Examples of these would include TEMP = 25°C; transistor type is *NPN*; the series resistance for a capacitor, or inductor is 0.1 ohm; etc.

Automatic Stress Equations. Device stress equations can be obtained automatically by entering the appropriate designator after the node connections. They include

Voltage and current sources—power and current

Transistors—power, IB, IC, and IE

Capacitors and inductors—current and voltage

Resistors, diodes, and FETs—power and current

Transformers—B, H data and winding current

Stress Checking. Automatic device stress checking and summary tables are available.

Automatic Device State Determination. At the start of every run the states of all active elements are indicated for the initial conditions, e.g., active, saturated, breakdown, etc. Also, any change of state from nominal is indicated in the DICAP worst-case option.

Output. Automatically provided are all nodal voltages and any auxiliary equations which were either written or requested. This can be reduced by selecting certain outputs to be tabulated or plotted. The output is plotted in either printer plots ($8\frac{1}{2} \times 11$ or continuous) or SC4020 plots. A variable or function of a variable may be plotted against time or frequency, another variable, a function of another variable, a component failure, or a dc parameter change. Equal increment transient outputs are also available. Bar graphs are provided for component overstresses. Summary tables are available for overstresses, secondary stresses, min-max values, and integration results.

Functions. The FUNC option consists of up to nine user-created FUNCTION subprograms which are included in the subroutine section. These subprograms allow the user the capability of FORTRAN statements to calculate a variable which is readily available for use in the main program. This permits component values to be set equal to a variable.

User Subroutines. User subroutines are available for special controls, special model convergence and acceptance criteria, and user-written equations. These subroutines permit a greater degree of control over the program operations but normally are not used.

Rerun. Multiple TRACAP and ALCAP reruns can be performed based on a single basic circuit. The user supplies only the changes that differ from the basic circuit.

Termination. A transient run can be terminated contingent on the behavior of specific network quantities. Also, runs can be restarted from the previous ending time.

Special Transient Plotting. Special transient plots are available with
1. Start and end time specified.
2. Threshold specified.
3. Start on a threshold and end on a specific time.

Nonlinear Switching. Nonlinear incremental switching as a function of another variable is provided.

MACRO Models. MACRO models allow the user to code the configuration once and then use the same configuration numerous times in the analysis. The required node renumbering and additional components generated are automatically accounted for by the program.

SYSCAP Application Areas

Stress Analysis. DICAP and TRACAP can be used together to perform an analysis of the electrical stresses on the components of a circuit. DICAP can be used to obtain dc stresses for nominal and worst-case conditions. The DICAP statistical option can be used to determine the distribution of component stresses. TRACAP can provide a simulation of the component stresses from transient and/or duty cycle circuit operation. The SOPSTO feature of DICAP can also be used to detect any potential overstress arising from multiple power supplies increasing from zero volts to rated voltage in a disproportionate manner.

Functional Analysis. DICAP can be used to determine circuit nominal, worst-case, and statistical functional steady-state performance. For circuits where the transient performance is the important function, the TRACAP program can provide nominal simulations. When ac linear gain, phase, and stability are the important functions, the ALCAP program is used. The user can obtain worst-case dynamic simulations by setting specific parameters to their limits. Selection of these parameters and the direction of their limits is made by the user on the basis of sensitivities calculated in the DICAP program and circuit engineering judgment.

Radiation Analysis. An ionizing radiation analysis determines the effects of ionizing radiation pulses on circuit functional operation and component stresses. TRACAP is used since the effects of ionizing radiation are time-dependent. The relative timing of the ionizing radiation pulse and the normal electrical signals are usually investigated. Neutron radiation is known to degrade semiconductor device properties such as forward current gain. When these "degraded" device parameters are used in the component part models, the extent of circuit degradation can be determined on either DICAP, ALCAP, or TRACAP. An EMP analysis determines the signal amplitude and frequency at which either the circuit ceases to perform its required system function or a component burn-out occurs. The TRACAP program is used for such an evaluation, with stimuli applied at one or multi ports. However, ALCAP can be used to determine the most critical frequency for the circuit.

Failure Mode and Effects Analysis. Failure mode and effects analysis are performed on the CFS and DCONE options of DICAP. A catalog of potential component part failures (opens or shorts) and the related secondary overstresses, secondary failures, and circuit failure symptoms which would result are obtained. This catalog is useful in fault isolation, equipment repair, and system level failure mode analysis. Selected use of TRACAP or ALCAP may be needed to form a complete catalog on some circuits (e.g., oscillators).

MOS/SOS MSI Analysis. Large-scale integrated circuits containing MOS and SOS field effect transistors are analyzed using TRACAP. Most design approaches involve the establishment of logic stage ground rules or other characterizations of standard circuit functions on an LSI chip (e.g., drivers, level shifters, time delay circuits, etc.). These basic circuits are to provide a simulation which takes the place of ordinary breadboard techniques (since breadboarding is difficult because of parasitic effects and loading effects of monitoring equipment). These models, once constructed, can then be used to examine the effects of variations in threshold voltage and other process parameters. Significantly increased yields are the results of these analyses. ALCAP is employed on linear circuits to determine gain and phase relationships. The DICAP program can be used to determine various dc effects, i.e., effects of W/C threshold voltage.

Advanced Design Techniques. With adequate turnaround time, each of the above types of analyses can be performed during the course of circuit design and development. Use of SYSCAP as an analytic breadboard provides a means of exploring untried design concepts and may eliminate test equipment interaction or deficiencies. Also such parameters as f_T can only be evaluated by an analytic breadboard.

SYSCAP Availability

The SYSCAP program is maintained, supported, and updated by the Rockwell International Corporation, Anaheim, California. If may be used in a batch or interactive mode at any Control Data Corporation Data Center in the United States (as shown in Fig. 10.3), Australia, Austria, Belgium, Canada, Denmark, Finland, Germany, Mexico, The Netherlands, Norway, and Sweden. If a SYSCAP user does not have walk-in access to a CDC data center, he may install in his own facility a terminal linked to CYBERNET® [9] through telephone lines.

The interactive version of SYSCAP can be accessed from any of the commonly available low-cost conversational terminals (Teletype ASR33, TI Silent 700, CDC 713 CRT, etc.) The program provides a quick response capability via the KRONOS [10] on-line system. Thus, data from a schematic

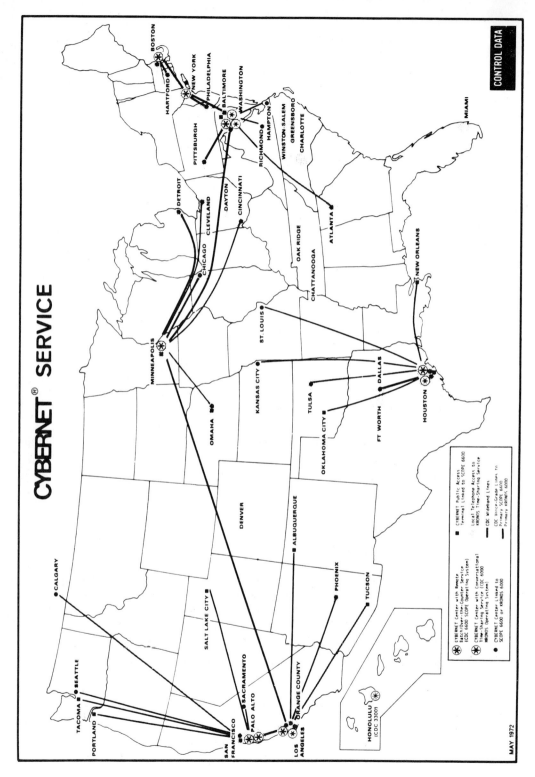

Figure 10.3 SYSCAP Domestic Availability.

can be entered at a keyboard to produce an immediate output. A complete set of documentation for SYSCAP is available [11] from Control Data Corporation.

10.2 PROGRAM STRUCTURE

SYSCAP consists of three separate programs: DICAP, ALCAP, and TRACAP. Each of these programs has a multioverlay structure, but all use a common data bank.

DICAP

DICAP is divided into overlays on a functional basis, as illustrated in Fig. 10.4. The main overlay (0, 0) is resident in core throughout execution of the program. All other overlays are transferred into the computer as required. Overlay (0, 0) contains the DICAP main program, which controls the overlay execution sequence as specified by the DICAP control cards. The data required by the program are stored dynamically in blank common for maximum use of available space.

Overlay (3, 0) processes the input data. Input processing is completed with reading End of File or a special card.

The nominal solution is then performed by overlay (0, 0). The rest of the solutions depend on the type of analyses. The analyses for DCONE, CFS, and statistical are performed by overlays (8, 0), (7, 0), and (2, 0), respectively. Overlay (1, 0) plots the results of the DCONE, CFS, and statistical analyses.

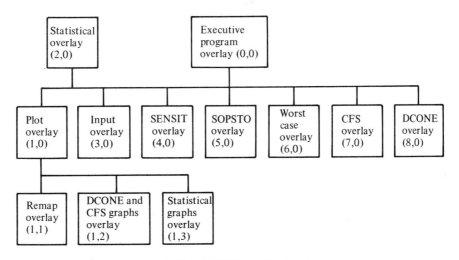

Figure 10.4 DICAP Overlay Structure.

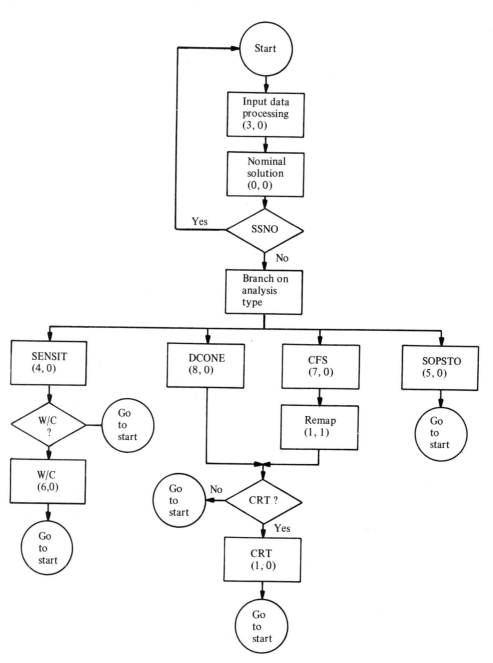

Figure 10.5 DICAP General Flow Chart.

The SENSIT analysis is performed in overlay (4, 0). If a worst-case analysis is specified, the sensitivities are calculated in the SENSIT overlay and stored for use in overlay (6, 0) for the worst-case analysis.

The SENSIT analysis assumes linear models around the operating point; therefore, convergence checking and nonlinear iteration are not performed. A general flow chart for DICAP is shown in Fig. 10.5. This flow chart illustrates the basic computation sequence.

ALCAP

ALCAP is divided into overlays on a functional basis, as illustrated in Fig. 10.6. Overlay (0, 0) contains the ALCAP main program, which controls the overlay execution sequence. Also present are various subroutines which, because of their function, must remain in core at all times.

Overlay (1, 0) processes the input data. Input processing is completed upon reading End of File or a special card.

Overlay (2, 0) controls the solution process by calling overlay (2, 1) for dc solution and then calling overlay (2, 2) to perform the ac solution. Overlay (2, 1) performs matrix stuffing, solving, convergence checking, and nonlinear iteration for the dc solution.

Overlay (2, 2), using the small-signal models obtained from the dc solution, performs stuffing of the complex matrix, solves the matrix, and stores

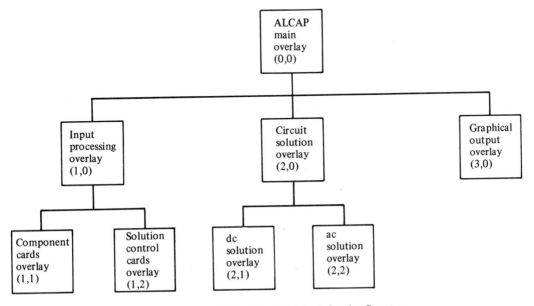

Figure 10.6 ALCAP Overlay Structure.

and prints the results as a function of frequency as specified by the solution control cards.

Overlay (3, 0) processes the data generated by overlay (2, 2) and produces graphical output in accordance with user-specified plot options.

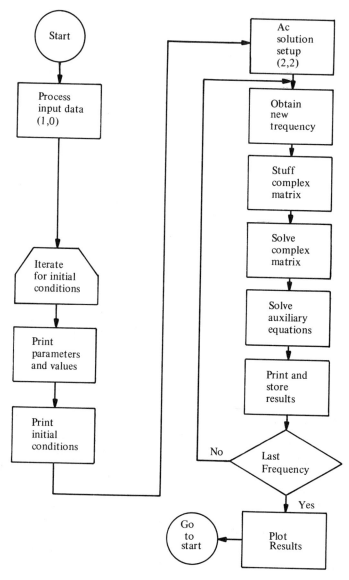

Figure 10.7 ALCAP General Flow Chart.

A general flow chart for ALCAP is shown in Fig. 10.7. This flow chart illustrates the basic computational sequence.

TRACAP

TRACAP is divided into overlays on a functional basis, as illustrated in Fig. 10.8. Overlay (0, 0) contains the TRACAP main program, which controls the overlay execution sequence. Also present are various subroutines which, because of their functions, must remain in core at all times.

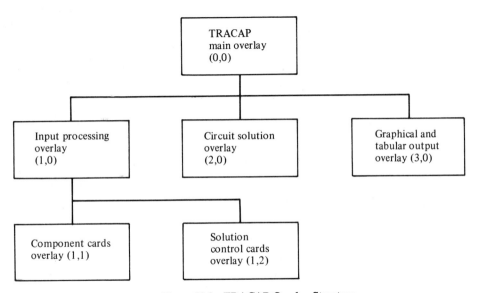

Figure 10.8 TRACAP Overlay Structure.

Overlay (1, 0) processes the input data. Input processing is completed upon reading End of File or a special card.

Overlay (2, 0) performs matrix stuffing, solving, convergence checking, and nonlinear iteration. In accordance with TRACAP control card options and solution control card specifications, overlay (2, 0) computes initial conditions and/or a transient solution.

Overlay (3, 0) processes the data generated by overlay (2, 0) and produces graphical and tabular output in accordance with user-specified plot and print options.

A general flow chart for TRACAP is shown in Fig. 10.9.

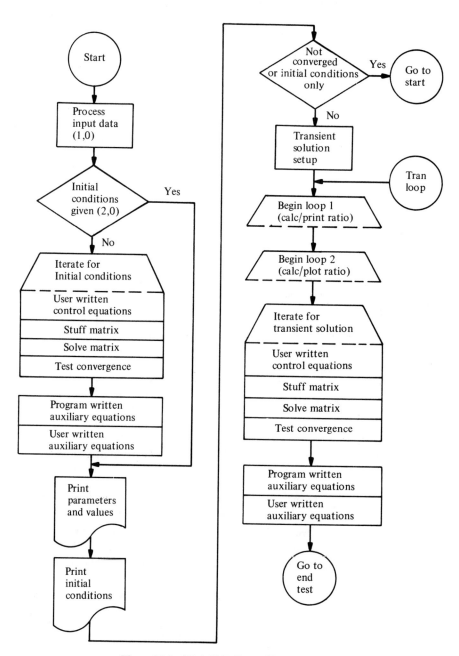

Figure 10.9 TRACAP General Flow Chart.

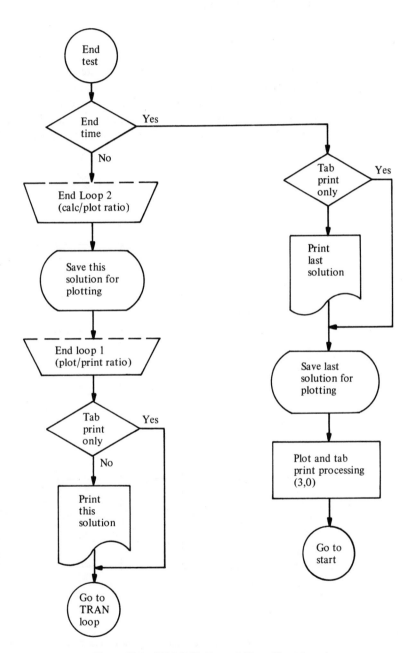

Figure 10.9 TRACAP General Flow Chart (*cont.*).

10.3 SYSCAP MODELS

Diode-Zener

The diode model in SYSCAP is an extension of the large-signal "ideal" diode. The ideal diode is modified by including a proportionality coefficient in the exponential term, a series bulk resistance, and zener (avalanche) effects. In the time and frequency domain, diffusion and depletion capacitance are added. To increase the frequency range of the diode model, deep storage

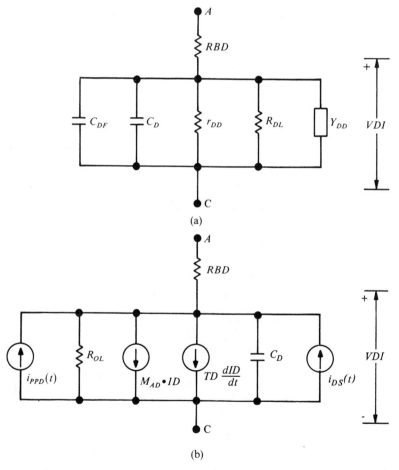

Figure 10.10 (a) Small Signal Diode Model; (b) Large Signal Diode Model.

effects [12] are included. For radiation applications short-term gamma effects can be simulated as described in Appendix 10.2.

The diode model is shown in Fig. 10.10. The parameters defining the model are listed in Table 10.1. These parameters are listed in the same order as used in the program.

Table 10.1 Diode/(Zener) Parameters

No.	Parameters	Description
1	RBD	Diode series bulk resistance
2	IDS	Diode reverse saturation current
3	MD	Diode proportionality constant
		$ID = IDS\left(\exp\dfrac{VDI}{MD\cdot\theta} - 1\right)$ = ideal diode current
		$\theta = kT/q$ ($= 0.026$ volts at 300°K)
		k = Boltzmann's constant (1.380×10^{-23} joule/°K)
		q = charge on electron (1.602×10^{-19} coulomb)
		T = temperature in degrees kelvin (T°C + 273)
		$r_{DD} = MD\cdot\theta/ID$ = dynamic resistance
4	RDL	Diode leakage resistance
5	BVR (VZ)	Breakdown voltage for diode (Zener voltage)
6	IBVR (IZ)	Current at breakdown voltage (Zener current)
7	RVBR (RZ)	Impedance at breakdown voltage (Zener voltage)
	ac and Transient Analysis Parameters	
8	CD	Diode depletion junction capacitance at VDI = 0
9	VDBI	Diode intrinsic built-in voltage
10	XDC	Diode depletion capacitance exponent
		$C_D = \dfrac{CO}{\left(1 - \dfrac{VDI}{VDBI}\right)^{XDC}}$ = depletion capacitance
11	TD	Storage time constant
		$C_{DF} = \dfrac{TD}{r_{DD}}$ = diode diffusion capacitance
12	IPPD	Diode steady-state primary photocurrent
13	DEPD	Depletion portion of ionization current
		$i_{PPD}(t) = f(IPPD, DEPD, \text{ionization pulse})$
		= diode primary photocurrent
14	T5D	Deep storage time constant
15	AD	Deep storage transport factor
		$Y_{DD} = \dfrac{AD}{1 + j\omega\cdot T5D} * \dfrac{1}{r_{DD}}$ = deep storage small-signal impedance
		$i_{DS}(t) = AD\cdot ID - T5D\dfrac{d(ids(t))}{dt}$
		= deep storage current

Transistor

The transistor model in SYSCAP (Fig. 10.11) is an extension of the large-signal Ebers and Moll transistor model. The two diodes are modified by including a proportionality coefficient in the exponential term and by breakdown effects. Series bulk resistance is included in the base, collector, and emitter. In the time and frequency domains, each diode includes diffusion and depletion capacitance. To increase the frequency range of the transistor model, deep storage effects [12] are included.

For radiation applications, short-term gamma effects can be simulated as described in Appendix 10.2. The parameters defining the model are listed in Table 10.2. These parameters are listed in the same order as used in the program. The parameters with a T after their number are used only in the ac and transient programs.

Table 10.2 *Transistor Parameters*

No.	Parameters	Description
1.	BN	Maximum beta normal
2	IP	Collector current for BN
3	BL	Beta at a current below IP
4	IL	Collector current for BL
5	BH	Beta at a current higher than IP
6	IH	Collector current for BH
7	BI	Maximum beta inverse
		α_N = normal current gain = $\frac{B}{1+B}$
		$B = (K_1 + K_2 \log IE) \frac{BN}{1 + K_3 \exp(K_4 * IE)}$
		IE = emitter current
		K_1, K_2, K_3, K_4 are determined from the above input data
		The inverse beta is calculated in a similar manner
8	RB	Bulk resistance base
9	RC	Bulk resistance collector
10	RE	Bulk resistance emitter
11	ICS	Collector diode reverse saturation current
12	MCS	Collector diode proportionality constant
		$IDC = ICS\left(\exp\frac{VDC}{MCS \cdot \theta} - 1\right)$ = collector diode current
		$\theta = kT/q$ (≈ 0.026 volts at 300°K)
		k = Boltzmann's constant (1.380×10^{-23} joule/°K)
		q = charge electron (1.602×10^{-19} coulomb)
		T = temperature in degrees kelvin (T°C + 273)
		$r_{DC} = MCS \cdot \theta/IDC$ = collector dynamic resistance
13	RCL	Collector diode leakage resistance
14	VCBO	Breakdown voltage collector diode
15	ICB	Current at breakdown
16	RCB	Impedance at breakdown voltage

Table 10.2—Cont.

No.	Parameters	Description
17T	CCB	Collector diode depletion junction capacitance at $V_{DC} = 0$
18T	VCBI	Collector diode intrinsic built-in voltage
19T	XCC	Collector diode capacitance exponent

$$C_C = \frac{CCB}{\left(1 - \dfrac{V_{DC}}{VCBI}\right)^{XCC}}$$

20T	TC	Collector diode storage time constant

$$C_{CF} = \frac{TC}{r_{DC}} = \text{collector diode diffusion capacitance}$$

21T	IPPC	Collector diode steady-state primary photocurrent
22T	DEPC	Depletion portion of ionization current

$i_{PPC}(t) = f(IPPC, DEPC, \text{ionization pulse})$
$\qquad = $ collector diode primary photocurrent

23T	T5C	Collector diode deep storage time constant
24T	AC	Collector diode deep storage transport factor

$$Y_{CD} = \frac{AC}{1 + j\omega T5C} \cdot \frac{1}{r_{DC}} = \text{deep storage small-signal impedance}$$

$$i_{CS}(t) = AC \cdot IDC - T5C \cdot \frac{di_{CS}(t)}{dt}$$
$$\qquad = \text{deep storage current}$$

25	IES	Emitter diode reverse saturation current
26	MES	Emitter diode proportionality constant

$$IDE = IES\left(\exp\frac{VDE}{MES \cdot \theta} - 1\right) = \text{emitter diode current}$$

$$r_{DE} = \frac{MES \cdot \theta}{IDE} = \text{emitter dynamic resistance}$$

27	REL	Emitter diode leakage resistance
28	VEBO	Breakdown voltage emitter diode
29	IEB	Current at breakdown
30	REB	Impedance at breakdown voltage
31T	CEB	Emitter diode depletion junction capacitance at $V_{DE} = 0$
32T	VEBI	Emitter diode intrinsic built-in voltage
33T	XEC	Emitter diode capacitance exponent

$$C_E = \frac{CEB}{\left(1 - \dfrac{V_{DE}}{VEBI}\right)^{XEC}}$$

34T	TE	Emitter diode storage time constant

$$C_{EF} = \frac{TE}{r_{DE}} = \text{emitter diode diffusion capacitance}$$

35T	IPPE	Emitter diode steady-state primary photocurrent
36T	DEPE	Depletion portion of ionization current

$i_{ppe}(t) = f(IPPE, DEPE, \text{ionization pulse})$
$\qquad = $ emitter diode primary photocurrent

37T	T5E	Emitter diode deep storage time constant
38T	AE	Emitter diode deep storage transport factor

$$Y_{ED} = \frac{AE}{1 + j\omega T5E} \cdot \frac{1}{r_{DE}} = \text{deep storage small-signal impedance}$$

$$i_{ES}(t) = AE \cdot IDE = T5E \cdot \frac{di_{ES}(t)}{dt} = \text{deep storage current}$$

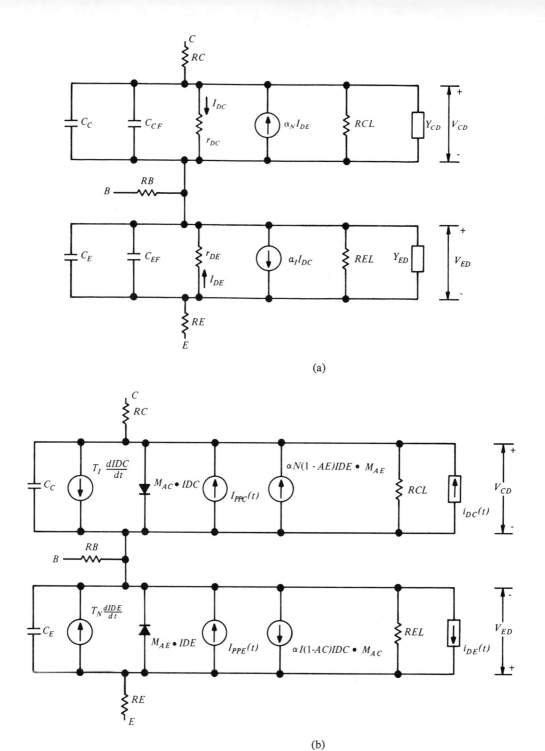

Figure 10.11 (a) Small Signal *PNP* Transistor Mode; (b) Large Signal *PNP* Transistor Mode.

Transformer

The transformer model in SYSCAP uses a phenomenological magnetic core model [13]. The model accounts for major and minor B-H loops, including initial conditions. This model also includes a single parameter TC, which is used to account for core dynamics. Provision has been made for the introduction of an airgap, ag., which permits analyses of circuits using gapped cores. Each coil has a series impedance but does not include leakage inductance. Table 10.3 presents the core data used to define the core model.

Table 10.3 *Transformer Core Data*

No.	Parameters	Description
1	CA	Core area (cm²)
2	CL	Core length (cm)
3	BS	
4	HS	
5	BM	
6	HM	
7	BR	
8	HC	
9	BK	
10	HK	
11	AG	Air gap (cm)
12	τ_C	Time constant (sec) controls growth of core as a function of frequency
13	IF	Initial flux density

Field Effect Transistor

The field effect transistor model in SYSCAP is a MOS/SOS enhancement or depletion mode field-effect N or P channel device. Figure 10.12 shows the entire model and the geometric location of the parameters. The built-in diodes which represents the *PN* junctions (Dsb, Ddb) are similar to the standard discrete diode models.

The parameters Cgs and Cgd are the result of the overlap of the gate metal and the source or drain diffusions. The parameters Rgs and Rgd represent the surface leakage between the metal patterns. The parameter Cgb exists in the cutoff region as a capacitance between the gate metal and the silicon substrate.

The dc characteristics of the channel are simulated by RD and RS (dynamic resistances) and IS and ID (variable current generators). The drain-

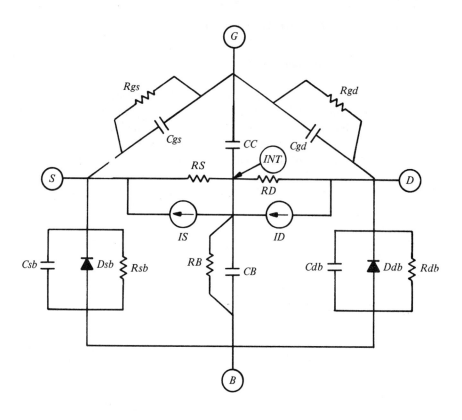

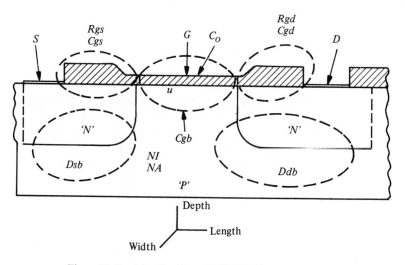

Figure 10.12 Identification of MOSNE Parameters.

source characteristics for an active device are determined primarily from the equations

$$RD = \frac{1}{8K1 \{VGB-VFB-\phi B-VDB-K_2 (VDB+\phi B)^{1/2}\}}$$

$$RS = \frac{1}{8K1 \{VGB-VFB-\phi B-VPB-K_2 (VPB+\phi B)^{1/2}\}}$$

$$ID = 4K1 \left\{(VGB-VFB-\phi B)\times(VDB-VPB) - \frac{VDB^2}{2}\right.$$
$$\left. + \frac{VPB^2}{2} - \frac{2}{3} K_2 [(VDB+\phi B)^{3/2} - (VPB+\phi B)^{3/2}]\right\}$$

$$IS = 4K1 \left\{(VGB-VFB-\phi B)\times(VPB-VSB) - \frac{VPB^2}{2}\right.$$
$$\left. + \frac{VSB^2}{2} - \frac{2}{3} K_2 [(VPB+\phi B)^{3/2} - (VSB+\phi B)^{3/2}]\right\}$$

$$K1 = \frac{W\mu C_o}{2L}$$

$$K2 = \frac{\sqrt{2\epsilon_R \epsilon_0 q N_A}}{C_o}$$

where

W = OPTOMASK width

μ = surface channel mobility

C_o = gate oxide capacitance per cm^2

L = OPTOMASK length

ϵ_R = relative dielectric constant

ϵ_0 = dielectric constant

q = charge of an electron

N_A = substrate impurity doping level per cm^3

IS = pseudonode to source current

ID = drain to pseudonode current

RS = dynamic source resistance

RD = dynamic drain resistance

VGB = voltage gate to body

VFB = flat band voltage (calculated below)

ϕB = 2∗ FERMI potential

VDB = voltage drain to body

VSB = voltage source to body

VPB = voltage pseudonode to body

These equations are modified to account for operation in the saturated region and the effect of body voltage.

The body effect is defined by the equation

$$\text{VTH} = \text{VFB} + \phi B + \text{VSB} + K_2 (\text{VSB} + \phi B)^{1/2}$$

where VTH = threshold voltage. This dc representation of the channel is shown in Fig. 10.13 for a fixed gate-source voltage. The model represents symmetrical operation in the normal and inverted modes. CB, RB, and CC are varied to obtain CG as defined in Fig. 10.14.

Two methods of entering data for the FET model are available in SYSCAP. The special model uses data (Table 10.4) obtained from the device characteristics. The process model uses data (Table 10.5) from the physical dimensions of the device.

Table 10.4 *Special Model Input Data*

No.	Parameter	Description
1	VG	Gate voltage
2	VPO (RON)*	Pinch-off voltage, VPO < VG − VTH. If fourth entry is negative (i.e., K_2 is entered), then entry is RON.
3.	VTH	Threshold voltage
4	DID ($-K_2$)*	Current at VPO. If zero or negative, then value of K_2 is entered directly.
5	RDsat	Dynamic resistance
6	θB	Two times Fermi potential (≈ 0.7)
7	Rgd	Leakage resistance of gate to drain
8	Rgs	Gate-to-source leakage resistance
9	Rdb	Drain-to-body leakage resistance
10	Rsb	Source-to-body leakage resistance
11	CG	Channel capacitance
12	Cgd	Gate-to-drain capacitance
13	Cgs	Gate-to-source capacitance
14	Cdbo	Depletion junction capacitance for drain-to-body substrate diode
15	Csbo	Depletion junction capacitance for source-to-body substrate diode

*Alternate method of entering data. Initiated by fourth entry being zero or negative.

Data for the MASK card is entered in either units of milliinches, microns, or centimeters as indicated after the key word 'MASK,' i.e., M, U, or C, respectively. A pictorial representation of the MASK data is shown in Fig. 10.15.

For either model a SUB card is required to define the ideal diode equations from source and drain to body and the corresponding leakage resistance

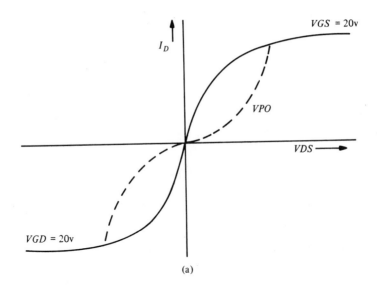

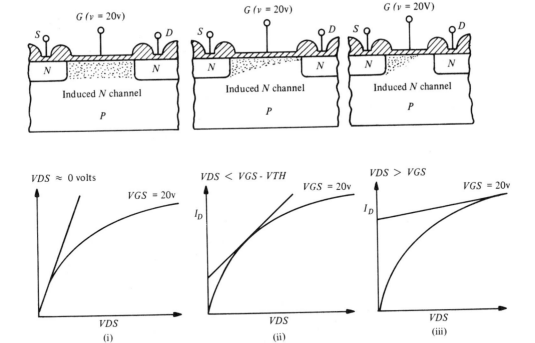

Figure 10.13 Model Operation and Relation to Simplified Physics of the Device: (a) Symmetrical Operation of the MOS/SOS Model; (b) Model Operation and Relation to Simplified Physics of the Device.

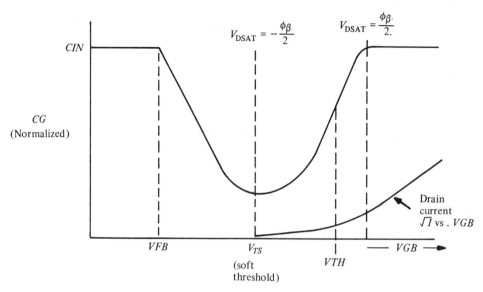

Figure 10.14 CG vs. VGB.

Table 10.5 *Process Model Input Data*

No.	Parameter	Description	Defaults
		MSI Data	
1	K'	Constant for unit effective gate	3U
2	VTH	Threshold voltage	3.6
3	C_O	Gate capacitance per cm^2	.0263U
4	N_A	Substrate impurity doping level per cm^3	.815E+15
5	VFB	Flat band voltage (enter as '0' and program will automatically caluate)	0
6	μV	Normalized mobility variation } See Fig. 10.14	1.0
7	$V\mu V$	Voltage at μV	25.
8	NVAR	0—Consider geometric effects 1—No geometric effects Used to account for shielding due to source and drain diffusion.	0
9	MULT	Multiplier of RD (SAT)/RON for unit device	10.
10	MOBIL	Surface channel mobility	228.
		Mask Data	
1	UNIT	OPTOMASK size of unit device	1.0
2	CENT	Distance between plotting centers	1.0
3	FW	Fringing factor for gate width	.01
4	FL	Fringing factor for source & drain diffusions	.062
5	PLEN	1/2 the average diffused length between gates	1.0
6	ROVRLP }	Resistance or capacitance for 1 unit of gate width	1T
7	COVRLP		.018P

and voltage-dependent capacitance. Entries on the SUB card are scaled according to diode junction area and periphery. The entries are as follows:

Entry	Units	Model Usage	Default
RBD	$\Omega - V^2$	Series bulk resistance	0
IDS	AMPS/V^2	Saturation current	.1P
MD	—	Proportionality constant	1.0
RDL	$\Omega - U$	Leakage resistance	.1G
BVR	VOLTS	Breakdown voltage	100
IBVR	AMPS	Current at BVR	10M
RVBR	Ω	Resistance at BVR	10
CD	FARADS/U^2	Depletion capacitance at 0 volts	7P
VDBI	VOLTS	Built-in-junction voltage	.75
XDC	—	Exponential factor	.333
TD	SEC	Recombination time of junction	3U
IPPD	AMPS/U^2	Steady state primary photocurrent	0
DEPD	–	Depletion portion of ionization current	0
T5D	SEC	Deep storage time constant	0
AD	–	Deep storage transport factor	0

The model parameters for the process model are calculated from user-entered data as follows:

$$Rgd = ROVLP/W'$$
$$Cgd = COVRLP*W'$$
$$Rgs = ROVLP/W'$$
$$Cgs = COVRLP*W'$$
$$CG = C_o*(W'*L')$$

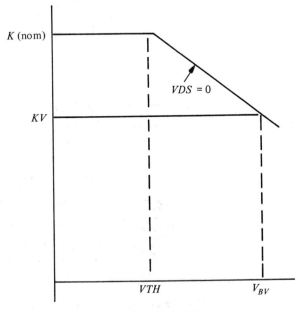

Figure 10.15 Normalized K Variation.

where

W1 = rounded value of W to the nearest unit
W2 = W − W1
L1 = rounded value of L to the nearest unit
L2 = L − L1
W′ = maximum of (W1 × CENT − CENT + UNIT, 0) + W2 × UNIT + 2F × W
L′ = maximum of (L1 × CENT − CENT + UNIT, 0) + L2 × UNIT − 2F × W

The variation of K′ for the process model is as shown in Fig. 10.16.

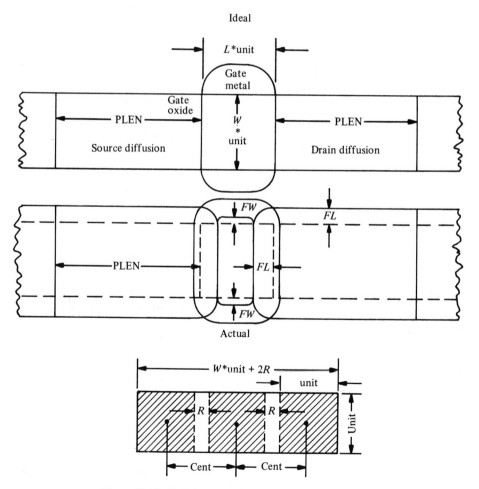

Figure 10.16 Definition of Mask-Card Data Entries (Ideal, Actual).

The first six entries for the special model are used to define the first six parameters of the FET model, i.e., K1, K2, relationship of RD_{SAT} to R_{ON}, K' variation, VFB and ϕB. For the special model K' is considered constant. The remainder of the entries are used directly in the model. Typical FET device characteristics from SYSCAP are shown in Fig. 10.17.

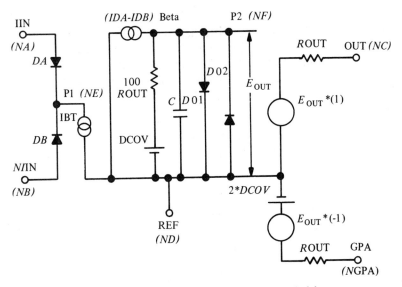

Figure 10.17 General Purpose Amplifier Model.

General-Purpose Amplifier (AMP/AMP WC Model)

The general-purpose amplifier is a simplified model which represents the operation of most general-purpose amplifiers and operational amplifiers in the active state and saturation state. The model is shown in Fig. 10.18.

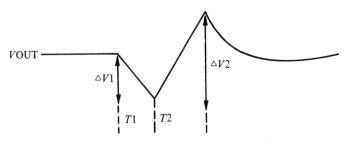

Figure 10.18 Gamma Radiation Parameters.

In Fig. 10.17 the following symbols are used:

IIN = inverted input node
NIIN = noninverted input node
OUT = output node
REF = reference node
P1, P2 = pseudonodes
GPA = GPA output (if 0, model is operational amplifier)

The input diode (DA) has M and IS selected to obtain the impedance (RIN/2) at a current (IBC) and a voltage drop of 0.6 volts. Input diode (DB) has the same M but IS is selected to obtain the impedance (RIN/2) at a current (IBC) and a voltage drop of 0.6 volts + DIVO. IBT is a current source equal to twice input current (IBC).

RIN = resistance looking into either input terminal with the other grounded
IBC = average of the two input currents
DIVO = differential input offset voltage

D01 and D02 are zener diodes which have M and IS adjusted so that at (0.8) times the clamp voltage (VUL, VLL) they will draw 10% of the generated current.

VUL = clamp voltage limit, upper
VLL = clamp voltage limit, lower

Beta is determined from the small-signal voltage gain (GAIN), RIN, and ROUT as

$$\text{BETA} = \frac{\text{RIN} \cdot \text{GAIN}}{200 \cdot \text{ROUT}}$$

where

GAIN = voltage gain of amplifier
ROUT = output resistance

CMOV is the common mode offset voltage for GPA (i.e., VNA = VNB, VNC − VND = DCOV with DIVO = 0). The transient model parameters which define the model transient responses are

FBR = frequency at which gain drops 3 dB
CIN = capacitance between input terminals
TIN = emitter storage time constant of input transistors

10.4 INPUT LANGUAGE

The SYSCAP input language has been designed to be as compatible as possible with the engineer/analyst's engineering background. The overall basic deck setup is shown in Fig. 10.19. This illustrates the card setup required for either a DICAP, ALCAP, or TRACAP simulation. Calls for special models, special plots, or the Fourier option would be performed in the relocatable SYSCAP control deck itself. Also, particular user auxiliary equations or models would be entered in this section.

The various SYSCAP analysis options are enumerated in Table 10.6 along with the relevant cards which should be considered when performing that type of analysis.

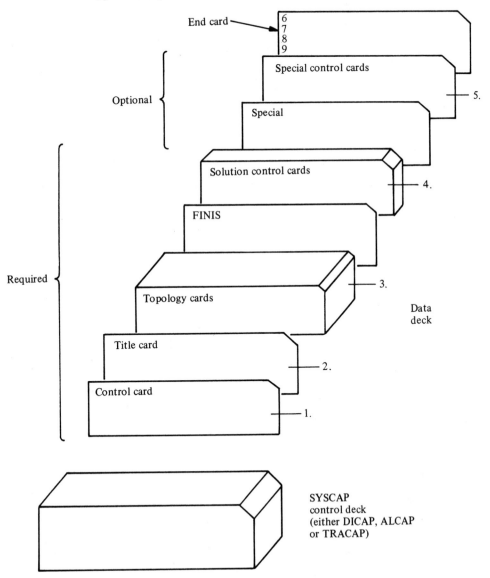

Figure 10.19 SYSCAP Deck Setup.

Table 10.6 SYSCAP Analyses

Deck Setup

Type of Analysis	Control Deck	Possible Control Card Entries	Title Card	Topology Cards	Possible Solution Control Cards	Possible Special Solution Control Cards
dc Nominal	DICAP	NOMINAL ITER AUX(REL) NODES(REL) TEMP	X	X (no transient sources)	STRESS LIMITS CHANGE	None
dc Parameter variation	DICAP	DCONE ITER CRT AUX(REL) NODES(REL) TEMP	X	X (no transient sources)	STRESS LIMITS* CHANGE PLOT*	PARM=
dc Failure mode and fault isolation	DICAP	CFS ITER CRT AUX(REL) NODES(REL) TEMP	X	X (no transient sources)	STRESS LIMITS* CHANGE PLOT*	NAME, NAME- - -
dc Power supply sequencing	DICAP	[SOPSTO1 SOPSTO2] ITER MINIPRINT AUX(REL) NODES(REL) TEMP	X	X (no transient sources)	STRESS LIMITS* CHANGE	EXXX, EXXX- - -
dc Sensitivity	DICAP	SENSIT ITER SENTOL AUX(REL) NODES(REL) TEMP	X	X (no transient sources)	CHANGE	None
dc Worst-case automatic	DICAP	[W/C W/C ALL] ITER SENTOL MATCH AUX(REL) NODES(REL) TEMP	X	X (no transient sources & component tolerances S/B added)	STRESS LIMITS* CHANGE	Parameters matched

dc Monte Carlo	DICAP	MONTECARLO
		  ⎡UNIFORM
		   NORMAL
		   MIN-MAX
		  ⎣TABLES
		MCARLO
		MATCH
		ITER
		AUX(REL)
		NODES(REL)
			X	STRESS LIMITS*
				CHANGE
				PLOT
					Parameters matched
					Table distribution

dc Worst-case selected	DICAP	W/C SEL ITER SENTOL MATCH AUX(REL) NODES(REL) TEMP	X (no transient sources and component tolerances S/B added)	STRESS LIMITS* CHANGE	Variables worst-cased Parameters matched
Transient	TRACAP	CALC/PLT PLT/PRT SAVE TEMP RAD 1/2/3 INCREMENT ITER AUX(REL) NODES(REL)	X (semiconductor must have transient data)	INITIAL TIME STOP GAMMA DOT PLOT XYPLOT PRINT INTEGRATE STORE STRESS SWITCH CHANGE MODIFY/RESOLVE	Fourier (REL) Extended plots (REL)
ac Gain and phase	ALCAP	CAL/PLT PLT/PRT TEMP ITER AUX(REL) NODES(REL)	X (no transient sources and add ac inputs)	INITIAL FREQ PLOT DBPLOT GAIN DBGAIN PRINT DBPRINT CHANGE MODIFY/RESOLVE	None

Notes: (1) Underlined items must be included.
      (2) *indicates items which S/B included.
      (3) **(REL)** indicates the relative version.

General Ground Rules for Data Entries

The following describes the general rules for entering input data:

1. Decimal points, if not used, are assumed to be after the last digit.
2. The sign is assumed to be positive unless indicated otherwise.
3. Scientific notation and the following suffixes may be used anywhere in the data deck:

    T: 10^{12} K: 10^3 N: 10^{-9}
    G: 10^9 M: 10^{-3} P: 10^{-12}
    MEG: 10^6 U: 10^{-6} E: FORTRAN SCIENTIFIC NOTATION

    Examples:

    $1000 = 1000. = 1K = 1.K = 1E + 03 = 1.E + 03$
    $.01 = 10M = 10.M = 1E - 02 = .1E - 01$

4. Brackets, [ ], imply that *any one option enclosed or none* may be specified. An option in brackets which is underlined is the default.
5. Braces, { }, imply that one option enclosed in the braces *must* be specified.
6. Capital letters indicate entries which go directly on the cards.
7. A small n is used for an integer. Appropriate small letters are used for real number entries.
8. Entries are not required to be tightly packed. They may be freely spaced from columns 1–80 provided they are correctly separated with the proper delimiters.
9. Nodes must be designated numerically. Zero is reserved for ground. Ground can also be designated as G.
10. A single delimiter (comma, parenthesis, or slash) must be used to separate each entry. Each section will indicate the appropriate type of delimiter to be used. A delimiter is never required at the end of a data entry.
11. The RS and RI designators for source resistance can be shortened to just R.
12. Component names are limited to four characters.

Input Coding

The input coding required for each of the five sections—(1) control card, (2) title card, (3) topology cards, (4) solution control cards, and (5) special control cards—is described in the following subsections.

Control Card

The initial control card is used to select the specific dc option, set the sensitivity and worst-case screen level, determine the Monte-Carlo sample size, reduce output data, select the appropriate radiation model, set a defined increment on the time base for output plots and tabulations, save output data for restart, indicate additional nonstandard circuit nodes and auxiliary equations, set temperature, and adjust the number of initial dc iterations. A single control card is required; however, all entries may be defaulted, resulting in a blank card. Entries may be in any order but must be separated by commas. The possible card entries and default values are defined in Table 10.7.

Table 10.7 *Control Card Entries*

Possible Entry	Definition	Default
	DC	
⎡NOMINAL⎤ DCONE CFS SOPSTO1 SOPSTO2 SENSIT W/C W/C SEL W/C ALL ⎣MONTECARLO⎦	NOMINAL: nominal solutions only DCONE: dc one-at-a-time parameter variation CFS: circuit failure simulation. SOPSTO1: simulation of power supply turn-on; off supplies at zero volts and nominal impedance SOPSTO2: simulation of power supply turn-on; off supplies at zero volts and 10 megohm SENSIT: sensitivity W/C: worst case, auxiliary equations max only W/C SEL: worst case for selected variables W/C ALL: worst case, min & max, for all nodes and auxiliaries MONTECARLO: dc Monte Carlo option	Nominal only
⎡UNIFORM⎤ NORMAL MIN-MAX ⎣TABLE⎦	The distribution for the dc Monte Carlo analysis	UNIFORM
MCARLO = integer	Monte Carlo sample size	100
SENTOL = real no.	The sensitivity tolerance product (in %) for worst-case analysis, and the sensitivity tolerance for sensitivity; a value of 0.01 would set SENTOL=.01 for sensitivity and SENTOL=.01% for W/C	SENTOL = .05

Table 10.7—*Cont.*

Possible Entry	Definition	Default
MATCH = integer	Number of matched parameter pairs (100 max) for W/C options, or Monte Carlo	MATCH = 0
CRT	CFS Bar Graph Stress Summary (SC4020)	No bar graph stress summary provided

ac-Transient

Possible Entry	Definition	Default
CALC/PLT = integer	Ratio of calculated points to plotted points or printed points for tabular print (PRINT option)	CALC/PLT = 1
PLT/PRT = integer	Ratio of plotted points to printed points	PLT/PRT = 1

Transient

Possible Entry	Definition	Default
⎡RAD1⎤ ⎢RAD2⎥ ⎣RAD3⎦	Radiation model to be used RAD1: exponential rise and fall time, dependent on TI RAD2: diffusion and depletion components RAD3: diffusion and voltage dependent depletion components	RAD1
INCREMENT = real no.	Forced output plots and tabular print points to be generated at equal increments; YYY seconds apart	Outputs as calculated intervals controlled by CALC/PLT ratio
SAVE	Punches output cards and prints parameter values at the completion of a run to allow restart	No output punch cards or printed final parameter values

General

Possible Entry	Definition	Default
NODES = integer	Total number of nodes in the circuit; used for I/C and/or nonstandard models with pseudonodes	No. of nodes in the circuit, excluding I/C and/or nonstandard models pseudo nodes
AUX = integer	Number of user-written auxiliary equations in USER subroutine; used only with relocatable version; max = 50	AUX = 0
TEMP = real no. ⎡F⎤ ⎢C⎥ ⎣K⎦	Temperature for simulation; F for Fahrenheit, C for centigrade, and K for Kelvin	TEMP = 25°C (no entry) TEMP = YYY°K (with TEMP = YYY)
ITER = integer	Maximum number of iteration for initial dc solution	ITER = 200
SEQ	Sequence number may be added to columns 73–80	Data can be entered from columns 1 to 80

Title Card

The title card is a single card on which some type of user-defined descriptive entry is made between column 1 and 80. This card is also required and may be blank.

Topology Cards

In the topology card section, the components in the electronic circuit under analysis are entered. A comment card, with an asterisk in column 1, can be placed anywhere in this section. Component ordering is left up to the user. The order in which the nodes are entered for nonpolarized devices determines the sign on the auxiliary equation. The component card general format is

$$\text{component name, device type (node connections)} \begin{Bmatrix} \text{auxiliary} \\ \text{equations} \end{Bmatrix}, \begin{Bmatrix} \text{part} \\ \text{data} \end{Bmatrix}$$

The component name consists of a key letter followed by alphanumeric characters. The key letters are

- R = resistors
- C = capacitors
- L = inductors
- E = voltage sources
- I = current sources
- VV = voltage-dependent voltage source
- VJ = voltage-dependent current source
- JJ = current-dependent current source
- JV = current-dependent voltage source
- T = transformers
- K = transformer cores
- D = diodes
- DZ = zeners
- Q = transistors
- F = field-effect transistors
- A = Operational amplifiers

Device types refer to *NPN* or *PNP* for transistors and MP, MN, SP or SN for FETs. The user, by entering the appropriate key (Table 10.8) automatically gets an auxiliary equation for that specific component. The name for the auxiliary equations which appear in the printout are also shown in Table 10.8. The component name, device type, node connection, auxiliary equations, and DATA BANK index or single data entry must be on one card. Subsequent card sets are used when multiple data entries are used (e.g.,

Table 10.8 *Possible Automatic Auxiliary Equations*

Component Name	Request Key	Solution Name	Definition
R name	P	R name PR	Power
	I	R name IR	Current
C name	V	C name VC	Voltage drop
	I	C name IC	Current
L name	V	L name VL	Voltage drop
	I	L name IL	Current
E name	P	E name PS	Source power
	I	E name IS	Source current
I name	P	I name PI	Source power
	I	I name II	Source current
T name	B	T name B	Flux density
	H	T name H	Magnetizing force
	I	T name IW winding #	Coil current
D name	P	D name PD	Power
	I	D name IDT	Current
DZ name	P	DZ name PDZ	Power
	I	DZ name IZT	Current
Q name	P	Q name PT	Power
	IB	Q name IB	Base current
	IC	Q name IC	Collector current
	IE	Q name IE	Emitter current
F name	P	F name PF	Power
		F name IDS	Drain-to-source current
		TFETPWR	Total FET Power.

transistors, diodes transformers, sources, and FETs). A FINIS card is used to end the topology card section.

Resistor $n_1 \underset{I}{\overset{Rxxx}{-\!\!\!-\!\!\!\!/\!\!\backslash\!/\!\!\backslash\!/\!\!\backslash\!-\!\!\!-}} n_2$

R name (n_1, n_2) [I,P,] Resistance Value $\left[\, , \begin{Bmatrix} +\text{tol (in \%)} \\ \text{or} \\ \text{max. value MAX} \end{Bmatrix}, \begin{Bmatrix} -\text{tol (in \%)} \\ \text{or} \\ \text{min. value MIN} \end{Bmatrix} \right]$

Example:

R12 (1,2) P, 10K, +10, 9K MIN

Tolerance information is required only for DICAP worst case and Monte Carlo.

Capacitor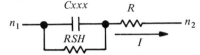

C name (n_1, n_2) [I,V] Capacitance Value [, R = series resistance, II = initial current, RSH = shunt resistance]	

Example:

$$C1\ (1,2)\ I0U,\ R = 1$$

The default value for the series resistance is 0.1 ohm, initial current zero, and shunt resistance 10^{12} ohms.

Inductor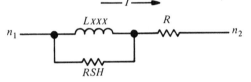

L name (n_1, n_2) [I,V,] Inductance Value [, R = series resistance value, II = initial current value, RSH = shunt resistance]

Example:

$$LB10\ (5,6)\ V,\ 1M,\ II = 1M$$

The default value for the series resistance is 0.1 ohm, initial current zero, and parallel resistance 10^{12} ohms.

DC Source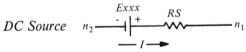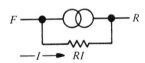

$\begin{Bmatrix} E\ name\ (+n_1, -n_2) \\ I\ name\ (Fn_1, Rn_2) \end{Bmatrix}$ [I,P,] dc value $\left[, \begin{Bmatrix} +\text{tol (in \%)} \\ \text{or} \\ \text{max. value MAX} \end{Bmatrix} \right]$,

$\left[\begin{Bmatrix} -\text{tol (in \%)} \\ \text{or} \\ \text{min. value MIN} \end{Bmatrix} \right]$ $\left[, \begin{matrix} RS = \text{series resistance for voltage,} \\ RI = \text{parallel resistance for current,} \end{matrix} \right]$ $\left[, \begin{Bmatrix} +\text{tol (in \%)} \\ \text{or} \\ \text{max. value MAX} \end{Bmatrix} \right]$,

$$\left\{ \begin{array}{c} -\text{tol (in \%)} \\ \text{or} \\ \text{min. value MIN} \end{array} \right\}$$

Example:

E01 (+1, −G) I, 10, +10, −10, RS = 1
I02 (F4, R16) 1

Tolerance information is required only for DICAP worst case and Monte Carlo. The default value for RI is 1 MEG. If the user omits the series resistance for a voltage source and no auxiliaries are requested, then zero ohms is used; otherwise 0.1 ohms is used. Zero ohms may not be entered.

Pulse Source

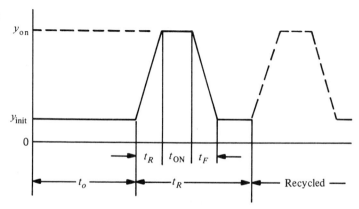

$$\left\{ \begin{array}{l} \text{E name } (+n_1, -n_2) \\ \text{I name } (Fn_1, Rn_2) \end{array} \right\} \text{[I,P,] PULSE,} \left[\begin{array}{l} RS = \text{series resistance} \\ \text{for battery,} \\ RI = \text{parallel resistance} \\ \text{for current,} \end{array} \right]$$

$$y_{init}, y_{on}, t_o, t_R, t_{ON}, t_F, t_R \left[\frac{SEC}{HZ} \right]$$

Example:

Single (assuming run time is less than 1):

E01 (+1, −0) I, PULSE, 0, 6, 0, 10N, 500N, 10, 1

Repetitive:

I02 (F3, R4) PULSE, −1M, 1M, 10N, 10N, 500N, 10N, 530N

Point by Point Source

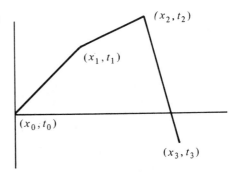

$\begin{cases} \text{E name } (+n_1, -n_2) \\ \text{I name } (Fn_1, Rn_2) \end{cases}$ [I,P,] POINTS,	$\begin{bmatrix} RS = \text{series resistance} \\ \text{for battery,} \\ RI = \text{parallel resistance} \\ \text{for current,} \end{bmatrix}$

$$x_0, t_0, x_1, t_1, x_2, t_2, x_3, t_3, \ldots$$

Example:

E1 (+5, −6) POINTS, RS = 10K, 0,0,1, 10U
I1 (R1, F2) P, POINTS, RI = MEG, 10M, 0, 15M, 1U,
25M, 2U, 25M, 1

Transient Sine Source

$$y(t) = y_0 + (mt + y_1) \text{ SIN } 2\pi F (t - t_0)$$

$\begin{cases} \text{E name } (+n_1, -n_2) \\ \text{I name } (Fn_1, Rn_2) \end{cases}$ [I,P] SINE,	$\begin{bmatrix} RS = \text{series resistance} \\ \text{for battery,} \\ RI = \text{parallel resistance} \\ \text{for current,} \end{bmatrix}$

$$y_0, y_1, t_0, F\left[\frac{HZ}{SEC}\right], m$$

Examples:

I01 (F2, R3) SINE, 0, 1, 0, 10K, 0
E01 (+1, −4) SINE, 0, 1, 01, 10USEC, 0

Exponential Source

$$y(t) = y_0 + (y_0 - y_1)e^{-(t-t_0)/\gamma}$$

$\begin{cases} \text{E name } (+n_1, -n_2) \\ \text{I name } (Fn_1, Rn_2) \end{cases}$ [I,P,] EXP, $\begin{bmatrix} RS = \text{series resistance} \\ \quad \text{for battery,} \\ RI = \text{parallel resistance} \\ \quad \text{for current,} \end{bmatrix}$ y_0, y_1, t_0, γ

Examples:
     E01 (+1, −2) I, EXP, RS = 10K, 0, 1, 0, 10U
     I02 (F4, R6) P, EXP, 0, 5, 0, 1U

AC Source (ALCAP)

$\begin{cases} \text{E name } (+n_1, -n_2) \\ \text{I name } (Fn_1, Rn_2) \end{cases}$ [I,P,] AC, dc voltage value, $\begin{bmatrix} RS = \text{series resistance for battery,} \\ RI = \text{parallel resistance for current,} \end{bmatrix}$ ac amplitude, ac phase $\begin{bmatrix} DEG \\ RAD \end{bmatrix}$

Examples:
     E1 (+1, −4) AC, 0, 1, 90
     I2 (F1, R4) AC, 1M, 1, 2 RAD

Voltage Dependent Voltage Source

VV name $(+n_1, -n_2)$ $k_1 * (+n_4, -n_5)$ RS = series resistance for battery
Example: VV1 (+5, −0) 5.*(+7, −1) RS = 1

Voltage Dependent Current Source

VJ name (Fn_1, Rn_2) $k_1 * (+n_4, -n_5)$ RI = parallel resistance for current
Example: VJ02 (F4, R3) 10K*(+6, −G) RI = 1 MEG

Current Dependent Current Source

$$\text{JJ name (Fn}_1, \text{Rn}_2) \, k_1 * \begin{Bmatrix} \text{R name} \\ \text{C name} \\ \text{L name} \\ \text{D name} \\ \text{I name} \\ \text{E name} \end{Bmatrix},$$

RI = parallel resistance for current

Examples:

JJ01 (F1, R2) 10*R1, RI = 1 MEG
JJ2 (R4, F3) 10*L1, RI = 100K

Current Dependent Voltage Source

$$\text{JV name } (+n_1, -n_2) \, k_1 * \begin{Bmatrix} \text{R name} \\ \text{C name} \\ \text{L name} \\ \text{D name} \\ \text{I name} \\ \text{E name} \end{Bmatrix},$$

RS = series resistance for current

Examples:

JV1 (+5, −0) 10*R01, RS = 1K
JVAA (+4, −6) L1, RS = 10K

Transformer

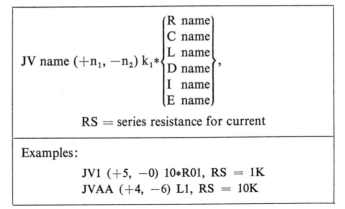

$$\text{T name}, \begin{Bmatrix} \text{K name} \\ \text{L = primary inductance} \end{Bmatrix}, \text{PSEUDO} = n_1 \, [,B,H]$$

/ turns in primary $(+n_2, -n_3)$ R = dc resistance [,I]/
/ turns in secondary $(+n_4, -n_5)$ R = dc resistance [,I]/

Examples:

T01, L = 1M, PSEUDO = 10, B / 10 (+3, −4) R = 10/
20 (+6, −7) R = 20 / 30 (+8, −9) R = 30/
T02, K1, PSEUDO = 15 / 20 (+1, −2) R = 1/
100 (+5, −6) R = 5.6/

If the K name option is used, then a core card must be entered. A unique PSEUDO node number is needed for each transformer. The value of the PSEUDO node is $d\phi/dt$, or volts per turn. More than one card may be used; however, the entire data set for any one winding must be on a single card.

Linear Core

K name, core area in centimeters squared, core length in centimeters, μ max, initial flux density (gauss)

Example:

K1, 1.1, 1.05, 1.4K, 0

Saturable Core

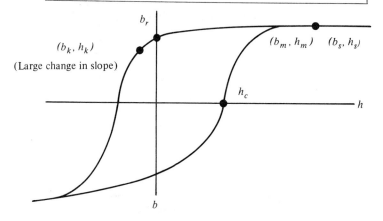

K name, core area in centimeters squared, core length in centimeters, b_s, h_s, b_m, h_m, b_r, h_c, b_k, h_k, air gap in centimeters, time constant, initial density (gauss)

Example:

K1, .25, 9.47, 15K, .6, 14.5K, .2, 14.3K, .11, 14.2K
−.1, 0, 1U, 0

Diode/Zener (*Data Bank*)

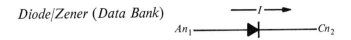

$$\begin{Bmatrix} \text{D name} \\ \text{DZ name} \end{Bmatrix} (An_1, Cn_2) \; [I,P,] \; \text{jedec} \; \# \; \begin{bmatrix} , \text{VDMIN} \\ , \text{VDMAX} \end{bmatrix} \begin{bmatrix} , \text{MIN} \\ , \text{MAX} \end{bmatrix}$$

Examples:

    D1 (A3, C4) I, 1N4151
    DZ02 (A4, C5) P, 1N4372, VDMAX

The VDMIN or VDMAX entry results in min. or max. junction data from the data bank. The MIN or MAX entry results in the entire set of either min or max data from the data bank.

Diode/Zener (*User-Supplied Data*)

$$\begin{Bmatrix} \text{D name} \\ \text{DZ name} \end{Bmatrix} (An_1, Cn_2) \; [I,P,] \; \begin{bmatrix} \text{D other name} \\ \text{DZ other name} \\ \text{NOM} \\ \hline \text{MIN} \\ \text{MAX} \\ \text{ALL} \end{bmatrix}$$

Next card: User-supplied data as defined in Sec. 10.3.

Examples:

    D1 (A1, C3) I, D02
    D2 (A4, C5)   } DICAP NOM.
    10., 1P, 1, 1G, 100, 10M, 10   DATA
    DZ 3 (A6, CG) ALL
    10., 1P. 1, 1G, 6, 1M, 5   } DICAP W/C
    10., 1P, 1, 1G, 5.7, 1M, 3   DATA
    10., 1P, 1, 1G, 6.3, 1M, 10
    D4 (A6, C3)
    5., .1P, 1.1, 1.1G, 100, 10M, 10, } ALCAP, TRACAP
    10P, .75, .5, 1N, 0, 0, 0, 0   NOM. DATA

The "D other name or DZ other name" are used when data are identical to another diode or zener for which the user has supplied his own data. In this case an additional card is not required. The NOM, MIN, MAX, or ALL, designator is used in DICAP when a string of data is to follow for either just nominal, or nominal and minimum, or nominal and maximum, or nominal, minimum, and maximum, respectively. The parameters in the data string which follows on a subsequent card are defined in Sec. 10.3. For ALCAP and TRACAP a blank is used, with the data string following on the next card. For DICAP, only the dc parameter data are required; however, the transient string is permissible for the NOM option.

Transistor (Data Bank)

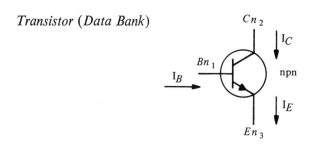

Q name, $\begin{bmatrix} \text{NPN} \\ \text{PNP} \end{bmatrix}$ (Bn_1, Cn_2, En_3) [IB, IC, IE, P,]

jedec # $\begin{bmatrix} \text{, BNBIMIN} \\ \text{, BNBIMAX} \end{bmatrix} \begin{bmatrix} \text{, VBECMIN} \\ \text{, VBECMAX} \end{bmatrix} \begin{bmatrix} \text{, RLEAKMIN} \\ \text{, RLEAKMAX} \end{bmatrix} \begin{bmatrix} \text{, FAST} \\ \text{, SLOW} \end{bmatrix} \begin{bmatrix} \text{, MIN} \\ \text{, MAX} \end{bmatrix}$

Examples:

   Q1, PNP (B3, C4, EG) IE, P, 2N2222
   Q3 (B4, C6, E10) P, 2N1722, BNBIMAX

The entries after the jedec # result in either min. or max. data from the DATA BANK.

Transistor (User-Supplied Data)

Q name, $\begin{bmatrix} \text{NPN} \\ \text{PNP} \end{bmatrix}$ (Bn_1, Cn_2, En_3)

[IB, IC, IE, P,] $\begin{bmatrix} \text{Q other name} \\ \text{NOM} \\ \text{MIN} \\ \text{MAX} \\ \text{ALL} \end{bmatrix}$

Next card: User-supplied data as defined in Sec. 10.3.

Example:

   Q1, NPN (B1, C4, EG) P, Q2
   Q4, PNP (B5, E6, CG) ALL

```
         100, 10M, .5, .1M, .6, 100M, 100, 1.0,  ⎫
          10, .1, 1P, 1, 1G, 100, 10M, 10, 1P,   ⎪
          1, 1G, 30, 10M, 10                      ⎪
         90, 10M, .5, .1M, .6, 100M, 100, 1.0,   ⎬  DICAP
          10, .1, 1P, 1, 1G, 80, 10M, 10, 1P,    ⎪  WORST-CASE
          1, 1G, 20, 10M 10                       ⎪  DATA
         150, 10M, .5, .1M, .6, 100M, 100, 1.0,  ⎪
          10, .1, 1P, 1, 1G, 150, 10M, 10, 1P,   ⎪
          1, 1G, 50, 10M, 10                     ⎭
         Q3 (B6, E7, CO) P, IB,                  ⎫
         100, 10M, .5, .1M, .6, 100M, 100, 1.0,  ⎪
          10, .1, 1P, 1, 1G, 100, 10M, 10, 10P,  ⎬  ALCAP
         .75, .5, 1U, 0, 0, 0, 0, 1P, 1, 1G, 20, ⎪  TRACAP
         10M, 10, 10P, .75, .5, 1N, 0, 0, 0, 0   ⎭  NOMINAL DATA
```

The "Q other name" is used when data are identical to a previously entered transistor, for which the user has supplied his own data. The NOM, MIN, MAX, and ALL designator is used in DICAP when a string of data is to follow for either just nominal, or nominal and minimum, or nominal and maximum, or nominal, minimum, and maximum, respectively. The parameters in the data string which follow on a subsequent card are defined in Sec. 10.3. For ALCAP and TRACAP a blank is used, with the data string following on the next card. For DICAP, only the dc parameter data are required; however, the transient string is permissible for the NOM option.

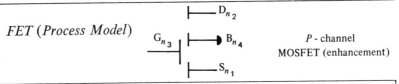

FET (Process Model) — P-channel MOSFET (enhancement)

F name, $\begin{Bmatrix} M \\ S \end{Bmatrix} \begin{Bmatrix} N \\ P \end{Bmatrix}$ [n](Sn_1, Dn_2, Gn_3, Bn_4, Pn_5)
[I,P,] OPTOMASK width, length [, K', threshold voltage]

Next cards: MSI [n],

MASK $\begin{Bmatrix} M \\ \overline{U} \\ C \end{Bmatrix}$ [n], {Data strings as defined in Sec. 10.3}

SUB [n],

Examples:

    F15, MP2(S20, D21, G25, B2, P3) P, 1, 4.5, 9, 4
    MSI2, X7 = .9, X 8 = 20

> MASK M2, X6 = .5T
> SUB2, 0, .1P, 1, 1G, 100, 10M, 10, 4P, .8, .5, 5N
> F16, MP (S25, D20, G33, B4, P5) P, 2, 3.4

The FET model is either MOS {M} or SOS {S}, N-channel {N} or P-channel {P}. The number n is used to correlate with the appropriate MSI MASK and SUB cards (Default=1). An entry of K' and/or threshold voltage changes the entry on the MSI card for that FET. A pseudo P node is required for each FET. Parameter and dimensional data are entered on a MSI, MASK, and SUB card, respectively. The $\left\{\begin{matrix}M\\U\\C\end{matrix}\right\}$ option on the MASK card allows the user to enter data in milli-inches {M}, microns {U}, or centimeters {C}. The default values (Sec. 10.3) on an MSI, MASK, or SUB card may be modified by using Xn = – – –. Also only a partial data string need be entered when the default values at the end suffices.

FET (*Special Model*)

> F name, $\left\{\begin{matrix}M\\S\end{matrix}\right\}$ $\left\{\begin{matrix}N\\P\end{matrix}\right\}$ [n](Sn_1, Dn_2, Gn_3, Bn_4, Pn_5)
> [I, P,] [F other name]

Next card: Data string as defined in Sec. 10.3.

Examples:

> F15, MP (S20, D21, G21, B2, P4) P
> 10, 5, 4, 30M, 2K, .7, 2T, 2T, 450MEG
> 450MEG, .01P, .007P, .007P, .02P, .02P
> SUB   , X3 = 1.5
> F16, MP (S40, D30, G41, B3, P4) I, P, F15

A pseudo P node is required for each FET. The parameter data for the FET are entered as a data string, on the next card/cards, if "F other name" has not been used, a SUB card is required if SUB default values are not adequate.

Operational Amplifiers

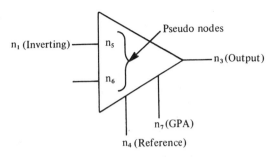

A name, n_1, n_2, n_3, n_4, n_5, n_6 [, n_7]
Second card:
rin, ibc, divo, gain, rout, vul, vll, cmov, fbc, cin, tin

Example:
A13(1, 4, 6, 0, 9, 10, 0)
1 MEG, .2U, 0, .1 MEG, 100, 12, −12, 0, 10, 5P, 1N

Two pseudo nodes (n_5 and n_6) are required for each operational amplifier.

Solution Control Cards

The possible solution control cards are enumerated in Table 10.9. For a DICAP run no solution control cards are required. However, to get overstress tables and stress checks, the STRESS LIMIT option must be used. For an ALCAP run one solution control card is required: the frequency card. For TRACAP the time step simulation end time card is required. In a TRACAP ionization analysis, the ionization pulse shape card is required. All other solution control cards are optional.

Special Control Cards

A card with the word SPECIAL must precede the special control card section. Special control cards are required only for the following:

    DICAP:

    DCONE

    CFS

    SOPSTO 1

    SOPSTO 2

    W/C (if MATCH is used, for Match Data)

Table 10.9 Solution Control Cards

Possible Entry	Definition
INITIAL = $\begin{Bmatrix} \text{COMPUTE} \\ \text{COMPUTE HALT} \\ \text{COMPUTE ONLY} \\ \text{ZERO} \\ \text{READ}, t_e, t_0, v_1, \ldots v_n \end{Bmatrix}$	Defines initial conditions for DICAP, TRACAP, or ALCAP (required)
FREQ = $f_1, \ldots f_n$	
FREQ = $\begin{Bmatrix} \text{MULT} \\ \text{ADD} \end{Bmatrix}$, initial frequency, multiplier or additive constant, total number of frequencies	Defines frequencies for ALCAP simulation (required)
TIME = $\Delta t_1, et_1, \ldots \Delta t_n, et_n$	Defines maximum delta time and end time for TRACAP (required)
[DB] PLOT $\begin{bmatrix} , \text{SC4020} \\ , \text{CPP} \\ , \text{CPPCS} \end{bmatrix}$ = $name_1, \ldots name_n$	Plots variables. CPP and CPPCS are used in TRACAP for continuous printer plot and continuous printer plot common scale, respectively. Upper and lower limits may be added after a name for TRACAP (UL = n_1, LL = n_2).
[DB]GAIN = $name_1/name_2, \ldots name_n/name_y$	Plots ratio of 2 variables in ALCAP
xyPLOT = $name_1/name_2, \ldots name_n/name_y$	Plots $name_1$ versus $name_2$ in TRACAP
PRINT = $name_1, \ldots name_n$	Provides tabular listing in TRACAP or ALCAP
STOP IF $\left(name_1 \begin{Bmatrix} .LT. \\ .GE. \end{Bmatrix} v_1 \right)$	Terminates TRACAP run when $name_1$ exceeds criteria
SWITCH IF $\left(name_1 \begin{Bmatrix} .LT. \\ .GE. \end{Bmatrix} v_1 \right)$ $name_2 = v_2$	Changes parameter in TRACAP run when criteria are achieved
GAMMA DOT, SCALE = scale factor, $\dot{\gamma}_1, t_1, \ldots \dot{\gamma}_n, t_n$	Ionization pulse shape
INTEGRATE [FROM t_1 TO t_2], $name_1, \ldots name_n$	TRACAP only
STORE [FROM t_1 TO t_2], $name_1 \begin{bmatrix} \text{MIN} \\ \text{MAX} \end{bmatrix}, \ldots name_n \begin{bmatrix} \text{MIN} \\ \text{MAX} \end{bmatrix}$	TRACAP only
STRESS, $name_1 = v_1, \ldots name_n = v_n$	Transient stress level determination
STRESS LIMITS $\quad name_1 = v_1 \begin{bmatrix} \text{MIN} \\ \text{MAX} \end{bmatrix}$ $\quad \vdots$ $\quad name_n = v_n \begin{bmatrix} \text{MIN} \\ \text{MAX} \end{bmatrix}$ STOP	dc Stress level determination
CHANGE $name_1 \begin{bmatrix} \text{MIN} \\ \text{MAX} \end{bmatrix} = v_1$	Change parameter prior to the run
MODIFY $name_1 = v_1$ $\quad \vdots$ MODIFY $name_n = v_n$ RESOLVE	Change parameters for subsequent TRACAP or ALCAP runs

W/C SEL

W/C ALL (if MATCH is used, for Match Data)

MONTE CARLO (tabled frequency)

MONTE CARLO (if MATCH is used, for Match Data)

TRACAP:

Fourier option.

Extended plot option.

If none of the above-listed models or options are desired, a SPECIAL card is not required. The special control cards, except for Fourier and the extended plot options, are described in Table 10.10. The reader is referred

Table 10.10 *Special Solution Control Cards*

Possible Entry	Definition
PARM = $name_1$, $\begin{cases} UL=v_1, US=v_2, LL=v_3, \\ DS=v_4 \\ VALUES=v_1 \ldots v_n \end{cases}$	Parameters to be varied for DCONE option
STOP	
$name_1, \ldots name_n$, STOP	Components to be failed for CFS option
$Ename_1, \ldots Ename_n$, STOP	Power supplies to be failed for SOPSTO1/2 options
NODES, $n_1, \ldots n_n$, STOP	Solutions to be worst-cased, min. and max.
MAX, $n_1, \ldots n_n$, STOP	Solutions to be worst-cased, max. only
MIN, $n_1, \ldots n_n$, STOP	Solutions to be worst-cased, min. only
$name_1$, $name_2$, max for $name_1$ when $name_2$ is min, min for $name_1$ when $name_2$ is max, max for $name_2$ when $name_1$ is min, min for $name_2$ when $name_1$ is max	Worst-case matching
$v_1, \ldots v_9$, 100	Tabled frequency distribution for Monte-Carlo
$name_1$, $name_2$, correlation coefficient, matching type	Monte-Carlo matching

to the SYSCAP manual on the input coding for the Fourier and extended plot options. ALCAP does not use any Special control cards.

10.5 OUTPUT OPTIONS

The output options which are available in SYSCAP are summarized in Table 10.11. All these outputs are in a form which allows them to be directly incorporated into report format without any additional typing. Examples of many of these outputs are shown in the next section.

10.6 SYSCAP EXAMPLES

The ease of data setup and component modeling using SYSCAP will be demonstrated on some actual circuit computer runs. Two simple single-stage circuits were run on the transient and ac programs, and a low-frequency linear amplifier was run on all the major options. The transient simulation on the single-stage circuit was performed interactively on the CDC KRONOS system. The ac analysis and all the linear amplifier simulations were run batch.

Transient Simulation

An initial dc and transient analysis was performed, by terminal on the logic circuit shown in Fig. 10.20. Characterization of the transistor was not required since it was called directly from the SYSCAP DATA BANK. A circuit file consisting of the 10 lines shown in Fig. 10.21 was created as a KRONOS file named LOGIC.

The interactive TRACAP program is initiated (Fig. 10.22) by TRACAP (D = LOGIC). The last entry is the previously created circuit file, LOGIC. The circuit topology, node connections, and solution control entries are the first output. Any diagnostic messages would have appeared in the topology output. After the request for 'NEXT CARD', the user indicates which solutions are to be saved (SAVE = NODE 1, NODE 2). After the next 'NEXT CARD' request, he indicates the program should proceed with the dc solution (RUN). The dc solution is then automatically outputted indicating Node 1 is -2.15 volts and Node 2 is 5.99 volts. The user then requests ITRACAP to go 50 time solutions (G, 50). When the program has completed the requested 50 solutions, it outputs the simulation elapsed time. In this example, the user requested an additional 20 solutions (G, 20). After 20 more solutions, the simulation time was out to the desired 200 NS (actually 203.2 NS). The user has now completed the calculations and is ready to examine the results.

Table 10.11 Output Options

		DICAP						MONTE-		
Analysis → Output ↓	NOM	DCONE	CFS	SOPSTO	SENSIT	W/C	CARLO	ALCAP	TRACAP
Data bank summary	A	A	A	A	A	A	A	A	A
State table	A	A	A	A	A	A	A	A	A
Nominal overstress summary	O	O	O	O	O	O	O		O
All nodal voltages & auxiliary answers at each solution	A	A	A	A	A	A		AR	AR
Reduced output in tabular form		O	O						
Overstresses at each solution	O	O	O	O		O	A	O	O
Min-max summary table						A	A		O
Overstress summary table		O	O	O		O	O		O
Plots		O	O				O	O	O
Integral values						O			O
Fourier coefficients									O
User-written auxiliary equations	O	O	O	O	O	O	O	O	O

A, automatic.
O, optional.
AR, automatic, but may be reduced by user control.

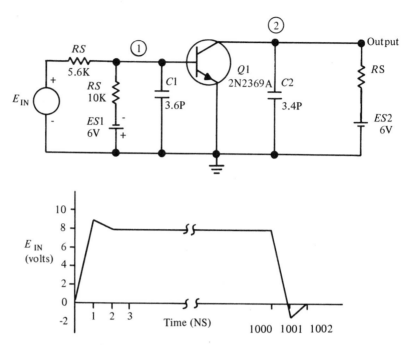

Figure 10.20 LOGIC Circuit and Input Signal.

```
OLD,LOGIC
READY.
LNH
00100 NOMINAL
00110 LOGIC INVERTER
00120 EIN(+1,-0)POINTS,RS=5.6K,0,0,9,1N,8,2N,8,.1U,-.1,101N,0,102N,0,1
00130 ES1(+1,-0)  -6,RS=10K
00140 ES2(+2,-0)   6,RS=1K
00150 Q1(B1,C2,E0) 2N2369A
00160 C1(1,0)   3.6P
00170 C2(2,0)   3.4P
00180 FINIS
00190 TIME=5N,200N
READY.
```

Figure 10.21 LOGIC Circuit File.

The first output is the value of Node 1 and Node 2 at the last time step, bottom of Figure 10.22. Next, the values of Node 1 and Node 2 are listed for every fifth time step, Figure 10.23. Finally, Node 1 and Node 2 are plotted versus time, Figure 10.24.

The required user inputs in Figures 10.21, 10.22, 10.23 and 10.24 are underlined.

```
-ITRACAP(D=LOGIC)
         TRACAP CONTROL CARD - NOMINAL

              LOGIC INVERTER
              EIN(+1,-0)POINTS,RS=5.6K,0,0,9,1N,8,2N,8,.1U,-.1,101N,0,102N,0,1
0             ES1(+1,-0)  -6,RS=10K
              ES2(+2,-0)   6,RS=1K
              Q1(B1,C2,E0)  2N2369A
              C1(1,0)   3.6P
              C2(2,0)   3.4P
              FINIS
1 NODE CONNECTIONS

  NODE   1          EIN ,     ES1 ,      Q1 ,     C1 ,

  NODE   2          ES2 ,     Q1  ,      C2 ,
SOLUTION CONTROL CARD(S)
 TIME=5N,200N
 NEXT CARD
? SAVE=NODE1,NODE2
  NEXT CARD
? RUN


0  TE  0.            STEP 1.00000E+37  MS    3  GDOT 0.
0    NODE VOLTAGE AND AUXILIARY ANSWERS

     1 NODE    1-2.1533E+00    2 NODE    2 5.9999E+00

          BUFFER SIZE =        935
   TE=  0.           MS=    3   BUF=    0       17.57.17.        .789
   RUN/OUTPUT
? 6,50
   TE=  1.13200E-07  MS=   75   BUF=   51       17.58.15.       2.238
   RUN/OUTPUT
? 6,20
   TE=  2.03200E-07  MS=   29   BUF=   69       17.58.40.       2.764
   RUN/OUTPUT
? A
0  TE  2.03200E-07  STEP 5.00000E-09  MS   29  GDOT 0.
0    NODE VOLTAGE AND AUXILIARY ANSWERS

     1 NODE    1-1.9516E+00    2 NODE    2 5.9892E+00
   RUN/OUTPUT
?
```

Figure 10.22 Interactive TRACAP Nominal and Transient Results.

For the input processing, dc initial solution, and transient solutions, it took approximately 3 system seconds on a CDC 6400 computer.

ac Simulation

An initial dc operating point calculation and ac frequency sweep from 1 KHz to 5×10^{11} Hz was performed on the ac inverter circuit shown in

```
          RUN/OUTPUT
        ? LR,5,1,2
           TIME              NODE    1       NODE    2
           0.               -2.1533E+00     5.9999E+00
           1.000000E-09     -2.0091E+00     6.0234E+00
           2.000000E-09     -1.7666E+00     6.0633E+00
           8.200000E-09     -6.2925E-01     6.1416E+00
           3.320000E-08      8.7013E-01     4.8502E+00
           5.820000E-08      9.3363E-01     1.4279E+00
           8.320000E-08      9.4593E-01     1.6472E-01
           1.002000E-07      9.4737E-01     1.5285E-01
           1.012000E-07      9.4452E-01     1.5514E-01
           1.022000E-07      9.4136E-01     1.6603E-01
           1.132000E-07      9.1212E-01     1.3590E+00
           1.382000E-07      5.3062E-01     5.5985E+00
           1.632000E-07     -1.0529E+00     5.9464E+00
           1.882000E-07     -1.7673E+00     5.9797E+00
          RUN/OUTPUT
```

Figure 10.23 LOGIC Tabulated Results.

Fig. 10.25. Typical data from the SYSCAP DATA BANK were used for transistor Q1. The initial dc operating point calculation resulted in a beta of 98 and a dynamic resistance of 50 ohms in the base and 10^{20} ohms in the collector. These values and the corresponding diffusion and depletion capacitance values were automatically included in the frequency domain calculations. The input coding is shown in Fig. 10.25, and the output Bode and Nichols plots are shown in Fig. 10.26. The circuit was rerun with the transistor diffusion and depletion capacitors set to zero. These results are shown in Fig. 10.27. The entire run, input processing, dc initial solution, frequency domain calculations, and output processing took less than 6 cpu seconds on a CDC 6600 computer.

General Simulation

The ease of using all the major options in SYSCAP is best demonstrated on a single circuit on which each of the options is exercised. The circuit selected was the linear amplifier shown in Fig. 10.28. The circuit was taken from the suggested schematic for a Fairchild μA702 High-Gain, Wide-Band dc Amplifier. No attempt was made, however, to characterize the semiconductors and parasitic elements so that the analysis would represent the operation of that device. Numbers were assigned each node consecutively from 1 to 16. Data for a high-speed 2N3251 transistor, off of the SYSCAP DATA BANK, were used for all eight transistors. Typical data (TYPDIODE) were used for the single diode. Power for all the semiconductors and the collector current of transistor Q1 were monitored on all dc analyses. The stress

```
RUN/OUTPUT
? PV=1,-3,1,2,0,7,5E-9,50
PLOT DELTA T=50.00E-10
1
A    -3.00000E+00           -1.00000E+00            1.00000E+00
B     0.                     3.50000E+00            7.00000E+00
0.          .........AAAAAA.............................B........
5.000E-09   .              .'        A A. A    .         .  BB      .
1.000E-08   .              .          .        A         .  B       .
1.500E-08   .              .          .         .        A .  B     .
2.000E-08   .              .          .         .        . AB       .
2.500E-08   .              .          .         .        . B     A  .
3.000E-08   .              .          .         .      B.         A .
3.500E-08   .              .          .         .  B    .         A .
4.000E-08   .              .          .       B .       .          A.
4.500E-08   .              .          .    B    .       .          A.
5.000E-08   .              .          . B.      .       .          A.
5.500E-08   .              .        . B          .       .          A.
6.000E-08   .              .        B  .         .       .          A.
6.500E-08   .           B  .         .          .       .          A.
7.000E-08   .        .  B  .         .          .       .          A.
7.500E-08   .B              .         .          .       .          A.
8.000E-08   .B              .         .          .       .          A.
8.500E-08   .B              .         .          .       .          A.
9.000E-08   .B              .         .          .       .          A.
9.500E-08   .B              .         .          .       .          A.
1.000E-07   .B              .         .          .       .          A.
1.050E-07   .BB             .         .          .       .          A.
1.100E-07   .         B     .         .          .       .          A.
1.150E-07   .              B.         .          .       .          A.
1.200E-07   .              .   B       .         .       .          A.
1.250E-07   .              .         . B         .       .         A .
1.300E-07   .              .          .         B.       .        A  .
1.350E-07   .              .          .         .       B.      A    .
1.400E-07   .              .          .         .       .  A B       .
1.450E-07   .              .          .         .       A .  B       .
1.500E-07   .              .          .         .     .A  .  B       .
1.550E-07   .              .          .         .  A    .  . B       .
1.600E-07   .              .          .         . A     .  . B       .
1.650E-07   .              .          .       . A       .  . B       .
1.700E-07   .              .          .      . A         .  .  B     .
1.750E-07   .              .          .     A           .  .  B      .
1.800E-07   .              .          .   A .            .  .  B     .
1.850E-07   .              .          .  A .             .  .  B     .
1.900E-07   .              .          .A  .              .  .  B     .
1.950E-07   .              .         . A  .              .  .  B     .
2.000E-07   .              .         . A  .              .  .  B     .
RUN/OUTPUT
? 0
```

Figure 10.24 LOGIC Plotted Results.

levels were arbitrarily set at 12.5 milliwatts and 12 milliamperes, respectively. Auxiliary equations were not requested for any other stresses.

A $\pm 3\%$ tolerance was assigned for all circuit resistors, and a $\pm 10\%$ tolerance was assigned the feedback resistor. The tolerance values are used

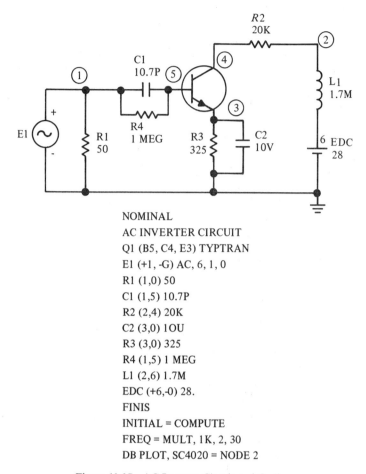

Figure 10.25 AC Inverter Circuit and Coding.

only in the worst-case and Monte Carlo options; all others will ignore them. The NOMINAL input deck is shown in Fig. 10.29. The analyses which were made and the results are as follows:

1. DICAP NOMINAL (Figs. 10.29 through 10.31). A NOMINAL dc analysis was performed to verify the circuit coding, transistor states, and dc bias levels. The state table (Fig. 10.30) indicates that all the semiconductors are active, as was expected for a linear amplifier. Also, the nodal solution (Fig. 10.31) at the output, node 16, was −2.94v, resulting in a gain of 98. This compared favorably with the expected ideal gain of 100. The basic deck used for the NOMINAL dc analysis was modified for each of the other options.

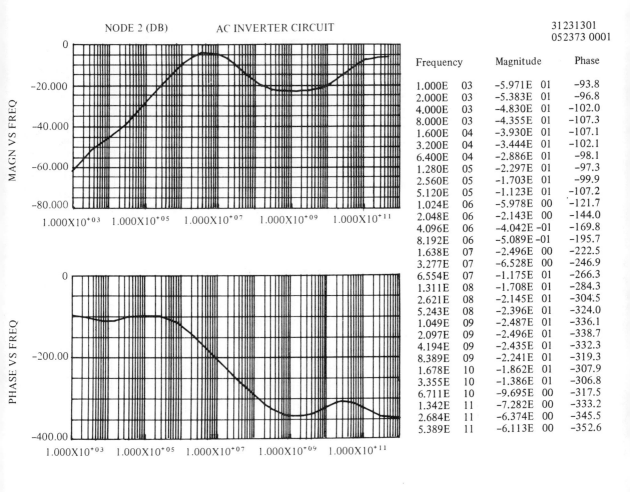

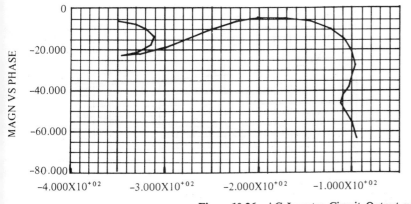

Figure 10.26 AC Inverter Circuit Output with **TYPTRAN**.

Frequency	Magnitude	Phase
1.000E 03	−5.971E 01	−93.8
2.000E 03	−5.383E 01	−96.8
4.000E 03	−4.830E 01	−102.0
8.000E 03	−4.355E 01	−107.3
1.600E 04	−3.930E 01	−107.1
3.200E 04	−3.444E 01	−102.1
6.400E 04	−2.886E 01	−98.1
1.280E 05	−2.297E 01	−97.3
2.560E 05	−1.703E 01	−99.9
5.120E 05	−1.123E 01	−107.2
1.024E 06	−5.978E 00	−121.7
2.048E 06	−2.143E 00	−144.0
4.096E 06	−4.042E −01	−169.8
8.192E 06	−5.089E −01	−195.7
1.638E 07	−2.496E 00	−222.5
3.277E 07	−6.528E 00	−246.9
6.554E 07	−1.175E 01	−266.3
1.311E 08	−1.708E 01	−284.3
2.621E 08	−2.145E 01	−304.5
5.243E 08	−2.396E 01	−324.0
1.049E 09	−2.487E 01	−336.1
2.097E 09	−2.496E 01	−338.7
4.194E 09	−2.435E 01	−332.3
8.389E 09	−2.241E 01	−319.3
1.678E 10	−1.862E 01	−307.9
3.355E 10	−1.386E 01	−306.8
6.711E 10	−9.695E 00	−317.5
1.342E 11	−7.282E 00	−333.2
2.684E 11	−6.374E 00	−345.5
5.389E 11	−6.113E 00	−352.6

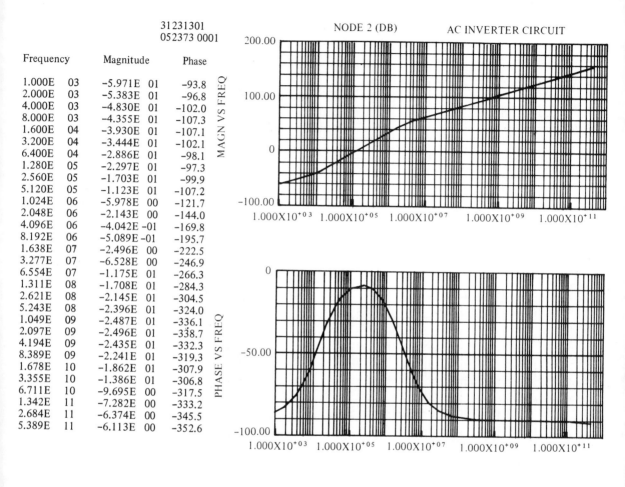

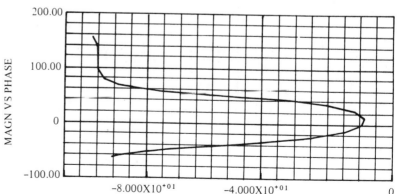

Figure 10.27 AC Inverter Circuit Output with TYPTRAN's Capacitors Set to Zero.

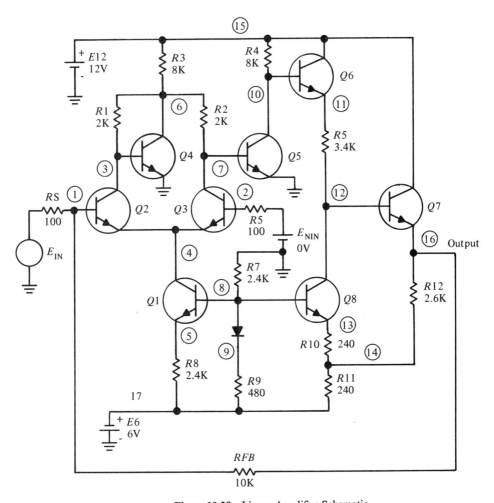

Figure 10.28 Linear Amplifier Schematic.

2. DICAP DCONE (Figs. 10.32 through 10.36). A DCONE dc analysis was performed to determine the maximum input level for which the circuit is essentially linear and the maximum output swing that can be obtained from the amplifier. The initial control card was modified from NOMINAL to DCONE. A plot card was added for the output nodal voltage and a sweep card for the input voltage (Fig. 10.32). EIN was varied from $-.12$ to $+.10$ in 10-millivolt steps. The correlation between step numbers and the value of EIN is shown in Fig. 10.33. This aids the user in the evaluation of the output plots. A typical output solution for one of the steps is shown in Fig. 10.34. At this particular step, Q6 had exceeded the allowed value of power. The output plot (Fig. 10.35) shows linearity from step 3,

```
DICAP
NOMINAL
LINEAR AMPLIFIER
EIN(+1,-0),03,RS=100
E6(+17,-0) -6
E12(+15,-0) 12
ENIN(+2,-0) 0 ,RS=100
Q1,NPN(B8,C4,E5 )P,IC,2N3251
Q2,NPN(B1,E4,C3)P,2N3251
Q3,NPN(C7,E4,B2)P,2N3251
Q4,NPN(B3,C6,E0)P,2N3251
Q5,NPN(B7,C10,E0)P,2N3251
Q6,NPN(B10,C15,E11)P,2N3251
Q7,NPN(B12,C15,E16)P,2N3251
Q8,NPN(B8,C12,E13)P,2N3251
D1(AB,C9)P,TYPDIODE
R1(3,6)2K,   +3,-3
R2(6,7)2K,   +3,-3
R3(6,15)8K,  +3,-3
R4(10,15)8K, +3,-3
R5(11,12)3.4K, +3,-3
R7(8,0)2.4K, +3,-3
R8(5,17)2.4K, +3,-3
R9(9,17)480, +3,-3
R10(13,14)240, +3,-3
R11(14,17)240, +3,-3
R12(14,16)2.6K, +3,-3
RFB(16,1)10K,+10,-10
FINIS
STRESS LIMITS
Q1PT=12.5M
Q1IC=12M
Q2PT=12.5M
Q3PT=12.5M
Q4PT=12.5M
Q5PT=12.5M
Q6PT=12.5M
Q7PT=12.5M
Q8PT=12.5M
D1PD=12.5M
STOP
```

Figure 10.29 DICAP NOMINAL Input Deck.

```
NOMINAL SOLUTION - SEMICONDUCTOR STATES

    TRANSISTOR STATE TABLE

    DEVICE  MODE
      Q1    ACTIVE NORMAL
      Q2    ACTIVE NORMAL
      Q3    ACTIVE NORMAL
      Q4    ACTIVE NORMAL
      Q5    ACTIVE NORMAL
      Q6    ACTIVE NORMAL
      Q7    ACTIVE NORMAL
      Q8    ACTIVE NORMAL

    DIODE STATE TABLE

    DEVICE  MODE
      D1    ON
```

Figure 10.30 DICAP NOMINAL State Table.

INITIAL CONDITIONS

NO STRESS LEVELS EXCEEDED FOR THIS SOLUTION

MATRIX SOLUTIONS= 35

NODE VOLTAGE AND AUXILIARY ANSWERS

```
1  NODE 1   4.34454E-04   10  NODE 10   1.83517E+00   19  Q1  IC   3.14443E-04
2  NODE 2  -1.08316E-04   11  NODE 11   1.18330E+00   20  Q2  PT   1.99199E-04
3  NODE 3   6.52640E-01   12  NODE 12  -2.30288E+00   21  Q3  PT   1.95592E-04
4  NODE 4  -6.05179E-01   13  NODE 13  -5.26944E+00   22  Q4  PT   1.03535E-03
5  NODE 5  -5.24043E+00   14  NODE 14  -5.51615E+00   23  Q5  PT   2.32563E-03
6  NODE 6   9.80561E-01   15  NODE 15   1.20000E+01   24  Q6  PT   1.10304E-02
7  NODE 7   6.57211E-01   16  NODE 16  -2.94504E+00   25  Q7  PT   1.03172E-02
8  NODE 8  -4.61773E+00   17  NODE 17  -6.00000E+00   26  Q8  PT   3.03350E-03
9  NODE 9  -5.08030E+00   18  Q1 PT     1.45880E-03   27  D1  PD   8.86297E-04
```

Figure 10.31 DICAP NOMINAL Nodal and Auxiliary Solutions.

```
DICAP
DCONE
LINEAR AMPLIFIER
EIN(+1,-0).03,RS=100
E6(+17,-0) -6
E12(+15,-0) 12
ENIN(+2,-0) 0 ,RS=100
Q1.NPN(B8,C4,E5 )P,IC,2N3251
Q2.NPN(B1,E4,C3)P,2N3251
Q3.NPN(C7,E4,B2)P,2N3251
Q4.NPN(B3,C6,E0)P,2N3251
Q5.NPN(B7,C10,E0)P,2N3251
Q6.NPN(B10,C15,E11)P,2N3251
Q7.NPN(B12,C15,E16)P,2N3251
Q8.NPN(B8,C12,E13)P,2N3251
D1(A8,C9)P,TYPDIODE
R1(3,6)2K,    +3,-3
R2(6,7)2K,    +3,-3
R3(6,15)8K,   +3,-3
R4(10,15)8K,  +3,-3
R5(11,12)3.4K, +3,-3
R7(8,0)2.4K,  +3,-3
R8(5,17)2.4K, +3,-3
R9(9,17)480,  +3,-3
R10(13,14)240, +3,-3
R11(14,17)240, +3,-3
R12(14,16)2.6K, +3,-3
RFB(16,1)10K,+10,-10
FINIS
STRESS LIMITS
Q1PT=12.5M
Q1IC=12M
Q2PT=12.5M
Q3PT=12.5M
Q4PT=12.5M
Q5PT=12.5M
Q6PT=12.5M
Q7PT=12.5M
Q8PT=12.5M
D1PD=12.5M
STOP
PLOT=NODE16
SPECIAL
PARM=EIN,UL=.1,US=10M,LL=-.12,DS=10M
STOP
```

Figure 10.32 DICAP DCONE Input Deck.

EIN = −.10 volts to step 18, EIN = +.05 volts. The output swing is from approximately +10 to −5 volts. Figure 10.36 summarizes the various overstresses which occurred at each of the step numbers.

3. DICAP CFS (Figs 10.37 through 10.42). A CFS dc analysis was performed to determine the effect of resistor failures on the output voltage and the resulting secondary overstresses. Also, any overstresses as a result of the output being shorted to ground were to be determined. The initial control card was modified from NOMINAL to CFS. A plot card was added for

TABLE
STEP NUMBER – PARAMETER VALUE CORRELATION CHART

STEP NUMBER	VARIABLE PARAMETER	PARAMETER VALUE	STEP NUMBER	VARIABLE PARAMETER	PARAMETER VALUE	STEP NUMBER	VARIABLE PARAMETER	PARAMETER VALUE
0	ALL PARAMETERS AT NOMINAL		8	EIN	-5.0000E-02	15	EIN	2.0000E-02
1	EIN	-1.2000E-01	9	EIN	-4.0000E-02	16	EIN	3.0000E-02
2	EIN	-1.1000E-01	10	EIN	-3.0000E-02	17	EIN	4.0000E-02
3	EIN	-1.0000E-01	11	EIN	-2.0000E-02	18	EIN	5.0000E-02
4	EIN	-9.0000E-02	12	EIN	-1.0000E-02	19	EIN	6.0000E-02
5	EIN	-8.0000E-02	13	EIN	-1.1102E-16	20	EIN	7.0000E-02
6	EIN	-7.0000E-02	14	EIN	1.0000E-02	21	EIN	8.0000E-02
7	EIN	-6.0000E-02				22	EIN	9.0000E-02
						23	EIN	1.0000E-01

NOTE – UNITS FOR PARAMETER VALUES ARE IN VOLTS FOR POWER SUPPLY VARIABLES AND IN OHMS FOR RESISTOR VARIABLES

Figure 10.33 DICAP DCONE Step Number Identification.

STEP NUMBER = 23
EIN = 1.00000E-01

LIST OF OVERSTRESSES

STRESS NAME	STRESS NUMBER	CALCULATED VALUE	ALLOWED VALUE	STRESS RATIO
Q6 PT MAX	24	1.603872E-02	1.250000E-02	1.28310

MATRIX SOLUTIONS= 3

NODE VOLTAGE AND AUXILIARY ANSWERS

1 NODE	1	5.50589E-02	10 NODE	10	3.61862E-02	19	Q1	IC	3.14213E-04
2 NODE	2	-2.62981E-05	11 NODE	11	-6.21137E-01	20	Q2	PT	3.41956E-04
3 NODE	3	6.45417E-01	12 NODE	12	-4.96528E+00	21	Q3	PT	3.77372E-05
4 NODE	4	-5.64905E-01	13 NODE	13	-5.27577E+00	22	Q4	PT	9.60329E-04
5 NODE	5	-5.24099E+00	14 NODE	14	-5.58418E+00	23	Q5	PT	2.17249E-04
6 NODE	6	1.21796E+00	15 NODE	15	1.20000E+01	24	Q6	PT	1.60387E-02
7 NODE	7	6.66974E-01	16 NODE	16	-4.42053E+00	25	Q7	PT	5.90486E-08
8 NODE	8	-4.61831E+00	17 NODE	17	-6.00000E+00	26	Q8	PT	4.01543E-04
9 NODE	9	-5.08084E+00	18 Q1 PT		1.47056E-03	27	D1	PD	8.85718E-04

Figure 10.34 DICAP DCONE Nodal Solutions (Available for each step).

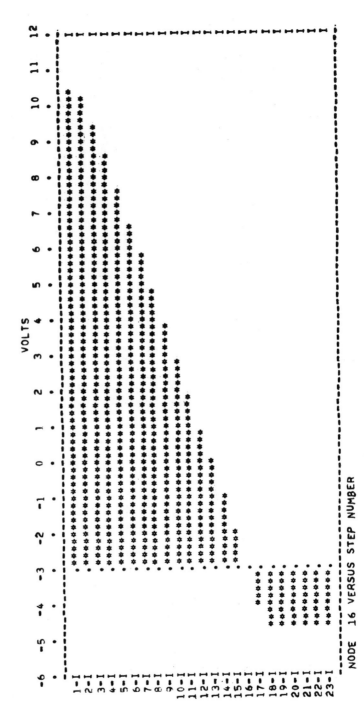

Figure 10.35 DICAP DCONE Graphic Output.

TABLE
DC ONE-AT-A-TIME OVERSTRESS TABLE

STEP NUMBER	VARIABLE PARAMETER	FAILURE POINT VALUE	OVERSTRESSED FUNCTION		ALLOWED LIMITS		CALCULATED VALUE
					MINIMUM	MAXIMUM	
3	EIN	-1.0000E-01	Q7	PT	0.00MW	12.50MW	15.91MW
4	EIN	-9.0000E-02	Q7	PT	0.00MW	12.50MW	20.15MW
5	EIN	-8.0000E-02	Q7	PT	0.00MW	12.50MW	23.85MW
6	EIN	-7.0000E-02	Q7	PT	0.00MW	12.50MW	26.87MW
7	EIN	-6.0000E-02	Q7	PT	0.00MW	12.50MW	29.13MW
8	EIN	-5.0000E-02	Q7	PT	0.00MW	12.50MW	30.59MW
9	EIN	-4.0000E-02	Q7	PT	0.00MW	12.50MW	31.23MW
10	EIN	-3.0000E-02	Q7	PT	0.00MW	12.50MW	30.98MW
11	EIN	-2.0000E-02	Q7	PT	0.00MW	12.50MW	29.81MW
12	EIN	-1.0000E-02	Q7	PT	0.00MW	12.50MW	27.73MW
13	EIN	-1.1102E-16	Q7	PT	0.00MW	12.50MW	24.74MW
14	EIN	1.0000E-02	Q7	PT	0.00MW	12.50MW	20.84MW
15	EIN	2.0000E-02	Q7	PT	0.00MW	12.50MW	16.03MW
17	EIN	4.0000E-02	Q6	PT	0.00MW	12.50MW	13.37MW
18	EIN	5.0000E-02	Q6	PT	0.00MW	12.50MW	15.99MW

Figure 10.36 DICAP DCONE Overstress Summary.

```
DICAP
CFS
LINEAR AMPLIFIER
EIN(+1,-0).03,RS=100
E6(+17,-0) -6
E12(+15,-0) 12
ENIN(+2,-0) 0 ,RS=100
Q1.NPN(B8,C4,E5 )P,IC,2N3251
Q2.NPN(B1,E4,C3)P,2N3251
Q3.NPN(C7,E4,B2)P,2N3251
Q4.NPN(B3,C6,E0)P,2N3251
Q5.NPN(B7,C10,E0)P,2N3251
Q6.NPN(B10,C15,E11)P,2N3251
Q7.NPN(B12,C15,E16)P,2N3251
Q8.NPN(B8,C12,E13)P,2N3251
D1(A8,C9)P,TYPDIODE
R1(3,6)2K,    +3,-3
R2(6,7)2K,    +3,-3
R3(6,15)8K,   +3,-3
R4(10,15)8K,  +3,-3
R5(11,12)3.4K, +3,-3
R7(8,0)2.4K,  +3,-3
R8(5,17)2.4K, +3,-3
R9(9,17)480,  +3,-3
R10(13,14)240, +3,-3
R11(14,17)240, +3,-3
R12(14,16)2.6K, +3,-3
RFB(16,1)10K,+10,-10
FINIS
STRESS LIMITS
Q1PT=12.5M
Q1IC=12M
Q2PT=12.5M
Q3PT=12.5M
Q4PT=12.5M
Q5PT=12.5M
Q6PT=12.5M
Q7PT=12.5M
Q8PT=12.5M
D1PD=12.5M
STOP
PLOT=NODE16
SPECIAL
RALL,N16/0,STOP
```

Figure 10.37 DICAP CFS Input Deck.

the output nodal voltage and a special control card indicating that all resistors were to be failed and that the output node (node 16) was to be shorted to ground (Fig. 10.37). The correlation between failure numbers and induced failure is shown in Fig. 10.38. A typical output solution for one of the induced failures is shown in Fig. 10.39. At this particular induced failure, RFB opened, transistor Q6 had exceeded the allowed value of power. Figure 10.40 provides a graphic display of the output as a function of induced failure. The numbers on the ordinate correspond to

TABLE
FAILURE IDENTIFICATION CHART

FAILURE NUMBER	FAILURE DESCRIPTION	
1	RFB	OPEN
2	RFB	SHORT
3	R10	OPEN
4	R10	SHORT
5	R11	OPEN
6	R11	SHORT
7	R12	OPEN
8	R12	SHORT
9	R1	OPEN
10	R1	SHORT
11	R2	OPEN
12	R2	SHORT
13	R3	OPEN
14	R3	SHORT
15	R4	OPEN
16	R4	SHORT
17	R5	OPEN
18	R5	SHORT
19	R7	OPEN
20	R7	SHORT
21	R8	OPEN
22	R8	SHORT
23	R9	OPEN
24	R9	SHORT
25	NODE 16 AND 0	SHORT

NOTE-SEE FIGURE   FOR CORRELATION

Figure 10.38 DICAP CFS Failure Number Identification.

induced failures as described in the failure number identification chart. Figure 10.41 summarizes the secondary overstresses grouped by type of overstress. A single asterisk (*) indicates greater than 150% stress, and a double asterisk (**) indicates greater than 500% stress. No secondary overstresses occurred as a result of the output being shorted to ground. The program also provides a summary of secondary overstresses grouped by induced failures (this was not included in the text). Figure 10.42 shows the SC4020 graphic stress table which can be obtained by adding CRT to the initial control card.

4. DICAP SOPSTO1 (Figs. 10.43 and 10.44). A SOPSTO1 dc analysis was performed to determine the overstresses as a result of either E6 or E12 being shorted to ground. The initial control card was modified from NOMINAL TO SOPSTO1 and a special control card was added indicating that E6 and E12 were to being shorted to ground (Fig. 10.43). A

```
FAILURE NUMBER  1
RFB  OPEN

STRESS NAME      STRESS NUMBER         CALCULATED VALUE      ALLOWED VALUE      STRESS RATIO

Q6 PT MAX             24                  1.708831E-02       1.250000E-02        1.36707

MATRIX SOLUTIONS=  11

                              NODE VOLTAGE AND AUXILIARY ANSWERS

1 NODE  1   2.98448E-02    10 NODE 10   4.71462E-02    19 Q1 IC    3.02427E-04
2 NODE  2  -5.34032E-05    11 NODE 11  -6.11767E-01    20 Q2 PT    2.86156E-04
3 NODE  3   6.48641E-01    12 NODE 12  -5.24351E+00    21 Q3 PT    8.68216E-05
4 NODE  4  -5.85226E-01    13 NODE 13  -5.30867E+00    22 Q4 PT    1.00729E-03
5 NODE  5  -5.26944E+00    14 NODE 14  -5.65433E+00    23 Q5 PT    1.75852E-04
6 NODE  6   1.12149E+00    15 NODE 15   1.20000E+01    24 Q6 PT    1.70883E-02
7 NODE  7   6.64954E-01    16 NODE 16  -5.65416E+00    25 Q7 PT    1.08835E-06
8 NODE  8  -4.64770E+00    17 NODE 17  -6.00000E+00    26 Q8 PT    1.40304E-04
9 NODE  9  -5.10883E+00    18 Q1 PT    1.41786E-03    27 D1 PD    8.56131E-04
```

Figure 10.39 DICAP CFS Nodal Solutions (Available for each induced failure).

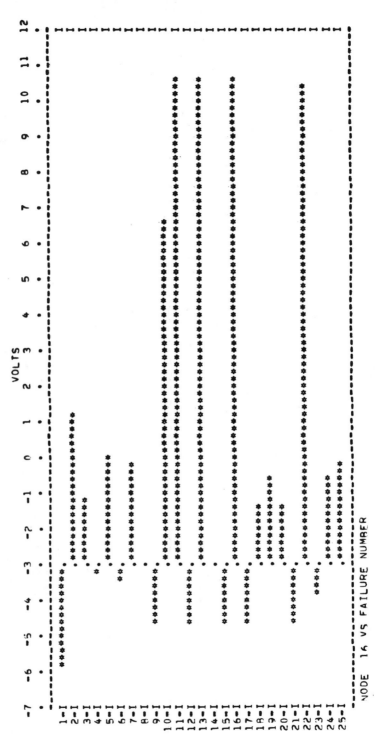

Figure 10.40 DICAP CFS Graphic Output.

TABLE
CIRCUIT FAILURE SIMULATION
(GROUPED BY TYPE OF OVERSTRESS)

LINEAR AMPLIFIER

FAILED PART	MODE OF INDUCED FAILURE	TYPE OF OVERSTRESS		VALUE OF STRESS		ALLOWABLE VALUE	
R8	SHORT	Q1	IC MAX	110.5	MA**	12.0	MA
R8	SHORT	Q1	PT MAX	18.2	MW	12.5	MW
R8	SHORT	Q2	PT MAX	40.0	MW*	12.5	MW
R8	SHORT	Q3	PT MAX	38.6	MW	12.5	MW
R8	SHORT	Q4	PT MAX	14.5	MW	12.5	MW
R8	SHORT	Q5	PT MAX	14.5	MW	12.5	MW
R4	SHORT	Q5	PT MAX	484.9	MW**	12.5	MW
RFB	OPEN	Q6	PT MAX	17.1	MW	12.5	MW
R4	OPEN	Q6	PT MAX	16.2	MW	12.5	MW
R10	SHORT	Q6	PT MAX	15.2	MW	12.5	MW
R8	OPEN	Q6	PT MAX	15.2	MW	12.5	MW
R1	OPEN	Q6	PT MAX	16.1	MW	12.5	MW
R2	SHORT	Q6	PT MAX	17.1	MW	12.5	MW
R12	OPEN	Q6	PT MAX	13.5	MW	12.5	MW
R11	SHORT	Q7	PT MAX	27.2	MW*	12.5	MW
RFB	SHORT	Q7	PT MAX	165.7	MW**	12.5	MW
R9	SHORT	Q7	PT MAX	23.1	MW*	12.5	MW
R5	SHORT	Q7	PT MAX	19.6	MW*	12.5	MW
R10	OPEN	Q7	PT MAX	20.7	MW*	12.5	MW
R7	OPEN	Q7	PT MAX	23.7	MW*	12.5	MW
NODE 16 AND	OSHORT	NONE					
R3	OPEN	NONE					
R5	OPEN	NONE					
R2	OPEN	NONE					
R11	OPEN	NONE					
R7	SHORT	NONE					
R9	OPEN	NONE					

* GREATER THAN 150 PERCENT STRESS--RELIABILITY DEGRADED.
** GREATER THAN 500 PERCENT STRESS--PROBABLE SECONDARY FAILURE.

Figure 10.41 DICAP CFS Overstress Summary Sorted by Type of Overstress.

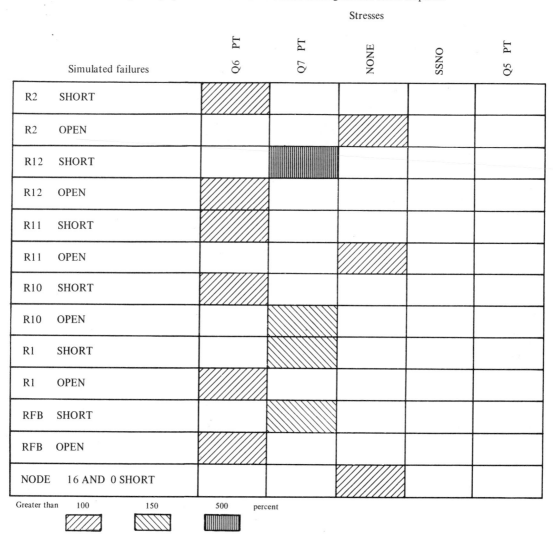

Figure 10.42 DICAP CFS Graphic Overstress Summary.

typical output solution for a power supply turn-on configuration is shown in Fig. 10.44. No overstresses were encountered for either failure; therefore, no overstress summary resulted. If overstresses had occurred, the program would have automatically provided a summary table similar to that in the DCONE and CFS options.

5. DICAP SENSIT (Figs. 10.45 and 10.46). A SENSIT analysis was performed to determine the sensitive dc parameters in the circuit. The initial

```
DICAP
SOPSTO1
LINEAR AMPLIFIER
EIN(+1,-0),03,RS=100
E6(+17,-0) -6
E12(+15,-0) 12
ENIN(+2,-0) 0 ,RS=100
Q1,NPN(B8,C4,E5 )P,IC,2N3251
Q2,NPN(B1,E4,C3)P,2N3251
Q3,NPN(C7,E4,B2)P,2N3251
Q4,NPN(B3,C6,E0)P,2N3251
Q5,NPN(B7,C10,E0)P,2N3251
Q6,NPN(B10,C15,E11)P,2N3251
Q7,NPN(B12,C15,E16)P,2N3251
Q8,NPN(B8,C12,E13)P,2N3251
D1(A8,C9)P,TYPDIODE
R1(3,6)2K,    +3,-3
R2(6,7)2K,    +3,-3
R3(6,15)8K,   +3,-3
R4(10,15)8K,  +3,-3
R5(11,12)3.4K, +3,-3
R7(8,0)2.4K,  +3,-3
R8(5,17)2.4K, +3,-3
R9(9,17)480,  +3,-3
R10(13,14)240, +3,-3
R11(14,17)240, +3,-3
R12(14,16)2.6K, +3,-3
RFB(16,1)10K,+10,-10
FINIS
STRESS LIMITS
Q1PT=12.5M
Q1IC=12M
Q2PT=12.5M
Q3PT=12.5M
Q4PT=12.5M
Q5PT=12.5M
Q6PT=12.5M
Q7PT=12.5M
Q8PT=12.5M
D1PD=12.5M
STOP
SPECIAL
E12,E6,STOP
```

Figure 10.43 DICAP SOPSTO1 Input Deck.

control card was modified from NOMINAL to SENSIT, SENTOL = .01 (Fig. 10.45). The entry of SENTOL = .01 will result in sensitivities down to 0.01 being outputed instead of the default value of 0.05. The sensitivity of every node and the auxiliary equations were calculated. A typical sensitivity output is shown in Fig. 10.46.

6. W/C SEL (Figs. 10.47 through 10.51). A W/C SEL analysis was performed to determine the minimum and maximum nodal voltages and whether any overstresses that existed. The stress calculations were not per se worst cased. The initial control card was modified from NOMINAL to W/C SEL, and a special control card was added to worst-case all nodes (Fig.

```
                    POWER SUPPLY TURN ON CONFIGURATION   2

             SOURCE VOLTAGE (VOLTS)      SOURCE IMPEDANCE (OHMS)

              E12         0.00            E12 RS    0.
              E6         -6.00            E6  RS    0.

      NO STRESS LEVELS EXCEEDED FOR THIS SOLUTION

MATRIX SOLUTIONS=   15

                     NODE VOLTAGE AND AUXILIARY ANSWERS

1 NODE  1  -2.38574E-02   10 NODE 10  -5.83899E-02   19 Q1 IC  3.14139E-04
2 NODE  2  -1.46891E-02   11 NODE 11  -7.15755E-01   20 Q2 PT  5.94390E-05
3 NODE  3  -6.28903E-01   12 NODE 12  -5.06726E+00   21 Q3 PT  9.09881E-05
4 NODE  4  -6.33635E-01   13 NODE 13  -5.27601E+00   22 Q4 PT  1.97855E-09
5 NODE  5  -5.24117E+00   14 NODE 14  -5.58504E+00   23 Q5 PT  2.06073E-09
6 NODE  6  -5.60251E-01   15 NODE 15   0.            24 Q6 PT  9.15637E-04
7 NODE  7  -6.31661E-01   16 NODE 16  -4.43750E+00   25 Q7 PT  7.11868E-09
8 NODE  8  -4.61849E+00   17 NODE 17  -6.00000E+00   26 Q8 PT  2.72276E-04
9 NODE  9  -5.08102E+00   18 Q1   PT   1.44868E-03   27 D1 PD  8.85533E-04
```

Figure 10.44 DICAP SOPSTO1 Nodal Solutions (Available for each power supply turn-on configuration).

```
DICAP
SENSIT,SENTOL=.01
LINEAR AMPLIFIER
EIN(+1,-0),.03,RS=100
E6(+17,-0) -6
E12(+15,-0) 12
ENIN(+2,-0) 0 ,RS=100
Q1,NPN(B8,C4,E5 )P,IC,2N3251
Q2,NPN(B1,E4,C3)P,2N3251
Q3,NPN(C7,E4,B2)P,2N3251
Q4,NPN(B3,C6,E0)P,2N3251
Q5,NPN(B7,C10,E0)P,2N3251
Q6,NPN(B10,C15,E11)P,2N3251
Q7,NPN(B12,C15,E16)P,2N3251
Q8,NPN(B8,C12,E13)P,2N3251
D1(A8,C9)P,TYPDIODE
R1(3,6)2K,   +3,-3
R2(6,7)2K,   +3,-3
R3(6,15)8K,  +3,-3
R4(10,15)8K, +3,-3
R5(11,12)3.4K, +3,-3
R7(8,0)2.4K,  +3,-3
R8(5,17)2.4K, +3,-3
R9(9,17)480,  +3,-3
R10(13,14)240, +3,-3
R11(14,17)240, +3,-3
R12(14,16)2.6K, +3,-3
RFB(16,1)10K,+10,-10
FINIS
STRESS LIMITS
Q1PT=12.5M
Q1IC=12M
Q2PT=12.5M
Q3PT=12.5M
Q4PT=12.5M
Q5PT=12.5M
Q6PT=12.5M
Q7PT=12.5M
Q8PT=12.5M
D1PD=12.5M
STOP
```

Figure 10.45 DICAP SENSIT Input Deck.

10.47). Figure 10.48 shows the W/C nodal results for a typical solution, node 6. The parameters which were set to their limits are noted, and any overstresses which occurred are printed out. Also, if any changes of state occurred, these would have been indicated. Figures 10.49 and 10.50 summarize the worst-case limits for the nodes and auxiliaries, respectively. The output node went from $+1.0$ to -4.3 volts. The overstresses which occurred are summarized in Fig. 10.51, and the solution which produced the overstress is noted.

7. MONTECARLO (Figs. 10.52 through 10.54). A MONTECARLO analysis was performed to determine the output voltage distribution assuming

SENSITIVITY OF NODE 14 WITH RESPECT TO A POSITIVE CHANGE IN THE FOLLOWING PARAMETERS

PARAMETER	SENSITIVITY
Q3 VBE	-.021
R2	-.022
EIN RS	-.024
R11	-.045
R9	-.061
E6	-.953
R7	.064
EIN	.025
RFB	.024
R12	.022
R1	.022
Q2 VBE	.021
R10	.020

Figure 10.46 DICAP SENSIT Sensitivity Output (Available for all nodes and auxiliaries).

```
DICAP
W/C SEL
LINEAR AMPLIFIER
EIN(+1,-0),03,RS=100
E6(+17,-0) -6
E12(+15,-0) 12
ENTN(+2,-0) 0 ,RS=100
Q1,NPN(B8,C4,E5 )P,IC,2N3251
Q2,NPN(B1,E4,C3)P,2N3251
Q3,NPN(C7,E4,B2)P,2N3251
Q4,NPN(B3,C6,E0)P,2N3251
Q5,NPN(B7,C10,E0)P,2N3251
Q6,NPN(B10,C15,E11)P,2N3251
Q7,NPN(B12,C15,E16)P,2N3251
Q8,NPN(B8,C12,E13)P,2N3251
D1(A8,C9)P,TYPDIODE
R1(3,6)2K,    +3,-3
R2(6,7)2K,    +3,-3
R3(6,15)8K,   +3,-3
R4(10,15)8K,  +3,-3
R5(11,12)3.4K, +3,-3
R7(8,0)2.4K,  +3,-3
R8(5,17)2.4K, +3,-3
R9(9,17)480,  +3,-3
R10(13,14)240, +3,-3
R11(14,17)240, +3,-3
R12(14,16)2.6K, +3,-3
RFB(16,1)10K,+10,-10
FINIS
STRESS LIMITS
Q1PT=12.5M
Q1IC=12M
Q2PT=12.5M
Q3PT=12.5M
Q4PT=12.5M
Q5PT=12.5M
Q6PT=12.5M
Q7PT=12.5M
Q8PT=12.5M
D1PD=12.5M
STOP
SPECIAL
  NODES,STOP
```

Figure 10.47 DICAP W/C SEL Input Deck.

that the components in the circuit were normally distributed. A sample size of 100 solutions was used. The initial control card was modified from NOMINAL to MONTECARLO, NORMAL, MCARLO = 100. A plot card was added to obtain the histogram for the output nodal voltage (Fig. 10.52). The nominal, minimum, maximum, average, standard deviation, and three sigma limits for all nodes and auxiliary equations are shown in Fig. 10.53. The histogram for the output is shown in Fig. 10.54. The three sigma limits of -2.63 and -3.3 for the output are within the worst-case limits calculated in the previous analysis.

```
WORST CASE MINIMUM COMPUTATION FOR NODE  6
THE FOLLOWING PARAMETERS HAVE BEEN SET TO WORST CASE.

         PARAMETER   PARAMETER    PARAMETER       SENSITIVITY
         NAME        MATCHED TO   VALUE           X TOLERANCE

         D1   VD                     MIN           2.519
         R7                       2.472000E+03     -.944
         R8                       2.472000E+03     -.935
         R9                       4.656000E+02      .878
         R1                       1.940000E+03      .498
         R2                       1.940000E+03      .481
         Q5   BNBI                   MAX           -.269
         Q4   BNBI                   MAX           -.232
         Q5   VBEC                   MAX           -.164
         Q4   VBEC                   MAX           -.154
         Q1   BNBI                   MIN            .087
         R3                       8.240000E+03    -.071
         R4                       8.240000E+03    -.057
         Q1   VBEC                   MIN            .051

NO STRESS LEVELS EXCEEDED FOR THIS SOLUTION

MATRIX SOLUTIONS=   5

                        NODE VOLTAGE AND AUXILIARY ANSWERS

1 NODE 1   5.39623E-04   10 NODE 10   1.05386E+00   19 Q1 IC   2.60005E-04
2 NODE 2  -9.13635E-05   11 NODE 11   4.08326E-01   20 Q2 PT   1.64040E-04
3 NODE 3   6.50190E-01   12 NODE 12  -2.29313E+00   21 Q3 PT   1.60528E-04
4 NODE 4  -6.00401E-01   13 NODE 13  -5.37562E+00   22 Q4 PT   9.85862E-04
5 NODE 5  -5.34922E+00   14 NODE 14  -5.56641E+00   23 Q5 PT   1.39813E-03
6 NODE 6   9.11453E-01   15 NODE 15   1.20000E+01   24 Q6 PT   9.15859E-03
7 NODE 7   6.54973E-01   16 NODE 16  -2.93615E+00   25 Q7 PT   1.06624E-02
8 NODE 8  -4.73007E+00   17 NODE 17  -6.00000E+00   26 Q8 PT   2.43900E-03
9 NODE 9  -5.11280E+00   18 Q1   PT  1.23673E-03   27 D1 PD   7.29291E-04
```

Figure 10.48 DICAP W/C SEL Nodal Solutions (Available for each node and auxiliary being W/C'd).

TABLE

SUMMARY OF WORST CASE NODE VOLTAGES

NODE NAME	NOMINAL VALUE (VOLTS)	MINIMUM VALUE (VOLTS)	MAXIMUM VALUE (VOLTS)
NODE 1	4.34454E-04	-1.78156E-02	3.85905E-02
NODE 2	-1.08316E-04	-5.38798E-05	-2.36981E-04
NODE 3	6.52640E-01	6.47762E-01	6.57953E-01
NODE 4	-6.05179E-01	-5.64161E-01	-6.32835E-01
NODE 5	-5.24043E+00	-5.16153E+00	-5.35092E+00
NODE 6	9.80561E-01	9.11453E-01	1.14923E+00
NODE 7	6.57211E-01	6.50957E-01	6.71380E-01
NODE 8	-4.61773E+00	-4.53638E+00	-4.73212E+00
NODE 9	-5.08030E+00	-5.02043E+00	-5.13120E+00
NODE 10	1.83517E+00	1.19458E+00	3.08958E+00
NODE 11	1.18330E+00	-6.46478E-01	2.43171E+00
NODE 12	-2.30288E+00	1.67851E+00	-5.18762E+00
NODE 13	-5.26964E+00	-5.18794E+00	-5.38269E+00
NODE 14	-5.51615E+00	-5.37303E+00	-5.59972E+00
NODE 15	1.20000E+01	1.20000E+01	1.20000E+01
NODE 16	-2.94508E+00	1.00413E+00	-4.31241E+00
NODE 17	-6.00000E+00	-6.00000E+00	-6.00000E+00

Figure 10.49 DICAP W/C SEL Nodal Voltage Summary.

TABLE

SUMMARY OF WORST CASE AUXILIARY SOLUTIONS

SOLUTION NAME	NOMINAL VALUE	MINIMUM VALUE	MAXIMUM VALUE	STRESS RATIO
Q1 PT	1.45880E-03	1.23634E-03	1.63350E-03	1.30680E-01
Q1 IC	3.14443E-04	2.59858E-04	3.58533E-04	2.98778E-02
Q2 PT	1.99199E-04	1.56659E-04	2.97703E-04	2.38162E-02
Q3 PT	1.95592E-04	1.44909E-04	2.31425E-04	1.85140E-02
Q4 PT	1.03535E-03	9.85862E-04	1.08903E-03	8.71227E-02
Q5 PT	2.32563E-03	1.03971E-04	3.53533E-03	2.82826E-01
Q6 PT	1.10304E-02	1.65824E-03	1.68000E-02	1.34400E+00
Q7 PT	1.03172E-02	6.29136E-08	2.77926E-02	2.22341E+00
Q8 PT	3.03350E-03	7.69834E-05	4.09503E-03	3.27602E-01
D1 PD	8.86297E-04	7.27499E-04	9.77319E-04	7.81855E-02

NOTE - N/A INDICATES STRESS RATIO IS NOT APPLICABLE.
       NSL INDICATES NO STRESS LIMITS WERE SPECIFIED.

Figure 10.50 DICAP W/C SEL Auxiliary Solution Summary.

TABLE
SUMMARY OF WORST-CASE OVERSTRESSES

STRESS NAME			STRESS NUMBER	SOLUTION BEING WORST-CASED	CALCULATED VALUE	ALLOWED VALUE	STRESS RATIO
Q6	PT	MAX	24	NODE 1 MIN	12.639 MW	12.500 MW	1.011
Q7	PT	MAX	25	NODE 1 MAX	12.989 MW	12.500 MW	1.039
Q7	PT	MAX	25	NODE 2 MIN	27.793 MW	12.500 MW	2.223
Q6	PT	MAX	24	NODE 2 MAX	16.800 MW	12.500 MW	1.344
Q6	PT	MAX	24	NODE 4 MAX	12.521 MW	12.500 MW	1.002
Q7	PT	MAX	25	NODE 10 MAX	13.266 MW	12.500 MW	1.061
Q7	PT	MAX	25	NODE 11 MAX	13.266 MW	12.500 MW	1.061
Q7	PT	MAX	25	NODE 12 MIN	15.097 MW	12.500 MW	1.208
Q7	PT	MAX	24	NODE 12 MAX	14.191 MW	12.500 MW	1.135
Q7	PT	MAX	25	NODE 14 MIN	12.681 MW	12.500 MW	1.014
Q7	PT	MAX	25	NODE 16 MIN	14.549 MW	12.500 MW	1.164
Q6	PT	MAX	24	NODE 16 MAX	14.139 MW	12.500 MW	1.131

NOTE: THIS TABLE DOES NOT INCLUDE POSSIBLE OVERSTRESSES AT NOMINAL.

NA INDICATES A STRESS LESS THAN MINIMUM OR A NODE VOLTAGE STRESS.

Figure 10.51 DICAP W/C SEL Overstress Summary.

```
DICAP
MONTECARLO,NORMAL,MCARLO=100
LINEAR AMPLIFIER
EIN(+1,-0).03,RS=100
E6(+17,-0),-6
E12(+15,-0) 12
ENIN(+2,-0) 0 ,RS=100
Q1,NPN(B8,C4,E5 )P,IC,2N3251
Q2,NPN(B1,E4,C3)P,2N3251
Q3,NPN(C7,E4,B2)P,2N3251
Q4,NPN(B3,C6,E0)P,2N3251
Q5,NPN(B7,C10,E0)P,2N3251
Q6,NPN(B10,C15,E11)P,2N3251
Q7,NPN(B12,C15,E16)P,2N3251
Q8,NPN(B8,C12,E13)P,2N3251
D1(A8,C9)P,TYPDIODE
R1(3,6)2K,  +3,-3
R2(6,7)2K,  +3,-3
R3(6,15)8K,  +3,-3
R4(10,15)8K,  +3,-3
R5(11,12)3.4K,  +3,-3
R7(8,0)2.4K,  +3,-3
R8(5,17)2.4K,  +3,-3
R9(9,17)480,  +3,-3
R10(13,14)240,  +3,-3
R11(14,17)240,  +3,-3
R12(14,16)2.6K,  +3,-3
RFB(16,1)10K,+10,-10
FINIS
STRESS LIMITS
Q1PT=12.5M
Q1IC=12M
Q2PT=12.5M
Q3PT=12.5M
Q4PT=12.5M
Q5PT=12.5M
Q6PT=12.5M
Q7PT=12.5M
Q8PT=12.5M
D1PD=12.5M
STOP
PLOT=NODE16
```

Figure 10.52 DICAP MONTECARLO Input Deck.

8. ALCAP (Figs. 10.55 through 10.57). An ALCAP analysis was performed to determine the frequency response of the circuit open loop; i.e., the feedback resistor RFB was removed. The input dc source was changed to an ac source, with 0 volt dc, 1 volt ac, and 0° phase. All stress-monitoring cards were removed. A dB plot of the output was requested on an SC4020. The initial dc conditions were to be automatically calculated and the semiconductor models linearized. The frequency points at which calculations are required were indicated. The completed deck setup is shown in Fig. 10.55. The nodal and auxiliary answers are calculated for each of the

PAGE 1

MONTECARLO OUTPUT VARIABLE
SUMMARY TABLES

SOLN NO.	OUTPUT VARIABLE	NOMINAL VALUE	MINIMUM VALUE	MAXIMUM VALUE	AVERAGE VALUE	STANDARD DEVIATION	3 SIGMA MINUS	3 SIGMA PLUS
1	NODE 1	4.34454E-04	-4.45861E-04	1.76859E-03	5.11527E-04	4.97043E-04	-9.79601E-04	2.00266E-03
2	NODE 2	-1.08316E-04	-1.87937E-04	-7.02446E-05	-1.09877E-04	2.00474E-05	-1.70020E-04	-4.97352E-05
3	NODE 3	6.52640E-01	6.51277E-01	6.55388E-01	6.52984E-01	1.00423E-03	6.49971E-01	6.55996E-01
4	NODE 4	-6.05179E-01	-6.06855E-01	-6.02552E-01	-6.05260E-01	7.64573E-04	-6.07554E-01	-6.02966E-01
5	NODE 5	-5.24043E+00	-5.30384E+00	-5.19736E+00	-5.24553E+00	2.00430E-02	-5.30566E+00	-5.18541E+00
6	NODE 6	9.80561E-01	9.55112E-01	9.97023E-01	9.78903E-01	8.52147E-03	9.53338E-01	1.00447E+00
7	NODE 7	-6.57211E-01	-6.52118E-01	-6.60732E-01	-6.57669E-01	1.29165E-03	-6.53794E-01	-6.61543E-01
8	NODE 8	-4.61718E+00	-4.68338E+00	-4.57285E+00	-4.62268E+00	2.05962E-02	-4.68446E+00	-4.56089E+00
9	NODE 9	-5.08030E+00	-5.10550E+00	-5.05103E+00	-5.07968E+00	1.19501E-02	-5.11553E+00	-5.04383E+00
10	NODE 10	1.83517E+00	1.39020E+00	2.26148E+00	1.80612E+00	1.61028E-01	1.32304E+00	2.28921E+00
11	NODE 11	1.18330E+00	7.40721E-01	1.60766E+00	1.15421E+00	1.60168E-01	6.73702E-01	1.63471E+00
12	NODE 12	-2.30288E+00	-2.55151E+00	-2.00724E+00	-2.29803E+00	1.07824E-01	-2.62150E+00	-1.97456E+00
13	NODE 13	-5.26964E+00	-5.33048E+00	-5.22724E+00	-5.27467E+00	1.95087E-02	-5.33320E+00	-5.21615E+00
14	NODE 14	-5.51616E+00	-5.54728E+00	-5.49491E+00	-5.51885E+00	1.13533E-02	-5.55291E+00	-5.48479E+00
15	NODE 15	1.20000E+01	1.20000E+01	1.20000E+01	1.20000E+01	0.	1.20000E+01	1.20000E+01
16	NODE 16	-2.94508E+00	-3.19014E+00	-2.65362E+00	-2.94065E+00	1.06106E-01	-3.25897E+00	-2.62234E+00
17	NODE 17	-6.00000E+00	-6.00000E+00	-6.00000E+00	-6.00000E+00	0.	-6.00000E+00	-6.00000E+00
18	Q1 PT	1.45880E-03	1.34216E-03	1.52347E-03	1.45012E-03	3.46992E-05	1.34603E-03	1.55422E-03
19	Q1 IC	3.14443E-04	2.85119E-04	3.31540E-04	3.12267E-04	8.75653E-06	2.85997E-04	3.38537E-04
20	Q2 PT	1.99199E-04	1.84347E-04	2.11773E-04	1.98111E-04	5.67445E-06	1.81088E-04	2.15135E-04
21	Q3 PT	1.95592E-04	1.72401E-04	2.06367E-04	1.94076E-04	5.89050E-06	1.76405E-04	2.11748E-04
22	Q4 PT	1.03535E-03	9.98816E-04	1.08454E-03	1.03715E-03	1.49056E-05	9.92438E-04	1.08187E-03
23	Q5 PT	2.32563E-03	1.83770E-03	2.73310E-03	2.29101E-03	1.67992E-04	1.78703E-03	2.79499E-03
24	Q6 PT	1.10304E-02	9.48677E-03	1.17954E-02	1.09279E-02	4.12316E-04	9.69099E-03	1.21649E-02
25	Q7 PT	1.03172E-02	8.88309E-03	1.20644E-02	1.03442E-02	5.77473E-04	8.61174E-03	1.20766E-02
26	Q8 PT	3.03350E-03	2.68594E-03	3.37656E-03	3.00595E-03	1.27211E-04	2.62432E-03	3.38759E-03
27	D1 PD	8.86297E-04	7.84146E-04	9.56062E-04	8.76933E-04	3.60673E-05	7.68731E-04	9.85135E-04

NO OVERSTRESSES WERE DETECTED IN THE MONTECARLO OUTPUT RESULTS.

...

Figure 10.53 DICAP MONTECARLO Solution Summary Table.

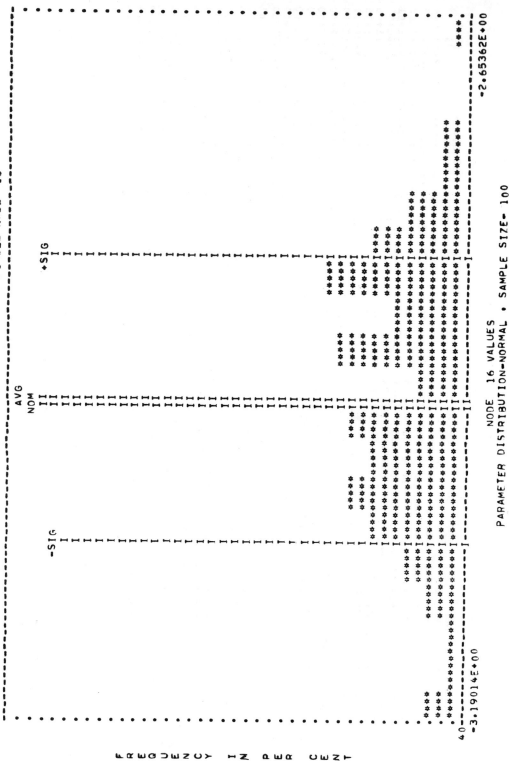

Figure 10.54 DICAP MONTECARLO Output Voltage Distribution.

```
ALCAP
NOMINAL
LINEAR AMPLIFIER
EIN(+1,-0)AC,0,RS=100,1,0
E6(+17,-0) -6
E12(+15,-0) 12
ENIN(+2,-0) 0 ,RS=100
Q1,NPN(B8,C4,E5 )P,IC,2N3251
Q2,NPN(B1,E4,C3)P,2N3251
Q3,NPN(C7,E4,B2)P,2N3251
Q4,NPN(B3,C6,E0)P,2N3251
Q5,NPN(B7,C10,E0)P,2N3251
Q6,NPN(B10,C15,E11)P,2N3251
Q7,NPN(B12,C15,E16)P,2N3251
Q8,NPN(B8,C12,E13)P,2N3251
D1(A8,C9)P,TYPDIODE
R1(3,6)2K,   +3,-3
R2(6,7)2K,   +3,-3
R3(6,15)8K,  +3,-3
R4(10,15)8K, +3,-3
R5(11,12)3.4K, +3,-3
R7(8,0)2.4K, +3,-3
R8(5,17)2.4K, +3,-3
R9(9,17)480, +3,-3
R10(13,14)240, +3,-3
R11(14,17)240, +3,-3
R12(14,16)2.6K, +3,-3
FINIS
DBPLOT=NODE16
INITIAL=COMPUTE
FREQ=100K,500K,1MEG,5MEG,10MEG,15MEG,20MEG,25MEG
```

Figure 10.55 ALCAP Input Deck.

entered frequencies. Semiconductors powers are zero for ac calculations. A typical output at 0.5 MHz is shown in Fig. 10.56. The Bode, Nichols plot for the output (or gain since the input is 1.0) is shown in Fig. 10.57. The gain roll off of this device was remarkably close to the 702 Fairchild specification.

9. TRACAP (Figs. 10.58 through 10.60). A TRACAP analysis was performed to determine the transient response of the circuit. The initial control card was modified from NOMINAL to PLT/PRT = 20. This resulted in a reduction of the normal output printing to one-twentieth of the calculated solutions. The input dc source was replaced by a single transient pulse from -100 to $+100$ millivolts with a 0 time delay, 100-nanosecond rise time, 5 microseconds on time and a 100-nanosecond fall time. All stress-monitoring cards were again removed. A printer plot of the output was requested. The initial dc conditions were to be automatically calculated.

$$ FREQUENCY = 5.00000E+05\ HZ - AC\ NODE\ VOLTAGES\ AND\ AUXILIARY\ ANSWERS:\ $$

SOLUTION NUMBER	SOLUTION NAME	REAL COMPONENT	IMAGINARY COMPONENT	MAGNITUDE	PHASE$ (DEGREES)
1	NODE 1	9.98440E-01	-3.62692E-03	9.98647E-01	-.208
2	NODE 2	4.65663E-03	2.04847E-03	5.08728E-03	23.745
3	NODE 3	-1.13502E-01	-2.63179E-02	1.16513E-01	-166.945
4	NODE 4	6.39320E-01	-7.03669E-02	6.43181E-01	-6.281
5	NODE 5	-4.29787E-01	1.95752E-02	4.72266E-01	-204.488
6	NODE 6	4.38979E+00	5.50497E-01	4.42418E+00	7.148
7	NODE 7	3.39981E-01	-7.74525E-01	8.43464E-01	-66.674
8	NODE 8	-4.43784E+00	2.02164E+00	4.87662E+00	-204.491
9	NODE 9	-4.22981E+00	1.92475E+00	4.64714E+00	-204.468
10	NODE 10	-1.71013E+01	2.73144E+02	2.73679E+02	93.583
11	NODE 11	-1.31024E+01	2.77538E+02	2.77847E+02	92.703
12	NODE 12	-1.92036E+02	5.03826E+02	5.39183E+02	69.135
13	NODE 13	-1.76961E+00	7.23108E+00	7.44447E+00	103.751
14	NODE 14	7.59450E+00	2.56011E+01	2.67038E+01	73.477
15	NODE 15	0.	0.	0.	0.000
16	NODE 16	1.91313E+02	5.01955E+02	5.37177E+02	69.136
17	NODE 17	0.	0.	0.	0.000
18	Q1	0.	0.	0.	0.000
19	Q1	-1.75187E-03	8.76342E-04	1.95883E-03	153.424
20	Q2	0.	0.	0.	0.000
21	Q3	0.	0.	0.	0.000
22	Q4	0.	0.	0.	0.000
23	Q5	0.	0.	0.	0.000
24	Q6	0.	0.	0.	0.000
25	Q7	0.	0.	0.	0.000
26	Q8	0.	0.	0.	0.000
27	D1	0.	0.	0.	0.000

Figure 10.56 ALCAP Output Solutions (Available at each frequency).

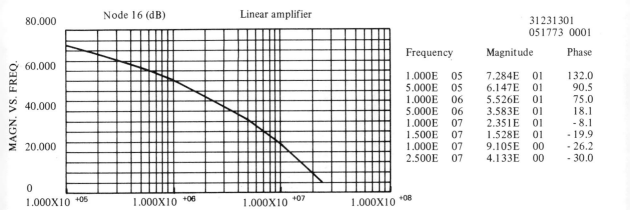

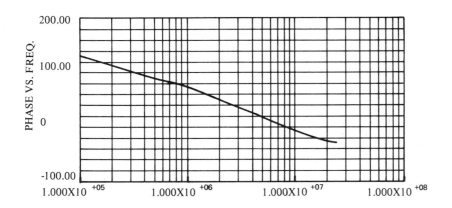

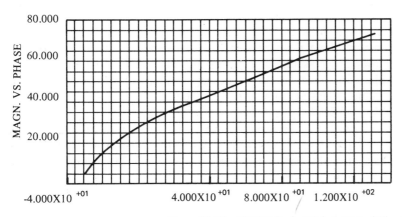

Figure 10.57 ALCAP Bode, Nichols Plot of Circuit Gain.

```
TRACAP
PLT/PRT=20
LINEAR AMPLIFIER
EIN(+1,-0)PULSE,RS=100,-.1,.1,0,.1U,5U,.1U,1
E6(+17,-0)  -6
E12(+15,-0) 12
ENIN(+2,-0) 0 ,RS=100
Q1.NPN(B8,C4,E5 )P,IC,2N3251
Q2.NPN(B1,E4,C3)P,2N3251
Q3.NPN(C7,E4,B2)P,2N3251
Q4.NPN(B3,C6,E0)P,2N3251
Q5.NPN(B7,C10,E0)P,2N3251
Q6.NPN(B10,C15,E11)P,2N3251
Q7.NPN(B12,C15,E16)P,2N3251
Q8.NPN(B8,C12,E13)P,2N3251
D1(A8,C9)P,TYPDIODE
R1(3,6)2K,    +3,-3
R2(6,7)2K,    +3,-3
R3(6,15)8K,   +3,-3
R4(10,15)8K,  +3,-3
R5(11,12)3.4K, +3,-3
R7(8,0)2.4K,  +3,-3
R8(5,17)2.4K, +3,-3
R9(9,17)480,  +3,-3
R10(13,14)240, +3,-3
R11(14,17)240, +3,-3
R12(14,16)2.6K, +3,-3
RFB(16,1)10K,+10,-10
FINIS
PLOT=NODE16
INITIAL=COMPUTE
TIME=.1U,10U
```

Figure 10.58 TRACAP Input Deck.

The time step was to have a maximum value of 100 nanoseconds for a total run of 10 microseconds. Since the input waveforms have a rise and fall time which is less than 10 times the entered time step, the program will automatically cut the time step to 10 nanoseconds during that period. The overall deck setup is shown in Fig. 10.58. The nodal and auxiliary solutions are shown for three time steps in Fig. 10.59. The printer plot of the output voltage is shown in Fig. 10.60.

A tabular listing of selected nodes and auxiliaries could have been requested for analyses 2, 3, 8, and 9. Also, either printer or SC4020 plots could have been obtained in analyses 2, 3, 7, 8, and 9. Transient overstresses could have been monitored in analysis 9.

```
SIMULATION TIME=  1.14000E-06     TIME STEP= 1.00000E-07  MATRIX SOLUTIONS= 133   GAMMADDOT= 0.

                                  NODE VOLTAGE AND AUXILIARY ANSWERS
 1  NODE   1    5.47801E-02       10  NODE 10    3.60337E-02      19  Q1  IC    3.10819E-04
 2  NODE   2   -2.60284E-05       11  NODE 11   -6.20393E-01      20  Q2  PT    3.37907E-04
 3  NODE   3    6.45680E-01       12  NODE 12   -4.81111E+00      21  Q3  PT    3.77052E-05
 4  NODE   4   -5.64884E-01       13  NODE 13   -5.28297E+00      22  Q4  PT    9.64611E-04
 5  NODE   5   -5.24815E+00       14  NODE 14   -5.58883E+00      23  Q5  PT    2.14380E-04
 6  NODE   6    1.20998E+00       15  NODE 15    1.00000E+01      24  Q6  PT    1.54686E-02
 7  NODE   7    6.66930E-01       16  NODE 16   -4.44801E+00      25  Q7  PT    9.98539E-05
 8  NODE   8   -4.62571E+00       17  NODE 17   -6.00000E+00      26  Q8  PT    6.05917E-04
 9  NODE   9   -5.08788E+00       18  Q1   PT    1.45717E-03      27  D1  PD    8.78244E-04

SIMULATION TIME=  3.14000E-06     TIME STEP= 1.00000E-07  MATRIX SOLUTIONS= 159   GAMMADDOT= 0.

                                  NODE VOLTAGE AND AUXILIARY ANSWERS
 1  NODE   1    5.50589E-02       10  NODE 10    3.61868E-02      19  Q1  IC    3.14213E-04
 2  NODE   2   -2.62982E-05       11  NODE 11   -6.21137E-01      20  Q2  PT    3.41956E-04
 3  NODE   3    6.45417E-01       12  NODE 12   -4.96527E+00      21  Q3  PT    3.77372E-05
 4  NODE   4   -5.64905E-01       13  NODE 13   -5.27577E+00      22  Q4  PT    9.60329E-04
 5  NODE   5   -5.24099E+00       14  NODE 14   -5.58418E+00      23  Q5  PT    2.17249E-04
 6  NODE   6    1.21796E+00       15  NODE 15    1.20000E+01      24  Q6  PT    1.60387E-02
 7  NODE   7    6.66974E-01       16  NODE 16   -4.42053E+00      25  Q7  PT    5.90690E-08
 8  NODE   8   -4.61831E+00       17  NODE 17   -6.00000E+00      26  Q8  PT    4.01544E-04
 9  NODE   9   -5.08084E+00       18  Q1   PT    1.47056E-03      27  D1  PD    8.85718E-04

SIMULATION TIME=  5.10000E-06     TIME STEP= 6.00000E-08  MATRIX SOLUTIONS= 179   GAMMADDOT= 0.

                                  NODE VOLTAGE AND AUXILIARY ANSWERS
 1  NODE   1    5.50589E-02       10  NODE 10    3.61867E-02      19  Q1  IC    3.14213E-04
 2  NODE   2   -2.62982E-05       11  NODE 11   -6.21137E-01      20  Q2  PT    3.41956E-04
 3  NODE   3    6.45417E-01       12  NODE 12   -4.96527E+00      21  Q3  PT    3.77372E-05
 4  NODE   4   -5.64905E-01       13  NODE 13   -5.27577E+00      22  Q4  PT    9.60329E-04
 5  NODE   5   -5.24099E+00       14  NODE 14   -5.58418E+00      23  Q5  PT    2.17249E-04
 6  NODE   6    1.21796E+00       15  NODE 15    1.20000E+01      24  Q6  PT    1.60387E-02
 7  NODE   7    6.66974E-01       16  NODE 16   -4.42053E+00      25  Q7  PT    5.90486E-08
 8  NODE   8   -4.61831E+00       17  NODE 17   -6.00000E+00      26  Q8  PT    4.01544E-04
 9  NODE   9   -5.08084E+00       18  Q1   PT    1.47056E-03      27  D1  PD    8.85718E-04
```

Figure 10.59 TRACAP Standard Output Print.

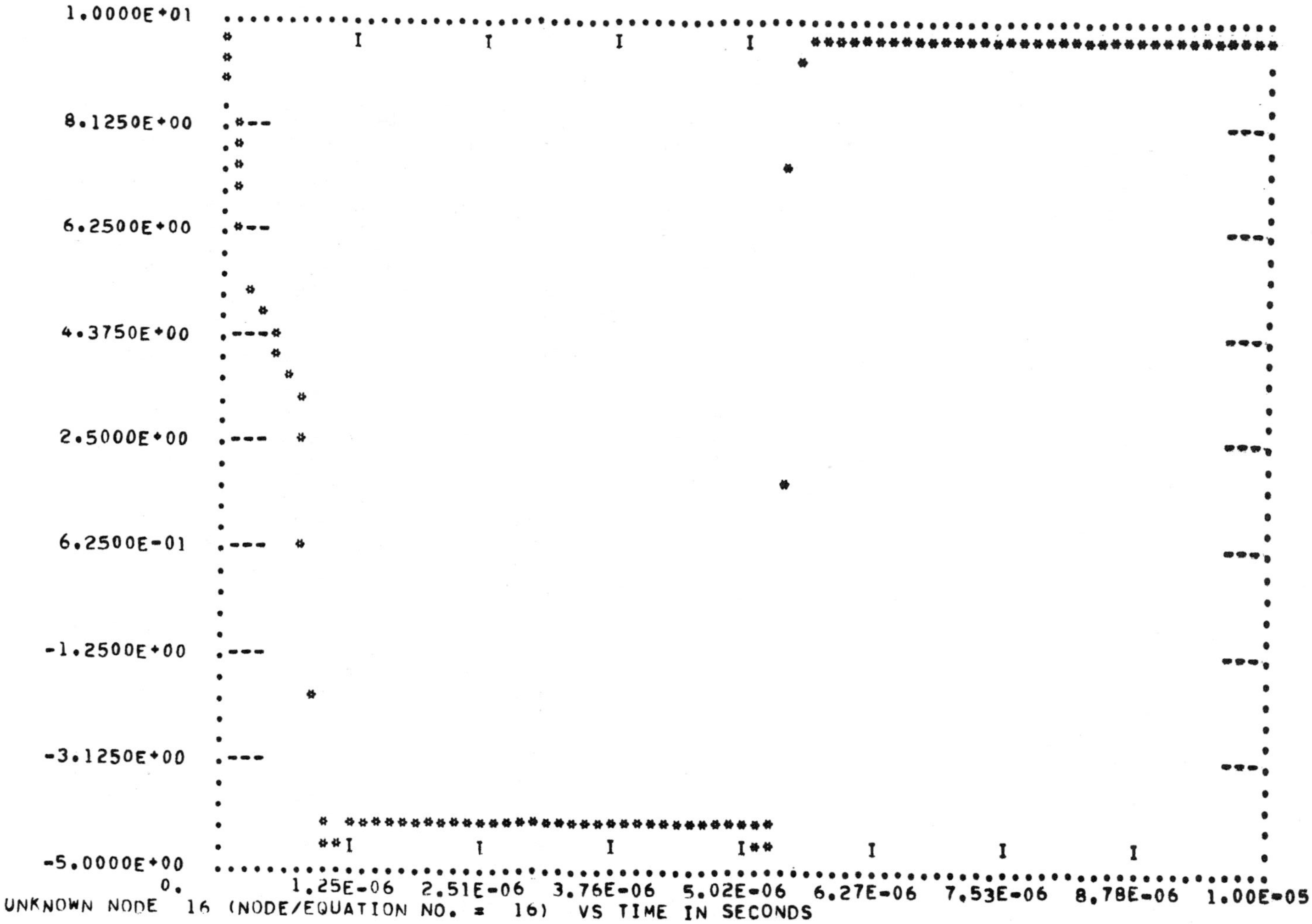

Figure 10.60 TRACAP Output as a Function of Time (Printer Plot).

10.7 SPECIAL USER MODELING

Special user modeling is available in SYSCAP primarily through the FUNC Option and MACRO MODELS.

The FUNC Option consists of up to nine user created FUNCTION subprograms. These subprograms allow the user the capability of FORTRAN statements to calculate a variable, which is readily available for use in the main program. The FUNC Option also permits either resistors, capacitors, inductors or sources to be set equal to a variable. Other parameters are set equal to variables through a CHANGE card.

MACRO MODELS are used when configuration of built-in models and/or functions are to be used repetitively, with minor changes in the analysis.

FUNC Option

The general format of the FUNCTION subprogram is as follows:

```
FUNCTION FUNC {1-9} (a1, a2, ..., a10)
    .
    .
    .
FUNC {1-9} =
    .
RETURN
END
```
} Included in the subroutine section

where

FUNC {1-9} is the FUNCTION name which is called in the topology section. a1, a2, ... a10 are distinct (within the same statement nonsubscripted variables. These are the dummy arguments. The name of the FUNCTION (FUNC {1-9}) must be assigned a value at least once within the subprogram.

The value of FUNC {1-9} is available for use within either the topology section or on a CHANGE card. In the topology section the general format is as follows:

$$name = FUNC\ \{1-9\}\ (S1, S2, \ldots, S10)$$

where

name is an alpha entry of up to four characters. (It must, however be a unique name.)

S1, S2, ..., S10 are any program solution names (e.g. NODE 1, RIIR, etc.) another *name* of a FUNC {1-9}, or a constant. For TRACAP, TE may be used for elapsed time and TO for delta time. In ALCAP, FREQ may be used for frequency (*Note*; FREQ = 0 for dc solution).

Four built-in model values may be set equal to name, or any other solution name or linear multiple of either. These include resistors, capacitors, inductors and/or dc sources (either dc value or resistance), e.g.,

    R12(1,2)   = A
    C14(5,2)   = A*5K
    L13(4,2)   = NODE 6
    E1 (+1,−4)   = BB, RS= NODE 5
    I1(F5,R3)   = BBCC,RI=NODE 10*5MEG

However, the linear multiplier must appear after *name* or the solution name. In TRACAP the dc resistances of the transient sources may also be set equal to a function, e.g.,

    E1(+1,−0)PULSE,RS=A,0,1,0,1N,1U,1N,1

The other model parameters are set equal to *name* through a CHANGE card, e.g.,

    CHANGE   DZ1VZ = AA
    CHANGE   Q1BN = ZI
    CHANGE   Q1RCL = NODE 6*1 K

Examples.

(1) In TRACAP, DICAP or the DC part of ALCAP, change a supply from 0 volts to 10 volts if node 1 and node 2 are greater than 6 volts.

```
FUNCTION FUNC1(A,B)
FUNC1 = 0
IF(A.GT.6..AND.B.GT.6.) FUNC1 = 10      } Subroutine coding
RETURN
END

A = FUNC1 (NODE1, NODE2)                 } Topology coding
E1 (+1,−0) = A
```

(2) In the AC part of ALCAP vary two resistors as the square root of frequency, and a third as the cube root of frequency.

```
FUNCTION FUNC1 (A,B,C)
FUNC1 = A*((B**C) +1.)                   } Subroutine coding
RETURN
END

A = FUNC1 (1E+3, FREQ,.5)
R1 (1,2) = A
B = FUNC1 (10E+3, FREQ,.5)               } Topology coding
R2 (3,4) = B
G = FUNC1 (5E+3, FREQ,.3)
R4 (5,6) = G
```

(3) In TRACAP, vary a capacitor as the square root of the sum of the voltage across it, divided by a constant and one. Node voltages are positive, and if B is less than C, then hold the value A for the capacitor.

```
FUNCTION FUNC2 (A,B,C,D)
IF (B.LE,C) B = C                        ⎫
FUNC2 = A/((1.+(B−C)/D)**.5)              ⎬  Subroutine coding
RETURN                                   
END                                      ⎭

AA = FUNC2 (10E−12,NODE4,NODE6,.8) ⎫  Topology
CAA (4,6) = AA                     ⎬  coding
```

MACRO Models

The MACRO model description is the first set of cards in the topology card section. The program automatically adds any additional nodes required, internal to the macro model, at the end of the regular circuit node numbers. The maximum number of models is ten. The general format for macro models is as follows:

Model Description

MODEL, *name* (A_1 nodeix, A_2 nodeiy, ..., A_n nodeiz) $S_1, S_2, ..., S_{10}$
.
.
.
TOPOLOGY ENTRIES
.
.
.
FINIS

Model Request

    name (A_1 nodex, A_2 nodey, ..., A_n nodez) $K_1, K_2, ..., K_{10}$

where

$A_1, A_2, ..., A_n$ are unique single alpha characters which are used prior to the node number, in the same way B, C & E are used for transistors.

nodeix, nodeiy, ..., nodeiz are the internal node numbers within the macro mode which are to interface externally.

$S_1, S_2, ..., S_{10}$ are single alpha characters which are used to pass data internally to the macro model from the model request (equal to $K_1, K_2, ..., K_{10}$).

nodex, nodey, ..., nodez are the eternal node numbers which interconnects the macro models and/or components in the actual circuit.

$K_1, K_2, \ldots, K_{10}$ are constants and/or solution names which are to be passed to the macro model.

Consider the following examples:

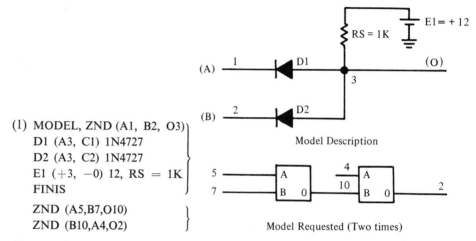

(1) MODEL, ZND (A1, B2, O3)
    D1 (A3, C1) 1N4727
    D2 (A3, C2) 1N4727
    E1 (+3, −0) 12, RS = 1K
    FINIS

    ZND (A5,B7,O10)
    ZND (B10,A4,O2)

Model Description

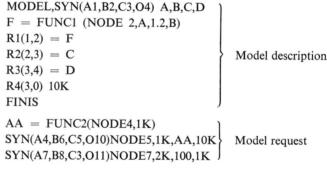

Model Requested (Two times)

(2) Functions may also be used with the macro model.

    MODEL,SYN(A1,B2,C3,O4) A,B,C,D
    F = FUNC1 (NODE 2,A,1.2,B)
    R1(1,2) = F
    R2(2,3) = C
    R3(3,4) = D
    R4(3,0) 10K
    FINIS

    Model description

    AA = FUNC2(NODE4,1K)
    SYN(A4,B6,C5,O10)NODE5,1K,AA,10K
    SYN(A7,B8,C3,O11)NODE7,2K,100,1K

    Model request

Components within a macro model can be changed through a CHANGE card using the following ground rules:

1. The name is the same as in the macro model definition.

2. After the name, a number is included between parentheses indicating the specific model request to be changed.

If for the previous example (1), E1's series resistance is to be changed to 5K on the first request and 100K on the second request, the appropriate CHANGE cards are:

    CHANGE E1 RS(1) = 5K
    CHANGE E1 RS(2) = 100K

10.8 LIMITATIONS

SYSCAP has been designed to make maximum use of computer space through dynamic storage of data, sparse admittance matrix techniques [14], etc. The following limits are therefore given as guidelines only.

Maximum number of nodes ≈ 255.

Maximum number of elements ≈ 500.

Circuit may not contain a floating ground.

Two nodes may not be shorted together (zero impedance).

Inductors, capacitors, and transformer windings may not have zero series resistance (default is 0.1 ohms).

TRACAP. The time step between solutions are controlled based on three items: user-specified limits, rate of convergence of nonlinear models, and variation of input sources. It is therefore possible in solution regions where nonlinear elements have little effect for numerical oscillation to occur in the capacitor [15] and inductor models if the user-specified time step limits exceed the time constants involving the capacitor or inductor model.

SYSCAP has included many warning and error messages. A complete listing of these outputs may be obtained from the SYSCAP users guide.

10.9 REFERENCES

1. HURST, R., "SYSCAP, TESS used for B-1 Bomber Radiation Effects Analysis," *CDC Scientific Services Newsletter*, No. 4, June 1972, p. 1.

2. HOCHWALD, W., "Computer-Aided Design Analysis of Modern Large-Scale Circuits and Subsystems," presented at the NATO Advisory Group for Aerospace Research and Development (AGARD) Conference in Denmark, May 1973.

3. GALLAGHER, F., "SYSCAP Solves Large and Small Circuits Economically," *IEEE Proc.*, Spring Seminar on Reliability, Boston, Mass., April 27, 1972.

4. RIVERA, E., et al., "Two-in-One Design Program Offers Big System Flexibility with Small Job Cost," *Electronics*, 43, No. 13, June 22, 1970, pp. 74–81.

5. JOHNSON, E. D., et al., *Transient Radiation Analysis by Computer Program (TRAC)*, Autonetics Division, North American Rockwell, Harry Diamond Laboratories, Anaheim, Cal., June 1968.

6. KLEINER, C. T., G. KINOSHITA, and E. D. JOHNSON, "Simulation and Verification of Transient Nuclear Radiation Effects on Semiconductor Electronics," *IEEE Trans. Nuclear Science*, NS-11, Nov. 1964, pp. 82–104.

7. CHERNEY, R. W., and R. RANALLI, "The Management of Computer-Aided Circuit Analysis," in *Proceedings January 10–12, 1967 Annual Symposium on Reliability*, Washington, D.C., pp. 759–765.

8. MORRIS, W. L., and R. RANALLI, "Today's Computer-Aided Design Programs and Application Capability," National Electronics Conference, Chicago, Dec. 9–11, 1968.

9. CYBERNET—A division of Control Data Corporation that provides remote batch and time-sharing computer services via a series of computer centers located throughout the United States.

10. KRONOS—Operating system of CYBERNET Service's CDC® 6400 Computer Systems.

11. *SYSCAP, User Information Manual*, Control Data Corporation, Minneapolis, Minn., Feb. 1975.

12. MALDONADO, C. D., and C. T. KLEINER, "Modification of the TRAC Ebers-Moll Model for Improved High Frequency and Collector Storage Time Predictions," IEEE Annual Conference on Nuclear and Space Radiation Effects, University of California at San Diego, July 21–23, 1970.

13. DIERKING, WILLIAM H., and C. T. KLEINER, "Phenomenological Magnetic Core Model for Circuit Analysis Programs," *IEEE Trans. Magnetics*, 8, No. 3, Sept. 1972, pp. 594–596.

14. SATO, NOBUO, and W. F. TINNEY, "Techniques for Exploiting the Sparsity of the Network Admittance Matrix," *IEEE Trans.*, on Power and Apparatus, 82, Dec. 1963, pp. 944–950.

15. JOHNSON, E. D., et al., *Transient Radiation Analysis by Computer Program (TRAC)*, Vol. II, Autonetics Division, North American Rockwell, Harry Diamond Laboratories, Anaheim, Cal., June 1968, pp. 49–50.

APPENDIX 10.1. MAJOR SYSCAP DEVELOPERS/CONTRIBUTORS

Name	Function
J. D. Bastian*	SYSCAP coordinator
R. Blankenhorn	Numerical techniques investigator
B. R. Bourbon	Originator of transient program structure
W. T. Clary	Program manager
H. L. Creech	Iteration techniques investigator
W. H. Dierking*	Technical leader
P. Elsaesser	Programmer
W. Hochwald*	Program manager
H. Jacobson	Special options programmer
E. D. Johnson*	New capability innovator
C. T. Kleiner*	Modeling/program consultant
S. A. Louie*	Lead programmer
J. H. Maenpaa	Programmer
L. R. McMurray*	Program consultant
J. H. Montero	Programmer
K. J. Nakamura	Programmer
J. H. Nelson*	Data reduction techniques
R. Ranalli	Program manager (originator of concept)
E. Rivera	Originator of dc program structure
W. C. Russell	Programmer
F. T. Suzuki*	Programmer
F. A. Vassallo*	Programmer
L. R. Young*	Lead ac programmer
L. M. White*	Computer specialist
P. D. Zehner*	Programmer

*Working on SYSCAP or related projects as of 1975.

APPENDIX 10.2. RADIATION SIMULATION

There are three radiation models in TRACAP: RAD1, RAD2, and RAD3.

RAD1

This model uses exponential rise and fall times (dependent on TC, TE, or TD) for the square radiation pulse as shown in Fig. 10.61. This represents all the ionization current being generated in the diffusion region.

RAD2

This model splits the ionization current into two components (diffusion and depletion) whose steady-state total equals that of the RAD1 model. The

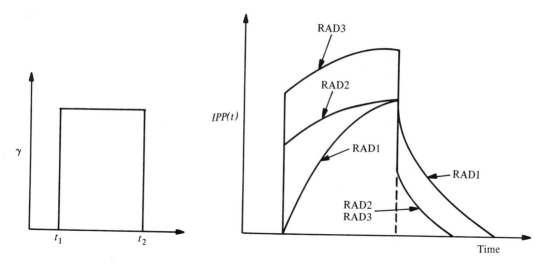

Figure 10.61 IPP (t) for a Square $\dot{\gamma}$ Pulse.

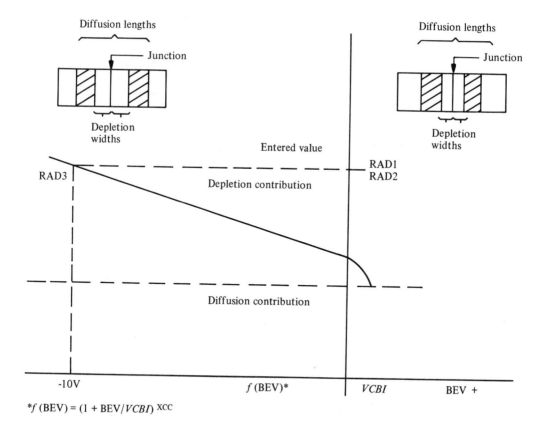

*f (BEV) = (1 + BEV/$VCBI$) ^{XCC}

Figure 10.62 RAD3 Voltage Dependence of the Generation Constant.

diffusion portion time constant is different from TC, TE, or TD, so an additional parameter TSC, TSE, or TSD, is used in both the RAD2 and RAD3 models. The depletion portion, DEPC, DEPE, or DEPD, is time-independent and varies directly as the radiation pulse varies. This is also shown in Fig. 10.61.

RAD3

This model is identical to the RAD2 model with one exception, the depletion portion, DEPC, DEPE, or DEPD, of the response is voltage-dependent. The response magnitude of the RAD3 model would be that shown in Fig. 10.61 for RAD2 if the base-collector voltage were held constant at 10 volts. The voltage dependence is the reciprocal function of the voltage dependence of the junction capacitance, which represents depletion width variation as a function of voltage. This variation is shown in Fig. 10.62. The insets represent the variation of the depletion width, which changes the depletion volume contribution.

Appendix A
Diode and Transistor Equivalent Circuits

RANDALL W. JENSEN
LAWRENCE P. McNAMEE

There are nearly as many diode and transistor equivalent circuits and variations of each basic form as there are circuit analysis programs to use them. Some of the more popular model configurations are those developed by Beaufoy and Sparkes, Linvill, Ebers and Moll, and, most recently, Gummel and Poon.

Except for the Gummel-Poon model, the equivalent circuits just mentioned are directly relatable to each other, as will be shown later in this appendix. Most of the diode and transistor models available to the circuit analyst/designer can be readily derived from the Ebers-Moll equivalent circuits. For this reason the basic diode and transistor equivalent circuits and their associated notations described in this appendix will be limited to the Ebers-Moll configuration.

A.1 DIODE EQUIVALENT CIRCUIT

The general diode equivalent circuit that we shall describe is shown in Fig. A.1. This configuration is an extension of the Ebers-Moll model in which JD is a perfect diode representing the diode equation. The current through the diode is given by the expression

$$i_d = I_s(\exp(\theta V_j) - 1) \tag{A.1}$$

where

I_s = diode saturation current

$\theta = q/\eta kT$

η = emission constant ($1 \leq \eta \leq 2$)

V_j = diode junction voltage

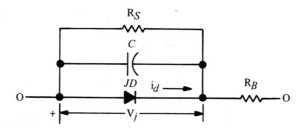

Figure A.1 General Diode Model.

The perfect diode is represented in many programs, such as SCEPTRE, as a voltage-dependent current source.

The resistive parameters (or elements) in the diode model are the series bulk resistance of the diode, R_B, and the shunt resistance, R_S. The capacitance shown in the model is actually the sum of the diode junction and diffusion capacitances as expressed by

$$C = C_j + C_d \qquad (A.2)$$

where the junction capacitance is

$$C_j = \frac{C_0}{(V_0 - V_j)^n} \qquad (A.3)$$

and where

C_0 = transition capacitance constant (approximately C_j at $V_j = 0$ for silicon diodes)

V_0 = junction contact potential

n = junction grading constant (depends on junction formation, $= \frac{1}{3}$ for linear grade junction, $= \frac{1}{2}$ for step junction)

and where the diffusion capacitance is given by

$$C_d = K(i_d + I_s) \qquad (A.4)$$

where

K = diffusion capacitance constant
$= \theta/2\pi F$

F = frequency parameter related to the diode storage time T_s by the expression

$$= \frac{\ln(1 + I_f/I_r)}{2\pi T_s} \qquad (A.5)$$

I_f = diode forward current in T_s test

I_r = diode reverse current in T_s test

i_d = current through diode JD

Since C_j is normally much greater than C_d at negative values of V_j, the capacitance term can be simplified to

$$C = \frac{C_0}{(V_0 - V_j)^n} + K * i_d \tag{A.6}$$

The user must be careful, however, to guarantee that $|C_j| > |K * i_d|$ for all negative values of V_j to assure himself that C is always a positive quantity. Negative capacitances never occur in the "real" world, but they certainly can in a circuit analysis.

A.2 TRANSISTOR EQUIVALENT CIRCUIT

The general bipolar transistor model for an *npn* transistor derived by Ebers-Moll is shown in Fig. A.2. This model is widely used because it provides a reasonably accurate representation of bipolar transistor operation in all four regions of operation with a minimum model complexity. The model

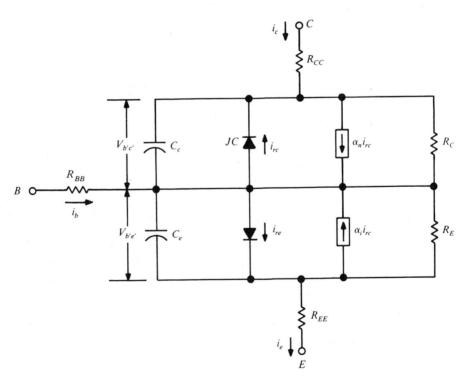

Figure A.2 Basic Ebers-Moll Bipolar Transistor Equivalent Circuit.

consists of two diodes connected together and coupled by two current-dependent current generators $\alpha_n i_{re}$ and $\alpha_i i_{rc}$, as shown in the figure.

The current, i_{re} and i_{rc}, through the perfect diodes, JE and JC, respectively, are defined by the expressions

$$i_{re} = I_{es}(\exp(\theta_e V_{b'e'}) - 1) \tag{A.7}$$

and

$$i_{re} = I_{cs}(\exp(\theta_c V_{b'c'}) - 1) \tag{A.8}$$

where

I_{es} = base-emitter reverse saturation current with base-collector short-circuited

I_{cs} = base-collector reverse saturation current with base-emitter short-circuited

$\theta_e = q/\eta_1 kT$ = emitter-base junction constant (volts^{-1})

$\theta_c = q/\eta_2 kT$ = collector-base junction constant (volts^{-1})

η_1, η_2 = constants depending on junction injection level ($1 \leq \eta \leq 2$)

The current-dependent current generators, $\alpha_n i_{re}$ and $\alpha_i i_{rc}$, are nonlinear functions of the diode currents. That is, α_n and α_i are not constant in the real world. Many of the circuit analysis programs have provisions to account for this nonlinearity, while other programs require that the α_n and α_i values be fixed. The phenomena that cause the nonlinearity at low currents are primarily transition region and surface recombination. Under high injection conditions, several other complex effects become dominant, which cause the values of α_n and α_i to decrease. For our purpose in this appendix, we shall consider the values to be fixed and independent of injection level.

The parasitic resistances will also be assumed constant. The series bulk resistances, R_{EE} and R_{CC}, and the shunt resistances, R_E and R_C, are essentially independent of current and cause no problem when assumed fixed. However, the base-spreading resistance R_{BB} is a nonlinear function of the junction voltages and can vary over several orders of magnitude depending on the bias conditions. Not only is the R_{BB} parameter nonlinear, but it is, unfortunately, somewhat a function of the technique used to measure it. The majority of analysis applications are not adversely affected by assuming that the value of R_{BB} is constant over the operating range. We shall also make this assumption here and in Appendix C, where tables of diode and transistor parameters are listed.

The total capacitances associated with the emitter and collector junctions are usually expressed as

$$C_e = C_{je} + C_{de}$$
$$= \frac{C_{oe}}{(V_{oe} - V_{b'e'})^{n_e}} + \theta_e T_e(i_{re} + I_{es}) \tag{A.9}$$

and
$$C_c = C_{jc} + C_{dc}$$
$$= \frac{C_{0c}}{(V_{0c} - V_{b'c'})^{n_c}} + \theta_c T_c (i_{rc} + I_{cs}) \quad (A.10)$$

where

C_{oe} = emitter transition capacitance constant

C_{oc} = collector transition capacitance constant

V_{oe} = emitter-base junction contact potential

V_{oc} = collector-base junction contact potential

n_e = emitter junction grading constant

n_c = collector junction grading constant

$T_e = (2\pi f_N)^{-1}$ = time constant for emitter diffusion capacitance equation

$T_c = (2\pi f_I)^{-1}$ = time constant for collector diffusion capacitance equation

f_N = average f_T normal mode

f_I = average f_T inverse mode

The capacitance equations are frequently written as

$$C_e = \frac{C_{0c}}{(V_{oe} - V_{b'e'})^{n_e}} + K_e(i_{re} + I_{es}) \quad (A.11)$$

and

$$C_c = \frac{C_{0c}}{(V_{0c} - V_{b'c'})^{n_c}} + K_c(i_{re} + I_{cs}) \quad (A.12)$$

for simplicity. As in the diode capacitance equation [Eq. (A.6)], the I_{es} and I_{cs} terms can be omitted if the user guarantees that $|C_j| > |C_d|$ for all negative bias voltage values.

A.3 REFERENCES

1. BEAUFOY, R., and J. J. SPARKES, "The Junction Transistor as a Charge-Controlled Device," *ATE J.* (*London*), 13, No. 4, Oct. 1957, pp. 310–324.

2. EBERS, J. J., and J. L. MOLL, "Large-Signal Behavior of Junction Transistors," *Proc. IRE*, 42, No. 12, Dec. 1954, pp. 1761–1772.

3. GUMMEL, H. K., and H. C. POON, "An Integral Charge Control Model of Bipolar Transistors," *Bell System Technical J.*, 49, No. 5, May–June 1970, pp. 827–852.

4. LINVILL, J. G., "Lumped Models of Transistors and Diodes," *Proc. IRE*, XLVI, June 1958, pp. 1141–1152.

Appendix B

Derivation of the Ebers-Moll and Charge–Control Transistor Model from the Two-lump Intrinsic Drift Model

RANDALL W. JENSEN
LAWRENCE P. McNAMEE

The Ebers-Moll and Beaufoy-Sparkes charge-control transistor models are equivalent to the basic two-lump intrinsic drift transistor model. In this appendix both models are derived from the two-lump approximation, and each of the lumped model parameters is related to externally measurable quantities or to manufacturing data. The *pnp* transistor and the model shown in Fig. B.1 are used for simplicity.

The transistor cross-sectional area is assumed to vary gradually across the base region so that the unidimensional approximation is valid, and the base region minority-carrier lifetime τ_p is assumed to vary with distance in the region. Then, $H_{C_1} \neq H_{C_2}$, $S_1 \neq S_2$, and the lumped element values are as specified in Fig. B.1, where the cross-sectional areas and lifetimes have appropriate average values.

B.1 EBERS-MOLL TRANSISTOR MODEL

The equations governing the two-lump drift model in Fig. B.1 are

$$I_e = p_e\left(H_{C_1} + S_1\frac{d}{dt} + H_D + mF\right) - p_c[H_D - (1-m)F] + C_{je}\frac{dV_e}{dt} \quad \text{(B.1)}$$

$$I_c = -p_e(H_D + mF) + p_c\left[H_{C_2} + S_2\frac{d}{dt} + H_D - (1-m)F\right] + C_{jc}\frac{dV_c}{dt} \quad \text{(B.2)}$$

$$I_b = -p_e\left(H_{C_1} + S_1\frac{d}{dt}\right) - p_c\left(H_{C_2} + S_2\frac{d}{dt}\right) - C_{je}\frac{dV_e}{dt} - C_{jc}\frac{dV_c}{dt} \quad \text{(B.3)}$$

The excess minority-carrier densities at the emitter and collector edges of the base region are related to the junction voltages by the expressions

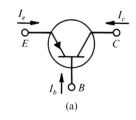

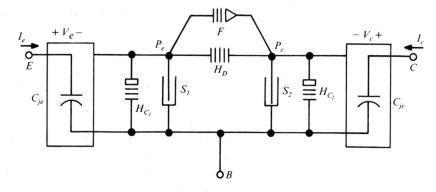

Figure B.1 (a) *pnp* Transistor and (b) Two-lump Intrinsic Drift Model.

$p_e = p_{ne}[\exp(\theta_e V_e) - 1]$ and $p_c = p_{nc}[\exp(\theta_c V_c) - 1]$. Substituting these expressions into Eqs. (B.1) and (B.2), the terminal current equations become

$$I_e = p_{ne}(H_{C_1} + H_D + mF)[\exp(\theta_e V_e) - 1] + S_1 \theta_e p_{ne}[\exp(\theta_e V_e)]\frac{dV_e}{dt}$$

$$- p_{nc}[H_D - (1-m)F][\exp(\theta_c V_c) - 1] + C_{je}\frac{dV_e}{dt} \quad (B.4)$$

and

$$I_c = -p_{ne}(H_D + mF)[\exp(\theta_e V_e) - 1]$$
$$+ p_{nc}[H_{C_2} + H_D - (1-m)F][\exp(\theta_c V_c) - 1]$$
$$+ S_2 \theta_c p_{nc}[\exp(\theta_c V_c)]\frac{dV_c}{dt} + C_{jc}\frac{dV_c}{dt} \quad (B.5)$$

From Eq. (B.4) the base-emitter reverse saturation current is defined as

$$I_{es}\Big|_{V_c=0} \equiv p_{ne}(H_{C_1} + H_D + mF) \quad (B.6)$$

Using Eq. (B.6) and $I_{be} = I_{es}[\exp(\theta_e V_e) - 1]$, $\exp(\theta_e V_e)$ can be expressed as

$$\exp(\theta_e V_e) = \frac{I_{be} + I_{es}}{I_{es}} \tag{B.7}$$

Similarly, from Eq. (B.5) the base-collector reverse saturation current is

$$I_{cs}\big|_{V_e=0} \equiv p_{nc}[H_{C_2} + H_D - (1-m)F] \tag{B.8}$$

and from $I_{bc} = I_{cs}[\exp(\theta_c V_c) - 1]$,

$$\exp(\theta_c V_c) = \frac{I_{bc} + I_{cs}}{I_{cs}} \tag{B.9}$$

Incorporating Eqs. (B.6)–(B.9) into Eqs. (B.4) and (B.5) the terminal current equations reduce to

$$I_e = I_{be} + \left[S_1 \theta_e p_{ne}\left(\frac{I_{be} + I_{es}}{I_{es}}\right) + C_{je}\right]\frac{dV_e}{dt}$$
$$- \frac{H_D - (1-m)F}{H_D + H_{C_2} - (1-m)F} I_{bc} \tag{B.10}$$

$$I_c = -\frac{H_D + mF}{H_C + H_D + mF} I_{be} + I_{bc}$$
$$+ \left[S_2 \theta_c p_{nc}\left(\frac{I_{bc} + I_{cs}}{I_{cs}}\right) + C_{jc}\right]\frac{dV_c}{dt} \tag{B.11}$$

The terminal current expressions for the nonlinear Ebers-Moll transistor model shown in Fig. B.2 are

$$I_e = I_{es}[\exp(\theta_e V_e) - 1] - \alpha_I I_{cs}[\exp(\theta_c V_c) - 1]$$
$$+ [CE + C_{je}]\frac{dV_e}{dt} \tag{B.12}$$

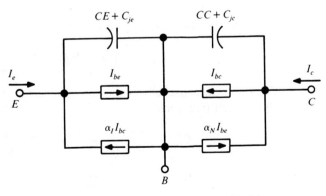

Figure B.2 Ebers-Moll Transistor Model.

and
$$I_c = -\alpha_N I_{es}[\exp(\theta_e V_e) - 1] + I_{cs}[\exp(\theta_c V_c) - 1]$$
$$+ [CC + C_{jc}]\frac{dV_c}{dt} \tag{B.13}$$

where

CE = emitter-junction diffusion capacitance

CC = collector-junction diffusion capacitance

The equivalence between Eqs. (B.10)–(B.13) is apparent. Formally,

$$I_{es} = p_{ne}(H_{C_1} + H_D + mF) \tag{B.14a}$$
$$I_{cs} = p_{nc}(H_{C_2} + H_D - (1-m)F) \tag{B.14b}$$
$$\alpha_N = \frac{H_D + mF}{H_{C_1} + H_D + mF} \tag{B.14c}$$
$$\alpha_I = \frac{H_D - (1-m)F}{H_D + H_{C_2} - (1-m)F} \tag{B.14d}$$
$$CE = S_1 \theta_e p_{ne}\left(\frac{I_{be} + I_{es}}{I_{es}}\right) \tag{B.14e}$$
$$CC = S_2 \theta_c p_{nc}\left(\frac{I_{bc} + I_{cs}}{I_{cs}}\right) \tag{B.14f}$$

The frequency dependence of the normal common-base current gain $\alpha_N(s)$ in the two-lump model is given by

$$\alpha_N(s) = \frac{H_D + mF}{H_{C_1} + H_D + mF + sS_1} = \frac{\alpha_{N_0}}{(1 + s/\omega_{\alpha N})} \tag{B.15}$$

The normal alpha cutoff frequency $\omega_{\alpha N}$ for the two-lump transistor model is

$$\omega_{\alpha N} = \frac{H_D + H_{C_1} + mF}{S_1} \tag{B.16}$$

Similarly, the inverse alpha cutoff frequency is

$$\omega_{\alpha I} = \frac{H_D + H_{C_2} - (1-m)F}{S_2} \tag{B.17}$$

The diffusion capacitance expressions in terms of the normal and inverse alpha cutoff frequencies are

$$CE = \frac{S_1 \theta_e (I_{be} + I_{es})}{H_D + H_{C_1} + mF} = \frac{\theta_e (I_{be} + I_{es})}{\omega_{\alpha N}} \tag{B.18}$$
$$CC = \frac{S_2 \theta_c (I_{bc} + I_{cs})}{H_{C_2} + H_D - (1-m)F} = \frac{\theta_c (I_{bc} + I_{cs})}{\omega_{\alpha I}} \tag{B.19}$$

It is obviously impossible to obtain independent solutions for the lumped-model parameters from the Ebers-Moll parameters in Eqs. (B.14), (B.16), and (B.17) or from terminal measurements alone. Additional manufacturing information (including the impurity profile) is required before a model can be constructed.

B.2 CHARGE-CONTROL TRANSISTOR MODEL

The Beaufoy-Sparkes charge-control transistor model can also be derived from the terminal-current expressions in Eqs. (B.1)–(B.3) if the relationship between the minority-carrier density at the edges of the base region (p_e, p_c) and the charge stored in the emitter and collector lumps (Q_N, Q_I) is understood. The relationships between p_e, Q_N, p_c, and Q_I are illustrated for the drift transistor in Fig. B.3.

The charge stored in the emitter lump is defined by*

$$Q_N = qAmwp_e = S_1 p_e \qquad (B.20)$$

and the charge stored in the collector lump is defined by

$$Q_I = qA(1-m)wp_c = S_2 p_c \qquad (B.21)$$

Furthermore, from the lumped-parameter definitions:

$$\frac{H_{C_1}}{S_1} = \frac{1}{\tau_{p_1}} \equiv \frac{1}{\tau_{BN}} \qquad (B.22)$$

$$\frac{H_{C_2}}{S_2} = \frac{1}{\tau_{p_2}} \equiv \frac{1}{\tau_{BI}} \qquad (B.23)$$

$$\tau_{tN} = \frac{\tau_{BN} H_C}{H_D + mF} = \frac{\tau_{BN}}{\beta_N} \equiv \tau_{CN} \qquad (B.24)$$

$$\tau_{tI} = \frac{\tau_{BI} H_{C_2}}{H_D(1-m)F} = \frac{\tau_{BI}}{\beta_I} \equiv \tau_{EI} \qquad (B.25)$$

where

τ_{BN} = minority-carrier lifetime in the emitter lump (normal operation)

τ_{BI} = minority-carrier lifetime in the collector lump (inverted operation)

τ_{CN} = minority-carrier base-transit time from the emitter to the collector junction

τ_{EI} = minority-carrier base-transit time from the collector to the emitter junction

*The coefficient m is the same as that used to compute the current through the driftance element.

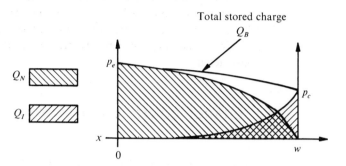

Total stored charge Q_B is proportional to the area underneath the line. $Q_B = Q_N + Q_I$

(a)

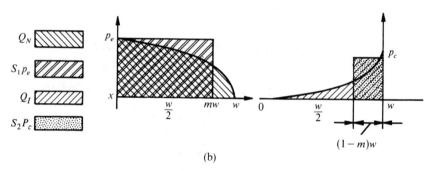

(b)

Figure B.3 Stored-charge Relationship Between Two-lump Drift and (a) Charge-control Transistor Models: Division of Stored Base Charge; (b) Equivalence between Q_N and $S_1 p_e$ and Q_I and $S_2 p_c$.

The terminal current equations can be rewritten in charge-control form by substituting Eqs. (B.22)–(B.25) into Eqs. (B.1)–(B.3). The charge-control equations are

$$I_e = Q_N \left(\frac{1}{\tau_{BN}} + \frac{1}{\tau_{CN}} \right) + \frac{d}{dt}(Q_N) - \frac{Q_I}{\tau_{EI}} \qquad (B.26)$$

$$I_c = -\frac{Q_N}{\tau_{CN}} + Q_I \left(\frac{1}{\tau_{BI}} + \frac{1}{\tau_{EI}} \right) + \frac{d}{dt}(Q_I) \qquad (B.27)$$

$$I_b = \frac{Q_N}{\tau_{BN}} + \frac{d}{dt}(Q_N) + \frac{Q_I}{\tau_{BI}} + \frac{d}{dt}(Q_I) \qquad (B.28)$$

The equivalent charge-control drift transistor model is shown in Fig. B.4.

The frequency dependence of the normal common-base current gain $\alpha_N(s)$ in the charge-control model is given by

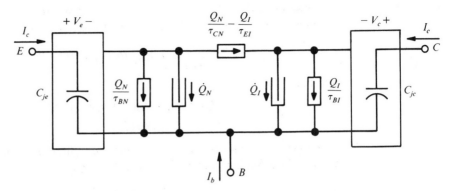

Figure B.4 Charge-control Transistor Model.

$$\alpha_N(s) = \frac{\dfrac{1}{\tau_{CN}}}{\dfrac{1}{\tau_{CN}} + \dfrac{1}{\tau_{BN}} + s} = \frac{1}{1 + \dfrac{1}{\beta_N} + s\tau_{CN}} \tag{B.29}$$

The normal alpha cutoff frequency from Eq. (B.29) is

$$\omega_{\alpha N} = \frac{1}{\alpha_N \tau_{CN}} \tag{B.30}$$

Similarly, the inverse alpha cutoff frequency is

$$\omega_{\alpha I} = \frac{1}{\alpha_I \tau_{EI}} \tag{B.31}$$

If two composite time constants, τ_{EN} and τ_{CI}, are defined as

$$\frac{1}{\tau_{EN}} \equiv \frac{1}{\tau_{BN}} + \frac{1}{\tau_{CN}} \tag{B.32a}$$

$$\frac{1}{\tau_{CI}} \equiv \frac{1}{\tau_{BI}} + \frac{1}{\tau_{EI}} \tag{B.32b}$$

the relationships between the alpha cutoff frequencies and the composite time constants become

$$\tau_{EN} = \frac{1}{\omega_{\alpha N}} \tag{B.33a}$$

and

$$\tau_{CI} = \frac{1}{\omega_{\alpha I}} \tag{B.33b}$$

Basically, the charge-control and two-lump models are identical, involving only a transformation of variables ($p_e \leftrightarrow Q_N$, $p_c \leftrightarrow Q_I$). Since it is easier to relate the charge parameters to the lumped-model parameters through the Ebers-Moll parameters, the Ebers-Moll to charge-control transformation is derived.

Assuming that Q_N is a function of V_e only and that Q_I is a function of V_c only, then by comparing Eqs. (B.12) and (B.26), we observe that

$$\frac{Q_N}{\tau_{EN}} = I_{es}[\exp(\theta_e V_e) - 1] \tag{B.34}$$

and that

$$Q_N = \alpha_N \tau_{CN} I_{es}[\exp(\theta_e V_e) - 1]$$
$$= Q_{BN}^*[\exp(\theta_e V_e) - 1] \tag{B.35}$$

where Q_{BN}^* is the effective charge stored in the emitter lump for large negative values of V_e. Similarly, from Eqs. (B.13) and (B.27),

$$Q_I = \alpha_I \tau_{EI} I_{cs}[\exp(\theta_e V_c) - 1]$$
$$= Q_{BI}^*[\exp(\theta_e V_c) - 1] \tag{B.36}$$

where Q_{BI}^* is the effective charge stored in the collector lump for large negative values of V_c.

The charge-control equations are frequently written in the form

$$I_e = \left(\frac{1}{\beta_N} + 1\right) I_N - I_I \tag{B.37}$$

$$I_c = -I_N + \left(\frac{1}{\beta_I} + 1\right) I_I \tag{B.38}$$

where

$$I_N = \frac{Q_{BN}^*}{\tau_{CN}}[\exp(\theta_e V_e) - 1] = I_{es}^{(c)}[\exp(\theta_e V_e) - 1] \tag{B.39}$$

$$I_I = \frac{Q_{BI}^*}{\tau_{BI}}[\exp(\theta_e V_c) - 1] = I_{cs}^{(c)}[\exp(\theta_e V_c) - 1] \tag{B.40}$$

where

$I_{es}^{(c)}$ = emitter reverse-saturation current (charge-control configuration)

$I_{cs}^{(c)}$ = collector saturation current (charge-control configuration)

By comparing the charge-control equations, Eqs. (B.37) and (B.38), and the Ebers-Moll equations specified in Eqs. (B.12) and (B.13), the relationships between the saturation currents for the two models are

$$I_{es}^{(c)} = \alpha_N I_{es} = \frac{\alpha_N I_{e0}}{1 - \alpha_N \alpha_I} \tag{B.41}$$

and

$$I_{cs}^{(c)} = \alpha_I I_{cs} = \frac{\alpha_I I_{c0}}{1 - \alpha_N \alpha_I} \tag{B.42}$$

The parameter conversions between Ebers-Moll, charge-control, and two-lump transistor models are summarized in Table B.1.

Table B.1 *Nonlinear Transistor Model Parameter Conversions*

Parameter	Two-Lump	Ebers-Moll	Charge-Control
Two-lump			
H_{D_1}	$H_D + mF$	$\dfrac{\alpha_N I_{es}}{p_{ne}}$	$\dfrac{I_{es}^{(c)}}{p_{ne}}$
H_{D_2}	$H_D - (1-m)F$	$\dfrac{\alpha_I I_{cs}}{p_{nc}}$	$\dfrac{I_{cs}^{(c)}}{p_{nc}}$
H_{C_1}		$\dfrac{I_{es}(1-\alpha_N)}{p_{nc}}$	$\dfrac{I_{es}^{(c)}}{\beta_N p_{ne}}$
H_{C_2}		$\dfrac{I_{cs}(1-\alpha_I)}{p_{nc}}$	$\dfrac{I_{cs}^{(c)}}{\beta_I p_{nc}}$
S_1		$\dfrac{I_{es}}{\omega_{\alpha N} p_{ne}}$	$\dfrac{\tau_{CN} I_{es}^{(c)}}{p_{ne}}$
S_2		$\dfrac{I_{cs}}{\omega_{\alpha I} p_{nc}}$	$\dfrac{\tau_{EI} I_{cs}^{(c)}}{p_{nc}}$
F	$H_{D_1} - H_{D_2}$	$\alpha_N I_{es}\left[\dfrac{1}{p_{ne}} - \dfrac{1}{p_{nc}}\right]$	$I_{es}^{(c)}\left[\dfrac{1}{p_{ne}} - \dfrac{1}{p_{nc}}\right]$
Ebers-Moll			
I_{es}	$p_{ne}(H_{C_1} + H_D + mF)$	$\dfrac{I_{eo}}{1 - \alpha_N \alpha_I}$	$\dfrac{I_{es}^{(c)}}{\alpha_N}$
I_{cs}	$p_{nc}[H_{C_2} + H_D - (1-m)F]$	$\dfrac{I_{co}}{1 - \alpha_N \alpha_2}$	$\dfrac{I_{cs}^{(c)}}{\alpha_I}$
α_N	$\dfrac{H_D + mF}{H_{C_1} + H_D + mF}$		$\dfrac{\beta_N}{1 + \beta_N}$
α_I	$\dfrac{H_D - (1-m)F}{H_{C_2} + H_D - (1-m)F}$		$\dfrac{\beta_I}{1 + \beta_I}$
$\omega_{\alpha N}$	$\dfrac{H_{C_1} + H_D + mF}{S_1}$		$\dfrac{1}{\tau_{EN}}$
$\omega_{\alpha I}$	$\dfrac{H_{C_2} + H_D - (1-m)F}{S_2}$		$\dfrac{1}{\tau_{CI}}$
Charge-control			
τ_{BN}	$\dfrac{S_1}{H_{C_1}}$	$\dfrac{\alpha_N}{(1-\alpha_N)\omega_{\alpha N}}$	
τ_{BI}	$\dfrac{S_2}{H_{C_2}}$	$\dfrac{\alpha_I}{(1-\alpha_I)\omega_{\alpha I}}$	
τ_{CN}	$S_1(H_D + mF)$	$\dfrac{1}{\alpha_N \omega_{\alpha N}}$	$\dfrac{\tau_{BN}}{\beta_N}$
τ_{CI}	$\dfrac{S_2}{H_{C_2} + H_D - (1-m)F}$	$\dfrac{1}{\omega_{\alpha I}}$	$\left[\dfrac{1}{\tau_{BI}} + \dfrac{1}{\tau_{EI}}\right]^{-1}$
τ_{EN}	$\dfrac{S_1}{H_{C_1} + H_D + mF}$	$\dfrac{1}{\omega_{\alpha N}}$	$\left[\dfrac{1}{\tau_{BN}} + \dfrac{1}{\tau_{CN}}\right]^{-1}$
τ_{EI}	$\dfrac{S_2}{H_D - (1-m)F}$	$\dfrac{1}{\alpha_I \omega_{\alpha I}}$	$\dfrac{\tau_{BI}}{\beta_I}$
$I_{es}^{(c)}$	$p_{ne}(H_D + mF)$	$\alpha_N I_{es}$	
$I_{cs}^{(c)}$	$p_{nc}(H_D - (1-m)F)$	$\alpha_I I_{cs}$	
β_N	$\dfrac{H_D + mF}{H_{C_1}}$	$\dfrac{\alpha_N}{1 - \alpha_N}$	$\dfrac{\tau_{BN}}{\tau_{CN}}$
β_I	$\dfrac{H_D - (1-m)F}{H_{C_2}}$	$\dfrac{\alpha_I}{1 - \alpha_I}$	$\dfrac{\tau_{BI}}{\tau_{EI}}$

Appendix C
Parameter Tables
Diode Model

RANDALL W. JENSEN
LAWRENCE P. McNAMEE

This appendix* contains parameter tables for diode models based on the Ebers-Moll formulation shown in Fig. A.1. The diode current i_d is given by the relationship

$$i_d = I_s[\exp(\theta V_i) - 1] \tag{C.1}$$

The diode capacitance C is given by the relationship

$$C = \frac{C_0}{(V_0 - V_i)^n} + K(i_d + i_s) \tag{C.2}$$

All the parameters conform to the high-speed units given in Table C.1.

Table C.1 *High-Speed Units*

Parameter	Unit
Time	Nanosecond
Resistance	Kilohm
Capacitance	Picofarad
Inductance	Microhenry
Current	Milliampere
Voltage	Volt

*Portions of the model parameter data has been derived from W. C. Cordwell, *Transistor and Diode Model Handbook*; USAF Technical Report AFWL-TR-69-44, Air Force Weapons Laboratory, Oct. 1969.

Table C.2 Diode Model Parameters

Diode Type	IS (mA)	θ (v^{-1})	RB (kΩ)	RS (kΩ)	C_0	V (v^0)	n	K (pF/mA)
1N63	2.8E−10	38.9	0.1	1000	324.0	0.5	0.5	6.9E3
1N93	2.4E−2	24.1	1.85E−4	4E6	19.4	0.486	1.31	1.74E5
1N100	2.5E−3	14.7	0.017	1000	0.354	0.5	0.5	23.4
1N140	2.87E−9	38.9	0.01	160	24.5	0.5	0.5	6.19E3
1N191	1.25E−12	38.9	0.05	400	0.3536	0.5	0.5	619.1
1N270	1.24E−3	23.1	0.0024	500	2.83	0.5	0.5	818.0
1N276	5.6E−4	32.0	5.E−4	800	0.5	0.6	0.29	130.0
1N279	1.24E−3	23.1	6.E−3	3800	24.5	0.5	0.5	737.0
1N457	7.4E−9	28.5	0.01	2.7E7	5.6	0.85	0.62	1.4E4
1N459	5.8E−8	27.4	0.014	2.E9	12.0	0.8	0.45	1.6E4
1N486A	2.27E−13	27.0	2.E−3	1.2E6	227.0	0.5	0.5	1.05E4
1N645	0.25E−6	26.1	0.53E−3	1.23E8	10.8	0.864	0.577	125E3
1N646	2.2E−7	27.0	1.E−4	3.1E7	16.0	1.0	0.5	2.6E4
1N647	1.6E−6	22.0	1.E−4	4.2E7	7.7	1.0	0.51	1.9E4
1N648	2.8E−6	21.0	1.E−4	5.8E7	6.6	1.0	0.48	1.4E5
1N649	2.51E−7	26.1	5.32E−4	1.23E8	10.8	0.864	0.577	125E3
1N658	3.9E−6	22.0	2.E−3	6.1E6	1.8	1.0	0.11	3.3E3
1N659	1.4E−8	33.7	45.E−3	5.6E5	19.0	0.8	0.4	3.E3
1N660	4.8E−6	21.3	45.E−3	9.9E5	3.4	0.9	0.32	6.9E3
1N661	8.24E−12	38.9	1.E−2	2.E4	5.91	0.5	0.333	124.0
1N695	7.8E−4	25.0	1.E−3	320.0	0.7	0.6	0.17	360.0
1N903	7.6E−9	28.4	0.1	1.2E7	1.3	0.8	0.24	1200.0
1N908	5.2E−6	20.2	7.E−3	8.9E5	1.2	0.8	0.18	1000.0
1N914	2.9E−6	21.5	2.E−3	1.1E3	24.0	0.9	0.5	1.81E4
1N914B	8.7E−7	24.0	2.E−3	1.1E3	2.4	0.9	0.19	3.2E3
1N970B	2.E−8	25.9	1.4E−4	3.E5	389.0	12.0	0.8	2.75E3
1N995	7.3E−4	27.0	0.015	4.E3	2.83	0.5	0.5	683.0
1N2199	2.E−6	22.6	0.01	1.E6	1.41	0.5	0.5	514.0
1N3070	4.23E−6	21.8	1.65E−3	8.64E6	1.76	1.0	0.166	628.0
1N3071	9.1E−6	20.0	1.6E−3	4.5E6	2.0	1.0	0.19	1.6E4
1N3600	5.06E−6	20.9	0.6E−3	1.36E6	0.9	0.85	0.05	3.98
1N3605	3.77E−6	19.8	0.777E−3	4.86E6	1.01	1.28	0.05	114.0
1N3669	1.3E−7	28.0	2.0E−3	1.1E3	23.0	0.8	0.46	1.3E4
1N4001	8.E−6	21.0	1.E−4	6.4E5	13.0	1.0	0.41	2.4E4
1N4003	4.2E−6	22.0	0.5E−3	2.E7	230.0	1.0	0.48	1.5E5
1N4005	6.2E−6	21.0	1.E−3	1.2E6	36.0	1.0	0.46	3.1E5
1N4406	8.0E−6	21.0	1.E−3	8.2E6	25.0	1.0	0.5	1.3E5
1N4572	7.42E−9	28.5	0.01	2.4E6	5.6	0.85	0.615	1.35E4
1N4610	2.52E−6	22.2	0.4E−3	1.37E6	1.4	1.0	0.493	51.3
FA2008	4.1E−6	20.9	0.01	1.5E6	1.4	1.0	0.12	91.0
FA2010	2.3E−6	22.5	5.E−4	2.9E6	1.2	1.0	0.1	6100.0
FD100	1.1E−5	18.1	6.E−3	2.E3	1.0	0.8	0.18	200.0
FD200	3.E−8	26.3	0.37E−3	9.71E8	4.41	0.96	0.389	2.59E4
FD300	3.E−8	26.3	0.37E−3	9.71E8	4.41	0.96	0.389	2.59E4
FD600	4.89E−6	21.1	0.47E−3	1.43E6	1.01	1.0	0.0088	48.8
FD624	1.E−4	20.0	2.E−3	1.6E3	20.0	1.0	0.0	0.0
FD700	5.15E−8	24.2	5.67E−3	5.E5	0.775	1.04	0.797	17.1

Table C.2—*Cont.*

Diode Type	IS (mA)	θ (v^{-1})	RB (kΩ)	RS (kΩ)	C_0	V (v^0)	n	K (pF/mA)
FD6666	5.36E−6	20.8	4.33E−4	1.2E6	1.05	1.0	0.102	90.0
FDA630	2.52E−6	22.2	0.39E−3	1.37E6	1.4	1.0	0.049	51.3
FDM1000	4.04E−6	19.7	4.5E−3	9.74E5	0.977	1.0	0.274	99.4
FSP220	3.2E−6	21.4	5.5E−3	1.4E6	1.6	1.0	0.15	160.0
H2969	0.1	9.7	2.E−3	1.6E6	4.0	0.8	0.08	58.2
HPA2001	4.6E−9	34.6	0.012	4.1E6	0.33	0.8	0.14	4.7E3
PS760	2.6E−6	24.9	4.5E−2	5.6E5	1.9	0.8	0.4	3000.0
PS4750	1.8E−7	24.9	5.E−3	9.9E6	2.3	0.8	0.17	7.2E3
PS4902	2.E−9	25.0	8.E−3	2.1E5	1.7	0.8	0.23	5.3E3
SD500	3.97E−11	38.9	1.E−3	—	3240.0	0.5	0.5	6.19E5
SG5250	3.2E−6	22.4	5.E−4	1.7E6	1.8	0.8	0.07	4.6E3
SG5270	1.4E−6	23.2	5.E−4	3.1E6	2.3	0.8	0.31	790.0
UT262	1.52E−6	24.0	1.E−4	3.9E6	24.0	1.0	0.41	4.E4
UT484	1.72E−8	29.4	2.5E−4	3.E5	190.0	0.75	0.5	655.0
UT4410	9.18E−8	35.3	2.E−4	1.E4	99.5	0.75	0.5	246.0

Appendix D
Transistor Model Parameter Tables

RANDALL W. JENSEN
LAWRENCE P. McNAMEE

This appendix* contains parameter tables for bipolar transistor models based on the Ebers-Moll formulation shown in Fig. A.2.

The diode currents in the model are defined by the relationships

$$i_{re} = I_{es}[\exp(\theta_e V_{b'e'}) - 1] \tag{D.1}$$

and

$$i_{rc} = I_{cs}[\exp(\theta_c V_{b'c'}) - 1] \tag{D.2}$$

The capacitance elements in the model are defined by the relationships

$$C_e = \frac{C_{0e}}{(V_{0e} - V_{b'e'})^{n_e}} + K_e(i_{re} + I_{es}) \tag{D.3}$$

and

$$C_c = \frac{C_{0c}}{(V_{0c} - V_{b'c'})^{n_c}} + K_c(i_{rc} + I_{cs}) \tag{D.4}$$

All the model parameters contained in the following parameter tables are specified in high-speed units as shown in Table C.1.

Because of the large number of parameters required in the transistor model, the transistor parameters have been divided into three tables: diode and dependent source table, capacitance table, and resistance table. The division was performed in this manner to isolate the dc and ac model parameters.

*Portions of the model parameter data has been derived from W. C. Cordwell, *Transistor and Diode Model Handbook*, USAF Technical Report AFWL-TR-69-44, Air Force Weapons Laboratory, Oct. 1969.

Table D.1 Transistor Diode Parameters

Transistor Type	Emitter I_{es} (mA)	Emitter θ_e (v^{-1})	α_I	Collector I_{cs} (mA)	Collector θ_c (v^{-1})	α_N
2N174	0.545	32.1	0.833	0.551	32.4	0.982
2N315	1.65E−5	38.9	0.833	1.65E−5	38.9	0.909
2N329A	2.16E−11	39.0	0.866	6.22E−8	29.0	0.972
2N335	4.83E−10	29.6	0.500	2.78E−10	29.6	0.981
2N336	2.91E−9	36.7	0.584	1.46E−9	28.2	0.991
2N343	4.72E−5	21.6	0.834	1.83E−5	21.6	0.973
2N356	1.83E−5	38.9	0.833	1.83E−5	38.9	0.937
2N384	3.3E−5	38.9	0.900	3.3E−5	38.9	0.909
2N385	4.29E−7	38.9	0.833	5.94E−8	38.9	0.909
2N393	5.72E−7	38.9	0.900	5.72E−7	38.9	0.909
2N398	4.26E−7	38.9	0.900	4.26E−7	38.9	0.909
2N404	2.03E−2	23.3	0.900	2.14E−2	21.9	0.984
2N414	4.44E−3	47.5	0.833	5.55E−3	45.8	0.984
2N457	2.58E−5	38.9	0.833	2.58E−5	38.9	0.968
2N585	5.05E−4	38.0	0.7777	4.11E−3	36.0	0.9895
2N597	5.22E−6	38.9	0.900	5.22E−6	38.9	0.986
2N598	1.05E−4	38.9	0.833	1.05E−4	38.9	0.993
2N657	1.E−4	20.7	0.500	1.96E−4	20.7	0.981
2N697	2.14	0.949	0.833	2.14E−12	38.9	0.976
2N705	2.86E−2	21.2	0.667	2.86E−2	21.2	0.976
2N706	1.03E−11	39.0	0.500	1.58E−10	35.0	0.968
2N706A	4.71E−10	31.6	0.833	7.79E−7	23.8	0.976
2N711A	4.95E−8	38.9	0.900	4.95E−8	38.9	0.962
2N718	3.3E−11	40.0	0.500	2.E−10	38.0	0.991
2N718A	4.08E−11	39.0	0.834	5.3E−11	39.0	0.988
2N720A	6.6E−11	40.0	0.500	1.86E−10	39.0	0.991
2N722	4.07E−11	37.0	0.500	3.6E−10	35.0	0.984
2N743	3.26E−11	36.0	0.800	1.49E−11	38.0	0.980
2N797	1.8E−6	38.9	0.833	1.8E−6	38.9	0.976
2N834	1.32E−12	41.0	0.878	1.14E−11	38.0	0.9843
2N835	3.58E−12	39.0	0.500	2.2E−11	37.0	0.984
2N910	4.07E−11	37.5	0.285	4.87E−11	37.5	0.985
2N914	6.4E−12	39.0	0.500	5.2E−11	37.0	0.988
2N915	6.65E−12	40.0	0.500	1.86E−10	27.0	0.992
2N916	4.91E−12	39.0	0.500	1.E−10	35.0	0.993
2N918	2.58E−12	38.0	0.492	2.12E−11	36.3	0.977
2N955A	1.86E−4	37.0	0.500	4.6E−3	36.0	0.967
2N964	9.41E−4	28.0	0.909	2.27E−3	28.9	0.980
2N976	3.33E−3	31.7	0.667	3.22E−3	31.6	0.988
2N995	2.74E−9	26.6	0.667	6.16E−8	26.1	0.986
2N1016B	2.73E−7	38.0	0.8333	8.76E−7	35.0	0.9896
2N1016E	6.44E−8	40.0	0.7368	2.2E−10	38.0	0.9783
2N1037	7.E−3	38.9	0.900	7.E−30	38.9	0.952
2N1039	3.21E−5	38.9	0.833	3.21E−5	38.9	0.976

Table D.1—Cont.

Transistor Type	Emitter			Collector		
	I_{es} (mA)	θ_e (v^{-1})	α_I	I_{cs} (mA)	θ_c (v^{-1})	α_N
2N1099	6.595	20.4	0.6875	2.23	29.7	0.8931
2N1131	3.56E−9	33.0	0.667	1.44E−6	23.9	0.952
2N1132	7.78E−9	27.2	0.500	7.86E−9	27.8	0.982
2N1184	8.12E−3	47.0	0.882	1.13E−2	40.0	0.9855
2N1225	1.23E−4	37.0	0.2593	5.4E−4	35.0	0.9722
2N1228	1.75E−7	23.0	0.667	5.33E−9	25.0	0.941
2N1301	5.72E−7	38.9	0.833	5.72E−7	38.9	0.986
2N1304	1.92E−3	25.7	0.833	1.96E−3	25.9	0.986
2N1306	1.96E−3	25.7	0.833	2.E−3	25.9	0.990
2N1307	1.96E−3	25.7	0.833	2.E−3	25.9	0.990
2N1308	1.99E−3	25.7	0.833	2.03E−3	25.9	0.993
2N1342	2.05E−11	39.0	0.3333	1.68E−10	37.0	0.9773
2N1483	3.38E−11	38.9	0.833	3.38E−11	38.9	0.952
2N1486	1.62E−8	37.0	0.787	4.57E−9	40.0	0.9876
2N1490	2.73E−11	37.3	0.833	1.23E−6	23.5	0.980
2N1499A	2.82E−3	29.3	0.667	2.82E−3	28.2	0.968
2N1506A	1.167E−10	38.4	0.257	3.82E−11	38.4	0.9662
2N1613	2.19E−7	32.0	0.833	2.26E−7	32.4	0.988
2N1709	2.08E−10	37.0	0.200	3.15E−9	35.0	0.9615
2N1711	2.25E−8	30.2	0.833	2.07E−8	30.3	0.993
2N1724	2.1E−9	27.7	0.800	2.08E−9	27.8	0.976
2N1893	7.2E−11	39.0	0.500	5.E−11	40.0	0.990
2N1900	1.6E−8	37.8	0.5782	2.32E−8	37.8	0.993
2N2048	3.4E−5	38.9	0.900	3.4E−5	38.9	0.980
2N2060	4.61E−12	35.8	0.286	2.24E−11	35.8	0.992
2N2087	3.99E−9	38.9	0.900	1.33E−8	38.9	0.976
2N2102	5.37E−7	24.6	0.714	8.41E−7	24.2	0.972
2N2126	1.03E−9	25.6	0.500	2.94E−8	20.5	0.962
2N2187	5.67E−11	38.9	0.3793	5.67E−11	36.1	0.9706
2N2188	8.19E−2	22.2	0.889	8.19E−2	22.2	0.988
2N2192	1.31E−10	42.0	0.875	9.49E−10	38.0	0.9949
2N2222	3.02E−11	40.1	0.697	1.2E−10	38.9	0.993
2N2223	1.08E−10	38.4	0.1667	1.78E−9	35.7	0.9868
2N2243A	5.63E−11	39.6	0.7586	7.53E−11	39.6	0.9890
2N2258	9.8E−9	35.4	0.500	9.81E−8	27.8	0.980
2N2369	2.3E−11	37.3	0.800	1.04E−6	23.5	0.978
2N2411	6.06E−13	29.4	0.2913	1.75E−13	39.4	0.9662
2N2453	2.85E−12	46.1	0.500	1.23E−11	36.0	0.995
2N2481	8.95E−12	41.0	0.0177	1.86E−9	37.0	0.983
2N2484	1.41E−6	21.5	0.627	1.04E−5	19.2	0.996
2N2538	1.2E−11	41.0	0.8276	6.E−10	36.0	0.9808
2N2656	4.85E−12	40.0	0.500	3.E−11	38.0	0.9896
2N2695	1.4E−11	39.0	0.500	6.4E−11	38.0	0.989
2N2708	1.61E−13	39.0	0.834	7.2E−12	34.0	0.991

Table D.1—*Cont.*

Transistor Type	Emitter			Collector		
	I_{es} (mA)	θ_e (v^{-1})	α_I	I_{cs} (mA)	θ_c (v^{-1})	α_N
2N2784	9.24E−11	35.2	0.142	4.24E−7	23.5	0.988
2N2801	3.33E−11	39.0	0.438	1.78E−10	39.0	0.9896
2N2808	4.17E−14	41.0	0.5650	9.02E−12	35.0	0.958
2N2845	2.34E−11	39.0	0.1300	2.45E−8	29.0	0.9804
2N2887	2.85E−10	37.3	0.2424	1.15E−9	37.3	0.9797
2N2894	6.85E−9	26.6	0.875	1.54E−8	26.1	0.986
2N2905	2.32E−11	40.6	0.583	6.13E−11	38.7	0.993
2N2907	2.21E−11	40.0	0.8077	1.48E−10	27.0	0.9954
2N3017	8.E−10	38.0	0.8210	8.64E−10	39.0	0.9877
2N3019	2.73E−10	37.3	0.833	1.23E−6	23.5	0.980
2N3021	1.01E−7	46.0	0.500	3.74E−7	46.0	0.976
2N3026	9.52E−8	46.0	0.666	2.14E−7	46.0	0.992
2N3055	9.27E−6	23.4	0.091	5.23E−8	24.5	0.500
2N3108	1.77E−8	30.2	0.800	1.63E−8	30.3	0.976
2N3117	3.13E−8	22.2	0.500	2.8E−8	23.8	0.998
2N3119	1.56E−9	31.4	0.500	9.76E−9	28.8	0.976
2N3227	2.86E−9	37.3	0.833	1.29E−6	23.5	0.990
2N3244	1.71E−10	38.0	0.677	1.02E−10	39.0	0.991
2N3251	5.37E−9	26.6	0.833	1.21E−8	26.1	0.990
2N3252	5.22E−11	40.0	0.500	4.2E−10	37.0	0.976
2N3287	9.48E−13	40.0	0.580	5.35E−11	34.0	0.960
2N3309	8.4E−12	40.0	0.500	8.4E−11	37.0	0.976
2N3468	2.26E−7	24.9	0.500	8.54E−7	24.9	0.972
2N3498	6.98E−12	43.7	0.666	1.49E−11	43.0	0.988
2N3499	5.35E−11	39.0	0.375	6.93E−10	37.0	0.9908
2N3501	3.99E−8	30.2	0.909	3.66E−8	30.3	0.993
2N3502	2.32E−8	29.7	0.833	2.34E−8	30.3	0.995
2N3503	2.32E−8	27.2	0.833	2.34E−8	27.8	0.995
2N3507	1.8E−6	24.3	0.100	5.53E−5	17.7	0.987
2N3600	9.18E−7	30.4	0.492	2.12E−9	28.4	0.977
2N3635	2.26E−8	27.2	0.833	2.29E−8	27.8	0.990
2N3737	3.96E−12	38.0	0.333	7.89E−12	38.0	0.986
2N3738	7.41E−8	40.0	0.500	7.32E−7	39.0	0.990
2N3738	7.29E−8	31.3	0.834	8.75E−7	32.0	0.990
2N3766	7.29E−8	31.3	0.834	8.75E−7	32.0	0.990
2N3792	4.1E−8	26.8	0.500	2.98E−7	26.8	0.993
2N3828	3.1E−11	38.0	0.6770	1.4E−10	38.0	0.960
2N3866	2.08E−9	38.4	0.500	8.04E−9	38.4	0.984
2N3904	2.97E−11	35.4	0.500	3.96E−9	27.8	0.990
2N3906	7.92E−11	33.0	0.500	1.19E−8	25.9	0.990
2N3913	8.24E−13	43.5	0.910	8.96E−13	43.5	0.990
2N3914	4.05E−13	43.5	0.953	4.23E−13	43.5	0.994
2N3915	2.08E−13	43.5	0.969	2.14E−13	43.5	0.997
2N3959	4.9E−12	35.1	0.500	1.96E−11	33.0	0.980

Table D.1—Cont.

Transistor Type	Emitter			Collector		
	I_{es} (mA)	α_I (v^{-1})	α_I	I_{cs} (mA)	θ_c (v^{-1})	α_N
2N3960	3.51E−13	38.1	0.834	1.22E−12	38.0	0.982
2N4125	4.75E−12	51.2	0.666	4.68E−12	51.2	0.990
2N4260	3.9E−12	36.6	0.500	3.9E−9	28.1	0.976
2N4923	2.93E−8	30.6	0.500	1.95E−7	23.6	0.976
2N10000	9.95E−10	35.36	0.500	3.98E−8	27.79	0.995

Table D.2 Transistor Capacitance Parameters

Transistor Type	Emitter				Collector			
	C_{0e}	V_{0e} (v)	K_e (pF/mA)	n_e	C_{0c}	V_{0c} (v)	K_c (pF/mA)	n_c
2N174	1.41E3	1.0	5.11E4	0.5	1.41E3	1.0	5.16E5	0.5
2N315	23.5	0.5	1.24E3	0.5	32.8	0.5	6.19E3	0.5
2N329A	28.0	0.8	3432.0	0.47	96.0	0.8	4.06E5	0.5
2N335	51.0	1.0	182.0	0.43	10.0	0.8	1.36E4	0.32
2N336	57.0	1.0	190.8	0.43	10.0	0.8	1.297E4	0.32
2N343	80.0	0.75	1.77E4	0.5	35.0	0.75	4.13E3	0.5
2N356	23.5	0.5	2.06E3	0.5	32.8	0.5	7.74E3	0.5
2N384	10.9	0.5	61.9	0.333	6.96	0.5	310.0	0.333
2N385	14.1	0.5	124.0	0.5	15.0	0.5	6.19E3	0.5
2N393	7.07	0.5	124.0	0.5	11.2	0.5	6.19E3	0.5
2N398	61.2	0.5	6.19E3	0.5	61.2	0.5	3.1E4	0.5
2N404	15.6	1.18	286.0	0.59	56.6	3.13	3.48E3	0.826
2N414	24.5	1.0	1.08E3	0.5	31.7	1.0	1.46E4	0.5
2N457	122.0	0.5	1.44E4	0.5	122.0	0.5	6.19E4	0.5
2N585	13.0	0.4	602.0	0.42	29.0	0.4	6159.0	0.41
2N597	20.8	0.5	2060.0	0.5	70.4	0.5	1.24E4	0.5
2N598	42.1	1.81	953.0	0.823	120.0	2.66	6.19E3	0.806
2N657	80.0	0.9	372.6	0.41	47.0	0.9	9190.8	0.395
2N697	50.0	0.5	1.51	0.333	76.6	0.5	68.8	0.333
2N705	3.5	0.3	22.5	0.7	11.9	0.3	1.28E3	0.25
2N706	5.4	0.9	7.41	0.35	4.5	1.0	4.2E3	0.14
2N706A	8.14	0.5	25.1	0.333	6.96	0.5	378.0	0.333
2N711A	11.4	0.5	41.3	0.333	14.1	0.5	619.0	0.5
2N718	37.0	0.9	30.8	0.41	25.0	0.8	9.5E4	0.38
2N718A	76.0	0.9	206.7	0.36	45.0	0.9	1.014E4	0.37
2N720A	78.0	1.0	60.0	0.41	32.0	0.9	1.64E5	0.39
2N722	29.0	1.0	28.86	0.42	15.0	0.8	8.4E3	0.45
2N743	4.6	0.9	15.88	0.39	5.3	0.8	1642.0	0.25
2N797	4.58	0.5	10.3	0.333	7.06	0.5	61.9	0.333

Table D.2—Cont.

Transistor Type	Emitter				Collector			
	C_{0e}	V_{0e} (v)	K_e (pF/mA)	n_e	C_{0c}	V_{0c} (v)	K_c (pF/mA)	n_c
2N834	9.0	0.8	18.56	0.5	7.3	0.9	1868.4	0.31
2N835	7.2	0.9	36.27	0.34	5.0	0.8	2.035E4	0.16
2N910	50.0	0.9	101.6	0.20	22.0	0.9	3468.7	0.31
2N914	5.9	0.8	14.82	0.32	4.7	0.9	4.44E3	0.13
2N915	7.1	0.9	20.8	0.31	4.6	0.8	4.86E4	0.41
2N916	7.3	0.9	19.89	0.30	6.7	0.9	5.25E4	0.34
2N918	2.0	0.5	8.284	0.15	1.7	0.8	461.0	0.12
2N955A	4.3	0.5	3.33	0.38	6.0	0.4	540.0	0.22
2N964	2.29	0.5	9.7	0.333	4.81	0.5	92.0	0.333
2N976	6.64	0.86	5.6	0.608	2.54	0.665	101.0	0.293
2N995	12.7	1.08	42.3	0.32	15.2	1.01	415.0	0.176
2N1016B	400.0	1.0	7897.0	0.2	1500.0	0.7	5.83E4	0.51
2N1016E	350.0	0.8	1.37E4	0.47	1600.0	0.7	8.12E4	0.5
2N1037	122.0	0.5	2.06E4	0.5	68.2	0.5	6.19E4	0.5
2N1039	1220.0	0.5	7.74E5	0.5	255.0	0.5	6.19E6	0.5
2N1099	1500.0	0.6	1.33E4	0.5	1200.0	0.6	3.06E4	0.5
2N1131	83.8	1.63	105.0	0.496	72.7	0.936	475.0	0.361
2N1132	83.8	1.63	217.0	0.496	72.7	0.936	885.0	0.361
2N1184	200.0	0.4	3.6E3	0.5	240.0	0.4	6.E3	0.48
2N1225	3.3	0.5	356.1	0.47	3.1	0.4	3902.5	0.19
2N1228	178.0	1.08	3050.0	0.32	137.0	1.01	3.32E4	0.176
2N1301	13.6	0.5	177.0	0.333	22.4	0.5	619.0	0.333
2N1304	21.0	1.4	510.0	0.4	42.8	1.4	2060.0	0.407
2N1306	21.0	1.4	340.0	0.4	42.8	1.4	2060.0	0.407
2N1307	14.7	1.4	340.0	0.4	42.8	1.4	2060.0	0.41
2N1308	21.0	1.4	204.0	0.4	42.8	1.4	2060.0	0.41
2N1342	37.0	0.8	60.98	0.34	17.0	0.8	12321.0	0.43
2N1483	309.0	0.5	4960.0	0.333	600.0	0.5	6.19E4	0.333
2N1486	230.0	0.9	2192.0	0.43	780.0	0.8	1.48E4	0.28
2N1490	3.31	1.06	11.9	0.31	2.93	1.1	74.7	0.079
2N1499A	5.72	0.5	233.0	0.333	5.6	0.5	449.0	0.333
2N1506A	155.0	0.9	31.17	0.39	46.0	0.8	12840.0	0.5
2N1613	54.6	0.713	56.5	0.323	2690.0	1.68	258.0	4.46
2N1709	420.0	1.0	56.92	0.37	102.0	0.8	3.85E4	0.47
2N1711	80.0	0.5	68.7	0.33	54.3	0.5	241.0	0.33
2N1724	2320.0	0.92	350.0	0.40	968.0	0.92	3510.0	0.40
2N1893	78.0	0.9	78.0	0.38	32.0	0.80	2.E5	0.39
2N1900	7000.0	0.8	63.79	0.38	1800.0	0.70	4.08E4	0.47
2N2048	5.72	0.5	23.8	0.333	1.5	0.50	619.0	0.333
2N2060	71.5	0.975	63.4	0.445	34.5	0.605	991.7	0.445
2N2087	70.0	0.5	41.3	0.383	26.3	0.50	124.0	0.333
2N2102	29.8	1.25	2.5E4	0.446	0.0	2.5	1.45E4	0.373
2N2126	1600.0	0.8	1.17E4	0.50	2410.0	0.8	1.48E5	0.41
2N2187	5.4	0.8	75.51	0.36	7.8	0.7	246.5	0.41

Table D.2—Cont.

Transistor Type	Emitter				Collector			
	C_{0e}	V_{0e} (v)	K_e (pF/mA)	n_e	C_{0c}	V_{0c} (v)	K_c (pF/mA)	n_c
2N2188	13.6	0.5	39.3	0.333	33.9	0.5	708.0	0.333
2N2192	39.0	0.9	34.26	0.45	21.0	0.8	3990.0	0.41
2N2222	22.0	0.9	2200.0	0.4	13.0	0.9	1.07E4	0.35
2N2223	59.0	0.8	60.63	0.41	33.0	0.9	2.2E4	0.41
2N2243A	67.0	0.8	37.99	0.41	30.0	0.8	1.05E4	0.41
2N2258	39.4	0.8	113.0	0.3	12.4	0.2	7370.0	0.2
2N2369	3.31	1.06	11.9	0.31	2.93	1.1	74.7	0.079
2N2411	5.6	0.9	26.65	0.28	5.3	0.8	2066.0	0.25
2N2453	4.5	0.7	128.6	0.414	5.6	0.7	1911.6	0.225
2N2481	5.5	0.8	16.52	0.32	4.0	0.8	124.4	0.46
2N2484	6.0	0.5	6830.0	0.33	10.5	0.5	3.06E4	0.33
2N2538	19.0	0.9	14.23	0.38	6.9	0.8	8640.0	0.26
2N2656	4.3	0.9	15.83	0.37	5.8	0.8	1.254E4	0.31
2N2695	45.0	0.8	35.88	0.35	19.0	0.9	2356.0	0.33
2N2708	1.2	0.9	6.24	0.14	1.5	0.9	1972.0	0.15
2N2784	1.58	1.0	5.6	0.347	1.73	1.0	0.701	0.229
2N2801	65.0	1.0	15.05	0.41	49.0	0.8	187.9	0.47
2N2808	0.9	1.0	5.5	0.08	1.3	1.0	870.1	0.23
2N2845	4.6	0.9	22.94	0.37	9.0	0.9	452.4	0.23
2N2887	440.0	0.9	10.6	0.39	103.0	0.8	1.805E4	0.49
2N2894	4.53	1.08	14.1	0.32	4.42	1.01	82.9	0.176
2N2905	18.0	1.0	1.96E3	0.42	13.0	0.8	3.08E3	0.38
2N2907	21.0	1.0	19.9	0.43	26.0	1.0	959.6	0.44
2N3017	480.0	0.9	48.8	0.39	170.0	0.9	3.5E5	0.42
2N3019	3.31	1.05	11.9	0.31	2.93	1.1	74.7	0.794
2N3021	715.0	1.22	11.5	0.475	332.0	0.614	1.47E4	0.475
2N3026	715.0	1.22	4.6	0.475	332.0	0.614	763.6	0.434
2N3055	1800.0	1.0	2.48E5	0.855	1.E4	1.0	2.E5	0.519
2N3108	80.0	0.5	50.1	0.33	43.5	0.5	502.0	0.33
2N3117	6.86	0.9	29.5	0.4	8.38	0.9	315.0	0.35
2N3119	130.0	0.8	1.72E4	0.29	65.5	0.23	9.18E4	0.11
2N3227	3.31	1.06	11.9	0.31	2.93	1.1	74.7	0.794
2N3244	55.0	1.0	18.84	0.45	43.0	0.8	1953.8	0.41
2N3251	10.1	1.08	14.1	0.32	9.15	1.0	82.9	0.176
2N3252	55.0	0.9	20.0	0.40	14.0	0.8	1850.0	0.23
2N3287	1.9	1.0	26.5	0.22	1.8	1.0	3155.2	0.23
2N3309	24.0	1.0	16.4	0.33	16.0	1.0	481.0	0.25
2N3468	118.0	0.9	29.88	0.49	61.7	0.9	4357.5	0.38
2N3498	74.4	1.28	87.4	0.518	19.0	0.9	2.75E4	0.378
2N3499	16.0	1.0	23.18	0.25	24.0	1.0	8325.0	0.33
2N3501	80.0	0.5	68.8	0.33	54.3	0.5	241.0	0.33
2N3502	21.8	1.63	237.0	0.496	10.7	0.936	966.0	0.361
2N3503	21.8	1.63	217.0	0.496	10.7	0.936	885.0	0.361
2N3507	300.0	1.0	7.096	0.1	40.0	0.8	3.45E4	0.25

Table D.2—Cont.

Transistor Type	Emitter				Collector			
	C_{0e}	V_{0e} (v)	K_e (pF/mA)	n_e	C_{0c}	V_{0c} (v)	K_c (pF/mA)	n_c
2N3600	2.0	0.5	6.63	0.15	1.7	0.8	352.2	0.12
2N3635	83.2	1.63	21.7	0.496	31.7	0.936	221.0	0.361
2N3737	71.0	0.75	1.193	0.37	9.1	0.75	269.8	0.27
2N3738	208.0	0.9	2544.0	0.378	46.0	0.9	3.35E5	0.34
2N3766	224.2	0.94	331.8	0.428	122.5	1.15	1.1E4	0.312
2N3792	4.8	0.75	1066.6	0.33	2.4	0.75	6860.8	0.30
2N3828	18.0	0.8	1094.0	0.31	6.8	0.75	1541.0	0.23
2N3866	7.6	1.1	7.642	0.52	5.4	1.0	940.8	0.366
2N3904	8.04	0.55	1130.0	0.16	6.1	0.219	681.0	0.08
2N3906	22.7	0.9	1310.0	0.2	15.3	0.2	2750.0	0.7
2N3913	21.0	0.8	219.7	0.5	21.0	0.8	2479.5	0.5
2N3914	21.0	0.8	112.7	0.5	21.0	0.8	1570.3	0.5
2N3915	21.0	0.8	59.16	0.5	21.0	0.8	1104.9	0.5
2N3959	6.34	1.2	6200.0	0.07	9.73	0.6	6820.0	0.05
2N3960	2.8	0.75	3.734	0.246	3.75	0.75	3.078	0.261
2N4125	3.7	0.75	14.54	0.188	3.2	0.75	384.0	0.266
2N4260	8.66	1.2	971.0	0.10	8.11	0.3	1490.0	0.05
2N4923	73.9	1.0	3250.0	0.3	83.5	0.4	7.51E4	0.32
2N10000	22.97	1.0	1414.0	0.2	16.27	0.5	5896.0	0.2

Table D.3 Transistor Resistance Parameters

Transistor Type	R_{BB} (kΩ)	R_{EE} (kΩ)	R_{CC} (kΩ)	R_E (kΩ)	R_C (kΩ)
2N174	3.E−3	0.0	2.E−5	400.0	160.0
2N315	0.02	0.0	0.001	800.0	800.0
2N329A	0.350	2.5E−3	0.032	5.E4	5.E4
2N335	0.595	0.004	0.065	5.E4	5.E4
2N336	1.0	2.5E−3	0.06	5.E4	5.E4
2N343	0.015	0.007	0.12	5.E4	5.E4
2N356	0.02	0.0	0.001	800.0	800.0
2N384	0.10	0.0	0.02	40.0	1000.0
2N385	0.07	0.0	0.005	1.5E3	2.5E3
2N393	0.115	0.0	0.010	500.0	1000.0
2N398	0.1	0.0	0.02	1000.0	180.0
2N404	1.9E−3	0.0	1.E−4	6.47E3	7.46E3
2N414	0.055	0.0	0.001	2.8E3	0.5E3
2N457	0.01	0.0	1.E−4	60.0	60.0
2N585	0.13	0.0	0.0023	—	—
2N597	0.006	0.0	2.5	3.E6	300.0
2N598	0.04	0.0	0.001	600.0	300.0

Table D.3—Cont.

Transistor Type	R_{BB} (kΩ)	R_{EE} (kΩ)	R_{CC} (kΩ)	R_E (kΩ)	R_C (kΩ)
2N657	0.018	4.5E−3	6.1E−3	5.E4	5.E4
2N697	0.01	0.0	0.002	3.E4	3.E4
2N705	0.05	0.0	0.012	1000.0	1000.0
2N706	3.5E−6	1.E−4	0.011	5.E4	5.E4
2N706A	0.013	0.0	1.E−4	1.05E7	3.59E6
2N711A	0.10	0.0	0.01	15.0	1.E4
2N718	0.03	1.E−4	0.0074	5.E4	5.E4
2N718A	0.06	5.E−4	0.0056	5.E4	5.E4
2N720A	0.05	0.001	0.0087	5.E4	5.E4
2N722	0.01	1.E−4	0.0054	5.E4	5.E4
2N743	0.20	0.0	0.0083	—	—
2N797	0.15	0.0	0.005	40.0	200.0
2N834	0.01	0.0	0.011	—	—
2N835	0.012	1.E−4	0.019	5.E4	5.E4
2N910	0.013	3.E−4	0.008	5.E4	5.E4
2N914	0.025	0.001	0.009	5.E4	5.E4
2N915	0.015	0.001	0.017	5.E4	5.E4
2N916	0.042	0.001	0.033	5.E4	5.E4
2N918	0.012	0.0085	0.012	5.E4	5.E4
2N955A	0.03	1.E−4	0.003	5.E4	5.E4
2N964	1.49E−3	0.0	4.E−5	1.83E3	7.29E4
2N976	0.05	0.0	0.002	127.0	1600.0
2N995	0.025	0.0	0.002	1.3E8	1.3E8
2N1016B	0.04	0.0	0.0036	—	—
2N1016E	0.03	0.0	0.005	—	—
2N1037	0.10	0.0	0.01	5.E4	5.E4
2N1039	0.01	0.0	2.5E−4	27.0	24.0
2N1099	0.0	0.0	5.E−5	—	—
2N1131	1.8E−3	0.0	1.E−4	2.E6	1.9E6
2N1132	6.5E−3	0.0	0.007	3.E4	3.E4
2N1184	0.02	0.0	1.2E−3	—	—
2N1225	0.02	0.0	5.8E−3	—	—
2N1228	0.002	0.0	1.E−4	1.35E7	8.09E7
2N1301	0.05	0.0	0.005	40.0	650.0
2N1304	0.04	0.0	0.001	1.5E4	4000.0
2N1306	0.04	0.0	0.001	1.5E4	4000.0
2N1307	0.04	0.0	0.001	1.5E4	4000.0
2N1308	0.04	0.0	0.001	1.5E4	4000.0
2N1342	0.012	0.0	0.0026	—	—
2N1483	0.012	0.0	2.7E−3	800.0	2000.0
2N1486	2.E−3	0.001	1.9E−3	—	—
2N1490	0.02	0.0	0.002	5.E5	5.E5
2N1499A	0.05	0.0	0.01	5.E2	1.6E3
2N1506A	0.023	0.0	0.001	—	—
2N1613	0.01	0.0	0.001	5.E5	6.E6
2N1709	5.E−3	0.001	6.E−4	—	—

Table D.3—*Cont.*

Transistor Type	R_{BB} (kΩ)	R_{EE} (kΩ)	R_{CC} (kΩ)	R_E (kΩ)	R_C (kΩ)
2N1711	1.44E−3	0.0	1.E−4	1.28E8	2.38E8
2N1724	0.10	0.0	0.1	3.9E7	9.47E8
2N1893	0.047	0.001	0.011	5.E4	5.E4
2N1900	5.E−3	0.0	6.E−4	—	—
2N2048	0.10	0.0	0.01	1000.0	5000.0
2N2060	0.094	0.006	0.002	5.E4	5.E4
2N2087	0.014	0.0	0.001	100.0	3.0E4
2N2102	1.2E−4	0.0	4.E−5	6.E10	4.E10
2N2126	3.5E−4	1.E−5	2.5E−5	5.E4	5.E4
2N2187	0.10	0.0	0.015	—	—
2N2188	5.E−3	0.0	0.002	2000.0	2000.0
2N2192	0.04	0.0	3.5E−3	—	—
2N2222	0.05	0.0	0.002	1.E6	1.E6
2N223	0.032	0.0	0.009	—	—
2N2243A	0.124	0.0	0.01	—	—
2N2258	1.5E−3	0.0	1.E−4	6.E6	5.E6
2N2369	0.02	0.0	0.002	5.E5	5.E5
2N2411	6.5E−2	0.001	1.3E−2	—	—
2N2453	0.10	0.01	0.02	5.E4	5.E4
2N2481	0.02	0.0	9.8E−3	—	—
2N2484	0.002	0.0	1.86E−4	6000.0	4000.0
2N2538	0.032	0.0	2.9E−3	—	—
2N2656	0.035	0.0	0.019	—	—
2N2695	0.04	1.E−4	2.7E−3	5.E4	5.E4
2N2708	0.08	0.001	0.015	5.E4	5.E4
2N2784	170.0	0.0	0.003	1.E5	5.E5
2N2801	0.055	0.001	6.1E−3	—	—
2N2808	0.08	0.0	0.036	—	—
2N2845	0.025	0.0	0.002	—	—
2N2887	0.013	0.0	3.E−4	—	—
2N2894	0.025	0.0	0.002	1.3E8	1.3E8
2N2905	0.022	0.0	0.002	1.E6	1.E6
2N2907	0.05	0.001	0.0089	—	—
2N3017	0.0045	0.0	2.E−4	—	—
2N3019	0.02	0.0	0.002	5.E5	5.E5
2N3021	0.0078	2.E−4	3.E−4	5.E4	5.E4
2N3026	0.0078	2.E−4	4.E−4	5.E4	5.E4
2N3055	0.002	0.0	3E−4	3.E4	4.5E5
2N3108	1.44E−3	0.0	1.E−4	1.28E7	2.38E7
2N3117	2.18E−3	0.0	1.4E−3	6.45E8	1.46E9
2N3119	6.E−4	0.0	1.E−6	4.E6	4.E7
2N3227	0.02	0.0	0.002	5.E5	5.E5
2N3244	0.02	0.0	1.4E−3	—	—
2N3251	0.025	0.0	0.002	1.3E8	1.3E8
2N3252	0.018	1.E−4	1.8E−3	5.E4	5.E4
2N3287	6.E−3	0.0	0.029	—	—

Table D.3—Cont.

Transistor Type	R_{BB} (kΩ)	R_{EE} (kΩ)	R_{CC} (kΩ)	R_E (kΩ)	R_C (kΩ)
2N3309	0.003	1.E−4	1.6E−3	5.E4	5.E4
2N3468	0.001	7.5E−4	2.2E−4	5.E4	5.E4
2N3498	0.0127	1.3E−3	1.66E−3	5.E4	5.E4
2N3499	0.003	0.0	1.6E−3	—	—
2N3501	1.44E−3	0.0	1.E−4	1.28E8	2.38E7
2N3502	6.5E−3	0.0	0.007	500.0	3.E6
2N3503	6.5E−3	0.0	0.007	500.0	3.E6
2N3507	1.E−4	2.E−4	2.E−4	5.E4	5.E4
2N3600	0.012	0.0085	0.0115	5.E4	5.E4
2N3635	0.0065	0.0	0.007	3.E4	3.E4
2N3737	0.02	4.2E−4	1.E−4	5.E4	5.E4
2N3738	0.0192	8.4E−4	4.9E−4	5.E4	5.E4
2N3766	0.023	4.1E−4	3.2E−4	5.E4	5.E4
2N3792	0.0024	5.E−4	0.008	5.E4	5.E4
2N3828	1.2E−3	0.0	0.017	—	—
2N3866	0.0045	5.E−4	0.008	5.E4	5.E4
2N3904	0.002	0.0	0.001	5.E6	1.E7
2N3906	0.001	0.0	7.E−4	2.E8	5.E6
2N3913	0.127	0.0058	0.034	5.E4	5.E4
2N3914	0.129	0.0038	0.021	5.E4	5.E4
2N3915	0.131	0.0021	0.018	5.E4	5.E4
2N3959	0.0028	0.0	8.E−4	5.E7	1.E8
2N3960	0.039	0.002	0.0015	5.E4	5.E4
2N4125	0.15	0.0014	8.E−4	5.E4	5.E4
2N4260	0.0042	0.0	2.E−4	5.E8	1.8E7
2N4923	4.E−4	0.0	3.E−5	5.E6	9.E6
2N10000	2.E−4	3.E−5	1.E−4	—	—

DIGITAL COMPUTER
LABORATORY
LIBRARY